How to use *Technical Communication*

Technical communicators know how to find and filter key information quickly. Here's an overview of some of the major components of this book to give you an edge as you navigate the exciting world of *Technical Communication*.

feature	function
Technical Communication in the News 	These short news items highlight recent examples of the impact and importance of technical communication. Learn about the human side of tech comm and see why it matters.
On the Job	On the Job boxes comprise honest, authentic comments from a wide range of communicators about the role that writing plays—sometimes unexpectedly—in the jobs they do.
Consider This	Consider This boxes summarize current research and offer concise insight into the topics discussed in each chapter. Today's scholarship at a glance.
Guidelines	Guidelines boxes give you direct, practical advice and accessible strategies you can use. Grasp the practical side of each chapter in an authoritative and handy format.
Checklists	Use Checklists to review and revise your documents. Key questions help focus on user needs and improve the effectiveness and usability of your documents.

Technical
Communication

TENTH EDITION

John M. Lannon
University of Massachusetts, Dartmouth

PEARSON
Longman

New York Boston San Francisco
London Toronto Sydney Tokyo Singapore Madrid
Mexico City Munich Paris Cape Town Hong Kong Montreal

Acquisitions Editor: Ginny Blanford
Senior Marketing Manager: Melanie Craig
Senior Supplements Editor: Donna Campion
Media Supplements Editor: Jenna Egan
Production Manager: Ellen MacElree
Project Coordination, Text Design, and Electronic Page Makeup: Nesbitt Graphics, Inc.
Cover Design Manager: John Callahan
Cover Designer: Maria Ilardi
Cover Photo: © International Stock Photography Ltd/ImageState
Photo Researcher: Julie Tesser
Manufacturing Buyer: Roy Pickering
Printer and Binder: R. R. Donnelley & Sons
Cover Printer: Phoenix Color Corps.

Photo Credits: Part I, AP/Wide World Photos; Part II, Nicholas Edwige/Photo
Researchers, Inc.; Part III, Richard Lord/PhotoEdit; Part IV, Dennis MacDonald/PhotoEdit;
Part V, Burger/Photo Researchers, Inc.

For permission to use copyrighted material, grateful acknowledgment is made to the copyright holders on pp. 779–94 which are hereby made part of this copyright page.

Between the time Web site information is gathered and published, some sites may have closed. Also, the transcription of URLs can result in typographical errors. The publisher would appreciate notification where these occur so that they may be corrected in subsequent editions.

Many of the designations used by manufacturers and sellers to distinguish their products are claimed as trademarks. Where these designations appear in this book, and Addison Wesley Longman was aware of a trademark claim, the designations have been printed in initial caps.

Library of Congress Catalog on file with the Library of Congress.

Please visit our website at http://www.ablongman.com/lannonweb

ISBN 0-321-27076-2

1 2 3 4 5 6 7 8 9 10—DOC—08 07 06 05

Brief Contents

PART V   Specific Documents and Applications   382

PART VI   A Brief Handbook with Additional Sample Documents   676

Detailed Contents

PART III

Structural and Style Elements 216

CHAPTER 12

Organizing for Users 218

CHAPTER 13

Editing for Readable Style 243

PART IV

Visual, Design, and Usability Elements   286

CHAPTER 14

Designing Visual Information   288

CHAPTER 15

Designing Pages and Documents   339

Preface

Whether handwritten, electronically mediated, or face-to-face, workplace communication is more than a value-neutral exercise in "information transfer"; it is a complex social transaction. Each rhetorical situation has its own specific interpersonal, ethical, legal, and cultural demands. Moreover, today's professional is not only a fluent communicator, but also a discriminating consumer of information, skilled in the methods of inquiry, retrieval, evaluation, and interpretation essential to informed decision making.

Designed in response to these issues, *Technical Communication,* Tenth Edition, addresses a wide range of interests for classes in which students from a variety of majors are enrolled. The text explains, illustrates, and applies rhetorical principles to an array of assignments, from brief memos and summaries to formal reports and proposals. To help students develop awareness of audience and accountability, exercises incorporate the problem-solving demands typical in college and on the job. Self-contained chapters allow for various course plans and customized assignments.

HALLMARKS OF THE TENTH EDITION

The hallmarks of the Tenth Edition of *Technical Communication* include:

- **A focus on applications beyond the classroom.** Clear ties to the workplace include more examples from everyday business situations and more sample documents, including on-the-job and internship documents written by students, new sections on career paths in Ch.1, and a new section on design skills for today's workplace in Ch. 15. Discussion about actual jobs held by technical communicators and about communication in various fields is supplemented by day-in-the-life observations in the new "On the Job" feature.
- **Updated technology coverage.** Chapters on Web design, document design, and usability, in particular, reflect changes in the technology and an increasing use of Web-based documents and platforms. Updated information on communicating electronically includes more on hypertext and mark-up language, expanded guidelines for creating a Web site, and new information on instant messaging, and electronic mail. Fully integrated computing advice is supplemented by "Consider This" discussions of technology and interpersonal issues that are shaping workplace communication.
- **Added coverage of international and global workplace issues.** This Tenth Edition includes additional samples, cases, and exercises premised on a multina-

tional intercultural workplace, foregrounding issues of cultural and social style differences. Many end of chapter exercises call for research into cultural differences in communication practices. Marginal globe icons identify this material.

- **A service-learning component.** End of chapter exercises include projects specifically intended for use in service-learning courses. A focus on nonprofit organizations has been added to supplement the corporate culture examples.
- **Expanded chapter on collaboration.** With new coverage of cross-cultural collaboration and an emphasis on computer-mediated and Internet collaboration, collaborative projects are featured throughout the text as well as in Ch. 6, Working in Teams.
- **Strong coverage of information literacy.** Information literate people are those who "know how knowledge is organized, how to find information, and how to use information in such a way that others can learn from them."* Critical thinking—the basis of information literacy—is covered intensively in Part II and integrated throughout the text.
- **Increased coverage of usability testing.** Usability receives consistent and explicit emphasis throughout, with an expanded chapter on usability (Chapter 17) and usability checklists at the end of relevant chapters in Parts IV and V.
- **Expanded treatment of ethical and legal issues.** Woven into the fabric of the communication process are legal and ethical considerations in word choice, product descriptions, instructions, and other forms of hard copy and electronic communication. Ch. 5, Weighing the Ethical Issues, also includes new information on recognizing unethical communication and day-to-day ethical dilemmas, as well as a new exercise on avoiding plagiarism. Chapters on collaboration and design also include extended discussion of ethical communication in the workplace.

ADDITIONAL FEATURES OF THE TENTH EDITION

Technical Communication, Tenth Edition, also includes the following:

- **"On the Job"** boxes demonstrate how the skills and strategies learned from this book are needed in real-world careers of all kinds.
- **"Tech Comm in the News"** underscores the book's current and real-world focus. Each part opener includes a 200-word boxed and illustrated summary of a current news story related to issues in technical communication (examples: the role of miscommunication in the *Columbia* shuttle crash; common flaws in Web site design; risks posed by the lack of health literacy among Americans).

---

*American Library Association Presidential Committee on Information Literacy: Final Report. Chicago: ALA, 1989.

- **"Consider This" boxes** provide interesting and topical applications of the important issues discussed in various chapters, such as collaborating, technology, and ethics.
- **Expanded Guidelines** help students apply and synthesize the information in the chapter and offer practical suggestions for real workplace situations.
- The **Companion Website** is fully integrated with the printed text. Marginal icons highlight sections where the Web site offers additional cases, sample documents, templates, and examples.
- **Exercises/Collaborative activities** at the end of each chapter help students apply what they've learned. More collaborative activities have been added to reflect the role of collaboration in all aspects of technical communication.
- **Usability Checklists** help students polish their writing by giving them points to consider and page cross-references so that they can refer to specific passages in the text to find more information on each point. All checklists have been expanded and redesigned for easier access.
- **Marginal annotations** highlight important concepts in the text. The "Note" annotation adds clarification or points out up-to-the-minute business and technological advances.

ORGANIZATION OF *TECHNICAL COMMUNICATION,* TENTH EDITION

The text begins with a brief overview of workplace communication in Chapter 1, followed by six major sections:

Part I: Communicating in the Workplace treats job-related communication as a problem-solving process. Students learn to think critically about the informative, persuasive, and ethical dimensions of their communications. They also learn how to adapt to the interpersonal challenges of collaborative work and to the various needs and expectations of global audiences.

Part II: The Research Process treats research as a deliberate inquiry process. Students learn to formulate significant research questions; to explore primary and secondary sources in hard copy and electronic form; to evaluate and interpret their findings; and to summarize for economy, accuracy, and emphasis.

Part III: Structural and Style Elements offers strategies for organizing and conveying messages that users can follow and understand. Students learn to control their material and to develop a readable style.

Part IV: Visual, Design, and Usability Elements treats the rhetorical implications of graphics and page design. Students learn to enhance a document's access, appeal, and visual impact for audiences who need to locate, understand, and use the information successfully.

Part V: Specific Documents and Applications applies earlier concepts and strategies to the preparation of print and electronic documents and oral presentations. Various letters, memos, reports, and proposals offer a balance of examples from the workplace and from student writing. Each sample document has been chosen so that students can emulate it easily.

Part VI: A Brief Handbook with Additional Sample Documents contains instructions for recording research findings and for documenting them in MLA or APA style, demonstrations of the writing process in three workplace settings, and a brief handbook of grammar, usage, and mechanics.

INSTRUCTIONAL SUPPLEMENTS

Accompanying *Technical Communication,* Tenth Edition is a wide array of supplements for both instructors and students. Specific to this text are:

- *Companion Website.* Find a wealth of resources at <www.ablongman.com/lannonweb>. The numbered icons printed in the text margins indicate topics and concepts that are illustrated in depth on the Companion Website. This Web site includes project-based individual and collaborative exercises, additional forms and document templates, and many sample Web and print documents for student response and class discussion. A comprehensive Instructor Resource section includes detailed strategies for incorporating Web resources and technology with the book; an extensive set of annotated links to resources in grammar and writing, technical communication organizations, and online journals and publications; chapter overviews and teaching notes; sample syllabi and downloadable *PowerPoint* slides for classroom use.
- *The Instructor's Resource Manual* for *Technical Communication,* Tenth Edition, supports both traditional and Web-based instruction. A guide to using the Companion Website with the book and an annotated index of the book's Web icons offer specific support for building connections between the textbook and its Companion Website, and helps teachers to effectively integrate the Web resources and projects into their teaching strategies. The manual also includes general suggestions and ideas for teaching technical communication from a composition standpoint; sample syllabi; transparency masters; and an annotated bibliography of resources for teaching technical communication. Resources for use in any technical communications include:
- *MyTechCommLab* is a comprehensive resource for students in technical communication. It offers the best multimedia resources for technical writing in one, easy-to-use place. Students will find guidelines, tutorials, and exercises for grammar, writing, and research, as well as a gallery of model documents, an online reference library, and Pearson's unique Research Navigator and Avoiding Plagiarism programs. Visit <www.mytechcommlab.com> for information about how to access this remarkable site.

- *Resources for Technical Communication.* This print supplement includes over forty sample documents in a variety of categories, as well as more than half a dozen case studies with exercises.
- *Visual Communication: A Writer's Guide,* examines the rhetoric and principles of visual design, with an emphasis throughout on audience and genre. Practical guidelines for incorporating graphics and visuals are featured along with sample planning worksheets and design samples and exercises. (Also available as part of MyTechCommLab.)

ACKNOWLEDGMENTS

Many of the refinements in this and earlier editions were inspired by generous and insightful suggestions from the following reviewers: Mary Beth Bamforth, Wake Technical Community College; Marian G. Barchilon, Arizona State University East; Christiana Birchak, University of Houston, Downtown; Susan L. Booker, Hampden-Sydney College; Gene Booth, Albuquerque Technical Vocational Institute; Alma Bryant, University of South Florida; Joanna B. Chrzanowski, Jefferson Community College; Jim Collier, Virginia Tech; Daryl Davis, Northern Michigan University; Charlie Dawkins, Virginia Polytechnical Institute and State University; Pat Dorazio, SUNY Institute of Technology; Julia Ferganchick-Neufang, University of Arkansas; Clint Gardner, Salt Lake Community College; Mary Frances Gibbons, Richland College; Lucy Graca, Arapahoe Community College; Roger Graves, DePaul University; Baotong Gu, Eastern Washington University; Gil Haroian-Guerin, Syracuse University; Susan Guzman-Trevino, Temple College; Wade Harrell, Howard University; Linda Harris, University of Maryland, Baltimore County; Michael Joseph Hassett, Brigham Young University; Cecilia Hawkins, Texas A & M University; Robert A. Henderson, Southeastern Oklahoma State University; TyAnna K. Herrington, Georgia Institute of Technology; Mary Hocks, Georgia State University; Robert Hogge, Weber State University; Glenda A. Hudson, California State University-Bakersfield; Gloria Jaffe, University of Central Florida; Bruce L. Janoff, University of Pittsburgh; Jeanette Jeneault, Syracuse University; Jack Jobst, Michigan Technical University; Kathleen Kincade, Stephen F. Austin State University; Susan E. Kincaid, Lakeland Community College; JoAnn Kubala, Southwest Texas State University; Karen Kuralt, Louisiana Tech University; Elizabeth A. Latshaw, University of South Florida; Lindsay Lewan, Arapahoe Community College; Sherry Little, San Diego State University; Linda Loehr, Northeastern University; Lisa J. McClure, Southern Illinois University-Carbondale; Devonee McDonald, Kirkwood Community College; James L. McKenna, San Jacinto College; Troy Meyers, California State University, Long Beach; Mohsen Mirshafiei, California State University, Fullerton; Thomas Murphy, Mansfield University; Thomas A. Murray, SUNY Institute of Technology; Shirley Nelson, Chattanooga State Technical

Community College; Gerald Nix, San Juan College; Megan O'Neill, Creighton University; Celia Patterson, Pittsburgh State University; Don Pierstorff, Orange Coast College; Carol Clark Powell, University of Texas at El Paso; Mark Rollins, Ohio University, Athens; Beverly Sauer, Carnegie Mellon University; Carol M. H. Shehadeh, Florida Institute of Technology; Sharla Shine, Terra Community College; Rick Simmons, Louisiana Technical University; Susan Simon, City College of the City University of New York; Tom Stuckert, University of Findlay; Terry Tannacito, Frostburg State University; Anne Thomas, San Jacinto College; Maxine Turner, Georgia Institute of Technology; Mary Beth VanNess, University of Toledo; Jeff Wedge, Embry-Riddle University; Kristin Woolever, Northeastern University; Carolyn Young, University of Wyoming; Stephanee Zerkel, Westark College; Beverly Zimmerman, Brigham Young University; and Don Zimmerman, Colorado State University.

Reviewers of the Tenth Edition

For this edition, I am grateful for the comments of the following people: Beth Camp, Linn-Benton Community College; Madelyn Flammia, University of Central Florida; Robert Henderson, Southeastern Oklahoma State University; Mitchell H. Jarosz, Delta College; Christopher Keller, University of Hawaii at Hilo; Thomas LaJeunesse, University of Minnesota Duluth; Lindsay Lewan, Arapahoe Community College; Tom Long, Thomas Nelson Community College; Michael McCord, Minnesota State University, Moorhead; Lisa McNair, Georgia Institute of Technology; Roxanne Munch, Joliet Junior College; Peter Porosky, Johns Hopkins University; Jan Schlegel, Tri-State University; Lauren Sewell Ingraham, University of Tennessee at Chattanooga; Christian Weisser, Florida Atlantic U. Honors College.

At the University of Massachusetts, Raymond Dumont was a steady source of help and ideas. Many other colleagues, graduate students, teaching assistants, and alumni offered helpful suggestions. As always, students gave me feedback and inspiration.

This Tenth Edition is the product of exceptional guidance and support from Ginny Blanford, Rebecca Gilpin, Michael Greer, Joe Opiela, Ellen MacElree, George Pullman, and Janet Nuciforo. Thank you all for sharing your talents so generously.

Special thanks to those who help me keep going: Chega, Daniel, Sarah, Patrick, and Zorro.

John M. Lannon

Introduction to Technical Communication

1

TECHNICAL COMMUNICATION IS USER-CENTERED

TECHNICAL COMMUNICATION IS EFFICIENT

TECHNICAL COMMUNICATION COMES IN ALL SHAPES AND SIZES

TECHNICAL COMMUNICATORS EMPLOY A BROAD ARRAY OF SKILLS

TECHNICAL COMMUNICATION IS PART OF MOST CAREERS

COMMUNICATION HAS BOTH AN ELECTRONIC AND A HUMAN SIDE

COMMUNICATION REACHES A DIVERSE AUDIENCE

CONSIDER THIS Twenty-First Century Jobs Require Portable Skills

T hanks to the revolution in electronic communication, the industrial age has given way to the *information age*. Instead of machines and physical goods, "information" and ideas have become our most prized commodities:

Information is the ultimate product

> The new source of wealth is not material; it is information, knowledge applied to work to create value. The pursuit of wealth is now largely the pursuit of information. (Wriston 8)

But information in itself has no value—unless it is *usable*. Usable information enables us to perform complex tasks, solve problems, make decisions, and create ideas. A particular company, for example, might need information like the following (Davenport 39, 61, 136):

Typical information needs in the workplace

1.1

For more on information needs related to bio-terrorism visit <www.ablongman.com/lannonweb>

- What is our competition doing and how should we respond?
- Are customer preferences changing, and if so, how?
- What new government regulations do we need to address?
- How can we design company operations to take advantage of the Internet?
- Should we build, rent, or buy?
- What new technology should our company be thinking about?

In the workplace and beyond, usable information not only helps us use all sorts of technical products, but also helps us understand and respond to complicated technical and societal issues:

Typical information needs in society

- Do the benefits of the Lyme disease vaccine outweigh its risks?
- How do I program my VCR?
- How safe are bioengineered foods?
- How can we protect individual privacy in the electronic age?
- Which brand of computer should I buy?
- How can our community service agency help prevent child abuse?

Usable information is not merely raw data

To answer these questions, Web sites, Intranets, and other resources—online or off—provide all sorts of *data* (measurements, observations, prices, statistics, and other raw facts). But to translate this data into usable information, we have to sift through it and figure out what it means and how it applies. Then we have to build a persuasive but honest case for our interpretation and recommendations. Finally—so that others can use this material—we shape it into some type of *document* (memo, letter, report, manual, online help, email, Web page, or script for an oral presentation). Often we do all this as a part of a team.

Whenever you convey usable information to various people in various situations, you work as a "technical communicator."

TECHNICAL COMMUNICATION IS USER CENTERED

What users
expect

Unlike poetry or fiction or essays, a technical document rarely focuses on the writer's personal thoughts and feelings. This doesn't mean that your document should have no personality (or *voice*), but only that the needs of your audience come first. Users typically are interested in "who you are" only to the extent that they want to know *what you have done, what you recommend,* or *how you speak for your company*. While a user-centered document never makes the writer "disappear," it does focus on what the audience considers most important.

TECHNICAL COMMUNICATION IS EFFICIENT

How workplace
and school
writing differ

Professors read to *test* our knowledge, but colleagues, customers, and supervisors read to *use* our knowledge. In fact, much of your own technical communication may involve translating specialized information for the use of nontechnical audiences, as the Figure 1.1 sample document illustrates.

Nontechnical readers expect a document that won't waste their time and energy. And so do all readers! In the United States especially, people using a technical document often go back and forth: instead of reading from beginning to end, they look up the information they need at that moment. For users to find the information and understand it, a document has to be easy to navigate and straightforward.

 NOTE

For any type of global communication, keep in mind that many cultures consider a direct, straightforward communication style offensive. For more detail, see page 37.

An efficient technical document is carefully designed to include the features such as those displayed in Figure 1.1.

Features of
efficient
documents

- *worthwhile content*—including all (and only) those details users need
- *sensible organization*—guiding the user and emphasizing important material
- *readable style*—promoting fluid reading and accurate understanding
- *effective visuals*—clarifying concepts and relationships, and substituting for words whenever possible
- *accessible design*—providing heads, lists, type styles, and other aids to navigation
- *supplements* (abstract, appendix, glossary, linked pages, and so on)—allowing users with different needs to read only those sections of a long document needed for their specific task

A
communicator's
legal
accountability

User-centered and efficient communication is no mere abstract notion: In the event of a lawsuit, faulty writing is treated like any other faulty product. If your inaccurate, unclear, or incomplete information leads to injury, damage, or loss, you and your company can be held legally responsible.

The topic and sponsoring agency are clearly identified

| United States Environmental Protection Agency | Office of Solid Waste and Emergency Response (5102G) | EPA 542-F-01-001 April 2001 www.epa.gov/superfund/sites www.cluin.org |

&EPA | # A Citizen's Guide to Bioremediation

The introduction offers a clear preview

The Citizen's Guide Series

EPA uses many methods to clean up pollution at Superfund and other sites. Some, like bioremediation, are considered new or innovative. Such methods can be quicker and cheaper than more common methods. If you live, work, or go to school near a Superfund site, you may want to learn more about cleanup methods. Perhaps they are being used or are proposed for use at your site. How do they work? Are they safe? This Citizen's Guide is one in a series to help answer your questions.

Headings aid navigation and are phrased as typical questions users would need answered

What is bioremediation?

Bioremediation allows natural processes to clean up harmful chemicals in the environment. Microscopic "bugs" or microbes that live in soil and groundwater like to eat certain harmful chemicals, such as those found in gasoline and oil spills. When microbes completely digest these chemicals, they change them into water and harmless gases such as carbon dioxide.

The illustration provides a clear visual referent for nonspecialists

Microbe eats oil | Microbe digests oil and changes it to water and harmless gases | Microbe releases water and harmless gases into soil or ground

How does it work?

Clear, direct writing explains a complex concept in terms geared to a general audience

In order for microbes to clean up harmful chemicals, the right temperature, nutrients (fertilizers), and amount of oxygen must be present in the soil and groundwater. These conditions allow the microbes to grow and multiply—and eat more chemicals. When conditions are not right, microbes grow too slowly or die. Or they can create more harmful chemicals. If conditions are not right at a site, EPA works to improve them. One way they improve conditions is to pump air, nutrients, or other substances (such as molasses) underground. Sometimes microbes are added if enough aren't already there.

The right conditions for bioremediation cannot always be achieved underground. At some sites, the weather is too cold or the soil is too dense. At such sites, EPA might dig up the soil to clean it above ground where heaters and soil mixing help improve conditions. After the soil is dug up, the proper nutrients are added. Oxygen also may be added by stirring the mixture or by forcing air through it. However, some microbes work better without oxygen. With the right temperature and amount of oxygen and nutrients, microbes can do their work to "bioremediate" the harmful chemicals.

FIGURE 1.1   An Efficient Technical Document   The text, organization, and design of this brief guide work together to make technical information accessible to a general audience.
Source: U.S. Environmental Protection Agency, April 2001. Information available at <www.epa.gov/superfund/sites> or <www.cluin.org>.

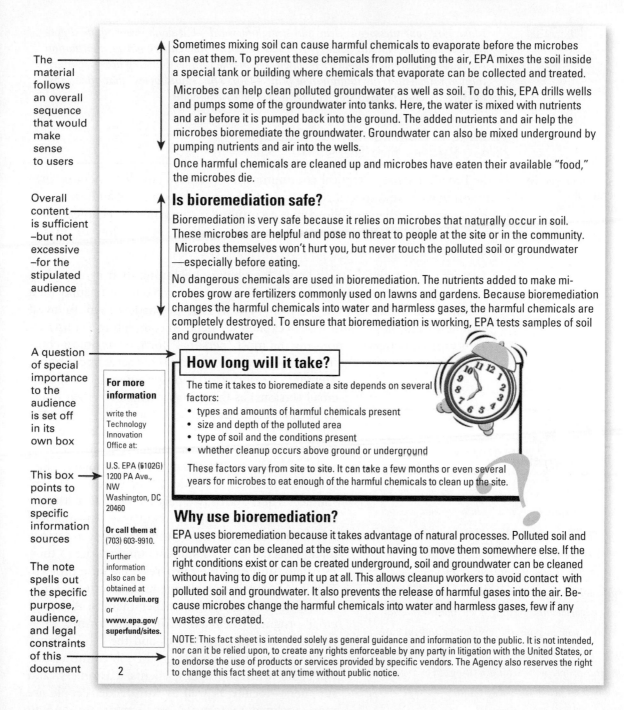

The material follows an overall sequence that would make sense to users

Sometimes mixing soil can cause harmful chemicals to evaporate before the microbes can eat them. To prevent these chemicals from polluting the air, EPA mixes the soil inside a special tank or building where chemicals that evaporate can be collected and treated.

Microbes can help clean polluted groundwater as well as soil. To do this, EPA drills wells and pumps some of the groundwater into tanks. Here, the water is mixed with nutrients and air before it is pumped back into the ground. The added nutrients and air help the microbes bioremediate the groundwater. Groundwater can also be mixed underground by pumping nutrients and air into the wells.

Once harmful chemicals are cleaned up and microbes have eaten their available "food," the microbes die.

Overall content is sufficient —but not excessive —for the stipulated audience

Is bioremediation safe?

Bioremediation is very safe because it relies on microbes that naturally occur in soil. These microbes are helpful and pose no threat to people at the site or in the community. Microbes themselves won't hurt you, but never touch the polluted soil or groundwater —especially before eating.

No dangerous chemicals are used in bioremediation. The nutrients added to make microbes grow are fertilizers commonly used on lawns and gardens. Because bioremediation changes the harmful chemicals into water and harmless gases, the harmful chemicals are completely destroyed. To ensure that bioremediation is working, EPA tests samples of soil and groundwater

A question of special importance to the audience is set off in its own box

How long will it take?

The time it takes to bioremediate a site depends on several factors:

- types and amounts of harmful chemicals present
- size and depth of the polluted area
- type of soil and the conditions present
- whether cleanup occurs above ground or underground

These factors vary from site to site. It can take a few months or even several years for microbes to eat enough of the harmful chemicals to clean up the site.

For more information

write the Technology Innovation Office at:

U.S. EPA (5102G) 1200 PA Ave., NW Washington, DC 20460

Or call them at (703) 603-9910.

Further information also can be obtained at **www.cluin.org** or **www.epa.gov/ superfund/sites.**

This box points to more specific information sources

The note spells out the specific purpose, audience, and legal constraints of this document

Why use bioremediation?

EPA uses bioremediation because it takes advantage of natural processes. Polluted soil and groundwater can be cleaned at the site without having to move them somewhere else. If the right conditions exist or can be created underground, soil and groundwater can be cleaned without having to dig or pump it up at all. This allows cleanup workers to avoid contact with polluted soil and groundwater. It also prevents the release of harmful gases into the air. Because microbes change the harmful chemicals into water and harmless gases, few if any wastes are created.

NOTE: This fact sheet is intended solely as general guidance and information to the public. It is not intended, nor can it be relied upon, to create any rights enforceable by any party in litigation with the United States, or to endorse the use of products or services provided by specific vendors. The Agency also reserves the right to change this fact sheet at any time without public notice.

2

FIGURE 1.1 An Efficient Technical Document *(continued)*

NOTE

*Make sure your message is clear and straightforward—but don't oversimplify. Information designer Nathan Shedroff reminds us that, while **clarity** makes information easier to understand, **simplicity** is "often responsible for the 'dumbing down' of information rather than the illumination of it" (280). The "sound bytes" that often masquerade as network news reports serve as a good case in point.*

TECHNICAL COMMUNICATION COMES IN ALL SHAPES AND SIZES

Common types of technical communication

In the broadest sense, technical communication includes any document or presentation that conveys specialized information for use by a specific audience—often a nonexpert audience. Following is a sampling of the kinds of technical communication you might encounter or prepare, either on the job or in the community.

- **Letters.**   Most people write letters long before beginning their careers. As a student, for example, you might write to request research data or to apply for a summer internship. On the job, you might write to persuade a client to invest in a new technology venture or to explain the delay in completing a construction project. Letters are not only the most "personal" form of technical communication, but they also provide written records and often serve as contracts.

- **Memos.**   Most organizations use memos (and their email versions) as the primary vehicle for their internal written communication. Unlike a conversation, a memoleaves a "paper trail" for future reference—requests, recommendations, directives, instructions, and so on. Memos are usually brief and follow a format that includes a header ("To," "From," "Date," "Subject") and one or two pages of body text. Memos cover just about any topic and purpose: An employee might write to her division head requesting assistance on a project; a team of students might write to their instructor outlining their progress on a term project.

- **Email.**   Email, basically a memo in electronic form, is used more widely than paper memos. A typical email program offers a built-in memo format, automatically inserting the "Date" and the "To," "From," and "Subject" lines. People on the job communicate via email with colleagues, clients, customers, and suppliers—locally as well as worldwide. People are more inclined to forward email messages, and to write more informally and hastily than they would with paper memos.

Types of writing

"Writing is essential to my work. Everything we do [at my company] results in a written product of some kind—a formal technical report, a summary of key findings, recommendations and submissions to academic journals or professional associations. We also write proposals to help secure new engagements. Writing is the most important skill we seek in potential employees and nurture and reward in current employees. It is very hard to find people with strong writing skills, regardless of their academic background."

—Paul Harder, President, mid-sized consulting firm

- **Brochures, Pamphlets, and Fact Sheets.** These brief documents are often designed for public consumption. For example, to market goods or services, companies produce brochures containing product descriptions. Professional organizations, such as the American Medical Association, produce brochures and pamphlets defining various medical conditions, explaining the causes and describing available treatments. Government agencies provide fact sheets (as in Figure 1.1) that offer technical definitions and descriptions on all sorts of topics, ranging from mad cow disease, to stem cells, to bioterrorism.

- **Instructional Material.** Instructions explain the sequence of steps or required course of action for completing a specific task, such as how to program a DVD player or how to install system software. Instructions come in various formats: Brief *reference cards* often fit on a single page; *brochures* can be mailed or handed out; book-length *manuals* accompany complex products; *online help* is built right into the computer application; *hyperlinked pages* offer various levels and layers of information. The more that people rely on complex technology, the more they need usable instructions.

- **Reports.** Reports, both short and long, provide a basis for informed decisions on matters ranging from the best recruit to hire for management training to the most economical cars to lease for the company's fleet. Some reports are strictly informative ("The Causes of Our Company's Network Crash"); other reports recommend solutions to urgent problems ("Recommended Security Measures for Airline Safety"); still others have an overtly persuasive goal, advocating a particular position or course of action ("Why Voters Should Reject the Nuclear Waste Storage Facility Proposed for Our County").

- **Proposals.** A proposal presents a strategy for solving a particular problem. Proposals attempt to persuade readers to improve conditions, accept a service or product, provide

ON THE JOB...

Types of writing

"I do 'Social History' assessments on all new patients that are admitted to the unit. This includes my written assessment (based on an interview of the patient) of why the patient has been admitted, a brief psychosocial history of the patient (previous hospitalizations, family history of mental illness or substance abuse, any history of physical or sexual abuse, and the family and cultural dynamics), a description of the patient's educational background, and a final evaluation of what issues I feel need to be treated or focused upon during the patient's stay (usually including aftercare plans, discharge planning, individual and group therapy)."

—Emma Bryant, social worker

Types of writing

"I generate emails continually. These include status reports, replies to queries, requests for missing information or outstanding materials due, jokes, etc. I sometimes have to write up instructions for media projects. I have to write bids. I write up invoices. I sometimes have to write for returns for equipment that isn't up to expectations. I sometimes have to write introductory messages to try to get new sales. Sometimes I have to write copy for Web sites."

—Lorraine Patsco, Director of Prepress
and Multimedia Production

research funding, or otherwise support a plan of action. Proposals are sometimes written in response to requests for proposals (RFPs). For example, a community may seek to expand its middle school or the Defense Department may wish to develop an intensive training program for airport baggage screeners. These organizations would issue RFPs, and each interested vendor would prepare a proposal that examines the problem, presents a solution, and stipulates a fee for carrying out the project.

This listing is by no means exhaustive. Each profession has its own specific formats for communicating, and many organizations use prepared forms for much of their internal communication. Moreover, most versions of paper-based communication can also be adapted to other media:

Other media for technical communications

- CD-ROM
- hyperlinked Web pages
- intranet Web pages (an organization's internal network)
- email attachments
- online help systems
- ebooks
- training sessions or oral presentations
- instructional videos

NOTE   *Despite stunning advances in electronic communication, paper is by no means disappearing from today's workplace. According to research firm IDC, the 1.49 trillion printed pages in 2002 will increase to 1.84 trillion pages in 2006 (Grimes).*

TECHNICAL COMMUNICATORS EMPLOY A BROAD ARRAY OF SKILLS

What technical communicators do

"Full-time" technical communicators serve many roles.[1] Trade and professional organizations employ technical communicators to produce newsletters, pamphlets, journals, and public relations material. Many work in business and industry, preparing instructional material, reports, proposals, and scripts for industrial

---

[1]My thanks to Pamela Herbert for this overview.

films. They also prepare sales literature, publicity releases, handbooks, catalogs, brochures, Web pages, intranet content, articles, speeches, and oral and multimedia presentations. To reduce costs and to speed production, technical communicators increasingly serve as "desktop publishers," using software such as *PageMaker* or *Quark* to design, illustrate, lay out, and print a finished publication—the kind of specialized tasks done in the past by compositors, typesetters, and professional illustrators.

Besides writing, technical communication specialists do other work. For example, they conduct research, help develop Web sites, and edit reports for punctuation, grammar, style, and logical organization. They oversee publishing projects, coordinating the efforts of writers, visual artists, graphic designers, content experts, and lawyers to produce a complex manual or proposal.

Types of writing

"Writing is probably 30 to 40 percent of my job, with editing taking up another 50 percent, and training and tutoring accounting for the remainder. I oversee semitechnical reports from my managers to upper managers and to our military sponsors—usually progress reports ranging from one to twenty pages; I also edit some sections of highly technical, engineer-to-engineer reports. The engineers and scientists write the body, and I handle the abstract, introduction, executive summary, acknowledgments, conclusion, and list of references, and I sometimes add figures and tables. I also write general reports—things like articles for the company newsletter—and training materials for engineers. I'm currently writing materials on grammar, audience analysis, and techniques for oral presentation."

—Bill Trippe, Communications Specialist
with military contract company

Related career paths

Given their broad range of skills, technical communicators often enter related fields such as publishing, magazine editing, radio, television, and college teaching.

TECHNICAL COMMUNICATION IS PART OF MOST CAREERS

Whatever your job description, expect to be evaluated, at least in part, on your communication skills. At one IBM subsidiary, for example, 25 percent of an employee's evaluation is based on how effectively that person shares information (Davenport 99). Even if you don't anticipate a "writing" career, expect to be a "part-time" technical communicator, who will routinely face situations like these:

How various professionals serve as part-time technical communicators

- As a medical professional, psychologist, social worker, or accountant, you will keep precise records that are crucial to patient or client welfare, and, increasingly, a basis for legal action.
- As a scientist, you will report on your research and explain its significance.

- As a manager, you will write memos, personnel evaluations, inspection reports, and give oral presentations.
- As a lab or service technician, you will keep daily activity records and help train coworkers in using, installing, or servicing equipment.
- As an attorney, you will research and interpret legal issues for clients.
- As an engineer or architect, you will collaborate with colleagues in related fields before presenting a proposal to your client. (For example, an architect's plans are reviewed by a structural engineer who certifies that the design is sound.)

The more you advance in your field, the more you share information and establish human contacts. Managers and executives, for example, spend much of their time negotiating, setting policies, and promoting their ideas—often among diverse cultures around the globe. In short, the higher your career goals, the more critical is your need to communicate.

NOTE   *Instead of joining the corporate ranks, you might decide to work in the nonprofit sector, say, for an environmental group such as the Sierra Club or a community service agency such as the United Way or Head Start, the preschool program for disadvantaged children. Or you might work as an intern or volunteer in these organizations. Whatever the setting, your writing will serve the community—say, in a brochure for public outreach, or a grant request for state funding, or a handbook for clients. In short, technical communication is not merely an instrument for financial profit: It can also serve the good of society. To explore this vital dimension, see the Service-Learning Project at the end of most chapters.*

1.2

For more on effective electronic collaboration visit <www.ablongman.com/lannonweb>

The rise of information technology

Limitations of information technology

COMMUNICATION HAS BOTH AN ELECTRONIC AND A HUMAN SIDE

Electronic mail, instant messaging, fax, teleconferencing, videoconferencing, Internet chat rooms, hypertext, multimedia—these and other resources, collectively known as *information technology* (IT)—enhance the speed, volume, and ways of transmitting information. Electronically mediated communication can reach a limitless audience instantly and globally, and can solicit immediate feedback.

Despite the tremendous advantages IT gives today's communicators, their information still needs to be *written*. Also, only humans can give *meaning* to all the information they convey and receive. Information technology, in short, is a tool, not a substitute for human interaction.

People make information meaningful by posing and answering questions no computer can answer.

QUESTIONS ONLY HUMANS CAN ANSWER

- *Which information is most relevant to this situation?*
- *Can I verify the accuracy of this source?*
- *What does this information mean?*
- *What action does it suggest?*
- *How does this information affect me or my colleagues?*
- *With whom should I share it?*
- *How might others interpret this information?*

Today more than ever, people who communicate on the job need to sort, organize, and interpret their material so users can understand it and act on it. With so much information required, and so much available, no one can afford to "let the data speak for themselves."

1.3

For more on global communication visit <www.ablongman.com/lannonweb>

COMMUNICATION REACHES A DIVERSE AUDIENCE

Electronically linked, our global community shares social, political, and financial interests. Multinational corporations often use parts manufactured in one country and shipped to another for assembly into a product to be marketed elsewhere. Cars may be assembled in the United States for a German automaker, or farm equipment manufactured in East Asia for a U.S. company. Research crosses national boundaries, and professionals transact across cultures with documents like these (Weymouth 143):

Documents that address global audiences

- scientific reports and articles on AIDS and other diseases
- studies of global pollution and industrial emissions
- specifications for hydroelectric dams and other engineering projects
- operating instructions for appliances and electronic equipment
- catalogs, promotional literature, and repair manuals
- contracts and business agreements

To connect diverse communities, any document must convey respect not only for language differences, but also for cultural differences:

How cultures shape communication styles

Our accumulated knowledge and experiences, beliefs and values, attitudes and roles—in other words, our cultures—shape us as individuals and differentiate us as a people. Our cultures, inbred through family life, religious training, and educational and work experiences . . . manifest themselves . . . in our thoughts and feelings, our actions and reactions, and our views of the world.

Most important for communicators, our cultures manifest themselves in our information needs and our styles of communication . . . our expectations as to how information should be organized, what should be included in its content, and how it should be expressed. (Hein 125)

For more on cross-cultural communication visit <www.ablongman.com/lannonweb>

How various cultures view U.S. communication style

Cultures differ over which behaviors seem appropriate for social interaction, business relationships, contract negotiation, and communication practices. An effective communication style in one culture may be offensive elsewhere. One survey of top international executives reveals the following attitudes toward U.S. communication style (Wandycz 22–23):

- *Latin America:* "Americans are too straightforward, too direct."
- *Eastern Europe:* "An imperial tone . . . It's always about how [Americans] know best."
- *Southeast Asia:* "To get my respect, American business[people] should know something about [our culture]. But they don't."
- *Western Europe:* "Americans miss the small points."
- *Central Europe:* "Americans tend to oversell themselves."

In short, global communication requires documents that achieve "efficiency" without being offensive. For more discussion, see pages 56, 59, 104, 282.

CONSIDER THIS   Twenty-First Century Jobs Require Portable Skills

A central theme in today's workplace is "nothing lasts forever." High-tech and dotcom companies emerge and vanish overnight. Even large, established companies expanding at one moment may be "downsizing" at the next.

To lower their costs and remain flexible amid rapidly shifting conditions, employers increasingly offer jobs that are temporary: for contract workers, part-timers, consultants, and the like (Jones 51). Instead of joining a company, climbing through the ranks, and retiring with a gold watch and a comfortable pension, today's college graduate can look forward to a series of employers—and careers.

UC-San Francisco researchers Yelin and Trupin recently found that only 33 percent of the California workforce held "traditional," permanent, full-time jobs, only 22 percent of them having held these jobs for three years or more (cited in Koretz 32). As of early 2002, workers nationwide had been with their current employer an average of 3.7 years, according to the Bureau of Labor Statistics.

No longer based on seniority, your job security in the twenty-first century will depend on skills you can carry from one employer to another—no matter what the job (Peters 172; Task Force 19):

- *Can you write and speak effectively?*
- *Can you research information, verify its accuracy, figure out what it means, and shape it for the user's specific purposes?*
- *Can you work on a team, with people from diverse backgrounds?*
- *Can you get along with, listen to, and motivate others?*
- *Are you flexible enough to adapt to rapid changes in business conditions and technology?*
- *Can you market yourself and your ideas persuasively?*
- *Are you ready to pursue lifelong learning and constant improvement?*

These, in short, are among the *portable skills* employers seek in today's college graduates—skills all related to communication.

EXERCISES For more exercises, visit
<www.ablongman.com/lannon>

1. Research the kinds of communicating you will do in your career. (Begin with the *Dictionary of Occupational Titles* in your library or on the Web.) You might interview a member of your chosen profession. What kinds of documents and presentations will you produce, and for what audiences and purposes? What types of global audiences can you expect? Explain in a memo to your instructor. (See pages 386, 387 for memo elements and format.)

2. Write a memo to your boss, justifying reimbursement for this course. Explain how the course will help you become more effective on the job.

3. Locate a Web site for a company or organization that hires graduates in your major. In addition to technical knowledge, what skills does this company seek in its job candidates? Discuss your findings in class. Also, trace the sequence of links you followed to reach your topic.

COLLABORATIVE PROJECT

Introducing a Classmate

Class members will work together often this semester. So that everyone becomes acquainted, your task is to introduce to the class the person seated next to you. (That person, in turn, will introduce you.) Follow this procedure:

a. Exchange with your neighbor whatever personal information you think the class needs: background, major, career plans, communication needs of your intended profession, and so on. Each person gets five minutes to tell her or his story.

b. Take careful notes; ask questions if you need to.

c. Take your notes home and select only what you think the class will find useful.

d. Prepare a one-page memo telling your classmates who this person is. (See pages 386, 387 for memo elements and format.)

e. Ask your neighbor to review the memo for accuracy; revise as needed.

f. Present the class with a two-minute oral paraphrase of your memo, and submit a copy of the memo to your instructor.

SERVICE-LEARNING PROJECT

Identify a community service agency in your area that needs to have one or more documents written. Start by looking in the yellow pages under "Social and Human Services" or "Environmental Organizations." Or look through your campus directory for campus service agencies such as the Writing and Reading Center, Health Services, International Student Services, Women's Resource Center, or Career Resources Center. Then narrow your list to one or two agencies that interest you. Explore the kinds of documents and publications that agency produces and then write a one-page memo reporting your findings to your classmates.

PART I

Communicating in the Workplace

Report Pinpoints Cause of *Columbia* Shuttle Crash

On February 1, 2003, near the end of mission STS-107, NASA Space Shuttle *Columbia* disintegrated on reentry into the Earth's atmosphere. All seven astronauts were killed in the accident. On August 26, 2003, the *Columbia* Accident Investigation Board (CAIB) presented to the press the conclusions of its seven-month investigation into the cause of the accident. In four words: "The foam did it."

As the Board members explained, a number of experiments concluded that a small piece of insulating foam, falling from the external fuel tank at about 500 mph, struck the leading edge of *Columbia*'s wing with sufficient force to cause a "breach" in the wing's surface. During reentry, superheated gases penetrated the hole in the wing and caused structural damage that eventually caused the entire shuttle to disintegrate.

The causes of the 2003 shuttle tragedy were vastly different from those behind the 1986 *Challenger* explosion. At the same time, a number of cultural, budgetary, and decision-making factors were outlined in the report. Similar foam incidents had happened on earlier flights. How were the signals missed? Like many organizations, NASA again finds itself needing to confront serious issues of culture and communications. ■

2

Preparing an Effective Technical Document

COMPLETE THE KEY TASKS

RELY ON CREATIVE AND CRITICAL THINKING

GUIDELINES for Writing with a Computer

MAKE PROOFREADING YOUR FINAL STEP

GUIDELINES for Proofreading

CONSIDER THIS Workplace Settings Are Increasingly "Virtual"

All professionals specialize in solving problems (how to repair equipment, how to improve a product, how to diagnose an ailment). But whatever your specialty, when you communicate on the job, your main problem is this: "How do I prepare the right document for this situation?"

COMPLETE THE KEY TASKS

To produce an effective document in a workplace setting, you typically need to complete four basic tasks (Figure 2.1):

What workplace communicators need to do

- **Deliver the essential information**—because different people in different situations have different information needs.
- **Make a persuasive case**—because people often disagree about what the information means and what action should be taken.
- **Weigh the ethical issues**—because the interests of your employer may conflict with the interests of other people involved.
- **Work in teams**—because this is how roughly 90 percent of U.S. workers spend some part of their day ("People" 57).

**FIGURE 2.1
How an
Effective
Document Is
Produced**

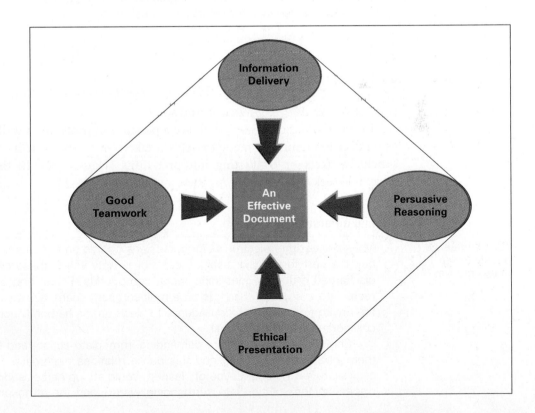

The scenarios that follow illustrate how a typical professional confronts these tasks in her own workplace communication.

Delivering the Essential Information

"Can I provide exactly what users need?"

Sarah Burnes was hired two months ago as a chemical engineer for Millisun, a leading maker of cameras, multipurpose film, and photographic equipment. Sarah's first major assignment is to evaluate the plant's incoming and outgoing water. (Waterborne contaminants can taint film during production, and the production process itself can pollute outgoing water.) Management wants an answer to this question: How often should we change water filters? The filters are expensive and hard to change, halting production for up to a day at a time. The company wants as much "mileage" as possible from these filters, without incurring government fines or tainting its film production.

Sarah will study endless printouts of chemical analysis, review current research and government regulations, do some testing of her own, and consult with her colleagues. When she finally decides on what all the data mean, Sarah will prepare a recommendation report for her bosses.

Later, she will collaborate with the company training manager and the maintenance supervisor to prepare a manual, instructing employees how to check and change the filters. To cut publishing costs, the company has asked Sarah to design and produce this manual using its desktop publishing system. ❏

Sarah's report, above all, needs to be accurate; otherwise, the company gets fined or lowers production. Once she has processed all the information, she faces the problem of giving users what they need: *How much explaining should I do? How will I organize? Do I need visuals?* And so on.

In other situations, Sarah will face a persuasion problem as well: for example, when decisions must be made or actions taken on the basis of incomplete or inconclusive facts or conflicting interpretations (Hauser 72). In these instances, Sarah will seek consensus for *her* view.

Making a Persuasive Case

"Can I influence people to see things my way?"

Millisun and other electronics producers are located on the shores of a small harbor, the port for a major fishing fleet. For twenty years, these companies have discharged effluents containing metal compounds, PCBs, and other toxins directly into the harbor. Sarah is on a multicompany team, assigned to work with the Environmental Protection Agency to clean up the harbor. Much of the team's collaboration occurs via email.

Enraged local citizens are demanding immediate action, and the companies themselves are anxious to end this public relations nightmare. But the team's analysis reveals that any type of cleanup would stir up harbor sediment, possibly dispersing the solution into surrounding waters and the atmosphere. (Many of the contaminants can be airborne.) Premature action might actually *increase* danger, but team members disagree on the degree of risk and on how to proceed.

Sarah's communication here takes on a persuasive dimension: She and her team members first have to resolve their own disagreements and produce an environmental impact report that reflects the team's consensus. If the report recommends further study, Sarah will have to justify the delays to her bosses and the public relations office. She will have to make people understand the dangers as well as she understands them. ❏

In the above situation, the facts are neither complete nor conclusive, and views differ about what these facts mean. Sarah will have to balance the various political pressures and make a case for *her* interpretation. Also, as company spokesperson, Sarah will be expected to protect her company's interests. Some elements of Sarah's persuasion problem: *Are other interpretations possible? Is there a better way? Can I expect political or legal fallout?*

Sarah also will have to reckon with the ethical implications of her writing, with the question of "doing the right thing." For instance, Sarah might feel pressured to overlook or sugarcoat or suppress facts that would be costly or embarrassing to her company.

Weighing the Ethical Issues

"Can I be honest and still keep my job?"

To ensure compliance with OSHA[1] standards for worker safety, Sarah is assigned to test the air purification system in Millisun's chemical division. After finding the filters hopelessly clogged, she decides to test the air quality and discovers dangerous levels of benzene (a potent carcinogen). She reports these findings in a memo to the production manager, with an urgent recommendation that all employees be tested for benzene poisoning. The manager phones and tells Sarah to "have the filters replaced," but says nothing at all about her recommendation to test for benzene poisoning. Now Sarah has to decide what to do about this lack of response: Assume the test is being handled, and bury the memo in some file cabinet? Raise the issue again, and risk alienating her boss? Send copies of her original memo to someone else who might take action? ❏

2.1
For more on ethics in technical communication visit <www.ablongman.com/ lannonweb>

Situations that compromise truth and fairness present the hardest choices of all: Remain silent and look the other way or speak out and risk being fired. Some elements of Sarah's ethics problem: *Is this fair? Who might benefit or suffer? What other consequences could this have?*

In addition to solving these various problems, Sarah has to reckon with the implications of working in a team setting: Much of her writing will be produced in collaboration with others (editors, managers, graphic artists), and her audience will extend beyond her own culture.

---

[1]Occupational Safety and Health Administration.

Working on a Team

Recent mergers have transformed Millisun into a multinational corporation with branches in eleven countries, all connected by an intranet. Sarah can expect to collaborate with coworkers from diverse cultures on research and development and with government agencies of the host countries on safety issues, patents and licensing rights, product liability laws, and environmental concerns.

In order to standardize the sensitive management of the toxic, volatile, and even explosive chemicals used in film production, Millisun is developing automated procedures for quality control, troubleshooting, and emergency response to chemical leakage. Sarah has been assigned to a team that is preparing computer-based training packages and instructional videos for all personnel involved in Millisun's chemical management worldwide. ❑

As a further complication, Sarah will have to develop working relationships with people she has never met face-to-face, people from other cultures, people she knows only via an electronic medium.

For Sarah Burnes, or any of us, writing is a process of *discovering* what we want to say, "a way to end up thinking something [we] couldn't have started out thinking" (Elbow 15). Throughout this process in the workplace, we rarely work alone, but instead collaborate with others for information, help in writing, and feedback (Grice, "Document" 29–30). We must satisfy not only our audience, but also our employer, whose goals and values ultimately shape the document (Selzer 46–47). Almost any document for people outside our organization will be *reviewed* for accuracy, appropriateness, usefulness, and legality before it is finally approved (Kleimann 521).

RELY ON CREATIVE AND CRITICAL THINKING

In *creative thinking*, we explore new ideas; we build on information; we devise better ways of doing things. (For example, "How do we get as much mileage as possible from our water filters?")

In *critical thinking*, we test the strength of our ideas or the worth of our information. Instead of accepting an idea at face value, we examine, evaluate, verify, analyze, weigh alternatives, and consider consequences—at every stage of that idea's development. We employ critical thinking to examine our evidence and our reasoning, to discover new connections and new possibilities, and to test the effectiveness and the limits of our solutions.

We apply creative and critical thinking throughout the four stages in the *writing process*:

1. Gather and evaluate ideas and information.
2. Plan the document.

3. Draft the document.

4. Revise the document.

One engineering professional describes how creative and critical thinking enrich every stage of the writing process:

Writing sharpens thinking

Good writing is a process of thinking, writing, revising, thinking, and revising, until the idea is fully developed. An engineer can develop better perspectives and even new technical concepts when writing a report of a project. Many an engineer, at the completion of a laboratory project, senses a new interpretation or sees a defect in the results and goes back to the laboratory for additional data, a more thorough analysis, or a modified design. (Franke 13)

As the arrows in Figure 2.2 indicate, no one stage of the writing process is complete until all stages are complete. Figure 2.3 lists the kinds of questions we answer at various stages. On the job, we must often complete these stages under deadline pressure. Like the exposed tip of an iceberg, the finished document provides the only visible evidence of our labor.

**FIGURE 2.2
The Writing
Process for
Technical
Documents**

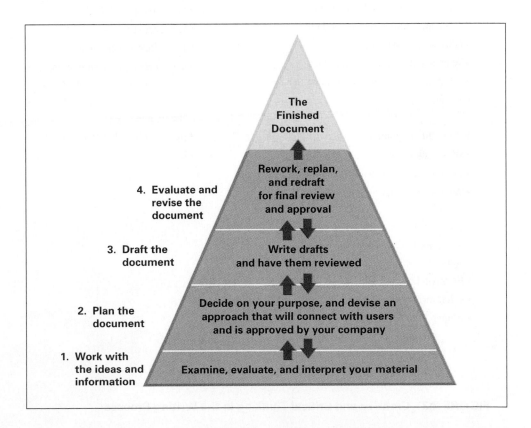

The
Finished
Document

Rework, replan,
and redraft
for final review
and approval

4. Evaluate and
revise the
document

Write drafts
and have them reviewed

3. Draft the
document

Decide on your purpose, and devise an
approach that will connect with users
and is approved by your company

2. Plan the
document

1. Work with
the ideas and
information

Examine, evaluate, and interpret your material

1. Work with the ideas and information:

• Have I defined the problem accurately?

• Is the information complete, accurate, reliable, and unbiased?

• Can it be verified?

• How much of it is useful?

• Do I need more information?

• What do these facts mean?

• What connections seem to emerge?

• Do the facts conflict?

• Are other interpretations or conclusions possible?

• Is a balance of viewpoints represented?

• What, if anything, should be done?

• Is the information honest and fair?

• Is there a better way?

• What are the risks and benefits?

• What other consequences might this have?

• Should I reconsider?

2. Plan the document:

• When is it due?

• What do I want it to do?

• Who is my audience, and why will they use it?

• What do they need to know?

• What are the "political realities" (feelings, egos, cultural differences, and so on)?

• How will I organize?

• What format and visuals should I use?

• Whose help will I need?

3. Draft the document:

• How do I begin, and what comes next?

• How much is enough?

• What can I leave out?

• Am I forgetting anything?

• How will I end?

• Who needs to review my drafts?

4. Evaluate and revise the document:

• Is the draft usable?

• Does it do what I want it to do?

• Is the content worthwhile?

• Is the organization sensible?

• Is the style readable?

• Is everything easy to find?

• Is the format appealing?

• Is everything accurate, complete, appropriate, and correct?

• Who needs to review and approve the final version?

• Does it advance my organization's goals?

• Does it advance my audience's goals?

FIGURE 2.3   Creative and Critical Thinking in the Writing Process

NOTE *Revising a draft doesn't always guarantee that you will improve it. Save each draft and then compare them to select the best material from each one.*

Computers, of course, are essential tools in the writing process. The following guidelines will help you capitalize on all the benefits a computer has to offer.

2.2

For more electronic writing resources visit <www.ablongman.com/lannonweb>

GUIDELINES for Writing with a Computer

1. *Beware of computer junk.* The ease of cranking out words on a computer can produce long, windy pieces that say nothing. Cut anything that fails to advance your meaning. (See pages 253–60 for ways to achieve conciseness.)

2. *Never confuse style with substance.* Laser printers and choices of typefaces, type sizes, and other design options can produce attractive documents. But not even the most engaging design can redeem a document with worthless or inaccessible content.

3. *Save and print often.* Save each paragraph as you write it; print out each page as you complete it; and keep a copy on a backup disk.

4. *Revise on hard copy.* Nothing beats scribbling on the printed page. The hard copy provides the whole text, right in front of you.

5. *Don't rely only on computerized writing aids.* A synonym found in an electronic thesaurus may distort your meaning. Spell checkers can root out incorrectly spelled words but not incorrectly *used* words such as "their," "they're," and "there" or "it's" versus "its." And grammar checkers often give bizarre or inaccurate advice. Page 284 summarizes the limitations of computerized aids. In the end, nothing substitutes for your own careful reading.

6. *Keep a different file for each draft.* Revision hardly ever occurs in a neat sequence ("good," "better," "best"). Sometimes parts of an earlier draft actually are better than something you've rewritten. Give each file a different name ("Draft #1," "Draft #2,"), in case you need to retrieve good, usable data.

7. *Select a design and a medium that your audience favors.* Should the document be primarily verbal, visual, or some combination? Should it travel by conventional mail, interoffice mail, or email? What would *these* users prefer in this situation—the solid feel of paper or the "hi-tech" lure of a computer screen? Younger audiences tend to like flashy graphics; older audiences prefer traditional text; and people in general trust printed text more than images (Horton, "Mix Media" 781).

MAKE PROOFREADING YOUR FINAL STEP

No matter how attractive and informative the document, basic errors annoy the user and make the writer look bad (including on various drafts that are being reviewed by colleagues). Proofreading detects easily correctable errors such as these:

Errors to look for during proofreading

- *Sentence errors* such as fragments, comma splices, or run-ons (see page 752)
- *Punctuation errors* such as missing apostrophes or excessive commas (see page 762)
- *Usage errors* such as "it's" for "its", "lay" for "lie," or "their" for "there" (see page 774)
- *Mechanical errors* such as misspelled words, inaccurate dates, or incorrect abbreviations (see page 773)
- *Format errors* such as missing page numbers, inconsistent spacing, or incorrect form of documenting sources (see page 344)
- *Typographical errors* (typos) such as repeated or missing words or letters, missing word endings (say, *-s* or *-ed* or *-ing*), or a left-out quotation mark or parenthesis (see page 284)

Refer to the page numbers in parentheses for advice on repairing these errors.

GUIDELINES for Proofreading

1. *Save it for the draft(s) others will read.* Proofreading the versions that only you will see might cause writer's block and distract you from the "rhetorical features" (content, organization, style, and design).

2. *Take a break before proofreading.* After you complete the piece, take a walk, take a nap, or whatever.

3. *Work from hard copy.* Research indicates that people read more perceptively (and become less tired) from a printed page than from a computer screen. Also, paper is easier to mark up and scribble on. Some people like to get comfortable or even lie down.

4. *Keep it slow.* Read each word—don't let yourself skim. Force yourself to slow down by sliding a ruler under each line or by moving backward throughout the document, sentence by sentence. For a long document, read only small chunks at one time.

5. *Be especially alert for problem areas in your writing.* Do you confuse commas with semicolons? Do you make typos? If punctuation is a problem, for example, make one final pass to check each punctuation mark.

6. *Proofread more than once.* The more you do it, the more errors you're likely to spot.

CONSIDER THIS Workplace Settings Are Increasingly "Virtual"

Office communication has evolved dramatically, as illustrated in the practices below.

- *Instead of being housed in one location, the virtual company may have branches across the state, the nation, or the world, to which many employees "commute" electronically. These telecommuters include freelance workers who are employed by other companies as well.*

- *Workplace discussions and document sharing occur via email, instant messaging, or videoconferencing. Networked employees worldwide collaborate and converse in real time. Email listservs announce daily developments for employees or readers worldwide such as price and inventory lists, changes or updates in policies or procedures, and press releases.*

- *Employees work and write collaboratively (as in developing a proposal or a marketing plan). Drafts circulated electronically allow colleagues to add comments directly on the manuscript. Multimedia systems present text, graphics, sound, and animated material retrieved from a computer file. Colleagues in any location work on the electronic document and comment on one another's "work in progress."*

- *Online databases store information from books, magazines, newspapers, journals,*

and so forth, and can be searched via the Web for the latest stock market quotations, trends in global weather patterns, sites of recent disease outbreaks, and so on. A single compact disc stores an entire encyclopedia, a medical dictionary, or interactive manuals and lessons.*

- *Instead of relying on secretaries, managers compose their own letters and memoranda for distribution via email to readers across the building or across the globe.*

- *On desktop publishing (DTP) networks, the composition, layout, graphics design, typesetting, and printing of external documents and Web pages are done in-house.*

- *Optical scanners take an electronic snapshot of any paper document produced or received, including incoming mail. Stored online, this image can be retrieved and edited, printed out, faxed or emailed, or posted on an electronic bulletin board or Web page. Company forms (requisitions, accident reports, etc.) can be produced, filled out, filed, updated, and distributed electronically.*

- *Unlike printed texts, which tend to be read front to back, electronic texts are often read nonsequentially (Grice and Ridgway 37). Readers navigate their own paths and choose various routes to explore.*

EXERCISES

For more exercises, visit
<www.ablongman.com/lannon>

1. Assume that a friend in your major thinks that computers have made writing skills obsolete and that anyone with the necessary hardware and software can write and design information without regard to the issues discussed in this chapter. Write your friend a memo based on this chapter explaining why you think these assumptions are mistaken. (See pages 386, 387 for details on memo format.)

Supplement this chapter's information with material from a brief search of Web sites that offer writing advice. Check out the Web sites listed below or use a search engine to locate other relevant sites. Trace the sequence of links you followed to reach your material, and cite each source. (See pages 695, 709 for citation formats for electronic sources.)

- <www.owl.english.purdue.edu> This online writing center offers all kinds of writing help.

- <www.inkspot.com> A good source of useful writing tips.

2. Locate a Web site that describes some form of multinational collaboration to address an environmental threat such as global warming, nuclear accident, deforestation, or species depletion. In a one-page memo, summarize how various cultures are working together to address the problem. (For example, to learn about international cooperation to save fish populations, go to the National Marine Fisheries site at <www.nmfs.noaa.gov>.) Trace the sequence of links you followed to reach your material, and cite each source.

3. As you respond to the following scenario,[2] carefully consider the information, persuasion, and ethical problems involved (and be prepared to discuss them in class).

> You are Manager of Product Development at High-Tech Toys, Inc. You need to send a memo to the Vice President of Information Services, explaining the following:
>
> a. The laser printer in your department is often out of order.
> b. The laser printer is seldom repaired satisfactorily.
> c. Either the machine is faulty or the repairperson is incompetent (but this person always appears promptly and cheerfully when summoned from Corporate Maintenance—and is a single parent raising three young children).
> d. It is difficult to get things done in your department without being able to use the laser printer.
> e. The members of your department share ideas and plans daily.
> f. You want the problem solved—but without getting the repairperson fired.

In your memo, recommend a solution, and briefly justify your recommendation.

COLLABORATIVE PROJECT

An Issue of Ethics

Working in small groups, analyze Sarah Burnes's ethical decisions (page 19). What could happen if Sarah follows her boss's orders? What could happen if she takes no further action? After discussing the issues involved and the possible consequences, try to reach a consensus about what action Sarah should take in this situation. Appoint one member to present your group's conclusion to the class.

SERVICE-LEARNING PROJECT

Social service agencies work toward varied goals. As you scan the list of United Way agencies and others researched by your classmates, can you identify agencies whose goals or values conflict with yours or your family's? If your instructor assigned you to work for one of these agencies, how might you respond?

In a one-page memo to your instructor, summarize the key values and goals of the agency you have researched, and explain how you would or would not make a good "fit" in working with that agency.

---

[2]My thanks to Teresa Pawelcyzk for the original version of this exercise.

3

Delivering the Essential Information

3.1

For more on under-
standing workplace
cultures visit
<www.ablongman.com/
lannonweb>

All technical communication is for people who will use and react to the information. Your task might be to *define* something—as in explaining what "variable
annuity" means for insurance clients; to *describe* something—as in showing an architectural client what a new office building will look like; to *explain* something—as in
telling a stereo technician how to eliminate bass flutter in your company's new line of
speakers. As depicted in Figure 3.1, usable information is based on **audience analysis,**
in which you learn all you can about those who will use your document.

Following are typical audiences you will encounter in the workplace. Granted,
these categories overlap to some extent, and it's impossible to speak of "all scientists" or "all engineers" without stereotyping. But a specific type of audience generally shares specific concerns and information needs (Gurak and Lannon 26).

**Different groups
have different
information
needs and
interests**

- *Scientists* search for knowledge to "understand the world as it is" (Petroski 2).
  Scientists look for at least 95 percent probability that chance played no role in
  a study's outcome. They want to know how well the study was designed and
  conducted and whether its findings can be replicated. Scientists know that
  their answers are never "final," but open-ended and ongoing: What seems
  probable today might be rendered improbable by tomorrow's research.
- *Engineers* rearrange "the materials and forces of nature" to improve the way
  things work (Petroski 1). Engineers solve problems like these: how to erect a
  suspension bridge that withstands high winds, how to design a lighter airplane
  or a smaller pacemaker, how to boost rocket thrust on a space shuttle. The engineer's concern is ultimately with practical applications, with structures and
  materials that are tested for safety and dependability.

- *Executives* focus on decision making. In a global business climate of overnight
  developments (in world markets, political strife, military conflicts, natural disasters) executives must often react on the spur of the moment. In such cases
  they rely on the best information immediately available—even when this information is incomplete or unverified (Seglin 54).

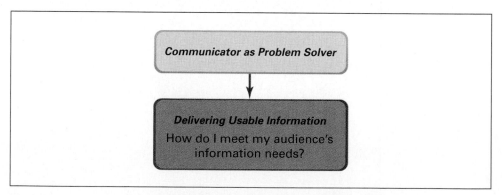

FIGURE 3.1   Communicators Begin by Considering Their Audience

- *Managers* oversee the day-to-day operations of their organization, focusing on problems like these: how to motivate employees, how to increase productivity, how to save money, how to avoid workplace accidents. They collaborate with colleagues and supervise various projects. To keep things running smoothly, managers rely on memos, reports, and other forms of information sharing.
- *Lawyers* focus on protecting the corporation from liability or corporate sabotage by answering questions like these: Do these instructions contain adequate warnings and cautions? Is there anything about this product or document that could generate a lawsuit? Have any of our trade secrets been revealed? Lawyers carefully review documents before approving their distribution outside the company.
- *The public* focuses on the big picture—on what pertains to them directly: What does this mean to me? How can I use this product safely and effectively? Why should I even read this? They rely on information for some immediate practical purpose: to complete a task (What do I do next?), to learn more about something (What are the facts and what do they mean?), to make a judgment (Is this good enough?).

Regardless of category, every audience expects a message tailored for its own specific interests and information needs.

ASSESS THE AUDIENCE'S INFORMATION NEEDS

Usable information connects with the audience by recognizing its unique background, needs, and preferences. The same basic message can be conveyed in different ways for different audiences. For instance, an article describing a new cancer treatment might appear in a medical journal for doctors and nurses. A less technical version might appear in a textbook for medical and nursing students. A more simplified version might appear in *Reader's Digest*. All three versions treat the same topic, but each meets the needs of a different audience.

Because your audience knows less than you, it will have questions.

TYPICAL AUDIENCE QUESTIONS ABOUT WORKPLACE DOCUMENTS

- *What is the purpose of this document?*
- *Why should I read it?*
- *What information can I expect to find here?*
- *What happened, and why?*
- *How should I perform this task?*
- *What action should be taken?*
- *How much will it cost?*
- *What are the risks?*
- *Do I need to respond to this document? If so, how?*

IDENTIFY LEVELS OF TECHNICALITY

When you write for a close acquaintance (coworker, engineering colleague, chemistry professor who reads your lab reports, or supervisor), you adapt your report to that person's knowledge, interests, and needs. But some audiences are larger and less defined (say, for a journal article, a computer manual, a set of first-aid procedures, or an accident report). When you have only a general notion about your audience's background, decide whether your document should be *highly technical,* *semitechnical,* or *nontechnical,* as depicted in Figure 3.2.

**FIGURE 3.2
Deciding on a
Document's
Level of
Technicality**

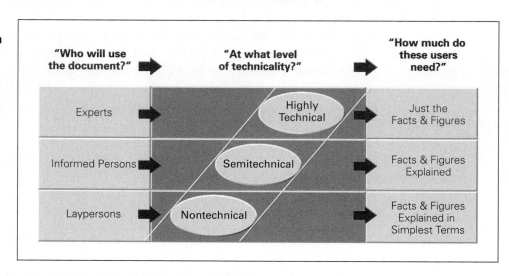

The Highly Technical Document

Users at a specialized level expect the facts and figures they need—without long explanations. In the following description of emergency treatment, an emergency-room physician reports to the patient's doctor, who needs an exact record of symptoms, treatment, and results.

A TECHNICAL VERSION

The patient was brought to the ER by ambulance at 1:00 A.M., September 27, 2002. The patient complained of severe chest pains, dyspnea, and vertigo. Auscultation and EKG revealed a massive cardiac infarction and pulmonary edema marked by pronounced cyanosis. Vital signs: blood pressure, 80/40; pulse, 140/min; respiration, 35/min. Lab: wbc, 20,000; elevated serum transaminase; urea nitrogen, 60 mg%. Urinalysis showed 4+ protein and 4+ granular casts/field, indicating acute renal failure secondary to the hypotension.

WWW

3.2

For more on the languages and cultures of expertise visit <www.ablongman.com/lannonweb>

Expert users need just the facts and figures, which they can interpret for themselves

The patient received 10 mg of morphine stat, subcutaneously, followed by nasal oxygen and 5% D & W intravenously. At 1:25 A.M. the cardiac monitor recorded an irregular sinus rhythm, indicating left ventricular fibrillation. The patient was defibrillated stat and given a 50 mg bolus of Xylocaine intravenously. A Xylocaine drip was started, and sodium bicarbonate administered until a normal heartbeat was established. By 3:00 A.M., the oscilloscope was recording a normal sinus rhythm.

As the heartbeat stabilized and cyanosis diminished, the patient received 5 cc of Heparin intravenously, to be repeated every six hours. By 5:00 A.M. the BUN had fallen to 20 mg% and vital signs had stabilized: blood pressure, 110/60; pulse, 105/min; respiration, 22/min. The patient was now conscious and responsive.

For her expert colleague, this physician defines no technical terms (*pulmonary edema, sinus rhythm*). Nor does she interpret lab findings (4+ *protein, elevated serum transaminase*). She uses abbreviations her colleague understands (*wbc, BUN, 5% D & W*). Because her colleague knows all about specific treatments and medications (*defibrillation, Xylocaine drip*), she doesn't explain their scientific bases. Her report answers concisely the main user questions she anticipates: *What was the problem? What was the treatment? What were the results?*

The Semitechnical Document

One broad class of users has some technical background, but less than the experts. For instance, first-year medical students have specialized knowledge, but less than advanced students. Yet all medical students could be considered semitechnical. Therefore, when you write for a semitechnical audience, identify the *lowest* level of understanding in the group, and write to that level. Too much explanation is better than too little.

The following partial version of the medical report might appear in a textbook for medical or nursing students, in a report for a medical social worker, or in a monthly report for the hospital administration.

Audiences

"Audience makes all the difference. I write for students, small groups of scholars, and general readers. I pitch grant proposals to larger groups of scholars, either nationally (as for the National Endowment for the Humanities) or locally (among colleagues throughout the disciplines at my university). I assume my audiences are happy enough to listen to me at first, but that to keep them reading I need to supply varying degrees of background and explanation pitched to their background and familiarity with the subject matter."

—John Bryant, Professor of English

A SEMITECHNICAL VERSION

**Informed but
nonexpert users
need enough
explanation to
understand what
the data mean**

Examination by stethoscope and electrocardiogram revealed a massive failure of the heart muscle along with fluid buildup in the lungs, which produced a cyanotic **discoloration of the lips and fingertips from lack of oxygen.**

The patient's blood pressure at 80 mm Hg (systolic)/40 mm Hg (diastolic) was **dangerously below its normal measure of 130/70.** A pulse rate of 140/minute was **almost twice the normal rate of 60–80.** Respiration at 35/minute was more than **twice the normal rate of 12–16.**

Laboratory blood tests yielded a white blood cell count of 20,000/cu mm (normal value: 5,000–10,000), **indicating a severe inflammatory response by the heart muscle.** The elevated serum transaminase enzymes (**produced in quantity only when the heart muscle falls**) confirmed the earlier diagnosis. A blood urea nitrogen level of 60 mg% (normal value: 12–16 mg%) **indicated that the kidneys had ceased to filter out metabolic waste products.** The 4+ protein and casts reported from the urinalysis (normal value: 0) **revealed that the kidney tubules were degenerating as a result of the lowered blood pressure.**

The patient immediately received **morphine to ease the chest pain,** followed by **oxygen to relieve strain on the cardiopulmonary system,** and an intravenous solution of **dextrose and water to prevent shock.**

This version explains the raw data (in boldface). Exact dosages are omitted because no one in this audience actually will be treating this patient. Normal values of lab tests and vital signs, however, help readers interpret. (Experts know these values.) Knowing what medications the patient received would be especially important to answering this audience's central question: *How is a typical heart attack treated?*

The Nontechnical Document

3.3

For more techniques for writing to a general audience visit <www.ablongman.com/lannonweb>

People with no training look for the big picture instead of complex details. They expect technical data to be translated into words most people understand. Laypersons are impatient with abstract theories but want enough background to help them make the right decision or take the right action. They are bored or confused by excessive detail, but frustrated by raw facts left unexplained or uninterpreted. They expect to understand the document after reading it only once.

The following nontechnical version of the medical report might be written for the patient's spouse who is overseas on business, or as part of a script for a documentary about emergency room treatment.

A NONTECHNICAL VERSION

Heart sounds and electrical impulses were both abnormal, **indicating a massive heart attack caused by failure of a large part of the heart muscle.** The lungs were swollen with fluid and the lips and fingertips showed **a bluish discoloration from lack of oxygen.**

Laypersons need everything translated into terms they understand

> Blood pressure was **dangerously low, creating the risk of shock.** Pulse and respiration were **almost twice the normal rate, indicating that the heart and lungs were being overworked** in keeping oxygenated blood circulating freely.
>
> **Blood tests** confirmed the heart attack diagnosis and **indicated that waste products usually filtered out by the kidneys were building up in the bloodstream. Urine tests showed that the kidneys were failing as a result of the lowered blood pressure.**
>
> The patient was given **medication to ease the chest pain, oxygen to ease the strain on the heart and lungs, and intravenous solution to prevent the blood vessels from collapsing and causing irreversible shock.**

Nearly all interpretation (in boldface), this version mentions no specific medications, lab tests, or normal values. It merely summarizes events and briefly explains them.

In a different situation, however (say, a malpractice trial), the nontechnical audience might need detailed information about medication and treatment. Such a report would, of course, be much longer—a short course in emergency coronary treatment.

Primary and Secondary Audiences

Whenever you prepare a single document for multiple users, classify your audience as *primary* or *secondary*. Generally, primary users are those who requested the document as a basis for decisions or actions. Secondary users are those who will carry out the project, who will advise the decision makers, or who will be affected by this decision in some way.

Primary and secondary audiences often differ in technical background. When you must write for audiences at different levels, follow these guidelines:

How to tailor a single document for multiple users

- If the document is short (a letter, memo, or anything less than two pages), rewrite it at various levels.
- If the document exceeds two pages, address the primary users. Then provide appendixes for secondary users. Transmittal letters, informative abstracts, and glossaries can also help nonexperts understand a highly technical report. (See pages 643–51 for use and preparation of appendixes and other supplements.)

The document in this next scenario must be tailored for both primary and secondary users.

Tailoring a Single Document for Different Users

Different users have different information needs

You are a metallurgical engineer in an automotive consulting firm. Your supervisor has asked you to test the fractured rear axle of a 2001 Delphi pickup truck recently involved in a fatal accident. Your assignment is to determine whether the fractured axle *caused* or *resulted from* the accident.

After testing the hardness and chemical composition of the metal and examining microscopic photographs of the fractured surfaces (fractographs), you conclude that the fracture resulted from stress that developed *during* the accident. Now you must report your procedure and your findings to a variety of readers.

"What do these findings mean?"

Because your report may serve as courtroom evidence, you must explain your findings in meticulous detail. But your primary users (the decision makers) will be nonspecialists (the attorneys who have requested the report, insurance representatives, possibly a judge and a jury), so you must translate your report, explaining the principles behind the various tests, defining specialized terms such as "chevron marks," "shrinkage cavities," and "dimpled core," and showing the significance of these features as evidence.

"How did you arrive at these conclusions?"

Secondary users will include your supervisor and outside consulting engineers who will be evaluating your test procedures and assessing the validity of your findings. Consultants will be focusing on various parts of your report, to verify that your procedure has been exact and faultless. For this group, you will have to include appendices spelling out the technical details of your analysis: *how* hardness testing of the axle's case and core indicated that the axle had been properly carburized; *how* chemical analysis ruled out the possibility that the manufacturer had used inferior alloys; *how* light-microscopic fractographs revealed that the origin of the fracture, its direction of propagation, and the point of final rupture indicated a ductile fast fracture, not one caused by torsional fatigue. ❏

In the previous situation, primary users need to know *what your findings mean,* whereas secondary users need to know *how you arrived at your conclusions.* Unless it serves the needs of each group independently, your information will be worthless.

Web-Based Documents for Multiple Audiences

Web pages are ideal for packaging and linking various levels of information. Notice how Figure 3.3 accommodates different users at different levels of technicality.

DEVELOP AN AUDIENCE AND USE PROFILE

Focus sharply on your audience by asking the questions below.

QUESTIONS ABOUT A DOCUMENT'S INTENDED AUDIENCE AND USE

- *Who wants the document? Who else will read it?*
- *Why do they want the document? How will they use it?*
- *What is the technical background of the primary audience? Of the secondary audience?*
- *How might cultural differences shape readers' expectations and interpretations?*

- *How much does the audience already know? What material will have informative value?*
- *What exactly does the audience need to know, and in what format? How much is enough?*
- *When is the document due?*

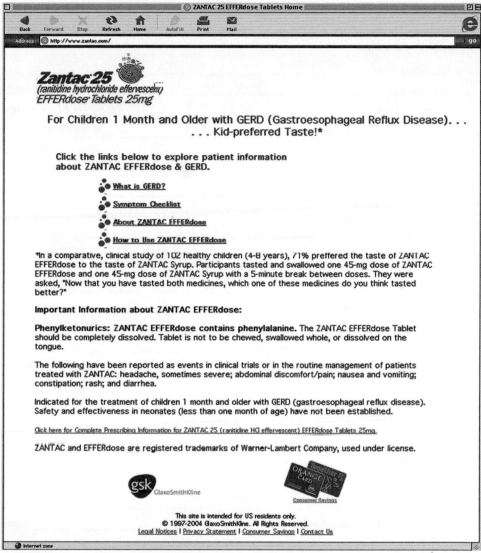

FIGURE 3.3   A Web Page Linked to Various User Needs This home page, designed for a primary audience of laypersons at its top layer, also provides a link titled *Prescribing Information* for a secondary audience for medical experts.

Source: From ZANTAC home page, <www.zantac.com>. Copyright GlaxoSmithKline. Used with permission.

3.4

For additional methods of audience analysis visit <www.ablongman.com/lannonweb>

To answer these questions, consider the suggestions that follow, and use a version of the Audience and Use Profile Sheet shown in Figure 3.4 for all your writing.

Audience Characteristics

Identify the primary audience by name, job title, and specialty (Martha Jones, Director of Quality Control, B.S. and M.S. in mechanical engineering). Are they su-

Audience Identity and Needs

Primary audience: _____ *(name, title)*

Secondary audience: _____

Relationship: _____ *(client, employer, other)*

Intended use of document: _____ *(perform a task, solve a problem, other)*

Prior knowledge about this topic: _____ *(knows nothing, a few details, other)*

Additional information needed: _____ *(background, only bare facts, other)*

Probable questions: _____ *?*

_____ *?*

_____ *?*

_____ *?*

Audience's Probable Attitude and Personality

Attitude toward topic: _____ *(indifferent, skeptical, other)*

Probable objections: _____ *(cost, time, none, other)*

Probable attitude toward this writer: _____ *(intimidated, hostile, receptive, other)*

Organizational climate: _____

Persons most affected by this document: _____ *(cautious, impatient, other)*

Temperament: _____ *(resistance, approval, anger, guilt, other)*

Probable reaction to document: _____

Risk of alienating anyone: _____

Audience Expectations about the Document

Reason document originated: _____ *(audience request, my idea, other)*

Acceptable length: _____ *(comprehensive, concise, other)*

Material important to this audience: _____ *(interpretations, costs*

_____ *conclusions, other)*

Most useful arrangement: _____ *(problem-causes-solutions, other)*

Tone: _____ *(businesslike, apologetic, enthusiastic, other)*

Cultural considerations: _____ *(level of detail or directness, other)*

Intended effect on this audience: _____ *(win support, change behavior, other)*

Due date: _____

FIGURE 3.4   Audience and Use Profile Sheet

periors, colleagues, or subordinates? Are they inside or outside your organization? What is their probable attitude toward this topic? Are they apt to accept or reject your conclusions and recommendations? Will your report convey good or bad news? How might cultural differences affect audience expectations and interpretations?

Identify others who might be interested in or affected by the document, or who will advise the primary audience.

Purpose of the Document

Learn why people want the document and how they will use it. Do they simply want a record of activities or progress? Do they expect only raw data, or conclusions and recommendations as well? Will people act immediately on the information? Do they need step-by-step instructions? Will the document be read and discarded, or filed, published, or distributed electronically? In your audience's view, what is most important about this document? Try asking them directly.

Audience's Technical Background

Colleagues who speak your technical language will understand raw data. Supervisors responsible for several technical areas may want interpretations and recommendations. Managers who have limited technical knowledge expect definitions and explanations. Clients with little or no technical background on your topic want to know what this information means to them, personally (to their health, pocketbook, financial prospects). However, none of these generalizations might apply to your situation. When in doubt, aim for low technicality.

Audience's Cultural Background

Some information needs are culturally determined. Germans, for example, tend to value thoroughness and complexity, with every detail included and explained in a businesslike tone. Japanese generally prefer multiple perspectives on the material, lots of graphics, and a friendly, encouraging tone (Hein 125–26).

Anglo-American business culture generally values plain talk that

Audiences

"I write procedures for technicians who install and service our company's photocopiers and other business machines. When the company comes out with a new machine or a better way of installing or servicing our equipment, all district offices get the technical information, the specifications, and a set of procedures written by the engineers who designed the equipment. I then rewrite the procedures to make them easier for technicians to follow. I also give follow-up training sessions to provide our technicians with hand-on experience."

—Leslie Jacobs, Service representative
and former technician

Audience

"I write for elected officials, public decision-making groups (commissions, boards of directors, etc.), community organizations, foundation staff and directors, and the general public. The audience makes a very important difference in the writing. We adjust the writing style and the content according to the interests of the audience, their technical knowledge, and the amount of detail needed to make a convincing case. Writing over the head of your audience (sometimes known as "talking down") is a big risk and a big mistake. We take a lot of time to understand our audiences before we start any writing assignment."

—Paul Harder, President, mid-sized consulting company

gets right to the point, but Asian cultures consider this rude, preferring indirect and somewhat ambiguous messages, which leave interpretation up to the reader (Leki 151; Martin and Chaney 276–77). To avoid seeming impolite, some readers might hesitate to ask for clarification or additional information. In Asian cultures even disagreement or refusal might be expressed as "We will do our best" or "This is very difficult," instead of "No"—to avoid offending and to preserve harmony (D. Rowland 47).

Also, many U.S. idioms ("breaking the bank," "cutthroat competition," "sticking your neck out") and cultural references ("the crash of '29," "Beantown") make no sense outside of U.S. culture (Coe "Writing" 17–19).[1]

Audience's Knowledge of the Subject

People expect to find something new and useful in your document. Writing has informative value[2] when it does at least one of these things:

A message with "informative value" does one or more of these things

- shares something new and significant
- reminds us about something we know but ignore
- offers fresh insight or perspective on something we already know

In short, informative writing gives people exactly what they need.

People approach most topics with some prior knowledge (or old information). They might need reminding, but they don't need a rehash of old information; they can "fill in the blanks" for themselves. On the other hand, readers don't need every bit of new information you can think of, either.

The more nonessential information people receive, the more likely they are to overlook or misinterpret the important material.

As a reader of this book, for example, you expect an introduction to technical communication, and my purpose is to provide that. Which of these statements would you find useful?

---

[1]For more information on how to analyze business practices in different cultures, consult one of the many available texts on this subject, for example, David A. Victor's *International Business Communication.*
[2]Adapted from James L. Kinneavy's assertion that discourse ought to be unpredictable, in *A Theory of Discourse* (Englewood Cliffs, NJ: Prentice, 1971).

**Not all facts are
equally useful**

1. Technical communication is hard work.
2. "The computerized version of the *Oxford English Dictionary* incorporates a modified SGML syntax" (Fawcett 379).
3. "Technical communication is a process of making deliberate decisions in response to a specific situation. In this process, you discover important meanings in your topic, and give your audience the information they need to understand your meanings" (Hogge 3).

Statement 1 offers no news to anyone who has ever picked up a pencil. You probably find Statement 2's information new but not relevant. Because Statement 3 offers new insight into a familiar process, then we can say it has informative value in this context. On the other hand, a technical communication professional may find Statement 2 more useful than 3.

 **ON THE JOB...**

Audiences

"I'm writing for psychiatrists, nurses, psychologists, and other staff on the unit (since we use a "medical model" I use standard "jargon" and abbreviations for medical staff."

—Emma Bryant, social worker

Appropriate Details, Format, and Design

The amount of detail in your document (*How much is enough?*) depends on what you can learn about your audience's needs. Were you asked to "keep it short" or to "be comprehensive"? Can you summarize, or does everything need spelling out? Are people more interested in conclusions and recommendations, or do they want all the details? Have they requested a letter; a memo; a short report; or a long, formal report with supplements (title page, table of contents, appendixes, and so on)? Can visuals and page design (charts, graphs, drawings, headings, lists) make the material more accessible?

NOTE *Although a detailed analysis can tell you a great deal, rarely is it possible to pin down an audience with absolute certainty—especially when the audience is large and diverse. Before submitting a final document, examine every aspect, trying to anticipate specific audience questions or objections. Better yet, ask selected readers for feedback on early drafts. (For more on usability testing, see Chapter 16.)*

3.5
For more on usability testing visit <www.ablongman.com/lannonweb>

Due Date and Timing

Does your document have a deadline? Workplace documents almost always do. Is there a best time to submit it? Due dates and timing are vital for competing effectively (for example, in submitting bids for a project). When possible, ask users to review an early draft and to suggest improvements.

CONSIDER THIS Communication Failure Can Have Drastic Consequences

Accidents that make headlines often result from human errors such as these:

- *The information is delivered "in the wrong form, at the wrong time, . . . to the wrong person" (Devlin 22).*
- *The information's complexity overwhelms the person receiving it (Wickens 2).*
- *The people involved aren't being attentive or assertive enough.*

Below is a sampling of "honest mistakes" in communication caused by human limitations.

Neglecting to Convey Vital Information

November 1973: The Vermont Yankee Nuclear Power Plant experienced near-meltdown "after day workers installing a closed-circuit television cut off power to a primary safety system and failed to inform the night shift" (Monmonier 209–10).

Not Being Assertive about Vital Information

January 1982: A Boeing 737 crashes on takeoff from Washington National Airport, killing 78 people. "The copilot had warned the captain of possible trouble several times—icy conditions were causing false readings on an engine-thrust gauge—but the copilot had not spoken forcefully enough, and the pilot ignored him" (Pool 44).

Downplaying Vital Information

September 11, 2001: Undetected by airport security, terrorists seize and crash four passenger planes, killing thousands. Investigators later find that, although it had no definite intelligence about suicide hijackings, the Federal Aviation Administration had considered such a possibility as early as 1998. But in its briefings and terrorism alerts to airlines, the FAA either discounted this possibility or failed to mention it at all, focusing instead on the threat of explosives inside of baggage (Yen 7).

Conveying the Wrong Information

December 1998: Nine months after launching, the spacecraft *Mars Climate Orbiter* disappeared just before reaching its orbit around the planet. This $2 billion loss was traced to the fact that metric units of measurement had been used by one NASA ground control team and English units by the other (e.g., meters versus yards). As a result, the confusing instructions to the spacecraft caused it to miss its orbit and burn up in the Martian atmosphere (M. Martin 3).

Underestimating Vital Information

December 1941: "Some people within the U.S. Army knew that a large group of airplanes was headed toward Pearl Harbor; others knew that six Japanese aircraft carriers weren't where they were supposed to be; yet nobody acted on that information" (Davenport 7).

Overlooking Vital Mistakes

August 1996: A 2-month-old boy with a heart defect is admitted to a major hospital in apparent heart failure. He is treated with the drug Digoxin and expected to recover fully. Within hours, his heart stops. All revival efforts fail.

A medical resident had mistakenly ordered a dose of 0.9 milligram—instead of the correct .09 mg. The mistake was then overlooked by the attending doctor, the pharmacy technician who filled the order, and a second resident physician who was asked by a nurse to check the dosage. Because of this error of one decimal point, the patient received 10 times the correct dose of Digoxin (Belkin 28+).

EXERCISES For more exercises, visit
<www.ablongman.com/lannon>

1. Using Internet or print sources, locate a short article from your field (or part of a long article or a selection from your textbook for an advanced course). Choose a piece written at the highest level of technicality you understand and then translate the piece for a layperson, as in the example on page 32. Exchange translations with a classmate from a different major. Read your neighbor's translation and write a paragraph evaluating its level of technicality. Submit to your instructor a copy of the original, your translated version, and your evaluation of your neighbor's translation.

2. Assume that a new employee is taking over your job because you have been promoted. Identify a specific problem in the job that could cause difficulty for the new employee. Write instructions for the employee for avoiding or dealing with the problem. Before writing, create an audience and use profile by answering (on paper) the questions on page 34.

3. The U.S. Immigration and Naturalization Service's Web site, at <www.ins.gov>, is designed for a truly global audience. After visiting the site, answer these questions:

• Would this site be easy for virtually any English speaker to navigate? List the features that accommodate users from diverse areas of the globe.

• Could improvements be made in the site's usability? What improvements would you recommend?

Print out relevant site pages, and be prepared to discuss your evaluation and recommendations in class.

4. Locate a Web site that accommodates various users at various levels of technicality. Sites for government agencies such as those listed below are good sources of both general and specialized information.

• Environmental Protection Agency (EPA)
<www.epa.gov>
• Nuclear Regulatory Commission (NRC)
<www.nrc.gov>

• National Institute of Health (NIH)
<www. nih.gov>
• Food and Drug Administration (FDA)
<www.fda.gov>

Examine one of these sites and find one example of (a) material aimed at a general audience, and (b) material on the same topic aimed at a specialized or expert audience. First, list the specific features that enabled you to identify each piece's level of technicality. Next, using the Audience and Use Profile Sheet, record the assumptions about the audience made by the author of the nontechnical version. Finally, evaluate how well that piece addresses a nontechnical user's information needs. (*Hint:* Check out, for instance, the MEDLINE link at the NIH site.)

Print out the site's home page as well as key linking pages, and be prepared to discuss your evaluation in class.

5. The 1979 release of radiation and near-meltdown at Pennsylvania's Three-Mile Island nuclear power plant; the 1984 explosion of an American-owned chemical plant in Bhopal, India; the November 2000 Singapore Airlines crash in Tapei, Taiwan—these disasters occurred because of human errors in communication. Search the Web for details on one of these disasters (or a similar one), and prepare a one- or two-page memo that summarizes the role of miscommunication in causing the disaster.

COLLABORATIVE PROJECTS

Form teams according to major (electrical engineering, biology, etc.), and respond to the following situation.

Assume that your team has received the following assignment from your major department's chairperson: An increasing number of first-year students are dropping out of the major because of low grades, stress, or inability to keep up with the work. Your task is to prepare a "Survival Guide," for distribution to incoming students. This one- or two-page memo should focus on the challenges and the pitfalls of the first year in a given major and should include a brief motivational section, along with whatever else team members think readers need.

Analyze Your Audience

Use Figure 3.4 as a guide for developing your audience and use profile.

Devise a Plan for Achieving Your Goal

From the audience traits you have identified, develop a plan for communicating your information. Express your goal and plan in a statement of purpose. For example:

> The purpose of this document is to explain the challenges and pitfalls of the first year in our major. We will show how dropouts have increased, discuss what seems to go wrong, give advice on avoiding some common mistakes, and emphasize the benefits of remaining in the program.

Plan, Draft, and Revise Your Document

Brainstorm for worthwhile content (for this exercise, make up some reasons for the dropout rate if you need to), do any research that may be needed, write a workable draft, and revise until it represents your team's best work.

Appoint a team member to present the finished document (along with a complete audience and use analysis) for class evaluation, comparison, and response.

Alternative Projects

a. In a one- or two-page memo to *all* incoming students develop a "First-Year Survival Guide." Spell out the least information anyone should know in order to get through the first year of college.

b. In one or two pages, describe the job outlook in your field (prospects for the coming decade, salaries, subspecialties, promotional opportunities, etc.). Write for high school seniors interested in your major. Your team's description will be included in the career handbook published by your college.

c. Identify an area or situation on campus that is dangerous or inconvenient or in need of improvement (endless cafeteria lines, poorly lit intersections or parking lots, noisy library, speeding drivers, inadequate dorm security, etc.). Observe the situation as a group during a peak-use period. Spell out the problem in a letter to a specified decision maker (dean, campus police chief, head of food service) who presumably will use your information as a basis for action.

SERVICE-LEARNING PROJECT

Create a one-page summary of the purpose, programs, and history of the agency you are planning to work with. Design your summary as an information flyer/fact sheet or as a brochure (page 524) to be distributed to first-time visitors to the agency, or to be included in grant applications or other mailings to request support.

4

Making a Persuasive Case

Persuasion means trying to influence someone's actions, thinking, or decision making (Figure 4.1). In the workplace, we rely on persuasion daily: to win coworker support, to attract clients and customers, to request funding. But changing someone's mind is never easy, and sometimes impossible. Your success will depend on who you are trying to persuade, what you are requesting, and how entrenched they are in their own views.

FIGURE 4.1
Informing and Persuading Both Require Audience Awareness

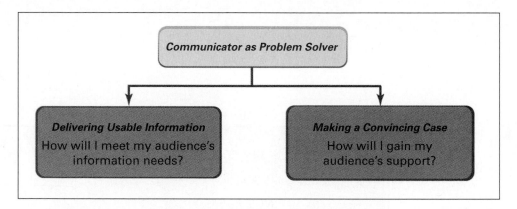

Persuasion is required whenever you tackle an issue about which people disagree. Assume, for example, that you are Manager of Employee Relations at Softbyte, a software developer whose recent sales have plunged. To avoid layoffs, the company is trying to persuade employees to accept a temporary cut in salary. As you plan your various memos and presentations on this volatile issue, you identify your major *claims*. (A claim is a statement of the point you are trying to prove.) For instance, you might want employees to recognize facts they've chosen to ignore:

A claim about what the facts are

> Because of the global recession, our software sales in two recent quarters have fallen nearly 30 percent, and earnings should remain flat all year.

Even when a fact is obvious, people often disagree about what it means or what action it indicates. And so you might want to influence their interpretation of the facts:

A claim about what the facts mean

> Reduced earnings mean temporary layoffs for roughly 25 percent of our staff. But we could avoid layoffs entirely if each of us at Softbyte would accept a 10-percent salary cut until the market improves.

And eventually you might want to ask for direct action:

A claim about what should be done

> Our labor contract stipulates that such an across-the-board salary cut would require a two-thirds majority vote. Once you've had time to examine the facts, we hope you'll vote "yes" on next Tuesday's secret ballot.

Whenever people disagree about what the facts are or what the facts mean or what should be done, you need to make the best case for your own view.

On the job, your memos, letters, reports, and proposals advance claims like these (Gilsdorf, "Executives' and Academics' Perception" 59–62):

Claims require support

> - We can't possibly meet this production deadline without sacrificing quality.
> - We're doing all we can to correct your software problem.
> - This hiring policy is discriminatory.
> - Our software is exactly what you need.
> - I deserve a raise.

Such claims, of course, are likely to be rejected—unless they are backed up by a convincing argument.

NOTE
"Argument," in this context, means "a process of careful reasoning in support of a claim"—and not "a quarrel or dispute." People who "argue skillfully" connect with others in a rational, sensible way, without animosity. But people who are merely "argumentative," on the other hand, simply make others defensive.

IDENTIFY YOUR SPECIFIC GOAL

"What do I want people to be doing or thinking?"

Arguments differ in how much they ask from people. Is your goal to (a) influence people's opinions, (b) seek their support, (c) induce direct action, or (d) alter people's behavior?

Arguing to Influence People's Opinions

Asking only for a change in thinking

Some arguments ask for minimal audience involvement. Maybe you want people to agree that the benefits of bioengineered foods outweigh the risks, or that your company's monitoring of employee emails is hurting morale. The goal here is merely to move readers to change their thinking, to say "I agree."

Arguing to Enlist People's Support

Asking for active support

Some arguments ask people to take a definite stand. Maybe you want readers to support a referendum to restrict cloning experiments, or to lobby for a daycare center where you work. The goal is to get people actively involved, to get them to ask "How can I help?"

Presenting a Proposal

Asking for direct action

Proposals offer specific plans for solving technical problems. The proposals we examine in Chapter 23 typically ask audiences to take—or to approve—some form of direct action (say, a plan for improving your firm's computer security or a Web-based orientation program for new employees). But before you can get people to act, you must complete these preliminary persuasive tasks:

A proposal
involves these
persuasive tasks

1. Spell out the problem (and its causes) in enough detail to convince readers of its importance.
2. Point out the benefits of solving the problem.
3. Offer a realistic, cost-effective solution.
4. Address anticipated objections to your plan.
5. Give reasons why your readers should be the ones to act.

Your proposal goal is achieved when people say "Okay, let's do this project."

Arguing to Change People's Behavior

Asking for
different behavior

Getting people to change their behavior is a huge challenge. Maybe you want a coworker to stop dominating your staff meetings, or to be more open about sharing information that you need to do your job. People take such arguments very personally. And the more personal the issue, the greater their resistance. After all, you're trying to get them to admit, "I was wrong. From now on, I'll do it differently."

The above categories can overlap, depending on your exact situation. But never launch an argument without a clear view of exactly *what* you want to see happen.

Try to Predict Audience Reaction

4.1

For more on power
dynamics in the work-
place visit
<www.ablongman.com/
lannonweb>

Any document can evoke different reactions depending on a user's temperament, interests, fears, biases, ambitions, or assumptions. Whenever peoples' views are challenged, they react with questions like these:

TYPICAL AUDIENCE QUESTIONS ABOUT ANY ATTEMPT TO PERSUADE

- *Says who?*
- *So what?*
- *Why should I?*
- *Why rock the boat?*
- *What's in this for me?*
- *What will it cost?*

- *What are the risks?*
- *What are you up to?*
- *What's in it for you?*
- *What does this really mean?*
- *Will it mean more work for me?*
- *Will it make me look bad?*

People read between the lines. Some might be impressed and pleased by your suggestions for increasing productivity or cutting expenses; some might feel offended or threatened. Some might suspect you of trying to undermine your boss. Furthermore, no one wants bad news; some people prefer to ignore it. Such are the "political realities" in any organization (Hays 19).

EXPECT AUDIENCE RESISTANCE

People who haven't made up their minds are more receptive to persuasion than those who have:

We rely on persuasion to help us make up our minds

> We need others' arguments and evidence. We're busy. We can't and don't want to discover and reason out everything for ourselves. We look for help, for short cuts, in making up our minds. (Gilsdorf, "Write Me" 12)

People who *have* decided what to think, however, naturally assume they're right, and they often refuse to budge. Whenever you question people's stance on an issue or try to change their behavior, expect resistance:

Why persuasion is so difficult

> By its nature, informing "works" more often than persuading does. While most people do not mind taking in some new facts, many people do resist efforts to change their opinions, attitudes, or behaviors. (Gilsdorf, "Executives' and Academics' Perception" 61)

For people to admit you might be right they must often admit they might be wrong, and the more strongly they identify with their position, the more resistance you can expect.

When people do yield to persuasion, they may respond grudgingly, willingly, or enthusiastically (Figure 4.2). Researchers categorize these responses as compliance, identification, or internalization (Kelman 51–60):

Some ways of yielding to persuasion are more productive than others

- *Compliance:* "I'm giving in to your demand in order to be rewarded or to avoid punishment. I really disagree, but I feel pressured, and so I'll go along to get along."

FIGURE 4.2 The Levels of Response to Persuasion

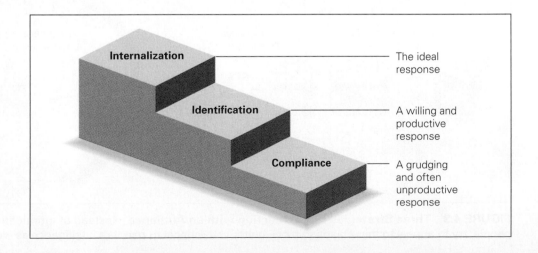

- *Identification:* "I'm going along with your appeal because I like and believe you, I want you to like me, and I believe we have something in common."
- *Internalization:* "I'm going along because what you're saying makes good sense and it fits my goals and values."

Although mere compliance is sometimes necessary (as in military orders or workplace safety regulations), nobody likes to be coerced. If people comply only because they feel intimidated and have no choice, you can expect to lose their loyalty and goodwill—and as soon as the threat or reward disappears, so will their compliance. Persuasion relies on identification and, especially, internalization.

KNOW HOW TO CONNECT WITH THE AUDIENCE

Persuasive people know when to simply declare what they want, when to reach out and create a relationship, or when to appeal to reason (Kipnis and Schmidt 40–46). These strategies can be labeled the *power connection*, the *relationship connection*, and the *rational connection*, respectively (Figure 4.3).

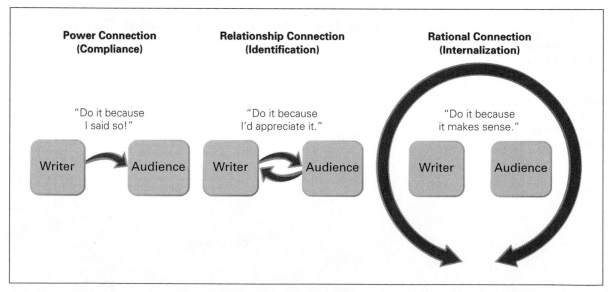

FIGURE 4.3 Three Strategies for Connecting with an Audience Instead of intimidating people, try to appeal to the relationship or—better yet—appeal to people's intelligence as well.

For illustration of these different connections, picture this situation: Your Company, XYZ Engineering, has just developed a fitness program, based on findings that healthy employees work better, take fewer sick days, and cost less to insure. This program offers clinics for smoking, stress reduction, and weight loss, along with group exercise. In your second month on the job you read this notice in your email:

POWER CONNECTION

MEMORANDUM
To: Employees at XYZ.com
FROM: GMaximus@XYZ.com
DATE: June 6, 20xx
SUBJECT: Physical Fitness

Orders readers to show up

On Monday, June 10, all employees will report to the company gymnasium at 8:00 a.m. for the purpose of choosing a walking or jogging group. Each group will meet for 30 minutes three times weekly during lunch time.

How would you react? Despite the reference to "choosing," people are given no real choice but simply ordered to show up. Typically used by bosses and other authority figures, the power connection does get people to comply but it almost always alienates them as well!

Suppose, instead, that you receive this next version of the memo. How would you react now?

RELATIONSHIP CONNECTION

MEMORANDUM
TO: Employees at XYZ.com
FROM: GMaximus@XYZ.com
DATE: June 6, 20xx
SUBJECT: An Invitation to Physical Fitness

Invites readers to participate

I realize most of you spend lunch hour playing cards, reading, or just enjoying a bit of well-earned relaxation in the middle of a hectic day. But I'd like to invite you to join our lunchtime walking/jogging club.

Leaves the choice to the reader

We're starting this club in hopes that it will be a great way for us all to feel better. Why not give it a try?

4.2

For more on the politics of memo writing visit <www.ablongman.com/lannonweb>

This second version conveys the sense that "we're all in this together." Instead of being commanded, readers are invited—offered a real choice.

How the audience perceives the writer is often more important than the writer's argument itself. People are more open to someone they like and trust. The relationship connection is especially vital in cross-cultural communication—as long as it is not too "chummy" and informal (see pages 59–61).

NOTE

Of course, you would be unethical in appealing to—or faking—the relationship merely to hide the fact that you have no evidence to support your claim (R. Ross 28). People need to find the claim believable ("Exercise will help me feel better") and relevant ("I, personally, need this kind of exercise").

Here is a third version of the memo. As you read, think about the ways it makes a persuasive case.

RATIONAL CONNECTION

MEMORANDUM
TO: Employees at XYZ.com
FROM: GMaximus@XYZ.com
DATE: June 6, 20xx
SUBJECT: Invitation to Join One of Our Jogging or Walking Groups

Presents authoritative evidence

I want to share a recent study from the *New England Journal of Medicine*, which reports that adults who walk two miles a day could increase their life expectancy by three years.

Other research shows that 30 minutes of moderate aerobic exercise, at least three times weekly, has a significant and long-term effect in reducing stress, lowering blood pressure, and improving job performance.

Offers alternatives

As a first step in our exercise program, XYZ Engineering is offering a variety of daily jogging groups: The One-Milers, Three-Milers, and Five-Milers. All groups will meet at designated times on our brand new, quarter-mile, rubberized clay track.

For beginners or skeptics, we're offering daily two-mile walking groups. For the truly resistant, we offer the option of a Monday–Wednesday–Friday two-mile walk.

Offers a compromise

Coffee and lunch breaks can be rearranged to accommodate whichever group you select.

Leaves the choice to the reader
Offers incentives

Why not take advantage of our hot new track? As small incentives, XYZ will reimburse anyone who signs up as much as $100 for running or walking shoes, and will even throw in an extra fifteen minutes for lunch breaks. And with a consistent turnout of 90 percent or better, our company insurer may be able to eliminate everyone's $200 yearly deductible in medical costs.

This version conveys respect for the reader's intelligence *and* for the relationship. With any reasonable audience, the rational connection stands the best chance of succeeding.

NOTE

Keep in mind that no cookbook formula exists, and in many situations, even the best persuasive attempts may be rejected.

ALLOW FOR GIVE-AND-TAKE

Reasonable people expect a balanced argument, with both sides of the issue considered evenly and fairly. Persuasion requires flexibility on your part. Instead of merely pushing your own case forward, consider other viewpoints. In advocating your position, for example, you need to (Senge 8):

How to promote your view

- explain the reasoning and evidence behind it
- invite people to find weak spots in your case, and to improve on it
- invite people to challenge your ideas (say, with alternative reasoning or data)

When others offer an opposing view, you need to:

How to respond to opposing views

- try to see things their way, instead of insisting on your way
- rephrase an opposing position in your own words, to be sure you understand it accurately
- try reaching agreement on what to do next, to resolve any insurmountable differences
- explore possible compromises others might accept

Perhaps some XYZ employees, for example, have better ideas for making the exercise program work for everyone.

ASK FOR A SPECIFIC RESPONSE

Unless you are giving an order, diplomacy is essential. But don't be afraid to ask for what you want:

Spell out what you want

> The moment of decision is made easier for people when we show them what the desired action is, rather than leaving it up to them. . . . No one likes to make decisions: there is always a risk involved. But if the writer asks for the action, and makes it look easy and urgent, the decision itself looks less risky. (Cross 3)

Tell people exactly what you want them to do or think.

 NOTE

Keep in mind that overly direct communication can offend audiences from other cultures. Don't mistake bluntness for clarity.

NEVER ASK FOR TOO MUCH

Stick with what is achievable

People never accept anything they consider unreasonable. And the definition of "reasonable" depends on the individual. Employees at XYZ, for example, differ as to which walking/jogging option they might accept. To the runner writing the memo, a daily five-mile jog might seem perfectly reasonable, but to most people

Persuasive challenges

"Argument and persuasion are crucial or you will never get done working."

—Lorraine Patsco, Director of Prepress and Multimedia Production

this would seem outrageous. XYZ's program, therefore, must offer something most people (except, say, couch potatoes and those in poor health) accept as reasonable.

Any request that exceeds its audience's "latitude of acceptance" (Sherif 39–59) is doomed. To get a clear sense of exactly how much is *achievable* in the given situation, ask around beforehand and feel out the people involved.

RECOGNIZE ALL COMMUNICATION CONSTRAINTS

Constraints are limits or restrictions imposed by the situation—who can say what to whom, and so on.

Communication constraints in the workplace

- What can I say around here, to whom, and how?
- Should I say it in person, by phone, in print, online?
- Could I be creating any ethical or legal problems?
- Is this the best time to say it?
- What's my relationship with the audience?
- Who are the personalities involved?
- Is there any peer pressure to overcome?
- How big an issue is this?

Organizational Constraints

Organizations announce their own official constraints: deadlines; budgets; guidelines for organizing, formatting, and distributing documents; and so on. But communicators also face *unofficial* constraints:

Decide carefully when to say what to whom

Most organizations have clear rules for interpreting and acting on statements made by colleagues. Even if the rules are unstated, we know who can initiate interaction, who can be approached, who can propose a delay, what topics can or cannot be discussed, who can interrupt or be interrupted, who can order or be ordered, who can terminate interaction, and how long interaction should last. (Littlejohn and Jabusch 143)

The exact rules vary among organizations, and most are unspoken, but anyone who breaks those rules (say, by going over a supervisor's head with a complaint or suggestion) invites disaster.

Airing even a legitimate gripe in the wrong way through the wrong medium to the wrong person can be fatal to your work relationships. The following email, for instance, is likely to be interpreted by the executive officer as petty and whining behavior, and by the maintenance director as a public attack.

Wrong way to the wrong person

> MEMORANDUM
> To: CEO@XYZ.com
> CC: MaintenanceDirector@XYZ.com
> FROM: Middle Manager@XYZ.com
> DATE: May 13, 20xx
> RE: Trash Problem
>
> Please ask the Maintenance Director to get his people to do their job for a change. I realize we're all understaffed, but I've gotten dozens of complaints this week about the filthy restrooms and overflowing wastebaskets in my department. If he wants us to empty our own wastebaskets, why doesn't he let us know?

Instead, why not address the message directly to the key person—or better yet, phone the person?

A better way to the right person

> MEMORANDUM
> To: MaintenanceDirector@XYZ.com
> FROM: MiddleManager@XYZ.com
> DATE: May 13, 20xx
> RE: Staffing Shortage
>
> I wonder if we could meet to exchange some ideas about how our departments might be able to support one another during these staffing shortages.

Can you identify the unspoken rules in companies where you have worked? What happens when such rules are ignored?

Legal Constraints

What you are allowed to say may be limited by contract or by laws that govern confidentiality or customers' rights or product liability:

Major legal constraints on communication

- In a collection letter for nonpayment, you can threaten to take legal action, but you cannot threaten to publicize the refusal to pay, nor can you pretend to be an attorney (Varner and Varner 31–40).
- If someone requests information on your employee, you can "respond only to specific requests that have been approved by the employee. Further, your comments should relate only to job performance which is documented" (Harcourt 64).
- If you write sales literature or manuals, you and your company can be sued over faulty information that causes injury or damage.

4.3

For more on public relations and legal liabilities visit <www.ablongman.com/lannonweb>

Whenever you prepare a document, be aware of possible legal problems. For instance, suppose an employee of XYZ Engineering (page 49) drops dead during the new exercise program you've marketed so persuasively. Could you and your company be liable? Should you require physical exams and stress tests (at company expense) for participants? When in doubt, always consult an attorney.

Ethical Constraints

Ethical constraints are defined not by law, but by honesty and fair play. For example, it may be perfectly legal to promote a new pesticide by emphasizing its effectiveness, while downplaying its carcinogenic effects; whether such action is ethical, however, is another issue. To earn people's trust, you will find that "saying the right thing" involves more than legal considerations. (See Chapter 5 for more on ethics.)

NOTE   *Persuasive skills carry tremendous potential for abuse. "Presenting your best case" does not mean deceiving others—even if the dishonest answer is the one people want to hear.*

ON THE JOB...
Persuasive challenges

"All writing is argument of one kind or another: even poetry. I make sure that my final written product has a clear line of argumentation, and that each paragraph is tied to that argument. I use comparisons, metaphors, humor, and other devices to augment my argument and make reading pleasurable, but for me no piece of writing is truly persuasive (or pleasing) without a clear argument."

—John Bryant, Professor of English

Time Constraints

Persuasion is often a matter of good timing. Should you wait for an opening, release the message immediately, or what? Let's assume that you're trying to "bring out the vote" among members of your professional society on some hotly debated issue, say, whether to refuse to work on any project related to biological warfare. You might prefer to wait until you have all the information you need or until you've analyzed the situation and planned a strategy. But if you delay, rumors or paranoia could cause people to harden their positions—and their resistance.

Social and Psychological Constraints

Too often, the human side of communication leads to misunderstanding, due to constraints like these:

"What's our relationship?"

- *Relationship with the audience:* Is your reader a superior, a subordinate, a peer? (Try not to dictate to subordinates or shield superiors from bad news.) How well do you know each other? Can you joke around or should you be serious? Do you get along or have a history of conflict or mistrust? What you say and how you say it—and how it is interpreted—will be influenced by the relationship.

"How receptive is this audience?"

- *Audience's personality:* Willingness to be persuaded depends largely on personality (Stonecipher 188-89). Does this person tend to be more open- or closed-minded, more skeptical or trusting, more bold or cautious, more of a conformist or a rugged individual? The less persuadable your audience, the harder you must work. For a totally resistant audience, you may want to back off—or give up altogether.

"How unified is this audience?"

- *Audience's sense of identity and affiliation as a group:* Does the group have a strong identity (union members, conservationists, engineers)? Will group loyalty or pressure prevent certain appeals from working? Address the group's collective concerns.

"Where are people coming from on this?"

- *Perceived size and urgency of the issue:* Does the audience see this as a cause for fear or for hope? Is trouble looming or has a great opportunity emerged? Has the issue been understated or overstated? Big problems often cause people to exaggerate their fears, loyalties, and resistance to change—or to seek quick solutions. Assess the problem realistically. Don't downplay a serious problem, but don't cause panic, either.

CONSIDER THIS People Often React Emotionally to Persuasive Appeals

We've all been on the receiving end of attempts to influence our thinking:

- *You need this product!*
- *This deal beats the competitor's deal!*
- *This candidate is the one to vote for!*
- *Try doing things this way!*
- *Donate to this good cause!*

How, exactly, do we decide which persuasive appeals to accept or reject?

One way is by evaluating the argument itself, by asking *Does the argument supporting this claim make good sense? Is it balanced and fair?*

But arguments rarely succeed or fail on their own merits. Researchers have identified common emotional factors that influence our receptiveness to persuasion.

Why We Say No

Management expert Edgar Schein outlines various fears that prevent people from trying or learning something new (34–39):

- Fear of the unknown: *Why rock the boat? (Change can be scary, and so we cling to old, familiar ways of doing things, even when those ways aren't working.)*

- Fear of disruption: *Who needs these headaches? (We resist change if it seems too complicated or troublesome.)*
- Fear of failure: *Suppose I screw up? (We worry about the shame or punishment that might result from making errors.)*

To overcome these basic fears, Schein explains, people need to feel "psychologically safe":

> . . . they have to see a manageable path forward, a direction that will not be catastrophic. They have to feel that a change will not jeopardize their current sense of identity and wholeness. They must feel that new habits are possible, that they can . . . try out new things without fear of punishment. (59)

Any effective argument for change must include such types of reassurance.

Why We Say Yes

Social psychologist Robert Cialdini pinpoints six subjective criteria that move people to accept a persuasive appeal (76–81):

- Reciprocation: *Do I owe this person a favor? (We feel obligated—and we look for the chance—to reciprocate, or return, a good deed.)*

(continues)

Consider This (continued)

- Consistency: *Have I made some earlier commitment along these lines? (We like to see ourselves as behaving consistently. Cialdini notes that people who have declared even minor support for a particular position [say by signing a petition], will tend to accept future requests for major support of that position [say, a financial contribution].)*
- Social validation: *What are other people saying about this argument? Are they agreeing or disagreeing? (We usually feel reassured by going along with our peers.)*
- Liking: *Do I like the person making the argument? (We are far more receptive to people we like—and often more willing to accept a bad argument from a likable person than a good one from an unlikable person!)*
- Authority: *How knowledgeable does this person seem about the issue? (We place confidence in experts and authorities.)*
- Scarcity: *Does this person know (or have) something that others don't? (The scarcer something seems, the more we value it [say, a hot tip about the stock market].)*

Sales professionals know very well how these criteria apply: A typical sales pitch, for example, might include a "free sample of our most popular brand, which is nearly sold out"

offered by a chummy salesperson full of "expert" details about the item itself.

While an ethical persuader's appeal to these criteria often enhances a legitimate argument, Cialdini warns of exploitation by unethical persuaders, as when a phony doctor in a TV commercial makes "authoritative" claims about some brand of medication.

Cross-Cultural Differences

To show how different cultures often weigh these various criteria differently, Cialdini (81) cites a recent survey of Citibank employees in four countries (United States, China, Spain, and Germany) by Stanford researchers Morris, Podolny, and Ariel. When asked by a coworker for help with a task, U.S. bank employees felt obligated to comply, or reciprocate, if they owed that person a favor. Employees in the Chinese bank were influenced mostly by the requester's status, or authority, while Spanish employees based their decision mainly on liking and friendship, regardless of the requester's status. German employees were motivated mainly by a sense of consistency in following the bank's official rules: If the rules stipulated they should help coworkers, they felt compelled to do so.

SUPPORT YOUR CLAIMS

The persuasive argument is the one that makes the best case—in the audience's view:

A persuasive case offers reasons that matter to the audience

When we seek a project extension, argue for a raise, interview for a job . . . we are involved in acts that require good reasons. Good reasons allow our audience and ourselves to find a shared basis for cooperating. . . . [Y]ou can use marvelous language, tell great stories, provide exciting metaphors, speak in enthralling tones, and even use your reputation to advantage, but what it comes down to is that you must speak to your audience with reasons they understand. (Hauser 71)

Imagine the following situation: As documentation manager for Bemis Software, a rapidly growing company, you supervise preparation and production of all user manuals. The present system for producing manuals is inefficient because three different departments are involved in (1) assembling the material, (2) word processing and designing, and (3) publishing the manuals. Much time and energy are wasted as a manual goes back and forth among software specialists, communication specialists, and the art and printing department. After studying the problem and calling in a consultant, you decide that greater efficiency would result if desktop pub-

Persuasive challenges

"I spend much of my time trying to persuade people that the information we offer can help them. Managers have to be persuaded that your recommendations are worthwhile and the best way to go. Clerical staff want to know, 'Will this help me keep my job?'"

—Blair Cordasco, training specialist
for an international bank

lishing software were installed in all computer terminals. Everyone involved could then contribute during all three phases. To sell this plan to bosses and coworkers you need reasons based on *evidence* and *appeals to everyone's needs and values* (Rottenberg 104–06).

Offer Convincing Evidence

Evidence (factual support from an outside source) is a powerful persuader—as long as it meets an audience's standards. A discerning audience evaluates evidence using these criteria (Perloff 157–58):

Criteria for worthwhile evidence

- *The evidence has quality.* People expect evidence that is strong, specific, new, different, and verifiable (provable).
- *The sources are credible.* People want to know where the evidence comes from, how it was collected, and who collected it.
- *The evidence is considered reasonable.* The evidence falls within the audience's "latitude of acceptance" (discussed on page 51–52).

Common types of evidence include factual statements, statistics, examples, and expert testimony.

FACTUAL STATEMENTS.   A *fact* is something that can be demonstrated by observation, experience, research, or measurement—and that your audience is willing to recognize.

Offer the facts

Most of our competitors already have desktop publishing networks in place.

Be selective. Decide which facts best support your case.

STATISTICS.   Numbers can be highly convincing. Many readers focus on the "bottom line": costs, savings, losses, profits.

Cite the numbers

> After a cost/benefit analysis, our accounting office estimates that an integrated desktop publishing network will save Bemis 30 percent in production costs and 25 percent in production time—savings that will enable the system to pay for itself within one year.

But numbers can mislead. Your statistics must be accurate, trustworthy, and easy to understand and verify (see pages 181–87). Always cite your source.

EXAMPLES.   Examples help people visualize and remember the point. For example, the best way to explain what you mean by "inefficiency" in your company is to show "inefficiency" occurring:

Show what you mean

The figure illustrates the inefficiency of Bemis's present system for producing manuals:

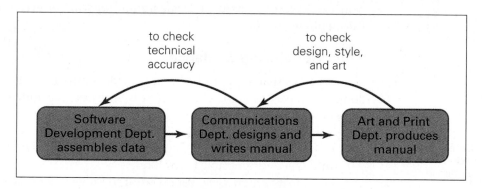

A manual typically goes back and forth through this cycle three or four times, wasting time and effort in all three departments.

Always explain how the example fits the point it is designed to illustrate.

EXPERT TESTIMONY.   Expert opinion—if it is unbiased and if people recognize the expert —lends authority and credibility to any claim.

Cite the experts

> Ron Catabia, nationally recognized networking consultant, has studied our needs and strongly recommends we move ahead with the integrated network.

See page 125 for the limits of expert testimony.

NOTE   *Finding evidence to support a claim often requires that we go beyond our own experience by doing some type of research. (See Part II.)*

Evidence alone may not be enough to change a person's mind. At Bemis, for example, the bottom line might be very persuasive for company executives but managers and employees will be asking: Does this threaten my authority? Will I have to work harder? Will I fall behind? Is my job in danger? This group expects some benefit beyond company profit.

Appeal to Common Goals and Values

Identify at least one goal you and your audience have in common: "What do we all want most?"

"What makes these people tick?"

Everyone at Bemis, for example, presumably wants job security and some control over their career. A persuasive recommendation will therefore take these goals into account:

Appeal to shared goals

> I'd like to show how desktop publishing skills, instead of threatening anyone's job, would only increase career mobility for all of us.

People's goals are shaped by their values (qualities they believe in, ideals they stand for): friendship, loyalty, honesty, equality, fairness, and so on (Rokeach 57–58).

At Bemis, you might appeal to the commitment to quality and achievement shared by the company and individual employees:

Appeal to shared values

> None of us needs reminding of the fierce competition in the software industry. The improved collaboration among networking departments will result in better manuals, keeping us on the front line of quality and achievement.

Give people reasons that have real meaning for *them* personally. For example, in a recent study of teenage attitudes about the hazards of smoking, respondents listed these reasons for not smoking: bad breath, difficulty concentrating, loss of friends, and trouble with adults. No respondents listed dying of cancer—presumably because this reason carries little meaning for young people personally (Baumann et al. 510–30).

NOTE *We are often tempted to emphasize anything that advances our case and to ignore anything that impedes it. But any message that prevents readers from making their best decision is unethical.*

 ## CONSIDER THE CULTURAL CONTEXT

Roughly 60 percent of business ventures between the United States and other countries fail (Isaacs 43) often because of obstacles posed by cultural differences.

Depending on social customs and values, people from different cultures react differently to persuasive appeals[1]:

How cultural
differences
govern a
persuasive
situation

4.4

For more on global
communication visit
<www.ablongman.com/
lannonweb>

- Some cultures hesitate to debate, criticize, or express disagreement. They prefer indirect, roundabout ways of approaching an issue—viewing it from all angles before declaring a position.
- Some cultures observe special formalities in communicating (say, expressions of concern for a person's family).
- Many cultures consider the source of a message as important as its content. Establishing trust and building a relationship are essential before doing business.
- Some cultures trust oral more than written communication.
- Cultures respond differently to different emotional pressures, such as feeling obliged to return favors or following the lead of their peers. (See Consider This, page 56.)
- Cultures differ in their attitudes toward big business, technology, competition, or women in the workplace. They might value delayed gratification more than immediate reward, stability more than progress, time more than profit, politeness more than candor, age more than youth.

Face saving is
every person's
bottom line

One especially volatile cause of clashes among different cultures is related to *face saving:* "the act of preserving one's prestige or outward dignity" (Victor 159–61). People lose face in situations like these:

How people lose
face

- *When they are offended or embarrassed by blatant criticism:* A U.S. businessperson in China decides to "tell it like it is," and proceeds to criticize the Tiananmen Square massacre and China's illegal contributions to American political parties (Stepanek 4).
- *When their customs are ignored:* An American female arrives to negotiate with older, Japanese males; Silicon Valley businesspeople show up in T-shirts and baseball caps to meet with hosts wearing suits.
- *When their values are trivialized:* An American in Paris greets his French host as "Pierre," slaps him on the back, and jokes that the "rich French food" on the Concorde flight had him "throwing up all the way over" (Isaacs 43).

4.5

For more on
ethnocentrism and
cultural difference visit
<www.ablongman.com/
lannonweb>

Anytime someone feels insulted, meaningful interaction is over.

Show respect for a country's cultural heritage by learning all you can about its history, landmarks, famous people, and especially its customs and values (Isaacs 43). The following questions can get you started.

---

[1]Adapted from Beamer 293–95; Gesteland 24; Hulbert, "Overcoming" 42; Jameson 9–11; Kohl et al. 65; Martin and Chaney 271–77; Nydell 61; Thatcher 193–94; Thrush 276–77; Victor 159–66.

QUESTIONS FOR ANALYZING CULTURAL DIFFERENCES*

What behavior does the culture consider acceptable?

- *Casual versus formal interaction*
- *Directness and plain talk versus indirectness and ambiguity*
- *Rapid decision making versus extensive analysis and discussion*
- *Willingness to request clarification*
- *Willingness to argue, criticize, or disagree*
- *Willingness to be contradicted*
- *Willingness to express emotion*

What values does the culture consider most important?

- *Attitude toward big business, competition, and U.S. culture*
- *Youth versus age*
- *Rugged individualism versus group loyalty*
- *Status of women in the workplace*
- *Feelings versus logic*
- *Candor versus face saving*
- *Progress and risk taking versus stability*
- *Importance of trust and relationship building*
- *Importance of time ("Time is money!" or "Never rush!")*
- *Preference for oral versus written communication*

*Adapted from Beamer 293–95; Gesteland 24; Hulbert, "Overcoming" 42; Jameson 9–11; Kohl et al. 65; Martin and Chaney 271–77; Nydell 61; Thatcher 193–94; Thrush 276–77; Victor 159–66.

NOTE *Violating a person's cultural frame of reference is offensive, but so is reducing individual complexity to a laundry list of cultural stereotypes. Any generalization about any group presents a limited picture and in no way accurately characterizes any or all members of the group.*

Shape Your Argument

Readers need to follow your reasoning; they expect to see how, exactly, you support your claim. The model in Figure 4.4 lays out a standard shape that can be adapted for most arguments. Select the elements appropriate to your situation and order them in a sequence that reveals a clear line of thought.

NOTE *People rarely change their minds quickly or without good reason. A truly resistant audience will dismiss even the best arguments and may end up feeling threatened and resentful. Even with a receptive audience, attempts at persuasion can fail. Often, the best you can do is avoid disaster and give people the chance to appreciate the merits of the case.*

Figure 4.5 shows a letter from a company that distributes systems for generating electrical power from recycled steam (cogeneration). President Tom Ewing persuasively answers a potential customer's question: "Why should I invest in the system you are proposing for my plant?" As you read, notice how the evidence and appeals support the opening claim. Notice also the focus on reasons important to the reader.

Figure 4.6 shows the audience and use profile for the writing situation in Figure 4.5.

Standard Shape for an Argument

Introduction: *Attract and Invite Your Audience and Provide a Forecast*

- Identify the issue clearly and immediately. Show that your argument deserves attention.

- Be clear about the points over which you and opponents disagree.

- Acknowledge the opposing viewpoint accurately and concede its merit.

- Offer at least one point of your own that your audience will agree with.

- Give enough background for people to understand your position accurately.

- State a clear, concrete, and definite claim (or thesis). Never delay your claim without good reason: If the issue is highly controversial or the audience is multicultural, for instance, you might offer convincing evidence and discussion first. (See page 388 for direct versus indirect approaches.)

- Keep the introduction short—no more than a few brief paragraphs.

Body: *Offer Support and Refutation*

- Focus on reasons that your audience will consider important.

- Organize your supporting points for best emphasis. If you think the audience has little interest, begin with the strongest material. Sometimes you can sandwich weaker points between stronger ones. But if all points are more-or-less equal, begin with the most familiar and acceptable to your audience—to elicit early agreement. In general, try to save strongest points for last.

- Reinforce each supporting point with verifiable evidence.

- String your supporting points and evidence together to show a definite line of reasoning.

- In at least one separate paragraph, refute opposing arguments (including any anticipated objections to your points).

Conclusion: *Sum Up Your Case and Make a Direct Appeal*

- Summarize your main points and refutation, emphasizing your strongest material. Offer a view of the Big Picture.

- Appeal directly to the audience for definite action (where appropriate).

- Let people know what they should do, think, or feel.

FIGURE 4.4   Shaping a Clear Line of Thought   Give the audience a clear and logical path, but remember that no argument rigidly follows the order of elements shown here.

EWING
POWER SYSTEMS
5 North Street   South Deerfield MA 01373

July 20, 20XX

Mr. Richard White, President
Southern Wood Products
Box 84
Memphis, TN 37162

Dear Mr. White:

The writer states his claim

In our meeting last week, you asked me to explain why we have such confidence in the project we are proposing. Let me outline what I think are excellent reasons.

Offers first reason
Gives examples

First, you and Don Smith have given us a clear idea of your needs, and our recent discussions confirm that we fully understand these needs. For instance, our proposal specifies an air-cooled condenser rather than a water-cooled condenser for your project because water in Memphis is expensive. And besides saving money, an air-cooled condenser will be easier to operate and maintain.

Offers second reason

Appeals to shared value (quality)

Gives example
Further examples

Second, we have confidence in our component suppliers and they have confidence in this project. We don't manufacture the equipment; instead, we integrate and package cogeneration systems by selecting for each application the best components from leading manufacturers. For example, Alias Engineering, the turbine manufacturer, is the world's leading producer of single-stage turbines, having built more than 40,000 turbines in 70 years. Likewise, each component manufacturer leads the field and has a proven track record. We have reviewed your project with each major component supplier, and each guarantees the equipment. This guarantee is of course transferable to you and is supplemented by our own performance guarantee.

Appeals to reader's goal (security)

Offers third reason

Third, we have confidence in the system design. We developed the CX Series specifically for applications like yours, in which there is a need for both a condensing and a backpressure turbine. In our last meeting, I pointed out the cost, maintenance, and performance benefits of the CX Series. And although the CX Series is an innovative design, all components are fully proven in many other applications, and our suppliers fully endorse this design.

Cites experts

Phone: (413)555-1767   Fax: (413)555-8791   Email: eps@valcom.com

FIGURE 4.5   Supporting a Claim with Good Reasons
Source: Reprinted with permission of Thomas S. Ewing, President, Ewing Power Systems, So. Deerfield, MA 01373.

Closes with best reason

Appeals to shared value
(trust) and shared goal
(success)

Richard White, July 20, 20XX, p. 2

Finally, and perhaps most important, you should have confidence in this project because we will stand behind it. As you know, we are eager to establish ourselves in Memphis-area industries. If we plan to succeed, this project must succeed. We have a tremendous amount at stake in keeping you happy.

If I can answer any questions, please phone me. We look forward to working with you.

Sincerely,

Tom Ewing

Thomas S. Ewing
President
EWING POWER SYSTEMS, INC.

FIGURE 4.5 Supporting a Claim with Good Reasons *(continued)*

Audience Identity and Needs

Primary audience: _Richard White, President of Southern Wood Products_

Secondary audience: _Don Smith, Plant Engineer; several plant managers_

Relationship: _A possible customer for a customized cogeneration system_

Intended use of document: _To provide information for a purchasing decision_

Prior knowledge about this topic: _Has compared various power-generation systems_

Additional information needed: _Seems to doubt the reliability of our system_

Probable questions: _How much money will your proposed system really save?_

How reliable is the equipment?

Can we depend on the innovative design you are proposing?

What quality of service can we expect?

Audience's Probable Attitude and Personality

Attitude toward topic: _Highly interested, but somewhat skeptical_

Probable objections: _This technology is too recent to have a solid track record_

Probable attitude toward this writer: _Receptive but cautious_

Organizational climate: _Open and flexible; lots of collaboration_

Persons most affected by this document: _White and other decision makers_

Temperament: _White takes a deliberate and conservative approach to decision making_

Probable reaction to document: _Readers should feel somewhat reassured_

Risk of alienating anyone: _No apparent risks_

Audience Expectations about the Document

Reason document originated: _Response to management concerns about reliability_

Cultural considerations: _None in particular_

Acceptable length: _Average business letter that gets right to the point_

Material important to this audience: _Solid evidence to back up our claim of "confidence" in this project_

Most useful arrangement: _A convincing list of reasons and appeals_

Tone: _Encouraging, friendly, and confident_

Intended effect on this audience: _To pave the way for a decision to purchase_

Due date: _ASAP—to illustrate our responsiveness to customer concerns_

FIGURE 4.6 Audience and Use Profile Sheet for Figure 4.5

GUIDELINES for Making Your Case

In any attempt at persuasion, the **audience** is your main focus. Whenever you set out to influence someone's thinking, remember this:

> **No matter how brilliant, any argument rejected by its audience is a failed argument.**

If people dislike you or decide that what you say has no meaning for them personally, they reject your argument outright. Instead of insisting that audiences see things your way, try to see things from their perspective.

Analyze the Situation

1. *Assess the political climate.* Whom will this affect? How will people react? How will they interpret your motives? Can you be outspoken? Could the argument cause legal problems? To avoid backlash:

   - Be aware of your status in the group; don't overstep.
   - Don't expect perfection from anyone—including yourself.
   - Don't make anyone look bad, lose face, or feel coerced.
   - Never fake certainty or make promises you can't keep.
   - Ask directly for support: "Is this idea worthy of your commitment?" (Senge 7).
   - Build support by inviting intended readers—especially the group opinion makers—to review early drafts.

   When reporting something that others do not want to hear, expect fallout. Decide beforehand whether you want to keep your job (or status) or your dignity. (See Chapter 5 for more on ethics.)

2. *Learn the unspoken rules.* Know the constraints (especially the legal ones) on what you can say to whom and how and when. Consider the cultural context.

3. *Decide on a connection (or combination of connections).* Should you require compliance, or appeal to a relationship or to common sense and reason?

4. *Anticipate audience reaction.* Will people be defensive, shocked, annoyed, angry? Try to neutralize major objections beforehand. Express your judgments on the issue ("We could do better") without blaming people ("It's all your fault").

Develop a Clear and Credible Plan

1. *Define your precise goal.* Develop the clearest possible view of what you want to see happen.

2. *Think your idea through.* Are there any holes in this argument? Will it stand up under scrutiny?

3. *Do your homework.* Be sure your facts are straight and your figures are accurate.

(continues)

Guidelines (continued)

4. *Never make a claim or ask for something you know people will reject outright.* Get a realistic sense about what is achievable in this particular situation by asking what people are thinking. Invite them to share in decision making. Offer real choices.

Prepare Your Case

1. *Be clear about what you want.* Diplomacy is important—especially in cross-cultural communication—but don't leave people guessing about your purpose.

2. *Project a likable and reasonable persona.* Persona is the image or impression you project in your tone and diction. Audiences are wondering "What do I think about the person making the argument?" "Do I like and trust this person?" "Does this person seem to know what he or she is talking about?" "Is this person trying to make me look stupid?"

   Audiences tune out aggressive people—no matter how sensible the argument. Resist the urge to preach, to "sound off," or to be sarcastic. Admit the imperfections or uncertainty in your case. Invite people to respond. A little humility never hurts.

3. *Find points of agreement with your audience.* What does everyone involved want? To reduce conflict, focus early on a shared value, goal, or experience. Emphasize your similarities. Pat deserving people on the back.

4. *Never distort the opposing position.* A sure way to alienate people is to cast the opponent in a more negative light than the facts warrant.

5. *Concede **something** to the opposing position.* Reasonable people respect an argument that is fair and balanced. Admit the merits of the opposing case before arguing for your own. Show empathy and willingness to compromise. Encourage people to air their own views.

6. *Don't merely criticize.* If you're arguing that something is wrong, be sure you can offer realistic suggestions for making it right.

7. *Stick to claims you can support.* Show people what's in it for them—but never distort the facts just to please the audience. Be honest about the risks.

8. *Stick to your best material.* Not all points are equal. Decide which material—from your audience's view—best advances your case.

Present Your Case

1. *Before releasing the document, get a second opinion.* Ask a reader you trust and who has no stake in the issue. If possible, have your company's legal department approve the document.

2. *Get the timing right.* When will your case most likely fly—or crash and burn? What else is going on that could influence people's reactions? Look for a good opening.

3. *Decide on the appropriate medium.* Given the specific issue and audience, should you communicate in person, via hard copy, phone, email, intranet, fax, newsletter, bulletin board? (See also page 52.) Should all recipients receive your message via the same medium? If your written message is likely to surprise readers, try to warn them face-to-face or by phone.

4. *Be sure everyone involved receives a copy.* People hate being left out of the loop—especially when any "change" that affects them is being discussed.

5. *Invite responses.* After people have had a chance to consider your argument, gauge their reactions by asking them directly.

6. *Don't be defensive about negative reactions.* Instead, admit mistakes, invite people to improve on your ideas, and try to build support (Bashein and Marcus 43).

7. *Know when to back off.* If you seem to be "hitting the wall," don't push. Consider trying again later or even dropping the whole effort. People who feel they have been bullied or deceived will likely become your enemies.

NOTE

Trying to change someone's opinion can be hard—but can also make a huge difference. For example, consider this study by the Rand Institute: Health insurers and HMOs often refuse to pay for costly medical treatment or additional services. Among patients who are denied benefits, only 3 or 4 per thousand ever make an appeal to the company, yet 42 percent of such appeals are successful (Fischman 50).

☑ CHECKLIST for Cross-Cultural Documents

Use this checklist* to verify that your documents respect audience diversity. (Page numbers in parentheses refer to the first page of discussion.)

☐ Does the document allow everyone to save face? (60)

☐ Is the document sensitive to the culture's customs and values? (60)

☐ Does the document conform to the safety and regulatory standards of the country? (00)

☐ Does the document provide the expected level of detail? (39)

☐ Does the document avoid possible misinterpretation? (38)

☐ Is the document organized in a way that readers will consider appropriate? (227)

☐ Does the document observe interpersonal conventions important to the culture (accepted forms of greeting or introduction, politeness requirements, first names, titles, and so on)? (59)

☐ Does the document's tone reflect the appropriate level of formality or casualness? (60)

☐ Is the document's style appropriately direct or indirect? (60)

☐ Is the document's format consistent with the culture's expectations? (360)

☐ Does the document embody universal standards for ethical communication? (81)

☐ Should the document be supplemented, if possible, by direct, face-to-face communication? (68)

*This list was largely adapted from Caswell-Coward 265; Weymouth 144; Beamer 293–95; Martin and Chaney 271–77; Thatcher 193–94; Victor 159–61.

For more exercises, visit
<www.ablongman.com/lannon>

1. You work for a technical marketing firm proud of its reputation for honesty and fair dealing. A handbook being prepared for new personnel includes a section titled "How to Avoid Abusing Your Persuasive Skills." All employees have been asked to contribute to this section by preparing a written response to the following:

   > Share a personal experience in which you or a friend were the victim of persuasive abuse in a business transaction. In a one- or two-page memo, describe the situation and explain exactly how the intimidation, manipulation, or deception occurred.

   Write the memo and be prepared to discuss it in class.

2. Recall an experience in which you accepted or rejected a persuasive appeal that involved some major decision in your life (selecting a college, buying your first car, supporting a political or environmental cause, joining the military, or the like). After reviewing pages 55–56, identify the major influences that caused you to say "yes" or "no." Be prepared to discuss your experience and its persuasion dynamics with the class.

3. Find an example of an effective persuasive letter. In a memo to your instructor, explain why and how the message succeeds. Base your evaluation on the persuasion guidelines, pages 66–68. Attach a copy of the letter to your evaluation memo. Be prepared to discuss your evaluation in class.

   Now, evaluate an ineffective document, explaining how and why it fails.

4. Think about some change you would like to see on your campus or at work. Perhaps you would like to promote something new, such as a campus-wide policy on plagiarism, changes in course offerings or requirements, more access to computers, a policy on sexist language, or a day-care center. Or perhaps you would like to improve something, such as the grading system, campus lighting, the system for student evaluation of teachers, or the promotion system at work. Or perhaps you would like to stop something from happening, such as noise in the library or sexual harassment at work.

   Decide whom you want to persuade and write a memo to that audience. Anticipate carefully your audience's implied questions, such as:

   - Do we really have a problem or need?
   - If so, should we care enough about it to do anything?
   - Can the problem be solved?
   - What are some possible solutions?
   - What benefits can we anticipate? What liabilities?

   Can you think of additional audience questions? Do an audience and use analysis based on the profile sheet, page 68.

   Don't think of this memo as the final word but as a consciousness-raising introduction that gets the reader to acknowledge that the issue deserves attention. At this early stage, highly specific recommendations would be premature and inappropriate.

5. Challenge an attitude or viewpoint that is widely held by your audience. Maybe you want to persuade your classmates that the time required to earn a bachelor's degree should be extended to five years or that grade inflation is watering down your school's reputation. Maybe you want to claim that the campus police should (or should not) wear guns. Or maybe you want to ask students to support a 10 percent tuition increase in order to make more computers and software available.

   Do an audience and use analysis based on the profile sheet, page 68. Write specific answers to the following questions:

   - What are the political realities?
   - What kind of resistance could you anticipate?
   - How would you connect with readers?
   - What about their latitude of acceptance?
   - Any other constraints?
   - What reasons could you offer to support your claim?

   In a memo to your instructor, submit your plan for presenting your case. Be prepared to discuss your plan in class.

6. Assess the political climate of an organization where you have worked—as an employee, a volunteer, an intern, a member of the military, or a member of a campus group (say, the school newspaper or the student senate). Analyze the decision-making culture of that organization:

   - Who are the key decision makers? How are decisions made?
   - How are policies primarily communicated (via power connection, relationship connection, rational connection)?
   - How much resistance occurs? How much give-and-take occurs?
   - What major constraints govern communication?
   - How would these considerations affect the way you would construct a persuasive case on an issue of importance to this organization?
   - How could the organizational structure be improved to encourage the sharing of new and constructive ideas?

   Prepare a memo reporting your findings and recommendations, addressed to a stipulated audience, and based on a thorough audience and use profile. (See pages 386, 387 for more on recommendation reports.)

   NOTE: This assignment might serve as the basis for the major term project (the formal proposal or analytical report, Chapters 23 and 24).

7. Use the questions on page 61 as a basis for interviewing a student from another country. Be prepared to share your findings with the class.

COLLABORATIVE PROJECTS

1. You work for an environmental consulting firm that is under contract with various countries for a range of projects, including these:

   - A plan for rain forest regeneration in Latin America and Sub-Saharan Africa
   - A plan to decrease industrial pollution in Eastern and Western Europe
   - A plan for "clean" industries in developing countries
   - A plan for organic agricultural development in Africa and India
   - A joint American/Canadian plan to decrease acid rain
   - A plan for developing alternative energy sources in Southeast Asia

   Each project will require environmental impact statements, feasibility studies, grant proposals, and a legion of other documents, often prepared in collaboration with members of the host country, and in some cases prepared by your company for audiences in the host country—from political, social, and industrial leaders to technical experts and so on.

   For such projects to succeed, people from different cultures have to communicate effectively and sensitively, creating goodwill and cooperation.

   Before your company begins work in earnest with a particular country, your coworkers will need to develop a degree of cultural awareness. Your assignment is to select a country and to research that culture's behaviors, attitudes, values, and social system in terms of how these variables influence the culture's communication preferences and expectations. What should your colleagues know about this culture in order to communicate effectively and diplomatically? Do the necessary research using the questions from page 61 as a guide.

   For a useful source of online information about world cultures, consult <http://lcweb2.loc.gov/frd/cs/cshome.html>. The Library of Congress Country Studies site is easy to navigate and provides a wealth of cultural information on over 100 countries. For an excellent source of information on Asian cultures, go to <www.asiasource.org>.

   Prepare a recommendation report in memo form (page 397). Be prepared to present your findings in class.

2. Often, workplace readers need to be persuaded to accept recommendations that are controversial or unpopular. This project offers practice in dealing with the persuasion problems of communicating within organizations.

Divide into teams. Assume that your team agrees strongly about one of these recommendations and is seeking support from classmates and instructors (and administrators, as potential readers) for implementing the recommendations.

Choose One Goal

a. Your campus Writing Center always needs qualified tutors to help first-year composition students with writing problems. On the other hand, students of professional writing need to sharpen their own skill in editing, writing, motivation, and diplomacy. All students in your class, therefore, should be assigned to the Writing Center during the semester's final half, to serve as tutors for twenty hours (beyond normal course time).

b. To prepare students for communicating in an automated work environment, at least one course assignment (preferably the long report) should be composed, critiqued, and revised online. Students not yet skilled in HTML will be required to develop the skill by midsemester.

c. This course should help individuals improve at their own level, instead of forcing them to compete with stronger or weaker writers. All grades, therefore, should be Pass/Fail.

d. To prepare for the world of work, students need practice in peer evaluation as well as self-evaluation. Because this textbook provides definite criteria and checklists for evaluating various documents, students should be allowed to grade each other and to grade themselves. These grades should count as heavily as the instructor's grades.

e. In preparation for writing in the workplace, no one should be allowed to limp along, cruising by with minimal performance. This course, therefore, should carry only three possible grades: A, B, or F. Those whose work would otherwise merit a C or D would instead receive an Incomplete and be allowed to repeat the course as often as needed to achieve a B grade.

f. To ensure that all graduates have adequate communication skills for survival in a world in which information is the ultimate product, each student in the college should pass a writing proficiency examination as a graduation requirement.

Analyze Your Audience

Your audience here consists of classmates and instructors (and possibly administrators). From your recent observation of this audience, what reader characteristics can you deduce?

Follow the model in Figure 4.6 for designing a profile sheet to record your audience and use analysis, and to duplicate for use throughout the semester. (Feel free to improve on the design and content of this model.)

Following is one possible set of responses to questions about audience identity and needs for goal (e) from the previous list.

- *Who is my audience?* Classmates and instructors (and possibly some college administrators).
- *How will readers use my information?* Readers will decide whether to support our recommendation for limiting possible grades in this course to three: A, B, or F.
- *How much is the audience likely to know already about this topic?* Everyone here is already a grade expert, and will need no explanation of the present grading system.
- *What else does the audience need to know?* Instructors should need no persuading; they know all about the quality of writing expected in the workplace. But some of our classmates probably will have questions like these: Why should we have to meet such high expectations? How can this grading be fair to the marginal writers? How will I benefit from these tougher requirements? Don't we already have enough work here?

You will have to answer questions by explaining how the issue boils down to "suffering now" or "suffering later," and that one's skill in communication will determine one's career advancement.

Devise a Plan for Achieving Your Goal

From the audience traits you have identified, develop a plan for justifying your recommendation. Express your goal and plan in a statement of purpose. Here is an example for goal (e):

> The purpose of this document is to convince classmates that our recommendation for an A/B/F grading system deserves your support. We will explain how skill in workplace writing affects career advancement, how higher standards for grading would help motivate students, and how our recommendation could be implemented realistically and fairly.

Plan, Draft, and Revise Your Document

Brainstorm (page 107) for worthwhile content, do any research that may be needed, write a draft, and revise as often as needed to produce a document that stands the best chance of connecting with your audience.

Appoint a member of your team to present the finished document (along with a complete audience and use analysis) for class evaluation and response.

SERVICE-LEARNING PROJECT

Just as the cultures of other nations have different values, so too do subcultures within the United States. Write a letter inviting neighborhood residents to an open house at a Latino community center in a predominantly Hispanic neighborhood. What factors influence how you shape and write your invitation? What language(s) would you use? Explain how your persuasive writing is influenced by cultural and linguistic differences.

5

Weighing the Ethical Issues

RECOGNIZE UNETHICAL COMMUNICATION IN THE WORKPLACE

KNOW THE MAJOR CAUSES OF UNETHICAL COMMUNICATION

UNDERSTAND THE POTENTIAL FOR COMMUNICATION ABUSE

RELY ON CRITICAL THINKING FOR ETHICAL DECISIONS

ANTICIPATE SOME HARD CHOICES

NEVER DEPEND ONLY ON LEGAL GUIDELINES

LEARN TO RECOGNIZE PLAGIARISM

DECIDE WHERE AND HOW TO DRAW THE LINE

CONSIDER THIS Ethical Standards Are Good for Business

GUIDELINES for Ethical Communication

CHECKLIST for Ethical Communication

An effective message (one that achieves its purpose) isn't necessarily an ethical message. Think of examples from advertising: "Our artificial sweetener is composed of proteins that occur naturally in the human body (amino acids)" or "Our potato chips contain no cholesterol." Such claims are technically accurate but misleading: amino acids in certain sweeteners can alter body chemistry to cause headaches, seizures, and possibly brain tumors; potato chips are loaded with saturated fat—which produces cholesterol.

Communication is unethical when it leaves recipients at a disadvantage or prevents them from making their best decisions (Figure 5.1). Ethical communication is measured by standards of honesty, fairness, and concern for everyone involved (Johannesen 1).

**FIGURE 5.1
Three Problems
Confronted by
Communicators**

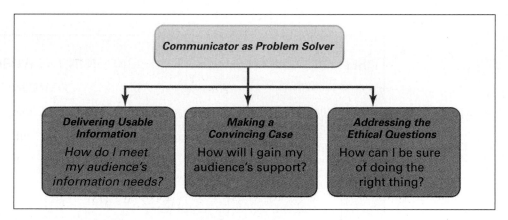

RECOGNIZE UNETHICAL COMMUNICATION IN THE WORKPLACE

Recent financial scandals reveal a growing list of corporations accused of boosting the value of company stocks by overstating profits and understating debt. As inevitable bankruptcy loomed, executives hid behind deceptive accounting practices; company officers quietly unloaded personal shares of inflated stock while employees and investors were kept in the dark and ended up losing billions. Small wonder that "opinion polls now place business people in lower esteem than politicians" (Merritt, "For MBAs" 64).

NOTE   *Among the 84 percent of college students surveyed who claim to be "disturbed" by corporate dishonesty, "59 percent admit to cheating on a test . . . and only 19 percent say they would report a classmate who cheated" (Merritt, "You Mean" 8).*

Corporate scandals make for dramatic headlines, but more routine examples of deliberate miscommunication rarely are publicized:

Routine instances of unethical communication

- A person lands a great job by exaggerating his credentials, experience, or expertise
- A marketing specialist for a chemical company negotiates a huge bulk sale of its powerful new pesticide by downplaying the carcinogenic hazards
- A manager writes a strong recommendation to get a friend promoted, while overlooking someone more deserving

BusinessWeek reports that 20 percent of employees surveyed claim to have witnessed fraud on the job. Common abuses range from falsifying expense accounts to overstating hours worked ("Crime Spree" 8).

Other instances of unethical communication are less black and white. Here is one engineer's description of the gray area in which debates over product safety versus quality often occur:

Ethical decisions are not always "black and white"

> The company must be able to produce its products at a cost low enough to be competitive. . . . To design a product that is of the highest quality and consequently has a high and uncompetitive price may mean that the company will not be able to remain profitable, and be forced out of business. (Burghardt 92)

Do you emphasize to a customer the need for scrupulous maintenance of a highly sensitive computer—and risk losing the sale? Or do you focus instead on the computer's positive features? The decisions we make in these situations are often influenced by the pressures we feel.

KNOW THE MAJOR CAUSES OF UNETHICAL COMMUNICATION

Well over 50 percent of managers surveyed nationwide feel "pressure to compromise personal ethics for company goals" (Golen et al. 75). To save face, escape blame, or get ahead, anyone might be tempted to say what people want to hear or to suppress bad news. But normally honest people usually break the rules only when compelled by an employer, coworkers, or their own bad judgment. Figure 5.2 depicts how workplace pressures can influence ethical values.

Yielding to Social Pressure

Sometimes, you may have to choose between doing what you know is right and doing what your employer or organization expects, as in this example:

Pressure to "look the other way"

> Just as your automobile company is about to unveil its new pickup truck, your safety engineering team discovers that the reserve gas tanks (installed beneath the truck but outside the frame) may, in rare circumstances, explode on impact from a side collision. You know that this information should be included in the owner's manual or, at a minimum, in a letter to the truck dealers, but the company has spent a fortune building this truck and doesn't want to hear about this problem.

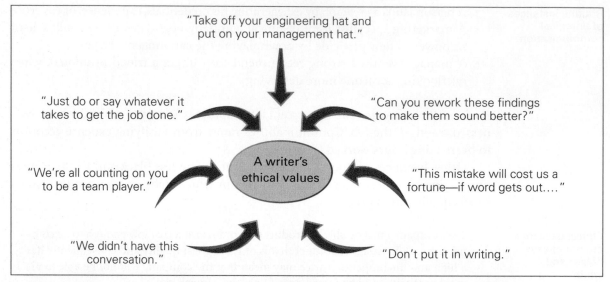

FIGURE 5.2   How Workplace Pressures Can Influence Ethical Values   A decision that is more efficient, profitable, or better for the company might overshadow a person's sense of what is right.

5.1

For more on group dynamics and collaboration visit <www.ablongman.com/lannonweb>

Companies often face the contradictory goals of *production* (producing a product and making money on it) and *safety* (producing a product but spending money to avoid accidents that may or may not happen). When productivity receives first priority, safety concerns may suffer (Wickens 434–36). In these circumstances, you need to rely on your own ethical standards. If, in the case of the reserve gas tanks, you decide to publicize the problem, expect to be fired for defying the company.

Mistaking Groupthink for Teamwork

Organizations rely on teamwork and collaboration to get a job done. For example, technical communicators often work as part of a larger team of writers, editors, designers, engineers, and other technical experts. Teamwork is important in these situations, but teamwork should not be confused with *groupthink*, which occurs when group pressure prevents individuals from questioning, criticizing, reporting bad news, or "making waves" (Janis 9). Group members may need to feel accepted by the team, often at the expense of making the right decision. Anyone who has ever given in to adolescent peer pressure has experienced a version of groupthink.

"I was only
following orders!"

5.2

View historical Nazi
technical documents at
<www.ablongman.com/
lannonweb>

Groupthink also can provide a handy excuse for individuals to deny responsibility. For example, because countless people work on a complex project (say, a new airplane), identifying those responsible for an error is often impossible—especially in errors of omission, that is, of overlooking something that should have been done (Unger 137).

> People commit unethical acts inside corporations that they never would commit as individuals representing only themselves. (Bryan 86)

Figure 5.3 depicts the kind of thinking that allows individuals to deny responsibility.[1]

**FIGURE 5.3
Hiding behind
Groupthink**

My job is to put together a persuasive ad for this brand of diet pill. It is someone else's job to make sure the claims are accurate and cause the customer no harm. It was someone else's job to make sure the stuff was safe. My job is only to promote the product. If someone does get hurt, it won't be my problem!

UNDERSTAND THE POTENTIAL FOR COMMUNICATION ABUSE

On the job, your effectiveness is judged by how well your documents speak for the company and advance its interests and agendas (Ornatowski 100–01). You walk the proverbial line between telling the truth and doing what your employer expects (Dombrowski 79).

Workplace communication influences the thinking, actions, and welfare of numerous people: customers, investors, coworkers, the public, policymakers—to name a few. These people are victims of communication abuse whenever we give them information that is less than the truth as we know it, as in the following situations.

Suppressing Knowledge the Public Needs

Pressures to downplay the dangers of technology can result in censorship of important information. Threat of a lawsuit, for instance, might cause a journal editor to suppress an article about adverse effects of a popular over-the-counter medication.

---

[1]My thanks to Judith Kaufman for this idea.

Here are additional examples:

Examples of
suppressed
information

- The biotech industry continues to resist any food labeling that would identify genetically modified ingredients (Raeburn 78).
    - Some prestigious science journals have refused to publish legitimate studies linking chlorine and fluoride in drinking water with cancer risk, and fluorescent lights with childhood leukemia (Begley, "Is Science" 63).
    - MIT's Arnold Barnett has found that information about airline safety lapses and near-accidents is often suppressed by air traffic controllers because of "a natural tendency not to call attention to events in which their own performance was not exemplary" or their hesitation to "squeal" about pilot error (qtd. in Ball 13).
    - Scientific breakthroughs, say, in cancer treatment, may be kept secret for months until lucrative patents can be obtained. One survey revealed this practice among 82 percent of biomedical companies and 50 percent of university researchers (Gibbs, "The Price" 16). Consider this notable example from the human genome project: Some companies seek to patent their particular genetic discoveries, thereby restricting public access and hindering development of vital medical treatments based on this information.

Ethical issues

"I have to be EXTREMELY careful of cultural bias. I also have to be conscious of as many related issues as possible when assessing the patient so as not to skew the evaluation. This documentation may also be seen by state agencies and case workers and if for some reason this child brings the hospital to court or vice versa, the document may be used in court."

—Emma Bryant, social worker

Hiding Conflicts of Interest

It's hard to expect fair and impartial information from scientists and other experts who have a financial stake in a particular issue:

Hidden conflicts
of interest

- In one analysis of 800 scientific papers, Tufts University's Sheldon Krimsky found that 34 percent of authors had "research-related financial ties," but none had been disclosed (King B1).
- *Los Angeles Times* medical writer Terence Monmaney recently investigated 36 drug review pieces in a prestigious medical journal and found "eight articles by researchers with undisclosed financial links to drug companies that market treatments evaluated in the articles" (qtd. in Rosman 100).
- Analysts on a popular TV financial program have recommended certain company stocks (thus potentially inflating the price of that stock) without disclosing that their investment firms hold stock in these companies (Oxfeld 105).

5.3

Learn more about recent conflicts of interest in medical research at <www.ablongman.com/lannonweb>

Exaggerating Claims about Technology

Organizations that have a stake in a particular technology (say, bioengineered foods) may be especially tempted to exaggerate its benefits, potential, or safety and downplay its risks. If your organization depends on outside funding (as in the defense or space industry), you might find yourself pressured to make unrealistic promises.

Falsifying or Fabricating Data

Research data might be manipulated or invented to support, say, a tobacco company facing lawsuits or by a scientist seeking grant money. Developments in fields such as biotechnology often occur too rapidly to allow for adequate peer review of articles before they are published ("Misconduct Scandal" 2).

More examples of distorted visuals and exaggerated claims at <www.ablongman.com/lannonweb>

Using Visual Images That Conceal the Truth

Generally more powerful than words, pictures can easily distort the real meaning of a message. For example, as required by law, TV commercials for prescription medications must inform consumers of a drug's side effects—which can often be serious. But the typical drug commercial, as commentator Rob Edelman notes, describes side effects while showing benign images of smiling people holding hands as they walk through the flowers or enjoy their grandchildren's play, and so on. In short, the happy image is designed to eclipse the sobering verbal message.

View a sample nondisclosure agreement at <www.ablongman.com/lannonweb>

Stealing or Divulging Proprietary Information

Information that originates in a specific company is the exclusive intellectual property of that company. Proprietary information includes insider financial information, employee records, product formulas, test and experiment results, surveys financed by clients, market research, plans, specifications, and minutes of meetings (Lavin 5). In theory, such information is legally protected, but it remains vulnerable to sabotage, theft, or exposure to the press. Fierce competition for the latest intelligence among rival companies leads to measures like these:

Examples of corporate espionage

> Companies have been known to use business school students to garner information on competitors under the guise of conducting "research." Even more commonplace is interviewing employees for slots that don't exist and wringing them dry about their current employer. (Gilbert 24)

The law prohibits employees who switch jobs from revealing proprietary information about a previous employer, or from soliciting its clients (page 86). A court recently ruled that even a collection of customer business cards can be a trade secret. The "doctrine of inevitable disclosure" asserts that an employee in a new job, sooner or later and deliberately or not, will disclose sensitive information. Acting on this doctrine, courts require some employees to remain idle between jobs until the information they possess has become outdated (Lenzer and Shook 100–02).

Misusing Electronic Information

With so much information stored in databases (by schools, employers, government, mail order retailers, credit bureaus, banks, credit card companies, insurance companies, pharmacies), how we combine, use, and share the information raises questions about personal privacy (Finkelstein 471). Also, compared with hard copy, a database is easier to alter; one simple command can wipe out or transform the facts. Private or inaccurate information can be sent from one database to countless others, but "correcting information in one database does not guarantee that it will be corrected in others" (Turner 5).

The proliferation of Web transactions, of course, creates broad opportunities for communication abuse. Notable examples include:

Web-based communication abuses

- Plagiarizing or republishing electronic sources without giving proper credit or obtaining permission.
- Copying digital files—music CDs, for example—without consent of the copyright holders.
- Failing to safeguard privacy. People worry about the ease with which Web sites can exchange personal information about visitor's health, finances, or buying habits. In one survey, "significant proportions of consumers" claimed that concerns for personal privacy prevent them from using health-related Web sites ("Online Health" 3–4).
- Publishing anonymous attacks, or smear campaigns, against people, products, or organizations. Anonymity is, of course, essential for people speaking out against a political dictatorship or for employees exposing corrupt business practices, but free speech is no excuse for threatening or defaming others (Gibbs, "Speech" 35–36).
- Selling prescription medications online without adequate patient screening or physician consultation. Terming online pharmacies "the true Wild West of the Web sites," journalists Claudia Kalb and Deborah Branscum point out that online pharmacies operating outside the United States can even market drugs never subjected to United States testing or approval (66).
- Offering inaccurate medical advice or information. University of Michigan researchers found erroneous information on a type of bone cancer at nearly one-third of 371 medical Web sites. On the basis of such information, laypersons could make disastrous decisions about treatment or choice of therapy ("Advisories" 2–3).

Withholding Information People Need for Their Jobs

Nowhere is the adage that "information is power" truer than among coworkers. One sure way to sabotage a colleague is to deprive that person of information about the task at hand. Studies show that employees withhold information for more benign reasons as well, such as fear that someone else might take credit for their work or might "shoot them down" (Davenport 90).

Exploiting Cultural Differences

Based on their business experience or their social values, people from certain cultures might be especially vulnerable to manipulation or deception. Some countries, for example, place greater reliance on interpersonal trust than on lawyers or legal wording, and a handshake can be worth more than the fine print of a legal contract. Other countries may tolerate abuse or destruction of their natural resources in order to generate much-needed income.

U.S. corporations, one researcher notes, "have had to address [ethical] questions about doing business with repressive governments, in environmentally sensitive areas, and with work forces whose low expectations can lead to exploitation" (Rivers 404). Two communication experts offer these examples of unethical behavior in a cross-cultural context: "[excluding] graphics of women in positions of power from presentation materials prepared for an Arabic audience" or "[saving] 20% on publishing costs by sending a document to a Third World sweat shop that exploits child labor" (Allen and Voss 59–60). If you know something about a culture's habits or business practices and if you use this information at their expense to get a sale or make a profit, you are behaving unethically.

Consider this recent attempt to use cultural differences as a basis for violating personal privacy: In a program of library surveillance, the FBI asked librarians to compile lists of materials being read by any "foreign national patron" or anyone with a "foreign sounding name." The librarians refused (Crumpton 8).

RELY ON CRITICAL THINKING FOR ETHICAL DECISIONS

Because of their effects on people and on your career, ethical decisions challenge your critical thinking skills:

Ethical decisions require critical thinking

> - How can I know the "right action" in this situation?
> - What are my obligations, and to whom, in this situation?
> - What values or ideals do I want to stand for in this situation?
> - What is likely to happen if I do X—or Y?

Ethical issues resist simple formulas, but the following criteria offer some guidance.

Reasonable Criteria for Ethical Judgment

Reasonable criteria (standards that most people consider acceptable) take the form of *obligations, ideals,* and *consequences* (Ruggiero, 3rd ed. 33–34; Christians et al. 17–18).

Obligations are the responsibilities you have to everyone involved:

Our obligations are varied and often conflicting

- *Obligation to yourself,* to act in your own self-interest and according to good conscience
- *Obligation to clients and customers,* to stand by the people to whom you are bound by contract—and who pay the bills
- *Obligation to your company,* to advance its goals, respect its policies, protect confidential information, and expose misconduct that would harm the organization
- *Obligation to coworkers,* to promote their safety and well-being
- *Obligation to the community,* to preserve the local economy, welfare, and quality of life
- *Obligation to society,* to consider the national and global impact of your actions

When the interests of these parties conflict—as they often do—you have to decide where your primary obligations lie.

Ideals are the values that you believe in or stand for: loyalty, friendship, compassion, dignity, fairness, and whatever qualities make you who you are (Ruggiero, 3rd ed. 33).

Consequences are the beneficial, or harmful, results of your actions. Consequences may be immediate or delayed, intentional or unintentional, obvious or subtle (Ruggiero, 3rd ed. 33). Some consequences are easy to predict; some are difficult; some are impossible.

Figure 5.4 depicts the relationship among these three criteria.

**FIGURE 5.4
Reasonable
Criteria for
Ethical
Judgment**

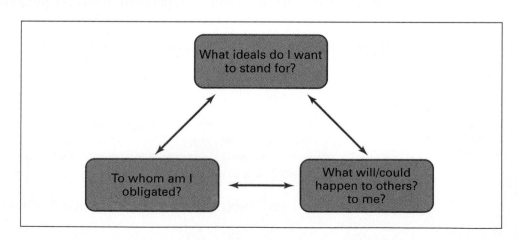

The above criteria help us understand why even good intentions can produce bad judgments, as in the following situation.

What seems like the "right action" might be the wrong one

Someone observes . . . that waste from the local mill is seeping into the water table and polluting the water supply. This is a serious situation and requires a remedy. But before one can be found, extremists condemn the mill for lack of conscience and for exploiting the community. People get upset and clamor for the mill to be shut down and its management tried on criminal charges. The next thing you know, the plant does close, 500 workers are without jobs, and no solution has been found for the pollution problem. (Hauser 96)

Because of their zealous dedication to the *ideal* of a pollution-free environment, the extremists failed to anticipate the *consequences* of their protest or to respect their *obligation* to the community's economic welfare.

Ethical Dilemmas

Ethical decisions are especially frustrating when no single choice seems acceptable (Ruggiero, 3rd ed. 35).

Ethical questions often resist easy answers

In private and public ways, such dilemmas are inescapable. For example, the announced intention of "welfare reform" includes freeing people from lifelong dependence. One could argue that dedication to this *consequence* would violate our *obligations* (to the poor and the sick) and our *ideals* (of compassion or fairness). On the basis of our three criteria, how else might the welfare-reform issue be considered?

Ethical dilemmas also confront the medical and scientific communities. For instance, in late 1992, a brain-dead, pregnant woman in Western Europe was kept alive for several weeks until her child could be delivered. Also in 1992, a California court debated the ethics of using the medical findings of Nazi doctors who had experimented on prisoners in concentration camps. More recently, scientists researching deadly organisms such as anthrax, smallpox, and other bioweapons are debating whether to censor their own findings. Censorship might slow the spread of dangerous information among terrorists, but it also would prevent the kind of information sharing among the scientific community that could lead to better protection against bioweapons (Green 104). In terms of our three criteria, how might these dilemmas be considered?

ANTICIPATE SOME HARD CHOICES

Communicators' ethical choices basically involve revealing or concealing information:

- What exactly do I report and to whom?
- How much do I reveal or conceal?
- How do I say what I have to say?
- Could misplaced obligation to one party be causing me to deceive others?

For illustration of a hard choice in workplace communication, consider the following scenario:

A Hard Choice

You are a structural engineer working on the construction of a nuclear power plant in a developing country. After years of construction delays and cost overruns, the plant finally has received its limited operating license from the country's Nuclear Regulatory Commission.

During your final inspection of the nuclear core containment unit, on February 15, you discover a ten-foot-long, hairline crack in a section of the reinforced concrete floor, twenty feet from where the cooling pipes enter the containment unit. (The especially cold and snowless winter likely has caused a frost heave beneath the foundation.) The crack has either just appeared or was overlooked by NRC inspectors on February 10.

The crack might be perfectly harmless, caused by normal settling of the structure; and this is, after all, a "redundant" containment system (a shell within a shell). But the crack might also signal some kind of serious stress on the entire containment unit, which could damage the entry and exit cooling pipes or other vital structures.

You phone your boss, who is just about to leave on vacation. He tells you, "Forget it; no problem," and hangs up.

You know that if the crack is reported, the whole start-up process scheduled for February 16 will be delayed indefinitely. More money will be lost; excavation, reinforcement, and further testing will be required—and many people with a stake in this project (from company executives to construction officials to shareholders) will be furious—especially if your report turns out to be a false alarm. Media coverage will be widespread. As the bearer of bad news—and bad publicity—you suspect that, even if you turn out to be right, your own career could be damaged by your apparent overreaction.

On the other hand, ignoring the crack could have unforeseeable consequences. Of course, no one would ever be able to implicate you. The NRC has already inspected and approved the containment unit—leaving you, your boss, and your company in the clear. You have very little time to decide. Start-up is scheduled for tomorrow, at which time the containment system will become intensely radioactive.

What would you do? Come to class prepared to justify your decision on the basis of the obligations, ideals, and consequences involved. ❑

> You may have to choose between the goals of your organization and what you know is right

Working professionals commonly face similar choices, which they must often make alone or on the spur of the moment, without the luxury of meditation or consultation.

NEVER DEPEND ONLY ON LEGAL GUIDELINES

Communication can be legal without being ethical

Can the law tell you how to communicate ethically? Sometimes. If you stay within the law, are you being ethical? Not always—as illustrated in this chapter's earlier section on communication abuses. In fact, even threatening statements made on the Web are considered legal by the Supreme Court as long as they are not likely to cause "imminent lawless action" (Gibbs, "Speech" 34).

Legal standards "sometimes do no more than delineate minimally acceptable behavior." In contrast, ethical standards "often attempt to describe ideal behavior, to define the best possible practices for corporations" (Porter 183).

Except for the instances listed below, lying is rarely illegal. Common types of legal lies are depicted in Figure 5.5. Later chapters cover other kinds of legal lying, such as page design that distorts the real emphasis or words that are deliberately unclear, misleading, or ambiguous.

**FIGURE 5.5
Lies That Are
Legal**

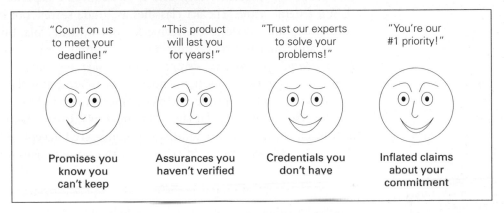

"Count on us to meet your deadline!"	"This product will last you for years!"	"Trust our experts to solve your problems!"	"You're our #1 priority!"
Promises you know you can't keep	Assurances you haven't verified	Credentials you don't have	Inflated claims about your commitment

What, then, are a communicator's legal guidelines? Among the laws regulating workplace communication are these:

Laws that govern workplace communication

- *Laws against deception* prohibit lying under oath, lying to a federal agent, lying about a product so as to cause injury, or breaking a contractual promise.
- *Libel law* prohibits any false written statement that maliciously attacks or ridicules anyone. A statement is considered libelous when it damages someone's reputation, character, career, or livelihood or when it causes humiliation or mental suffering. Material that is damaging but truthful would not be considered libelous unless it were used intentionally to cause harm. In the event of a libel suit, a writer's ignorance is no defense; even when the damaging material has been obtained from a presumably reliable source, the writer (and publisher) are legally accountable.[2]
- *Laws protecting employee privacy* impose strict limits on information employers are allowed to give out regarding an employee's job references, disciplinary

---

[2]Thanks to my colleague Peter Owens for the material on libel.

problems, health problems, or reasons for being fired. (See page 284 for more on this topic.)

- *Copyright law* (pages 138, 144) protects the ownership rights of authors—or of their employers, in cases where the writing was done as part of their employment.

- *Law against software theft* prohibits unauthorized duplication of copyrighted software. A first offense carries up to five years in prison and fines up to $250,000. The Software Publisher's Association estimates that software piracy costs the industry more than $2 billion yearly ("On Line" A29).

- *Law against electronic theft* prohibits unauthorized distribution of copyrighted material via the Internet as well as possession of ten or more electronic copies of any material worth $2,500 or more (Evans, "Legal Briefs" 22).

- *Laws against stealing or revealing trade secrets.* According to FBI estimates, roughly $25 billion of proprietary information (trade secrets and other exclusive intellectual property) is stolen yearly. The 1996 Economic Espionage Act makes such theft a federal crime. The act classifies as "trade secret" not only items such as computer source code or the recipe for our favorite cola, but even a listing of clients and contacts brought from a previous employer (Farnham 114, 116).

- *Laws against fraudulent, deceptive, or misleading advertising* prohibit false claims or suggestions, say, that a product or treatment will cure disease, or representation of a used product as new. Fraud is defined as lying that causes another person monetary damage (Harcourt 64). Even a factual statement such as "our cigarettes have no additives" is considered deceptive because it implies that a cigarette with no additives is safer than other cigarettes (Savan 63). The Federal Trade Commission offers the following legal guidelines:

When ads break the law

> A claim can be misleading if relevant information is left out or if the claim implies something that's not true. For example, a lease advertisement for an automobile that promotes "$0 Down" may be misleading if significant and undisclosed charges are due at lease signing. In addition, claims must be substantiated. . . . If your ad specifies a certain level of support for a claim—"tests show X"— you must have at least that level of support. . . . Advertising agencies or Web-site designers also may be liable for making or disseminating deceptive representations . . . [and they] are responsible for reviewing the information used to substantiate ad claims. ("Advertising and Marketing")

- *Liability laws* define the responsibilities of authors, editors, and publishers for damages resulting from incomplete, unclear, misleading, or otherwise defective information. The misinformation might be about a product (such as failure to warn about the toxic fumes from a spray-on oven cleaner) or a procedure (such as misleading instructions in an owner's manual for using a tire jack). Even if misinformation is given out of ignorance, the writer is liable (Walter and Marsteller 164–65).

Legal standards for product literature vary from country to country. A document must satisfy the legal standards for safety, health, accuracy, or language for the country in which it is distributed. For example, instructions for assembly or operation must carry warnings stipulated by the country in which the product is sold. Inadequate documentation, as judged by that country's standards, can result in a lawsuit (Caswell-Coward 264–66; Weymouth 145).

NOTE

Laws regulating communication practices are few because such laws have traditionally been seen as threats to freedom of speech (Johannesen 86). Large companies typically have legal departments to advise you about a document's legal aspects. Also, most professions have their own ethics guidelines. If your field has its own formal code, obtain a copy.

LEARN TO RECOGNIZE PLAGIARISM

Ethical communication includes giving proper credit to the work of others. In both workplace and academic settings, plagiarism (representing the words, ideas, or perspectives of others as your own) is a serious breach of ethics.

Examples of plagiarism

Blatant cases of plagiarism occur when a writer consciously lifts passages from another work (print or online) and incorporates them into his or her own work without quoting or documenting the original source. As most students know, this can result in a failing grade and potential disciplinary action. More often, writers will simply fail to cite a source being quoted or paraphrased, often because they misplaced the original source and publication information, or forgot to note it during their research (Anson and Schwegler 633–36). Whereas this more subtle, sometimes unconscious form of misrepresentation is less blatant, it still constitutes plagiarism and can undermine a technical communicator's credibility, or worse. Whether the infraction is intentional or unintentional, writers, researchers, and other professionals accused of plagiarism can lose their reputation and be sued or fired.

Plagiarism and the Internet

The rapid development of Internet resources has spawned a wild array of misconceptions about plagiarism. Some people mistakenly assume that because material posted on a Web site is free, it can be paraphrased or copied without citation. Despite the ease of cutting and pasting from Web sites, the fact remains: Any time you borrow someone else's words, ideas, perspectives, or images—regardless of the medium used in the original source—you need to document the original source accurately.

5.6

For more on plagiarism and corporate document "reuse" visit <www.ablongman.com/lannonweb>

Whatever your job, learning to gather, incorporate, and document authoritative source material is an absolutely essential career skill. By properly citing a range of sources in your work, you bolster your own credibility and demonstrate your skills as a researcher and a writer. (For more on incorporating and documenting sources, see Appendix A. For more on recognizing plagiarism, go to <www.indiana.edu/~wts/wts/plagiarism.html>.)

NOTE

Plagiarism and copyright infringement are not the same. You can plagiarize someone else's work without actually infringing copyright. These two issues are frequently confused, but plagiarism is primarily an ethical issue, whereas copyright infringement is a legal and economic issue. (For more on copyright and related legal issues, see Chapter 8.)

CONSIDER THIS   Ethical Standards Are Good for Business

People look for companies they can trust. And businesses have discovered that trustworthiness creates goodwill. To earn public trust, for example, companies involved in human cloning, bioengineered foods, and other controversial technologies increasingly are hiring "professional ethical advisors" to help them "sort out right from wrong when it comes to developing, marketing, and talking about new technology" (Brower 25). Here are some notable instances of good standards that pay off.

By Telling Investors the Truth, a Company Can Increase Its Stock Value

Publicly traded companies that tell the truth about profits and losses are tracked by more securities analysts than companies that use "accounting smoke screens" to hide bad economic news. "All it takes is the inferential leap that more analysts touting your stock means a higher stock price" (Fox 303).

High Standards Earn Customer Trust

Wetherill Associates, Inc., a car parts supply company, was founded on the principle of honesty and "taking the right action," instead of trying to maximize profits and minimize losses. Among Wetherill's policies:

- *Employees are given no sales quotas, so that no one will be tempted to camouflage disappointing sales figures.*
- *Employees are required to be honest in all business practices.*
- *Lies (including "legal lies") to colleagues or customers are grounds for being fired.*
- *Employees who gossip or backbite are penalized.*

From a $50,000 start-up budget and 45 people who shared this ethical philosophy in 1978, the company has grown to 480 employees and $160 million in yearly sales and $16 million in profit—and continues growing at 25 percent annually (Burger 200–01).

Socially Responsible Action Earns International Goodwill

John Brown, CEO of British Petroleum Co., is committed to the economic and social prosperity of all locations in which his global corporation operates. Some of BP's efforts to earn employee loyalty and community goodwill:

- *Building schools, providing job training for local employees, and supporting small business development*
- *Providing medical equipment and assorted technology*
- *Repairing environmental damage from forest fires and other natural disasters*
- *Keeping detailed, open records of workplace and environmental accidents, and working constantly to eliminate such accidents*
- *Listening to the concerns of local residents and seeking their feedback*

Rather than occurring at shareholder expense, BP's investment in social responsibility apparently improves its bottom line: Annual returns of 33 percent have outperformed the Dow-Jones industrial average by more than 50 percent since 1992 (Garten 26).

Sharing Information with Coworkers Leads to a Huge Invention

Information expert Keith Devlin describes this result of one company's "strong culture of sharing ideas":

> The invention of Post-it Notes by 3M's Art Fry came about as a result of a memo from another 3M scientist who described the new glue he had developed. The new glue had the unusual property of providing firm but very temporary adhesion. As a traditional bonding agent, it was a failure. But Fry was able to see a novel use for it, and within a short time, Post-it Notes could be seen adorning every refrigerator door in the land. (179–80)

DECIDE WHERE AND HOW TO DRAW THE LINE

Suppose your employer asks you to cover up fraudulent Medicare charges or a violation of federal pollution standards. If you decide to resist, your choices seem limited: resign or go public (i.e., blow the whistle).

Walking away from a job isn't easy, and whistle-blowing can spell career disaster. Many organizations refuse to hire anyone blacklisted as a whistle-blower. Even if you aren't fired, expect animosity on the job. Consider this excerpt from a study of sixty-eight whistle-blowers by the Research Triangle Institute:

Consequences of
whistle-blowing

> "More than two-thirds of all whistle-blowers reported experiencing at least one negative outcome. . . . " Those most likely to experience adverse consequences were "lower ranking [personnel]." Negative consequences included pressure to drop their allegations, [ostracism] by colleagues, reduced research support, and threatened or actual legal action. Interestingly, . . . three-fourths of these whistle-blowers experiencing "severe negative consequences" said they would definitely or probably blow the whistle again. (qtd. in "Consequences of Whistle Blowing" 2)

Despite the negative consequences, few people surveyed regretted their decision to go public.

Employers are generally immune from lawsuits by employees who have been dismissed unfairly but who have no contract or union agreement specifying length of employment (Unger 94). Current law, however, offers some protection for whistle-blowers.

Limited legal
protections for
whistle-blowers

- The Federal False Claims Act allows an employee to sue, in the government's name, a contractor who defrauds the government (say, by overcharging for military parts). The employee receives up to 25 percent of money recovered by the suit. Also, this law allows employees of government contractors to sue when they are punished for whistle-blowing (Stevenson A7).
- Anyone punished for reporting employer violations to a regulatory agency (Federal Aviation Administration, Nuclear Regulatory Commission, Occupational Health and Safety Administration, and so on) can request a Labor Department investigation. A claim ruled valid leads to reinstatement and reimbursement for back pay and legal expenses.[3]
- Laws in several states protect employees who report discrimination or harassment on the basis of sexual orientation (Fisher, "Can I Stop" 205).
- Beyond requiring greater accuracy and clarity in the financial reports of publicly traded companies, The Sarbanes-Oxley Act of 2002 imposes criminal penalties for executives who retaliate against employees who blow the whistle on corporate misconduct. This legislation also requires companies to establish confidential hotlines for reporting ethical violations.

[3]Although employees are legally entitled to speak confidentially with OSHA inspectors about health and safety violations, one survey reveals that inspectors themselves believe such laws offer little protection against company retribution (Kraft 5).

• One Web-based service, Ethicspoint.com, allows employees to file their reports anonymously and then forwards this information to the company's ethics committee.

5.7

Compare ethical codes for various professions at <www.ablongman.com/lannonweb>

Even with such protections, an employee who takes on a company without the backing of a labor union or other powerful group can expect lengthy court battles, high legal fees (which may or may not be recouped), and disruption of life and career.

Before accepting a job offer, do some discreet research about the company's reputation. (Of course you can learn only so much before actually working there.) Learn whether the company has *ombudspersons* (who help employees file complaints) or hotlines for advice on ethics problems or for reporting misconduct. Ask whether the company or organization has a formal code for personal and organizational behavior (Figure 5.6). Finally, assume that no employer, no matter how ethical, will tolerate any public statement that makes the company look bad.

NOTE   *Sometimes the right choice is obvious, but often it is not so obvious. No one has any sure way of always knowing what to do. This chapter is only an introduction to the inevitable hard choices that, throughout your career, will be yours to make and to live with. For further guidance and case examples, go to The Online Ethics Center for Engineering and Science at <www.onlineethics.org>.*

☑ CHECKLIST for Ethical Communication

Use this checklist* for any document you prepare or for which you are responsible. Numbers in parentheses refer to the first page of discussion)

Accuracy

☐ Have I explored all sides of the issue and all possible alternatives? (121)

☐ Do I provide enough information and interpretation for recipients to understand the facts as I know them? (92)

☐ Do I avoid exaggeration, understatement, sugarcoating, or any distortion or omission that would leave recipients at a disadvantage? (74)

☐ Do I state the case clearly, instead of hiding behind jargon and generalities? (266)

Honesty

☐ Do I make a clear distinction between "certainty" and "probability"? (79)

☐ Are my information sources valid, reliable, and relatively unbiased? (165)

☐ Do I actually believe what I'm saying, instead of being a mouthpiece for groupthink or advancing some hidden agenda? (76)

☐ Would I still advocate this position if I were held publicly accountable for it? (77)

☐ Do I inform people of the consequences or risks (as I am able to predict) of what I am advocating? (75)

☐ Do I give candid feedback or criticism, if it is warranted? (76)

Fairness

☐ Am I reasonably sure this document will harm no innocent persons or damage their reputations? (85)

☐ Am I respecting all legitimate rights to privacy and confidentiality? (80)

☐ Am I distributing copies of this document to every person who has the right to know about it? (81)

☐ Do I credit all contributors and sources of ideas and information? (87)

*Adapted from Brownell and Fitzgerald 18; Bryan 87; Johannesen 21–22; Larson 39; Unger 39–46; Yoos 50–55.

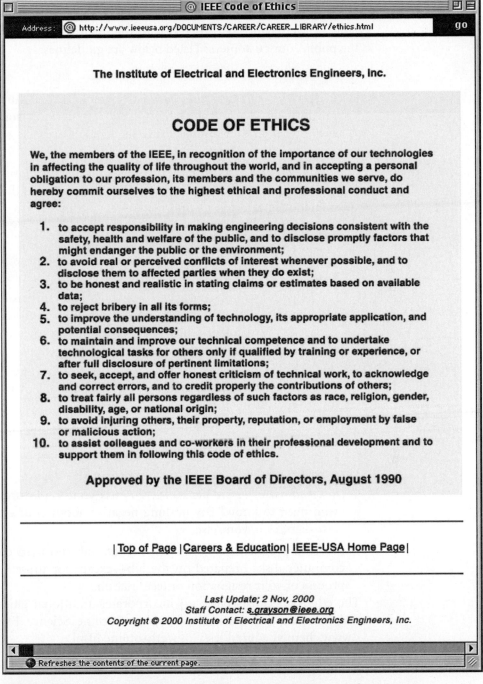

FIGURE 5.6   A Sample Code of Ethics.   Notice the many references to *communication* in this engineering association's code of professional conduct.

Source: The Institute of Electrical and Electronics Engineers, Inc., <www.ieee.org>. Copyright © 2004 IEEE. Reprinted with permission of the IEEE.

GUIDELINES for Ethical Communication*

How do we balance self-interest with the interests of others —our employers, the public, our customers? Listed below are guidelines:

Satisfying the Audience's Information Needs

1. *Give the audience everything it needs to know.* To see things as accurately as you do, people need more than just a partial view. Don't bury readers in needless details, but do make sure they get all the facts straight. If you're at fault, admit it and apologize immediately.

2. *Give the audience a clear understanding of what the information means.* Facts can be misinterpreted. Ensure that readers understand the facts as you do. If you're not certain about your own understanding, say so.

Taking a Stand versus the Company

1. *Get your facts straight, and get them on paper.* Don't blow matters out of proportion, but do keep a paper (and digital) trail in case of legal proceedings.

2. *Appeal your case in terms of the company's interests.* Instead of being pious and judgmental ("This is a racist and sexist policy, and you'd better get your act together"), focus on what the company stands to gain or lose ("Promoting too few women and minorities makes us vulnerable to legal action").

3. *Aim your appeal toward the right person.* If you have to go to the top, find someone who knows enough to appreciate the problem and who has enough clout to make something happen.

4. *Get professional advice.* Contact an attorney and your union or professional society for advice about your legal rights.

Leaving the Job

1. *Make no waves before departure.* Discuss your departure only with people "who need to know." Say nothing negative about your employer to clients, coworkers, or anyone else.

2. *Leave all proprietary information behind.* Take no hard copy documents or computer disks prepared on the job—except for those records tracing the process of your resignation or termination.

The ethics checklist (page 90) incorporates additional guidelines from other chapters. For additional advice, go to "Online Science Ethics Resources" at <www.chem.vt.edu/ethics/vinny/ethxonline.html>.

*Adapted from G. Clark 194; Unger 127–30; Lenzer and Shook 102.

EXERCISES For more exercises, visit
<www.ablongman.com/lannon>

1. Prepare a memo (one or two pages) for distribution to first-year students in which you introduce the ethical dilemmas they will face in college. For instance:

   - If you received a final grade of A by mistake, would you inform your professor?
   - If the library lost the record of books you've signed out, would you return them anyway?
   - Would you plagiarize—and would that change in your professional life?
   - Do you support lowering standards for student athletes if the team's success was important for the school's funding and status?
   - Would you allow a friend to submit a paper you've written for some other course?
   - What other ethical dilemmas can you envision? Tell your audience what to expect, and give them some *realistic* advice for coping. No sermons, please.

2. In your workplace communications, you may face hard choices concerning what to say, how much to say, how to say it, and to whom. Whatever your choice, it will have definite consequences. Be prepared to discuss the following cases in terms of the obligations, ideals, and consequences involved. Can you think of similar choices you or someone you know has faced? What happened? How might the problems have been avoided?

   - While traveling on an assignment that is being paid for by your employer, you visit an area in which you would really like to live and work, an area in which you have lots of contacts but never can find time to visit on your own. You have five days to complete your assignment, and then you must report on your activities. You complete the assignment in three days. Should you spend the remaining two days checking out other job possibilities, without reporting this activity?
   - As a marketing specialist, you are offered a lucrative account from a cigarette manufacturer; you are expected to promote the product. Should you accept the account? Suppose instead the account were for beer, junk food, suntanning parlors, or ice cream. Would your choice be different? Why, or why not?
   - You have been authorized to hire a technical assistant, so you are about to prepare an advertisement. This is a time of threatened cutbacks for your company. People hired as "temporary," however, have never seemed to work out well. Should your ad include the warning that this position could be only temporary?
   - You are one of three employees being considered for a yearly production bonus, which will be awarded in six weeks. You've just accepted a better job, at which you can start any time in the next two months. Should you wait until the bonus decision is made before announcing your plans to leave?
   - You are marketing director for a major importer of coffee beans. Your testing labs report that certain African beans contain roughly twice the caffeine of South American varieties. Many of these African varieties are big sellers, from countries whose coffee bean production helps prop otherwise desperate economies. Should your advertising of these varieties inform the public about the high caffeine content? If so, how much emphasis should this fact be given?
   - You are research director for a biotechnology company working on an AIDS vaccine. At a national conference, a researcher from a competing company secretly offers to sell your company crucial data that could speed discovery of an effective vaccine. Should you accept the offer?

3. Visit a Web site for a professional association in your major or career (American Psychological Association, Society for Technical Communi-

cation, American Nursing Association, or the like) and locate its code of ethics. How often are communication-related issues mentioned? Print out a copy of the code for sharing in a class discussion of the role of ethical communication in different fields.

4. Prepare a brief presentation for your classmates or coworkers in which you answer these questions: *What is plagiarism? How do I avoid it?* Start by exploring the following sites:

- *Plagiarism: What It Is and How to Avoid It,* from the Indiana University Writing Tutorial Services at <http://www.indiana.edu/~wts/wts/plagiarism.html>
- *Avoiding Plagiarism,* from the Purdue Online Writing Lab at <http://owl.english.purdue.edu/handouts/research/r_plagiar.html>"

Find at least one additional Web source on plagiarism (you may need to do a search).

For your class presentation, your goal is to summarize in one page or less a practical, working definition of plagiarism, and a list of strategies for avoiding it. (See page 199 for guidelines for summarizing information.) Attach a copy of the relevant Web page(s) to your presentation. Be sure to credit each source of information (page 685).

5. Examine Web sites that make competing claims about a controversial topic such as bioengineered foods and crops, nuclear power, and herbal medications or other forms of alternative medicine. For example, compare claims about biotech foods from the Biotechnology Council <www.whybiotech.org>, the Sierra Club <www.sierraclub.org>, American Growers Foods <www.americangrowers.com>, and the Food and Drug Administration <www.fda.gov>. Or compare claims about nuclear energy from the Nuclear Energy Institute <www.nei.org> with claims from the Sierra Club <www.sierraclub.org>, the American Council on Science <www.acsh.org>, and the Nuclear Regulatory Commission <www.nrc.gov>. Do you find possible examples of unethical communication, such as conflicts of interest or exaggerated claims? Refer to pages 77–79 and the Checklist for Ethical

Communication (page 90) as a basis for evaluating the various claims. Report your findings in a memo to your instructor and classmates.

6. In July 1999 the American Telemedicine Association (ATA) issued the following advice to consumers who use the Internet for health-related information and services. ATA's criteria for a quality site include the following ("Advisories" 2–3):

- The site is sponsored by a reputable healthcare organization (American Cancer Society, American Medical Association, nationally recognized medical college, or the like). Information from a commercial interest such as a drug company should include assurances that the material is reasonably balanced and objective, and does not merely promote the company's own products.
- Each information source is clearly documented.
- A site providing online diagnosis or prescribing treatment and medication avoids any direct sales of the treatments or medications being prescribed.
- The professionals offering medical consultation are fully licensed, and their credentials are clearly posted.
- The site clearly describes its policies and procedures for maintaining records of the consultation and safeguarding patient privacy.

Visit a health-related Web site and evaluate it according to the above criteria. Focus on sites that cover alternative health such as <www.vicus.com>, sites that create specific recommendations based on the information you provide such as <www.webmd.com>, sites that offer specialized consultation about specific medical conditions such as <www.mediconsult.com>, or discussion sites for people with a specific medical condition such as <www.cancersurvivors.network.com>. Assume that you are a Web site consultant, and prepare a memo for the site's Webmaster pointing out specific problems and recommending changes to improve the site's credibility.

7. Assume that you are a training manager for ABC Corporation, which is in the process of overhauling its policies on company ethics. Developing the company's official Code of Ethics will require several months of research and collaboration with attorneys, ethics consultants, editors, and the like. Meanwhile, your boss has asked you to develop a brief but practical set of "Guidelines for Ethical Communication," as a quick and easy reference for all employees until the official code is finalized. Using the material in this chapter, prepare a two-page memo for employees, explaining how to avoid major ethical pitfalls in corporate communication.

COLLABORATIVE PROJECT

After dividing into groups, study the following scenario and complete the assignment: You belong to the Forestry Management Division in a state whose year-round economy depends almost totally on forest products (lumber, paper, etc.) but whose summer economy is greatly enriched by tourism, especially from fishing, canoeing, and other outdoor activities. The state's poorest area is also its most scenic, largely because of the virgin stands of hardwoods. Your division has been facing growing political pressure from this area to allow logging companies to harvest the trees. Logging here would have positive and negative consequences: for the foreseeable future, the area's economy would benefit greatly from the jobs created; but traditional logging practices would erode the soil, pollute waterways, and decimate wildlife, including several endangered species—besides posing a serious threat to the area's tourist industry. Logging, in short, would give a desperately needed boost to the area's standard of living, but would put an end to many tourist-oriented businesses and would change the landscape forever.

Your group has been assigned to weigh the economic and environmental effects of logging, and prepare recommendations (to log or not to log) for your bosses, who will use your report in making their final decision. To whom do you owe the most loyalty here: the unemployed or underemployed residents, the tourist businesses (mostly owned by residents), the wildlife, the land, future generations? The choices are by no means simple. In cases like this, it isn't enough to say that we should "do the right thing," because we are sometimes unable to predict the consequences of a particular action—even when it seems the best thing to do. In a memo to your supervisor, tell what action you would recommend and explain why. Be prepared to defend your group's ethical choice in class on the basis of the obligations, ideals, and consequences involved.

SERVICE-LEARNING PROJECT

Identify a service agency or advocacy group whose goals and values you support: for example, an environmental group or one that opposes the use of animals in laboratory experiments. What is the main ethical argument advanced by this group? What are two or three major objections that opponents offer to justify a different position? After reviewing Chapter 4, prepare a one- or two-page memo responding to these objections for distribution to group members as "Arguing Points."

6

Working in Teams

The power of collaboration

Electronic Communication allows more and more writing to be done collaboratively, by teams who share information, expertise, ideas, and responsibilities. Successful collaboration brings together the best that each team member has to offer. It promotes feedback, new perspectives, group support, and the chance to test one's ideas in group discussion.

EXAMPLES OF SUCCESSFUL COLLABORATION

Our notion of a solitary engineer, scientist, or businessperson laboring in some quiet corner is quickly disappearing. For inspiration and feedback, workplace professionals collaborate with peers and coworkers—across the company, the nation, or the globe.

Companywide collaboration to design a new refrigerator

> Various components and aspects must be considered once the [refrigerator's] size is determined. These include the compressor, . . . the structure of the motor that drives the compressor, . . . the control system, . . . aesthetic considerations [and so on]. The point is that various individuals work on each of the components or subsystems, and then share information as they design the entire refrigerator system. (Burghardt 209)

Nationwide collaboration to restore the Everglades

> A team of civil engineers, ecologists, and biologists is designing new flowways and levees to transport unpolluted water into the Everglades ecosystem. Team members might work from various sites, meeting electronically to share and refine design ideas, compare research findings, and edit reports and proposals. (Boucher 32–33)

Worldwide collaboration to build the International Space Station

> To study the effects of prolonged space travel on the human mind and body, and on plants and animals, American astronauts lived with Russian cosmonauts on the Russian space station, Mir. This was the first phase of the International Space Station, being built with the collaboration of 15 countries. This ambitious project requires experts worldwide to collaborate on problems in designing, building, staffing, and operating a multi-national space station ("From Mir to Mars").

WWW

6.1

Read "Does the File Cabinet Have a Sex Life?" at <www.ablongman.com/lannonweb>

Uniting each of these projects are the countless documents that must be produced: proposals, specifications, progress reports, feasibility reports, operating manuals, and so on. And, like other project assignments, the related documents are completed jointly.

How a collaborative document is produced

Not all members of a collaborative team do the actual "writing"; the refrigerator *Owner's Manual*, for instance, is produced by writers, engineers, graphic artists, editors, reviewers, marketing personnel, and lawyers (Debs, "Recent Research" 477). Others might research, edit, proofread, or test the document's *usability* (Chapter 16).

THE ROLE OF PROJECT MANAGEMENT IN SUCCESSFUL COLLABORATION

The previous examples of collaboration all rely on one attribute: *teamwork*, a cooperative effort toward a shared goal. But to interact as a cohesive unit, a team needs to agree on what its goals are and on what process it will use to achieve them. This is where project management comes in.

A *project*, simply stated, is an organized effort to get something done. Whether it's a college research paper or the construction of the Brooklyn Bridge, no successful project unfolds haphazardly; instead, it needs to be managed in a systematic way—especially when a team is involved. The following guidelines explain how to capitalize on a team's strength by establishing a shared sense of purpose and direction.

GUIDELINES for Managing a Collaborative Project*

1. *Appoint a group manager.* The manager assigns tasks, enforces deadlines, conducts meetings and keeps them on track, consults with supervisors, and generally "runs the show."

2. *Define a clear and definite goal.* Compose a purpose statement (page 42) that spells out the project's goal and the group's plan for achieving it. Be sure each member understands the goal.

3. *Decide on the type of document required.* Will this be a report, proposal, manual, brochure, or pamphlet? Are visuals and supplements (abstract, appendices, and so on) needed? Will the document be in hard copy or electronic form, or both?

4. *Decide how the group will be organized.* Here are two possibilities:
   a. The group researches and plans together, but each person writes a different part of the document.
   b. Some members plan and research; one person writes a complete draft; others review, edit, revise, and produce the final version.

   NOTE *The final version should display one consistent style throughout—as if written by one person only.*

5. *Divide the task.* Who will be responsible for which parts of the document or which phases of the project? Should one person alone do the final revision? Which jobs are the hardest? Who is best at doing what (writing, editing, layout, design and graphics, oral presentation)? Who will make final decisions?

   NOTE *Spell out—in writing—clear expectations for each team member.*

6. *Establish a timetable.* Specific completion dates for each phase keep everyone focused on what is due, and when. Charts for planning and scheduling, as in Figure 6.1, help the team visualize the whole project and each part, along with start-up and completion dates for each phase (Horn 41, 45). See pages 312, 314 for more on Gantt charts.

7. *Decide on a meeting schedule and format.* How often will the group meet, and where and for how long? Who will take notes (or minutes)? Set a strict time

6.2

Learn more about using project-planning software and tools at <www.ablongman.com/lannonweb>

limit for each discussion topic. Distribute copies of the meeting agenda and timetable beforehand, and stick to it. Meetings work best when each member prepares a specific contribution beforehand.

8. *Establish a procedure for responding to the work of other members.* Will reviewing and editing be done in writing, face-to-face, as a group, one-on-one, or online? Will this process be supervised by the project manager?

9. *Develop a file-naming system for various drafts.* When working with multiple drafts, it's too easy to save over a previous version and lose something important.

10. *Establish procedures for dealing with group problems.* How will gripes and disputes be aired (to the manager, the whole group, the offending individual)? How will disputes be resolved (by vote, the manager)? How will irrelevant discussion be avoided or curtailed? Expect some conflict but try to use it positively, and try to identify a natural peacemaker in the group.

11. *Select a group decision-making style beforehand.* Will decisions be made alone by the group manager or be based on group input or majority vote?

12. *Appoint a different "observer" for each meeting.* At Charles Schwab & Co., the designated observer keeps a list of what worked or didn't work during the meeting. The list is then added to that meeting's minutes (Matson, "The Seven Sins" 31).

13. *Decide how to evaluate each member's contribution.* Will the manager assess each member's performance and, in turn, be evaluated by each member? Will members evaluate each other? What are the criteria? Figure 6.2 shows one possible form for a manager's evaluation of members. Equivalent criteria for evaluating the manager include open-mindedness, ability to organize the team, fairness in assigning tasks, ability to resolve conflicts, or ability to motivate. (Members might keep a journal of personal observations for overall evaluation of the project.)

14. *Prepare a project management plan.* Figure 6.3 shows a sample planning form. Distribute completed copies to members.

15. *Submit progress reports regularly.* Progress reports (page 388) enable everyone to track activities, problems, and progress.

*Adapted from Debs, "Collaborative Writing" 38–41; Hill-Duin 45–50; Hulbert, "Developing" 53–54; McGuire 467–68; Morgan 540–41.

SOURCES OF CONFLICT IN COLLABORATIVE GROUPS

Workplace surveys show that people view meetings as "their biggest waste of time" (Schrage 232). This fact alone accounts for the boredom, impatience, or irritability

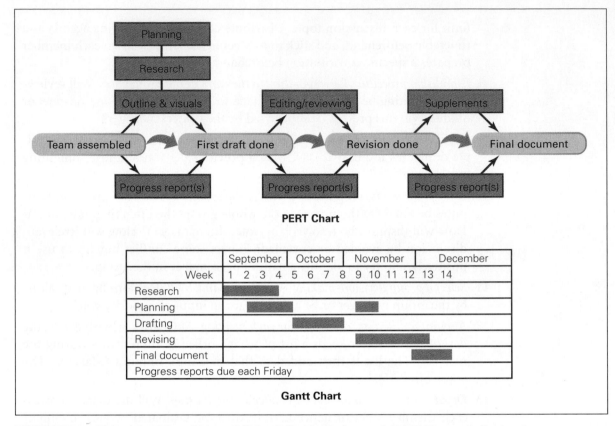

FIGURE 6.1   Charts for Planning and Scheduling a Project   A PERT (Program Evaluation and Review Technique) chart maps out the major activities (rectangles) and events (ovals) for a complex project. Heavy arrows indicate the *critical path* (straightest line) through the project. A Gantt chart depicts specific beginning and ending dates for each phase of the project and shows overlapping phases as well. Note that these are simplified versions; your own charts may need to include additional activities such as "updating," "usability testing," "legal checks," and so on.

6.3

Read about a team assessment controversy at Microsoft at <www.ablongman.com/lannonweb>

that might crop up in any meeting. But even the most dynamic group setting can produce conflict because of differences like the following.

Interpersonal Differences

People might clash because of differences in personality, working style, commitment, standards, or ability to take criticism. Some might disagree about exactly what or how much the group should accomplish, who should do what, or who should have the final say. Some might feel intimidated or hesitant to speak out.[1]

---

[1]Adapted from Bogert and Butt 51; Burnett 533–34; Debs, "Collaborative Writing" 38; Hill-Duin 45–46; Nelson and Smith 61.

Performance Appraisal for J. Fishkill

(Rate each element as [superior], [acceptable], or [unacceptable] and use the "Comment" section to explain each rating briefly.)

- *Cooperation:* [___superior___]

    Comment: works extremely well with others; always willing to help out; responds positively to constructive criticism

- *Dependability:* [___acceptable___]

    Comment: arrives on time for meetings; completes all assigned work

- *Effort:* [___acceptable___]

    Comment: does fair share of work; needs no prodding

- *Quality of work produced:* [___superior___]

    Comment: produces work that is carefully researched, well documented, and clearly written

- *Ability to meet deadlines:* [___superior___]

    Comment: delivers all assigned work on or before the deadline; helps other team members with last-minute tasks

R. P. Ketchum

Project manager's signature

FIGURE 6.2 Sample Form for Evaluating Team Members Any evaluation of strengths and weaknesses should be backed up by comments that explain the ratings. A group needs to decide beforehand what constitutes "effort," "cooperation," and so on.

www

6.4

For more on gender and workplace culture visit <www.ablongman.com/lannonweb>

These interpersonal conflicts can actually worsen when the group interacts exclusively online (page 111).

Gender and Cultural Differences

Collaboration involves working with peers—those of equal status, rank, and expertise. But gender and cultural differences can cause some participants to feel less than equal.

GENDER CODES AND COMMUNICATION STYLE.   Research on ways women and men communicate in meetings indicates a gender gap. Communication specialist Kathleen Kelley-Reardon offers the following assessment of gender differences in workplace communication:

Project Planning Form

Project title:

Audience:

Project manager:

Team members:

Purpose of the project:

Type of document required:

Specific Assignments   **Due Dates**

   Research: Research due:

   Planning: Plan and outline due:

   Drafting: First draft due:

   Revising: Reviews due:

   Preparing final document: Revision due:

   Presenting oral briefing: Progress report(s) due:

                                                Final document due:

Work Schedule

Team meetings:	Date	Place	Time	Note taker
#1				
#2				
#3				
etc.				

Mtgs. w/instructor

   #1

   #2

   etc.

Miscellaneous

   How will disputes and grievances be resolved?

   How will performances be evaluated?

   Other matters (Internet searches, email routing, computer conferences, etc.)?

FIGURE 6.3   Sample Project Planning Form for Managing a Collaborative Project
To manage a team project you need to (a) spell out the project goal, (b) break the entire task down into manageable steps, (c) create a climate in which people work well together, and (d) keep each phase of the project under control.

How gender
codes influence
communication

Women and men operate according to communication rules for their gender, what experts call "gender codes." They learn, for example, to show gratitude, ask for help, take control, and express emotion, deference, and commitment in different ways. (88–89)

Kelley-Reardon explains how women tend to communicate during meetings: Women are more likely to take as much time as needed to explore an issue, build consensus and relationship among members, use tact in expressing views, use care in choosing their words, consider the listener's feelings, speak softly, and allow interruptions. Women generally issue requests instead of commands (*Could I have the report by Friday?* versus *Have this ready by Friday.*) and qualify their assertions in ways that avoid offending (*I don't want to seem disagreeable here, but . . .*).

One study of mixed-gender peer interaction indicates that women, in contrast to men, tend to: be agreeable, solicit and admit the merits of other opinions, ask questions, and admit uncertainty (say, with qualifiers such as *maybe, probably, it seems as if*) (Wojahn 747).

None of these traits is gender specific. People of either gender can be soft-spoken and reflective. But such traits often attach to the "feminine" stereotype. Supreme Court Justice Sandra Day O'Connor, for example, recalls the difficulty of getting male colleagues to pay attention to what she had to say (Hugenberg, LaCivita, and Lubanovic 215).

Any woman who breaches the gender code, say, by being assertive, may be seen as "too controlling" (Kelley-Reardon 6). Studies suggest women have less freedom than male peers to alter their communication strategies: less assertive males often are still considered persuasive, whereas more assertive females often are not (Perloff 273). In the

Collaborative writing

"Most of our writing happens in teams. Typically, our project teams (or sometimes just the Project Director) develop a detailed outline for a written product. The team then goes over it together and writing assignments are made (intentional use of the passive voice there). The assignments usually match the content areas for which team members have been responsible. The Project Director (generally me) is responsible for the introduction that describes the purpose or importance of the work and the conclusions/implications. The middle part is "technical" and I usually review those sections without getting into the details. We have a writing style that is consistent across most of our products. It takes about six months for team members to learn to write in that style. One of the biggest team writing challenges we have is the tendency for recent graduates to overwrite. They have learned in their academic programs to write very formally, with ponderous vocabulary and lots of passive voice. Our stuff has to get the point in a hurry."

—Paul Harder, President,
mid-sized consulting company

words of one researcher, fitting into the workplace culture requires that women decide "to be quiet and popular, or speak out, and not be accepted" (Jones 50).

As one consequence of these gender differences, males tend "to become leaders of task-oriented groups, whereas women emerge as social leaders more frequently than men" (Dillard, Solomon, and Samp 709).

Entrenched attitudes about gender in the U.S. workplace do appear to be changing: Ratings by peers, subordinates, and bosses indicate that women excel in a variety of interpersonal areas such as motivating others, listening to others, being flexible, keeping people informed, coaching, and team building. Not only are such "people skills" now being recognized as legitimate business skills, but they are also in high demand at a time in which teamwork is more and more essential (Sharpe 75+). As a result, books,[2] seminars, and training programs in "qualities typically associated with women" are becoming increasingly popular among male managers (Gogoi 84).

 NOTE   *Despite apparent changes in U.S. attitudes, negative views toward women in the workplace persist in many cultures. Surveys show that female executives in U.S. company branches outside this country continue to be underrepresented ("The Big Picture" 14).*

 CULTURAL CODES AND COMMUNICATION STYLE. International business expert David A. Victor describes cultural codes that influence interaction in group settings:

How cultural codes influence communication

- Some cultures value silence more than speech, intuition and ambiguity more than hard evidence or data, politeness, and personal relationships more than business relationships (145–46).
- Cultures differ in their perceptions of time. Some are "all business" and in a big rush; others take as long as needed to weigh the issues, engage in small talk and digressions, chat about family, health, and other personal matters (233).
- Cultures differ in willingness to express disagreement, question or be questioned, leave things unstated, touch, shake hands, kiss, hug, or backslap (209–11).
- Direct eye contact is not always a good indicator of listening. In some cultures it is offensive. Other eye movements, such as squinting, closing the eyes, staring away, staring at legs or other body parts, are acceptable in some cultures but insulting in others (206).

For detailed and up-to-date information about any of 181 cultures around the world, go to *Culturgrams* at <www.culturgrams.com>. Here you can access "free country pages" or, for a modest fee, purchase a full report on a particular country.

---

[2]See, for example, Daniel P. Goleman's best-selling management book, *Emotional Intelligence.*

MANAGING GROUP CONFLICT

No team can afford to assume that all members share one viewpoint, one communication style, one approach to problem solving. Conflicts must be expressed and addressed openly. Pointing out that "conflict can be good for an organization—as long as it's resolved quickly," management expert David House offers these strategies for overcoming personal differences (Warshaw 48):

How to manage group conflict

- Give everyone a chance to be heard.
- Take everyone's feelings and opinions seriously.
- Don't be afraid to disagree.
- Offer and accept constructive criticism.
- Find points of agreement with others who hold different views.
- When the group does make a decision, support it fully.

Business etiquette expert Ann Marie Sabath offers these additional suggestions for maintaining civility during meetings (108–10):

How to reduce animosity

- If someone is overly aggressive or insists on wandering off track, respond politely and try to acknowledge valid and constructive reasons for this person's behavior: "I understand your concern or frustration about this, and it's probably something we should look at more closely." If you think the point might have some value, suggest a later meeting: "Why don't we take some time to think about this and schedule another meeting to discuss it?"
- Never attack or point the finger by using "aggressive 'you' talk": "You should," "You haven't," or "You need to realize," for example, imply blame, and only increase animosity. See page 252 for ways to avoid a blaming tone.

Ultimately, collaboration requires compromise and consensus: In order for people with different

Collaborative writing

"Unfortunately in my experience, whenever I've had to create documents with a team it's been a small agony. I am good at this process, but it only takes one clueless person on the team to grind the process to a halt. Usually these documentations or reports are created IN ADDITION TO your normal workload, so everyone involved is under pressure because every minute they're collaborating they're not doing the backlog of their normal workload. Then you get the weirdo or two who has a job of no timely responsibility and they find the meetings to be the high point of their careers to date, and they want to shine out and prolong the talking as much as possible, and then nothing gets done. I would much rather create these types of documents alone or with one or two sensible colleagues, and then put the shaped-up document out for review."

—Terry Vilante, Chief Financial Officer, small public relations company views to reach agreement, each person has to be willing to give a little. Before any meeting, review the persuasion guidelines on page 66, and try really *listening* to what other people have to say.

OVERCOMING DIFFERENCES BY ACTIVE LISTENING

Listening is key to getting along, building relationships, and learning. Information expert Keith Devlin points out that "managers get around two-thirds of their knowledge from face-to-face meetings or telephone conversations and only one-third from documents and computers" (163). In a recent manager survey, the ability to listen was ranked second (after the ability to follow instructions) among thirteen communication skills sought in entry-level graduates (cited in Goby and Lewis 42).

Nearly half our time communicating at work is spent listening (Pearce, Johnson, and Barker 28). But poor listening behaviors cause us to retain only a fraction of what we hear. How effective are your listening behaviors? Assess them by using the questions below.

QUESTIONS FOR ASSESSING YOUR LISTENING BEHAVIORS
- *Do I remember people's names after being introduced?*
- *Do I pay close attention to what is being said, or am I easily distracted?*
- *Do I make eye contact with the speaker, or stare off elsewhere?*
- *Do I actually appear interested and responsive, or bored and passive?*
- *Do I allow the speaker to finish, or do I interrupt?*
- *Do I tend to get the message straight, or misunderstand it?*
- *Do I remember important details from previous discussions, or forget who said what?*
- *Do I ask people to clarify complex ideas, or just stop listening?*
- *Do I know when to keep quiet, or do I insist on being heard?*

When you communicate, are you "listening or just talking" (Bashein and Markus 37)? Many of us seem more inclined to speak, to say what's on our minds, than to listen. We often hope someone else will do the the listening. Effective listening requires *active* involvement—not just passive reception.

THINKING CREATIVELY

Today's rapidly changing workplace demands new and better ways of doing things:

Creativity is a vital asset

> More than one-fourth of U.S. companies employing more than 100 people offer some kind of creativity training to employees. (Kiely 33)

Creative thinking is especially effective in group settings, using the following techniques.

GUIDELINES for Active Listening*

1. *Don't dictate.* If you are the group moderator, don't express your view until everyone else has their chance.

2. *Be receptive.* Instead of resisting different views, develop a "learner's" mindset: take it all in first, and evaluate it later.

3. *Keep an open mind.* Judgment stops thought (Hayakawa 42). Reserve judgment until everyone has had their say.

4. *Be courteous.* Don't smirk, roll your eyes, whisper, fidget, or wisecrack.

5. *Show genuine interest.* Eye contact is vital, and so is body language (nodding, smiling, leaning toward the speaker). Make it a point to remember everyone's name.

6. *Hear the speaker out.* Instead of "tuning out" a message you find disagreeable, allow the speaker to continue without interruption (except to ask for clarification). Delay your own questions, comments, and rebuttals until the speaker has finished. Instead of blurting out a question or comment, raise your hand and wait to be recognized.

7. *Focus on the message.* Instead of thinking about what you want to say next, try to get a clear understanding of the speaker's position.

8. *Ask for clarification.* If anything is unclear, say so: "Can you run that by me again?" To ensure accuracy, paraphrase the message: "So what you're saying is. . . . Is that right?" Whenever you respond, try repeating a word or phrase that the other person has just used.

9. *Be agreeable.* Don't turn the conversation into a contest, and don't insist on having the last word.

10. *Observe the 90/10 rule.* You rarely go wrong spending 90 percent of your time listening, and 10 percent speaking. President Calvin Coolidge claimed that "Nobody ever listened himself out of a job." Some historians would argue that "Silent Cal" listened himself right into the White House.

*Adapted from Armstrong 24+; Bashein and Markus 37; Cooper 78–84; Dumont and Lannon 648–51; Pearce, Johnson, and Barker 28–32; Sittenfeld 88; Smith 29.

Multiple ideas are better than one

6.5

For more on other writing processes visit <www.ablongman.com/lannonweb>

Brainstorming

When we begin working with a problem, we search for useful material: insights, facts, statistics, opinions, images—anything that sharpens our view of the problem and potential solutions ("How can we increase market share for Zappo software?"). *Brainstorming* is a technique for coming up with useful material. Its aim is to produce as many ideas as possible (on paper, screen, whiteboard, or the like). Although brainstorming can be done individually, it is especially effective in a group setting.

A procedure for
brainstorming

1. *Choose a quiet setting and agree on a time limit.*
2. *Decide on a clear and specific goal for the session.* For instance, "We need at least five good ideas about why we are losing top employees to other companies."
3. *Focus on the issue or problem.*
4. *As ideas begin to flow, record every one.* Don't stop to judge relevance or worth and don't worry about spelling or grammar.
5. *If ideas are still flowing at session's end, keep going.*
6. *Take a break.*
7. *Now confront your list.* Strike out what is useless and sort the remainder into categories. Include any new ideas that pop up.

Limitations of
brainstorming

Because of intimidation, groupthink, and other social pressures (page 75), group brainstorming often fails to achieve its "nonjudgmental ideal" (Kiely 34). Lower-status members, for instance, might feel reluctant to express their ideas or criticize others.

Brainstorming online, using email or asynchronous "chat" software, can relieve social pressure on participants. Some software allows participants to create pseudonyms and mask their identities.

Brainwriting

An alternative to brainstorming, *brainwriting*, enables group members to record their ideas—anonymously—on slips of paper or on a networked computer file. Ideas are then exchanged or posted on a large screen for comment and refinement by other members (Kiely 35).

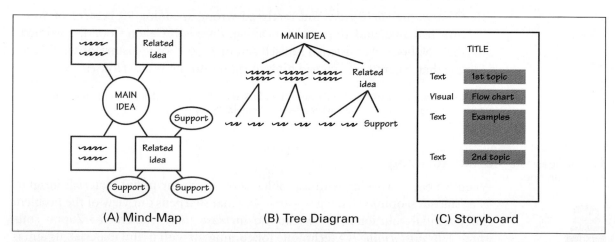

FIGURE 6.4   Visual Techniques for Thinking Creatively   Each of these techniques helps participants develop a concrete mental image of an otherwise abstract process (i.e., the thinking process).

Mind-Mapping

A more structured version of brainstorming, *mind-mapping* (Figure 6.4A) helps visualize relationships. Group members begin by drawing a circle around the main issue or concept, centered on the paper or whiteboard. Related ideas are then added, each in its own box, connected to the circle by a ruled line (or "branch"). Other branches are then added, as lines to some other distinct geometric shape containing supporting ideas. Unlike a traditional outline, a mind-map does not require sequential thinking: as each idea pops up, it is connected to related ideas by its own branch. Mind-mapping software such as *Mindjet* automates this process of visual thinking.

A simplified form of mind-mapping is the *tree diagram* (Figure 6.4B), in which major topic, minor topics, and subtopics are connected by branches that indicate their relationships. Page 120 shows a sample tree diagram for a research project.

Collaborative writing

"*My work in preparing user manuals is almost entirely collaborative. The actual process of writing takes maybe 30 percent of my time. I spend more time consulting with my information sources such as the software designers and field support people. I then meet with the publication and graphics departments to plan the manual's structure and format. As I prepare various drafts, I have to keep track of which reviewer has which draft. Because I rely on others' feedback, I circulate materials often. And so I write email memos on a regular basis. One major challenge is getting everyone involved to agree on a specific plan of action and then to stay on schedule so we can meet our publication deadline.*"

—Pam Herbert, technical writer, software firm

Storyboarding

A technique for visualizing the shape of an entire process (or a document) is *storyboarding* (Figure 6.4C). Group members write each idea and sketch each visual on a large index card. Cards are then displayed on a wall or bulletin board so that others can comment or add, delete, refine, or reshuffle ideas, topics, and visuals (Kiely 35–36). Page 231 shows a final storyboard for a long report.

REVIEWING AND EDITING OTHERS' WORK

Documents produced collaboratively are reviewed and edited extensively. *Reviewing* means evaluating how well a document connects with its intended audience and meets its intended purpose. Reviewers typically examine a document for these specific qualities:

What reviewers
look for

- accurate, appropriate, useful, and legal content
- material organized for the reader's understanding
- clear, easy to read, and engaging style
- effective visuals and page design
- a document that is safe, dependable, and easy to use

In reviewing, you explain to the writer how you respond as a reader. This commentary helps a writer think about ways of revising. Criteria for reviewing various documents appear in checklists throughout this book. (See also Chapter 16, on Usability).

Editing means actually "fixing" the piece by making it more precise and readable. Editors typically suggest improvements like these:

Ways in which
editors "fix"
writing

- rephrasing or reorganizing sentences
- clarifying a topic sentence
- choosing a better word or phrase
- correcting spelling, usage, or punctuation, and so on

Criteria for editing appear in Chapter 13 and Appendix C.

NOTE   *Your job as a reviewer or editor is to help clarify and enhance a document—but without altering its intended meaning.*

GUIDELINES for Peer Reviewing and Editing

1. *Read the entire piece at least twice before you comment.* Develop a clear sense of the document's purpose and its intended audience. Try to visualize the document as a whole before you evaluate specific parts or features.

2. *Remember that mere correctness does not guarantee effectiveness.* Poor usage, punctuation, or mechanics do distract readers and harm the writer's credibility. However, a "correct" piece of writing might still contain faulty rhetorical elements (inferior content, confusing organization, or unsuitable style).

3. *Understand the acceptable limits of editing.* In the workplace, editing can range from fine-tuning to an in-depth rewrite (in which case editors are cited prominently as consulting editors or coauthors). In school, however, rewriting a piece to the extent that it ceases to belong to its author may constitute plagiarism (pages 87, 682, 684).

4. *Be honest but diplomatic.* Most of us are sensitive to criticism—even when it is constructive—and we all respond more favorably to encouragement. Begin with something positive before moving to material that needs improvement. Be supportive instead of judgmental.

5. *Always explain why something doesn't work.* Instead of "this paragraph is confusing," say "because this paragraph lacks a clear topic sentence, I had trouble discovering the main idea." (See page 379 for sample criteria.) Help the writer identify the cause of the problem.

6. *Focus first on the big picture.* Begin with the content and the shape of the document. Is the document appropriate for its audience and purpose? Is the supporting material relevant and convincing? Is the discussion easy to follow? Does each paragraph do its job? Then discuss specifics of style and correctness (tone, word choice, sentence structure, and so on).

7. *Make specific recommendations for improvements.* Write out suggestions in enough detail for the writer to know what to do. Provide brief reasons for your suggestions.

8. *Be aware that not all feedback has equal value.* Even professional editors can disagree. If different readers offer conflicting opinions of your own work, seek your instructor's advice.

6.6

For more on effective electronic collaboration visit <www.ablongman.com/lannonweb>

When face-to-face meetings are preferred

When electronic meetings are preferred

FACE-TO-FACE VERSUS ELECTRONICALLY MEDIATED COLLABORATION

Should groups meet in the same physical space or in virtual space? Face-to-face collaboration seems preferable when people don't know each other, when the issue is sensitive or controversial, or when people need to interact on a personal level (Munter 81).

Electronically mediated (virtual) collaboration is preferable when people are in different locations, have different schedules, or when it is important to avoid personality clashes, to encourage shy participants, or to prevent intimidation by dominant participants (Munter 81, 83).

Here are some technologies that erase distance and enable people to work together in the same virtual space:

Tools for electronic collaboration

- *basic email,* for exchanging ideas as schedules permit
- *text messaging and chat systems,* for communicating in real time
- *groupware,* for group authoring and editing
- *digital whiteboard,* a large screen that allows participants to write, sketch, and erase in real time, from their own computers
- *Web conferencing,* using a password-protected site or company intranet
- *teleconferencing,* using speakerphones.
- *videoconferencing,* for live online meetings in which participants at different sites can see each other

Benefits and drawbacks of electronic meetings

Research indicates that electronic meetings are more productive than face-to-face meetings (Tullar, Kaiser, and Balthazard 54). Written ideas can be more carefully considered and expressed, and they provide a durable record for feedback and reference. On the negative side, equipment crashes are disruptive. And the lack of personal contact (say, a friendly grin or handshake or small talk) makes it hard for trust to develop. Even worse, opposing but anonymous participants might engage in open hostility, as in email flaming (Clark, "Teaching" 49–50).

Some experts argue that computer-based meeting tools eliminate equality issues that crop up in face-to-face meetings. Thus, some people feel more secure about saying "what they really think" (Matson, "The Seven Sins" 30). Also, "status cues" such as age, gender, appearance, or ethnicity are invisible online (Wojahn 747–48).

 NOTE   *Many cultures value the social (or relationship) function of communication as much as its informative function (Archee 41). Therefore, a recipient might consider certain communications media more appropriate than others, preferring, say, a phone conversation to text messaging or email.*

ETHICAL ABUSES IN WORKPLACE COLLABORATION

Our "lean" and "downsized" corporate world spells fierce competition among coworkers, often creating this dilemma:

Teamwork versus survival of the fittest

> Many companies send mixed signals . . . saying they value teamwork while still rewarding individual stars, so that nobody has any real incentive to share the glory. (Fisher, "My Team Leader" 291)

WWW
6.7
For more on power dynamics in the workplace visit <www.ablongman.com/lannonweb>

The resulting mistrust interferes with genuine teamwork and promotes the following kinds of unethical behavior.

Intimidating One's Peers

A dominant personality may intimidate peers into silence or agreement (Matson, "The Seven Sins" 30). Intimidated employees resort to "mimicking"—merely repeating what the boss says (Haskin, "Meetings without Walls" 55).

Claiming Credit for Others' Work

Workplace plagiarism occurs when the team or project leader claims all the credit. Even with good intentions, "the person who speaks for a team often gets the credit, not the people who had the ideas or did the work" (Nakache 287–88). Team expert James Stern describes one strategy for avoiding plagiarism among coworkers:

How to ensure that the deserving get the credit

> Some companies list "core" and "contributing" team members, to distinguish those who did most of the heavy lifting from those who were less involved. (qtd. in Fisher, "My Team Leader" 291)

Stern advises groups to decide beforehand—and in writing—what credit will be given for which contributions.

Hoarding Information

Surveys reveal that the biggest obstacle to workplace collaboration is people's "tendency to hoard their own know-how" (Cole-Gomolski 6) when confronted with questions like these:

Information
people need to
do their jobs

- *Whom do we contact for what?*
- *Where do we get the best price, the quickest repair, the most dependable service?*
- *What's the best way to do X?*

Despite all the technology available for information sharing, fewer than 10 percent of companies succeed in persuading employees to share ideas on a routine basis (Koudsi 233).

People hoard information when they think it gives them power or self-importance, or when having exclusive knowledge might provide job security (Devlin 179). In a worse case, they withhold information in order to sabotage peers.

CONSIDER THIS How You Speak Shows Where You Rank*

Popular discussion of communication style in recent years has centered on differences between the sexes. The subject has been fodder for TV talk shows, corporate seminars, and bestsellers, notably Deborah Tannen's You Just Don't Understand *and John Gray's* Men Are From Mars, Women Are From Venus. *But Sarah McGinty, a teaching supervisor at Harvard University's school of education, believes language style is based more on power than on gender—and that marked differences distinguish the powerful from the powerless loud and clear. As a consultant, she is often called on to help clients develop more effective communication styles. FORTUNE's* Justin Martin *spoke with McGinty about her ideas:*

What Style of Speaking Indicates That Someone Possesses Power?

A person who feels confident and in control will speak at length, set the agenda for a conversation, stave off interruptions, argue openly, make jokes, and laugh. Such a person is more inclined to make statements, less inclined to ask questions. They are more likely to offer solutions or a program or a plan. All this creates a sense of confidence in listeners.

What about People Who Lack Power? How Do They Speak?

The power deficient drop into conversations, encourage other speakers, ask numerous questions, avoid argument, and rely on gestures such as nodding and smiling that suggest agreement. They tend to offer empathy rather than solutions. They often use unfinished sentences. Unfinished sentences are a language staple for those who lack power.

How Do You Figure Out What Style of Communication You Lean Toward?

It's quite hard to do. We're often quite ignorant about our own way of communicating. Everyone comes home at night occasionally and says, "I had that idea, but no one heard me, and everyone thinks it's Harry's idea." People like to pin that on gender and a lot of other things as well. But it's important to find out what really did happen. Maybe it was the volume of your voice, and you weren't heard. Maybe you overex-

(continues)

Consider This (continued)

plained, and the person who followed up pulled out the nugget of your thought.

But it's important to try to get some insight into what your own language habits are so that you can be analytical about whether you're shooting yourself in the foot. You can tape your side of phone calls, make a tape of a meeting, or sign up for a communications workshop. That's a great way to examine how you conduct yourself in conversations and in meetings.

Does Power Language Differ from Company to Company?

Certainly. The key is figuring out who gets listened to within your corporate culture. That can make you a more savvy user of language. Try to sit in on a meeting as a kind of researcher, observing conversational patterns. Watch who talks, who changes the course of the discussion, who sort of drops in and out of the conversation. Then try to determine who gets noticed and why.

One very effective technique is to approach the person who ran the meeting a couple of days after the fact and ask for an overall impression. What ideas were useful? What

ideas might have a shot at being implemented?

How Can You Get More Language Savvy?

You can start by avoiding bad habits, such as always seeking collaboration in the statements you make. Try to avoid "as Bob said" and "I pretty much agree with Sheila." Steer clear of disclaimers such as "I may be way off base here, but. . . . " All these serve to undermine the impact of your statements.

The amount of space you take up can play a big part in how powerful and knowledgeable you appear. People speaking before a group, for instance, should stand with their feet a little bit apart and try to occupy as much space as possible. Another public-speaking tip: Glancing around constantly creates a situation in which nobody really feels connected to what you're saying.

Strive to be bolder. Everyone tends to worry that they will offend someone by stating a strong opinion. Be bold about ideas, tentative about people. Saying "I think you're completely wrong" is not a wise strategy. Saying "I have a plan that I think will solve these problems" is perfectly reasonable. You're not attacking people. You're being bold with an idea.

*"How You Speak Shows Where You Rank," interview by Justin Martin with Sarah McGinty, from FORTUNE, February 2, 1998, (c) 1998 Time Inc. All rights reserved.

EXERCISES   For more exercises, visit <www.ablongman.com/lannon>

1. Describe the role of collaboration in a company, organization, or campus group where you have worked or volunteered. Among the questions: What types of projects require collaboration? How are teams organized? Who manages the projects? How are meetings conducted? Who runs the meetings? How is conflict managed? Summarize your findings in a one- or two-page memo.
   Hint: If you have no direct experience, interview a group representative, say a school administrator or faculty member or editor of the campus newspaper. (See page 151 for interview guidelines.)

 2. On the Web, examine the role of global collaboration in building the International Space Station. Summarize your findings in a memo to be shared with the class. Trace the sequence of links you followed to reach your material, and cite each source.
   Hint: You might begin by exploring the Personal Space link at <www.nasa.gov> for profiles of the people behind the Space Station. Also explore the International Space Station link at <www.pbs.org>.

COLLABORATIVE PROJECTS

1. **Gender Differences:** Divide into small groups of mixed genders. Review pages 101–04 and

113–14. Then test the hypothesis that women and men communicate differently in the workplace.

Each group member prepares the following brief messages—without consulting with other members:

- A thank-you note to a coworker who has done you a favor.
- A note asking a coworker for help with a problem or project.
- A note asking a collaborative peer to be more cooperative or to stop interrupting or complaining.
- A note expressing impatience, frustration, confusion, or satisfaction to members of your group.
- A recommendation for a friend who is applying for a position with your company.
- A note offering support to a good friend and coworker.
- A note to a new colleague, welcoming this person to the company.
- A request for a raise, based on your hard work.
- The meeting is out of hand, so you decide to take control. Write what you would say.
- Some members of your group are dragging their feet on a project. Write what you would say.

As a group, compare messages, draw conclusions about the original hypothesis, and appoint one member to present findings to the class.

2. As an "observer" (page 99), keep a journal during a collaborative project, noting what succeeded and what did not, what interpersonal conflicts developed and how they were resolved, what other issues contributed to progress or delay, the role and effectiveness of electronic tools, and so on. In a memo report to classmates and instructor, summarize the achievements and setbacks in your project and recommend improvements.

Avoid attacking, blaming, or offending anyone. Offer constructive suggestions for improving collaborative work *in general*.

3. *Listening Competence:* Use the questions on page 106 to:

a. assess the listening behaviors of one member in your group during collaborative work,

b. have some other member assess your behaviors, and

c. do a self-assessment.

Record the findings and compare each self-assessment with the corresponding outside assessment. Discuss findings with the class.

4. Use your listserv (or email instant messaging network) to confer electronically on all phases of the collaborative project, including peer review (page 109) and usability testing (Chapter 16).

When your project is complete, write an explanation telling how electronic conferencing eased or hampered the group's efforts and how it improved or detracted from the overall quality of your document.

5. Form teams of three to six people and draw up a one-page set of guidelines, or ground rules (in memo form), for helping your team meetings run efficiently and with minimum conflict. Supplement material from this chapter with ideas from your own group brainstorming. Compare your memo with those from other groups.

6. Hold two brainstorming sessions for Exercise 5: one face-to-face and one via email. Decide on the benefits and drawbacks of each version and record your findings in a memo to be shared with the class.

7. Web pages from The Writer's Block <www.writersblock.ca/spring95/team.htm> describe the relationship between writers and editors in the workplace. Prepare a one-page summary of this information, in your own words, for class discussion. (Page 199 offers guidelines for summarizing.) Attach copies of relevant Web pages to your summary.

SERVICE-LEARNING PROJECT

Plan a group visit to one of the agencies or organizations your class is working with. Include in your planning document instructions detailing who will be in charge of note-taking, leading interviews or conversations, research, and photographing the site. Review and edit the planning document until all of your team members feel comfortable and knowledgeable about their role in the agency visit.

PART II

The Research Process

Surgeons Rely on Video Game Skills

A research study conducted by Beth Israel Medical Center and the National Institute on Media and the Family, released in April 2004, reported that video game skills translate into surgical prowess. Doctors who spent at least three hours a week playing video games made about 37 percent fewer mistakes in laparoscopic surgery than doctors who did not play games. They were also significantly faster and more efficient in surgery than their non-game-playing peers. As study coauthor Paul J. Lynch comments, "The study heralds the arrival of Generation X into medicine."

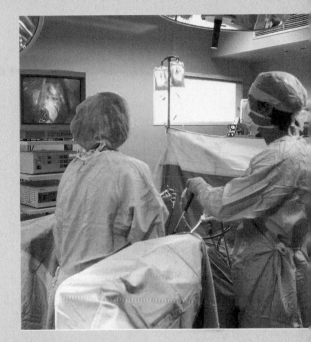

Researchers are now considering ways to implement the findings of their study. Dr. James Rosser, one of the researchers, is developing a training course called "Top Gun" in which surgical trainees "warm up" with a video game, honing their coordination, agility, and accuracy, before entering the operating room. In laparoscopic surgery, surgeons use a tiny camera and instruments controlled by joysticks and a video monitor to perform minimally intrusive surgery on body parts ranging from gallbladders to knees. The process is very delicate. As Dr. Rosser describes, "It's like tying your shoelaces with 3-foot-long chopsticks" (Dobnik). ■

7

Thinking Critically about the Research Process

ASKING THE RIGHT QUESTIONS

EXPLORING A BALANCE OF VIEWS

ACHIEVING ADEQUATE DEPTH IN YOUR SEARCH

EVALUATING YOUR FINDINGS

INTERPRETING YOUR FINDINGS

CONSIDER THIS   Expert Opinion Is Not Always Reliable

GUIDELINES   for Evaluating Expert Information

Major decisions in the workplace typically are based on careful research, with the findings recorded in a written report. Managers spend an estimated 17 percent of their time searching for information (Davenport 157).

Research is a deliberate form of inquiry, a process of problem solving in which certain procedures follow a recognizable sequence (Figure 7.1). But research is not simply a numbered set of procedures. The procedural stages depend on the many decisions that accompany any legitimate inquiry (Figure 7.2).[1]

FIGURE 7.1
The Procedural Stages of the Research Process

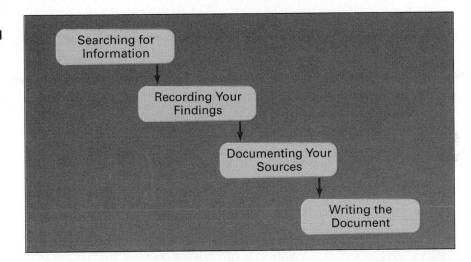

FIGURE 7.2
The Inquiry Stages of the Research Process

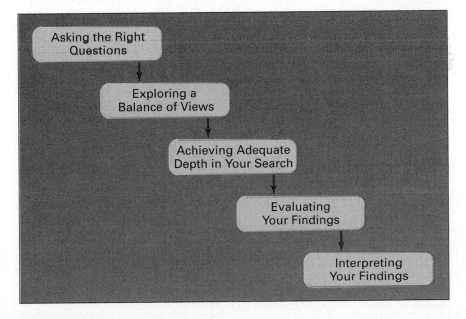

[1]My thanks to University of Massachusetts Dartmouth librarian Shaleen Barnes for inspiring this entire chapter.

7.1

Sometimes questions are more important than answers. Learn more at <www.ablongman.com/lannonweb>

ASKING THE RIGHT QUESTIONS

The answers you uncover will depend on the questions you ask. Assume, for instance, that you are faced with the following scenario:

Defining and Refining a Research Question

You are the public health manager for a small, New England town in which high-tension power lines run within one hundred feet of the elementary school. Parents are concerned about danger from electromagnetic radiation (EMR) emitted by these power lines in energy waves known as electromagnetic fields (EMFs). Town officials ask you to research the issue and prepare a report to be distributed at the next town meeting in six weeks. ❑

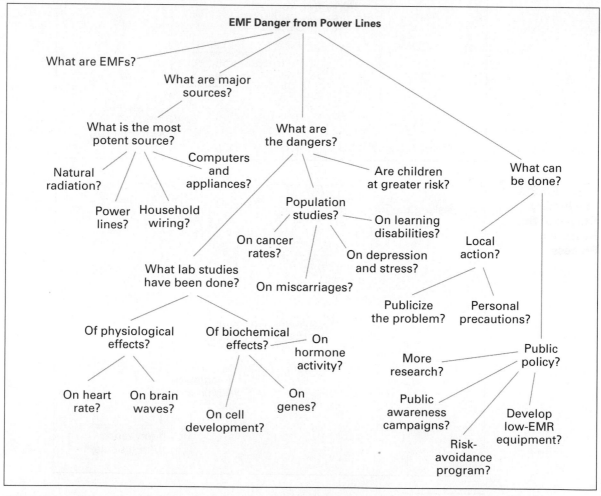

FIGURE 7.3   How the Right Questions Help Define a Research Problem   You cannot begin to solve a problem until you have defined it clearly.

First, you need to identify your exact question or questions. Initially, the main question might be: *Do the power lines pose any real danger to our children?* After phone calls around town and discussions at the coffee shop, you discover that townspeople actually have three main questions: *What are electromagnetic fields? Do they endanger our children? If so, then what can be done?*

To answer these questions, you need to consider a range of subordinate questions, like those in the Figure 7.3 tree chart. Any one of the questions could serve as topic of a worthwhile research report on such a complicated issue. As research progresses, this chart will grow. For instance, after some preliminary reading, you learn that electromagnetic fields radiate not only from power lines but from *all* electrical equipment, and even from the Earth itself. So you face this additional question: *Do power lines present the greatest hazard as a source of EMFs?*

You now wonder whether the greater hazard comes from power lines or from other EMF sources. Critical thinking, in short, has helped you define and refine the essential questions.

Exploring a Balance of Views

Instead of settling for the most comforting or convenient answer, pursue the *best* answer. Even "expert" testimony may not be enough, because experts can disagree or be mistaken. To answer fairly and accurately, consider a balance of perspectives from up-to-date and reputable sources (Figure 7.4).

FIGURE 7.4
Effective Research Considers Multiple Perspectives
Try to consider all the angles.

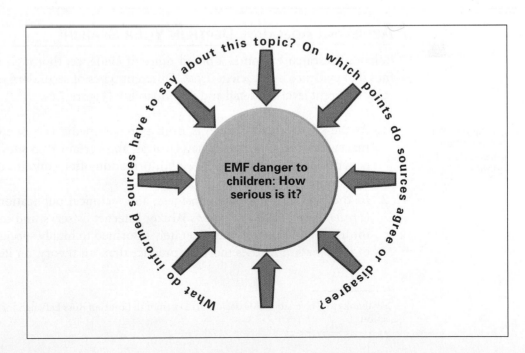

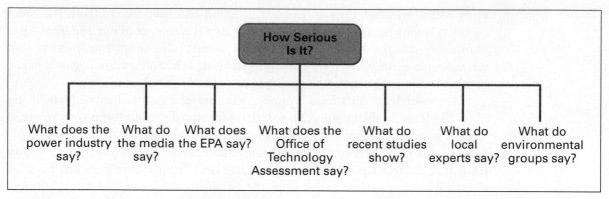

FIGURE 7.5   A Range of Essential Viewpoints   No single source is likely to offer "the final word." Ethical researchers rely on evidence that represents a fair balance of views.

Let's say you've chosen this question: *Do electromagnetic fields from various sources endanger our children?* Now you can consider sources to consult (journals, interviews, reports, Internet sites, database searches, and so on). Figure 7.5 illustrates some likely sources for information on the EMF topic.

NOTE   *Recognize the difference between "balance" (sampling a full range of opinions) and "accuracy" (getting at the facts). Government or industry spokespersons, for example, might present a more positive view (or "spin") than the facts warrant. Not every source is equal, nor should we report points of view as if they were equal (Trafford 137).*

ACHIEVING ADEQUATE DEPTH IN YOUR SEARCH[2]

Balanced research examines a broad *range* of evidence; thorough research examines that evidence in sufficient *depth*. Different types of secondary information occupy different levels of detail and dependability (Figure 7.6).

The depth of a source often determines its quality

1. At the surface level are items from popular media (newspapers, radio, TV, magazines, certain Internet newsgroups, and certain Web sites). Designed for general consumption, this layer of information often contains more journalistic interpretation than factual detail.
2. At the next level are trade, business, and technical publications or Web sites (*Frozen Food World, Publisher's Weekly,* Internet listservs, and so on). Designed for users who range from moderately informed to highly specialized, this layer of information focuses more on practice than on theory, on items considered

---

[2]My thanks to University of Massachusetts Dartmouth librarian Ross LaBaugh for inspiring this section.

**FIGURE 7.6
Effective
Research
Achieves
Adequate
Depth**

7.2

Find out more about
when to skim and
when to drill down at
<www.ablongman.com/
lannonweb>

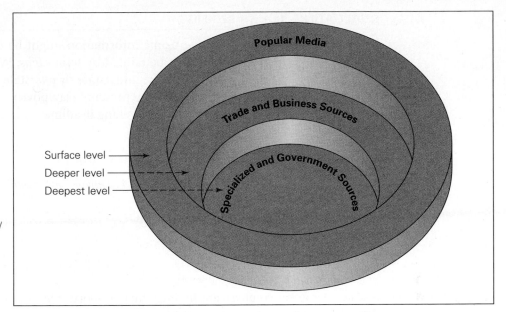

newsworthy to group members, on issues affecting the field, on public relations, and on viewpoints that tend to reflect a field's particular biases.

3. At a deeper level is the specialized literature (journals from professional associations—academic, medical, legal, engineering). Designed for practicing professionals, this layer of information focuses on theory as well as practice, on
descriptions of the latest studies (written by the researchers themselves and
scrutinized by peers for accuracy and objectivity), on debates among scholars
and researchers, and on reviews and critiques of prior studies.

Also at this deepest level are government sources (reports by NASA, EPA,
FAA, the Defense Department, Congress) and corporate documents available
through the Freedom of Information Act (page 143). Designed for anyone
willing to investigate its complex resources, this information layer offers facts
and highly detailed and (in many instances) *relatively* impartial views.

NOTE   *Web pages, of course, offer links to increasingly specific levels of detail. But the actual
"depth" and quality of a Web site's information depend on the sponsorship and reliability of that site (page 166).*

How deep is deep enough? This depends on your purpose, your audience, and
your topic. But the real story and the hard facts most likely reside at deeper levels.
Research on the EMF issue, for example, would need to look beneath popular
"headlines" and biased special interests (say, electrical industry or environmental
groups), focusing instead on studies by experts.

EVALUATING YOUR FINDINGS

Not all findings have equal value. Some information might be distorted, incomplete, or misleading. Material might be tainted by *source bias*: With an emotional issue involving children, a source might understate or overstate certain facts, depending on whose interests that source represents (say power company, government agency, parent group, or a reporter seeking headlines).

QUESTIONS FOR EVALUATING A PARTICULAR FINDING

- *Is this information accurate, reliable, and relatively unbiased?*
- *Do the facts verify the claim?*
- *How much of the information is useful?*
- *Is this the whole or the real story?*
- *Does something seem to be missing?*
- *Do I need more information?*

Instead of merely emphasizing findings that support their own biases or assumptions, ethical researchers seek out the most *accurate* answer. Only near the end of your inquiry can you settle on a *definite* conclusion—based on what the facts suggest.

INTERPRETING YOUR FINDINGS

Once you have decided which of your findings seem legitimate, you need to decide what they all mean.

QUESTIONS FOR INTERPRETING YOUR FINDINGS

- *What are my conclusions and do they address my original research question?*
- *Do any findings conflict?*
- *Are other interpretations possible?*
- *Should I reconsider the evidence?*
- *What, if anything, should be done?*

Perhaps you will reach a definite conclusion: For example, "The evidence about EMF dangers seems persuasive enough for us to be concerned and to take the following actions." Perhaps you will not.

Even the best research can produce contradictory or indefinite conclusions. For instance, critics point out that some EMF studies indicate increased cancer risk while others indicate beneficial health effects. Some scientists claim that stronger EMFs are emitted by natural sources, such as Earth's magnetic field, than by electrical sources (McDonald, "Some Physicists" 5). An accurate conclusion would have to come from your analyzing all views and then deciding that one outweighs the others—or that only time will tell.

CONSIDER THIS Expert Opinion Is Not Always Reliable

An expert is someone capable of doing the right thing at the right time.
     (Holyoak, qtd. in Woodhouse and Nieusma 23).

What We Expect from Experts

Whenever we face uncertainty, we consult experts to help us make informed decisions about complex issues:

- *What should we do about global warming?*
- *Where and how should we store nuclear waste?*
- *How promising is the newly announced "cancer cure"?*
- *What is causing the massive die-off of frogs worldwide?*

How We Confer Expert Status

As information consumers, we confer expert status onto someone who seems to know more than we do—based on credentials, and relevant skills, experience, and knowledge. But researchers point out that "expert status is . . . in the eye of the beholder"—not necessarily based on a person's knowledge or *analytical skills,* but instead on that person's *linguistic skills:* use of technical language and persuasion strategies (Rifkin and Martin 31–36).

The Limits of Expert Opinion

Even though experts tend to consider themselves neutral, in controversial issues, outside influences often cause neutrality to disappear (Rifkin and Martin 31–33):

- *The expert might have a financial stake in the issue—say an environmental researcher who receives financial support from nuclear power companies. So, even though we might recognize this person's knowledge and skill, we might have cause to mistrust her recommendations or conclusions.*
- *The expert might have an extreme point of view, radically different from mainstream,*

accepted, scientific opinion—for example, about the risks or benefits of human cloning experiments.
- *The expert might be venturing in areas beyond his expertise—for example, a real estate lawyer dabbling in copyright law.*
- *The expert can be mistaken—say, a meteorologist predicting the weather.*

A Typical Case of Dueling Experts

For years, scientists and engineers have debated whether Yucca Mountain, Nevada, is an appropriate site for burying high-level nuclear waste deep underground. Some $3 billion worth of technical studies have assessed risks and benefits to health and safety.

Supporting Arguments

- *Storing nuclear waste in one secure facility is safer and cheaper than storing it at the various power plants.*
- *A number of power plants are already running out of storage space.*
- *The Yucca Mountain site is remote, has a dry climate and stable geology, and abuts a desert already contaminated by nuclear-weapons testing over forty years ago.*

Opposing Arguments

- *Some scientists claim that earthquake possibilities have been greatly underestimated, pointing out that tremors occur periodically in this area.*
- *Leaking waste could contaminate ground water.*
- *Some of this material will remain highly dangerous for at least ten thousand years—a period longer than any written language (for warning) or "human-made edifice" has lasted.*

Both sides of the argument are based on *expert* opinion or analysis. In short, "when it comes to Yucca Mountain, scientists do not have the answers" (Gross 134+).

NOTE

Never force a simplistic conclusion on a complex issue. Sometimes the best you can offer is an indefinite conclusion: "Although controversy continues over the extent of EMF hazards, we all can take simple precautions to reduce our exposure." A wrong conclusion is far worse than no definite conclusion at all.

7.3

For more on the languages and cultures of expertise visit <www.ablongman.com/lannonweb>

GUIDELINES for Evaluating Expert Information

To use expert information effectively, follow these suggestions:

1. *Look for common ground.* When opinions conflict, consult as many experts as possible and try to identify those areas in which they agree (Detjen 175).

2. *Consider all reasonable opinions.* Science writer Richard Harris notes that "Often [extreme views] are either ignored entirely or given equal weight in a story. Neither solution is satisfying. . . . Putting [the opinions] in balance means . . . telling . . . where an expert lies on the spectrum of opinion. . . . The minority opinion isn't necessarily wrong—just ask Galileo" (170).

3. *Be sure the expert's knowledge is relevant in this context.* Don't seek advice about a brain tumor from a podiatrist.

4. *Don't expect certainty.* In complex issues, experts cannot *eliminate* uncertainty; they can only help us cope with it.

5. *Expect special interests to produce their own experts to support their position.*

6. *Learn all you can about the issue before accepting anyone's final judgment.*

EXERCISES

 For more exercises, visit <www.ablongman.com/lannon>

1. Students in your major want a listing of one or two discipline-specific information sources from different depths of specialization:

   a. the popular press (newspaper, radio, TV, magazines)
   b. trade/business publications (newsletters and trade magazines)
   c. professional literature (journals)
   d. government sources (corporate data, technical reports, etc.)

   Prepare the list (in memo form) and include a one-paragraph description of each source.

2. Select an issue from science or technology (for example, possible health hazards from cellular phones). Survey expert opinions by consulting Web sources such as these:

- *Ask Jeeves* at <www.askjeeves.com>
- *Webhelp* at <www.webhelp.com>
- *AskMe.com* at <www.askme.com>
- *Scientific American's* "Ask the Experts" link at <www.sciam.com>

Locate one example of each of the following:

- a point on which most experts agree
- a point on which many experts disagree
- an opinion that seems influenced by financial or political motives
- an opinion that resides on the radical end of the spectrum

Report your findings in a memo to share with the class. Be sure to document clearly each source that you are quoting or paraphrasing (see Appendix A).

3. Assume that as communications director for XYZ, an international corporation, you oversee intercultural training of native U.S. employees who will be working in various company branches worldwide and collaborating routinely with members of different cultures. To enhance employee training, you decide to compile a short list of Web sites that provide up-to-date information on various cultures.

After reviewing the following sites, provide a brief description of each, and rank the sites in terms of the depth of information each provides on a given culture (say, Pakistani, Saudi Arabian, and so on). In a memo to all employees, recommend which site(s) to visit for general or specific information and for certain types of information (history, behaviors, values, and so on).

- *Culturegrams* at <www.culturegrams.com/culturegram2000.htm>
- *CIA World Factbook Online* at <www.odci.gov/cia/publications/factbook>
- *U.S. Department of State Electronic Research Collection* at <dosfan.lib.uic.edu>
- *AsiaSource* at <www.asiasource.org>

4. Begin researching for the analytical report (Chapter 24) due at semester's end. Complete these steps. (Your instructor might establish a timetable.)

Phase One: Preliminary Steps

a. Choose a topic of immediate practical importance, something that affects you, your workplace, or your community directly.

b. Identify a specific audience and its intended use of your information. Complete an audience and use profile (page 36).

c. Narrow your topic, and check with your instructor for approval.

d. Make a working bibliography to ensure sufficient primary and secondary sources. Don't delay this step!

e. List the things you already know about your topic.

f. Write a clear statement of purpose and submit it in a proposal memo (page 42) to your instructor.

g. Develop a tree chart of possible questions (as on page 120).

h. Make a working outline.

Phase Two: Collecting Data (Read Chapters 8–10 in preparation for this phase.)

a. In your research, move from general to specific; begin with general works for an overview.

b. Skim your material, looking for high points.

c. Take selective notes. Use notecards or electronic file software.

d. Plan and administer questionnaires, interviews, and inquiries.

e. Whenever possible, conclude your research with direct observation.

f. Evaluate and interpret your findings.

g. Use the checklist on page 194 to reassess your research methods and reasoning.

Phase Three: Organizing Your Data and Writing the Report

a. Revise your working outline as needed.

b. Fully document all sources of information.

c. Write your final draft according to the checklist on page 194.

d. Proofread carefully and add all needed supplements (title page, letter of transmittal, abstract, summary, appendix, glossary).

Due Dates: To Be Assigned by Your Instructor

List of possible topics due:
Final topic due:
Proposal memo due:
Working bibliography and working outline due:
Notecards due:
Copies of questionnaires, interview questions, and inquiry letters due:
Revised outline due:
First draft of report due:
Final draft with supplements and documentation due:

8

Exploring Electronic and Hard Copy Sources

8.1

For more on
disciplinary research
domains online visit
<www.ablongman.com/
lannonweb>

Although electronic searches for information are becoming the norm, a thorough search often requires careful examination of hard copy sources as well. Advantages and drawbacks of each search medium (Table 8.1) often provide good reason for exploring both.

**TABLE 8.1
Hard Copy
versus
Electronic
Sources:
Benefits and
Drawbacks**

	Benefits	**Drawbacks**
Hard Copy Sources	• discovered and organized by librarians • easier to preserve and keep secure	• time-consuming and inefficient to search • hard to update
Electronic Sources	• more current, efficient, and accessible • searches can be narrowed or broadened • can offer material that has no hard copy equivalent	• access to recent material only • not always reliable; sources may be very biased • user might get lost • material might disappear

HARD COPY VERSUS ELECTRONIC SOURCES

Benefits of hard copy sources

Hard copy libraries offer the judgment and expertise of librarians who organize and search for information. Compared with electronic files (on disks, tapes, hard drives), hard copy is easier to protect from tampering and to preserve from aging. (An electronic file's life span can be as brief as ten years.)

Drawbacks of hard copy sources

Manual searches (flipping pages by hand), however, are time-consuming and inefficient: books can get lost; relevant information has to be pinpointed and retrieved, or "pulled" (page 130), by the user. Also, hard copy cannot be updated easily (Davenport 109–11).

Benefits of electronic sources

Compared with hard copy, electronic sources are more current, efficient, and accessible. Sources can be updated rapidly. Ten or fifteen years of an index can be reviewed in minutes. Searches can be customized: for example, narrowed to specific dates or topics. They can also be broadened: a keyword search (page 135) can uncover material that a hard copy search might overlook; Web pages can link to all sorts of material—much of which exists in no hard copy form.

Drawbacks of electronic sources

Drawbacks of electronic sources include the fact that databases rarely contain entries published before the mid-1960s and that material, especially on the Internet, can change or disappear overnight or be highly unreliable. Also, given the researcher's potential for getting lost in cyberspace, a thorough electronic search calls for a preliminary conference with a trained librarian.

CONSIDER THIS Information Can Be "Pushed" or "Pulled"

The conventional strategy for distributing information requires recipients to *pull* the information—retrieve what they need when they need it, say, from the corporate library, the company files, or the Internet. An alternate strategy pushes hard copy or electronic information directly to selected recipients. "The best argument for [push] strategy is that people don't know what they don't know" (Davenport 148).

How Push Technology Works

Specialized software (such as *Headliner Professional* or *Pointcast Network*) allows users to stipulate the types of information they want. After searching, retrieving, and highlighting information tailored to a user's needs, push software automatically downloads the material to the client's desktop.

For Internet searches, users stipulate categories of information (say, municipal bond prices) or specific Web sites to be searched, with updates as requested. Intranet data or Internet material, links included, can be downloaded and then browsed at will.

A push-type search of the company intranet might combine field data from various salespeople (say, about orders or competition), or track the number of hits to the company Web site, or analyze email questions to the Web site about specific products. The user can then view search results (displayed in charts and graphs, spreadsheets, or reports organized by product category) via his or her email or browser (Baker 65; Cortese 152; Cronin, "Using the Web" 254; Haskin, "A Push" 75+).

Workplace Applications of Push Technology

The ultimate promise is that push technology will enable any company to become "event driven." "Every employee will know everything he or she needs to know as soon as the information becomes available" (Desmond 149). Examples:

- *announcements of new developments, breaking news, weather conditions, and so on*
- *internal company information—say, new policies, sales updates, market events*
- *product updates, news about competing companies or research breakthroughs, conference proceedings*
- *electronic copies of relevant articles in journals or trade publications*

Drawbacks of Push Technology

- *No electronic search agent can discriminate like a human reviewer.*
- *Push software might introduce viruses, clutter one's hard drive with worthless material, and breach system security.*

Examples of Push Technology

- *Comprehensive retrieval systems, such as* SavvySearch, *simultaneously employ dozens of Internet search engines* (Infoseek, WebCrawler), *develop a customized search plan, and rank each source on the basis of usefulness to the researcher.*
- *"Bozo filters" sift through email messages, weeding out the nonessential and give priority to others on the basis of particular names or other keywords.*
- *"Personalized newspaper" programs monitor hundreds of news and information sources, select the news most relevant to the individual subscriber (e.g., about a particular company, industry, or medical treatment), and assemble and deliver the document via fax or email (Hafner 77).*

A coming generation of electronic "mentors will search databases for information useful to a particular individual, and . . . spark the user's creativity with questions and facts like those from a human consultant" ("Electronic" 56).

With many automated searches, a manual search of hard copy is usually needed as well. One recent study found greater than 50 percent inconsistency among database indexers. Thus, even an electronic search by a trained librarian can miss improperly indexed material (Lang and Secic 174–75). In contrast, a manual search provides the whole "database" (the bound index or abstracts). As you browse, you often randomly discover something useful.

8.2

Learn more about database networks on the Web at <www.ablongman.com/lannonweb>

INTERNET SOURCES

Internet service providers (ISPs) including Compu-Serve, America Online, and Microsoft Network provide access to the Internet via "gateways," along with aids for navigating its many resources.

Usenet

Usenet is a worldwide system for online discussions via newsgroups, an electronic bulletin board on which users post, share, and discuss information via email.

Moderated newsgroups try to filter out unreliable material

Newsgroups are either *moderated* or *unmoderated*. In a moderated group, each contribution is evaluated by a reviewer who must approve the material before it is posted. In an unmoderated group, all contributions are posted. Most newsgroups are unmoderated, and thus generally less reliable than moderated groups.

For a wealth of reliable information, consult *newsfeed* newsgroups, which post news items from wire services such as the Associated Press. Whereas newspapers can print only a fraction of the information received from wire services, newsfeed groups provide all of it online.

Newsgroups typically publish answers to "frequently asked questions" (FAQ lists) about a topic (acupuncture, sexual harassment, and so on). Although potentially informative, FAQs reflect the biases of those who contribute to and edit them (Maeglin 5). A group's particular convictions might politicize information and produce all sorts of inaccuracies (Snyder 90).

To locate newsgroups on any topic, go to <www.liszt.com/news>.

Listservs

Listservs usually are more specialized than newsgroups

A listserv is a computer-operated mailing list. Like newsgroups, listservs are special-interest groups for email discussion and information sharing. In contrast to newsgroups, listservs usually focus on specialized topics, with discussions usually among experts (say, cancer researchers), often before their findings or opinions appear in published form. Many listservs include a FAQ listing.

Listserv access is available to subscribers who receive mailings automatically via email. Like a newsgroup, a listserv may be moderated or unmoderated, but subscribers are expected to observe proper Internet etiquette and to stick to the topic, without digressions, "flaming" (attacking someone), or "spamming" (posting irrele-

The role of research

"If I have to research I look everywhere—books, online, databases, newsgroups, library, periodicals, etc. And I have to say—if you ever get stuck for information, do check out a newsgroup or mailing list about the subject. Everyone interested in the subject will be there and someone will be able to get you started."

—Lorraine Patsco, Director of Prepress and Multimedia Production

vant messages). Some lists allow anyone to subscribe, while others require approval by the list owner.

To find listservs on any particular topic, go to <www.liszt.com> or <http://tile.net.lists>.

Library Chatrooms

Major libraries offer the research expertise of reference librarians on a round-the-clock basis via live chat. In response to a researcher query, the librarian locates the answer or guides the researcher to the appropriate sources (Kinik 38). See, for example, the *Santa Monica Public Library* site (Figure 8.1) at <www.smpl.org>.

Electronic Magazines (E-zines)

E-zines offer information available only in electronic form. Despite the broad differences in quality among e-zines, this online medium offers certain benefits over hard copy magazines:

Benefits of online magazines

- links to related information
- immediate access to earlier magazine issues
- interactive forums for discussions among readers, writers, and editors
- rapid updating and error correction

Major news publications and television and radio news programs also offer interactive editions online. Examples include *ABCNews.Com* <www.abcnews.go.com>, *National Public Radio* <www.npr.org>, and *PBS Online NewsHour* <www.pbs.org/newshour>.

Email Inquiries

Email is excellent for contacting knowledgeable people in any field. Email addresses are increasingly accessible via locator programs that search the Internet. However, unsolicited and indiscriminate email inquiries might annoy the recipient.

Online reference librarians in the Library of Congress at <http://lcWeb.loc.gov> respond directly to email queries or they forward the question to a member library (Kinik 38).

World Wide Web

The Web is a global network of databases, documents, images, and sounds. All types of information can be accessed and explored through navigation programs such as *Netscape Navigator* or *Microsoft Internet Explorer*, known as "browsers." Hypertext links (page 461) enable users to explore information along different paths by clicking on keywords or icons.

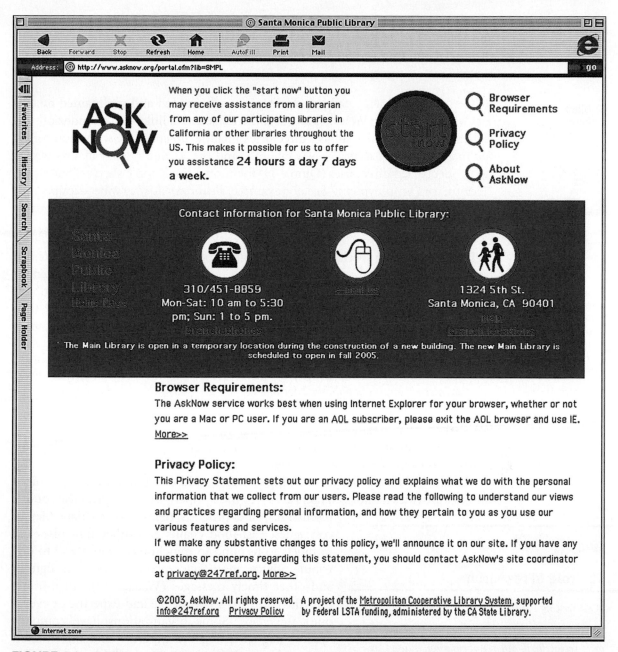

FIGURE 8.1   A Library Chatroom   An on-call librarian can respond to questions via the Interactive Reference link even when the library is closed.

Source: "Ask Now" Web page from Santa Monica Public Library, <http://www.asknow.org/portal.cfm?lib=SMPL>, or <http://www.smpl.org>, Ask Now Link. Used by permission of Metropolitan County Library System.

Each Web site has its own *home page* that serves as an introduction to the site and is linked to additional "pages" that individual users can explore according to their needs. To find various sites on the Web, we use two basic tools: *subject directories and search engines.*

Subject directories are maintained by humans

8.3

For more on evaluating the output of search engines visit <www.ablongman.com/lannonweb>

Most search engines are maintained by computers and not people

NOTE

SUBJECT DIRECTORIES.   Subject directories are compiled and maintained by editors who sift through Web sites and sort the most useful links into an index of subject categories. Popular subject directories include *Yahoo!* at <www.yahoo.com>, the *Internet Public Library* <www.ipl.org>, and the *Virtual Library* at <www.vlib.org>. Specialized directories (Quible 59) focus on a single topic such as "Software," "Health," or "Employment." See, for example, *Beaucoup* at <www.beaucoup.com> for a listing of specialized directories (and search engines) organized by category.

SEARCH ENGINES.   Search engines such as *AltaVista* <www.altavista.com> scan for Web sites containing specific keywords. Because most search engines are maintained by computers instead of people, none of the information gets filtered, evaluated, or organized. So even though search engines yield a lot more information than subject directories, much of it can be irrelevant. Some search engines are more selective than others, and some, such as *SearchIQ* <www.searchiq.com>, focus on specialized topics.

> *Assume that any material obtained from the Internet is protected by copyright. Document all online sources in college papers, and obtain written permission any time you use Internet sources in published documents.*

Intranets and Extranets

An *intranet* is an in-house computer network that employs Internet technology for information access within a company.

Invaluable for on-the-job training, research, and collaboration, a customized intranet provides authorized users access to the company's library, price lists, online discussions, and progress reports—even data files of colleagues. Fast-food chains and other franchises use intranets to respond to franchise questions and to distribute advice, industry/company news, sales figures, and other timely messages (Wallace 12). An intranet puts the company's knowledge and expertise at everyone's fingertips. Some organizations have company "yellow pages," listing the expertise and information possessed by each employee.

An *extranet* integrates a company's intranet with the global Internet. External users (customers, subcontractors, outside vendors) with an Internet connection and a password can browse nonrestricted areas of a company's Web site and download selected information, including customized reports. Extranets elimi-

The role of research

"Our writing is all about research. We hardly ever use the library. Mostly, the writing is about our own research. If other information is used, research assistants will find it on the Web. The Internet has changed my professional life dramatically for the better."

—Paul Harder, President,
mid-sized consulting company

nate the need for the traditional printing and mailing of information to clients or suppliers. They also enable collaboration among organizations. At Caterpillar Tractors, Inc., for example, when a customer's equipment breaks down, the entire company can mobilize immediately, contact in-house and outside experts, access records of previous solutions to similar problems, and collaborate on a solution—for example, designing an improved mechanical part for the equipment (Haskin, "The Extranet" 57–60).

The extranet blend of in-house information and Internet access leaves an organization's network vulnerable to hackers, spies, or saboteurs. Therefore, each extranet site has its own *firewall* (software that keeps out uninvited users and that controls the data they can access). Firewalls offer password protection and encryption (coding) of sensitive information.

KEYWORD SEARCHES USING BOOLEAN OPERATORS

Most engines that search by keyword allow the use of Boolean[1] operators (commands such as "AND," "OR," "NOT," and so on) to define relationships among various keywords. Table 8.2 shows how these commands can expand a search or narrow it by generating more or fewer "hits."

**TABLE 8.2
Using Boolean
Operators to
Expand or Limit
a Search**

If you enter these terms . . .	The computer searches for . . .
electromagnetism *AND* health	Only entries that contain both terms
electromagnetism *OR* health	All entries that contain either term
electromagnetism *NOT* health	Only entries that contain term 1 and do not contain term 2
electromag* = truncated search word	All entries that contain this root within other words

Boolean commands can also be combined, as in

(electromagnetic **OR** radiation) **AND** (fields **OR** tumors)

The hits produced from this query would contain any of these combinations:

electromagnetic fields, electromagnetic and tumors, radiation and fields, radiation and tumors

Using *truncation* (cropping a word to its root and adding an asterisk), as in *electromag**, would produce a broad array of hits, including these:

electromagnet, electromagnetic energy, electromagnetic impulse, electromagnetic wave

---

[1]British mathematician and logician George Boole (1815–1864) developed the system of symbolic logic (Boolean logic) now widely used in electronic information retrieval.

Different search engines use Boolean operators in slightly different ways; many include additional options (such as NEAR, to search for entries that contain search terms within ten or twenty words of each other). Click on the HELP option of your search engine to see which strategies it supports.

GUIDELINES for Researching on the Internet*

1. *Try to focus your search beforehand.* The more precisely you identify the information you seek, the lower your chance of wandering through cyberspace.

2. *Select keywords or search phrases that are varied and technical, rather than general.* Some search terms generate better hits than others. In addition to "electromagnetic radiation," for example, try "electromagnetic fields," "power lines and health," or "electrical fields." Specialized terms (for example, "vertigo" versus "dizziness") offer the best access to sites that are reliable, professional, and specific. Always check your spelling.

3. *Look for Web sites that are specific.* Compile a hotlist of sites most relevant to your needs and interests. Specialized newsletters and trade publications are good sources.

4. *Set a time limit for searching.* Set a ten- to fifteen-minute time limit, and avoid tangents.

5. *Expect limited results from any search.* Each search engine (*AltaVista, Excite, HotBot, Infoseek,* etc.) has its own strengths and weaknesses. Some are faster and more thorough while others yield more targeted and updated hits. Some search titles only—instead of the full text—for keywords. No single search engine can index more than a fraction of ever-increasing Web content. Broaden your coverage by using multiple engines.

6. *Use bookmarks and hotlists for quick access to favorite Web sites.* Mark a useful site with a bookmark, which you can then add to your hotlist.

7. *Download or print out files.* Site addresses can change overnight; material is rapidly updated or discarded—especially from a newsgroup or listserv. When you find something you need, download or print it before it changes or disappears. Be sure to accurately record the URL where you found the information and the date on which you accessed the site.

8. *Be selective about what you download.* Download only what you need. Unless they are crucial to your research, omit graphics, sound, and video files. Focus on text files only.

9. *Never download copyrighted material without written authorization from the copyright holder (page 144).* It is a federal crime to possess or give out electronic copies of copyrighted material worth over $1,000 without permission (Grossman 37). Only material in the public domain (page 145) is exempted.

Before downloading *anything* from the Internet, ask yourself: "Am I violating someone's privacy (as in forwarding an email or a newsgroup entry)?" or "Am I decreasing, in any way, the value of this material for the person who owns it?" For any type of commercial use, obtain permission beforehand and credit the source exactly as directed by the owner.

10. *Consider using information retrieval services.* An electronic service such as Inquisit or DIALOG protects copyright holders by selling access to materials in its database. For a monthly or per-page fee, users can download full texts of articles. Subscribers to these Internet-accessible databases include companies and educational institutions. Check with your library.

    Although these services effectively filter out a lot of Internet junk, they do not catalog material that exists only in electronic form (say, e-zines or newsgroup and listserv entries). Therefore, these databases exclude potentially valuable material (such as research studies not yet available in hard copy) accessible only through a general Web search.

*Guidelines adapted from Baker 57+; Branscum 78; Busiel and Maeglin 39–40, 76; Fugate, "Mastering" 40–41; Kawasaki, "Get Your Facts" 156; Matson, "(Search) Engines" 249–52.

OTHER ELECTRONIC SOURCES

In addition to the Internet, other electronic technologies are used for storing and retrieving information. These technologies are accessible at libraries and, increasingly, via the Web.

Compact Discs

A single CD-ROM can store the equivalent of an entire encyclopedia and serves as a portable database, usually searchable via keyword. One useful CD-ROM for business information is *ProQuest*™: its *ABI/INFORM* database offers full text of more than 600 management, marketing, and business journals published since 1989, along with indexes and abstracts of roughly 1,500 journals from the 1970s onward; its *UMI* database indexes major U.S. newspapers. A useful CD-ROM for information about psychology, nursing, education, and social policy is *SilverPlatter*™.

NOTE   *In many cases, CD-ROM access via the Internet is restricted to users who have passwords for entering a particular library or information system.*

Online Retrieval Services

Libraries and corporations subscribe to three types of mainframe database services that are highly specialized and current, and often updated daily (Lavin 14).

Types of online databases

- *Bibliographic databases* list publications in a particular field and sometimes include abstracts of each work listed.
- *Full-text databases* display the entire article or document (usually excluding graphics), and also allow the article to be printed.
- *Factual databases* provide global and up-to-the-minute stock quotations, weather data, and credit ratings of major companies—among facts of all kinds.

CONSIDER THIS Information in Electronic Form Is Copyright Protected

Unanswered Questions

Copyright and fair use law (page 144) is quite specific for printed works or works in other tangible form (paintings, photographs, music). But how do we define "fair use" (page 144) of intellectual property in electronic form? How does copyright protection apply? How do fair use restrictions apply to material used in multimedia presentations or to text or images that have been altered or reshaped to suit the user's specific needs?

Information obtained via email or discussion groups presents additional problems: Sources often do not wish to be quoted or named or to have early drafts made public. How do we protect source confidentiality? How do we avoid infringing on works in progress that have not yet been published? How do we quote and cite this material without violating ownership and privacy rights (Howard 40–41)?

Present Status of Electronic Copyright Law

Subscribers to commercial online services such as DIALOG pay fees, and copyholders in turn receive royalties. But few specific legal protections exist for noncommercial types of electronic information.

Since April 1989, however, most works are considered copyrighted as soon as they are produced in *any* tangible form—even if they carry no copyright notice. Fair use of electronic information is generally limited to brief excerpts that serve as a basis for response—for example, in a discussion group. Except for certain government documents, no Internet posting is in the "public domain" (page 145) unless it is expressly designated as such by its author (Templeton).

Until specific laws are enacted, the following uses of copyrighted material in electronic form can be considered copyright violations (Communication Concepts, Inc. 13; Elias 85, 86; Templeton):

- *Downloading a work from the Internet and forwarding copies to other readers.*
- *Editing, altering, or incorporating an original work as part of your own document or multimedia presentation.*
- *Placing someone else's printed work online without the author's written permission.*
- *Reproducing and distributing original software or material from a privately owned database.*
- *Copying and forwarding an email message without the sender's authorization. The exact wording of an email message is copyrighted, but its content may legally be revealed—except for "proprietary information" (page 79).*

Some copyright violations (say, reproducing and distributing trade secrets) may exceed the boundaries of civil law and be prosecuted as felonies (Templeton).

When in doubt, assume the work is copyrighted, and obtain written permission from the owner.

For more on the open source software and copy-leftist movement visit <www.ablongman.com/lannonweb>.

Database networks such as DIA-LOG are accessible via the Internet, for a fee. Specialized databases such as MEDLINE or ENVIROLINE offer free bibliographies and abstracts, and for a fee, copies of the full text. Ask your librarian for help searching online databases.

NOTE

Never assume that computers yield the best material. Database specialist Charles McNeil points out that "the material in the computer is what is cheapest to put there." Reference librarian Ross LaBaugh warns of a built-in bias: "The company that assembles the database often includes a disproportionate number of its own publications." Like any collection of information, a database can reflect the biases of its assemblers.

ON THE JOB...
The role of research

"As a freelance researcher, I search online databases and Web sites for any type of specialized information needed by my clients. For example, yesterday I did a search for a corporate attorney who needed the latest information on some specific product-liability issues, plus any laws or court decisions involving specific products. For the legal research I accessed LEXIS, the legal database that offers full-text copies of articles and cases. For the liability issue I began with Dow-Jones News/Retrieval and then doubled-checked by going into the Dialog database."

—Martha Casamonte, freelance researcher

HARD COPY SOURCES

Where you begin your search of hard copy sources depends on whether you are looking for background and basic facts or for the latest information. Library sources are shown in Figure 8.2.

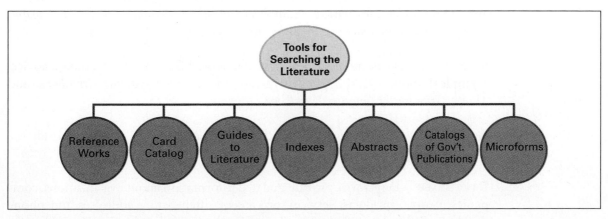

FIGURE 8.2 Ways of Searching the Literature Many of these resources are now accessible via the Web.

If you are an expert in the field, you might simply do a computerized database search or browse through specialized journals and listservs. If you have limited knowledge, you probably want to begin with general reference sources.

NOTE   *Librarian Ross LaBaugh suggests beginning with the popular, general literature, then working toward journals and other specialized sources: "The more accessible the source, the less valuable it is likely to be."*

Reference Works

Reference works provide background that can lead to more specific information. Make sure the work is current by checking the most recent copyright date.

Reference sources provide basic facts

BIBLIOGRAPHIES.   These lists of publications on a given subject are generally issued yearly, or even weekly. However, some quickly become dated. To locate bibliographies in your field, begin with the *Bibliographic Index,* a list (by subject) of major bibliographies. For a list of everything published by the government in scientific and technical fields, consult *A Guide to U.S. Government Scientific and Technical Resources.* You can also find bibliographies on highly focused topics, such as *Health Hazards of Video Display Terminals: An Annotated Bibliography.*

ENCYCLOPEDIAS.   Encyclopedias provide basic information (which might be outdated). Examples include the *Encyclopedia of Building and Construction Terms* and the *Encyclopedia of Food Technology.* The *Encyclopedia of Associations* lists over 30,000 professional organizations worldwide (Institute of Electrical and Electronics Engineers, American Society of Women Accountants, and so on). Most of these organizations now have Web sites.

DICTIONARIES.   Dictionaries can be generalized or focus on specific disciplines or give biographical information. Examples include the *Dictionary of Engineering and Technology* and the *Dictionary of Scientific Biography.*

HANDBOOKS.   These research aids offer condensed facts (formulas, tables, advice, examples) about a field. Examples include the *Civil Engineering Handbook,* and *The McGraw-Hill Computer Handbook.*

ALMANACS.   Almanacs contain factual and statistical data. Examples include the *World Almanac and Book of Facts,* and the *Almanac for Computers.*

DIRECTORIES.   Directories provide updated information about organizations, companies, people, products, services, or careers, often listing addresses and phone numbers. Examples include *The Career Guide: Dun's Employment Opportunities*

Directory, and the *Directory of American Firms Operating in Foreign Countries.* For electronic versions, ask your librarian about *Hoover's Company Capsules* (for basic information on more than 13,000 companies) and *Hoover's Company Profiles* (for detailed information on roughly 3,400 companies).

NOTE   *Many of the reference works listed above are accessible free via the Internet. Go to the Internet Public Library at <www.ipl.org>.*

Card Catalog

All materials held by a library usually are listed in its card catalog under author, title, and subject. Instead of being made up of actual cards, most card catalogs are now electronic and can be accessed through the Internet or through terminals in the library. Visit the library's Web site or ask a librarian for help.

To search catalogs from many different libraries, go to the *Library of Congress Gateway* at <http://lcweb. loc.gov/Z3950/gateway.html#other> or *LibrarySpot* at <www.libraryspot.com>, a gateway to over 5,000 libraries worldwide.

Guides to Literature

If you simply don't know which books, journals, indexes, and reference works are available for your topic, consult a guide to literature. For a list of books in various disciplines, see Walford's *Guide to Reference Material* or Sheehy's *Guide to Reference Books.* For scientific and technical literature, see Malinowsky and Richardson's *Science and Engineering Literature: A Guide to Reference Sources.* Ask a librarian about guides for your discipline.

Indexes

Indexes list current works

Indexes are lists of books, newspaper articles, journal articles, or other works on a particular subject. Most are now searchable by computer.

BOOK INDEXES.   All books currently published (up to a set date) are listed in book indexes by author, title, or subject. Sample indexes include *Scientific and Technical Books and Serials in Print* (an annual listing) and *New Technical Books: A Selective List with Descriptive Annotations* (issued ten times yearly).

NOTE   *No book is likely to offer the very latest information, because of the time required to publish a book manuscript (from months to over one year).*

NEWSPAPER INDEXES.   Most indexes list articles by subject. Sample titles include the *New York Times Index* and the *Wall Street Journal Index.* Most newspapers and news magazines are searchable via their Web sites, and they usually charge a fee for searches of past issues.

PERIODICAL INDEXES.   A periodical index lists articles from magazines and journals. For general information, consult the *Magazine Index* (a subject listing on microfilm) and the *Reader's Guide to Periodical Literature* (updated every few weeks).

For specialized information, consult indexes that list articles by discipline, such as the *General Science Index* or *Business Periodicals Index*. Specific disciplines have their own indexes such as the *Index to Legal Periodicals* and the *International Nursing Index*. The *Expanded Academic Index*, on CD-ROM, lists some 1,200 journals and often provides full text and images. Ask a librarian about indexes for your topic.

CITATION INDEXES.   Citation indexes help answer this question: *Who else has said what about this published idea?* Using a citation index, a researcher can track down the publications in which the original material has been cited, quoted, critiqued, verified, or otherwise amplified (Garfield 200). Examples include the *Social Science Citation Index* and *Web of Science (Science Citation Index Expanded)*, which cross-references articles from more than 5,700 journals.

TECHNICAL REPORT INDEXES.   Government and private-sector reports written worldwide offer specialized and highly current information. Sample indexes include *Scientific and Technical Aerospace Reports* and the *Government Reports Announcements and Index*. Proprietary or security restrictions, of course, restrict public access to certain corporate or government documents.

PATENT INDEXES.   Patents are issued to protect rights to new inventions, products, or processes. Information experts Schenk and Webster point out that patents are often overlooked as sources of current information: "Since . . . complete descriptions of the invention [must be] included in patent applications, one can assume that almost everything that is new and original in technology can be found in patents" (121). Sample indexes include the *Index of Patents Issued from the United States Patent and Trademark Office* and the *World Patents Index*. Patents in various technologies are searchable through databases such as *Hi Tech Patents* and PI *(World Patents Index)*.

What happens when you are locked out of a source? Learn more at <www.ablongman.com/lannonweb>

INDEXES TO CONFERENCE PROCEEDINGS.   Many of the papers presented at the thousands of professional conferences are collected and then indexed in listings such as the *Index to Scientific and Technical Proceedings* and *Engineering Meetings*. The latest ideas or advances in a field often are unveiled at conferences, before appearing as journal publications.

Abstracts

Beyond indexing various works, abstracts summarize each article. The abstract can save you from going all the way to the journal to decide whether to read the article. Abstracts usually are titled by discipline: *Biological Abstracts, Computer Abstracts,* and so on. For some current research, you might consult abstracts of doctoral dissertations in *Dissertation Abstracts International*. Most abstracts are searchable by computer.

Access Tools for U.S. Government Publications

The federal government publishes maps, periodicals, books, pamphlets, manuals, research reports, and all sorts of other information. A few of the countless titles include *Electromagnetic Fields in Your Environment, Major Oil and Gas Fields of the Free World,* and the *Journal of Research of the National Bureau of Standards.* Much of this information can be searched online. Your best bet for tapping this complex resource is to request assistance from the librarian in charge of government documents. If your library does not hold the publication you seek, it can be obtained through electronic access or interlibrary loan.

Here are the basic access tools for documents issued or published at government expense and access tools for privately sponsored documents.

- *The Monthly Catalog of the United States Government,* the major pathway to government publications and reports.
- *Government Reports Announcements and Index,* a listing (with summaries) of more than one million federally sponsored research reports and patents since 1964.
- *The Statistical Abstract of the United States,* updated yearly, offers statistics on population, health, employment, and the like. It can be accessed via the Web. CD-ROM versions are now available.

Many unpublished documents are available under the Freedom of Information Act. The FOIA grants public access to all federal agency records except for classified documents, trade secrets, certain law enforcement files, records protected by personal privacy law, and similar types of exempted information.

Publicly
accessible
government
records

Suppose you have heard that a certain toy has been recalled as a safety hazard and you want to know the details. In this case, the Consumer Product Safety Commission could help you. Perhaps you want to read the latest inspection report on conditions at a nursing home certified for Medicare. Your local Social Security office keeps such records on file. Or you might want to know if the Federal Bureau of Investigation has a file that includes you. In all these examples, you may use the FOIA to request information. (U.S. General Services Administration 1)

Contact the agency that would hold the records you seek: say, for workplace accident reports, the Department of Labor; for industrial pollution records, the Environmental Protection Agency.

Much government information is posted to the Web. For example, the Food and Drug Administration maintains a bulletin board at <www.fda.gov> on experimental AIDS drugs, on drug recalls, and related items; the Department of Energy at <www.doe.gov> offers information on human radiation experiments. One gateway to government sites: the Library of Congress page at <http://lcWeb.loc.gov>.

Microforms

Microform technology allows vast quantities of printed information to be stored in rolls of microfilm or packets of microfiche. (This material is read on machines that magnify the reduced image.) Ask your librarian for assistance.

CONSIDER THIS Frequently Asked Questions about Copyright of Hard Copy Information

Copyright laws ultimately have an ethical purpose: to balance the reward for intellectual labors with the public's right to use information freely.

1. *What is a copyright?*

   A copyright is the exclusive legal right to reproduce, publish, and sell a literary, dramatic, musical, or artistic work. Written permission must be obtained to use all copyrighted material.

2. *What are the limits of copyright protection?*

   Copyright protection covers the exact wording of the original, but not the ideas, concepts, theories, or factual information that it conveys. For example, Einstein's theory of relativity has no copyright protection but the exact wording in announcing or explaining the theory does (Abelman 33; Elias 3). Also, anyone paraphrasing Einstein's ideas but failing to cite him as the source of that paraphrase would be guilty of plagiarism.

3. *How long does copyright protection last?*

   Works published before January 1, 1978, are generally protected for ninety-five years. Works published on or after January 1, 1978, are copyrighted for the author's life plus seventy years.

4. *Must a copyright be officially registered in order to protect a work?*

   No. Protection begins as soon as a work is created. But an author suing for copyright infringement would need to register the work before proceeding.

5. *Must a work be published in order to receive copyright protection?*

   No. In fact, a legal ruling of "fair use" for any unpublished work is less likely than for a published work because unauthorized use violates the author's right to decide if, when, and how to publish the work (Elias 108).

6. *What is "fair use"?*

   "Fair use" is the legal and limited use of copyrighted material without permission. The source of any material that is not your own should, of course, be acknowleged. Fair use does not ordinarily apply to case studies, charts and graphs, author's notes, or private letters ("Copyright Protection" 30).

7. *How is fair use determined?*

   In determining whether the use of copyrighted material is fair, the courts ask these questions:

   • *Is the material being used for commercial or for nonprofit purposes?* For example, nonprofit educational use is viewed more favorably than for-profit use.
   • *Is the copyrighted work published or unpublished?* Use of published work, say, a news article, is viewed more favorably than use of unpublished essays, correspondence, or the like.
   • *How much, and which part, of the origi-*

nal work is being used? The smaller the part of the original, the more favorably its use will be viewed. Never considered fair, however, is the use of a part that "forms the core, distinguishable, creative effort of the work being cited" *(Author's Guide 30).*

- *How will the economic value of the original work be affected?* Any use that reduces the potential market value of the original will be viewed unfavorably.

8. *What is the exact difference between copyright infringement and fair use?*

Although using ideas from an original work is considered fair use, a paraphrase that incorporates too much of the original expression can be considered infringement—even when the source is cited (Abelman 41). The reproduction of a government document that includes material previously protected by copyright (graphs, images, company logos, slogans, or other material) is considered infringement. The United States Copyright Office offers this general caution:

> The distinction between "fair use" and infringement may be unclear and not easily defined. There is no specific number of words, lines, or notes that may safely be taken without permission.
>
>   Acknowledging the source of the copyrighted material does not substitute for obtaining permission. ("Fair Use" 1–2)

When in doubt, obtain written permission from the copyright holder.

For updated information about fair use issues, especially pertaining to works in electronic form, visit the Stanford University Library Web site at: <http://www.fairuse.Stanford.edu>.

9. *What is material in the "public domain"?*

"Public domain" refers to material not protected by copyright or material on which copyright has expired. Works published in the United States ninety-five years before the current year are in the public domain. Most government publications and commonplace information, such as height and weight charts or a [metric conversion] table are in the public domain. These works occasionally contain copyrighted material (used with permission and properly acknowledged). A new translation or version of a work in the public domain can be protected by copyright; if you are not sure whether something is in the public domain, request permission. ("Copyright Protection" 31)

10. *What about international copyright protection?*

Copyright protection varies among individual countries, and some countries offer virtually no protection for foreign works (say, one produced in the United States):

> There is no such thing as an "international copyright" that will automatically protect an author's writings throughout the world . . . . An author who wishes copyright protection . . . in a particular country should first determine the extent of protection available to works of foreign authors in that country . . . . ("International Copyright" 1–2)

In the United States all foreign works that meet certain requirements are protected by copyright (Abelman 36).

For more on international copyright issues, go to The World Intellectual Property Organization at <www.wipo.org>.

11. *Who owns the copyright to a work prepared as part of one's employment?*

A work prepared in the service of one's employer or under written contract for a client is a "work made for hire." The employer or client is legally considered the author and therefore holds the copyright (Abelman 33–34). For example, a manual researched, designed, and written as part of one's employment would be a work made for hire.

12. *What guidelines cover material downloaded on a computer?*

In recent rulings on electronic copyright issues, courts have followed traditional copyright laws (Graybill 27). For the latest developments, visit the Web site of the United States Copyright Office at: <http://www.loc.gov/copyright>. For more on electronic copyright issues, see page 138, and visit *The Copyright Website* at <http://www.benedict.com> and the University of Texas legal site at <http://www.utsystem.edu/OGC>.

For links to articles about intellectual property issues, go to the *Publishing Law Center* at <www.publaw.com>.

EXERCISES   For more exercises, visit <www.ablongman.com/lannon>

1. Using the printed or electronic card catalog, locate and record the full bibliographic data for five books in your field or on your semester report topic, all published within the past year.
2. Consult the *Library of Congress Subject Headings* for alternative headings under which you might find information in the card catalog for your semester report topic.
3. List five major reference works in your field or on your topic by consulting Sheehy, Walford, or a more specific guide to literature.
4. List the title of each of these specialized reference works in your field or on your topic: a bibliography, an encyclopedia, a dictionary, a handbook, an almanac (if available), and a directory.
5. Identify the major periodical index in your field or on your topic. Locate a recent article on a specific topic (e.g., use of artificial intelligence in medical diagnosis). Photocopy the article and write an informative abstract.
6. Consult the appropriate librarian and identify two databases you would search for information on the topic in Exercise 2.
7. Identify the major abstract collection in your field or on your topic. Using the abstracts, locate a recent article. Photocopy the abstract and the article.
8. Using technical report indexes, locate abstracts of three recent reports on one specific topic in your field. Provide complete bibliographic information.

9. Using patent indexes, locate and describe three recently patented inventions in your field, and provide complete bibliographic information.
10. Using indexes of conference proceedings, locate abstracts of three recent conference papers on *one* specific topic in your field. Provide complete bibliographic information.
11. Using the *Monthly Catalog* or *Government Reports Announcements and Index,* locate and photocopy (or download and print) a recent government publication in your field or on your topic.
12. Using OCLC, RLIN, *ProQuest, SilverPlatter,* or similar online or CD-ROM services, locate and copy the bibliographic record (including abstracts, if available) of four current books and four current articles in your field or on your topic.
13. If your library offers students a free search of commercial database networks such as DIALOG, ask your librarian for help in preparing an electronic search for your semester report.
14. Explore Internet databases via Prodigy, Pathfinder, the Microsoft Network, or a similar access provider or "on-ramp." Prepare a list of promising database resources for your report topic.
15. Locate an Internet discussion list related to your report topic. Download and print out the group's FAQ list. For a directory of listservs, consult <http://www.liszt.com>.
16. Using *Netscape Navigator, Internet Explorer,* or a similar browsing program, search Web sites to locate resources for your report topic.

17. If your library belongs to a consortium of electronically networked libraries, search the holdings of other libraries on the network for topic resources not available in your library. Prepare a list of promising possibilities.

18. Students in your major want a listing of at least *two* of each of the following discipline-specific sources: the main reference books; indexes; periodicals; government publications; commercial, Internet, and CD-ROM databases; online newsgroups and discussion groups. Prepare the list (in memo form) and include a one-paragraph description of each source. Be prepared to discuss your list in class.

19. Most Web browsers allow you to do keyword searches (page 135) by using a search engine such as *AltaVista* or *Infoseek*. Each engine has its own guidelines and peculiarities; these usually are explained in a "help" file or user's guide. Learn to use at least one search engine; for your colleagues, write instructions for designing and conducting a Web search using that engine.

    URLs: <www.searchiq.com>
    <www.lycos.com>
    <www.altavista.com>
    <www.infoseek.com>

COLLABORATIVE PROJECTS

1. Group yourselves according to major. For other students in your major, prepare a guide, in the form of a brochure, to your library's electronic resources (CD-ROM services and commercial database services, electronic catalogs, network consortium, Internet gateways, and so on). Describe discipline-specific types of resources available via each electronic medium. Early in this project, arrange for a group tour and demonstration of your library's resources by a librarian. (In conjunction with this project, your instructor may assign Chapters 15 and 22.)

2. Divide into groups and prepare a comparative evaluation of literature search media. Each group member will select *one* of the resources listed below and create an individual bibliography (listing at least twelve recent and relevant works on a topic of interest selected by the group):

   - conventional print media
   - electronic catalogs
   - CD-ROM services
   - a commercial database service such as DIALOG
   - the Internet and World Wide Web
   - an electronic consortium of local libraries, if applicable

   After recording the findings and keeping track of the time spent in each search, compare the ease of searching and quality of results obtained from each type of search on your group's selected topic. Which medium yielded the most current sources (page 165)? Which provided abstracts and full texts as well as bibliographic data? Which consumed the most time? Which provided the most dependable sources (page 166)? The most diverse or varied sources (page 121)? Which cost the most to use? Finally, which yielded the greatest *depth* of resources (page 122)?

   Prepare a report and present your findings to the class. (In conjunction with this project, your instructor may assign Chapter 24.)

3. Divide into groups and decide on a campus or community issue or some other topic worthy of research. Elect a group manager to assign and coordinate tasks. At project's end, the manager will provide a performance appraisal by summarizing, in writing, the contribution of each team member. Assigned tasks will include planning, information gathering from primary and secondary sources, document preparation (including visuals) and revision, and classroom presentation. (See page 98 for collaboration guidelines.)

   Do the research, write the report, and present your findings to the class. (In conjunction with this project, your instructor may assign Chapter 24.)

4. Group yourselves according to major. Assume that several employers in your field are holding a job fair on campus next month and will be interview-

ing entry-level candidates. Each member of your group is assigned to develop a profile of *one* of these companies or organizations by researching its history, record of mergers and stock value, management style, financial condition, price/earnings ratio of its stock, growth prospects, products and services, multinational affiliations, ethical record, environmental record, employee relations, pension plan, employee stock options or profit-sharing plans, commitment to affirmative action, number of women in upper management, or any other features important to a prospective employee. The entire group will then edit each profile and assemble them in one single document to be used as a reference for students in your major.

SERVICE-LEARNING PROJECT

Two sites that provide information specific to service-learning and grant writing are the *Foundation Center* at <www.fdncenter.org> and *Donor's Forum* at <www.donorsforum.org>. Go to the *Foundation Center* site and click on "finding funders." Enter a search term that describes an issue your agency addresses (for example, "addiction" or "affordable housing"). Develop a list of ten foundations that look like primary sources of support for your agency. Detail these ten foundations in a one-page summary memo to the agency staff. (For more on grants and grant writing, see Chapter 23.)

9

Exploring Primary Sources

**FIGURE 9.1
Sources for
Primary
Research**

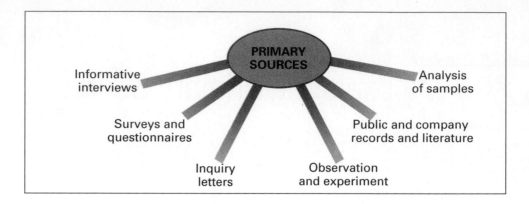

The role of research

"When researching a topic for a manual, I focus on usability, by trying to anticipate my audience's needs and asking the technical source person (usually a programmer or a systems analyst) specific questions keyed to my audience's needs: Who performs the task? What materials are required? What does the task accomplish? What can go wrong? Otherwise, I would waste time soaking up like a sponge any and all information the source person feels like rattling off, whether it's important or not."

—Pam Herbert, technical writer,
computer firm

Workplace decisions often rely on primary research—an original, firsthand study of the topic, involving sources like those in Figure 9.1.

INFORMATIVE INTERVIEWS

An excellent primary source for information unavailable in any publication is the personal interview. Much of what an expert knows may never be published (Pugliano 6). Also, a respondent might refer you to other experts or sources of information.

Of course, an expert's opinion can be just as mistaken or biased as anyone else's (page 125). As medical patients, for example, we would seek second opinions about serious medical conditions. As researchers we should seek a balanced range of expert opinions about a complex problem or controversial issue—not only from a company engineer and environmentalist, for example, but also from independent and presumably more objective third parties such as a professor or journalist who has studied the issue.

Selecting the Best Interview Medium

Once you decide whom to interview about what, select your medium carefully:

How interview
media compare

- *In-person interviews* are most productive because they allow human contact (Hopkins-Tanne 24).
- *Phone interviews* are convenient, but they lack the human contact of in-person interviews—especially when the interviewer and respondent have not met.

- *Email interviews* are convenient and inexpensive, and they allow plenty of time for respondents to consider their answers.
- *Fax interviews* are highly impersonal, and using them is generally a bad idea.

Whatever your medium, obtain a respondent's approval *beforehand*—instead of waylaying this person with an unwanted surprise.

GUIDELINES for Informative Interviews*

Planning the Interview

1. *Focus on your purpose.* Determine exactly what you hope to learn from this interview. Write out your purpose.

> I will interview Anne Hector, Chief Engineer at Northport Electric, to ask her about the company's approaches to EMF risk avoidance—within the company as well as in the community.

2. *Do your homework.* Learn all you can about the topic beforehand. If the respondent has published anything relevant, read it before the interview. Don't waste time asking questions you could have answered yourself.

3. *Contact the intended respondent.* Do this by phone, letter, or email, and be sure to introduce yourself and your purpose. (See Karen Granger's letter on page 439.)

4. *Request the interview at your respondent's convenience.* Give the respondent ample notice and time to prepare, and ask whether she/he objects to being quoted or taped. If you use a tape recorder, insert fresh batteries and a new tape and set the recording volume loud enough. If possible, submit a list of questions well before the actual interview.

Preparing the Questions

1. *Make each question clear and specific.* Vague, unspecific questions elicit vague, unspecific answers.

> How is this utility company dealing with the problem of electromagnetic fields?

> Which problem—public relations, potential liability, danger to electrical workers, to the community, or what?

> What safety procedures have you developed for risk avoidance by electrical work crews?

2. *Avoid questions that can be answered with "yes" or "no."*

*Several guidelines are adapted from Blum 88; Dowd 13–14; Hopkins-Tanne 23, 26; Kotulak 147; Lambe 32; McDonald, "Covering Physics" 190; Rensberger 15; Young 114, 115, 116.

Cla
qu

Fo
qu

Purpose
statement

9.1

For more resources
on interviewing
techniques visit
<www.ablongman.com/
lannonweb>

Co
qu

A vague question

A clear and
specific question

(continues)

2. *Design an engaging introduction and opening questions.* Persuade respondents that the survey relates to their concerns, that their answers matter, and that their anonymity is assured. Explain how respondents will benefit from your findings, or offer an incentive (say, a copy of your final report).

A survey
introduction

> Your answers will help our school board to speak accurately for your views at our next town meeting. Results of this survey will appear in our campus newspaper. Thank you.

Researchers often include a cover letter with the questionnaire.

Begin with the easiest questions, which usually are the closed-ended ones. Once respondents commit to these, they are likely to complete more difficult questions later.

3. *Make each question unambiguous.* All respondents should be able to interpret identical questions identically. An ambiguous question leaves room for misinterpretation.

An ambiguous
question

> Do you favor weapons for campus police? YES_____ NO_____

"Weapons" might mean tear gas, clubs, handguns, all three, or two out of three. Consequently, responses to the above question would produce a misleading statistic, such as "Over 95 percent of students favor handguns for campus police" when the accurate conclusion might be "Over 95 percent of students favor some form of weapon." Moreover, the limited "yes/no" format reduces an array of possible opinions to an either/or choice.

A clear and
incisive question

> Do you favor (check all that apply):
> _____ Having campus police carry mace and a club?
> _____ Having campus police carry nonlethal "stun guns"?
> _____ Having campus police store handguns in their cruisers?
> _____ Having campus police carry small-caliber handguns?
> _____ Having campus police carry large-caliber handguns?
> _____ Having campus police carry no weapons?
> _____ Don't know

To ensure a full range of possible responses, include options such as "Other," "Don't know," "Not Applicable," or an "Additional Comments" section.

4. *Make each question unbiased.* Avoid *loaded questions* that invite or advocate a particular viewpoint or bias:

A loaded
question

> Should our campus tolerate the needless endangerment of innocent students by lethal weapons?
> YES_____ NO_____

- *Email interviews* are convenient and inexpensive, and they allow plenty of time for respondents to consider their answers.
- *Fax interviews* are highly impersonal, and using them is generally a bad idea.

Whatever your medium, obtain a respondent's approval *beforehand*—instead of waylaying this person with an unwanted surprise.

GUIDELINES for Informative Interviews*

Planning the Interview

1. *Focus on your purpose.* Determine exactly what you hope to learn from this interview. Write out your purpose.

   > I will interview Anne Hector, Chief Engineer at Northport Electric, to ask her about the company's approaches to EMF risk avoidance—within the company as well as in the community.

2. *Do your homework.* Learn all you can about the topic beforehand. If the respondent has published anything relevant, read it before the interview. Don't waste time asking questions you could have answered yourself.

3. *Contact the intended respondent.* Do this by phone, letter, or email, and be sure to introduce yourself and your purpose. (See Karen Granger's letter on page 439.)

4. *Request the interview at your respondent's convenience.* Give the respondent ample notice and time to prepare, and ask whether she/he objects to being quoted or taped. If you use a tape recorder, insert fresh batteries and a new tape and set the recording volume loud enough. If possible, submit a list of questions well before the actual interview.

Preparing the Questions

1. *Make each question clear and specific.* Vague, unspecific questions elicit vague, unspecific answers.

   > How is this utility company dealing with the problem of electromagnetic fields?

   > Which problem—public relations, potential liability, danger to electrical workers, to the community, or what?

   > What safety procedures have you developed for risk avoidance by electrical work crews?

2. *Avoid questions that can be answered with "yes" or "no."*

*Several guidelines are adapted from Blum 88; Dowd 13–14; Hopkins-Tanne 23, 26; Kotulak 147; Lambe 32; McDonald, "Covering Physics" 190; Rensberger 15; Young 114, 115, 116.

Purpose statement

9.1

For more resources on interviewing techniques visit <www.ablongman.com/lannonweb>

A vague question

A clear and specific question

(continues)

Guidelines (continued)

An unproductive question

> In your opinion, can technology find ways to decrease EMF hazards?

Instead, phrase your question to elicit a detailed response:

A productive question

> Of the various technological solutions being proposed or considered, which do you consider most promising?

This is one instance in which your earlier homework pays off.

3. *Avoid loaded questions.* A loaded question invites or promotes a particular bias:

A loaded question

> Wouldn't you agree that EMF hazards have been overstated?

An impartial question does not lead the interviewee to respond in a certain way.

An impartial question

> In your opinion, have EMF hazards been accurately stated, overstated, or understated?

4. *Save the most difficult, complex, or sensitive questions for last.* Leading off with your toughest questions might annoy respondents, making them uncooperative.

5. *Write out each question on a separate blank page.* Use a three-ring binder with $8\frac{1}{2}" \times 11"$ pages to arrange your questions in logical order. You can then flip to a new page for each question and record responses easily.

Conducting the Interview

1. *Make a good start.* Dress appropriately and arrive on time. Thank your respondent; restate your purpose; explain why you believe he/she can be helpful; explain exactly how you will use the information .

2. *Be sensitive to cultural differences.* If the respondent belongs to a culture different from your own, then consider the level of formality, politeness, directness, relationship building, and other behaviors seen as appropriate in that culture. (See pages 59–61.)

3. *Let the respondent do most of the talking.* Keep opinions to yourself.

4. *Be a good listener.* Don't doodle or let your eyes wander. People reveal more when their listener seems genuinely interested. (For advice about active listening, see pages 106–07.)

5. *Stick to your interview plan.* If the respondent wanders, politely nudge the conversation back on track (unless the added information is useful).

6. *Ask for clarification or explanation.* If you don't understand an answer, say so. Request an example, an analogy, or a simplified version—and keep asking until you understand.

Clarifying
questions

- Could you go over that again?
- Is there a simpler explanation?

Science writer Ronald Kotulak argues that "[no] question is dumb if the answer is necessary to help you understand something. . . . Don't pretend to know more than you do" (144).

7. *Keep checking on your understanding.* Repeat major points in your own words and ask if the technical details are accurate and if your interpretation is correct. But don't put words into the respondent's mouth.

8. *Be ready with follow-up questions.* Some answers may reveal new directions for the interview.

Follow-up
questions

- Why is it like that?
- Could you say something more about that?
- What more needs to be done?
- What happened next?

9. *Keep note-taking to a minimum.* Record statistics, dates, names, and other precise data, but not every word. Jot key terms or phrases that later can refresh your memory.

Concluding the Interview

1. *Ask for closing comments.* Perhaps the respondent can lead you to additional information.

Concluding
questions

- Would you care to add anything?
- Is there anyone else I should talk to?
- Is there anyone who has a different point of view?
- Are there any other sources you are aware of that might help me better understand this issue?

2. *Request permission to follow up.* If additional questions arise, you might need to contact the respondent again, perhaps by phone, email, or fax—depending on the complexity of the questions and on the respondent's preference.

3. *Invite the respondent to review your version.* If the interview is to be published, ask the respondent to check your final draft (for misspelled names, inaccurate details, misquotations, and so on) and to approve it. Offer to provide copies of any document in which this information appears.

4. *Thank your respondent and leave promptly.*

5. *As soon as you leave the interview, write a complete summary (or record one verbally).* Do this while responses are fresh in your memory.

A Sample Interview

Figure 9.2 shows the partial text of an interview on persuasive challenges in the workplace. Notice how the interviewer probes, seeks clarification, and follows up on responses from XYZ's Director of Corporate Relations.

SURVEYS AND QUESTIONNAIRES

Surveys help you to develop profiles and estimates about the concerns, preferences, attitudes, beliefs, needs, or perceptions of a large, identifiable group (a *target population*) by studying representatives of that group (a *sample*).

> - Do consumers prefer brand A or brand B?
> - What percentage of students feel safe on our campus?
> - Is public confidence in technology increasing or decreasing?
> - Are people able to use this product safely and efficiently?

NOTE   *A "census" is a survey of an entire target population.*

The tool for conducting surveys is the questionnaire. While interviews allow for greater clarity and depth, questionnaires offer an inexpensive way to survey a large group. Respondents can answer privately and anonymously—and often more candidly than in an interview.

Questionnaires carry certain limitations, though:

Limitations of survey research

- *A low rate of response* (*often less than 30 percent*). People refuse to respond to a questionnaire that seems too long, too complicated, or in some way threatening. They might be embarrassed by the topic or afraid of how their answers could be used.
- *Responses that might be non-representative.* A survey will get responses from the people who want to respond, but you will know nothing about the people who didn't respond. Those who responded might have extreme views, a particular stake in the outcome, or some other motive that represents inaccurately the population being surveyed (Plumb and Spyridakis 625–26).
- *Lack of follow-up.* Survey questions do not allow for the kind of follow-up and clarification possible with interview questions.

Even surveys by professionals carry potential for error. As consumers of survey research, we need to understand how surveys are designed, administered, and interpreted, and what can go wrong in the process. The following is an introduction to creating surveys and to avoiding pitfalls along the way.

Probing and following up

Q. *Would you please summarize your communication responsibilities?*

A. The corporate relations office oversees three departments: customer service (which handles claims, adjustments, and queries), public relations, and employee relations. My job is to supervise the production of all documents generated by this office.

Q. *Isn't that a lot of responsibility?*

A. It is, considering we're trying to keep some people happy, getting others to cooperate, and trying to get everyone to change their thinking and see things in a positive light. Just about every document we write has to be persuasive.

Seeking clarification

Q. *What exactly do you mean by "persuasive"?*

A. The best way to explain is through examples of what we do. The customer service department responds to problems like these: Some users are unhappy with our software because it won't work for a particular application, or they find a glitch in one of our programs, or they're confused by the documentation, or someone wants the software modified to meet a specific need. For each of these complaints or requests, we have to persuade our audience that we've resolved the problem or that we're making a genuine effort to resolve it quickly.

The public relations department works to keep up our reputation through links outside the company. For instance, we keep in touch with this community, with consumers, the general public, government and educational agencies.

Seeking clarification

Q. *Can you be more specific? "Keeping in touch" doesn't sound much like persuasion.*

A. Okay, right now we're developing programs with colleges and universities, in which we offer heavily discounted software, backed up by an extensive support network (regional consultants, an 800 phone hotline, and workshops). We're hoping to persuade them that our software is superior to our well-entrenched competitor's. And locally we're offering the same kind of service and support to business clients.

Following up

Q. *What about employee relations?*

A. Day to day we face the usual kinds of problems: trying to get 100 percent employee contributions to the United Way, or persuading employees to help out in the community, or getting them to abide by new company regulations restricting smoking or to limit personal phone calls. Right now, we're facing a real persuasive challenge. Because of market saturation, software sales have flattened across the board. This means temporary layoffs for roughly 28 percent of our employees. Our only alternative is to persuade *all* employees to accept a 10-percent salary and benefit cut until the market improves.

Probing

Q. *How, exactly, do you persuade employees to accept a cut in pay and benefits?*

A. Basically, we have to make them see that by taking the cut, they're really investing in the company's future—and, of course, in their own.

[The interview continues.]

FIGURE 9.2 Partial Text of an Informative Interview

Defining the Survey's Purpose and Target Population

Why is this survey being performed? What, exactly, is it measuring? How much background research do you need? How will the survey findings be used?

Who is the exact population being studied (the chronically unemployed, part-time students, computer users)? For example, in its research on science and technology activity, the *Statistical Abstract of the United States* differentiates "scientists and engineers" from "technicians":

Target populations clearly defined

> Scientists and engineers are defined as persons engaged in scientific and engineering work at a level requiring a knowledge of sciences equivalent at least to that acquired through completion of a 4-year college course. Technicians are defined as persons engaged in technical work at a level requiring knowledge acquired through a technical institute, junior college, or other type of training less extensive than 4-year college training. Craftspersons and skilled workers are excluded. (U. S. Department of Commerce 603)

Identifying the Sample Group

How will intended respondents be selected? How many respondents will there be? Generally, the larger the sample surveyed, the more dependable the results (assuming a well-chosen and representative sample). Will the sample be randomly chosen? In the statistical sense, "random" does not mean "haphazard": a random sample means that each member of the target population stands an equal chance of being in the sample group.

Even a sample that is highly representative of the target population carries a measure of *sampling error*.

A type of survey error

> The particular sample used in a survey is only one of a large number of possible samples of the same size which could have been selected using the same sampling procedures. Estimates derived from the different samples would, in general, differ from each other. (U. S. Department of Commerce 949)

The larger the sampling error (usually expressed as the *margin of error*, page 185), the less dependable the survey findings.

9.2

For more qualitative and quantitative survey techniques visit <www.ablongman.com/lannonweb>

Defining the Survey Method

What type of data (opinions, ideas, facts, figures) will be collected? Is timing important? How will the survey be administered—in person, by mail, by phone? How will the data be collected, recorded, analyzed, and reported (Lavin 277)?

Phone, email, and in-person surveys yield fast results and high response rates, but respondents consider phone surveys annoying and, without anonymity, people tend to be less candid. Mail surveys are more confidential.

Electronic surveys conducted, via a Web form or an email message, are the least expensive. But these methods can have pitfalls. Computer connections can fail and you have less control over how many times the same person completes the survey. Also, because online text usually takes longer to read, people may give up or refuse to complete the survey.

GUIDELINES for Developing a Questionnaire

1. *Decide on the types of questions* (Adams and Schvaneveldt 202–12; Velotta 390). Questions can be *open-ended* or *closed-ended*. Open-ended questions allow respondents to answer in any way they choose:

Open-ended
questions

- How much do you know about electromagnetic radiation at our school?
- What do you think should be done about electromagnetic fields (EMFs) at our school?

It is more time-consuming to measure the data gathered, but open-ended questions provide a rich source of information.

When you want to measure exactly where people stand on an issue, choose closed-ended questions:

Closed-ended
questions

Are you interested in joining a group of concerned parents?
YES _____ NO _____

Rate your degree of concern about EMF's at our school.
HIGH _____ MODERATE _____ LOW _____ NO CONCERN _____

Circle the number that indicates your view about the town's proposal to spend $20,000 to hire its own EMF consultant.

1 . . . . 2 . . . . 3 . . . . 4 . . . . 5 . . . . 6 7
Strongly   No   Strongly
Disapprove   Opinion   Approve

Respondents may be asked to *rate* one item on a scale (from high to low, best to worst), to *rank* two or more items (by importance, desirability), or to select items from a list. Other questions measure percentages or frequency:

How often do you ...?
ALWAYS_____ OFTEN_____ SOMETIMES_____ RARELY_____ NEVER_____

Although they are easy to answer, tabulate, measure, and analyze, closed-ended questions might elicit biased responses. Some people, for instance, automatically prefer items near the top of a list or the left side of a rating scale (Plumb and Spyridakis 633). Also, respondents are more prone to agree than to disagree with assertions in a questionnaire (Sherblom, Sullivan, and Sherblom 61).

www

9.3

Can any question be free of bias? Find out more at
<www.ablongman.com/lannonweb>

(continues)

2. *Design an engaging introduction and opening questions.* Persuade respondents that the survey relates to their concerns, that their answers matter, and that their anonymity is assured. Explain how respondents will benefit from your findings, or offer an incentive (say, a copy of your final report).

A survey introduction

> Your answers will help our school board to speak accurately for your views at our next town meeting. Results of this survey will appear in our campus newspaper. Thank you.

Researchers often include a cover letter with the questionnaire.

Begin with the easiest questions, which usually are the closed-ended ones. Once respondents commit to these, they are likely to complete more difficult questions later.

3. *Make each question unambiguous.* All respondents should be able to interpret identical questions identically. An ambiguous question leaves room for misinterpretation.

An ambiguous question

> Do you favor weapons for campus police? YES_____ NO_____

"Weapons" might mean tear gas, clubs, handguns, all three, or two out of three. Consequently, responses to the above question would produce a misleading statistic, such as "Over 95 percent of students favor handguns for campus police" when the accurate conclusion might be "Over 95 percent of students favor some form of weapon." Moreover, the limited "yes/no" format reduces an array of possible opinions to an either/or choice.

A clear and incisive question

> Do you favor (check all that apply):
> _____ Having campus police carry mace and a club?
> _____ Having campus police carry nonlethal "stun guns"?
> _____ Having campus police store handguns in their cruisers?
> _____ Having campus police carry small-caliber handguns?
> _____ Having campus police carry large-caliber handguns?
> _____ Having campus police carry no weapons?
> _____ Don't know

To ensure a full range of possible responses, include options such as "Other," "Don't know," "Not Applicable," or an "Additional Comments" section.

4. *Make each question unbiased.* Avoid *loaded questions* that invite or advocate a particular viewpoint or bias:

A loaded question

> Should our campus tolerate the needless endangerment of innocent students by lethal weapons?
> YES_____ NO_____

Emotionally loaded and judgmental words ("endangerment," "innocent," "tolerate," "needless," "lethal") in a survey are unethical because their built-in judgments manipulate people's responses (Hayakawa 40).

5. *Make it brief, simple, and inviting.* Long questionnaires usually get very few replies. And when people do reply to a long survey, they tend to give less thought to their answers.

Try to limit questions and their response spaces to two sides of one page. Include a stamped, return-addressed envelope, and give a specific return date. Address each respondent by name; sign your letter or your introduction; and give your title.

A Sample Questionnaire

The student-written letter and questionnaire in Figures 9.3 and 9.4, sent to presidents of local companies, is designed to elicit responses that can be tabulated easily. (For a usability questionnaire, see page 372.)

Written reports of survey findings often include an appendix (page 650) that contains a copy of the questionnaire as well as the tabulated responses.

NOTE *For excellent, in-depth advice about planning and conducting surveys and interviews, go to William M. K. Trochims' "Survey Research" page at <http://trochim.human.cornell.edu/KB/survey.htm>, and click on the various links.*

INQUIRY LETTERS, PHONE CALLS, AND EMAIL INQUIRIES

Letters, phone calls, or email inquiries to experts listed in Web pages are handy for obtaining specific information from government agencies, legislators, private companies, university research centers, trade associations, and research foundations.

NOTE *Keep in mind that unsolicited inquiries, especially by phone or email, can be intrusive and offensive.*

PUBLIC RECORDS AND ORGANIZATIONAL PUBLICATIONS

The Freedom of Information Act and state public record laws grant access to an array of government, corporate, and organizational documents. Obtaining these documents (from state or federal agencies) takes time, but in them you can find answers to questions like these (Blum 90–92):

Public records may hold answers to tough questions

- Which universities are being investigated by the USDA (Dept. of Agriculture) for mistreating laboratory animals?
- Are IRS auditors required to meet quotas?
- What are the results of state and federal water-quality inspections in this region?

Organization records (reports, memos, computer printouts, and so on) are good primary sources. Most organizations also publish pamphlets, brochures, annual reports, or prospectuses for consumers, employees, investors, or voters.

NOTE *Be alert for bias in company literature. If you were evaluating the safety measures at a local nuclear power plant, you would want the complete picture. Along with the company's literature, you would also want studies and reports from government agencies and publications from environmental groups.*

April 5, 20xx

House 10
University of Massachusetts, Dartmouth
North Dartmouth, MA 02747

Name, Title
Company Name
Address

Dear _____:

I am exploring ways to enhance relationships between UMD's Professional Communication Program and the local business community.

Specific areas of inquiry:
1. the communication needs of local companies and industries
2. the feasibility of on-campus and in-house seminars for employees
3. the feasibilty of expanding communication course offerings at UMD

Please take a few minutes to complete the attached survey. Your response will provide an important contribution to my study.

All respondents will receive a copy of my report, scheduled to appear in the fall issue of The Business and Industry Newsletter. Thank you.

Sincerely,

L.S. Taylor
Technical Communication Student

FIGURE 9.3   A Questionnaire Cover Letter

Communication Questionnaire

1. Describe your type of company (e.g., manufacturing, high tech)

   _____

2. Number of employees (Please check one.)

   _____ 0–4   _____ 25–50   _____ 100–150   _____ 300–450
   _____ 5–25   _____ 50–100   _____ 150–300   _____ 450+

3. What types of written communication occur in your company? (Label by frequency: daily, weekly, monthly, never.)

   _____ memos   _____ letters   _____ advertising
   _____ manuals   _____ reports   _____ newsletters
   _____ procedures   _____ proposals   _____ other (Specify.)
   _____ email   _____ catalogs   _____

4. Who does most of the writing? (Pls. give titles.) _____

5. Please characterize your employees' writing effectiveness.

   _____ good   _____ fair   _____ poor

6. Does your company have formal guidelines for writing?

   _____ no   _____ yes (Pls. describe briefly.) _____

7. Do you offer in-house communication training?

   _____ no   _____ yes (Pls. describe briefly.) _____

8. Please rank the usefulness of the following areas in communication training (from 1–10, 1 being most important).

   _____ organizing information   _____ audience awareness
   _____ summarizing information   _____ persuasive writing
   _____ editing for style   _____ grammar
   _____ document design   _____ researching
   _____ email etiquette   _____ Web page design
   _____ other (Pls. specify.) _____

9. Please rank these skills in order of importance (from 1–6, 1 being most important).

   _____ reading   _____ listening   _____ speaking to groups
   _____ writing   _____ collaborating   _____ speaking face-to-face

10. Do you provide tuition reimbursement for employees?

    _____ no   _____ yes

11. Would you consider having UMD communication interns work for you part-time?

    _____ no   _____ yes

12. Should UMD offer Saturday seminars in communication?

    _____ no   _____ yes

    Additional comments/suggestions: _____

    _____

FIGURE 9.4   A Questionnaire

9.4

Learn techniques for ethnographic observations of work-place cultures at <www.ablongman.com/lannonweb>

PERSONAL OBSERVATION AND EXPERIMENT

Observation should be your final step because you now know what to look for. Have a plan. Know how, where, and when to look, and jot down observations immediately. You might even take photos or make drawings.

Informed observations can pinpoint real problems. Here is an excerpt from a report investigating low morale at an electronics firm. This researcher's observations and interpretation are crucial in defining the problem:

Direct observation is often essential

> Our on-site communications audit revealed that employees were unaware of any major barriers to communication. More than 75 percent of employees claimed they felt free to talk to their managers, but the managers, in turn, estimated that fewer than 50 percent of employees felt free to talk to them.
>
> The problem involves misinterpretation. Because managers don't ask for complaints, employees are afraid to make them, and because employees never ask for an evaluation, they never get one. Each side has inaccurate perceptions of what the other side expects, and because of ineffective communications, each side fails to realize that its perceptions are wrong.

NOTE   *Even direct observation is not foolproof: for instance, you might be biased about what you see (focusing on the wrong events or ignoring something important); or, instead of behaving normally, people being observed might behave in ways they think observers expect. (Adams and Schvaneveldt 244)*

An experiment is a controlled form of observation designed to verify an assumption (e.g., the role of fish oil in preventing heart disease) or to test something untried (the relationship between background music and worker productivity). Each field has its own guidelines for experiment design.

ANALYSIS OF SAMPLES

Workplace research can involve collecting and analyzing samples: water or soil or air, for contamination and pollution; foods, for nutritional value; ore, for mineral value; or plants, for medicinal value. Investigators analyze material samples to find the cause of an airline accident. Engineers analyze samples of steel, concrete, or other building materials to determine their load-bearing capacity. Medical specialists analyze tissue samples for disease.

EXERCISES For more exercises, visit
<www.ablongman.com/lannon>

1. Revise these questions to make them appropriate for inclusion in a questionnaire:

   a. Would a female president do the job as well as a male?
   b. Don't you think that euthanasia is a crime?
   c. Do you oppose increased government spending?
   d. Do you think welfare recipients are too lazy to support themselves?
   e. Are teachers responsible for the decline in literacy among students?
   f. Aren't humanities studies a waste of time?
   g. Do you prefer Rocket Cola to other leading brands?
   h. In meetings, do you think men are more interruptive than women?

2. Identify and illustrate at least six features that enhance the effectiveness of the questionnaire in Figure 9.4. (Review pages 157–59 for criteria.) Be prepared to discuss your evaluation in class.

3. Arrange an interview with someone in your field. Decide on general areas for questioning: job opportunities, chances for promotion, salary range, requirements, outlook for the next decade, working conditions, job satisfaction, and so on. Compose specific interview questions; conduct the interview, and summarize your findings in a memo to your instructor.

4. Return to Exercise 6 in Chapter 4 (page 70) and arrange an interview with a respondent who can provide the information you need.

- campus codes prohibiting hate speech or offensive language in general
- campus alcohol policy
- campus safety
- facilities for disabled students
- campus racial or gender issues
- access to computers

Once you have identified your survey's exact purpose and your target population, follow these steps:

   a. Decide on the size and makeup of a randomly selected sample group.
   b. Try to identify all sources of potential error.
   c. Develop a questionnaire that will measure accurately what your survey intends to measure. Design questions that are engaging, unambiguous, unbiased, and easy to answer and tabulate.
   d. Administer the survey to a representative sample group.
   e. Tabulate, analyze, and interpret the responses.
   f. Prepare a written report summarizing your survey purpose, process, findings, and conclusions. Discuss any survey limitations. Include a copy of the questionnaire as well as the tabulated responses.
   g. Appoint one group member to present your findings to the class.

In addition to reviewing pages 157–59, look over Chapter 10, especially the section on validity and reliability (page 188).

COLLABORATIVE PROJECT

Divide into small groups and decide on a survey of views, attitudes, preferences, or concerns about some issue affecting your campus or the community. Expand on this short list of possible survey topics:

SERVICE-LEARNING PROJECT

Plan and conduct an on-site interview at the agency you are working with. Write a memo report to your instructor summarizing what you learned from the interview.

10

Evaluating and Interpreting Information

Not all information is equally valuable. Not all interpretations are equally valid. For instance, if you really want to know how well the latest innovation in robotic surgery works, you need to check with other sources besides, say, its designer (from whom you could expect an overly optimistic or insufficiently critical assessment).

Whether you work with your own findings or those of other researchers, you need to decide if the findings are valid and reliable. Then you need to decide what your information means. Figure 10.1 outlines this challenge.

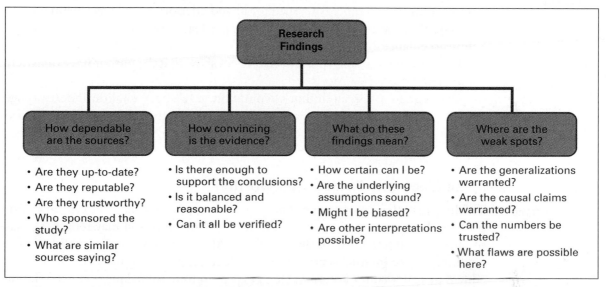

FIGURE 10.1   Decisions in Evaluating and Interpreting Information   Collecting information is often the easiest part of the research process. Your larger challenge is in getting the exact information you need, making sure it's accurate, and figuring out what it means. Throughout this process, there is much room for the types of errors covered in this chapter.

EVALUATE THE SOURCES

Not all data sources are equally dependable. A source might offer information that is out of date, inaccurate, incomplete, mistaken, or biased.

Is the Source Up-to-Date?

Even newly published books contain information that can be more than a year old, and journal articles often undergo a lengthy process of peer review.

"How current is the information?"

Certain types of information become outdated more quickly than others. For topics that focus on *technology* (multimedia law, superconductivity, alternative cancer treatments), information more than a few months old may be outdated. But for topics that focus on *people* (business ethics, management practices, workplace gender equality), historical perspectives often help.

The most recent information is not always the most reliable—especially in scientific re-search, a process of ongoing inquiry in which what seems true today may be proven false tomorrow. Consider, for example, the recent discoveries of fatal side effects from some of the latest "miracle" weight-loss drugs.

Is the Printed Source Dependable?

"What is the source's reputation?"

One way to assess a publication's reputation is to check its copyright page. Is the work published by a university, professional society, museum, or respected news organization? Do members of the editorial and advisory board have distinguished titles and degrees? Is the publication *refereed* (all submissions reviewed by experts before acceptance)?

Does the bibliography or list of references show that the author has extensively researched the issue (Barnes)?

One way to assess an author's reputation is to check citation indexes (page 142) to see what other experts have said about this research. Many periodicals also provide brief biographies or descriptions of authors' earlier publications and other achievements.

Is the Electronic Source Trustworthy?

"Can the source be trusted?"

The Internet offers information that may never appear in other sources, for example from listservs and newsgroups. But much of this material may reflect the bias of the special-interest groups that provide it. Moreover, anyone can publish almost anything on the Internet—including misinformation—without having it verified, edited, or reviewed for accuracy. Don't expect to find everything you need on the Internet. (Pages 170–71 offer suggestions for evaluating sources on the Web.)

Even in a commercial database, such as DIALOG, decisions about what to include and what to leave out depend on the biases, priorities, or interests of those who assemble the database.

Because it advocates a particular point of view, a special-interest Web site (as in Figure 10.2) can provide useful clues about the ideas and opinions of its sponsors; however, don't rely on special-interest sites for factual information—unless the facts can be verified elsewhere ("Evaluating Internet-Based Information").

Is the Information Relatively Unbiased?

"Who sponsored the study, and why?"

Much of today's research is paid for by private companies or special-interest groups that have their own social, political, or economic agendas (Crossen 14, 19). Medical research may be sponsored by drug or tobacco companies; nutritional research, by food manufacturers; environmental research, by oil or chemical companies. Public policy research (on gun control, school prayer, endangered species) may be sponsored by opposing groups (environmentalists versus the logging industry) producing opposing results.

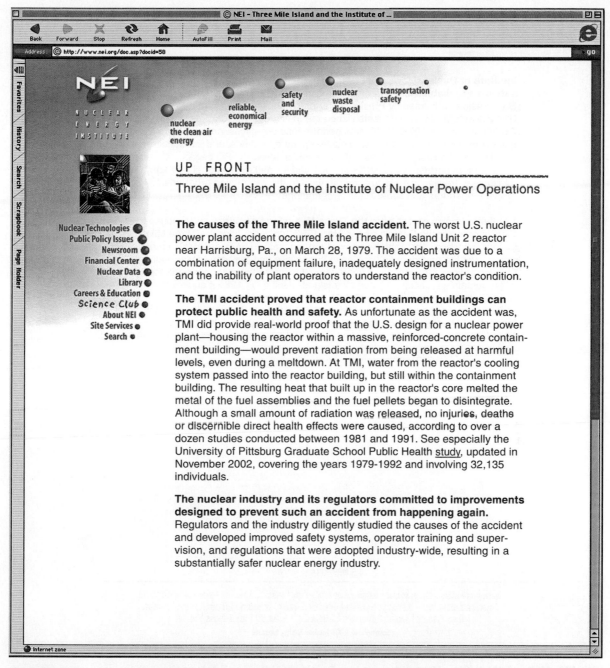

nuclear
the clean air
energy

reliable,
economical
energy

safety
and
security

nuclear
waste
disposal

transportation
safety

UP FRONT

Three Mile Island and the Institute of Nuclear Power Operations

Nuclear Technologies
Public Policy Issues
Newsroom
Financial Center
Nuclear Data
Library
Careers & Education
Science Club
About NEI
Site Services
Search

The causes of the Three Mile Island accident. The worst U.S. nuclear power plant accident occurred at the Three Mile Island Unit 2 reactor near Harrisburg, Pa., on March 28, 1979. The accident was due to a combination of equipment failure, inadequately designed instrumentation, and the inability of plant operators to understand the reactor's condition.

The TMI accident proved that reactor containment buildings can protect public health and safety. As unfortunate as the accident was, TMI did provide real-world proof that the U.S. design for a nuclear power plant—housing the reactor within a massive, reinforced-concrete containment building—would prevent radiation from being released at harmful levels, even during a meltdown. At TMI, water from the reactor's cooling system passed into the reactor building, but still within the containment building. The resulting heat that built up in the reactor's core melted the metal of the fuel assemblies and the fuel pellets began to disintegrate. Although a small amount of radiation was released, no injuries, deaths or discernible direct health effects were caused, according to over a dozen studies conducted between 1981 and 1991. See especially the University of Pittsburg Graduate School Public Health study, updated in November 2002, covering the years 1979-1992 and involving 32,135 individuals.

The nuclear industry and its regulators committed to improvements designed to prevent such an accident from happening again. Regulators and the industry diligently studied the causes of the accident and developed improved safety systems, operator training and supervision, and regulations that were adopted industry-wide, resulting in a substantially safer nuclear energy industry.

FIGURE 10.2   A Web Site That Advocates a Particular Viewpoint   What concerns or issues are reflected in the content of this page? What public attitudes does this government agency appear to be advocating?

Source: Home page from Nuclear Energy Institute, <www.nei.org>. Reprinted by permission of Nuclear Energy Institute.

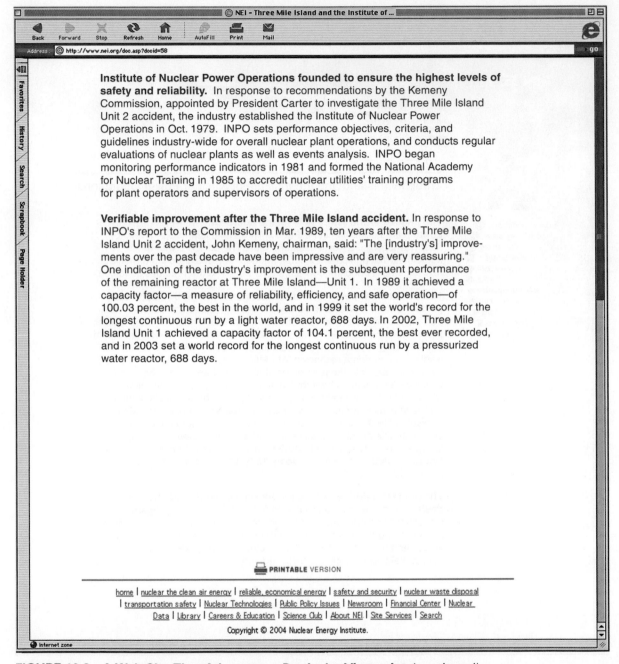

FIGURE 10.2   A Web Site That Advocates a Particular Viewpoint *(continued)*

Instead of a neutral and balanced inquiry, this kind of "strategic research" is designed to support one special interest or another (Crossen 132–34). Furthermore, those who pay for strategic research are not likely to publicize findings that contradict their original claims or opinions (profits lower than expected, losses or risks greater than expected). As consumers of research, we should try to determine exactly what the sponsors of a particular study stand to gain or lose from the results (234).

NOTE   *Keep in mind that any research ultimately stands on its own merits. Thus, funding by a special interest should not automatically discredit an otherwise valid and reliable study.*

Also, financing from a private company often sets the stage for beneficial research that might otherwise be unaffordable, as when research funded by Quaker Oats led to other studies proving that oats can lower cholesterol (Raloff, "Chocolate Hearts" 189).

How Does This Source Measure Up to Others?

"What are similar sources saying?"

Most studies have some type of flaw (page 189). Therefore, instead of relying on a single source of study, seek a consensus among various respected sources (Cohn 106).

NOTE   *Some issues (the need for defense spending or the causes of inflation) are always controversial and will never be resolved. Although we can get verifiable data and can reason persuasively on some subjects, no close reasoning by any expert and no supporting statistical analysis will "prove" anything about a controversial subject. Some problems are simply more resistant to solution than others, no matter how dependable the sources.*

EVALUATE THE EVIDENCE

Evidence is any finding used to support or refute a particular claim. While evidence can serve the truth, it can also create distortion, misinformation, and deception. For example:

Questions that invite distorted evidence

- How much money, material, or energy does recycling really save?
- How well are public schools educating children?
- Which investments or automobiles are safest?

Competing answers to such questions often rest on evidence that has been stacked to support a particular view or agenda.

Is the Evidence Sufficient?

"Is there enough evidence?"

Evidence is sufficient when nothing more is needed to reach an accurate judgment or conclusion. A study of the stress-reducing benefits of low-impact aerobics, for example, would require a broad survey sample: people who have practiced aerobics for a long time; people of both genders, different ages, different

10.1

For more on evaluating
questionable web
sources visit
<www.ablongman.com/
lannonweb>

GUIDELINES for Evaluating Sources on the Web*

1. *Consider the site's domain type and sponsor.* In this typical address, <http://www.umassd.edu>, the site or domain information follows the *www*. The *.edu* signifies the type of organization from which the site originates. Standard domain types in the United States:

   .com = business/commercial organization
   .edu = educational institution
   .gov = government organization
   .mil = military organization
   .net = any group or individual with simple software and Internet access
   .org = nonprofit organization

   The domain type might signal a certain bias or agenda that could skew the data. For example, at a .com site, you might find accurate information, but also some type of sales pitch. At a .org site, you might find a political or ideological bias (say, The Heritage Foundation's conservative ideology versus the Brookings Institution's more liberal slant). A tilde (~) in the address usually signifies a personal home page. Knowing about a site's sponsor can help you evaluate the credibility of its postings.

2. *Identify the purpose of the page or message.* Decide whether the message is intended to merely relay information, to sell something, or to promote a particular ideology or agenda.

3. *Look beyond the style of a site.* Fancy graphics, video, and sound do not always translate into dependable information. Sometimes the most reliable material resides in the less attractive, text-only sites. People can design flashy-looking pages without necessarily knowing what they are talking about. Even something written in clear, plain English instead of in difficult scientific terms (as in medical information) might be inaccurate.

4. *Assess the site's material's currency.* An up-to-date site should indicate when the material was created or published, and when it was posted and updated.

5. *Assess the author's credentials.* Learn all you can about the author's reputation, level of expertise on this topic, and institutional affiliation (a university, a Fortune 500 company, a reputable environmental group). Do this by following links to other sites that mention the author, by using search engines to track the author's name, or by consulting a citation index (page 142). Newsgroup postings often contain "a *signature file* that includes the author's name, location, institutional or organizational affiliation, and often a quote that suggests something of the writer's personality, political leanings, or sense of humor" (Goubil-Gambrell 229–30).

*Guidelines adapted from Barnes; Busiel and Maeglin 39; Elliot; Fackelmann 397; Grassian; Hall 60–61; Hammett; Harris, Robert; Kapoun 4; Stemmer.

NOTE *Don't confuse an author (the person who wrote the material) with a Webmaster (the person who created and maintains the Web site).*

6. *Decide whether the assertions/claims make sense.* Based on what you know about the issue, decide where, on the spectrum of informed opinion and accepted theory, this author's position resides. How well is each assertion supported? Never accept any claim that seems extreme without verifying it through other sources, such as a professor, a librarian, or a specialist in the field. Also, consider whether your own biases might predispose you to automatically accept certain ideas.

7. *Compare the site with other sources.* Check related sites and publications to compare the quality of information and to discover what others might have said about this site, author, or topic. Comparing many similar sites helps you create a benchmark, a standard for evaluating any particular site based on the criteria in these guidelines). Ask a librarian for help.

8. *Look for other indicators of quality.*

   - *Worthwhile content:* The material is technically accurate. All sources of data presented as "factual" are fully documented. (See Appendix A.)
   - *Sensible organization:* The material is organized for the user's understanding, with a clear line of reasoning.
   - *Readable style:* The material is well written (clear, concise, easy to understand) and free of typos, misspellings, and other mechanical errors.
   - *Relatively objective coverage:* Debatable topics are addressed in a balanced and impartial way, with fair, accurate representation of opposing views. The tone is reasonable, with no "sounding off" or name-calling ("radicals," "extremists," "fringe groups," "fear mongers").
   - *Expertise:* The author refers to related theory and other work in the field and uses specialized terminology accurately and appropriately.
   - *Peer review:* The material has been evaluated and verified by related experts.
   - *Up-to-date links to reputable sites:* The site offers a gateway to related sites that meet quality criteria.
   - *Follow-up option:* The material includes a signature block or a link for contacting the author or organization for clarification or verification.

For more on evaluating Web-based sources, go to <www.webcredibility.org> and <www.library.cornell.edu/okuref/research/webeval.html>.

occupations, and different lifestyles before they began aerobics; and so on. Even responses from hundreds of practitioners might be insufficient unless those responses were supported by laboratory measurements of metabolic and heart rates, blood pressure, and so on.

NOTE   *Although anecdotal evidence ("This worked great for me!") might be a good starting point for an investigation, your personal experience rarely provides enough evidence from which to generalize. No matter how long you might have practiced aerobics, for instance, you cannot tell whether your experience is representative.*

Is the Presentation of Evidence Balanced and Reasonable?

How evidence can be misused

Misuse of evidence in a courtroom often makes headlines. But evidence routinely is misused beyond the courtroom as well, as in the following instances.

"Is this claim too good to be true?"

OVERSTATEMENT.   Consumers are offered a daily menu of cures for ailments ranging from insomnia to cancer. Overzealous researchers might exaggerate their achievements, without mentioning the limitations of their study.

"Is there a downside?"

OMISSION OF VITAL FACTS.   Aspirin is widely promoted for *decreasing* heart attack and stroke risk caused by clotting, but far less emphasized is its role in *increasing* stroke risk caused by brain bleeding (Lewis 222). Acetaminophen products are advertised as a "safe" alternative pain reliever, without aspirin's side effects (stomach irritation, Reye's syndrome), but even small overdoses of acetaminophen have caused liver failure. Moreover, acetaminophen is the leading cause of U.S. drug fatalities (Easton and Herrara 42–44).

Is the glass "half full" or "half empty"?

DECEPTIVE FRAMING OF THE FACTS.   A *frame of reference* is a set of ideas, beliefs, or views that influences our interpretation of other ideas. For example, people with a fundamentalist view of the Bible might reject the concept of evolution. Or consider the well-known optimist/pessimist test: Is the glass half full (a positive frame of reference) or half empty (a negative frame of reference)? In medical terms, is a "90-percent survival rate" more acceptable than a "10-percent mortality rate"? Framing sways our perception (Lang and Secic 239–40). For example, what we now term a "financial recession" used to be a "financial depression," which was a euphemism for "financial panic" (P. Bernstein 183). For more on euphemisms, see page 268.

How framing of a survey question can affect responses

The framing of survey questions can manipulate responses. For example, in a survey of attitudes toward abortion, consider how responses might differ depending on the following phrasing: "Do you approve of abortion on demand?" versus "Do you approve of abortion under any circumstances?" (Phillips 192).

Even unintentional framing can have major consequences. Researchers Kahneman and Tversky describe this situation at one hospital:

How framing can influence a life-and-death decision

[D]octors were concerned that they might be influencing patients who had to choose between the life-or-death risks in different forms of treatment. The choice was between radiation and surgery in the treatment of lung cancer. Medical data . . . showed that no patients die during radiation but have a shorter life expectancy than patients who survive the risk of surgery; the overall difference in life expectancy was not great enough to provide a clear choice between the two forms of treatment. When the question was put in terms of risk of death during treatment, more than 40% of the patients favored radiation. When the question was put in terms of life expectancy, only about 20% favored radiation. (cited in P. Bernstein 276)

Whether the language is provocative ("rape of the environment"), euphemistic ("teachable moment" versus "mistake"), or demeaning to opponents ("bureaucrats," "tree huggers"), deceptive framing obscures the real issues. Common "spin" strategies of politicians, for example, include painting situations as rosier than they are or calling people names.

Can the Evidence Be Verified?

"Is the evidence hard or soft?"

Hard evidence consists of factual statements, expert opinion, or statistics that can be verified. *Soft evidence* consists of uninformed opinion or speculation obtained

"One question: If this is the Information Age, how come nobody knows anything?"

Source: © The New Yorker Collection 1998 Robert Mankoff from cartoonbank.com. All Rights Reserved.

or analyzed unscientifically, and findings that have not been replicated or reviewed by experts. Reputable news organizations employ fact-checkers to verify information before it appears in print.

INTERPRET YOUR FINDINGS

10.2

What is truth? Can there be more than one truth? More at <www.ablongman.com/ lannonweb>

Interpreting means trying to reach the truth of the matter: an overall judgment about what the findings mean and what conclusion or action they suggest.

Unfortunately, research does not always yield answers that are clear or conclusive. Instead of settling for the most *convenient* answer, we pursue the most reasonable answer by critically examining a full range of possible meanings.

What Level of Certainty Is Warranted?

As possible outcomes of research we can identify three distinct and very different levels of certainty:

1. The ultimate truth—*the conclusive answer:*

A practical definition of "truth:

> Truth is what is so about something, the reality of the matter, as distinguished from what people wish were so, believe to be so, or assert to be so. . . . In the words of Harvard philosopher Israel Scheffler, truth is the view "which is fated to be ultimately agreed to by all who investigate." The word *ultimately* is important. Investigation may produce a wrong answer for years, even for centuries. . . .
>
> One easy way to spare yourself any further confusion about truth is to reserve the word *truth* for the final answer to an issue. Get in the habit of using the words *belief, theory,* and *present understanding* more often. (Ruggiero, 3rd ed. 21–22)

Conclusive answers are the research outcome we seek, but often we have to settle for answers that are less than certain.

2. The *probable answer:* the answer that stands the best chance of being true or accurate—given the most we can know at this particular time. Probable answers are subject to revision in light of new information. This is especially the case with *emergent* science, which one expert defines as "science whose truth has not yet been settled by consensus" (Hornig-Priest 97). Examples include risks versus benefits of cloning, food irradiation, or genetically modified crops.
3. The *inconclusive answer:* the realization that the truth of the matter is more elusive, ambiguous, or complex than we expected.

"Exactly how certain are we?"

To ensure an accurate outcome, we must decide what level of certainty our findings warrant. For example, we are *highly certain* about the perils of smoking or sunburn, *reasonably certain* about the health benefits of fruits and vegetables and moderate exercise, and *less certain* about the perils of coffee drinking or electro-magnetic waves, or the benefits of vitamin supplements.

The "truth" never changes; however, our notions about "truthfulness" do, as in these examples:

Some changing notions about "Truth"

- *"The earth is the center of the universe."* Though dead wrong, Ptolemy's cosmology was based on the best information available in the second century A.D. And this certainty survived thirteen centuries—even after new information had discredited Ptolemy's theory. When Copernicus and Galileo proposed more truthful views in the fifteenth century, they were labeled heretics.
- *"Brush your teeth and blow your nose with asbestos."* Considered a "miracle fiber" for two thousand years, asbestos—soft, flexible, and fire resistant—was used in countless products ranging from asbestos handkerchiefs in the first century to tablecloths, toothpaste, and cigarette filters in the twentieth century. Not until the 1970s was the truth about the long-suspected role of asbestos in lung disease publicized (Alleman and Mossman 70–74).
- *"Fat is bad, Carbs are good."* This was the basic message of The Food Guide Pyramid, introduced in 1992 by the U.S. government. The Pyramid has largely been turned upside-down by research indicating that certain carbohydrates may actually promote chronic disease while certain fats may help prevent it (Willett and Stampfer 66).

Communication expert Katherine Rowan reminds us just how elusive certainty can be, in any context:

In science, uncertainty is a fact of life

. . . because more refined, more precise, better explanation is always possible, no phenomenon, no matter how thoroughly studied, is ever fully explained or understood. In this sense, all scientific knowledge is uncertain. (204–05)

Ethical violations in communicating uncertainty

Of course, unethical communicators can downplay certainty, as tobacco companies did for decades. Or they can exaggerate certainty, as in recent claims about the absolute safety of genetically modified foods. Each of these ploys is a type of *argument from ignorance,* in which the absence of evidence to the contrary is offered as "proof" of something: "X is true because it has not been proven false" (or vice versa). For example, "Since no one has yet demonstrated harmful side effects from genetically modified food, it must be safe."

The vast majority of scientific, technical, and social controversies are open-ended. Therefore, no irrefutable presentation of "the facts" will likely settle the controversy over questions like these:

Questions that invite controversy

- How rapidly is global warming progressing?
- What is causing the death and disfigurement of frogs worldwide?
- Can vitamins prevent cancer or heart disease?
- Should Affirmative Action programs be expanded or discontinued?

Does this mean that such questions should be ignored? Of course not. Even though some claims cannot be proven, we can still reach reasonable conclusions on the basis of the evidence.

Are the Underlying Assumptions Sound?

Assumptions are notions we take for granted, things we often accept without proof. The research process rests on assumptions like these: that a sample group accurately represents a larger target group, that survey respondents remember certain facts accurately, that mice and humans share enough biological similarities for meaningful research. For a particular study to be *valid* (page 188), the underlying assumptions have to be accurate.

How underlying assumptions affect research validity

Consider this example: You are an education consultant evaluating the accuracy of IQ testing as a predictor of academic performance. Reviewing the evidence, you perceive an association between low IQ scores and low achievers. You then verify your statistics by examining a cross-section of reliable sources. Can you then conclude that IQ tests *do* predict performance accurately? This conclusion might be invalid unless you could verify the following assumptions:

1. That neither parents, teachers, nor children had seen individual test scores, which could produce biased expectations.
2. That, regardless of score, each child had completed an identical curriculum, instead of being "tracked" on the basis of his or her score.

NOTE   *Assumptions are often easier to identify in someone else's thinking and writing than in our own. During collaborative discussions, ask group members to help you identify your own assumptions (Maeglin).*

10.3

Where does bias come from? Learn more at <www.ablongman.com/lannonweb>

To What Extent Has Personal Bias Influenced the Interpretation?

Personal bias is a fact of life

To support a particular version of the truth, our own bias might cause us to overestimate (or deny) the certainty of our findings.

Unless you are perfectly neutral about the issue, an unlikely circumstance, at the very outset . . . you will believe one side of the issue to be right, and that belief will incline you to . . . present more and better arguments for the side of the issue you prefer. (Ruggiero, 3rd ed. 134)

How bias can outweigh evidence

The following example illustrates how *cognitive bias* (seeing what we expect to see) can blind us to the most compelling evidence. The 1989 spill from the oil tanker *Exxon Valdez* polluted more than 1,000 miles of Alaskan shoreline and led to a massive recovery effort that included using high-pressure hot water to clean oil from the beaches. But a respected study shows that the *uncleaned* beaches are now healthier than those sterilized by the hot water: "Whatever [beach life] survived the oiling did not survive the cure" (Holloway 109). But this finding—that cleaning up is more harmful than helpful—remains highly unpopular with the Alaskan public, who continue to insist on the removal of virtually every drop of oil (109–12).

Because personal bias is hard to transcend, rationalizing often becomes a substitute for reasoning:

Rationalizing versus reasoning

> You are reasoning if your belief follows the evidence—that is, if you examine the evidence first and then make up your mind. You are rationalizing if the evidence follows your belief—if you first decide what you'll believe and then select and interpret evidence to justify it. (Ruggiero, 3rd ed. 44)

Personal bias often is subconscious until we examine our own value systems: attitudes long held but never analyzed, assumptions we've inherited from our own backgrounds, and so on. Recognizing our own biases is the crucial first step in managing them.

Are Other Interpretations Possible?

"What else could this mean?"

Perhaps other researchers disagree with the meaning of these findings. Settling on a final meaning can be hard. For instance, how should we interpret the reported increase in violent crime on U.S. college campuses—especially in light of statistics that show violent crime decreasing in general (Lederman 5)? Should we conclude (a) that college students are becoming more violent, (b) that some drugs and guns in high schools end up on campuses, (c) that off-campus criminals see students as easy targets, or (d) all of these? Or could these findings mean something else entirely? For example, (a) that increased law enforcement has led to more campus arrests—and thus, greater recognition of the problem or (b) that campus crimes really haven't increased, but more are being reported? Depending on our interpretation, we might conclude that the situation is worsening—or improving!

NOTE   *Not all interpretations are equally valid. Never assume that any interpretation that is possible is also allowable—especially in terms of its ethical consequences. Certain interpretations in the college crime example, for instance, might justify an overly casual or overly vigilant response—either of which could have disastrous consequences.*

AVOID ERRORS IN REASONING

Finding the truth, especially in a complex issue or problem, often is a process of elimination, of ruling out or avoiding errors in reasoning. As we interpret, we

---

CONSIDER THIS Standards of Proof Vary for Different Audiences and Cultural Settings

How much evidence is enough to "prove" a particular claim? The answer often depends on whether the inquiry occurs in the science lab, the courtroom, or the boardroom, as well as on the specific cultural setting:

- *The scientist demands evidence that indicates at least 95 percent certainty. A scientific finding must be evaluated and replicated by other experts. Good science looks at the entire picture. Findings are reviewed before they are reported. Inquiries and answers in science are never "final," but open-ended and ongoing: What seems probable today may be shown improbable by tomorrow's research.*

- *The juror demands evidence that indicates only 51 percent certainty (a "preponderance of the evidence"). Jurors are not scientists. Instead of the entire picture, jurors get only the information made available by lawyers and witnesses. A jury bases its opinion on evidence that exceeds "reasonable doubt"*

(Monastersky, "Courting" 249; Powell 32+). Based on such evidence, courts have to make decisions that are final.

- *The corporate executive demands immediate (even if insufficient) evidence. In a global business climate of overnight developments (in world markets, political strife, military conflicts, natural disasters), important business decisions are often made on the spur of the moment. On the basis of incomplete or unverified information—or even hunches—executives must make quick decisions to react to crises and capitalize on opportunities (Seglin 54).*

- *Specific cultures may have their own standards for authentic, reliable, and persuasive evidence. "For example, African cultures rely on storytelling for authenticity. Arabic persuasion is dependent on universally accepted truths. And Chinese value ancient authorities over recent empiricism" (Byrd and Reid 109).*

---

make *inferences:* We derive conclusions about what we don't know by reasoning from what we do know (Hayakawa 37). For example, we might infer that a drug that boosts immunity in laboratory mice will boost immunity in humans, or that a rise in campus crime statistics is caused by the fact that young people have become more violent. Whether a particular inference is on target or dead wrong depends largely on our answers to one or more of these questions:

Questions for testing inferences

- To what extent can these findings be generalized?
- Is *Y* really caused by *X*?
- How much can the numbers be trusted, and what do they mean?

Following are three major reasoning errors that can distort our interpretations.

Faulty Generalization

We engage in faulty generalization when we jump from a limited observation to a sweeping conclusion. Even "proven" facts can invite mistaken conclusions.

Factual
observations

> 1. "Some studies have shown that gingko [an herb] improves mental functioning in people with dementia [mental deterioration caused by maladies such as Alzheimer's Disease]" (Stix 30).
> 2. "For the period 1992–2005, two thirds of the fastest-growing occupations [called] for no more than a high-school degree" (Harrison 62).
> 3. "Adult female brains are significantly smaller than male brains—about 8% smaller, on average" (Seligman 74).

Invalid
conclusions

> 1. Gingko is food for the brain!
> 2. Higher education . . . Who needs it?!
> 3. Women are the less intelligent gender.

"How much can
we generalize
from these
findings?"

When we accept findings uncritically and jump to conclusions about their meaning (as in points 1 and 2, above) we commit the error of *hasty generalization*. When we overestimate the extent to which the findings reveal some larger truth (as in point 3, above) we commit the error of *overstated generalization*.

NOTE *We often need to generalize, and we should. For example, countless studies support the generalization that fruits and vegetables help lower cancer risk. But we ordinarily limit general claims by inserting qualifiers such as "usually," "often," "sometimes," "probably," "possibly," or "some."*

10.4

What are some consequences of relativism? Learn more at <www.ablongman.com/lannonweb>

Faulty Causal Reasoning

Causal reasoning tries to explain why something happened or what will happen, often very complex questions. Sometimes a *definite cause is apparent* ("The engine's overheating is caused by a faulty radiator cap"). We reason about definite causes when we explain why the combustion in a car engine causes the wheels to move, or why the moon's orbit makes the tides rise and fall.

But causal reasoning often explores *causes that are not so obvious, but only possible or probable*. In these cases, much analysis is needed to isolate a specific cause.

Suppose you want to answer this question: "Why does our college campus not have daycare facilities?" Brainstorming produces these possible causes:

"Did *X* possibly,
probably, or
definitely cause
Y?"

- lack of need among students
- lack of interest among students, faculty, and staff
- high cost of liability insurance
- lack of space and facilities on campus
- lack of trained personnel
- prohibition by state law
- lack of government funding for such a project

Say you proceed with interviews, questionnaires, and research into state laws, insurance rates, and availability of personnel. First, you rule out some items, and others appear as probable causes. Specifically, you find a need among students, high campus interest, an abundance of qualified people for staffing, and no state laws prohibiting such a project. Three probable causes remain: lack of funding, high insurance rates, and lack of space. Further inquiry shows that lack of funding and high insurance rates *are* issues. These obstacles, however, could be eliminated through new sources of revenue: charging a fee for each child, soliciting donations, diverting funds from other campus organizations, and so on. Finally, after examining available campus space and speaking with school officials, you arrive at one definite cause: lack of space and facilities.

When you report on your research, be sure readers can draw conclusions identical to your own on the basis of the evidence. The process might be diagrammed like this:

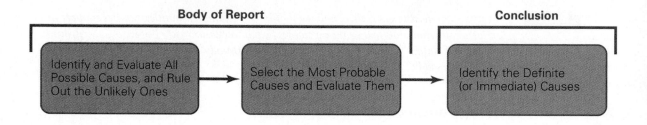

The persuasiveness of your causal argument will depend on the quality of evidence you bring to bear, as well as on your ability to clearly explain the links in the chain. Also, you must convince your audience that you haven't overlooked important alternative causes.

NOTE   *Anything but the simplest effect is likely to have multiple causes. You have to make the case that the cause you have isolated is the real one. In the daycare scenario, for example, you might argue that lack of space and facilities somehow is related to funding. And the college's inability to find funds or space might be related to student need or interest, which is not high enough to exert real pressure. Lack of space and facilities, however, appears to be the "immediate" cause.*

Here are common errors that distort or oversimplify cause-effect relationships:

Ignoring other causes	Investment builds wealth. [Ignores the roles of knowledge, wisdom, timing, and luck in successful investing.]
Ignoring other effects	Running improves health. [Ignores the fact that many runners get injured, and that some even drop dead while running.]

Inventing a causal sequence	Right after buying a rabbit's foot, Felix won the state lottery. [Posits an unwarranted causal relationship merely because one event follows another.]
Confusing correlation with causation	Women in Scandinavian countries drink a lot of milk. Women in Scandinavian countries have a high incidence of breast cancer. Therefore, milk must be a cause of breast cancer. [The association between these two variables might be mere coincidence and might obscure other possible causes, such as environment, fish diet, and genetic predisposition (Lemonick 85).]
Rationalizing	My grades were poor because my exams were unfair. [Denies the real causes of one's failures.]

Media Researcher Robert Griffin identifies three criteria for demonstrating a causal relationship:

> Along with showing correlation [say, an association between smoking and cancer], evidence of causality requires that the alleged causal agent occurs prior to the condition it causes (e.g., that smoking precedes the development of cancers) and—the most difficult task—that other explanations are discounted or accounted for (240).

For example, epidemiological studies found that people who eat lots of broccoli, cauliflower, and other "cruciferous" vegetables have lower rates of some cancers. But other explanations (say, that big veggie eaters might have many other healthful habits as well) could not be ruled out until lab studies showed how a special protein in these vegetables actually protects human cells (Wang 182). For more examples of causal claims, see Consider This, page 182.

AVOID STATISTICAL FALLACIES

How numbers can mislead

The purpose of statistical analysis is to determine the meaning of a collected set of numbers. In primary research, our surveys and questionnaires often lead to some kind of numerical interpretation ("What percentage of respondents prefer X?" "How often does Y happen?"). In secondary research, we rely on numbers collected by primary researchers.

Numbers seem more precise, more objective, more scientific, and less ambiguous than words. They are easier to summarize, measure, compare, and analyze. But numbers can be totally misleading. For example, radio or television phone-in surveys often produce distorted data: Although "90 percent of callers" might express support for a particular viewpoint, callers tend to be those with the greatest anger or extreme feelings—representing only a fraction of overall attitudes (Fineman 24). Mail-in surveys can produce similar distortion because only people with certain attitudes might choose to respond.

Before relying on any set of numbers, we need to know exactly where they come from, how they were collected, and how they were analyzed.

CONSIDER THIS Correlation Does not Equal Causation

Causal reasoning often relies on statistical analysis. The following definitions help us evaluate a statistical analysis of causal relationships.

- *Correlation:* a numerical measure of the strength of the relationship between two variables—say, smoking and lung cancer risk, or education and income (Black 513).

    We depict the strength of a correlation by plotting data in *scatter diagrams* (Lang and Secic 94):

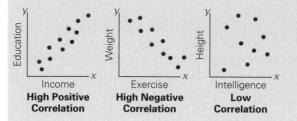

| High Positive Correlation | High Negative Correlation | Low Correlation |

As correlation increases, all data points approach the pattern of a perfect slanting line. In a *positive correlation,* as one variable (say, education) increases, so does the other (say, income). In a *negative correlation,* one variable (say, exercise) increases, while the other (say, weight) decreases.

But instead of "proving" that one thing "causes" another, correlation merely signals a possible relationship.

- *Causation:* the demonstrable production of a specific effect (smoking causes lung cancer). Correlations between smoking and lung cancer or between education and income signal a causal relationship that has been demonstrated by many studies.

Causation Actually Demonstrated

Multiple studies (especially in labs) have demonstrated how specific causes produce specific effects: say, how asbestos fibers invade lung tissue to cause a cancer known as mesothelioma, or how ionizing radiation alters cellular DNA (Harris, Richard 172).

Causation Strongly Indicated

Consider this: "7 percent of Americans eat at McDonald's on any given day, and the average child watches 10,000 food commercials on television a year" (In Brief 28). Can we speculate about the relationship between these eating and viewing habits and the fact that the U.S. obesity rate has risen by one-third since 1980 ("Fat Chance" 3)?

Possible Causation

Between 1971 and 2001, forearm fractures among adolescents during sports doubled. One possible cause, according to Mayo Clinic researchers: the increasing consumption of calcium-free soft drinks instead of milk (Seppa 221).

Improbable Causation

A recently discovered correlation between moderate alcohol consumption and decreased heart disease risk offers no sufficient proof that moderate drinking "causes" less heart disease. Only detailed research might answer this question.

No Demonstrable Causation

A 2001 study found that people who win Academy Awards live an average of 4 years longer than people who don't (Herper 12). But is Oscar a "cause" of longer life? Of course not.

- *Coincidence:* the random but simultaneous occurrence of two or more events.
- *Confounding (or confusing) factors:* other reasons or explanations for a particular outcome. For instance, studies indicating that regular exercise improves health might be overlooking this confounding factor: healthy people tend to exercise more than those who are unhealthy ("Walking" 3–4).

Common Statistical Fallacies

Faulty statistical reasoning produces conclusions that are unwarranted, inaccurate, or downright deceptive. Here are some typical fallacies.

"Exactly how well are we doing?"

THE SANITIZED STATISTIC.   Numbers can be manipulated (or "cleaned up") to obscure the facts. For example, the College Board's 1996 recentering of SAT scores has raised the "average" math score from 478 to 500 and the average verbal score from 424 to 500 (a boost of almost 5 and 18 percent, respectively) although actual student performance remains unchanged (Samuelson, "Merchants" 44).

THE MEANINGLESS STATISTIC.   Exact numbers can be used to quantify something so inexact, vaguely defined, or hard to count that it should only be approximated (Huff 247; Lavin 278):

"How many rats was that?"

- Boston has 3,247,561 rats.
- Zappo detergent makes laundry 10 percent brighter.

An exact number looks impressive, but certain subjects (child abuse, cheating in college, virginity, drug and alcohol abuse on the job, eating habits) cannot be quantified exactly because respondents don't always tell the truth (because of denial, embarrassment, or merely guessing). Or they respond in ways they think the researcher expects.

Overly precise numbers are also harder to read:

Needlessly precise and hard to read

Computer engineers' salaries increased from $34,717 to $41,346, while programmers' salaries increased from $26,807 to $32,112.

Easier to decipher and compare

Computer engineers' salaries increased roughly from $35,000 to $41,000, while programmers' salaries increased from $27,000 to $32,000.

Unless greater precision is required, round your numbers to a maximum of two significant digits ("0.00") (Lang and Secic 40).

THE UNDEFINED AVERAGE.   The mean, median, and mode are confused in determining an average (Huff 244; Lavin 279).

Three ways of reporting an "average"

- The *mean* is the result of adding up the value of each item in a set of numbers, and then dividing by the number of items.
- The *median* is the result of ranking all the values from high to low, and then identifying the middle value (or the 50th percentile, as in calculating SAT scores).
- The *mode* is the value that occurs most often in a set of numbers.

Each of these measurements represents some kind of average, but unless we know which average is being presented, we cannot interpret the figures accurately.

"Why is everyone griping?"

Assume, for instance, that we are calculating the "average" salary among female managers at XYZ Corporation:

Manager	Salary
A	$90,000
B	90,000
C	80,000
D	65,000
E	60,000
F	55,000
G	50,000

In the above example, the "mean salary" (total salaries divided by number of salaries) equals $70,000; the "median salary" (middle value) equals $65,000; the "mode" (most frequent value) equals $90,000. Each is legitimately an average, and each could be used to support or refute a particular assertion (for example, "Women managers are paid too little" or "Women managers are paid too much").

Research expert Michael Lavin sums up the potential for bias in the reporting of averages:

Unethical uses of "averages"

Depending on the circumstances, any one of these measurements [*mean, median, or mode*] may describe a group of numbers better than the other two. . . . [But] people typically choose the value which best presents their case, whether or not it is the most appropriate. (279)

What the "mean" doesn't tell us

Although the mean is the most commonly computed average, this measurement can be misleading when values on either end of the scale (*outliers*) are extremely high or low. Suppose, for instance, that Manager A (above) was paid a $200,000 salary. Because this figure deviates so far from the normal range of salary figures for B through G, it distorts the average for the whole group—increasing the mean salary by more than 20 percent (Plumb and Spyridakis 636). In this instance, the median figure—unaffected by outlying values—would present a more realistic picture.

What the "median" and the "mode" don't tell us

One drawback of the "median" figure is that it ignores extreme values. But the "mode" focuses only on the "typical" values, ignoring the data's *range* (distance between highest and lowest values), as well as their variability. In short, the clearest picture of a data set may require two of these measures—if not all three.

NOTE

Failure to report "outliers" (the small percentage of the results occurring at a great distance from the mean) in calculating the mean or secretly leaving the outliers out of the calculation is deceptive. An ethical approach is to calculate the results with and without the outliers and to report both figures (Lang and Secic 31).

THE DISTORTED PERCENTAGE FIGURE.   Percentages are often reported without explanation of the original numbers used in the calculation (Adams and Schvaneveldt 359; Lavin 280): "Seventy-five percent of respondents prefer our brand over the competing brand"—without mention that only four people were surveyed. Three out of four respondents is a far less credible number than, say, 3,000 out of 4,000—yet each of these ratios equals "75 percent."

NOTE   *In small samples, percentages can mislead because the percentage size can dwarf the number it represents: "In this experiment, 33% of the rats lived, 33% died, and the third rat got away." (Lang and Secic 41.) When your sample is small, report the actual numbers: "Five out of ten respondents agreed . . . ."*

"How much of a majority is 51 percent?"

Another fallacy in reporting percentages occurs when the *margin of error* is ignored. This is the margin within which the true figure lies, based on estimated sampling errors in a survey. For example, a claim that "the majority of people surveyed prefer Brand X" might be based on the fact that 51 percent of respondents expressed this preference; but if the survey carried a 2 percent margin of error, the true figure could be as low as 49 or as high as 53 percent. In a survey with a high margin of error, the true figure might be so uncertain that no definite conclusion can be drawn.

"Which car should we buy?"

THE BOGUS RANKING.   Items are compared on the basis of ill-defined criteria (Adams and Schvaneveldt 212; Lavin 284): "Last year, the Batmobile was the number-one selling car in Gotham City"—without mentioning that some competing car makers actually sold *more* cars to private individuals, and that the Batmobile figures were inflated by hefty sales—at huge discounts—to rental car companies and corporate fleets. Unless we know how the ranked items were chosen and how they were compared (the *criteria*), a ranking can produce a scientific-seeming number based on a completely unscientific method.

The Limitations of Number Crunching

Computers are great for comparing, synthesizing, and predicting—because of their speed, processing power, and "what if" capabilities. But limitless capacity to crunch numbers cannot guarantee accurate results.

DRAWBACKS OF DATA MINING.   Many highly publicized correlations are the product of *data mining*: in this process computers randomly compare one set of variables (say, eating habits) with another set (range of diseases). From these countless comparisons, certain relationships, or associations, are revealed (say, between coffee drinking and pancreatic cancer risk). At one retail company, data mining revealed a correlation between diaper sales and beer sales (presumably because young fathers go out at night to buy diapers). The retailer then displayed the diapers next to the beer and reportedly sold more of both (Rao 128).

Because it detects hidden relationships and has predictive power, data mining is a popular business tool. Companies assemble their own *data warehouses* (databases of information about customers, research and development, market conditions, legal matters). Much of what is "mined," however, can turn out to be trivial or absurd:

Bizarre correlations uncovered by data mining

- Venereal disease rates correlate with air pollution levels (Stedman, "Data Mining" 28).
- The hourly wages of self-proclaimed "natural blondes" (both male and female) averaged 75 cents more than the wages of nonblonde people in 1993 (Sklaroff and Ash 85).

NOTE

Despite its limitations, data mining is invaluable for "uncovering correlations that require computers to perceive but that thinking humans can evaluate and research further" (Maeglin).

THE BIASED META-ANALYSIS.   In a meta-analysis, researchers look at a whole range of studies that have been done on one topic (say, the role of high-fat diets in cancer risk). The purpose of this "study of studies" is to decide on the overall meaning of these collected findings.

As "objective" as it may seem, meta-analysis does carry potential for error (Lang and Secic):

Potential errors in meta-analysis

- *Selection bias.* Because results ultimately depend on which studies have been included and which omitted, a meta-analysis can reflect the biases of the researchers who select the material (174).
- *Publication bias.* Small studies have less chance of being published than large ones (175–76).
- *"Head counting."* This questionable method tallies the studies that have positive, negative, or insignificant results, and announces the winning category based on sheer numbers—without accounting for a study's relative size, design, or other differences.

THE FALLIBLE COMPUTER MODEL.   Computer models process complex assumptions to produce impressive but often inaccurate statistical estimates about costs, benefits, risks, or probable outcomes.

Computer models to predict global warming levels, for instance, are based on differing assumptions about wind and weather patterns, cloud formations, ozone levels, carbon dioxide concentrations, sea levels, or airborne sediment from volcanic eruptions. Despite their seemingly scientific precision, different global warming models generate 50-year predictions of sea-level rises that range from a few inches to several feet (Barbour 121). Other models suggest that warming effects could be offset by evaporation of ocean water and by clouds reflecting sunlight back to outer space (Monatersky 69). Still other models suggest that the 1-degree Fahrenheit warming over the last 100 years might not be the result of the greenhouse effect at all, but of "random fluctuations in global temperatures" (Stone 38).

How assumptions influence a computer model

Choice of assumptions might be influenced by researcher bias or the sponsors' agenda. For example, a prediction of human fatalities from a nuclear reactor meltdown might rest on assumptions about availability of safe shelter, evacuation routes, time of day, season, wind direction, and structural integrity of the containment unit. But the assumptions could be manipulated to produce an overstated or understated estimate of risk (Barbour 228). For computer-modeled estimates of accident risk (oil spill, plane crash) or of the costs and benefits of a proposed project or policy (a space station, welfare reform), consumers rarely know the assumptions behind the numbers.

According to risk expert Peter Bernstein, the major limitation of computer modeling is that it tries to predict the future on the basis of information from the past (334).

Misleading Terminology

The terms used to interpret statistics sometimes hide their real meaning.

"Do we all agree on what these terms mean?"

- *People treated for cancer have a "50 percent survival rate."* This widely publicized figure is misleading in two ways: (1) "survival," to laypersons, means "staying alive," but to medical experts, staying alive for only five years after diagnosis qualifies as survival; (2) the "50 percent" survival figure covers *all* cancers, including certain skin or thyroid cancers that have extremely high *cure rates*, as well as other cancers (such as lung or ovarian) that rarely are curable and have extremely low *survival rates* ("Are We" 5; *Facts and Figures* 2).
- *"The BODCARE Health Plan is proud of achieving 99 percent customer satisfaction."* The rate of customer "satisfaction," in this context, provides little real information about the quality of the health plan itself. Because most HMO customers are reasonably healthy, they rarely visit a doctor and so have little cause for dissatisfaction. Only those in poor health would be able to evaluate "quality of service." Also, customer service is not the same as "quality of treatment"—which is much harder for laypersons to evaluate. Finally, are we talking about customers who are "satisfied," "somewhat satisfied," or "very satisfied"—all combined into one result (Spragins 77)?

INTERPRET THE REALITY BEHIND THE NUMBERS

Even the most valid and reliable statistics require that we interpret the reality behind the numbers:

"Is this news good, bad, or insignificant?"

- *"Rates for certain cancers double after prolonged exposure to electromagnetic radiation."* Does the cancer rate actually increase from "1 in 10,000" to "2 in

10,000," or from "1 in 50" to "2 in 50"? Without knowing the *base rate* (original rate) we cannot possibly decide how alarming this "doubling" of risk actually is.

- *"Saabs and Volvos are involved in 75 percent fewer fatal accidents than average."* Is this only because of superior engineering or also because Saab and Volvo owners tend to drive very carefully ("The Safest" 72)?
- *"Eating charbroiled meat may triple the risk of stomach cancer."* The actual added risk of dying from a weekly charbroiled steak is 1 in 4 million—lower than the risk of drowning in the bathtub (Lee 280).

The above numbers may be "technically accurate" and may seem highly persuasive in the interpretations they suggest. But the actual "truth" behind these numbers is far more elusive. Any interpretation of statistical data carries the possibility that other, more accurate interpretations have been overlooked or deliberately excluded (Barnett 45).

10.5

What are some consequences of relativism? Learn more at <www.ablongman.com/lannonweb>

ACKNOWLEDGE THE LIMITS OF RESEARCH

Legitimate researchers live with uncertainty. They expect to be wrong far more often than right. Experimentation and exploration often produce confusion, mistakes, and dead ends. Following is a brief list of things that go wrong with research and interpretation.

Obstacles to Validity and Reliability

Validity and *reliability* determine the dependability of any research (Adams and Schvaneveldt 79–97; Burghardt 174–75; Crossen 22–24; Lang and Secic 154–55; Velotta 391). *Valid research* produces correct findings. A survey, for example, is valid when (1) it measures what you want it to measure, (2) it measures accurately and precisely, and (3) its findings can be generalized to the target population. A valid survey question helps each respondent to interpret the question exactly as the researcher intended, and it asks for information respondents are qualified to provide.

What makes a survey valid

Survey validity depends largely on trustworthy responses. Even clear, precise, and neutral questions can produce mistaken, inaccurate, or dishonest answers for these reasons:

Why survey responses can't always be trusted

- People often see themselves as more informed, responsible, or competent than they really are. For example, in surveys of job skills, respondents consistently rank themselves in the top 25th percentile (Fisher, "Can I Stop" 206). Also, 90 percent of adults rate their driving ability as "above average." And a mere 2 percent of high school students rate their leadership ability "below average," while 25 percent rate themselves in the top 1 percent (Baumeister 21).

- Respondents might suppress information that reflects poorly on their behavior, attitudes, or will power when asked questions like: "How often do you take needless sick days?" "Would you lie to get ahead?" "How much TV do you watch?" When asked about going to church, for example, "respondents typically overstate their attendance by 70 percent" ("Sunday" 26).
- Respondents might exaggerate or invent facts or opinions that reveal a more admirable picture when asked questions like: "How much do you give to charity?" "How many books do you read?" "How often do you hug your children?"
- Even when respondents don't know, don't remember, or have no opinion, they tend to guess in ways designed to win the researcher's approval.

What makes a survey reliable

Reliable research produces findings that can be replicated. A survey is reliable when its results are consistent, for instance, when a respondent gives identical answers to the same survey given twice or to different versions of the same questions. A reliable survey question can be interpreted identically by all respondents. Factors that compromise survey reliability include the following:

Why surveys aren't always reliable

- Each sample group has its own peculiarities (in distribution of age, gender, religious background, educational level, and so on).
- Various observers might interpret identical results differently.
- A single observer might interpret the same results differently at different times.

Much of your communication will be based on the findings of other researchers, so you will need to assess the validity and reliability of their research as well as your own.

Flaws in Study Design

While some types of studies are more reliable than others, each type has limitations (Cohn 106; Harris, Richard 170–72; Lang and Secic 8–9; Murphy 143).

EPIDEMIOLOGIC STUDIES. Observing various populations (human, animal, or plant), epidemiologists search for correlations (say, between computer use and cataracts). Conducted via observations, interviews, surveys, or review of records, these studies involve no controlled experiments. Their major limitations:

Common flaws in epidemiologic studies

- Faulty sampling techniques (page 156) may distort results.
- Observation bias, or cognitive bias, (seeing what one wants to see) may occur.
- Coincidence can easily be mistaken for correlation.
- Confounding factors (other explanations) often affect results.

Even with a correlation that is 99 percent certain, an epidemiological study alone doesn't "prove" anything. (The larger the study, however, the more credible.)

LABORATORY STUDIES.   Although laboratories offer controlled conditions, these studies also have certain limitations:

Common flaws in laboratory studies

- The reaction of an isolated group of cells does not always predict the reaction of the entire organism.
- The reactions of experimental animals to a treatment or toxin often are not generalizable to humans. For example, massive, short-term doses given to animals differ vastly from the lower, long-term doses taken by humans.
- Faulty lab technique may distort results. For example, a recent study linking Vitamin C pills to genetic damage created panic. Experts have since concluded that "researchers themselves may have created 90 to 99% of the genetic damage when they ground up the cells to examine the DNA" ("Vitamin C" 1).

HUMAN EXPOSURE STUDIES (CLINICAL TRIALS).   These studies compare one group of people receiving medication or treatment with an untreated group, the *control group*. Major limitations of human exposure studies:

Common flaws in clinical trials

- The study group may be non-representative or too different from the general population in overall health, age, or ethnic background. (For example, the fact that gingko might slow memory loss in sick people doesn't mean it will boost the memory of healthy people.)
- Anecdotal reports are unreliable. Respondents often invent answers to questions like: "How often do you eat ice cream?"
- Lack of objectivity may distort results. A drug's effectiveness is tested in a *randomized trial*, which compares a treated group of patients with a control group, who are given *placebos* (substitutes containing no actual medicine). These studies are *masked (or blind)*: Specific group assignment is concealed from patients. But doctors sometimes sneak their sickest patients into the treatment group hoping they might possibly benefit from new treatments. Despite good intentions, this practice subverts the random selection vital to such trials, making precise assessment of treatment impossible (Wallich 20+).

It is best to look for some consensus among a combination of studies.

Sources of Measurement Error

How scientific measurement can go wrong

All measurements (of length, time, temperature, weight, population characteristics) are prone to error (Taylor 3–4). The two basic types of measurement error are discussed below.

RANDOM ERROR.   Anything that causes variations in values from measure to measure is a random error (Taylor 46, 94):

Causes of
random error in
measurement

- Technique can vary from measure to measure: in running a stopwatch, reading markings on a scale, locating two points for measuring distance, changing body position in reading an instrument (say, a thermometer).
- Each observation differs with different observers or with the same observer in repeated observations.
- Each sample in a survey differs. Because of this sampling error, the same computation applied to multiple random samples drawn from the same given population produces varying results.

Researchers help neutralize random errors by repeating the measurements and averaging the range of values.

SYSTEMATIC ERROR. Any consistent bias that causes researchers to overestimate or underestimate a measurement's true value is a systematic error (Taylor 94, 97):

Causes of
systematic error
in measurement

- A measuring device can be faulty: for example, a watch that runs slow or an improperly calibrated instrument.
- The measuring device (for example, a scale) might be positioned improperly.

Systematic errors are harder to neutralize than random errors because *all* systematically flawed measurements are inaccurate.

Sources of Deception

One problem in reviewing scientific findings is "getting the story straight." Deliberately or not, consumers are often given a distorted picture.

"Has bad or
embarrassing
news been
suppressed?"

UNDERREPORTED HAZARDS. Although twice as many people in the United States are killed by medications than by auto accidents—and countless others harmed—doctors rarely report adverse drug reactions. One Rhode Island study showed some 26,000 such reactions noted in doctors' files, of which only 11 had been reported to the Food and Drug Administration (Freundlich 14).

"Is the topic 'too
weird' for
researchers?"

10.6

Why are fringe science
and bad logic thriving?
Learn more at
<www.ablongman.com/
lannonweb>

THE UNTOUCHABLE RESEARCH TOPIC. Paranormal phenomena and extraterrestrials are rarely the topics of respectable research. Neither is alternative medicine—despite the fact that one U.S. citizen in three uses some type of alternative treatment (chiropractic, acupuncture, acupressure, herbal therapy) at least once a year. Little or no scientific evidence suggests that any or all of these therapies have actual medical benefits. Why? "Partly because few 'respectable' scientists are willing to risk their reputations to do the testing required, and partly because few firms would be willing to pay for it if they were." Drug companies have little interest because "herbal medicines, not being new inventions, cannot be patented" ("Any Alternative?" 83).

"Does this report get the story straight?"

A "GOOD STORY" BUT BAD SCIENCE. Spectacular claims that are even remotely possible are more appealing than spectacular claims that have been disproven. This is why we hear plenty about claims like the following—but little about their refutation:

Even bad science makes good news

- Giant comet headed for earth!
- Tabletop device achieves cold fusion—producing more energy than it consumes!
- Insects may carry the AIDS virus!

Does all this mean we can't believe anything? Of course not. But we should be very selective about what we do choose to believe.

GUIDELINES for Evaluating and Interpreting Information

Evaluate the Sources

1. *Check the source's date of posting or publication.* Although the latest information is not always the best, it's important to keep up with recent developments.

2. *Assess the reputation of each printed source.* Check the copyright page, for background on the publisher; the bibliography, for the quality and extent of research; and (if available) the author's brief biography, for credentials.

3. *Assess the quality of each electronic source.* See page 170 for evaluating Internet sources. Don't expect comprehensive sources on any single database.

4. *Identify the study's sponsor.* If the study acclaims the crash-worthiness of the Batmobile, but is sponsored by the Batmobile Auto Company, be skeptical.

5. *Look for corroborating sources.* Usually, no single study produces dependable findings. Learn what other sources say, why they might agree or disagree with your source, and where most experts stand on this topic.

Evaluate the Evidence

1. *Decide whether the evidence is sufficient.* Evidence should surpass personal experience, anecdote, or news reports. It should be substantial enough for reasonable and informed observers to agree on its value, relevance, and accuracy.

2. *Look for a reasonable and balanced presentation of evidence.* Suspect any claims about "breakthroughs," or "miracle cures," as well as loaded words that invite emotional response or anything beyond accepted views on a topic. Expect a discussion of drawbacks as well as benefits.

3. *Do your best to verify the evidence.* Examine the facts that support the claims. Look for replication of findings. Go beyond the study to determine the direction in which the collective evidence seems to be leaning.

Interpret Your Findings

1. *Don't expect "certainty."* Most complex questions are open-ended and a mere accumulation of "facts" doesn't "prove" anything. Even so, the weight of solid evidence usually points toward some reasonable conclusion.

2. *Examine the underlying assumptions.* As opinions taken for granted, assumptions are easily mistaken for facts.

3. *Identify your personal biases.* Examine your own assumptions. Don't ignore evidence simply because it contradicts your way of seeing, and don't focus only on evidence that supports your assumptions.

4. *Consider alternate interpretations.* Consider what else this evidence might mean.

Check for Weak Spots

1. *Scrutinize all generalizations.* Decide whether the "facts" are indeed facts or assumptions, and whether the evidence supports the generalization. Suspect any general claim not limited by some qualifier ("often," "sometimes," "rarely," or the like).

2. *Treat causal claims skeptically.* Differentiate correlation from causation, as well as possible from probable or definite causes. Consider confounding factors (other explanations for the reported outcome).

3. *Look for statistical fallacies.* Determine where the numbers come from, and how they were collected and analyzed—information that legitimate researchers routinely provide. Note the margin of error.

4. *Consider the limits of computer analysis.* Data mining often produces intriguing but random correlations; meta-analysis might oversimplify relationships among various studies; a computer model is only as accurate as the assumptions and data programmed into it.

5. *Look for misleading terminology.* Examine terms that beg for precise definition in their specific context: "survival rate," "success rate," "customer satisfaction," "average increase," "risk factor," and so on.

6. *Interpret the reality behind the numbers.* Consider the possibility of alternative, more accurate, interpretations of these numbers.

7. *Consider the study's possible limitations.* Small, brief studies are less reliable than large, extended ones; epidemiologic studies are less reliable than laboratory studies (which also carry flaws); animal or human exposure studies are often not generalizable to larger human populations; "masked" (or blind) studies are not always as objective as they seem; measurements are prone to error.

8. *Look for the whole story.* Consider whether bad news may be underreported; good news, exaggerated; bad science, camouflaged and sensationalized; or research on promising but unconventional topics (say, alternative energy sources) ignored.

✓ CHECKLIST for the Research Process

(Use this checklist to assess your research process. Numbers in parentheses refer to the first page of discussion.)

Methods

☐ Did I ask the right questions? (120)
☐ Are the sources appropriately up-to-date? (165)
☐ Is each source reputable, trustworthy, relatively unbiased, and borne out by other, similar sources? (166)
☐ Does the evidence clearly support all of the conclusions? (124)
☐ Is a fair balance of viewpoints represented? (121)
☐ Can all of the evidence be verified? (173)
☐ Has the research achieved adequate depth? (122)
☐ Has the entire research process been valid and reliable? (188)

Reasoning

☐ Am I reasonably certain about the meaning of these findings? (174)
☐ Can I discern assumption from fact? (176)
☐ Am I reasoning instead of rationalizing? (177)
☐ Can I discern correlation from causation? (183)
☐ Is this the most reasonable conclusion (or merely the most convenient)? (174)
☐ Can I rule out other possible interpretations or conclusions? (177)

☐ Have I accounted for all sources of bias, including my own? (176)
☐ Are my generalizations warranted by the evidence? (178)
☐ Am I confident that my causal reasoning is accurate? (179)
☐ Can I rule out confounding factors? (182)
☐ Can all of the numbers, statistics, and interpretations be trusted? (181).
☐ Have I resolved (or at least acknowledged) any conflicts among my findings? (124)
☐ Can I rule out any possible error or distortion? (189)
☐ Am I getting the whole story, and getting it straight? (124)
☐ Should the evidence be reconsidered? (124)

Documentation

☐ Is the documentation consistent, complete, and correct? (Appendix A)
☐ Is all quoted material clearly marked throughout the text? (682)
☐ Are direct quotations used sparingly and appropriately? (682)
☐ Are all quotations accurate and integrated grammatically? (683)
☐ Are all paraphrases accurate and clear? (684)
☐ Have I documented all sources not considered common knowledge? (685)

EXERCISES   For more exercises, visit
<www.ablongman.com/lannon>

1. Assume you are an assistant communications manager for a new organization that prepares research reports for decision makers worldwide. (A sample topic: "What effect has the North American Free Trade Agreement had on the U.S. computer industry?") These clients expect answers based on the best available evidence and reasoning.

Although your recently hired coworkers are technical specialists, few have experience in the kind of wide-ranging research required by your clients. Training programs in the research process are being developed by your communications division but will not be ready for several weeks.

Meanwhile, your manager directs you to prepare a one- or two-page memo that introduces employees to major procedural and reasoning errors that affect validity and reliability in the research process. Your manager wants this memo to be comprehensive but not vague.

2. From print or broadcast media, personal experience, or the Internet, identify an example of each of the following sources of distortion or of interpretive error:

- a study with questionable sponsorship or motives
- reliance on insufficient evidence
- unbalanced presentation
- deceptive framing of facts
- overestimating the level of certainty
- biased interpretation
- rationalizing
- unexamined assumptions
- faulty causal reasoning
- hasty generalization
- overstated generalization
- sanitized statistic
- meaningless statistic
- undefined average
- distorted percentage figure
- bogus ranking
- fallible computer model
- misinterpreted statistic
- deceptive reporting

Hint: For examples of faulty (as well as correct) statistical reasoning in the news, check out Dartmouth College's *Chance Project* at <www.dartmouth.edu/~chance>.

Submit your examples to your instructor along with a memo explaining each error, and be prepared to discuss your material in class.

3. Referring to the list in Exercise 2, identify the specific distortion or interpretive error in these examples:

a. *The federal government excludes from unemployment figures an estimated 5 million people who remain unemployed after one year* (Morgenson 54).

b. *Only 38.268 percent of college graduates end up working in their specialty.*

c. *Sixty-six percent of employees we hired this year are women and minorities, compared to the national average of 40 percent.* No mention is made of the fact that only three people have been hired this year, by a company that employs 300 (mostly white males).

d. *Are you pro-life (or pro-choice)?*

4. Identify confounding factors (page 182) that might have been overlooked in the following interpretations and conclusions:

a. *The overall cancer rate today is higher than in 1910* ("Are We" 4). Does this mean the actual incidence of disease has increased or are other explanations for this finding possible?

b. *One out of every five patients admitted to Central Hospital dies* (Sowell 120). Does this mean that the hospital is bad?

c. *In a recent survey, rates of emotional depression differed widely among different countries—far lower in Asian than in Western countries* (Horgan 24+). Are these differences due to culturally specific genetic factors, as many scientists might conclude? Or is this conclusion *confounded* by other variables?

d. *"Among 20-year-olds in 1979, those who said that they smoked marijuana 11 to 50 times in the past year had an average IQ 15 percentile points higher than those who said they'd only smoked once"* (Sklaroff and Ash 85). Does this indicate that pot increases brain power or could it mean something else?

e. *Teachers are mostly to blame for low test scores and poor discipline in public schools.* How is our assessment of this claim affected by the following information? *From age 2 to 17, children in the U.S. average 12,000 hours in school, and 15,000 to 18,000 hours watching TV* ("Wellness Facts" 1).

5. Uninformed opinions are usually based on assumptions we've never really examined. Examples of popular assumptions that are largely unexamined:

- "Bottled water is safer and better for us than tap water."
- "Forest fires should always be prevented or suppressed immediately."

- "The fewer germs in their environment, the healthier the children."
- "The more soy we eat, the better."

Identify and examine one popular assumption for accuracy. For example, you might tackle the bottled water assumption by visiting the FDA Web site <www.FDA.gov> and the Sierra Club site <www.sierra.org>, for starters. (Unless you get stuck, work with an assumption not listed above.) Trace the sites and links you followed to get your information, and write up your findings in a memo to be shared with the class.

COLLABORATIVE PROJECTS

1. Exercises from the previous section may be done as collaborative projects.
2. Assess the findings below and then describe how you would rework the comparison to arrive at a meaningful conclusion.

   A 1998 article used colorful graphics to underscore the "gap" between teachers' wages and the "average wages" of other workers. State-by-state, teachers' wages exceeded the "average wage" by a figure ranging from 2.9 percent (in the District of Columbia) to 65.2 percent (in Pennsylvania)—the gap in most states ranging from 20 to 60 percent. These figures were offered as evidence for the claim that teachers are indeed handsomely paid. (Brimelow 51)

SERVICE-LEARNING PROJECT

Divide into groups and identify a controversial environmental or technology issue (for example, the need to drill for oil and natural gas in the Alaskan Wildlife Refuge or the feasibility of the Star Wars Defense Initiative). Assume that your public-interest group publishes a monthly newsletter designed to give readers an accurate assessment of opposing claims about such issues. Using this chapter as a guide, review and evaluate the main arguments and counterarguments about the issue you've chosen. Prepare the text for a 1,500-word article that will appear in the newsletter, pointing out specific examples of questionable sources, interpretive error, or distorted reasoning.

11

Summarizing and Abstracting Information

A *summary* is a concise statement of the main points and conclusions in a longer document. For those readers who are interested only in the big picture, the entire report may not be relevant, so most long reports are commonly preceded by some type of summary.

PURPOSE OF SUMMARIES

11.1

Learn about summaries and hierarchies of power at <www.ablongman.com/lannonweb>

On the job, you have to write concisely about your work. You might report on meetings or conferences, describe your progress on a project, or propose a money-saving idea. A routine assignment for many new employees is to provide superiors (decision makers) with summaries of the latest developments in their field.

Researchers and people who must act on information need to identify quickly what is most important in a long document. An abstract is a type of summary that does three things: (1) shows what the document is all about; (2) helps users decide whether to read all of it, parts of it, or none of it; and (3) gives users a framework for understanding what follows.

An effective summary communicates the *essential message* accurately and in the fewest words. Consider the following passage:

The original passage

> The lack of technical knowledge among owners of television sets leads to their suspicion about the honesty of television repair technicians. Although television owners might be fairly knowledgeable about most repairs made to their automobiles, they rarely understand the nature and extent of specialized electronic repairs. For instance, the function and importance of an automatic transmission in an automobile are generally well known; however, the average television owner knows nothing about the flyback transformer in a television set. The repair charge for a flyback transformer failure is roughly $150—a large amount to a consumer who lacks even a simple understanding of what the repairs accomplished. In contrast, a $450 repair charge for the transmission on the family car, though distressing, is more readily understood and accepted.

Three ideas make up the essential message: (1) television owners lack technical knowledge and are suspicious of repair technicians; (2) an owner usually understands even the most expensive automobile repairs; and (3) owners do not understand or accept expenses for television repairs. A summary of the above passage might read like this:

A summarized version

> Because television owners lack technical knowledge about their sets, they are often suspicious of repair technicians. Although consumers may understand expensive automobile repairs, they rarely understand or accept repair and parts expenses for their television sets.

NOTE   *For letters, memos, or other short documents that can be read quickly, the only summary needed is usually an opening thesis or topic sentence that previews the contents.*

Summaries are vital to key executives and other decision makers who have no time to read in detail everything that crosses their desks. For example one recent U.S. president required that all significant world news for the last twenty-four hours be condensed to one printed page and placed on his desk first thing each morning. Another president employed a full-time writer/researcher who summarized articles from relevant and reputable magazines.

GUIDELINES for Summarizing Information

1. *Be considerate of later readers.* Unless you own the book, journal, or magazine, work from a photocopy.

2. *Read the entire original.* When summarizing someone else's work, get a complete picture before writing a word.

3. *Reread and underline.* Identify the issue or need that led to the article or report. Focus on the essential message: thesis and topic sentences, findings, conclusions, and recommendations.

4. *Pare down your underlined material.* Omit technical details, examples, explanations, or any background that readers won't need in order to understand the original's main idea. In summarizing another's work, avoid quotations; if you must quote a crucial word or phrase directly, use quotation marks.

5. *Rewrite in your own words.* Include all essential material in the first draft; even if it is too long, you can trim later. Be sure to add no personal comments to the original, except for a brief, clarifying definition, if needed.

6. *Edit for conciseness.* When you have everything readers need, trim the word count (page 253).

   a. Cross out needless words—without harming clarity or grammar. Use complete sentences:

   > As far as artificial intelligence is concerned, the technology is only in its infancy.

   b. Cross out needless prefaces:

   > The writer argues . . . .
   > Also discussed is . . . .

   c. Combine related ideas (page 261) and rephrase to emphasize relationships:

   > A recent study emphasized job opportunities in the computer field. Fewer of tomorrow's jobs will be for programmers and other people who know how to create technology. More jobs will be for people who can use technology—as in marketing and finance (P. Ross, "Enjoy" 206).

Needless words omitted

Needless prefaces omitted

Disconnected and rambling

(continues)

Guidelines (continued)

Connected and concise

Compare this connected and more concise version:

> A recent study predicts fewer jobs for programmers and other creators of technology, and more jobs for users of technology—as in marketing and finance (P. Ross, "Enjoy" 206).

    d. Use numerals for numbers, except to begin a sentence.

7. *Check your version against the original.* Verify that you have preserved the essential message. Add no personal comments—unless you are preparing an executive abstract (page 208).

8. *Rewrite your edited version.* In this final version, strive for readability and conciseness. Add transitional expressions (page 772) to reinforce the connection between related ideas. Respect any stipulated word limit.

9. *Document your source.* Cite the full source below any summary not accompanied by its original (Appendix A).

WHAT USERS EXPECT FROM A SUMMARY

Whether you summarize your own documents (like the sample on page 649) or someone else's, users will have these expectations:

Elements of a usable summary

- *Accuracy:* Users expect a summary to precisely sketch the content, emphasis, and line of reasoning from the original.
- *Completeness:* Users expect to consult the original document only for more detail—but not to make sense of the main ideas and their relationships.
- *Readability:* Users expect a summary to be clear and straightforward—easy to follow and understand.
- *Conciseness:* Users expect a summary to be informative yet brief, and they may stipulate a word limit (say, two hundred words).
- *Nontechnical style:* Unless they are all experts, users expect plain English.

Although the summary is written last, it is read first. Take the time to do a good job.

A SITUATION REQUIRING A SUMMARY

Assume that you work in the information office of your state's Department of Environmental Management (DEM). In the coming election, citizens will vote on a referendum proposal for constructing the state's first nuclear power plant. Referendum supporters argue that nuclear power would help solve the growing problem of acid rain and global warming from burning fossil fuels. Opponents argue that nuclear power is expensive and unsafe.

To clarify the economic, environmental, and safety issues for voters, the DEM is preparing a newsletter that will be mailed to each registered voter. You have been assigned the task of researching the recent data on nuclear power and summarizing them for newsletter readers. Here is one of the articles marked up and then summarized according to the guidelines on page 199.

AN ARTICLE TO BE SUMMARIZED

U.S. Nuclear Power Industry: Background and Current Status

Combine as orienting statement (controlling idea)

Omit background details

Include causes of problem

The U.S. nuclear power industry, while currently generating more than 20 percent of the Nation's electricity, faces an uncertain future. No nuclear power plants have been ordered since 1978, and more than 100 reactors have been cancelled, including all ordered after 1973. No units are currently under active construction; the Tennessee Valley Authority's Watts Bar I reactor, ordered in 1970 and licensed to operate in 1996, was the last U.S. nuclear unit to be completed. The nuclear power industry's troubles include a slowdown in the rate of growth of electricity demand, high nuclear power plant construction costs, public concern about nuclear safety and waste disposal, and a changing regulatory environment.

Obstacles to Expansion

Include major cause

Omit nonvital details

Include key comparison

Omit speculation

Include key facts

High construction costs are perhaps the most serious obstacle to nuclear power expansion. Construction costs for reactors completed within the last decade have ranged from $2 billion to $6 billion, averaging about $3,000 per kilowatt of electric generating capacity (in 1995 dollars). The nuclear industry predicts that new plant designs could be built for about half that amount, but construction costs would still substantially exceed the projected costs of coal- and gas-fired plants.

Of more immediate concern to the nuclear power industry is the outlook for existing nuclear reactors in a deregulated electricity market. Electric utility restructuring, which is currently under way in several States, could increase the competition faced by existing nuclear plants. High operating costs and the need for costly improvements and equipment replacements have resulted in the permanent shutdown during the past decade of 10 U.S. commercial reactors before completion of their 40-year licensed operating periods. Several more reactors are currently being considered for early shutdown.

Include key facts and comparisons

Nevertheless, all is not bleak for the U.S. nuclear power industry, which currently comprises 109 licensed reactors at 68 plant sites in 38 States, Electricity production from U.S. nuclear power plants is greater than that from oil, natural gas, and hydropower, and behind only coal, which accounts for approximately 55 percent of U.S. electricity generation. Nuclear plants generate more than half the electricity in six states.

Include key fact

Omit nonvital details

Average operating costs of U.S. nuclear plants have dropped during the 1990s, and costly downtime has been steadily reduced. Licensed commercial reactors generated electricity at an average of 75 percent of their total capacity in 1996, slightly below the previous year's record.

Omit visual

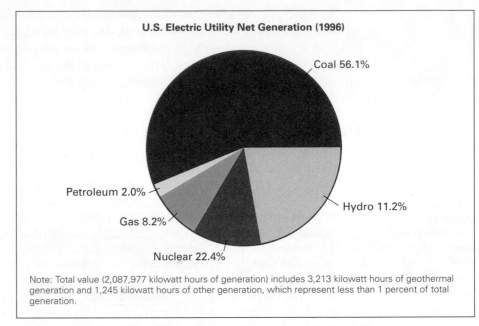

U.S. Electric Utility Net Generation (1996)

Coal 56.1%

Petroleum 2.0%

Gas 8.2%

Hydro 11.2%

Nuclear 22.4%

Note: Total value (2,087,977 kilowatt hours of generation) includes 3,213 kilowatt hours of geothermal generation and 1,245 kilowatt hours of other generation, which represent less than 1 percent of total generation.

Source: U.S. Department of Energy, Energy Information Administration.

Include key claim

Omit explanation

Include key fact

Omit nonvital fact

Include key fact

   Global warming that may be caused by fossil fuels—the "greenhouse effect"—is cited by nuclear power supporters as an important reason to develop a new generation of reactors. But the large obstacles noted above must still be overcome before electric utilities will order new nuclear units.

   Reactor manufacturers are working on designs for safer, less expensive nuclear plants, and the Nuclear Regulatory Commission (NRC) has approved new regulations to speed up the nuclear licensing process, consistent with the Energy Policy Act of 1992. Even so, the Energy Information Administration forecasts that no new U.S. reactors will become operational before 2010, if any are ordered at all.

Safety Concerns

Include key facts
Omit examples

Include key claims

Controversy over safety has dogged nuclear power throughout its development, particularly following the Three Mile Island accident in Pennsylvania and the April 1986 Chernobyl disaster in the former Soviet Union. In the United States, safety-related shortcomings have been identified in the construction quality of some plants, plant operation and maintenance, equipment reliability, emergency planning, and other areas. In addition, mishaps have occurred in which key safety systems have been disabled. NRC's oversight of the nuclear industry is an ongoing issue: nuclear utilities often complain that they are subject to overly rigorous and inflexible regulation, but nuclear critics charge that NRC frequently relaxes safety standards when compliance may prove difficult or costly to the industry.

Include key fact

Include striking exception

Omit long explanation

In terms of public health consequences, <u>the safety record of the U.S. nuclear power industry has been excellent. In more than 2,000 reactor-years</u> of operation in the United States, <u>the only incident</u> at a commercial power plant <u>that might lead to any deaths or injuries to the public</u> has been the <u>Three Mile Island</u> accident, in which more than half the core melted. Public exposure to radioactive materials released during that accident is expected to cause fewer than five deaths (and perhaps none) from cancer over the following 30 years. An independent study released in September 1990 found no "convincing evidence" that the Three Mile Island accident had affected cancer rates in the area around the plant. However, a study released in February 1997 concluded that much higher levels of radiation may have been released during the accident than was previously believed.

Omit speculation

Include key issue

Omit explanation

The relatively small amounts of radioactivity released by nuclear plants during normal operation are not generally believed to pose significant hazards. Documented public exposure to radioactivity from nuclear power plant waste has also been minimal, although the potential long-term hazard of waste disposal remains controversial. <u>There is substantial scientific uncertainty about</u> the level of <u>risk posed by low levels</u> of <u>radiation</u> exposure; as with many carcinogens and other hazardous substances, health effects can be clearly measured only at relatively high exposure levels. In the case of radiation, the assumed risk of low-level exposure has been extrapolated mostly from health effects documented among persons exposed to high levels of radiation, particularly Japanese survivors of nuclear bombing.

Include key claim

Omit nonvital details

Include key claim

The <u>consensus among most safety experts is that a severe nuclear power plant accident in the United States is likely to occur less frequently than one every 10,000 reactor-years</u> of operation. These experts believe that most severe accidents would have small public health impacts and that accidents causing as many as 100 deaths would be much rarer than once every 10,000 reactor-years. On the other hand, <u>some experts challenge the complex calculations that go into predicting such accident frequencies,</u> contending that accidents with serious public health consequences may be more frequent.

Regulation

For many years, a top priority of the nuclear industry was to modify the process for licensing new nuclear plants. No electric utility would consider ordering a nuclear power plant, according to the industry, unless licensing became quicker and more predictable, and designs were less subject to mid-construction safety-related changes ordered by NRC. The Energy Policy Act of 1992 largely implemented the industry's goals.

Omit long explanation

Nuclear plant licensing under the Atomic Energy Act of 1954 had historically been a two-stage process. NRC first issued a construction permit to build a plant and then, after construction was finished, an operating permit to run it. Each stage of the licensing process involved complicated proceedings. Environmental impact statements also are required under the National Environmental Policy Act.

Over the vehement objections of nuclear opponents, the Energy Policy Act provides a clear statutory basis for one-step nuclear licenses, allowing completed plants to operate without delay if construction criteria are met. NRC would hold preoperational hearings on the adequacy of plant construction only in specified circumstances.

Include key claims

A fundamental concern in the nuclear regulatory debate is the performance of NRC in issuing and enforcing nuclear safety regulations. The nuclear industry and its supporters have regularly complained that unnecessarily stringent and inflexibly enforced nuclear safety regulations have burdened nuclear utilities and their customers with excessive costs. But many environmentalists, nuclear opponents, and other groups charge NRC with being too close to the nuclear industry, a situation that they say has resulted in lax oversight of nuclear power plants and routine exemptions from safety requirements.

Omit explanation

Primary responsibility for nuclear safety compliance lies with nuclear utilities, which are required to find any problems with their plants and report them to NRC. Compliance is monitored directly by NRC, which maintains at least two resident inspectors at each nuclear power plant. The resident inspectors routinely examine plant systems, observe the performance of reactor personnel, and prepare regular inspection reports. For serious safety violations, NRC often dispatches special inspection teams to plant sites.

Decommissioning and Life Extension

Include key fact

When nuclear power plants end their useful lives, they must be safely removed from service, a process called decommissioning. NRC requires nuclear utilities to make regular contributions to special trust funds to ensure that money is available to remove all radioactive material from reactors after they closed. Because no full-sized U.S. commercial reactor has yet been completely decommissioned, which can take several decades, the cost of the process can only be estimated. Decommissioning cost estimates cited by a 1996 Department of Energy report, for one full-sized commercial reactor, ranged from about $150 million to $600 million in 1995 dollars.

Omit nonvital details

Include key fact

Include striking cost figure

Omit speculation

It is assumed that U.S. commercial reactors could be decommissioned at the end of their 40-year operating licenses, although several plants have been retired before their licenses expired and others could seek license renewals to operate longer. NRC rules allow plants to apply for a 20-year license extension, for a total operating time of 60 years. Assuming a 40-year lifespan, more than half of today's 109 licensed reactors could be decommissioned by the year 2016.

Source: *Congressional Digest* Jan. 1998: 7+.

Assume that in two early drafts of your summary, you rewrote and edited; for coherence and emphasis, you inserted transitions and combined related ideas. Here is your final draft.

A SUMMARY

U.S. Nuclear Power Industry: Background and Current Status

Although nuclear power generates more than 20 percent of U.S. electricity, no plants have been ordered since 1978, orders dating to 1973 are cancelled, and no units are now being built. Cost, safety, and regulatory concerns have led to zero growth in the industry.

Nuclear plant construction costs far exceed those for coal- and gas-fired plants. Also, high operating and equipment costs have forced permanent, early shutdown of 10 reactors, and the anticipated shutdown of several more.

On the positive side, the 109 licensed reactors in 38 states produce roughly 22 percent of the nation's electricity—more than oil, natural gas, and hydropower combined, and second only to coal, which produces roughly 55 percent. Moreover, nuclear power is cleaner than fossil fuels. Yet, despite declining costs and safer, less expensive designs, no new reactors could come online earlier than 2010—if any were ordered.

Safety concerns persist about plant construction, operation, and maintenance, as well as equipment reliability, emergency planning, and NRC's (Nuclear Regulatory Commission) oversight of the industry. Scientists disagree over the extent of long-term hazards from low-level emissions during plant operation and from waste disposal.

Except for the 1979 partial meltdown at Three Mile Island, however, the U.S. nuclear power industry has an excellent safety record for more than 2,000 reactor-years of operation. Most experts estimate that a severe nuclear accident in the United States will occur less than once every 10,000 reactor-years, but other experts are less optimistic.

Central to the nuclear power controversy is the NRC's role in policing the industry and enforcing safety regulations. Industry supporters claim that overregulation has created excessive costs. But opponents charge the NRC with lax oversight and enforcement.

One final unknown involves "decommissioning": safely closing down an aging power plant at the end of its 40-year operating life, a lengthy process expected to cost $150 million to $600 million per reactor.

Source: *Congressional Digest* Jan. 1998: 7+.

The version above is trimmed, tightened, and edited: word count is reduced to less than 20 percent of original length. A summary this long serves well in many situations, but other audiences might want a briefer and more compressed summary—say, roughly 15 percent of the original:

A MORE COMPRESSED SUMMARY

U.S. Nuclear Power Industry: Background and Current Status

Although nuclear power generates more than 20 percent of U.S. electricity, cost, safety, and regulatory concerns have led to zero growth in the industry. Moreover, operating and equipment costs are forcing many permanent, early shutdowns.

On the positive side, nuclear reactors generate more of the nation's electricity than all other fossil fuels except coal—and with far less pollution. Yet, despite declining operating costs and safer, less expensive designs, no new reactors could come online earlier than 2010—if any were ordered.

Safety concerns persist about plant construction, operation, and maintenance as well as equipment reliability, emergency planning, and NRC (Nuclear Regulatory Commission) oversight. Scientists disagree over the probability of a severe accident and the long-term hazards from normal, low-level emissions or from waste disposal. Except for the 1979 partial meltdown at Three Mile Island, however, the U.S. industry's safety record remains excellent.

Also controversial is the NRC's role in policing and enforcement. Industry supporters claim that excessive regulation has created excessive costs. But opponents charge the NRC with lax oversight and enforcement.

Finally, "decommissioning," safely closing down an aging power plant at the end of its operating life, is a lengthy and costly process.

Source: *Congressional Digest* Jan. 1998: 7+.

Notice that the essential message remains intact; related ideas are again combined and fewer supporting details are included. Clearly, length is adjustable according to your audience and purpose.

11.2

Learn more about summaries in online documentation at <www.ablongman.com/lannonweb>

FORMS OF SUMMARIZED INFORMATION

In preparing a report, proposal, or other document, you might summarize works of others as part of your presentation. But you will often summarize your own material as well. For instance, if your document extends to several pages, it might include different forms of summarized information, in different locations, with different levels of detail: *closing summary, informative abstract, descriptive abstract,* or *executive abstract*[1] (Figure 11.1).

[1] Adapted from David Vaughan. Although I take liberties with his classification, Vaughan helped clarify my thinking about the overlapping terminology that blurs these distinctions.

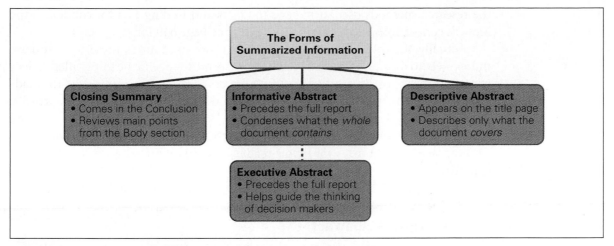

FIGURE 11.1   Summarized Information Assumes Various Forms

The Closing Summary

A *closing summary* appears at the beginning of a Conclusion section or at the end of a report's Body sections. It enables readers to review and remember the preceding main points or major findings. This look back at "the big picture" also helps readers appreciate and understand the conclusions and recommendations that will follow. (See pages 627 and 638 for examples.)

The Informative Abstract ("Summary")

Readers often appreciate condensed versions of reports. Some of these readers like to see a capsule version of the report before reading the complete document; others simply want to know basically what a report says without having to read the full document.

In order to meet reader needs, the *informative abstract* appears just after the title page. This type of summary tells the reader essentially what the full document says: It identifies the need or issue that has prompted the report; it describes

The importance of summaries

"Every time I run a training session in corporate communication, participants tell horror stories about working weeks or months on a report, only to have it disappear somewhere up the management chain. We use copies of those "invisible" reports as case studies, and invariably, the summary turns out to have been poorly written, providing readers few or no clues as to the report's significance. I'll bet companies lose millions because new ideas and recommendations get relegated to that stockpile of reports unread yearly in corporate America."

—Frank Sousa, communications consultant

the research methods used; it reviews the facts and findings; and it condenses the report's conclusions and recommendations. (See page 649 for an example.)

Actually, the title "Informative Abstract" is not used much these days. You are more likely to encounter the title "Summary." A more specific heading titled "Executive Summary" (or "Executive Abstract") refers to material summarized for readers who may not understand all the technical details contained in the report (See page 209). By contrast, a "Technical Summary" (or "Technical Abstract") is aimed at readers at the same technical level as the author of the report. You may need two or three levels of summary for report readers who have different levels of technical expertise.

See Chapter 25 for more discussion of the Summary section in a report.

The Descriptive Abstract

Another, more compressed form of summarized information can precede the full document (usually on its title page): a *descriptive abstract* merely describes what the report is about—its nature and extent. This type of abstract helps potential readers decide whether to read the document. It presents the broadest view and offers no major facts from the original. Compare, for example, the abstract that follows with the article summary on page 206:

A DESCRIPTIVE ABSTRACT

U.S. Nuclear Power Industry: Background and Current Status

The track record of the U.S. nuclear power industry is examined and reasons for its lack of growth are identified and assessed.

Because they tend to focus on methodology rather than results, descriptive abstracts are used most often in the sciences and social sciences.[2]

On the job, you might prepare informative abstracts for a boss who needs the information but who has no time to read the original. Or you might write descriptive abstracts to accompany a bibliography of works you are recommending to colleagues or clients (an annotated bibliography).

The Executive Abstract

A special type of informative abstract, the *executive abstract* (or "executive summary") essentially "replaces" the entire report. Aimed at decision makers rather than technical audiences, an executive abstract generally has more of a persuasive emphasis: to convince readers to act on the information. Executive abstracts are

---

[2]My thanks to Daryl Davis for this clarifying distinction.

crucial in cases when readers have no time to read the entire original document and when they expect the writer to help guide their thinking. ("Tell me how to think about this," instead of, "Help me understand this.") Unless the user stipulates a specific format, organize your executive abstract to answer these questions:

Users of an executive abstract have these questions

- *What did you find?*
- *What does it mean?*
- *What should be done?*

The following executive abstract addresses the problem of falling sales for a leading company in the breakfast cereal industry (Grant 223+).

AN EXECUTIVE ABSTRACT

Status Report: Market Share for Goldilocks Breakfast Cereals, Inc. (GBC)

In response to a request from GBC's Board of Directors, the accounting division analyzed recent trends in the company's sales volume and profitability.

FINDINGS

"What did you find?"

- Even though GBC is the cereal industry leader, its sales for the past four years increased at a mere average of 2.5 percent annually, to $5.2 billion, and net income decreased 12 percent overall, to $459 million.
- This weak sales growth apparently results from consumer resistance to retail price increases for cereal, totaling 91 percent in slightly more than a decade, the highest increase of any processed-food product.
- GBC traditionally offers discount coupons to offset price increases, but consumers seem to prefer a lower everyday price.
- GBC introduces an average of two new cereal products annually (most recently, "Coconut Whammos" and "Spinach Crunchies"), but such innovations do little to increase consumer interest.
- A growing array of generic cereal brands have been underselling GBC's products by more than $1 per box, especially in giant retail outlets.
- This past June, GBC dropped its cereal prices by roughly 20 percent, but by this time, the brand had lost substantial market share to generic cereal brands.

CONCLUSIONS

"What does it mean?"

- Slow but progressive loss of market share threatens GBC's dominance as industry leader.
- GBC must regain consumer loyalty to reinvigorate its market base.
- Not only have discount coupon promotions proven ineffective, but the manufacturer's cost for such promotions can total as much as 20 percent of sales revenue.

• New cereal products have done more to erode than to enhance GBC's brand image.

RECOMMENDATIONS

To regain lost market share and ensure continued dominance, GBC should implement the following recommendations:

1. Eliminate coupon promotions immediately.
2. Curtail development of new cereal products, and invest in improving the taste and nutritional value of GBC's traditional products.
3. Capitalize on GBC's brand recognition with an advertising campaign to promote GBC's "best-sellers" as an "all-day" food (say, as a healthful snack or lunch or an easy and inexpensive alternative to microwave dinners).
4. Examine the possibility of high-volume sales at discounted prices through giant retail chains.

ETHICAL CONSIDERATIONS IN SUMMARIZING INFORMATION

11.3

For more on the ethics of summaries visit <www.ablongman.com/ lannonweb>

Information in a summary format is increasingly attractive to today's readers, who often feel bombarded by more information than they can handle. Consider, for example, the popularity of the *USA Today* newspaper, with its countless news items offered in brief snippets for overtaxed readers. In contrast, the *New York Times* offers lengthy text that is information rich but more time-consuming to digest.

A summary format is especially adaptable to the hypertext-linked design of Web-based documents. Instead of long blocks of text, Web users expect pages with concise modules, or "chunks," of information that stand alone, are easy to scan, and require little or no scrolling. (See Chapter 19 for more on Web page design.) Moreover, magazine Web sites such as *Forbes* or *The Economist* offer email summaries of their hard copy editions. And while capsules or "digests" of information are an efficient way to stay abreast of new developments, the abbreviated presentation carries potential pitfalls, as media critic Ilan Greenberg points out (650):

Ways in which summarized information can be unethical

• A condensed version of a complicated issue or event may provide a useful overview, but this superficial treatment can rarely communicate the issue's full complexity—that is, the complete story.
• Whoever summarizes a lengthy piece makes decisions about what to leave out and what to leave in, what to emphasize, and what to ignore. During the selection process, the original message could very well be distorted.
• In a summary of someone else's writing, the tone or "voice" of the original author disappears—along with that writer's way of seeing. In some cases, this can be a form of plagiarism.

A summary's tip-of-the-iceberg view can alter any reader's accurate interpretation of the issue or event, as in the following headlines that summarize the story but distort the facts:

Summaries that fail to capture the real story

- "Study: Cannabis Makes Drivers More Cautious" This headline from the August 21, 2000 *Ottawa Citizen* is accompanied by the following summary on page A1: "Driving while high is less dangerous than while fatigued or drunk." Unless they turn to page A2, readers never encounter the essential fact that "Experts agree that driving while high is not as safe as driving while sober."
- "Chocolate: The New Heart-Healthy Food" Various forms of this claim have made headlines, as, for example, in the March 18, 2000 *Science News*: "Chocolate Hearts: Yummy and Good Medicine?" While the main ingredient in chocolate (cocoa) is rich in antioxidants that prevent arterial plaque buildup, most chocolate treats also contain high concentrations of sugar, caffeine, and cholesterol-laden butter fat or tropical oils (palm or coconut)—thus offsetting any apparent health benefits.

11.4

For more on usability testing visit <www.ablongman.com/lannonweb>

Informed decisions about countless science and technological controversies (human cloning, bioengineered foods, global warming, estrogen therapy) require an informed public. And while summaries do have their place in our busy world, scanning headlines or abstracts is no substitute for detailed reading and careful weighing of the facts. The more complex the topic, the more readers need the whole story.

☑ CHECKLIST for Usability of Summaries

Use this checklist to refine your summaries. (Page numbers in parentheses refer to first page of discussion.)

Content

- ☐ Does the summary contain only the essential message? (198)
- ☐ Does the summary make sense as an independent piece? (200)
- ☐ Is the summary accurate when checked against the original? (200)
- ☐ Is the summary free of any additions to the original? (199)
- ☐ Is the summary free of needless details? (199)
- ☐ Is the summary economical yet clear and comprehensive? (200)
- ☐ Is the source documented? (200)

- ☐ Does the descriptive abstract tell what the original is about? (208)

Organization

- ☐ Is the summary coherent? (233)
- ☐ Are there enough transitions to reveal the line of thought? (234)

Style

- ☐ Is the summary's level of technicality appropriate for its audience? (200)
- ☐ Is the summary free of needless words? (199)
- ☐ Are all sentences clear, concise, and fluent? (244)
- ☐ Is the summary written in correct English? (Appendix C)

EXERCISES 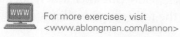 For more exercises, visit <www.ablongman.com/lannon>

1. Read each of these two paragraphs, and then list the significant ideas comprising each essential message. Write a summary of each paragraph.

   In recent years, ski-binding manufacturers, in line with consumer demand, have redesigned their bindings several times in an effort to achieve a noncompromising synthesis between performance and safety. Such a synthesis depends on what appear to be divergent goals. Performance, in essence, is a function of the binding's ability to hold the boot firmly to the ski, thus enabling the skier to rapidly change the position of his or her skis without being hampered by a loose or wobbling connection. Safety, on the other hand, is a function of the binding's ability both to release the boot when the skier falls, and to retain the boot when subjected to the normal shocks of skiing. If achieved, this synthesis of performance and safety will greatly increase skiing pleasure while decreasing accidents.

   Contrary to public belief, sewage treatment plants do not fully purify sewage. The product that leaves the plant to be dumped into the leaching (sievelike drainage) fields is secondary sewage containing toxic contaminants such as phosphates, nitrates, chloride, and heavy metals. As the secondary sewage filters into the ground, this conglomeration is carried along. Under the leaching area develops a contaminated mound through which groundwater flows, spreading the waste products over great distances. If this leachate reaches the outer limits of a well's drawing radius, the water supply becomes polluted. And because all water flows essentially toward the sea, more pollution is added to the coastal regions by this secondary sewage.

2. Attend a campus lecture and take notes on the significant points. Write a summary of the lecture's essential message.

3. Find an article about your major field or area of interest and write both an informative abstract and a descriptive abstract of the article.

4. Select a long paper you have written for one of your courses; write an informative abstract and a descriptive abstract of the paper.

5. After reading the article in Figure 11.2 prepare a descriptive abstract and an informative abstract, using the guidelines on page 199. Identify a specific audience and use for your material.

   A possible scenario: You are assistant communications manager for a leading software development company. Part of your job involves publishing a monthly newsletter for employees. After coming across this article, you decide to summarize it for the upcoming issue. (Aspirin is a popular item in this company, given the headaches, stiff necks, and other medical problems that often result from prolonged computer work.) You have 350–375 words of newsletter space to fill. Consider carefully what this audience does and doesn't need. In this situation, what information is most important?

   Bring your abstracts to class and exchange them with a classmate for editing according to the revision checklist. Revise your edited copies before submitting them to your instructor.

COLLABORATIVE PROJECTS

1. Organize into small groups and choose a topic for discussion: an employment problem, a campus problem, plans for an event, suggestions for energy conservation, or the like. (A possible topic: Should employers have the right to require lie detector tests, drug tests, or AIDS tests for their employees?)

   Discuss the topic for one class period, taking notes on significant points and conclusions. Afterward, organize and edit your notes in line with the directions for writing summaries. Next, write a summary of the group discussion in no more than 200 words. Finally, as a group, compare your individual summaries for accuracy, emphasis, conciseness, and clarity.

2. In class, form teams of students who have similar majors or interests. As a team, decide on a related topic that is currently in the news. Appoint a manager who will assign each team member a specific task. Using a combination of Web-based and hard copy versions of news coverage, compare summarized versions with more detailed coverage. For example:

   - a *USA Today* hard copy version versus one from the *New York Times*
   - a headline summary from the *New York Times'* "Quick News" and "Page One Plus" links <www.nytimes.com> versus the full-text hard copy version
   - summarized Web versions from *Forbes* <www.forbes.com> or *The Economist* <www.economist.com> versus the whole story in hard copy
   - a summarized cover story from "The Daily News Info" link on *Newsweek's* Web site <www.newsweek.com> versus the entire story in hard copy

   (Ask your reference librarian for additional suggestions.)

Each team member should compare the benefits and drawbacks of the story's shorter and longer versions, making a copy of each. Are there instances in which a summary version simply is ethically inadequate as a sole source of information? (Consult the Checklist for Ethical Communication, page 90.) Using your sample documents, explain and illustrate.

As a full team, assemble and discuss the collected findings, and appoint one member to present the findings to the class in a 15-minute oral report, showing overhead transparencies (pages 662, 665) of selected documents on the overhead projector.

SERVICE-LEARNING PROJECT

Obtain a copy of the Annual Report or other public document describing the activities and mission of the agency for which you are working. Write an informative abstract of the report for a general, public audience.

ASPIRIN
A New Look at an Old Drug
by Ken Flieger

Americans consume an estimated 80 billion aspirin tablets a year. The *Physician's Desk Reference* lists more than 50 over-the-counter drugs in which aspirin is the principal active ingredient. Yet, despite aspirin's having been in routine use for nearly a century, both scientific journals and the popular media are full of reports and speculation about new uses for this old remedy.

Almost a century after its development aspirin is the focus of extensive laboratory research and some of the largest clinical trials ever carried out in conditions ranging from cardiovascular disease and cancer to migraine headache and high blood pressure in pregnancy.

How Does It Work?

The mushrooming interest in aspirin has come about largely because of fairly recent advances in understanding how it works. What is it about this drug that, at small doses, interferes with blood clotting, at somewhat higher doses reduces fever and eases minor aches and pains, and at comparatively large doses combats pain and inflammation in rheumatoid arthritis and several other related diseases?

The answer is not yet fully known, but most authorities agree that aspirin achieves some of its effects by inhibiting the production of prostaglandins. Prostaglandins are hormone-like substances that influence the elasticity of blood vessels, control uterine contractions, direct the functioning of blood platelets that help stop bleeding, and regulate numerous other activities in the body.

In the 1970s, a British pharmacologist, John Vane, Ph.D., noted that many forms of tissue injury were followed by the release of prostaglandins. In laboratory studies, he found that two groups of prostaglandins caused redness and fever, common signs of inflammation. Vane and his co-workers also showed that by blocking the synthesis of prostaglandins, aspirin prevented blood platelets from aggregating, one of the initial steps in the formation of blood clots.

This explanation of how aspirin and other non-steroidal anti-inflammatory drugs (NSAIDs) produce their intriguing array of effects prompted laboratory and clinical scientists to form and test new ideas about aspirin's possible value in treating or preventing conditions in which prostaglandins play a role. Interest quickly focused on learning whether aspirin might prevent the blood clots responsible for heart attacks.

A heart attack or myocardial infarction (MI) results from the blockage of blood flow not *through* the heart, but *to* heart muscle. Without an adequate blood supply, the affected area of muscle dies and the heart's pumping action is either impaired or stopped altogether.

The most common sequence of events leading to an MI begins with the gradual build-up of plaque (atherosclerosis) in the coronary arteries. Circulation through these narrowed arteries is restricted, often causing the chest pain known as angina pectoris.

An acute heart attack is believed to happen when a tear in plaque inside a narrowed coronary artery causes platelets to aggregate, forming a clot that blocks the flow of blood. About 1,250,000 persons suffer heart attacks each year in the United States, and some 500,000 of them die. Those who survive a first heart attack are at greatly increased risk of having another.

Could Aspirin Help?

To learn whether aspirin could be helpful in preventing or treating cardiovascular disease, scientists have carried out numerous large randomized controlled clinical trials. In these studies, similar groups of hundreds or thousands of people are randomly assigned to receive either aspirin or a placebo, an inactive, lookalike tablet. The participants—and in double-blind trials the investigators, as well—do not know who is taking aspirin and who is swallowing a placebo.

Over the last two decades, aspirin studies have been conducted in three kinds of individuals: persons with a history of coronary artery or cerebral vascular disease, patients in the immediate, acute phases of a heart attack, and healthy men with no indication of current or previous cardiovascular illness.

The results of studies of people with a history of coronary artery disease and those in the immediate phases of a heart attack have proven to be of tremendous importance in the prevention and treatment of cardiovascular disease. The studies showed that aspirin substantially reduces the risk of death and/or non-fatal heart attacks in patients with a previous MI or unstable angina pectoris, which often occurs before a heart attack.

On the basis of such studies, these uses for aspirin (unstable angina, acute MI, and survivors of an MI) are described in the professional labeling of aspirin products, information provided to physicians and other health professionals. Aspirin labeling intended for the general public does not discuss its use in arthritis or cardiovascular disease because treatment of these serious conditions—even with a common over-the-counter drug—has to be medically supervised. The consumer labeling contains a general warning about

---

FIGURE 11.2   An Article To Be Summarized

Source: Excerpt from FDA Consumer *Jan./Feb. 1994: 19–21.*

excessive or inappropriate use of aspirin, and specifically warns against using aspirin to treat children and teenagers who have chickenpox or the flu because of the risk of Reye's syndrome, a rare but sometimes fatal condition.

Aspirin for Healthy People?

Once aspirin's benefits for patients with cardiovascular disease were established, scientists sought to learn whether regular aspirin use would prevent a first heart attack in healthy individuals. The findings regarding that critical question have thus far been equivocal. The major American study designed to find out if aspirin can prevent cardiovascular deaths in healthy individuals was a randomized, placebo-controlled trial involving just over 22,000 male physicians between 40 and 84 with no prior history of heart disease. Half took one 325-milligram aspirin tablet every other day, and half took a placebo.

The trial was halted early, after about four-and-a-half years, and the findings quickly made public in 1988 when investigators found that the group taking aspirin had a substantial reduction in the rate of fatal and non-fatal heart attacks compared with the placebo group. There was, however, no significant difference between the aspirin and placebo groups in number of strokes (aspirin-treated patients did slightly worse) or in overall deaths from cardiovascular disease.

A similar study in British male physicians with no previous heart disease found no significant effect nor even a favorable trend for aspirin on cardiovascular disease rates. The British study of 5,100 physicians, while considerably smaller than the American study, reported three-quarters as many vascular "events." FDA scientists believe the results of the two studies are inconsistent.

The U.S. Preventive Services Task Force, a panel of medical-scientific authorities in health promotion and disease prevention, is one of many groups looking at new information on the role of aspirin in cardiovascular disease. In its Guide to Clinical Preventive Services, issued in 1989, the task force recommended that low-dose aspirin therapy "should be considered for men aged 40 and over who are at significantly increased risk for myocardial infarction and who lack contraindications" to aspirin use. A revised Guide, scheduled for publication in the fall of 1994, is expected to include a slightly revised recommendation concerning aspirin and cardiovascular disease but no major change in advice to physicians about aspirin's possible role in preventing heart attacks.

Better understanding of aspirin's myriad effects in the body has led to clinical trials and other studies to assess a variety of possible uses: preventing the severity of migraine headaches, improving circulation to the gums thereby arresting periodontal disease, preventing certain types of cataracts, lowering the risk of recurrence of colorectal cancer, and controlling the dangerously high blood pressure (called preeclampsia) that occurs in 5 to 15 percent of pregnancies.

None of these uses for aspirin has been shown conclusively to be safe and effective, and there is concern that people may be misusing aspirin on the basis of unproven notions about its effectiveness. Last October, FDA proposed a new labeling statement for aspirin products advising consumers to consult a doctor before taking aspirin for new and long-term uses. The proposed statement would read, "IMPORTANT: See your doctor before taking this product for your heart or for other new uses of aspirin because serious side effects could occur with self-treatment."

The Other Side of the Coin

While examining new possibilities for aspirin in disease treatment and prevention, scientists do not lose sight of the fact that even at low doses aspirin is not harmless. A small subset of the population is hypersensitive to aspirin and cannot tolerate even small amounts of the drug. Gastrointestinal distress—nausea, heartburn, pain—is a well-recognized adverse effect and is related to dosage. Persons being treated for rheumatoid arthritis who take large daily doses of aspirin are especially likely to experience gastrointestinal side effects.

Aspirin's antiplatelet activity apparently accounts for hemorrhagic strokes, caused by bleeding into the brain, in a small but significant percentage of persons who use the drug regularly. For the great majority of occasional aspirin users, internal bleeding is not a problem. But aspirin may be unsuitable for people with uncontrolled high blood pressure, liver or kidney disease, peptic ulcer, or other conditions that might increase the risk of cerebral hemorrhage or other internal bleeding.

New understanding of how aspirin works and what it can do leaves no doubt that the drug has a far broader range of uses than imagined [nearly a century ago]. The jury is still out, however, on a number of key questions about the best and safest ways to use aspirin. And until some critical verdicts are handed down, consumers are well-advised to regard aspirin with appropriate caution.

FIGURE 11.2   An Article To Be Summarized *(continued)*

PART III

Structural and Style Elements

Top Ten Web Design Mistakes

Web design and usability expert Jakob Nielsen publishes an annual list of *Top Ten Web Design Mistakes* each year. In 2004, he reported that many Web sites are getting better at using minimalist design, maintaining archives, and offering comprehensive services. But for every two steps forward, there's at least one step backward.

At the top of Nielsen's widely publicized list this year: "unclear statement of purpose." As Nielsen writes: "Many companies, particularly in the high tech industry, use vague or generic language to describe their purpose. Obscuring this basic fact makes it much harder for users to interpret a websites's information and services."

Other significant Web design flaws include undated content; small thumbnail images of large, detailed photos; and long lists that can't be "winnowed by attributes." Nielsen concludes that Web design is improving as designers learn to account for users and their needs. But technology and design concerns often override user needs and rhetorical contexts, and the result is that users get annoyed and quickly click over to another site. ∎

12

Organizing for Users

Our thinking rarely occurs in a neat, predictable sequence, but we cannot report our ideas in the same random order in which they occur. Instead of forcing users to organize unstructured information themselves, we shape this material for their understanding. In setting out to organize, we face questions like these:

TYPICAL QUESTIONS IN ORGANIZING FOR USERS

- *What relationships do the collected data suggest?*
- *What should I emphasize?*
- *In which sequence will users approach this material?*

- *What belongs where?*
- *What do I say first? Why?*
- *What comes next?*
- *How do I end?*

To answer these questions, writers rely on the organizing strategies discussed below.

PARTITIONING AND CLASSIFYING

12.1

For more on "chunking" in electronic media visit <www.ablongman.com/lannonweb>

Partition and classification are both strategies for sorting things out. *Partition* deals with *one thing only*. It separates that thing into parts, chunks, sections, or categories for closer examination (say, a report separated into introduction, body, and conclusion).

Partition answers these user questions

- *What are its parts?*
- *What is it made of?*

Classification deals with *an assortment of things* that share certain similarities. It groups these things systematically (for example, grouping electronic documents into categories—reports, memos, Web pages).

Classification answers these user questions

- *What relates to what?*
- *What belongs where?*

Whether you choose to apply partition or classification depends on your purpose. For example, to describe a personal computer system to a novice, you might partition the system into *CPU, keyboard, printer, power cord,* and so on; for a seasoned user who wants to install an expansion card, you might partition the system into *processor-direct slot, video-in slot, communication slot,* and so on. On the other hand, if you have twenty-five software programs to arrange so you can easily locate the one you want, you will need to group them into smaller categories. You might want to classify programs according to function (*word processing, graphics, database management*) or according to expected frequency of use or relative ease of use—or some other basis.

Close examination of any complex problem usually requires both partition and classification, as in this example:

Data in Random Form

While researching the health effects of electromagnetic fields (EMFs), you encounter information about various radiation sources; ratio of risk to level of exposure; workplace studies; lab studies of cell physiology, biochemistry, and behavior; statistical studies of diseases in certain populations; conflicting expert views; views from local authorities, and so on. ❑

Figure 12.1 shows how classification might organize this random collection of EMF data into manageable categories. (Note that many of the Figure 12.1 categories might be divided further into subcategories, such as *kitchen sources, workshop sources, bedroom sources,* and so on.) Figure 12.2 shows how partition might reveal the parts of a single concept (the electromagnetic spectrum).

In organizing documents, writers use partition and classification routinely, in a process we know as *outlining*.

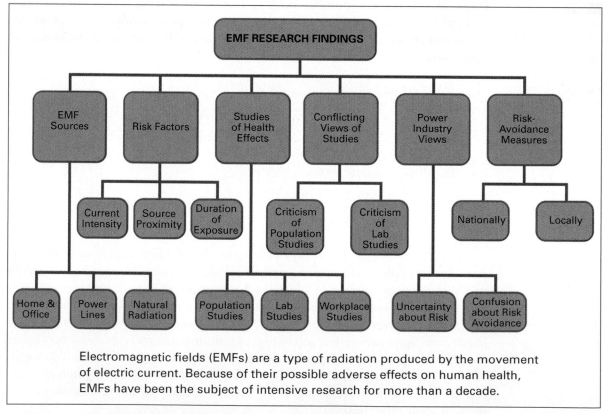

FIGURE 12.1   Assorted Items Classified by Category

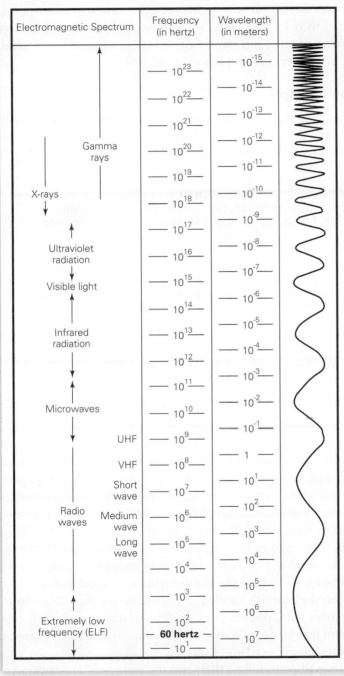

The Electromagnetic Spectrum

Electromagnetic fields can be characterized by either their **wavelength** or their **frequency**, which are related. The amount of energy an electric or magnetic field can carry depends on the frequency and wavelength of the field. The wave-length describes how far it is between one peak on the wave and the next peak. The frequency, measured in hertz, describes how many wave peaks pass by in one second of time.

If you take all the different kinds of electromagnetic fields we know about and place them on a chart, from the lowest frequency (i.e., lowest energy) to the highest, you have a chart of the electro-magnetic spectrum. The low end of the spectrum includes electric and magnetic fields produced by everyday electrical appliances. At the top of the spectrum are X-rays and gamma-rays.

When you hear about "EMFs" in the news media, the term usually refers to electric and magnetic fields at the extremely low frequency (or ELF) end of the spectrum, such as those associated with our use of electric power.

◄

This illustrates the point that the higher the frequency, the shorter the wavelength. The wavelengths are infinitely long at the bottom and infinitesimally short at the top of the spectrum so, obviously, the drawing cannot be done to scale.

FIGURE 12.2   One Item Partitioned into Its Components

Source: U.S. Environmental Protection Agency, EMF in Your Environment. *Washington, DC: GPO, 1992: 4–5.*

OUTLINING

When material is left in its original, unstructured form, people waste time trying to understand it. With an outline, you move from a random collection of ideas to an organized list that helps users follow your material.

A Document's Basic Shape

How should you organize to make the document logical from the user's point of view? Begin with the basics. Useful writing of any length—a book, chapter, news article, letter, or memo—typically follows this pattern:

Workplace documents often display this basic shape

> **INTRODUCTION**
> The introduction attracts attention, announces the viewpoint, and previews what will follow. All good introductions invite readers into the text.
>
> **BODY**
> The body explains and supports the viewpoint, achieving *unity* by remaining focused on the viewpoint. It achieves *coherence* by carrying a line of thinking from sentence to sentence in logical order.
>
> **CONCLUSION**
> The conclusion sums up the meaning of the piece or points toward other meanings to be explored. Good conclusions give readers a clear perspective on what they have just read.

The shape of useful writing

- The *introduction* provides orientation by doing any of these things: explaining the topic's origin and significance and the document's purpose; briefly identifying your intended audience and your information sources; defining specialized terms or general terms that have special meanings in your document; accounting for limitations such as incomplete or questionable data; previewing the major topics to be discussed in the body section.

  Some introductions need to be long and involved; others are better short and to the point. If you don't know the users well enough to give only what they need, use subheadings so they can choose what they want to read.
- The *body* delivers on the promise implied in your introduction ("Show me!"). Here you present your data, discuss your evidence, lay out your case, or tell users what to do and how to do it. Body sections come in all different sizes, depending on how much users need and expect.

  Body sections are titled to reflect their specific purpose: "Description and Function of Parts," for a mechanism description; "Required Steps," for a set of instructions; "Collected Data," for a feasibility analysis.
- The *conclusion* of a document has assorted purposes: It might evaluate the significance of the report, reemphasize key points, take a position, predict an out-

come, offer a solution, or suggest further study. If the issue is straightforward, the conclusion might be brief and definite. If the issue is complex or controversial, the conclusion might be lengthy and open-ended.

Conclusions vary with the document. You might conclude a mechanism description by reviewing the mechanism's major parts and then briefly describing one operating cycle. You might conclude a comparison or feasibility report by offering judgments about the facts you've presented and then recommending a course of action.

NOTE   *All readers expect a definite beginning, middle, and ending that provide orientation, discussion, and review. But specific people want these sections tailored to their expectations. Identify your readers' expectations by (1) anticipating their probable questions (pages 29, 46), and (2) visualizing the sequence in which users would want these questions answered.*

The computer is especially useful for rearranging outlines until they reflect the sequence in which you expect users to approach your message.

The Formal Outline

A simple list usually suffices for organizing short documents or as a tentative outline for longer documents. An author or team rarely begins by developing a formal outline when planning a manuscript. But at some stage (often a *later* stage) in the writing process, a long, complex document usually calls for a more systematic, formal outline. Here is a formal outline for the report examining the health effects of electromagnetic fields (pages 617–28):

A formal outline using alphanumeric notation

Children Exposed to EMFs: A Risk Assessment

I. INTRODUCTION
   A. Definition of electromagnetic fields
   B. Background on the health issues
   C. Description of the local power line configuration
   D. Purpose of this report
   E. Brief description of data sources
   F. Scope of this inquiry
II. DATA SECTION [Body]
   A. Sources of EMF exposure
      1. power lines
      2. home and office
         a. kitchen
         b. workshop [and so on]
      3. natural radiation
      4. risk factors
         a. current intensity
         b. source proximity
         c. duration of exposure

B. Studies of health effects
    1. population surveys
    2. laboratory measurements
    3. workplace links
C. Conflicting views of studies
    1. criticism of methodology in population studies
    2. criticism of overgeneralized lab findings
D. Power industry views
    1. uncertainty about risk
    2. confusion about risk avoidance
E. Risk-avoidance measures
    1. nationally
    2. locally
III. CONCLUSION
    A. Summary and overall interpretation of findings
    B. Recommendations

NOTE

Long reports often begin directly with a statement of purpose. For the intended audience (i.e., generalists) of this report, however, the technical topic must first be defined so that users understand the context. Also, each level of division yields at least two items. If you cannot divide a major item into at least two subordinate items, retain only your major heading.

A formal outline easily converts to a table of contents for the finished report, as shown in Chapter 24.

NOTE

Because they serve mainly to guide the writer, minor outline headings (such as items [a] and [b] under II.A.1 above) may be omitted from the table of contents or the report itself. Excessive headings make a document seem fragmented.

In technical documents, the alphanumeric notation shown above often is replaced by decimal notation:

Decimal notation in a technical document

2.0 DATA SECTION
    2.1 Sources of EMF Exposure
        2.1.1 home and office
            2.1.1.1 kitchen
            2.1.1.2 workshop [and so on]
        2.1.2 power lines
        2.1.3 natural radiation
        2.1.4 risk factors
            2.1.4.1 current intensity
            2.1.4.2 source proximity
            2.1.4.3 [and so on]

The decimal outline makes it easier to refer users to specifically numbered sections of the document ("See 2.1.2"). While both systems achieve the same organizing objective, decimal notation usually is preferred in business, government, and industry.

In some cases, you may wish to expand your *topic outline* into a *sentence outline*, in which each sentence serves as a topic sentence for a paragraph in the report:

A sentence
outline

> 2.0 DATA SECTION
>   2.1 Although the 2 million miles of power lines crisscrossing the United States have been the focus of the EMF controversy, potentially harmful waves also are emitted by household wiring, appliances, electric blankets, and computer terminals.
>   2.1.1 [and so on]

Sentence outlines are used mainly in collaborative projects in which various team members prepare different sections of a long document.

NOTE

A survey of the influence of computers on workplace writing found that traditional, formal outlining was giving way to outlining in the form of "notes on audience, purpose, direction, key content points, tone." These outlines were "flexible, sketchy, punctuated by arrows, numbers, or exclamation points; they looked more like lists" (Halpern 179). Outlines needn't be pretty, as long as they help you control your material.

*Also, the neat and ordered outlines in this book show the final **products** of writing and organizing, not the **process,** which is often initially messy and chaotic.*

Some writers don't start out with an outline at all! Instead, they scratch and scribble with pencil and paper or click away at the keyboard, making lots of false starts as they hammer out some kind of acceptable draft; only then do they outline to get their thinking straight.

Not until the final draft of a long document do you compose the finished outline, which serves as a model for your table of contents, as a check on your reasoning, and as a way of revealing to users a clear line of thinking.

Outlining and Reorganizing on a Computer

Most word-processing programs enable you to work on your document and your outline simultaneously. An "outline view" of the document helps you to see relationships among ideas at various levels, to create new headings, to add text beneath headings, and to move headings and their subtext (Figure 12.3). You also can *collapse* the outline view to display the headings only.

Switch between "normal view" (to work on your text) and "outline view" (to examine the arrangement of material). You can add or delete headings or text and reorganize whole sections of your document (*Microsoft Word* 504–05).

As a visual alternative to traditional outlining, many computer graphics programs enable you to display prose outlines as tree diagrams (page 314).

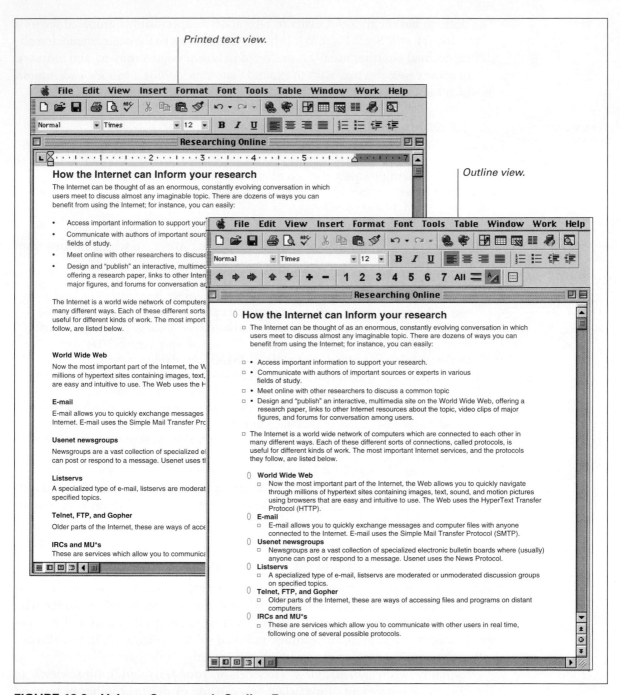

FIGURE 12.3   Using a Computer's Outline Features

Source: From Researching Online. *2nd ed. by David Munger, et al., 1988.*

NOTE *No single form of outline should be followed slavishly by any writer. The organization of any document ultimately is determined by the user's needs and expectations. In many cases, specific requirements about a document's organization and style are spelled out in a company's style guide.*

12.2

Learn more about cross-cultural awareness at <www.ablongman.com/lannonweb>

Organizing for Cross-Cultural Audiences

Different cultures have different expectations for how information should be organized. For instance, a paragraph in English typically begins with a main idea directly expressed as a topic or orienting sentence and followed by specific support; any digression from this main idea is considered harmful to the paragraph's *unity.* Some cultures, however, consider digression a sign of intelligence or politeness. To native readers of English, the long introductions and digressions in certain Spanish or Russian documents might seem tedious and confusing, but a Spanish or Russian reader might view the more direct organization of English as overly abrupt and simplistic (Leki 151).

Expectations differ even among same-language cultures. British correspondence, for instance, typically expresses the bad news directly up front, instead of taking the indirect approach preferred in the United States. A bad news letter or memo appropriate for a U.S. audience could be considered evasive by British readers (Scott and Green 19).

The Report Design Worksheet

As an alternative to the audience and use profile sheet (page 36), the worksheet in Figure 12.4 can supplement your outline and help you focus on your audience and purpose.[1]

STORYBOARDING

As you prepare a long document, one useful organizing tool is the *storyboard,* a sketch of the finished document. Much more specific and visual than an outline, a storyboard maps out each section (or module) of your outline, topic by topic, to help you see the shape and appearance of the entire document in its final form. Working from a storyboard, you can rearrange, delete, and insert material as needed—without having to wrestle with a draft of the entire document.

---

[1]This version is based on a worksheet developed by Professor John S. Harris of Brigham Young University.

REPORT DESIGN WORKSHEET

Preliminary Information

What is to be done?   *A report on the health effects of electromagnetic radiation*

Whom is it to be presented to, and when?   *Town Meeting, April 1*

Audience Analysis	Primary Audience	Secondary Audience
Position and title:	*town manager*	*selectpersons, various town officials, school board, parents, colleagues, friends*
Relationship to author or organization:	*employer*	
Technical expertise:	*nontechnical (for this topic)*	*nontechnical*
Personal characteristics:	*highly efficient; expects results*	*most have strong views on this issue*
Attitude toward author or organization:	*is preparing my annual performance review*	*friendly and respectful; selectboard will vote on my contract renewal and pay raise*
Attitude toward subject:	*extremely concerned*	*same*
Effect of report on audience or organization:	*will be read closely and acted upon*	*will be discussed at town meeting*

User's Purpose

Why has audience requested it?	*wants to address any potential hazards without delay*	
What does audience plan to do with it?	*use the data to make an informed decision about action*	*confer with the town manager about the decision*
What should audience know beforehand to understand it as written?	*nothing special; history of the issue is reviewed in report*	*same*
What does audience already know?	*has read and heard very general information about the issue*	*same*
What amount and kinds of detail will audience find significant?	*clear description of the issue and careful review of the evidence*	*same*
What should audience know and/or be able to do after reading it?	*make a decision based on the best evidence available*	*advise the town manager about her decision*

FIGURE 12.4   Report Design Worksheet

Writer's Purpose

Why am I writing? *to communicate my research findings*

What effect(s) do I wish to achieve? *to have my audience conclude that, while we await further research, we should take immediate and inexpensive steps toward risk avoidance and continue to assess the extent of EMF hazards throughout the school*

Design Specifications

Sources of data: *recently published research, including online and Internet sources; interviews with local authorities*

Tone: *semiformal*

Point of view: *mostly third person (except for recommendations)*

Needed visuals and supplements: *title page, letter of transmittal, table of contents, informative abstract, charts, graphs, and tables*

Appropriate format (letter, memo, etc.): *formal report format with full heading system*

Basic organization (problem-causes-solution, intro-instructions-summary, etc.): *causes-possible effects-conclusions and recommendations*

Main items in introduction: *Definition of electromagnetic fields*
Background on the health issue
Description of the local power line configuration
Purpose of report, and intended audience
Data sources
Scope of this inquiry

Main items in body: *Sources of EMF exposure*
Risk factors
Studies of health effects
Conflicting views of studies
Local power company views
Risk-avoidance measures

Main items in conclusion: *Summary of findings*
Overall interpretation of findings
Recommendations

Other Considerations: *no frills or complex technical data; this audience is interested in the "bottom line" as far as what action they should take*

FIGURE 12.4   Report Design Worksheet   *(Continued)*

Storyboarding is especially helpful in collaborative writing, in which various team members prepare various parts of a document and then get together to edit and assemble their material. In such cases, storyboard modules may be displayed on whiteboards, posterboards, or flip charts.

NOTE   *Try creating a storyboard after writing a full draft, for a bird's-eye view of the document's organization.*

Figure 12.5 displays one storyboard module based on Section II.A of the outline on page 223. Notice how the module begins with the section title, describes each text block and visual block, and includes a list of suggestions about special considerations.

PARAGRAPHING

¶   Users look for orientation, for shapes they can recognize. But a document's shape (introduction, body, conclusion) depends on the smaller shapes of each paragraph.

Although paragraphs can have various shapes and purposes (paragraphs of introduction, conclusion, or transition), our focus here is on *support paragraphs*. Although part of the document's larger design, each of these middle blocks of thought can usually stand alone in meaning and emphasis.

The Support Paragraph

All the sentences in a support paragraph relate to the main point, which is expressed as the *topic sentence:*

Topic sentences

> Computer literacy has become a requirement for all "educated" people.
> A video display terminal can endanger the operator's health.
> Chemical pesticides and herbicides are both ineffective and hazardous.

Each topic sentence introduces an idea, judgment, or opinion. But in order to grasp the writer's exact meaning, users need explanation. Consider the third statement:

> Chemical pesticides and herbicides are both ineffective and hazardous.

Imagine that you are a researcher for the Epson Electric Light Company, assigned this question: Should the company (1) begin spraying pesticides and herbicides under its power lines, or (2) continue with its manual (and nonpolluting) ways of minimizing foliage and insect damage to lines and poles? If you simply responded with the preceding assertion, your employer would have further questions:

> • *Why, exactly, are these methods ineffective and hazardous?*
> • *What are the problems?*
> • *Can you explain?*

**FIGURE 12.5
One Module
from a
Storyboard**

section title   **Sources of EMF Exposure**

text

Discuss milligauss measurements as indicators of cancer risk

text

Brief lead-in to power line emissions

visual

EPS table comparing power line emissions at various distances

text

Discuss EMF sources in home and office

visual

Table comparing EMF emissions from common sources

text

Discuss major risk factors: Voltage versus current; proximity versus duration of exposure; sporadic, high-level exposure versus constant, low-level exposure

visual

Line graph showing strength of exposure in relation to distance from electrical appliances

text

Focus on the key role of <u>proximity</u> to the EMF source in risk assessment

<u>Special considerations:</u>
- Define all specialized terms (<u>current</u>, <u>voltage</u>, <u>milligauss</u>, and so on) for a general audience.
- Emphasize that no "safe" level of EMF exposure has been established.
- Emphasize that even the earth's magnetic field emits significant electromagnetic radiation.

To answer these questions and to support your assertion, you need a fully developed paragraph:

Intro. (topic sent.)
Body (2–6)

Conclusion (7–8)

> [1]**Chemical pesticides and herbicides are both ineffective and hazardous.** [2]Because none of these chemicals has permanent effects, pest populations invariably recover and need to be resprayed. [3]Repeated applications cause pests to develop immunity to the chemicals. [4]Furthermore, most of these products attack species other than the intended pest, killing off its natural predators, thus actually increasing the pest population. [5]Above all, chemical residues survive in the environment (and living tissue) for years, often carried hundreds of miles by wind and water. [6]This toxic legacy includes such biological effects as birth deformities, reproductive failures, brain damage, and cancer. [7]Although intended to control pest populations, these chemicals ironically threaten to make the human population their ultimate victims. [8]I therefore recommend continuing our present control methods.

Most standard paragraphs in technical writing have an introduction-body-conclusion structure. They begin with a clear topic (or orienting) sentence stating a generalization. Details in the body support the generalization.

The Topic Sentence

Users look to a paragraph's opening sentences for a framework. When the paragraph's main point is missing, users struggle to grasp your meaning. Read this next paragraph once only.

A paragraph
without a topic
sentence

> Besides containing several toxic metals, it percolates through the soil, leaching out naturally present metals. Pollutants such as mercury invade surface water, accumulating in fish tissues. Any organism eating the fish—or drinking the water—in turn faces the risk of heavy metal poisoning. Moreover, acidified water can release heavy concentrations of lead, copper, and aluminum from metal plumbing, making ordinary tap water hazardous.

Can you identify the paragraph's main idea? Probably not. Without the topic sentence, you have no framework for understanding. You don't know where to place the emphasis: on polluted fish, on metal poisoning, on tap water?

Now, insert the following sentence at the beginning and reread the paragraph:

The missing topic
sentence

> Acid rain indirectly threatens human health.

With this orientation, the message's exact meaning becomes obvious.

The topic sentence should appear *first* (or early) in the paragraph, unless you have good reason to place it elsewhere. Think of your topic sentence as "the one sentence you would keep if you could keep only one" (U.S. Air Force Academy 11). In some instances, a paragraph's main idea may require a "topic statement" consisting of two or more sentences, as in this example:

A topic statement can have two or more sentences

> The most common strip-mining methods are open-pit mining, contour mining, and auger mining. The specific method employed will depend on the type of terrain that covers the coal.

The topic sentence or topic statement should focus and forecast. Don't write *Some pesticides are less hazardous and often more effective than others* when you mean *Organic pesticides are less hazardous and often more effective than their chemical counterparts.* The first topic sentence leads everywhere and nowhere; the second helps us focus, tells us what to expect from the paragraph. Don't write *acid rain poses a danger,* leaving readers to decipher what you mean by *danger.* If you mean that *Acid rain is killing our lakes and polluting our water supplies,* say so.

Paragraph Unity

A paragraph is unified when every word, phrase, and sentence directly supports the topic sentence.

A unified paragraph

> **Solar power offers an efficient, economical, and safe solution to the Northeast's energy problems.** To begin with, solar power is highly efficient. Solar collectors installed on fewer than 30 percent of roofs in the Northeast would provide more than 70 percent of the area's heating and air-conditioning needs. Moreover, solar heat collectors are economical, operating for up to twenty years with little or no maintenance. These savings recoup the initial cost of installation within only ten years. Most important, solar power is safe. It can be transformed into electricity through photovoltaic cells (a type of storage battery) in a noiseless process that produces no air pollution—unlike coal, oil, and wood combustion. In sharp contrast to its nuclear counterpart, solar power produces no toxic waste and poses no catastrophic danger of meltdown. Thus, massive conversion to solar power would ensure abundant energy and a safe, clean environment for future generations.

One way to destroy unity in the paragraph above would be to veer from the focus on *efficient, economical,* and *safe* by introducing topics such as the differences between active and passive solar heating, or manufacturers of solar technology, or the advantages of solar power over wind power.

Every topic sentence has a key word or phrase that carries the meaning. In the pesticide-herbicide paragraph (page 232), the key words are *ineffective* and *hazardous.* Anything that fails to advance the meaning of *ineffective* and *hazardous* throws the paragraph—and the users—off track.

Paragraph Coherence

¶coh

In a coherent paragraph, everything not only belongs, but also sticks together: Topic sentence and support form a *connected line of thought,* like links in a chain.

Paragraph coherence can be damaged by (1) short, choppy sentences; (2) sentences in the wrong order; (3) insufficient transitions and connectors

(Appendix C) for linking related ideas; or (4) an inaccessible line of reasoning. Here is how the solar energy paragraph might become incoherent:

An incoherent paragraph

>   Solar power offers an efficient, economical, and safe solution to the Northeast's energy problems. Unlike nuclear power, solar power produces no toxic waste and poses no danger of meltdown. Solar power is efficient. Solar collectors could be installed on fewer than 30 percent of roofs in the Northeast. These collectors would provide more than 70 percent of the area's heating and air-conditioning needs. Solar power is safe. It can be transformed into electricity. This transformation is made possible by photovoltaic cells (a type of storage battery). Solar heat collectors are economical. The photovoltaic process produces no air pollution.

In the above paragraph, the second sentence, about safety, belongs near the end. Also, because of short, choppy sentences and insufficient links between ideas, the paragraph reads more like a list than like a flowing discussion. Finally, a concluding sentence is needed to complete the chain of reasoning and to give readers a clear perspective on what they've just read.

Here, in contrast, is the original, coherent paragraph with sentences numbered for later discussion and with transitions and connectors shown in boldface. Notice how this version reveals a clear line of thought:

A coherent paragraph

>   [1]Solar power offers an efficient, economical, and safe solution to the Northeast's energy problems. [2]**To begin with,** solar power is highly efficient. [3]Solar collectors installed on fewer than 30 percent of roofs in the Northeast would provide more than 70 percent of the area's heating and air-conditioning needs. [4]**Moreover**, solar heat collectors are economical, operating for up to twenty years with little or no maintenance. [5]**These savings** recoup the initial cost of installation within only ten years. [6]**Most important,** solar power is safe. [7]**It** can be transformed into electricity through photovoltaic cells (a type of storage battery) in a noiseless process that produces no air pollution—unlike coal, oil, and wood combustion. [8]**In sharp contrast** to its nuclear counterpart, solar power produces no toxic waste and poses no danger of catastrophic meltdown. [9]**Thus,** massive conversion to solar power would ensure abundant energy and a safe, clean environment for future generations.

The line of thinking in this paragraph seems easy enough to follow:

1.   The topic sentence establishes a clear direction.
2–3.   The first reason is given and then explained.
4–5.   The second reason is given and explained.
6–8.   The third and major reason is given and explained.
9.   The conclusion reemphasizes the main point and completes the chain of reasoning.

To reinforce the logical sequence, related ideas are combined in individual sentences, and transitions and connectors signal clear relationships. The whole paragraph sticks together.

Paragraph Length

Paragraph length depends on the writer's purpose and the user's capacity for understanding. Writing that contains highly technical information or complex instructions may use short paragraphs or perhaps a list. In writing that explains concepts, attitudes, or viewpoints, support paragraphs generally run from 100 to 300 words.

But word count really means very little. What matters is *how thoroughly the paragraph makes your point.* A flabby paragraph buries users in needless words and details; but just skin-and-bones leaves readers looking for the meat.

Try to avoid too much of anything. A clump of short paragraphs can make some writing seem choppy and poorly organized, but a stretch of long ones is tiring. Occasional paragraphs of only one or two sentences can focus the user's attention and highlight important ideas.

NOTE   *In writing displayed on computer screens, short paragraphs and lists are especially useful because they allow for easy scanning and navigation.*

SEQUENCING

12.3

Learn about associational or nonlinear structures at <www.ablongman.com/lannonweb>

Items in logical sequence follow some pattern that reveals a relationship: cause-and-effect, comparison-contrast, and so on. For instance, a progress report usually follows a *chronological* sequence (events in order of occurrence). An argument for a companywide exercise program would likely follow an *emphatic* sequence (benefits in order of importance—least to most, or vice versa).

A single paragraph usually follows one particular sequence. A longer document may use one particular sequence or a combination of sequences. Some common sequences are described below.

Spatial Sequence

A spatial sequence begins at one location and ends at another. It is most useful in describing a physical item or a mechanism. Describe the parts in the sequence in which users would actually view the parts or in the order in which each part functions: (left to right, inside to outside, top to bottom). This description of a hypodermic needle proceeds from the needle's base (hub) to its point:

"What does it look like?"

> A hypodermic needle is a slender, hollow steel instrument used to introduce medication into the body (usually through a vein or muscle). It is a single piece composed of three parts, all considered sterile: the hub, the cannula, and the point. The hub is the lower, larger part of the needle that attaches to the necklike

opening on the syringe barrel. Next is the cannula (stem), the smooth and slender central portion. Last is the point, which consists of a beveled (slanted) opening, ending in a sharp tip. The diameter of a needle's cannula is indicated by a gauge number; commonly, a 24–25 gauge needle is used for subcutaneous injections. Needle lengths are varied to suit individual needs. Common lengths used for subcutaneous injections are $\frac{3}{8}$, $\frac{1}{2}$, $\frac{5}{8}$, and $\frac{3}{4}$ inch. Regardless of length and diameter, all needles have the same functional design.

Product and mechanism descriptions almost always have some type of visual to amplify the verbal description.

Chronological Sequence

A chronological sequence follows the actual sequence of events. Explanations of how to do something or how something happened generally are arranged according to a strict time sequence: first step, second step, and so on.

"How is it done?"

Instead of breaking into a jog too quickly and risking injury, take a relaxed and deliberate approach. Before taking a step, spend at least ten minutes stretching and warming up, using any exercises you find comfortable. (After your first week, consult a jogging book for specialized exercises.) When you've completed your warmup, set a brisk pace walking. Exaggerate the distance between steps, taking long strides and swinging your arms briskly and loosely. After roughly one hundred yards at this brisk pace, you should feel ready to jog. Immediately break into a very slow trot: lean your torso forward and let one foot fall in front of the other (one foot barely leaving the ground while the other is on the pavement). Maintain the slowest pace possible, just above a walk. *Do not bolt out like a sprinter!* The biggest mistake is to start fast and injure yourself. While jogging, relax your body. Keep your shoulders straight and your head up, and enjoy the scenery—after all, it is one of the joys of jogging. Keep your arms low and slightly bent at your sides. Move your legs freely from the hips in an action that is easy, not forced. Make your feet perform a heel-to-toe action: land on the heel; rock forward; take off from the toe.

The paragraph explaining how acid rain endangers human health (page 232) offers another example of chronological sequence.

Effect-to-Cause Sequence

Problem-solving analyses typically use a sequence that first identifies a problem and then traces its causes.

"How did this happen?"

Modern whaling techniques nearly brought the whale population to the threshold of extinction. In the nineteenth century, invention of the steamboat increased hunters' speed and mobility. Shortly afterward, the grenade harpoon was invented so that whales could be killed quickly and easily from the ship's deck. In 1904, a whaling station opened on Georgia Island in South America. This station became the gateway to Antarctic whaling for the nations of the world. In 1924, factory ships were designed that enabled round-the-clock whale tracking and processing. These ships could reduce a ninety-foot whale to its by-products in roughly thirty minutes. After World War II, more powerful boats with remote sensing devices gave a final boost to the whaling industry. The number of kills had now increased far beyond the whales' capacity to reproduce.

Cause-to-Effect Sequence

A cause-to-effect sequence follows an action to its results. Below, the topic sentence identifies the causes, and the remainder of the paragraph discusses its effects.

"What will happen if I do this?"

Some of the most serious accidents involving gas water heaters occur when a flammable liquid is used in the vicinity. The heavier-than-air vapors of a flammable liquid such as gasoline can flow along the floor—even the length of a basement—and be explosively ignited by the flame of the water heater's pilot light or burner. Because the victim's clothing frequently ignites, the resulting burn injuries are commonly serious and extremely painful. They may require long hospitalization, and can result in disfigurement or death. *Never, under any circumstances, use a flammable liquid near a gas heater or any other open flame.* (Consumer Product Safety Commission)

Emphatic Sequence

Emphasis makes important things stand out. Reasons offered in support of a specific viewpoint or recommendation often appear in workplace writing, as in the pesticide-herbicide paragraph on page 232 or the solar energy paragraph on page 233. For emphasis, the reasons or examples are usually arranged in decreasing or increasing order of importance.

"What should I remember about this?"

Although strip mining is safer and cheaper than conventional mining, it is highly damaging to the surrounding landscape. Among its effects are scarred mountains, ruined land, and polluted waterways. Strip operations are altering our country's land at the rate of 5,000 acres per week. An estimated 10,500 miles of streams have been poisoned by silt drainage in Appalachia alone. If strip mining continues at its present rate, 16,000 square miles of U.S. land will eventually be stripped barren.

In this paragraph, the most dramatic example appears last, for greatest emphasis.

Problem-Causes-Solution Sequence

The problem-solving sequence proceeds from description of the problem to diagnosis to solution. After outlining the cause of the problem, this next paragraph explains how the problem has been solved:

"How was the
problem solved?"

> On all waterfront buildings, the unpainted wood exteriors had been severely damaged by the high winds and sandstorms of the previous winter. After repairing the damage, we took protective steps against further storms. First, all joints, edges, and sashes were treated with water-repellent preservative to protect against water damage. Next, three coats of nonporous primer were applied to all exterior surfaces to prevent paint from blistering and peeling. Finally, two coats of wood-quality latex paint were applied over the nonporous primer. To keep coats of paint from future separation, the first coat was applied within two weeks of the priming coats, and the second within two weeks of the first. Two weeks after completion, no blistering, peeling, or separation has occurred.

Comparison-Contrast Sequence

Workplace writing often requires evaluation of two or more items on the basis of their similarities or differences.

"How do these
items compare?"

> The ski industry's quest for a binding that ensures good performance as well as safety has led to development of two basic types. Although both bindings improve performance and increase the safety margin, they have different release and retention mechanisms. The first type consists of two units (one at the toe, another at the heel) that are spring-loaded. These units apply their retention forces directly to the boot sole. Thus the friction of boot against ski allows for the kind of ankle movement needed at high speeds over rough terrain, without causing the boot to release. In contrast, the second type has one spring-loaded unit at either the toe or the heel. From this unit a boot plate travels the length of the boot to a fixed receptacle on its opposite end. With this plate binding, the boot has no part in release or retention. Instead, retention force is applied directly to the boot plate, providing more stability for the recreational skier, but allowing for less ankle and boot movement before releasing. Overall, the double-unit binding performs better in racing, but the plate binding is safer.

For comparing and contrasting more specific data on these bindings, two lists would be most effective.

> The Salomon 555 offers the following features:
> 1. upward release at the heel and lateral release at the toe (thus eliminating 80 percent of leg injuries)
> 2. lateral antishock capacity of 15 millimeters, with the highest available return-to-center force
> 3. two methods of reentry to the binding: for hard and deep-powder conditions
> 4. five adjustments
> 5. (and so on)

The Americana offers these features:

1. upward release at the toe as well as upward and lateral release at the heel
2. lateral antishock capacity of 30 millimeters, with moderate return-to-center force
3. two methods of reentry to the binding
4. two adjustments, one for boot length and another for comprehensive adjustment for all angles of release and elasticity
5. (and so on)

Instead of this block structure (in which one binding is discussed and then the other), the writer might have chosen a point-by-point structure (in which points common to both items, such as "Reentry Methods" are listed together). The point-by-point comparison works best in feasibility and recommendation reports because it offers readers a meaningful comparison between common points.

CHUNKING

Each of the organizing techniques discussed in this chapter is a way of *chunking* information: breaking the message down into discrete, digestible units, based on the users' needs and the document's purpose. Well-chunked material generally is easier to follow and more visually appealing.

Chunking enables us to show which pieces of information belong together and how the various pieces are connected (Horn 187). For instance, the opening page of Chapter 7 divides the research process into two chunks:

A major topic chunked into subtopics

- Procedural Stages
- Inquiry Stages

Each of these units then divides into smaller, somewhat parallel chunks:

Subtopics chunked into smaller topics

- Procedural Stages
    Searching for Information
    Recording Your Findings
    Documenting Your Sources
    Writing the Report
- Inquiry Stages
    Asking the Right Questions
    Exploring a Balance of Views
    Achieving Adequate Depth in Your Search
    Evaluating Your Findings
    Interpreting Your Findings

If any of these segments become too long to read and understand easily, they might be subdivided again.

NOTE

Chunking requires careful decisions about exactly how much is enough and what constitutes sensible proportions among the parts. Don't overdo it by creating such tiny segments that your document ends up looking fragmented and disconnected.

Using visuals for chunking

In addition to chunking information verbally we can chunk it visually:

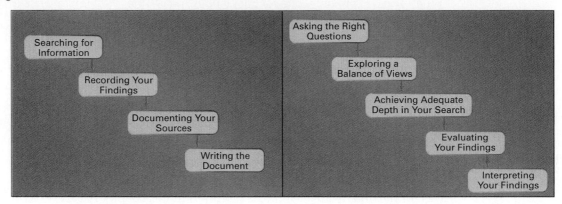

Notice how the visual display makes relationships immediately apparent. For more on visual design, see Chapter 14.

Finally, we can chunk information by using white space, shading, or other forms of page design:

Using page design for chunking

A well-designed page provides immediate cues about where to look and how to proceed. For more on page design, see Chapter 15.

Chunking is particularly useful in designing hypertext documents and Web pages (Chapter 19).

CREATING AN OVERVIEW

Once you've settled on a final organization for your document, give readers an immediate preview of its contents by answering their intial questions:

What readers want to know immediately

- What is the purpose of this document?
- Why should I read it?
- What information can I expect to find here?

Readers will have additional, more specific questions as well (see page 29), but first they want to know what the document is all about and how it relates to them.

Overviews come in various shapes. The overview for this book, for example, appears on page xxi, under the heading "Organization of Technical Communication, Tenth Edition." An informative abstract of a long document also provides an overview, as on page 649. An overview for an oral presentation appears on page 659 as an introduction to that presentation. Whatever its shape, a good overview gives readers the "big picture" to help them navigate the document and understand its details (Gurak and Lannon 43).

EXERCISES

 For more exercises, visit <www.ablongman.com/lannon>

1. Locate, copy, and bring to class a paragraph that has the following features:

   - an orienting topic sentence
   - adequate development
   - unity
   - coherence
   - a recognizable sequence
   - appropriate length for its purpose and audience

   Be prepared to identify and explain each of these features in a class discussion.

2. For each of the following documents, indicate the most logical sequence. (For example, a description of a proposed computer lab would follow a spatial sequence.)

   - a set of instructions for operating a power tool
   - a campaign report describing your progress in political fund-raising
   - a report analyzing the weakest parts in a piece of industrial machinery
   - a report analyzing the desirability of a proposed oil refinery in your area
   - a detailed breakdown of your monthly budget to trim excess spending
   - a report investigating the reasons for student apathy on your campus
   - a report evaluating the effects of the ban on DDT in insect control
   - a report on any highly technical subject, written for a general audience
   - a report investigating the success of a no-grade policy at other colleges
   - a proposal for a no-grade policy at your college

COLLABORATIVE PROJECTS

1. Organize into small groups. Choose *one* of these topics, or one your group settles on, and then brainstorm to develop a formal outline for the body of a report. One representative from your group can write the final draft on the board, for class revision. (In a computer classroom, your group's representative might

type revisions of the draft in *Microsoft Word* outline view.)

- job opportunities in your career field
- a physical description of the ideal classroom
- how to organize an effective job search
- how the quality of your higher educational experience can be improved
- arguments for and against a formal grading system
- an argument for an improvement you think this college needs most

2. Assume your group is preparing a report titled "The Negative Effects of Strip Mining on the Cumberland Plateau Region of Kentucky." After brainstorming and researching your subject, you all settle on these four major topics:

- economic and social effects of strip mining
- description of the strip-mining process
- environmental effects of strip mining
- description of the Cumberland Plateau

Arrange these topics in the most sensible sequence.

When your topics are arranged, assume that subsequent research and further brainstorming produce this list of subtopics:

- method of strip mining used in the Cumberland Plateau region
- location of the region
- permanent land damage
- water pollution
- lack of educational progress
- geological formation of the region
- open-pit mining
- unemployment
- increased erosion
- auger mining
- natural resources of the region
- types of strip mining
- increased flood hazards
- depopulation
- contour mining

Arrange these subtopics (and perhaps some sub-subtopics) under appropriate topic headings. Use decimal notation to create the body of a formal outline. Appoint one group member to present the outline in class.

Hint: Assume that your thesis is: "Decades of strip mining (without reclamation) in the Cumberland Plateau have devastated this region's environment, economy, and social structure."

13

Editing for Readable Style

You might write for a diverse or specific audience, or for experts or nonexperts. But no matter how technically appropriate your document, audience needs are not served unless your style is *readable*.

A definition of style

What is *writing style*, and how does it influence user reaction to a document? Your particular writing style is a blend of these elements:

What determines your style

- the way in which you construct each sentence
- the length of your sentences
- the way in which you connect sentences
- the words and phrases you choose
- the tone you convey

Readable sentences require correct grammar, punctuation, and spelling. But correctness alone is no guarantee of readability. For example, this response to a job applicant is mechanically correct but inefficient:

Inefficient style

> We are in receipt of your recent correspondence indicating your interest in securing the advertised position. Your correspondence has been duly forwarded for consideration by the personnel office, which has employment candidate selection responsibility. You may expect to hear from us relative to your application as the selection process progresses. Your interest in the position is appreciated.

Notice how hard you have worked to extract information that could be expressed this simply:

More efficient

> Your application for the advertised position has been forwarded to our personnel office. As the selection process moves forward, we will be in touch. Thank you for your interest.

Inefficient style makes readers work harder than they should.

Style can be inefficient for many reasons, but it is especially inept when it

Ways in which style goes wrong

- makes the writing impossible to interpret
- takes too long to make the point
- reads like a Dick-and-Jane story from primary school
- uses imprecise or needlessly big words
- sounds stuffy and impersonal

Regardless of the cause, inefficient style results in writing that is less informative, less persuasive. Moreover, inefficient style can be unethical—by confusing or misleading the audience.

To help your audience spend less time reading, you must spend more time revising for a style that is *clear, concise, fluent, exact,* and *likable*.

EDITING FOR CLARITY

A clear sentence requires no more than a single reading. The following suggestions will help you edit for clarity.

amb   **AVOID AMBIGUOUS PHRASING.**   Workplace writing ideally has *one* meaning only and allows for *one* interpretation. Does a person's "suspicious attitude" mean that he is "suspicious" or "suspect"?

AMBIGUOUS PHRASING	All managers are not required to submit reports. *(Are some or none required?)*
REVISED	Managers are not all required to submit reports.

or

Managers are not required to submit reports.

AMBIGUOUS PHRASING	Most city workers strike on Friday.
REVISED	Most city workers **are planning to strike** on Friday.

or

Most city workers **typically strike** on Friday.

ref   **AVOID AMBIGUOUS PRONOUN REFERENCES.**   Pronouns (*he, she, it, their,* and so on) must clearly refer to the noun they replace.

AMBIGUOUS REFERENT	Our patients enjoy the warm days while **they** last. *(Are the patients or the warm days on the way out?)*

Depending on whether the referent (or antecedent) for *they* is *patients* or *warm days,* the sentence can be clarified.

CLEAR REFERENT	While these warm days last, our patients enjoy them.

or

Our terminal patients enjoy the warm days.

AMBIGUOUS REFERENT	Jack resents his assistant because **he** is competitive. *(Who's the competitive one—Jack or his assistant?)*
CLEAR REFERENT	Because his assistant is competitive, Jack resents him.

or

Because Jack is competitive, he resents his assistant.

(See Appendix C for more on pronoun references, and page 281 for avoiding sexist bias in pronoun use.)

AVOID AMBIGUOUS PUNCTUATION. A missing hyphen, comma, or other punctuation mark can obscure meaning.

MISSING HYPHEN	Replace the trailer's inner wheel bearings. *(The inner-wheel bearings or the inner wheel-bearings?)*
MISSING COMMA	Does your company produce liquid hydrogen? If so, how[,] and where do you store it? *(Notice how the meaning changes with a comma after "how.")*
	Police surrounded the crowd[,] attacking the strikers. *(Without the comma, the crowd appears to be attacking the strikers.)*

A missing colon after *kill* yields the headline "Moose Kill 200." A missing apostrophe after *Myers* creates this gem: "Myers Remains Buried in Portland."

EXERCISE 1

Edit each sentence below to eliminate ambiguities in phrasing, pronoun reference, or punctuation.

  a. Call me any evening except Tuesday after 7 o'clock.
  b. The benefits of this plan are hard to imagine.
  c. I cannot recommend this candidate too highly.
  d. Visiting colleagues can be tiring.
  e. Janice dislikes working with Claire because she's impatient.
  f. Our division needs more effective writers.
  g. Tell the reactor operator to evacuate and sound a general alarm.
  h. If you don't pass any section of the test, your flying days are over.
  i. Dial "10" to deactivate the system and sound the alarm.

---

tel

AVOID TELEGRAPHIC WRITING. *Function words* signal relationships between the *content words* (nouns, adjectives, verbs, and adverbs) in a sentence. Some examples of function words:

Function words

• articles (*a, an, the*)
• prepositions (*in, of, to*)
• linking verbs (*is, seems, looks*)
• relative pronouns (*who, which, that*)

Some writers mistakenly try to compress their writing by eliminating these function words.

AMBIGUOUS	Proposal to employ retirees almost dead.
REVISED	The proposal to employ retirees **is** almost dead.

AMBIGUOUS	Uninsulated end pipe ruptured. *(What ruptured? The pipe or the end of the pipe?)*
REVISED	**The** uninsulated end **of the** pipe is ruptured.

or

The uninsulated pipe **on the** end ruptured.

AMBIGUOUS	The reactor operator told management several times she expected an accident. *(Did she tell them once or several times?)*
REVISED	The reactor operator told management several times **that** she expected an accident.

or

The reactor operator told management **that** several times she expected an accident.

mod AVOID AMBIGUOUS MODIFIERS. Modifiers explain, define, or add detail to other words or ideas. If a modifier is too far from the words it modifies, the message can be ambiguous.

MISPLACED MODIFIER	**Only** press the red button in an emergency. *(Does **only** modify **press** or **emergency?**)*
REVISED	Press **only** the red button in an emergency.

or

Press the red button in an emergency **only.**

Position modifiers to reflect your meaning.

Another problem with ambiguity occurs when a modifying phrase has no word to modify.

DANGLING MODIFIER	**Being so well known in the computer industry,** I would appreciate your advice.

The writer meant to say that the *reader* is well known, but with no word to connect to, the modifying phrase dangles. Eliminate the confusion by adding a subject:

REVISED	Because **you** are so well known in the computer industry, I would appreciate your advice.

See Appendix C for more on modifiers.

EXERCISE 2

Edit each sentence below to repair telegraphic writing or to clarify ambiguous modifiers.

    a. Replace main booster rocket seal.
    b. The president refused to believe any internal report was inaccurate.
    c. Only use this phone in a red alert.
    d. After offending our best client, I am deeply annoyed with the new manager.
    e. Send memo to programmer requesting explanation.
    f. Do not enter test area while contaminated.

`st mod` **UNSTACK MODIFYING NOUNS.**   One noun can modify another (as in "software development"). But when two or more nouns modify a noun, the string of words becomes hard to read (as in "nuclear core containment unit technician safety protection procedures" versus "procedures for protecting the safety of technicians in the nuclear core containment unit"). Besides being confusing, stacked modifiers can also be ambiguous.

> **AMBIGUOUS**   Be sure to leave enough time for a **training session participant** evaluation. *(Evaluation of the session or of the participants?)*

Stacked nouns also deaden your style. Bring your style *and* your audience to life by using action verbs (*complete, prepare, reduce*) and prepositional phrases.

> **REVISED**   Be sure to leave enough time **for** participants **to evaluate** the training session.
>
>   *or*
>
> Be sure to leave enough time **to evaluate** participants in **the** training session.

No such problem with ambiguity occurs when *adjectives* are stacked in front of a noun.

> **CLEAR**   He was a **nervous, angry, confused,** but **dedicated** employee.

The adjectives clearly modify *employee*.

`wo` **ARRANGE WORD ORDER FOR COHERENCE AND EMPHASIS.**   In coherent writing, everything sticks together; each sentence builds on the preceding sentence and looks ahead to the following sentence. Sentences generally work best when the beginning looks back at familiar information and the end provides the new (or unfamiliar) information:

Effective word order

Familiar		Unfamiliar
My dog	has	fleas.
Our boss	just won	the lottery.
This company	is planning	a merger.

Besides helping a message stick together, the familiar-to-unfamiliar structure emphasizes the new information. Just as every paragraph has a key sentence, every sentence has a key word or phrase that sums up the new information. That key word or phrase usually is emphasized best when it appears at the end of the sentence.

> **FAULTY EMPHASIS**   We expect a **refund** because of your error in our shipment.
>
> **CORRECT**   Because of your error in our shipment, we expect a **refund.**

> **FAULTY EMPHASIS**   In a business relationship, **trust** is a vital element.
>
> **CORRECT**   A business relationship depends on **trust.**

One exception to placing key words last occurs with an imperative statement (a command, an order, an instruction), with the subject [*you*] understood. For instance, each step in a list of instructions should contain an action verb (*insert, open, close, turn, remove, press*). To give readers a forecast, place the verb in that instruction at the beginning.

> **CORRECT**   **Insert** the diskette before activating the system.
> **Remove** the protective seal.

With the key word at the beginning of the instruction, users know immediately what action they need to take.

EXERCISE 3

Edit each sentence below to unstack modifying nouns or to rearrange the word order for clarity and emphasis.

a. Develop online editing system documentation.
b. We need to develop a unified construction automation design.
c. Install a hazardous materials dispersion monitor system.
d. I recommend these management performance improvement incentives.
e. Our profits have doubled since we automated our assembly line.
f. Education enables us to recognize excellence and to achieve it.
g. In all writing, revision is required.
h. We have a critical need for technical support.
i. Sarah's job involves fault analysis systems troubleshooting handbook preparation.

USE ACTIVE VOICE OFTEN.   A verb's *voice* signals whether a sentence's subject acts or is acted on. The active voice ("I did it") is more direct, concise, and persuasive than the passive voice ("It was done by me"). In active voice sentences, a clear agent performs a clear action on a recipient:

13.1

Consider the ethics of active and passive voice at <www.ablongman.com/lannonweb>

ACTIVE VOICE	*Agent*	*Action*	*Recipient*
	Joe	lost	your report.
	Subject	*Verb*	*Object*

Passive voice reverses this pattern, placing the recipient of the action in the subject slot.

PASSIVE VOICE	*Recipient*	*Action*	*Agent*
	Your report	was lost	by Joe.
	Subject	*Verb*	*Prepositional phrase*

Sometimes the passive eliminates the agent altogether:

> **PASSIVE VOICE**   Your report was lost. *(Who lost it?)*

Passive voice is unethical if it obscures the person or other agent who performed the action when that responsible person or agent should be identified.

Some writers mistakenly rely on the passive voice because they think it sounds more objective and important. But the passive voice often makes writing wordy, indecisive, and evasive.

CONCISE AND DIRECT (ACTIVE)	**I underestimated** labor costs for this project. *(7 words)*
WORDY AND INDIRECT (PASSIVE)	Labor costs for this project **were underestimated by me.** *(9 words)*
EVASIVE (PASSIVE)	Labor costs for this project were underestimated.

Do not evade responsibility by hiding behind the passive voice:

EVASIVE (PASSIVE)	A **mistake was made** in your shipment. *(By whom?)*
	It was decided not to hire you. *(Who decided?)*
	A **layoff is recommended.**

In reporting errors or bad news, use the active voice, for clarity and sincerity. The passive voice creates a weak and impersonal tone.

WEAK AND IMPERSONAL	An offer **will be made** by us next week.

STRONG AND PERSONAL	**We will make** an offer next week.

Use the active voice when you want action. Otherwise, your statement will have no power.

WEAK PASSIVE	If my claim is not settled by May 15, the Better Business Bureau **will be contacted,** and their advice on legal action **will be taken.**
STRONG ACTIVE	If you do not settle my claim by May 15, **I will contact** the Better Business Bureau for advice on legal action.

Notice how this active version emphasizes the new and significant information by placing it at the end.

Ordinarily, use the active voice for giving instructions.

PASSIVE	The bid **should be sealed.** Care **should be taken** with the dynamite.
ACTIVE	**Seal** the bid. **Be careful** with the dynamite.

Avoid shifts from active to passive voice in the same sentence.

FAULTY SHIFT	During the meeting, project members **spoke** and **presentations were given.**
CORRECT	During the meeting, project members **spoke** and **gave** presentations.

Unless you have a deliberate reason for choosing the passive voice, prefer the *active* voice for making forceful connections like the one described here:

Why the active voice is preferable

By using the active voice, you direct the reader's attention to the subject of your sentence. For instance, if you write a job application letter that is littered with passive verbs, you fail to achieve an important goal of that letter: to show the readers the important things you have done, and how prepared you are to do important things for them. That strategy requires active verbs, with clear emphasis on *you* and what you have done/are doing. (Pugliano 6)

EXERCISE 4

Convert these passive voice sentences to concise, forceful, and direct expressions in the active voice.

- a. The evaluation was performed by us.
- b. Unless you pay me within three days, my lawyer will be contacted.
- c. Hard hats should be worn at all times.

    d. It was decided to reject your offer.

    e. It is believed by us that this contract is faulty.

    f. Our test results will be sent to you as soon as verification is completed.

---

pv

USE PASSIVE VOICE SELECTIVELY.   Passive voice is appropriate in lab reports and other documents in which the agent's identity is immaterial to the message.

    Use the passive when your audience has no need to know the agent.

CORRECT PASSIVE	Mr. Jones **was brought** to the emergency room.
	The bank failure **was publicized** statewide.

Use the passive voice to focus on events or results when the agent is unknown, unapparent, or unimportant.

CORRECT PASSIVE	All memos in the firm **are filed** in a database.
	Fred's article **was published** last week.
	All policy claims **are kept** confidential.

Prefer the passive when you want to be indirect or inoffensive (as in requesting the customer's payment or the employee's cooperation, or to avoid blaming someone—such as your supervisor) (Ornatowski 94).

ACTIVE BUT	**You have not paid** your bill.
OFFENSIVE	**You need to overhaul** our filing system.
INOFFENSIVE	This bill **has not been paid.**
PASSIVE	Our filing system **needs to be overhauled.**

Use the passive voice if the person behind the action needs to be protected.

CORRECT PASSIVE	The criminal **was identified.**
	The embezzlement scheme **was exposed.**

EXERCISE 5

The sentences below lack proper emphasis because of improper use of the active voice. Convert each to passive voice.

    a. Joe's company fired him.

    b. Someone on the maintenance crew has just discovered a crack in the nuclear core containment unit.

    c. A power surge destroyed more than two thousand lines of our new applications program.

    d. You are paying inadequate attention to worker safety.

    e. You are checking temperatures too infrequently.

    f. You did a poor job editing this report.

OS

AVOID OVERSTUFFED SENTENCES.   A sentence crammed with ideas makes details hard to remember and relationships hard to identify.

> **OVERSTUFFED**   Publicizing the records of a private meeting that took place three weeks ago to reveal the identity of a manager who criticized our company's promotion policy would be unethical.

Clear things up by sorting out the relationships.

> **REVISED**   In a private meeting three weeks ago, a manager criticized our company's policy on promotion. It would be unethical to reveal the manager's identity by publicizing the records of that meeting. *(Other versions are possible, depending on the writer's intended meaning.)*

Give your readers no more information in one sentence than they can retain and process.

EXERCISE 6

Unscramble this overstuffed sentence by making shorter, clearer sentences.

> A smoke-filled room causes not only teary eyes and runny noses but also can alter people's hearing and vision, as well as creating dangerous levels of carbon monoxide, especially for people with heart and lung ailments, whose health is particularly threatened by secondhand smoke.

EDITING FOR CONCISENESS

Two kinds of wordiness

Writing can suffer from two kinds of wordiness: one kind occurs when readers receive information they don't need (think of an overly detailed weather report during local television news). The other kind of wordiness occurs when too many words are used to convey information readers *do* need (as in saying "a great deal of potential for the future" instead of "great potential").

Every word in the document should advance your meaning:

> Writing improves in direct ratio to the number of things we can keep out of it that shouldn't be there. (Zinsser 14)

Concise writing conveys the most information in the fewest words. But it does not omit details necessary for clarity.

Use fewer words whenever fewer will do. But remember the difference between *clear writing* and *compressed writing* that is impossible to decipher.

> **COMPRESSED**   Give new vehicle air conditioner compression cut-off system specifications to engineering manager advising immediate action.
>
> **CLEAR**   The cut-off system for the air conditioner compressor on our new vehicles is faulty. Give the system specifications to our engineering manager so they will be modified.

First drafts are rarely concise. Trim the fat.

AVOID WORDY PHRASES.   Each phrase here can be reduced to one word.

Avoiding wordy phrases

print out	= print
at a rapid rate	= rapidly
due to the fact that	= because
the majority of	= most
readily apparent	= obvious
a large number	= many
prior to	= before
aware of the fact that	= know
in close proximity	= near
boot up	= boot

ELIMINATE REDUNDANCY.   A redundant expression says the same thing twice, in different words, as in *fellow colleagues.*

completely eliminate	**end** result
basic essentials	cancel **out**
enter **into**	consensus **of opinion**
mental awareness	**utter** devastation
mutual cooperation	**the month of** August

AVOID NEEDLESS REPETITION.   Unnecessary repetition clutters writing and dilutes meaning.

> **REPETITIOUS**   In trauma victims, breathing is restored by **artificial respiration**. Techniques of **artificial respiration** include mouth-to-mouth **respiration** and mouth-to-nose **respiration.**

Repetition in the above passage disappears when sentences are combined.

> **CONCISE**   In trauma victims, breathing is restored by artificial respiration, either mouth-to-mouth or mouth-to-nose.

NOTE   *Don't hesitate to repeat, or at least rephrase, material (even whole paragraphs in a longer document) if you feel that users need reminders. Effective repetition helps avoid cross-references like these: "See page 23" or "Review page 10."*

EXERCISE 7

Make these sentences more concise by eliminating needless phrases, redundancy, and needless repetition.

- a. I have admiration for Professor Jones.
- b. Due to the fact that we made the lowest bid, we won the contract.
- c. On previous occasions we have worked together.
- d. She is a person who works hard.
- e. We have completely eliminated the bugs from this program.
- f. This report is the most informative report on the project.
- g. Through mutual cooperation, we can achieve our goals.
- h. I am aware of the fact that Sam is trustworthy.
- i. This offer is the most attractive offer I've received.

---

th   **AVOID *THERE* SENTENCE OPENERS.**   Many *There is* or *There are* sentence openers can be eliminated.

WEAK	**There is** a coaxial cable connecting the antenna to the receiver.
REVISED	A coaxial cable connects the antenna to the receiver.
WEAK	**There is** a danger of explosion in Number 2 mineshaft.
REVISED	Number 2 mineshaft is in danger of exploding.

Dropping these openers places the key words at the end of the sentence, where they are best emphasized.

NOTE   *Of course, in some contexts, proper emphasis would call for a* There *opener.*

CORRECT	People have often wondered about the rationale behind Boris's sudden decision. There are several good reasons for his dropping out of the program.

It   **AVOID SOME *IT* SENTENCE OPENERS.**   Avoid beginning a sentence with *It*—unless the *It* clearly points to a specific referent in the preceding sentence: "This document is excellent. It deserves special recognition."

WEAK	**It** was his bad attitude that got him fired.
REVISED	His bad attitude got him fired.

WEAK	**It** is necessary to complete both sides of the form.
REVISED	Please complete both sides of the form.

pref

DELETE NEEDLESS PREFACES. Instead of delaying the new information in your sentence, get right to the point.

WORDY	**I am writing this letter because** I wish to apply for the position of copy editor.
CONCISE	Please consider me for the position of copy editor.
WORDY	**As far as artificial intelligence is concerned,** the technology is only in its infancy.
CONCISE	Artificial intelligence technology is only in its infancy.

EXERCISE 8

Make these sentences more concise by eliminating *There* and *It* openers and needless prefaces.

   a. There was severe fire damage to the reactor.
   b. There are several reasons why Jane left the company.
   c. It is essential that we act immediately.
   d. It has been reported by Bill that several safety violations have occurred.
   e. This letter is to inform you that I am pleased to accept your job offer.
   f. The purpose of this report is to update our research findings.

wv

AVOID WEAK VERBS. Prefer verbs that express a definite action: *open, close, move, continue, begin.* Avoid weak verbs that express no specific action: *is, was, are, has, give, make, come, take.*

NOTE

In some cases, such verbs are essential to your meaning: "Dr. Phillips is operating at 7 a.m." "Take me to the laboratory."

All forms of *to be* (*am, are, is, was, were, will, have been, might have been*) are weak. Substitute a strong verb for conciseness:

WEAK	My recommendation **is** for a larger budget.
STRONG	I **recommend** a larger budget.

Don't disappear behind weak verbs and their baggage of needless nouns and prepositions.

WEAK AND WORDY	Please **take into consideration** my offer.
CONCISE	Please **consider** my offer.

Strong verbs, or action verbs, suggest an assertive, positive, and confident writer. Here are some weak verbs converted to strong:

Revising weak verbs

has the ability to	= can
give a summary of	= summarize
make an assumption	= assume
come to the conclusion	= conclude
take action	= act
make a decision	= decide
make a proposal	= propose

EXERCISE 9

Edit each of these wordy and vague sentences to eliminate weak verbs.

a. Our disposal procedure is in conformity with federal standards.
b. Please make a decision today.
c. We need to have a discussion about the problem.
d. I have just come to the realization that I was mistaken.
e. Your conclusion is in agreement with mine.
f. This manual gives instructions to end users.

---

to be **DELETE NEEDLESS *TO BE* CONSTRUCTIONS.**   Sometimes the *to be* form itself mistakenly appears behind such verbs as *appears, seems,* and *finds.*

> **WORDY**   Your product seems **to be** superior.
> I consider this employee **to be** highly competent.

prep   **AVOID EXCESSIVE PREPOSITIONS.**   Needless prepositions create wordiness.

> **WORDY**   The recommendation first appeared **in** the report written **by** the supervisor in January **about** that month's productivity.
>
> **CONCISE**   The recommendation first appeared in the supervisor's productivity report for January.

Each prepositional phrase here can be reduced.

Avoiding needless prepositions

with the exception of	= except for
in reference to	= about (*or* regarding)
in order that	= so
in the near future	= soon
in the event of	= if
at the present time	= now
in the course of	= during
in the process of	= during (*or* in)

FIGHT NOUN ADDICTION.   Nouns manufactured from verbs (nominalizations) often accompany weak verbs and needless prepositions.

WEAK AND WORDY	We ask for the **cooperation** of all employees.
STRONG AND CONCISE	We ask that all employees **cooperate**.

WEAK AND WORDY	Give **consideration** to the possibility of a career change.
STRONG AND CONCISE	**Consider** a career change.

Besides causing wordiness, nominalizations can be vague—by hiding the agent of an action. Verbs are generally easier to read because they signal action.

WORDY AND VAGUE	A **valid requirement** for immediate action exists. *(Who should take the action? We can't tell.)*
PRECISE	We **must act** immediately.

Here are nominalizations restored to their action verb forms:

Trading nouns for verbs

conduct an investigation of	=	investigate
provide a description of	=	describe
conduct a test of	=	test
make a discovery of	=	discover

Nominalizations drain the life from your style. In cheering for your favorite team, you wouldn't say "Blocking of that kick is a necessity!" instead of "Block that kick!"

NOTE   *Also avoid excessive economy. For example, "Employees must cooperate" would not be an acceptable alternative to the first example in this section. But, for the final example, "Block that kick" would be.*

EXERCISE 10

Make these sentences more concise by eliminating needless prepositions, *to be* constructions, and nominalizations.

a. Igor seems to be ready for a vacation.
b. Our survey found 46 percent of users to be disappointed.
c. In the event of system failure, your sounding of the alarm is essential.
d. These are the recommendations of the chairperson of the committee.
e. Our acceptance of the offer is a necessity.
f. Please perform an analysis and make an evaluation of our new system.
g. A need for your caution exists.
h. Power surges are associated, in a causative way, with malfunctions of computers.

MAKE NEGATIVES POSITIVE. A positive expression is easier to understand than a negative one.

INDIRECT AND WORDY	Please do not be late in submitting your report.
DIRECT AND CONCISE	Please submit your report on time.

Readers work even harder to translate sentences with multiple negative expressions:

CONFUSING AND WORDY	Do **not** distribute this memo to employees who have **not** received a security clearance.
CLEAR AND CONCISE	Distribute this memo only to employees who have received a security clearance.

Besides directly negative words (*no, not, never*), some indirectly negative words (*except, forget, mistake, lose, uncooperative*) also force readers to translate.

CONFUSING AND WORDY	**Do not neglect** to activate the alarm system. My diagnosis was **not inaccurate**.
CLEAR AND CONCISE	**Be sure** to activate the alarm system. My diagnosis was **accurate**.

The positive versions are more straightforward *and* persuasive.

Some negative expressions, of course, are perfectly correct, as in expressing disagreement.

CORRECT NEGATIVES	This is **not** the best plan. Your offer is **unacceptable**. This project will **never** succeed.

Prefer positives to negatives, though, whenever your meaning allows:

Trading negatives for positives

did not succeed	=	failed
does not have	=	lacks
did not prevent	=	allowed
not unless	=	only if
not until	=	only when
not absent	=	present

CLEAN OUT CLUTTER WORDS. Clutter words stretch a message without adding meaning. Here are some of the most common: *very, definitely, quite, extremely, rather, somewhat, really, actually, currently, situation, aspect, factor.*

CLUTTERED	**Actually**, one **aspect** of a business **situation** that could definitely make me **quite** happy would be to have a **somewhat** adventurous partner who **really** shared my **extreme** attraction to risks.
CONCISE	I seek an adventurous business partner who enjoys risks.

DELETE NEEDLESS QUALIFIERS.   Qualifiers such as *I feel, it seems, I believe, in my opinion,* and *I think* express uncertainty or soften the tone and force of a statement.

APPROPRIATE QUALIFIER	Despite Frank's poor grades last year he will, **I think,** do well in college.
	Your product **seems** to meet our needs.

But when you are certain, eliminate the qualifier so as not to seem tentative or evasive.

NEEDLESS QUALIFIERS	**It seems that** I've made an error.
	We **appear to** have exceeded our budget.
	In my opinion, this candidate is outstanding.

NOTE

In communicating across cultures, keep in mind that a direct, forceful style might be considered offensive (page 282).

EXERCISE 11

Make these sentences more concise by changing negatives to positives and by clearing out clutter words and needless qualifiers.

- a. Our design must avoid nonconformity with building codes.
- b. Never fail to wear protective clothing.
- c. Do not accept any bids unless they arrive before May 1.
- d. I am not unappreciative of your help.
- e. We are currently in the situation of completing our investigation of all aspects of the accident.
- f. I appear to have misplaced the contract.
- g. Do not accept bids that are not signed.
- h. It seems as if I have just wrecked a company car.

EDITING FOR FLUENCY

Fluent sentences are easy to read because of clear connections, variety, and emphasis. Their varied length and word order eliminate choppiness and monotony. Fluent sentences enhance *clarity*, emphasizing the most important ideas. Fluent sentences enhance *conciseness,* often replacing several short, repetitious sentences with one longer, more economical sentence. To write fluently, use the following strategies.

comb **COMBINE RELATED IDEAS.** A series of short, disconnected sentences is not only choppy and wordy, but also unclear.

DISCONNECTED	Jogging can be healthful. You need the right equipment. Most necessary are well-fitting shoes. Without this equipment you take the chance of injuring your legs. Your knees are especially prone to injury. *(5 sentences)*
CLEAR, CONCISE, AND FLUENT	Jogging can be healthful if you have the right equipment. Shoes that fit well are most necessary because they prevent injury to your legs, especially your knees. *(2 sentences)*

Most sets of information can be combined to form different relationships, depending on what you want to emphasize. Imagine that this set of facts describes an applicant for a junior management position with your company.

- Roy James graduated from an excellent management school.
- He has no experience.
- He is highly recommended.

Assume that you are a personnel director, conveying your impression of this candidate to upper management. To convey a negative impression, you might combine the facts in this way:

STRONGLY NEGATIVE EMPHASIS	Although Roy James graduated from an excellent management school and is highly recommended, **he has no experience.**

The *independent* idea (in boldface) receives the emphasis. (See also page 757 on subordination.) But if you are undecided, yet leaning in a negative direction, you might write:

STRONGLY NEGATIVE EMPHASIS	Roy James graduated from an excellent management school and is highly recommended, **but** he has no experience.

In this sentence, the ideas before and after *but* are both independent. Joining them with the coordinating word *but* suggests that both sides of the issue are equally important (or "coordinate"). Placing the negative idea last, however, gives it slight emphasis. (See also page 756 on coordination.)

Finally, to emphasize strong support for the candidate, you could say:

> Although Roy James has no experience, **he graduated from an excellent management school and is highly recommended.**

Here, the earlier idea is subordinated by *although*, leaving the two final ideas independent.

Caution: Combine sentences only to simplify the reader's task. Overstuffed sentences with too much information and too many connections can be hard for readers to sort out.

OVERSTUFFED	Our night supervisor's verbal order from upper management to repair the overheated circuit was misunderstood by Leslie Kidd, who gave the wrong instructions to the emergency crew, thereby causing the fire within 30 minutes.
CLEARER	Upper management issued a verbal order to repair the overheated circuit. When our night supervisor transmitted the order to Leslie Kidd, it was misunderstood. Kidd gave the wrong instruction to the emergency crew, and the fire began within 30 minutes.

Research indicates that most people can retain between fifteen and twenty-five words at one time (Boyd 18). But even short sentences can be hard to interpret if they include too many details.

OVERSTUFFED	Send three copies of Form 17-e to all six departments, unless Departments A or B or both request Form 16-w instead.

`var` **VARY SENTENCE CONSTRUCTION AND LENGTH.** Related ideas often need to be linked in one sentence, so that readers can grasp the connections:

DISCONNECTED	The nuclear core reached critical temperature. The loss-of-coolant alarm was triggered. The operator shut down the reactor.
CONNECTED	As the nuclear core reached critical temperature, triggering the loss-of-coolant alarm, the operator shut down the reactor.

But an idea that should stand alone for emphasis needs a whole sentence of its own:

CORRECT	Core meltdown seemed inevitable.

However, an unbroken string of long or short sentences can bore and confuse readers, as can a series with identical openings:

BORING AND REPETITIVE	There are some drawbacks about diesel engines. **They** are difficult to start in cold weather. **They** cause vibration. **They** also give off an unpleasant odor. **They** cause sulfur dioxide pollution.
VARIED	Diesel engines have some drawbacks. Most obvious are their noisiness, cold-weather starting difficulties, vibration, odor, and sulfur dioxide emission.

Similarly, when you write in the first person, overusing *I* makes you appear self-centered. (Some organizations require use of the third person, avoiding the first person completely, for all manuals, lab reports, specifications, product descriptions, and so on.)

Do not, however, avoid personal pronouns if they make the writing more readable (say, by eliminating passive constructions).

short   **USE SHORT SENTENCES FOR SPECIAL EMPHASIS.**   All this talk about combining ideas might suggest that short sentences have no place in good writing. Wrong. Short sentences (even one-word sentences) provide vivid emphasis. They stick in a reader's mind.

EXERCISE 12

Combine each set of sentences below into one fluent sentence that provides the requested emphasis.

SENTENCE SET	John is a loyal employee. John is a motivated employee. John is short-tempered with his colleagues.
COMBINED FOR POSITIVE EMPHASIS	Even though John is short-tempered with his colleagues, he is a loyal and motivated employee.
SENTENCE SET	This word processor has many features. It includes a spelling checker. It includes a thesaurus. It includes a grammar checker.
COMBINED TO EMPHASIZE THESAURUS	Among its many features, such as spelling and grammar checkers, this word processor includes a thesaurus.

a. The job offers an attractive salary.
   It demands long work hours.
   Promotions are rapid.
   (*Combine for negative emphasis.*)

  b. The job offers an attractive salary.
   It demands long work hours.
   Promotions are rapid.
   (*Combine for positive emphasis.*)
  c. Our office software is integrated.
   It has an excellent database management program.
   Most impressive is its word-processing capability.
   It has an excellent spreadsheet program.
   (*Combine to emphasize the word processor.*)
  d. Company X gave us the lowest bid.
   Company Y has an excellent reputation.
   (*Combine to emphasize Company Y.*)
  e. Superinsulated homes are energy efficient.
   Superinsulated homes create a danger of indoor air pollution.
   The toxic substances include radon gas and urea formaldehyde.
   (*Combine for a negative emphasis.*)
  f. Computers cannot *think* for the writer.
   Computers eliminate many mechanical writing tasks.
   They speed the flow of information.
   (*Combine to emphasize the first assertion.*)

13.2

See a list of online word finder resources at <www.ablongman.com/lannonweb>

Situations in which people often hide behind language

FINDING THE EXACT WORDS

Too often, language can *camouflage* rather than communicate. People see many reasons to hide behind language, as when they

- speak for their company but not for themselves
- fear the consequences of giving bad news
- are afraid to disagree with company policy
- make a recommendation some readers will resent
- worry about making a bad impression
- worry about being wrong
- pretend to know more than they do
- avoid admitting a mistake, or ignorance

Inflated and unfamiliar words, borrowed expressions, and needlessly technical terms camouflage meaning. Poor word choices produce inefficient and often unethical writing that resists interpretation and frustrates the audience.

  Following are strategies for finding words that are *convincing, precise,* and *informative.*

USE SIMPLE AND FAMILIAR WORDING. Don't replace technically precise words with nontechnical words that are vague or imprecise. Don't write *a part that makes*

the computer run when you mean *central processing unit*. Use the precise term, and define it in a glossary for nontechnical readers:

GLOSSARY ENTRY   **Central processing unit:** the part of the computer that controls information transfer and carries out arithmetic and logical instructions.

Certain technical words may be indispensable in certain contexts, but the nontechnical words usually can be simplified. Instead of *answering in the affirmative,* use *say yes;* instead of *endeavoring to promulgate* a new policy, *try to announce* it.

UNFAMILIAR WORDS   Acoustically attenuate the food consumption area.
REVISED   Soundproof the cafeteria.

Don't use three syllables when one will do. Generally, trade for less:

Trading multiple
syllables for
fewer

demonstrate	=	show
endeavor	=	effort, try
frequently	=	often
initiate	=	begin
is contingent upon	=	depends on
multiplicity of	=	many
subsequent to	=	after
utilize	=	use

Whenever possible, choose words you use and hear in everyday speaking.

Don't write *Keep me apprised* instead of *Keep me informed, I concur* instead of *I agree, securing employment* instead of *finding a job,* or *it is cost prohibitive* instead of *we can't afford it.*

Besides being annoying, needlessly big or unfamiliar words can be *ambiguous.*

AMBIGUOUS   Make an improvement in the clerical situation.

Should we hire more clerical personnel or better personnel or should we train the personnel we have? Words chosen to impress readers too often confuse them instead. A plain style is more persuasive because "it leaves no one out" (Cross 6).

PLAIN ENGLISH   Avoid prolix nebulosity.
NEEDED
REVISED   Don't be wordy and vague.

Of course, now and then the complex or more elaborate word best expresses your meaning. For instance, we would not substitute *end* for *terminate* in referring to something with an established time limit.

| **CORRECT** | Our trade agreement terminates this month. |

If a complex word can replace a handful of simpler words—and can sharpen your meaning—use it.

WEAK	Six rectangular grooves **around the outside edge** of the steel plate **are needed for** the pressure clamps **to fit into.**
INFORMATIVE AND PRECISE	Six rectangular grooves on the steel plate **perimeter accommodate** the pressure clamps.
WEAK	We need a **one-to-one exchange of ideas and opinions.**
INFORMATIVE AND PRECISE	We need a **dialogue.**

EXERCISE 13

Edit these sentences for straightforward and familiar language.

a. May you find luck and success in all endeavors.
b. I suggest you reduce the number of cigarettes you consume.
c. Within the copier, a magnetic reed switch is utilized as a mode of replacement for the conventional microswitches that were in use on previous models.
d. A good writer is cognizant of how to utilize grammar in a correct fashion.
e. I will endeavor to ascertain the best candidate.
f. In view of the fact that the microscope is defective, we expect a refund of our full purchase expenditure.
g. I wish to upgrade my present employment situation.

---

jarg

When jargon is appropriate

AVOID USELESS JARGON.   Every profession has its own shorthand and accepted phrases and terms. For example, *stat* (from the Latin "statim" or "immediately") is medical jargon for *Drop everything and deal with this emergency.* For computer buffs, a *glitch* is a momentary power surge that can erase the contents of internal memory; a *bug* is an error that causes a program to run incorrectly. Such useful jargon conveys clear meaning to a knowledgeable audience.

When jargon is inappropriate

Jargon can be useful when you are communicating with specialists. But some jargon is useless in any context. In the world of useless jargon, people don't *cooperate* on a project; instead, they *interface* or *contiguously optimize their efforts.* Rather than *designing a model,* they *formulate a paradigm.* Instead of *observing limits* or *boundaries,* they *function within specific parameters.*

A popular form of useless jargon is adding -*wise* to nouns, as shorthand for *in reference to* or *in terms of.*

| **USELESS JARGON** | **Expensewise** and **schedulewise,** this plan is unacceptable. |
| **REVISED** | In terms of expense and scheduling, this plan is unacceptable. |

Writers create another form of useless jargon when they invent verbs from nouns or adjectives by adding an *-ize* ending: Don't invent *prioritize* from *priority*; instead use *to rank priorities*.

Useless jargon's worst fault is that it makes the person using it seem stuffy and pretentious:

PRETENTIOUS   Unless all parties interface synchronously within given parameters, the project will be rendered inoperative.

POSSIBLE TRANSLATION   Unless we coordinate our efforts, the project will fail.

Beyond reacting with frustration, readers often conclude that useless jargon is camouflage for a writer with something to hide.

Before using any jargon, think about your specific audience and ask yourself: "Can I find an easier way to say exactly what I mean?" Use jargon only if it *improves* your communication.

NOTE *If your employer insists on needless jargon or elaborate phrasing, then you have little choice. What is best in matters of style is not always what some people consider appropriate. Use the style your employer or organization expects, but remember that most documents that achieve superior results use plain English.*

acr USE ACRONYMS SELECTIVELY.   Acronyms are another form of specialized shorthand, or jargon. They are formed from the first letters of words in a phrase (as in *LOCA* from *l*oss *o*f *c*oolant *a*ccident) or from a combination of first letters and parts of words (as in *bit* from *b*inary dig*it* or *pixel* from *pic*ture *el*ement).

Computer technology has spawned countless acronyms, including:

ISDN = Integrated Services Digital Network
Telnet = Telephone Network
URL   = Uniform Resource Locator

Acronyms *can* communicate concisely—but only when the audience knows their meaning, and only when you use the term often in your document. Whenever you first use an acronym, spell out the words from which it is derived.

An acronym defined

Modem ("modulator + demodulator"): a device that converts, or "modulates," computer data in electronic form into a sound signal that can be transmitted via phone line and then reconverted, or "demodulated," into electronic form for the receiving computer.

NOTE *To identify virtually any acronym, consult the acronym-finder site at Norway's University of Oslo <www.habrok.uio.no/cgi-bin/acronyms>.*

AVOID TRITENESS. Writers who rely on worn-out phrases (clichés) such as the following come across as either too lazy or too careless to find exact, unique ways of saying what they mean.

Worn out phrases

make the grade	the chips are down
in the final analysis	not by a long shot
close the deal	last but not least
hard as a rock	welcome aboard
water under the bridge	over the hill
holding the bag	bite the bullet
up the creek	work like a dog

EXERCISE 14

Edit these sentences to eliminate useless jargon and triteness.

    a. For the obtaining of the X–33 word processor, our firm will have to accomplish the disbursement of funds to the amount of $6,000.

    b. To optimize your financial return, prioritize your investment goals.

    c. The use of this product engenders a 50-percent repeat consumer encounter.

    d. We'll have to swallow our pride and admit our mistake.

    e. We wish to welcome all new managers aboard.

    f. Managers who make the grade are those who can take daily pressures in stride.

---

euph **AVOID MISLEADING EUPHEMISMS.** A form of understatement, euphemisms are expressions aimed at politeness or at making unpleasant subjects seem less offensive. Thus, *we powder our nose* or *use the boys' room* instead of *using the bathroom; we pass away* or *meet our Maker* instead of *dying.*

When a euphemism is appropriate

When euphemisms avoid offending or embarrassing people, they are perfectly legitimate. Instead of telling a job applicant he or she is *unqualified,* we might say, *Your background doesn't meet our needs.* In addition, there are times when friendliness and interoffice harmony are more likely to be preserved with writing that is not too abrupt, bold, blunt, or emphatic (MacKenzie 2).

Euphemisms, however, are unethical if they understate the truth when only the truth will serve. In the sugarcoated world of misleading euphemisms, bad news disappears:

When a euphemism is deceptive

- Instead of being *laid off or fired,* workers are *surplused* or *deselected,* or the company is *downsized.*
- Instead of *lying* to the public, the government *engages in a policy of disinformation.*
- Instead of *wars* and *civilian casualties,* we have *conflicts* and *collateral damage.*

Language loses all meaning when *criminals* become *offenders,* when *mistakes* become *teachable moments,* and when people who are just plain *lazy* become *underachievers.* Plain talk is always better than deception. If someone offers you a job *with limited opportunity for promotion,* expect a *dead-end job.*

AVOID OVERSTATEMENT.   Exaggeration sounds phony. Be cautious when using words such as *best, biggest, brightest, most,* and *worst.*

OVERSTATED	**Most** businesses have **no** loyalty toward their employees.
REVISED	**Some** businesses have **little** loyalty toward their employees.
OVERSTATED	You will find our product to be the **best.**
REVISED	You will **appreciate the high quality** of our product.

Be aware of the vast differences in meaning among these words:

Qualify your generalizations

few	rarely
some	sometimes
many	often
most	usually
all	always

Unless you specify *few, some, many,* or *most,* people can interpret your statement to mean *all.*

MISLEADING	Assembly-line employees are doing shabby work.

Unless you mean *all,* qualify your generalization with *some,* or *most*—or even better, specify *20 percent.*

EXERCISE 15
Edit these sentences to eliminate euphemism, overstatement, or unsupported generalizations.

a. I finally must admit that I am an abuser of intoxicating beverages.
b. I was less than candid.
c. This employee is poorly motivated.
d. Most entry-level jobs are boring and dehumanizing.
e. Clerical jobs offer no opportunity for advancement.
f. Because of your absence of candor, we can no longer offer you employment.

AVOID IMPRECISE WORDING.   Even words listed as synonyms can carry different shades of meaning. Do you mean to say *I'm slender, You're slim, She's lean,* or *He's scrawny?* The wrong choice could be disastrous.

**TABLE 13.1
Commonly
Confused
Words**

Find online lists of of-
ten misused words at
<www.ablongman.
com/lannonweb>

Similar Words	Used Correctly in a Sentence
Affect means "to have an influence on."	Meditation positively *affects* concentration.
Affect can also mean "to pretend."	Boris likes to *affect* a French accent.
Effect used as a noun means "a result."	Meditation has a positive *effect* on concentration.
Effect used as a verb means "to make happen" or "to bring about."	Meditation can *effect* an improvement in concentration.
Already means "before this time."	Our new laptops are *already* sold out.
All ready means "prepared."	We are *all ready* for the summer tourist season.
Among refers to three or more.	The prize was divided *among* the four winners.
Between refers to two.	The prize was divided *between* the two winners.
Continual means "repeated at intervals."	Our lower field floods *continually* during the rainy season.
Continuous means "without interruption."	His headache has been *continuous* for three days.
Differ from refers to unlike things.	This plan *differs* greatly *from* our earlier one.
Differ with means "to disagree."	Mary *differs with* John about the feasibility of this project.
Disinterested means "unbiased" or "impartial."	Good science calls for *disinterested* analysis of research findings.
Uninterested means "not caring."	Junior high school students are often *uninterested* in science.
Eminent means "famous" or "distinguished."	Dr. Ostroff, the *eminent* physicist, is lecturing today.
Imminent means "about to happen."	A nuclear meltdown seemed *imminent*.
Farther refers to physical distance (a measurable quantity).	The station is 20 miles *farther*.
Further refers to extent (not measurable).	*Further* discussion of this issue is vital.
Fewer refers to things that can be counted.	*Fewer* than fifty students responded to our survey.
Less refers to things that can't be counted.	This survey had *less* of a response than our earlier one.
Imply means "to hint at" or "to insinuate."	This report *implies* that a crime occurred.
Infer means "to reason from evidence."	From this report, we can *infer* that a crime occurred.
It's stands for "it is."	*It's* a good time for a department meeting.
Its stands for "belonging to."	The cost of the project has exceeded *its* budget.

Similar Words	Used Correctly in a Sentence
Lay means "to place or set something down." It always takes a direct object.	Please *lay* the blueprints on the desk.
Lie means "to recline." It takes no direct object.	This patient needs to *lie* on his right side all night.
(Note that the past tense of *lie* is *lay*.)	The patient *lay* on his right side all last night.
Phenomena is the plural form of *phenomenon*.	Many scientific *phenomena* remain unexplained.
Phenomenon is the singular form.	Tiger Woods continues to be a golfing *phenomenon*.
Precede means "to come before."	Audience analysis should *precede* a written report.
Proceed means "to go forward."	If you must wake the cobra, *proceed* carefully.
Principle is always a noun that means "basic rule or standard."	Ethical *principles* should govern all our communications.
Principal, used as a noun, means "the major person(s)."	All *principals* in this purchase must sign the contract.
Principal, used as an adjective, means "leading."	Martha was the *principal* negotiator for this contract.

Be on the lookout for imprecisely phrased (and therefore illogical) comparisons.

> **IMPRECISE**   Your bank's interest rate is higher than BusyBank. (Can a rate be higher than a bank?)
>
> **PRECISE**   Your bank's interest rate is higher than BusyBank's.

Imprecision can create ambiguity. For instance, is *send us more personal information* a request for more information that is personal or for information that is more personal? Does your client expect *fewer* or *less* technical details in your report?

Precision ultimately enhances conciseness, when one exact word replaces multiple inexact words.

> **WORDY AND LESS EXACT**   I have **put together** all the financial information.
>   **Keep doing** this exercise for ten seconds.
>
> **CONCISE AND MORE EXACT**   I have **assembled** all the. . . .
>   **Continue** this exercise. . . .

spec   **BE SPECIFIC AND CONCRETE.**   General words name broad classes of things, such as *job, computer,* or *person.* Such terms usually need to be clarified by more specific ones.

General terms
traded for
specific terms

job   = senior accountant for Softbyte Press
computer = Macintosh PowerBook G3
person   = Sarah Jones, production manager

The more specific your words, the sharper your meaning.

How the level of
generality affects
writing's visual
quality

General ↑ structure

dwelling

vacation home

log cabin

Specific │ log cabin in Vermont

↓ a three-room log cabin on the banks of the Battenkill River

Notice how the picture becomes more vivid as we move to lower levels of generality. To visualize your way of seeing, and your exact meaning, readers need specifics.

Abstract words name qualities, concepts, or feelings (*beauty, luxury, depression*) whose exact meaning has to be nailed down by *concrete* words—words that name things we can visualize.

Abstract terms
traded for
concrete terms

a **beautiful** view   = snowcapped mountains, a wilderness lake, pink granite ledge, ninety-foot birch trees

a **luxurious** condominium = imported tiles, glass walls, oriental rugs

a **depressed** worker   = suicidal urge, insomnia, feelings of worthlessness, no hope for improvement

Informative writing *tells* and *shows*.

> GENERAL   One of our **workers** was **injured** by a **piece of equipment recently**.

The boldface words only *tell* without showing.

> SPECIFIC   **Alan Hill** suffered a **broken thumb** while working on a **lathe yesterday**.

Choose informative words that express exactly what you mean. Don't write *thing* when you mean *lever, switch, micrometer,* or *disk*.

NOTE

In some instances, of course, you may wish to generalize for the sake of diplomacy. Instead of writing "Bill, Mary, and Sam have been tying up the office phones with personal calls," you might prefer to generalize: "Some employees . . . have been. . . ." The second version gets your message across without pointing the finger.

When you can, provide solid numbers and statistics that get your point across:

<table>
<tr><td>GENERAL</td><td>In 1972, thousands of people were killed or injured on America's highways. Many families had at least one relative who was a casualty. After the speed limit was lowered to 55 miles per hour in late 1972, the death toll began to drop.</td></tr>
<tr><td>SPECIFIC</td><td>In 1972, 56,000 people died on America's highways; 200,000 were injured; 15,000 children were orphaned. In that year, if you were a member of a family of five, chances are that someone related to you by blood or law was killed or injured in an auto accident. After the speed limit was lowered to 55 miles per hour in late 1972, the death toll dropped steadily to 41,000 in 1975.</td></tr>
</table>

NOTE *Most good writing offers both general and specific information. The most general material appears in the topic statement and sometimes in the conclusion because these parts, respectively, set the paragraph's direction and summarize its content.*

EXERCISE 16

Edit these sentences to make them more precise and informative.

a. Our outlet does more business than Chicago.
b. Anaerobic fermentation is used in this report.
c. Loan payments are due bimonthly.
d. Your crew damaged a piece of office equipment.
e. His performance was admirable.
f. This thing bothers me.

Analogy versus comparison

USE ANALOGIES TO SHARPEN THE IMAGE.   Ordinary comparison shows similarities between two things *of the same class* (two computer keyboards, two methods of cleaning dioxin-contaminated sites). Analogy, on the other hand, shows some essential similarity between two things of *different classes* (writing and computer programming, computer memory and post office boxes).

Analogies are good for emphasizing a point (*Some rain is now as acidic as vinegar*). They are especially useful in translating something abstract, complex, or unfamiliar, as long as the easier subject is broadly familiar to readers. Analogy therefore calls for particularly careful analyses of audience.

Analogies can save words and convey vivid images. *Collier's Encyclopedia* describes the tail of an eagle in flight as "spread like a fan." The following sentence from a description of a trout feeder mechanism uses an analogy to clarify the positional relationship between two working parts:

> **ANALOGY**   The metal rod is inserted (and centered, **crosslike**) between the inner and outer sections of the clip.

Without the analogy *crosslike*, we would need something like this to visualize the relationship:

> **MISSING ANALOGY**   The metal rod is inserted, **perpendicular to the long plane and parallel to the flat plane,** between the inner and outer sections of the clip.

Besides naming things, analogies help *explain* things. This next analogy helps clarify an unfamiliar concept (dangerous levels of a toxic chemical) by comparing it to something more familiar (human hair).

> **ANALOGY**   A dioxin concentration of 500 parts per trillion is lethal to guinea pigs. One part per trillion is roughly equal to the thickness of a human hair compared to the distance across the United States. *(Congressional Research Report 15)*

tone ADJUSTING YOUR TONE

How tone is created

Your tone is your personal stamp—the personality that takes shape between the lines. The tone you create depends on (1) the distance you impose between yourself and the reader, and (2) the attitude you show toward the subject.

Assume, for example, that a friend is going to take over a job you've held. You're writing your friend instructions for parts of the job. Here is your first sentence:

Informal

> Now that you've arrived in the glamorous world of office work, put on your track shoes; this is no ordinary manager-trainee job.

This sentence imposes little distance between you and the reader (it uses the direct address, *you,* and the humorous suggestion to *put on your track shoes*). The ironic use of *glamorous* suggests just the opposite: that the job holds little glamor.

For a different reader (say, the recipient of a company training manual), you would have chosen some other opening:

Semiformal

> As a manager trainee at GlobalTech, you will work for many managers. In short, you will spend little of your day seated at your desk.

The tone now is serious, no longer intimate, and you express no distinct attitude toward the job. For yet another audience (clients or investors who will read an annual report), you might alter the tone again:

Formal

> Manager trainees at GlobalTech are responsible for duties that extend far beyond desk work.

Here the businesslike shift from second- to third-person address makes the tone too impersonal for any document addressed to the trainees themselves.

We already know how tone works in speaking. When you meet someone new, for example, you respond in a tone that defines your relationship:

Tone announces interpersonal distance

> Honored to make your acquaintance. [*formal tone—greatest distance*]
> How do you do? [*formal*]
> Nice to meet you. [*semiformal—medium distance*]
> Hello. [*semiformal*]
> Hi. [*informal—least distance*]
> What's happening? [*informal—slang*]

Each of these greetings is appropriate in some situations, inappropriate in others. Whichever tone you decide on, be consistent throughout your document.

> **INCONSISTENT TONE**   My office isn't fit for a pig [*too informal*]: it is ungraciously unattractive [*too formal*].
>
> **REVISED**   The shabbiness of my office makes it an unfit place to work.

In general, lean toward an informal tone without using slang.

In addition to setting the distance between writer and reader, your tone implies your *attitude* toward the subject and the reader.

Tone announces attitude

> We dine at seven.
> Dinner is at seven.
> Let's eat at seven.
> Let's chow down at seven.
> Let's strap on the feedbag at seven.
> Let's pig out at seven.

The words you choose tell readers a great deal about where you stand.

One problem with tone occurs when your attitude is unclear. Say *I enjoyed the fiber optics seminar* instead of *My attitude toward the fiber optics seminar was one of high approval.* Say *Let's liven up our dull relationship* instead of *We should inject some rejuvenation into our lifeless liaison.*

In writing a memo about an upcoming meeting to review the reader's job evaluation, would you invite this person to *discuss* the evaluation, *talk it over, have a chat,* or *chew the fat*? Decide how casual or serious your attitude should be. Use the following strategies for making your tone conversational and appropriate.

> **GUIDELINES for Deciding about Tone**
>
> 1. Use a formal or semiformal tone in writing for superiors, professionals, or academics (depending on what you think the reader expects).
>
> 2. Use a semiformal or informal tone in writing for colleagues and subordinates (depending on how close you feel to your reader).
>
> 3. Use an informal tone when you want your writing to be conversational, or when you want it to sound like a person talking.
>
> 4. Above all, find out what tone your particular readers prefer.

USE AN OCCASIONAL CONTRACTION.   Unless you have reason to be formal, use (but do not overuse) contractions. Balance an *I am* with an *I'm*, a *you are* with a *you're*, and *it is* with an *it's*.

MISSING CONTRACTION	Do not be wordy and vague.
REVISED	Don't be wordy and vague.

Use contractions only with pronouns, not with nouns or proper nouns (names).

AWKWARD CONTRACTIONS	Barbara'll be here soon. Health's important.
AMBIGUOUS CONTRACTIONS	The dog's barking. Bill's skiing.

These ambiguous contractions could be confused with possessive constructions.

NOTE

*The contracted version often sounds less emphatic than the two-word version—for example, "**Don't** handle this material without protective clothing" versus "**Do not** handle this material without protective clothing." If your message requires emphasis, you should not use a contraction.*

ADDRESS READERS DIRECTLY.   Use the personal pronouns *you* and *your* to connect with readers.

IMPERSONAL TONE	Students at our college will find the faculty always willing to help.
PERSONAL TONE	As a student at our college, **you** will find the faculty always willing to help.

Readers relate better to something addressed to them directly.

NOTE

*Use **you** and **your** only to correspond directly with the reader, as in a letter, memo, instructions, or some form of advice, encouragement, or persuasion. By using **you** and **your** in a situation that calls for first or third person, you might write something like this:*

WORDY AND AWKWARD	When **you** are in northern Ontario, **you** can see wilderness lakes everywhere around **you**.
APPROPRIATE	Wilderness lakes are everywhere in northern Ontario.

EXERCISE 17

The sentences below suffer from pretentious language, unclear expression of attitude, missing contractions, or indirect address. Adjust the tone.

    a. Further interviews are a necessity to our ascertaining the most viable candidate.

    b. Do not submit the proposal if it is not complete.

    c. Employees must submit travel vouchers by May 1.

    d. Persons taking this test should use the HELP option whenever they need it.

    e. I am not unappreciative of your help.

    f. My disapproval is far more than negligible.

USE *I* AND *WE* WHEN APPROPRIATE.   Instead of disappearing behind your writing, use *I* or *We* when referring to yourself or your organization.

DISTANT	The writer would like a refund.
REVISED	**I** would like a refund.

A message becomes doubly impersonal when both writer and reader disappear.

IMPERSONAL	The requested report will be sent next week.
PERSONAL	**We** will send the report **you** requested next week.

PREFER THE ACTIVE VOICE.   Because the active voice is more direct and economical than the passive voice, it generally creates a less formal tone. (Review pages 250–52 for use of active and passive voice.)

PASSIVE AND IMPERSONAL	Travel expenses cannot be reimbursed unless receipts are submitted.
ACTIVE AND PERSONAL	We cannot reimburse your travel expenses unless you submit receipts.

EXERCISE 18

These sentences have too few *I* or *We* constructions or too many passive constructions. Adjust the tone.

    a. Payment will be made as soon as an itemized bill is received.

    b. You will be notified.

    c. Your help is appreciated.

Revising a document

"Deadlines affect how we approach the writing process. With plenty of time, we can afford the luxury of the whole process: careful decisions about audience, purpose, content, organization, and style—and plenty of revisions. At times, we have to take shortcuts."

—Blair Cordasco, Training Specialist
for an international bank

---

d. Our reply to your bid will be sent next week.

e. Your request will be given our consideration.

f. My opinion of this proposal is affirmative.

g. This writer would like to be considered for your opening.

EMPHASIZE THE POSITIVE.   Whenever you offer advice, suggestions, or recommendations, try to emphasize benefits rather than flaws.

CRITICAL TONE	Because of your division's lagging productivity, a management review may be needed.
ENCOURAGING TONE	A management review might help boost productivity in your division.

How tone can be too informal

AVOID AN OVERLY INFORMAL TONE.   We generally do not write in the same way we would speak to friends at the local burger joint or street corner. Achieving a conversational tone does not mean lapsing into substandard usage, slang, profanity, or excessive colloquialisms. *Substandard usage* ("He ain't got none," "I seen it today," "She brang the book") ignores standards of educated expression. *Slang* ("hurling," "belted," "bogus," "bummed") usually has specific meaning only for members of a particular in-group. *Profanity* ("This idea sucks," "pissed off," "What the hell") not only displays contempt for the audience but also triggers contempt for the person using it. *Colloquialisms* ("O.K.," "a lot," "snooze," "in the bag") are understood more widely than slang, but tend to appear more in speaking than in writing.

How tone can offend

Tone is offensive when it violates the reader's expectations: when it seems disrespectful, tasteless, distant and aloof, too "chummy," casual, or otherwise inappropriate for the topic, the reader, and the situation.

When to use an academic tone

A formal or academic tone is appropriate in countless writing situations: a research paper, a job application, a report for the company president. In a history essay, for example, you would not refer to George Washington and Abraham Lincoln as "those dudes, George and Abe." Whenever you begin with rough drafting or brainstorming, your tone might be overly informal and is likely to require some adjustment during subsequent drafts.

But while slang usually is inappropriate in school or workplace writing, some situations (say, certain email messages) call for a measure of informality. The occasional colloquial expression helps soften the tone of any writing.

bias   **AVOID PERSONAL BIAS.**   If people expect an impartial report, try to keep your own biases out of it. Imagine, for instance, that you have been sent to investigate the causes of an employee-management confrontation at your company's Omaha branch. Your initial report, written for the New York central office, is intended simply to describe what happened. Here is how an unbiased description might read:

A factual account

> At 9:00 a.m. on Tuesday, January 21, eighty women employees set up picket lines around the executive offices of our Omaha branch, bringing business to a halt. The group issued a formal protest, claiming that their working conditions were repressive, their salary scale unfair, and their promotional opportunities limited. The women demanded that the company's hiring and promotional policies and wage scales be revised. The demonstration ended when Garvin Tate, vice president in charge of personnel, promised to appoint a committee to investigate the group's claims and to correct any inequities.

Notice the absence of implied judgments; the facts are presented objectively. A less impartial version of the event, from a protestor's point of view, might read like this:

A biased version

> Last Tuesday, sisters struck another blow against male supremacy when eighty women employees paralyzed the company's repressive and sexist administration for more than six hours. The timely and articulate protest was aimed against degrading working conditions, unfair salary scales, and lack of promotional opportunities for women. Stunned executives watched helplessly as the group organized their picket lines, determined to continue their protest until their demands for equal rights were addressed. An embarrassed vice president quickly agreed to study the group's demands and to revise the company's discriminatory policies. The success of this long-overdue confrontation serves as an inspiration to oppressed women employees everywhere.

Judgmental words (*male supremacy, degrading, paralyzed, articulate, stunned, discriminatory*) inject the writer's attitude, even though it isn't called for. In contrast to this bias, the following version patronizingly defends the status quo:

A biased version

> Our Omaha branch was the scene of an amusing battle of the sexes last Tuesday, when a group of irate feminists, eighty strong, set up picket lines for six hours at the company's executive offices. The protest was lodged against supposed inequities in hiring, wages, working conditions, and promotion for women in our company. The radicals threatened to surround the building until their demands for "equal rights" were met. A bemused vice president responded to this carnival demonstration with patience and dignity, assuring the militants that their claims and demands—however inaccurate and immoderate—would receive just consideration.

Again, qualifying adjectives and superlatives slant the tone.

Being unbiased, of course, doesn't mean remaining "neutral" about something you know to be wrong or dangerous (Kremers 59). If, for instance, you conclude that the Omaha protest was clearly justified, say so.

sexist   **AVOID SEXIST USAGE.**   Sexist usage refers to doctors, lawyers, and other professionals as *he* or *him,* while referring to nurses, secretaries, and homemakers as *she* or *her*. In this traditional stereotype, males do the jobs that really matter and that pay higher wages, whereas females serve only as support and decoration. When females do invade traditional "male" roles, we might express our surprise at their boldness by calling them *female executives, female sportscasters, female surgeons,* or *female hockey players*. Likewise, to demean males who work in occupations that were traditionally seen as "female," we sometimes refer to *male secretaries, male nurses, male flight attendants,* or *male models*.

 offen   **AVOID OFFENSIVE USAGE OF ALL TYPES.**   Enlightened communication respects all people in reference to their specific cultural, racial, ethnic, and national background; sexual and religious orientation; age or physical condition. References to individuals and groups should be as neutral as possible; no matter how inadvertent, any expression that seems condescending or judgmental or that violates the reader's sense of appropriateness is offensive. Detailed guidelines for reducing biased usage appear in these two works:

Schwartz, Marilyn, et al. *Guidelines for Bias-Free Writing*. Bloomington: Indiana UP, 1995.
Publication Manual of the American Psychological Association, 5th ed. Washington, DC: American Psychological Association, 2001.

EXERCISE 19

The sentences below suffer from negative emphasis, excessive informality, biased expressions, or offensive usage. Adjust the tone.

   a. If you want your workers to like you, show sensitivity to their needs.
   b. By not hesitating to act, you prevented my death.
   c. The union has won its struggle for a decent wage.
   d. The group's spokesman demanded salary increases.
   e. Each employee should submit his vacation preferences this week.
   f. While the girls played football, the men waved pom-poms.
   g. Aggressive management of this risky project will help you avoid failure.
   h. The explosion left me blind as a bat for nearly an hour.
   i. This dude would be an excellent employee if only he could learn to chill out.

GUIDELINES for Nonsexist Usage

1. Use neutral expressions.

chair, or **chairperson**	rather than	**chairman**
businessperson	rather than	**businessman**
supervisor	rather than	**foreman**
postal worker	rather than	**postman**
homemaker	rather than	**housewife**
humanity, or **humankind**	rather than	**mankind**
actor	rather than	**actor vs. actress**

2. Rephrase to eliminate the pronoun, but only if you can do so without altering your original meaning.

> SEXIST   A writer will succeed if **he** revises.
> REVISED   A writer who revises succeeds.

3. Use plural forms.

> SEXIST   A writer will succeed if **he** revises.
> REVISED   Writers will succeed if **they** revise (but *not* A writer will succeed if **they** revise).

When using a plural form, don't create an error in pronoun-referent agreement by having the *plural* pronoun *they* or *their* refer to a *singular* referent (as in **Each writer** *should do* **their** *best*).

4. When possible (as in direct address) use *you:* **You** *will succeed if* **you** *revise.* But use this form *only* when addressing someone directly. (See page 276.)

5. Use occasional pairings (*him* or *her, she* or *he, his* or *hers*): *A writer will succeed if* **she or he** *revises.* But note that overuse of such pairings can be awkward: *A writer should do* **his or her** *best to make sure that* **he or she** *connects with* **his or her** *readers.*

6. Drop diminutive endings such as *-ess* and *-ette* used to denote females (*poetess, drum majorette, actress*).

7. Use *Ms.* instead of *Mrs.* or *Miss,* unless you know that person prefers one of the traditional titles. Or omit titles completely: *Roger Smith and Jane Kelly; Smith and Kelly.*

8. In quoting sources that have ignored present standards for nonsexist usage, consider these options:

   - Insert [*sic*] ("thus" or "so") following the first instance of sexist terminology in a particular passage.
   - Use ellipses to omit sexist phrasing.
   - Paraphrase instead of quoting directly.

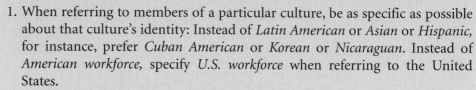

13.4

Politically correct or necessary sensitivity? Consider the debate at <www.ablongman. com/lannonweb>

Below is a sampling of suggestions adapted from the works listed above:

1. When referring to members of a particular culture, be as specific as possible about that culture's identity: Instead of *Latin American* or *Asian* or *Hispanic,* for instance, prefer *Cuban American* or *Korean* or *Nicaraguan.* Instead of *American workforce,* specify *U.S. workforce* when referring to the United States.

   Avoid judgmental expressions: Instead of *third-world* or *undeveloped nations* or the *Far East,* prefer *developing* or *newly industrialized nations* or *East Asia.* Instead of *nonwhites,* refer to *people of color.*

2. When referring to someone who has a disability, avoid terms that could be considered pitying or overly euphemistic, such as *victims, unfortunates, challenged,* or *differently abled.* Focus on the individual instead of the disability: Instead of *blind person* or *amputee,* refer to a *person who is blind* or a *person who has lost an arm.*

   Avoid expressions that demean those who have medical conditions: *retard, mental midget, insane idea, lame excuse, the blind leading the blind, able-bodied workers,* and so on.

3. When referring to members of a particular age group, prefer *girl* or *boy* for people of age fourteen or under; *young person, young adult, young man,* or *young woman* for those of high-school age; and *woman* or *man* for those of college age. (*Teenager* or *juvenile* carries certain negative connotations.) Instead of *the elderly,* prefer *older persons.*

EXERCISE 20

Find examples of overly euphemistic language (such as "chronologically challenged") or of insensitive language. Discuss your examples in class.

CONSIDERING THE CULTURAL CONTEXT

13.5

For more on cross-cultural and global communication visit <www.ablongman. com/lannonweb>

The style guidelines in this chapter apply specifically to standard English in North America. But technical communication is a global process: Practices and preferences differ widely among various cultures.

Certain cultures might prefer long sentences and technical language, to convey an idea's full complexity. Other cultures value expressions of respect, politeness, praise, and gratitude more than clarity or directness (Hein 125–26; Mackin 349–50).

Writing in non-English languages tends to be more formal than in English, and may rely heavily on the passive voice (Weymouth 144). French readers, for example, may prefer an elaborate style that reflects sophisticated and complex modes of thinking. In contrast, our "plain English," conversational style might connote simple-mindedness, disrespect, or incompetence (Thrush 277).

Documents may originate in English but then be translated into other languages. In such cases, writers must be careful to use English that is easy to translate. Analogies, idioms, and humor are often difficult for translators.

Also, in translation or in a different cultural context, certain words carry offensive or unfavorable connotations. For example, certain cultures use "male" and "female" in referring only to animals (Coe, "Writing for Other Cultures" 17). Other notable disasters (Gesteland 20; Victor 44):

ON THE JOB...
Revising a document

"For short pieces, I outline in my head, draft, and revise. On my first draft of a short piece, I spend 40 to 50 percent of the time on the first one or two paragraphs and crank out the rest quickly. Then I revise two or three times, and tinker with the mechanics and format right up until printing. I usually don't bother to make a printout until I'm pretty close to a final product."

—Bill Trippe, Communications Specialist
for a military contract company

- The Chevrolet *Nova*—meaning "doesn't go" in Spanish
- The Finnish beer *Koff*—for an English-speaking market
- Colgate's *Cue* toothpaste—an obscenity in French
- A brand of bicycle named *Flying Pigeon*—imported for a U.S. market

Idioms ("strike out," "ground rules") hold no logical meaning for other cultures. Slang ("bogus," "fat city") and colloquialisms ("You bet," "Gotcha") can strike readers as being too informal and crude.

Offensive writing (including inappropriate humor) can alienate audiences—toward you *and* your culture (Sturges 32).

LEGAL AND ETHICAL IMPLICATIONS OF WORD CHOICE

13.6

For more on ethics in technical communication visit <www.ablongman.com/lannonweb>

Situations in which word choice has ethical or legal consequences

Chapter 5 (page 85) discusses how workplace writing is regulated by laws against libel, deceptive advertising, and defective information. One common denominator among these violations resides in imprecise or inappropriate word choice. We are all accountable for the words we use—intentionally or not—in framing the audience's perception and understanding.

- *Assessing risk.* Is the investment you are advocating "a sure thing" or merely "a good bet," or even "risky"? Are you announcing a "caution," a "warning," or a "danger"? Should methane levels in mineshaft #3 "be evaluated" or do "they pose a definite explosion risk"? Never underestimate the risks.

- *Offering a service or product.* Are you proposing to "study the problem" or "explore solutions to the problem" or "eliminate the problem"? Do you "stand behind" your product or do you "guarantee" it? Never promise more than you can deliver.
- *Giving instructions.* Before inserting the widget between the grinder blades, should I "switch off the grinder" or "disconnect the grinder from its power source" or "trip the circuit breaker," or do all three? Triple-check the clarity of your instructions.
- *Comparing your product with competing products.* Instead of referring to a competitor's product as "inferior" or "second-rate" or "substandard," talk about your own "first-rate product" that "exceeds (or meets) standards." Never run down the competition.
- *Evaluating an employee* (T. Clark, "Teaching Students" 75–76). In a personnel evaluation, don't refer to the employee as a "troublemaker" or as "unprofessional," or "too abrasive" or "too uncooperative" or "incompetent" or "too old" for the job. Focus on the specific requirements of this job, and offer *factual* instances in which these requirements, or standards, have been violated: "Our monitoring software recorded five visits by this employee to X-rated Web sites during working hours." Or "This employee arrives late for work on average twice weekly, has failed to complete assigned projects on three occasions, and has difficulty working with others." Instead of expressing personal judgments, offer the facts. Be sure everyone involved knows exactly what the standards are well beforehand. Otherwise, you risk violating federal laws against discrimination and you invite costly libel or antidiscrimination suits.

ON THE JOB...
Revising a document

"I usually have to just send it out there due to the fast pace of working in a crisis setting. Can't even use white out . . . have to cross things out so that they can be seen, because they are legal documents."

—Emma Bryant, social worker

USING AUTOMATED EDITING TOOLS EFFECTIVELY

Many of the strategies in this chapter could be executed rapidly with word-processing software. By using the *global search and replace function,* you can command the computer to search for ambiguous pronoun references, overuse of passive voice, *to be* verbs, *There* and *It* sentence openers, negative constructions, clutter words, needless prefaces and qualifiers, overly technical language, jargon, sexist language, and so on. With an online dictionary or thesaurus, you can check definitions or see a list of synonyms for a word you have written.

The limits of automation   But these editing aids can be extremely imprecise. For example, both "its" and "it's" are spelled correctly, but only one of them means "it is." Your spell checker is great for words that are spelled incorrectly—but not for words that are *used* incorrectly such as "their," "they're," and "there" or for typos that create the wrong but correctly spelled word such as "howl" or "fort" instead of "how" or "for." Likewise, grammar checkers are great for helping you spot a possible problem, but don't rely only on what the software tells you. For example, not every sentence that the grammar checker flags as "long" should be shortened. These tools simply can't eliminate the writer's burden of *choice*.

Also, none of the rules offered in this chapter applies universally. Ultimately, your own sensitivity to meaning, emphasis, and tone—the human contact—will determine the effectiveness of your writing style.

EXERCISE 21

Try the grammar/style function of your word-processing program. First, look for problems with clarity, conciseness, and fluency yourself. Then compare your changes with those the computer suggests. If the computer contradicts your own judgment, ask a classmate for feedback. If the computer suggests changes that seem ungrammatical or incorrect, consult a good handbook for confirmation. Try to assess when and how the grammar function can be useful and when you can revise best on your own.

For class discussion, prepare a list of the advantages and disadvantages of your automated grammar checker. Use your grammar/style checker on the first two sentences from Exercises 1–10 in this chapter. Are the suggested changes correct? Which of the topics covered in this chapter does the checker miss?

13.7

For more online writing labs and resource links visit <www.ablongman.com/lannonweb>

EXERCISE 22

Do a Web search to find an online style and grammar source and, in a one-page memo for classmates, describe the major types of help the site offers.

SAMPLE SITES (Do not limit yourself to these):

Purdue Online Writing Lab—Offers all sorts of writing help. <www.owl.english.purdue.edu/introduction.html>.

Grammar and Style Notes—Articles cover usage and style. <www.andromeda.rutgers.edu/~jlynch/writing/>.

Writer's Workshop Online Writing Guide—Includes a basic grammar and usage handbook. <www.english.uiuc.edu/cws/wworkshop/mainmenu.html>.

Elements of Style—An online version of the classic text. <www.columbia.edu/acis/bartleby/strunk/>.

Plain English Network—Advice for achieving user-friendly style. <www.plainlanguage.gov>.

EXERCISE 23

Go to the *University of Victoria's Writer's Guide* <http://web.uvic.ca/wguide/>. Use the Table of Contents page to locate the section on *Audience and Tone*. Locate one item of information about audience and tone (or voice) not covered in this chapter. Take careful notes for a brief discussion of this information in class. Attach a copy of the relevant Web page(s) to your written notes.

To *PowerPoint* or Not to *PowerPoint?*

Is Microsoft's ubiquitous presentation software *PowerPoint* a useful tool for designing and delivering presentations? Or is it the latest example of the dumbing-down of professional discourse? Led by critics like Edward Tufte, a debate about the power of *PowerPoint* has reached a fever pitch. The *New York Times* comments that "*PowerPoint* has a tendency to turn any information into a dull recitation of look-alike factoids" (Schwartz 3). *Fast Company* commentator Seth Godin argues, simply, that "*PowerPoint* is a disaster" (Godin).

For a delightful example of the stylistic and design effects of *PowerPoint,* check out a parody of Abraham Lincoln's famous Gettysburg Address at <http://www.norvig.com/Gettysburg/index.htm>.

Musician and artist David Byrne presents a different perspective: "Although I began by making fun of the medium, I soon realized I could actually create things that were beautiful. I could bend the program to my own whim and use it as an artistic agent. "Byrne recently published a book including a DVD of *PowerPoint* artworks ("Learning to Love *PowerPoint*"). ■

14

Designing Visual Information

A visual is any pictorial representation used to clarify a concept, emphasize a particular meaning, illustrate a point, or analyze ideas or data. Besides saving space and words, visuals help people process, understand, and remember information. Because they offer powerful new ways of looking at data, visuals reveal trends, problems, and possibilities that otherwise might remain buried in lists of facts and figures.

WHY VISUALS ARE IMPORTANT

In printed or online documents, in oral presentations or multimedia programs, visuals are a staple of communication today. Compare, for example, a typical textbook or newspaper from the late 1980s or early 1990s with a recently published edition.

Visuals are easier than ever to produce

With graphics software, color printers, digital cameras, and video manipulation devices, anyone can create charts, graphs, diagrams, and digitally altered photographs. People routinely download clip art from the Internet, create their own Web pages, or give visual presentations at business meetings.

This doesn't mean that verbal messages have become obsolete. Instead, words integrate with shapes and images to create what design expert Robert Horn calls *visual language:*

Visual language provides the basis of modern communication

> As the world increases in complexity, as the speed at which we need to solve business and social problems increases, as it becomes increasingly critical to have the "big picture" as well as multiple levels of detail immediately accessible, visual language will become more and more prevalent in our lives. (15)

 Visuals make for efficient communication

In the workplace, our reliance on visuals is driven by the fact that communication is global and instantaneous, that people work in teams—often scattered worldwide, and that success depends on harnessing an increasing volume of complex information.

Our audiences expect visuals

Ultimately, we rely on visuals because our audiences expect them. Bombarded with information from all sources, people want to find what they need quickly and easily, and they want it to be understandable. People like to feel intelligent, to understand the message at a glance. Visuals help answer many of the questions users ask as they process information:

TYPICAL AUDIENCE QUESTIONS IN PROCESSING INFORMATION

- *Which information is most important?*
- *Where, exactly, should I focus?*
- *What do these numbers mean?*
- *What should I be thinking or doing?*

- *What should I remember about this?*
- *What does it look like?*
- *How is it organized?*
- *How is it done?*
- *How does it work?*

All You Need to Know About Metric
(For Your Everyday Life)

10

Metric is based on the Decimal system

The metric system is simple to learn. For use in your everyday life you need to know only ten units. You also need to get used to a few new temperatures. Of course, there are other units which most persons will not need to learn. There are even some metric units with which you are already familiar; those for time and electricity are the same as you use now.

BASIC UNITS

METER: a little longer than a yard (about 1.1 yards)
LITER: a little larger than a quart (about 1.06 quarts)
GRAM: a little more than the weight of a paper clip

(comparative sizes are shown)

1 METER

1 YARD

COMMON PREFIXES
(to be used with basic units)

milli: one-thousandth (0.001)
centi: one-hundredth (0.01)
kilo: one-thousand times (1000)
For example
1000 millimeters = 1 meter
100 centimeters = 1 meter
1000 meters = 1 kilometer

1 LITER

1 QUART

25 DEGREES FAHRENHEIT

OTHER COMMONLY USED UNITS

millimeter:	0.001 meter	diameter of a paper clip wire
centimeter:	0.01 meter	a little more than the width of a paper clip (about 0.4 inch)
kilometer:	1000 meters	somewhat farther than 1/2 mile (about 0.6 mile)
kilogram:	1000 grams	a little more than 2 pounds (about 2.2 pounds)
milliliter:	0.001 liter	five of them make a teaspoon

OTHER USEFUL UNITS
hectare: about 2 1/2 acres
metric ton: about one ton

25 DEGREES CELSIUS

WEATHER UNITS: **FOR TEMPERATURE**   **FOR PRESSURE**
degrees celsius   kilopascals are used
100 kilopascals = 29.5 inches of Hg (14.5 psi)

°C	−40	−20	0	20	37	60	80	100
°F	−40	0	32	80	98.6	160		212

water freezes   body temperature   water boils

1 POUND

1 KILOGRAM

FIGURE 14.1   Visuals That Clarify and Simplify   Notice how the words and numbers are enhanced and clarified by familiar images and shapes.

Source: National Institute of Standards and Technology, 1992.

NOTE   *Visual design is a complex art in which plenty can go wrong. No amount of visual technology can substitute for careful decisions about your audience, the purpose of your communication, and the design options you have available.*

HOW VISUALS WORK

More receptive to images than to words, most people resist unbroken pages of printed text. Visuals help diminish this resistance in several ways:

How visuals help
neutralize reader
resistance

- *Visuals enhance comprehension by displaying abstract concepts in concrete, geometric shapes.* "How does the metric system work?" (Figure 14.1). "How do lasers work?" (Figure 14.35).
- *Visuals make meaningful comparisons possible.* "How do industries compare in terms of the toxic chemicals they release into the environment?" (Figure 14.2). "How does one pound compare with one kilogram?" (Figure 14.1).
- *Visuals depict relationships.* "How does seasonal change affect the rate of construction in our county?" (Figure 14.10). "What is the relationship between Fahrenheit and Celsius temperature?" (Figure 14.1).

- *Visuals serve as a universal language.* In the global workplace, carefully designed visuals can transcend cultural and language differences, and thus facilitate international communication (Figure 14.1).

As one expert points out, "the visual elements of a text affect how readers interact with the words" (Hilligoss 9).

**FIGURE 14.2
A Visual
Displaying the
"Big Picture"**

*Source: Data
from U.S. Environmental Protection Agency.
See Table 14.3.*

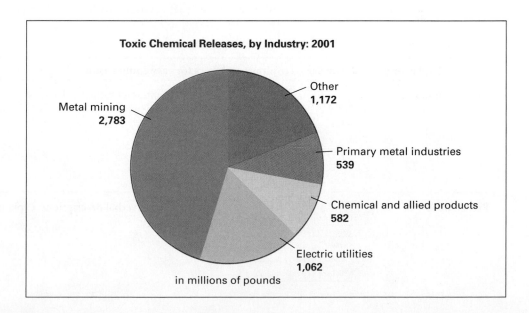

Toxic Chemical Releases, by Industry: 2001

Other
1,172

Metal mining
2,783

Primary metal industries
539

Chemical and allied products
582

Electric utilities
1,062

in millions of pounds

14.1
See examples of
effective visuals at
<www.ablongman.com/
lannonweb>

**Use visuals in
situations like
these**

WHEN TO USE A VISUAL

In general, use visuals whenever they make your point more clearly than the text. Use visuals to direct the audience's focus or to help them remember something, as in the following situations (Dragga and Gong 46–48):

- when you want to instruct or persuade
- when you want to draw attention to something immediately important
- when you expect the document to be consulted randomly or selectively (e.g., a manual or other reference work) instead of being read in its original sequence (e.g., a memo or letter)
- when you expect the audience to be relatively less educated, less motivated, or less familiar with the topic
- when you expect the audience to be distracted

There may be organizational reasons for using visuals; for example, some companies may always expect a chart or graph as part of their annual report. Certain industries, such as the financial sector, often use graphs and charts (such as the graph of the daily Dow Jones Industrial Average).

NOTE   *Use visuals to clarify and enhance your discussion, and not merely to decorate your document.*

WHAT TYPES OF VISUALS TO CONSIDER

Different types of visuals serve different functions. The following overview sorts visual displays into four categories: tables, graphs, charts, and graphic illustrations. Each type of visual offers a new way of seeing, a different perspective.

Types of Visuals and Their Uses

TABLES display organized data across columns and rows for easy comparison.		
Numerical tables	TABLE 1 Charting the Lesson Lesson | Page | Page | Page A | 1 | 3 | 6 B | 2 | 2 | 5 C | 3 | 1 | 4	Use to present exact numerical values.
Prose tables	TROUBLESHOOTING Problem | Cause | Solution • power | • cord | • plug-in • light | • bulb | • replace • flicker | • tube | • replace	Use to organize verbal descriptions, explanations, or instructions.

GRAPHS translate numbers into shapes, shades, and patterns.

Bar graphs	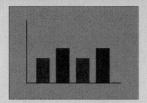	Use to show comparisons.
Line graphs	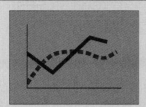	Use to show changes over time, cost, or other variables.

CHARTS depict relationships via geometric, arrows, lines, and other design elements.

Pie charts	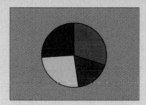	Use to relate parts or percentages to the whole.
Organization charts	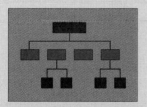	Use to show the links among departments, management structures, or other elements of a company.
Flowcharts	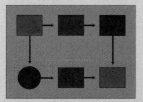	Use to trace the steps (or decisions) in a procedure or stages in a process.
Gantt charts	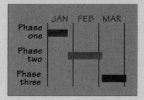	Use to depict how the phases of a project relate to each other.

(continues)

Types of Visuals and Their Uses *(continued)*

CHARTS *(continued)*

Tree charts

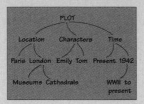

Use to show how the parts of an idea or a concept inter-relate.

Pictograms

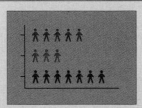

Use to symbolize the items being displayed or measured via icons (or isotypes).

GRAPHIC ILLUSTRATIONS rely on pictures rather than on data or words.

Representational diagrams

Use to present a realistic but simplified view, usually with essential parts labeled.

Exploded diagrams

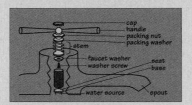

Use to explain how an item is put together or how a user should assemble a product.

Cutaway diagrams

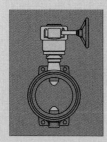

Use to show what is inside of a device or to help explain how a device works.

GRAPHIC ILLUSTRATIONS *(continued)*

Schematic diagrams	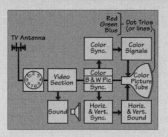	Use to present the conceptual elements of a principle, process, or system to depict *function* instead of appearance.
Maps	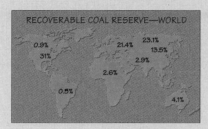	Use to help readers to visualize a specific location or to comprehend data about a specific geographic region.
Photographs		Use to show exactly what something looks like.
Symbols and icons		Use to make concepts understandable to broad audiences, including international audiences, children, or people who may have difficulty reading.

HOW TO SELECT VISUALS FOR YOUR PURPOSE AND AUDIENCE

To select the most effective display, consider your specific purpose and the abilities and preferences of your audience.

QUESTIONS ABOUT A VISUAL'S PURPOSE AND INTENDED AUDIENCE

What is my purpose?

- *What do I want the audience to do or think (know facts and figures, follow directions, make a judgment, understand how something works, perceive a relationship, identify something, see what something looks like, pay attention)?*
- *Do I want users to focus on one or more exact values, compare two or more values, or synthesize a range of approximate values?*

Who is my audience?

- *What is their technical background on this topic?*
- *What is their level of interest in this topic?*

- *Would they prefer the raw data or interpretations of the data?*
- *Are they accustomed to interpreting visuals?*
- *What is their cultural background?*

Which type of visual might work best in this situation?

- *What forms of information should this visual depict (numbers, shapes, words, pictures, symbols)?*
- *Which visual display would be most compatible with the type of judgment, action, or understanding I seek from this audience?*
- *Which visual display would this audience find most accessible?*

Here are a few examples of the choices you must consider in selecting visuals:

Choices to consider in selecting visuals

- If you merely want the audience to know facts and figures, a table might suffice, but if you want them to make a particular judgment about these data, a bar graph, line graph, or pie chart might be preferable.
- To depict the operating parts of a mechanism, an exploded or cutaway diagram might be preferable to a photograph.
- Expert audiences tend to prefer numerical tables, flowcharts, schematics, and complex graphs or diagrams they can interpret for themselves.
- General audiences tend to prefer basic tables, graphs, diagrams, and other visuals that direct their focus and interpret key points extracted from the data.

Although several alternatives might be possible, one particular type of visual (or a combination) usually is superior. The best option, however, may not always be available to you. Your particular audience or organization may express its own preferences. Or your choices may be limited by lack of equipment (software, scanners, digitizers), insufficient personnel (graphic designers, technical illustrators), or insufficient budget. In any case, your basic task is to enable the audience to interpret the visual correctly.

 NOTE   *Although visual communication has global appeal, certain visual displays can be inappropriate in certain cultures. For example, not all cultures read from left to right, so a chart designed to be read from left to right that is read in the opposite direction could be misunderstood.*

PREFERRED DISPLAYS FOR SPECIFIC VISUAL PURPOSES

Purpose . . . . Preferred Visual
- *Organize numerical data* . . . .*Table*
- *Show comparative data* . . . .*Table, bar graph, line graph*
- *Show a trend* . . . .*Line graph*
- *Interpret or emphasize data* . . . .*Bar graph, line graph, pie chart, map*
- *Introduce an unfamiliar object* . . . .*Photo, representational diagram*
- *Display a project schedule* . . . .*Gantt chart*
- *Show how parts are assembled* . . . . . .*Photo, exploded diagram*
- *Show how something is organized* . . . .*Organization chart, map*
- *Give instructions* . . . .*Prose table, photo, diagrams, flowchart*
- *Explain a process* . . . .*Flowchart, block diagram*
- *Clarify a concept or principle* . . . .*Block or schematic diagram, tree chart*
- *Describe a mechanism* . . . .*Photo, representational diagram, or cutaway diagram*

TABLES

A table is a powerful way to display dense textual information such as specifications, comparisons or conditions. Assume, for example, that you are researching recent death rates for heart disease and cancer. From various sources, you collect these data:

Technical data in printed form can be hard to interpret

1. In 1970, 419 males and 309 females per 100,000 people died of heart disease; 172 males and 135 females died of cancer.
2. In 1980, 369 males and 305 females per 100,000 people died of heart disease; 205 males and 164 females died of cancer.
3. In 1990 . . . .

In the textual form above, numerical information is repetitious and hard to interpret. As the numerical data increase, so does our difficulty in processing this material. In Table 14.1 (constructed via the "Table" command in a word-processing program), the above statistics are easier to compare and analyze.

NOTE *Include a caption with your visual, to analyze or interpret the trends or key points you want readers to recognize.*

Numerical tables such as Table 14.1 present *quantitative information* (data that can be measured). Prose tables present *qualitative information* (prose descriptions, explanations, or instructions). Table 14.2, for example, combines numerical data, probability estimates, comparisons, and instructions—all organized for the smoker's understanding of radon gas risk in the home.

No table should be overly complex for its audience. Although impressive-looking, Table 14.3 is hard for nonspecialists to interpret because it presents too much information at once. We can see how an unethical writer might use a complex table to bury numbers that are questionable or embarrassing (Williams 12). For laypersons, use fewer tables and keep them simple.

TABLE 14.1
Data Displayed in a Table

Organizes data in columns and rows, for easy comparison.

Death Rates for Heart Disease and Cancer 1970–2000				
	Number of Deaths (per 100,000)			
	Heart Disease		Cancer	
Year	Male	Female	Male	Female
1970	419	309	248	163
1980	369	305	272	167
1990	298	282	280	176
2000	256	260	257	206
% change, 1970–2000	−38.9	−15.8	+0.36	+26.3

A caption explaining the numerical relationships

Note: Both male and female death rates from heart disease decreased from 1970 to 2000, but males showed a sizably larger decrease. Cancer deaths during this period increased for both groups, with females showing a much larger increase.
Source: Adapted from *Statistical Abstract of the United States: 2003 (123rd ed.).* Washington: GPO. 96,97.

TABLE 14.2
A Prose Table

Displays complicated numerical and verbal information.
Source: Home Buyer's and Seller's Guide to Radon. *Washington: GPO, 1993.*

Radon Risk if You Smoke			
Radon level	If 1,000 people who smoked were exposed to this level over a lifetime ...	The risk of cancer from radon exposure compares to ...	WHAT TO DO: Stop Smoking and ...
20 pCi/L[a]	About 135 people could get lung cancer	←100 times the risk of drowning	Fix your home
10 pCi/L	About 71 people could get lung cancer	←100 times the risk of dying in a home fire	Fix your home
8 pCi/L	About 57 people could get lung cancer		Fix your home
4 pCi/L	About 29 people could get lung cancer	←100 times the risk of dying in an airplane crash	Fix your home
2 pCi/L	About 15 people could get lung cancer	←2 times the risk of dying in a car crash	Consider fixing between 2 and 4 pCi/L
1.3 pCi/L	About 9 people could get lung cancer	(Average indoor radon level)	(Reducing radon levels below 2 pCi/L is difficult)
0.4 pCi/L	About 3 people could get lung cancer	(Average outdoor radon level)	

Note: If you are a former smoker, your risk may be lower.
[a]picocuries per liter

TABLE 14.3   A Complex Table   Can cause information overload for nontechnical audiences.
Source: U.S. Environmental Protection Agency, Annual Toxics Release Inventory.

Toxic Chemical Releases by Industry: 2001

[In millions of pounds (6,158.0 represents 6,158,000,000), except as indicated.] *

Industry	1987 SIC[1] code	Total facilities (number)	Total on- and off-site releases	Total air emissions	Surface water discharges	On-site land release Total[2]	On-site land release Surface impoundments	Total on-site releases
Total[3]	(X)	24,896	6,158.0	5,580.0	1,679.4	220.8	3,680.1	577.7
Metal mining	10	89	2,782.6	2,782.0	2.9	0.4	2,778.7	0.5
Coal mining	12	88	16.1	16.1	0.8	0.8	14.6	-
Food and kindred products	20	1,688	125.1	118.9	56.1	55.2	7.6	6.2
Tobacco products	21	31	3.6	3.2	2.5	0.5	0.2	0.3
Textile mill products	22	289	7.0	6.2	5.7	0.2	0.3	0.7
Apparel and other textile products	23	16	0.4	0.3	0.3	-	-	0.1
Lumber and wood products	24	1,006	31.4	30.9	30.5	-	0.4	0.5
Furniture and fixtures	25	282	8.0	7.8	7.8	-	-	0.2
Paper and allied products	26	507	195.7	189.9	157.2	16.5	16.2	5.8
Printing and publishing	27	231	19.7	19.3	19.3	-	-	0.4
Chemical and allied products	28	3,618	582.6	501.3	227.8	57.6	215.9	81.3
Petroleum and coal products	29	542	71.4	68.1	48.2	17.1	2.8	3.3
Rubber and misc. plastic products	30	1,822	88.5	78.1	77.1	0.1	0.9	10.5
Leather and leather products	31	60	2.6	1.3	1.2	0.1	-	1.3
Stone, clay, glass products	32	1,027	40.5	35.4	31.3	0.2	4.0	5.1
Primary metal industries	33	1,941	558.6	286.8	57.6	44.7	184.5	271.8
Fabricated metals products	34	2,959	64.0	42.8	40.4	1.7	0.6	21.2
Industrial machinery and equipment	35	1,143	15.4	10.7	8.3	-	2.5	4.6
Electronic, electric equipment	36	1,831	23.9	16.4	12.7	2.9	0.7	7.6
Transportation equipment	37	1,348	80.6	67.7	66.7	0.2	0.8	13.0
Instruments and related products	38	375	9.4	8.6	7.2	1.4	-	0.8
Miscellaneous	39	312	8.4	6.8	6.8	-	-	1.6
Electric utilities	49	732	1,062.2	989.2	717.6	3.5	268.1	73.1
Chemical wholesalers	5169	475	1.5	1.3	1.3	-	-	0.2
Petroleum bulk terminals	5171	596	21.3	21.2	21.2	-	-	0.2
RCRA/solvent recovery	4953/ 7369	223	219.9	168.4	1.0	-	167.4	51.4

* Represents or rounds to zero. X Not applicable. [1]Standard Industrial Classification, see text, Section 12. [2]Includes underground injection for Class I and Class II to V wells and land releases. [3]Includes industries with no specific industry identified, not shown separately.

NOTE   *Like all other parts of a document, visuals are designed with audience and purpose in mind (Journet 3). An accountant doing an audit might need a table listing exact amounts, whereas the average public stockholder reading an annual report would prefer the "big picture" in an easily grasped bar graph or pie chart (Van Pelt 1). Similarly, scientists might find the complexity of data shown in Table 14.3 perfectly appropriate, but a nonexpert audience (say, environmental groups) might prefer the clarity and simplicity of a chart like Figure 14.2.*

How to use and
display a table

For displaying information in a table, follow these general guidelines:

- *Try to limit the table to one page.* Otherwise, write "continues" at the bottom, and begin the second page with the full title, "continued," and the original column headings.
- *If the table is too wide for the page, turn it 90 degrees so that the left-hand side faces the bottom of the page.* Or divide the data into two tables. (Few readers may bother rotating the page to read the table broadside.)
- *In your discussion, refer to the table by number, and explain what readers should be looking for.* Or include a prose caption. In short, introduce the table, present it, and then interpret it.

For specific information about creating tables, see How to Construct a Table and the accompanying Table 14.4 on the next page.

Tables work well for displaying exact values, but readers find graphs or charts easier to interpret. Geometric shapes (bars, curves, circles) are generally easier to remember than lists of numbers (Cochran et al. 25).

NOTE   *Any visual other than a table is usually categorized as a figure, and so titled* (Figure 1 Aerial View of the Panhandle Site).

GRAPHS

Graphs translate numbers into shapes, shades, and patterns. A graph displays, at a glance, the approximate values, the point being made about those values, and the relationship being emphasized. Graphs are especially useful for depicting comparisons, changes over time, patterns, or trends.

A graph's horizontal axis shows categories (the independent variables) to be compared, such as years within a period (1980, 1990, 2000). The vertical axis shows the range of values (the dependent variables) for comparing or measuring the categories, such as the number of deaths from heart failure in a given year. A dependent variable changes according to activity in the independent variable (for example, a decrease in quantity over a set time, as in Figure 14.3).

Bar Graphs

Generally easy to understand, bar graphs show discrete comparisons, such as year-by-year or month-by-month. Each bar represents a specific quantity. You can use bar graphs to focus on one value or to compare values over time.

SIMPLE BAR GRAPH.   A simple bar graph displays one trend or theme. The graph in Figure 14.3 shows one trend extracted from Table 14.1, male deaths from heart disease. To aid interpretation, you can record exact values above each bar—but only if the audience needs exact numbers.

HOW TO CONSTRUCT A TABLE

① **TABLE 14.4 ■ Federal Student Financial Assistance: 1999 – 2003**

STUB HEAD ② Number of Awards ③ (1000)[a]	1999	2000	COLUMN HEADS 2001	2002	2003[b]
Total	④14,567 ⑤	15,043	16,154	17,256	⑦18,858
⑥ Federal Pell Grant	3,764	3,899	4,341	4,812	4,884
Fed. Opportunity Grant	1,170	1,175	1,295	1,189	1,189
Federal Work-Study	733	713	740	1,073	1,073
Federal Perkins Loan	655	639	660	⑧ X	707
Fed. Direct Student Loan	2,891	2,739	2,763	2,908	3,086
Fed. Family Educ. Loan	5,354	5,878	6,355	7,274	7,919

ROW HEADS

NOTE ⑨ [a]As of June 30.   [b]Estimate. (X) Not available.

SRC ⑩ Source: *U.S. Department of Education, Office of Postsecondary Education, unpublished data. Statistical Abstract of the United States: 2003* (123rd Edition). Washington: GPO. 187.

1. Number the table in its order of appearance and provide a title that describes exactly what is being compared or measured.
2. Label stub, column, and row heads (*Number of Awards; 1999, 2000, and so on; Federal Pell Grant*) so readers know what they are looking at.
3. Stipulate all units of measurement or use familiar symbols and abbreviations ($, hr., no.). Define specialized symbols or abbreviations (Å for *angstrom, db* for *decibel*) in a footnote.
4. Compare data vertically (in columns) instead of horizontally (in rows). Columns are easier to compare than rows. Try to include row or column averages or totals, as reference points for comparing individual values.
5. Use horizontal rules to separate headings from data. In a complex table, use vertical rules to separate columns. In a simple table, use as few rules as clarity allows.

6. List the items in a logical order (alphabetical, chronological, decreasing cost). Space listed items for easy comparison. Keep prose entries as brief as clarity allows.
7. Convert fractions to decimals, and align decimals and all numerical values vertically. Keep decimal places for all numbers equal. Round insignificant decimals to the nearest whole number.
8. Use *x, NA*, or a dash to signify any omitted entry, and explain the omission in a footnote ("Not available," "Not applicable").
9. Use footnotes to explain entries, abbreviations, or omissions. Label footnotes with lowercase letters so readers do not confuse the notation with the numerical data.
10. Cite data sources beneath any footnotes. When adapting or reproducing a copyrighted table for a work to be published, obtain written permission from the copyright holder.

**FIGURE 14.3
A Simple Bar
Graph**

Shows a single
relationship in
the data.

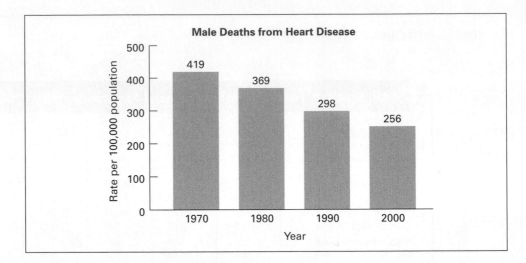

MULTIPLE-BAR GRAPH.   A bar graph can display two or three relationships simultaneously. Figure 14.4 contrasts two sets of information, allowing readers to see two trends. When you create a multiple-bar graph, be sure to use a different pattern or color for each bar, and include a key (or *legend*) so your audience knows which color or pattern corresponds with which bar.

The more relationships you include on a graph, the more complex the graph becomes, so try not to include more than three on any one graph.

**FIGURE 14.4
A Multiple-Bar
Graph**

Shows two or
more relation-
ships simultane-
ously.

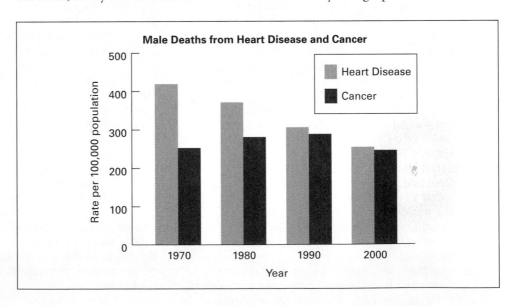

HORIZONTAL-BAR GRAPH.   Horizontal-bar graphs are good for displaying a large series of bars arranged in order of increasing or decreasing value, as in Figure 14.5. This format leaves room for labeling the categories horizontally (*Doctorate,* and so on).

FIGURE 14.5
A Horizontal-Bar Graph

Accommodates lengthy labels.
Source: Bureau of Labor Statistics.

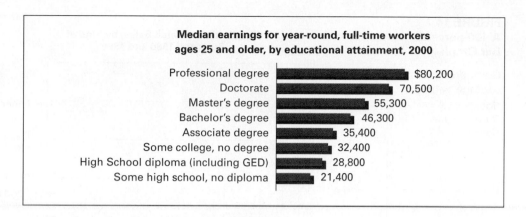

Median earnings for year-round, full-time workers ages 25 and older, by educational attainment, 2000

Professional degree	$80,200
Doctorate	70,500
Master's degree	55,300
Bachelor's degree	46,300
Associate degree	35,400
Some college, no degree	32,400
High School diploma (including GED)	28,800
Some high school, no diploma	21,400

STACKED-BAR GRAPH. Instead of displaying bars side-by-side, you can stack them. Stacked-bar graphs are especially useful for showing how much each item contributes to the whole. Figure 14.6 displays other comparisons from Table 14.1. To avoid confusion, don't display more than four or five sets of data in a single bar.

FIGURE 14.6
A Stacked-Bar Graph

Compares the parts that make up each total.

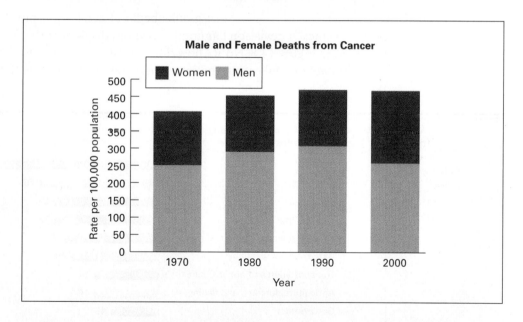

Male and Female Deaths from Cancer

Women Men

Rate per 100,000 population

Year

100-PERCENT BAR GRAPH. A type of stacked-bar graph, the 100-percent bar graph shows the value of each part that makes up the 100-percent value, as in Figure 14.7. Like any bar graph, the 100-percent graph can have either horizontal or vertical bars.

Notice how bar graphs become harder to interpret as bars and patterns increase. For a general audience, the data from Figure 14.7 might be displayed in pie charts (page 310).

FIGURE 14.7
A 100-percent
Bar Graph

Compares per-
centage values.
Source: U.S.
Bureau of the
Census.

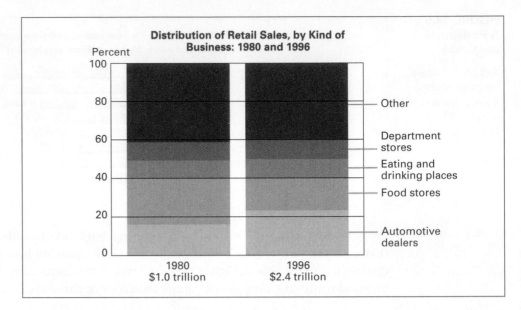

DEVIATION BAR GRAPH.   Most graphs begin at a zero axis point, displaying only positive values. A deviation bar graph, however, displays both positive and negative values, as in Figure 14.8. Notice how the vertical axis extends to the negative side of the zero baseline, following the same incremental division as the positive side of the graph.

FIGURE 14.8
A Deviation
Bar Graph

Displays both
positive and
negative values.
Source: Bureau
of Labor
Statistics.

14.2

Microsoft Excel can
generate many kinds of
charts. Find out
how at
<www.ablongman.com/
lannonweb>

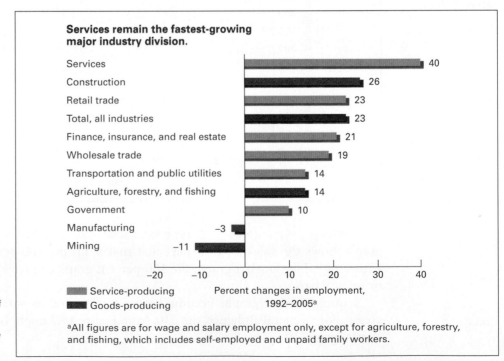

3-D BAR GRAPH.   Graphics software makes it easy to shade and rotate images for a three-dimensional view. The 3-D perspectives in Figure 14.9 engage our attention and visually emphasize the data.

NOTE   *Although 3-D graphs can enhance and dramatize a presentation, an overly complex graph can be misleading or hard to interpret. Use 3-D only when a two-dimensional version will not serve as well. Never sacrifice clarity and simplicity for the sake of visual effect.*

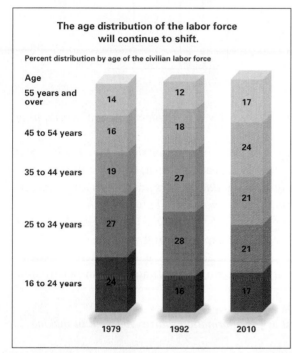

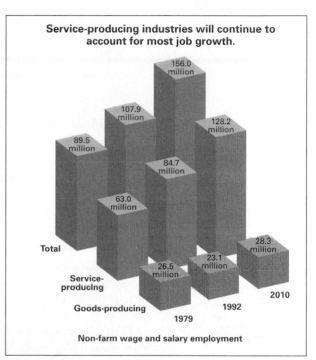

FIGURE 14.9   3-D Bar Graphs   Adding a third axis creates the appearance of depth.
Source: Bureau of Labor Statistics.

HOW TO DISPLAY A BAR GRAPH.   Once you have decided on a type of bar graph, follow these suggestions for achieving a user-friendly display.

How to display a bar graph

• *Use a bar graph only to compare values that are noticeably different.* Small value differences will yield bars that look too similar to compare.
• *Keep the graph simple and easy to read.* Don't plot more than three types of bars in each cluster. Avoid needless visual details.
• *Number your scales in units familiar to the audience.* Units of 1 or multiples of 2, 5, or 10 are best (Lambert 45). Space the numbers equally.
• *Label both scales to show what is being measured or compared.* If space allows, keep all labels horizontal for easier reading.

- *Label each bar or cluster of bars at its base.*
- *Use* tick marks *to show the points of division on your scale.* If the graph has many bars, extend the tick marks into *grid lines* to help readers relate bars to values.

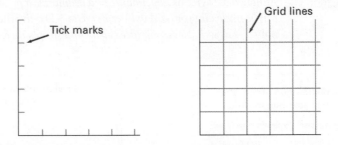

- *To avoid confusion, make all bars the same width (unless you are overlapping them).*
- *In a multiple-bar graph, use a different pattern, color, or shade for each bar in a cluster.* Provide a key, or legend, identifying each pattern, color, or shade.
- *If you are trying for emphasis, be aware that darker bars are seen as larger, closer, and more important than lighter bars of the same size (Lambert 93).*
- *In your discussion, refer to the graph by number ("Figure 1"), and explain what the user should look for.* Or include a prose caption along with the graph.
- *Cite data sources beneath the graph.* When adapting or reproducing a copyrighted graph for a work to be published, you must obtain written permission from the copyright holder.

NOTE   *Failure to cite the creator of a visual or the information sources you used in making your own visual is plagiarism.*

Computer graphics programs automatically employ most of these design features. Anyone producing visuals, however, should know the conventions.

Line Graphs

A line graph can accommodate many more data points than a bar graph (for example, a twelve-month trend, measured monthly). Line graphs help readers synthesize large bodies of information in which exact quantities don't need to be emphasized.

SIMPLE LINE GRAPH.   A simple line graph, as in Figure 14.10, uses one line to plot time intervals (or categories) on the horizontal scale and values on the vertical scale.

FIGURE 14.10
A Simple Line
Graph

Displays one
relationship.

14.3

What is "chartjunk?"
Find out more at
<www.ablongman.com/
lannonweb>

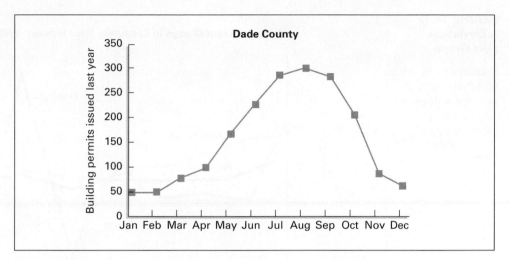

MULTILINE GRAPH. A multiline graph displays several relationships simultaneously, as in Figure 14.11. Include a caption to explain the relationships readers are
supposed to see.

FIGURE 14.11
A Multiple-Line
Graph

Displays multi-
ple relation-
ships.
*Source: Bureau
of Labor
Statistics.*

A caption
explaining the
visual
relationships

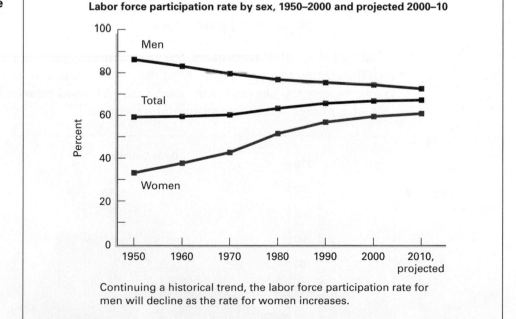

DEVIATION LINE GRAPH. Extend your vertical scale below the zero baseline to display positive and negative values in one graph, as in Figure 14.12. Mark values below the baseline in intervals parallel to those above it.

**FIGURE 14.12
A Deviation
Line Graph**

Displays both
negative and
positive values.
*Source: Chart
prepared by U.S.
Bureau of the
Census.*

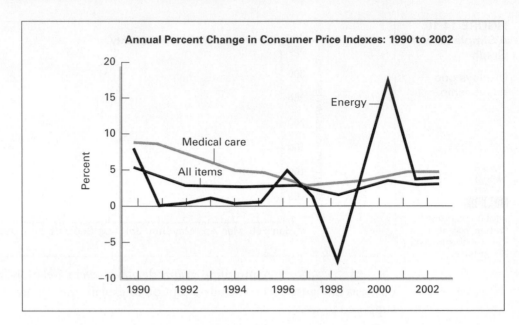

BAND OR AREA GRAPH.   By shading in the area beneath the main plot lines, you
can highlight specific features. Figure 14.13 is another version of the Figure 14.10
line graph.

The multiple bands in Figure 14.14 depict relationships among sums instead
of the direct comparisons depicted in the Figure 14.11 line graph. Despite their
visual appeal, multiple-band graphs are easy to misinterpret: In a multiline graph,
each line depicts its own distance from the zero baseline. But in a multiple-*band*
graph, the very top line depicts the *total*, with each band below it being a part of
that total. Always clarify these relationships for users.

**FIGURE 14.13
A Simple
Band Graph**

Shading adds
emphasis.

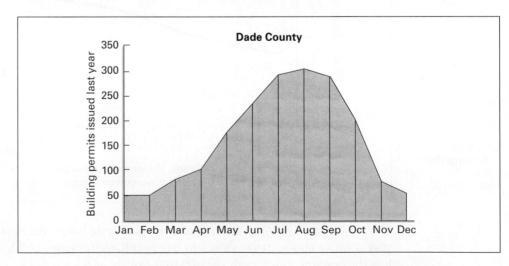

**FIGURE 14.14
A Multiple-Band
Graph**

Each item is
added to the one
below it.

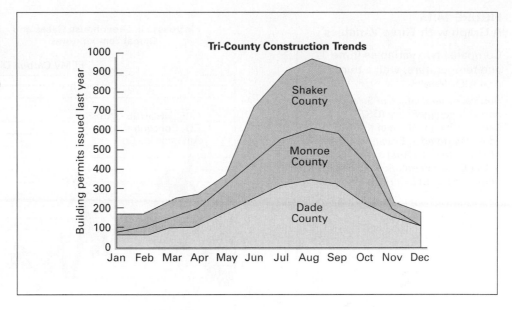

Tri-County Construction Trends

Shaker
County

Monroe
County

Dade
County

Building permits issued last year

Jan Feb Mar Apr May Jun Jul Aug Sep Oct Nov Dec

How to display a
line graph

HOW TO DISPLAY A LINE GRAPH. Follow the suggestions for displaying a bar graph (page 305), with these additions:

- *Display no more than three or four lines on one graph.*
- *Mark each individual data point used in plotting each line.*
- *Make each line visually distinct (using color, symbols, and so on).*
- *Label each line so users know what each one represents.*
- *Avoid grid lines that users could mistake for plotted lines.*

Graphs with Three Variables

As discussed on page 300, graphs usually depict a relationship between one independent variable (say, *time*) and one dependent variable (say, *global temperature fluctuations*). To provide additional perspective on the data, a graph might display two different but related dependent variables on parallel vertical axes. For example, Figure 14.15 plots the relationship between global temperatures and carbon dioxide levels over time, showing how one dependent variable (*temperature*) changes in respect to another one (CO_2 *levels*). Notice how the correlation between CO_2 levels and temperature becomes apparent.

NOTE   *Even though it may signal a possible causal relationship, correlation alone does not "prove" a direct causal link. See page 182 for more discussion.*

FIGURE 14.15
A Graph with Three Variables

Compares two variables (time and temperature) with a third one (CO_2 levels).

Source: Federal Office of Science and Technology Policy (OSTP). Climate Change: State of Knowledge. *Reprinted in Executive Office of Environmental Affairs.* The State of Our Environment. *Boston: State of Massachusetts, April 2000.*

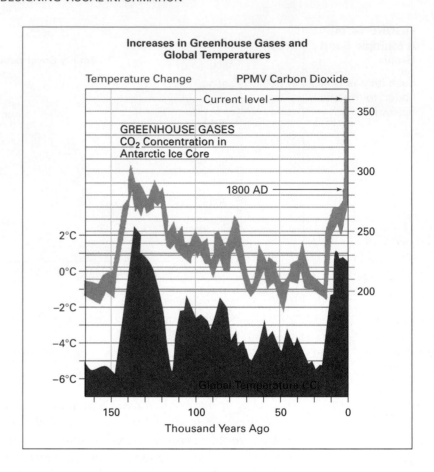

CHARTS

The terms *chart* and *graph* often are used interchangeably. But a chart displays relationships (quantitative or cause-and-effect) that are not plotted on a coordinate system (*x* and *y* axes).

Pie Charts

Easy for almost anyone to understand, a pie chart displays the relationship of parts or percentages to the whole. Readers can compare the parts to each other as well as to the whole (to show how much was spent on what, how much income comes from which sources, and so on). Figure 14.16 shows a simple pie chart. Figure 14.17 is an exploded pie chart. Exploded pie charts help highlight various pieces of the pie.

For displaying pie charts, follow these suggestions:

How to display a pie chart

- *Make sure the parts add up to 100 percent.*
- *Differentiate each slice clearly.* Use different colors or shades, or differentiate by "exploding" out various pie slices.

- *Include a key, or legend, to help readers identify each slice, or label each slice directly.*
- *Include no less than two and no more than eight segments.* A pie chart containing more than eight segments can be hard to interpret, especially if the segments are small (Hartley 96).
- *Combine very small segments under the heading "Other."*
- *For easy reading, keep all labels horizontal.*

**FIGURE 14.16
A Simple Pie
Chart**

Shows the
relationships
of parts or
percentages
to the whole.

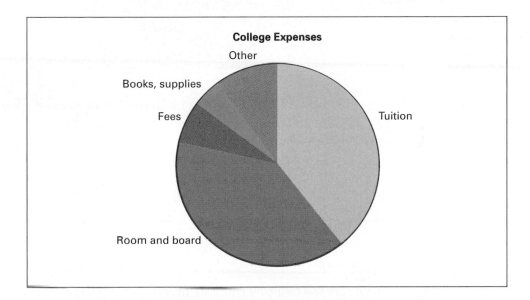

College Expenses

Other

Books, supplies

Fees

Tuition

Room and board

**FIGURE 14.17
An Exploded Pie
Chart.**
*Source: Bureau
of Labor Statis-
tics.*

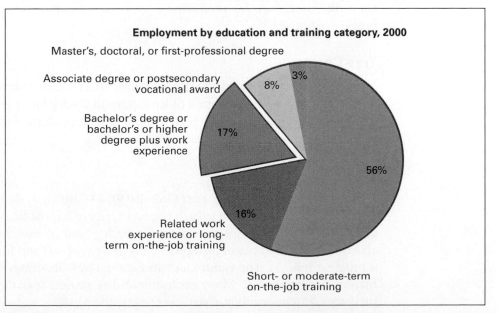

Employment by education and training category, 2000

Master's, doctoral, or first-professional degree

Associate degree or postsecondary
vocational award

Bachelor's degree or
bachelor's or higher
degree plus work
experience

8%   3%

17%

56%

16%

Related work
experience or long-
term on-the-job training

Short- or moderate-term
on-the-job training

Organization Charts

An organization chart shows the hierarchy and relationships between different departments and other units in an organization, as in Figure 14.18.

FIGURE 14.18
An Organization Chart

Shows how different people or departments are ranked and related.

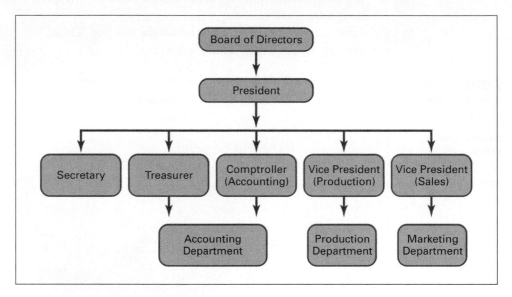

Flowcharts

A flowchart traces a procedure or process from beginning to end. Figure 14.19 traces the procedure for producing a textbook. (Another flowchart example appears on page 582.)

Tree Charts

Whereas flowcharts display the steps in a process, tree charts show how the parts of an idea or concept relate to each other. Figure 14.20 displays part of an outline for this chapter so that users can better visualize relationships. The tree chart seems clearer and more interesting than the prose listing.

Gantt and PERT Charts

Named for engineer H. L. Gantt (1861–1919), a Gantt chart depicts how the parts of an idea or concept relate to each other. A series of bars or lines (time lines) indicates start-up and completion dates for each phase or task in a project. Gantt charts are useful for planning a project (as in a proposal) and for tracking it (as in a progress report). The Gantt chart in Figure 14.21 illustrates the schedule for a manufacturing project. Many professionals use project management software to produce Gantt and similar charts (see pages 100, 314).

**FIGURE 14.19
A Flowchart for
Producing a
Textbook**

Depicts a
sequence of
events, activi-
ties, steps, or
decisions.

*Source: Adapted
from* Harper &
Row Author's
Guide.

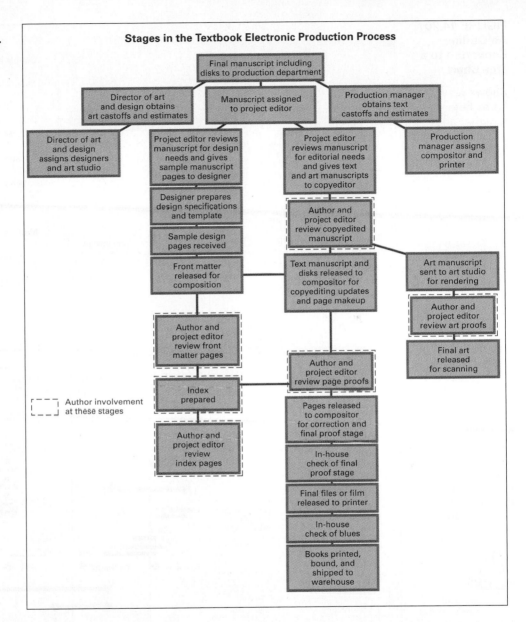

A related type of chart used for scheduling activities on a project is the PERT chart. See page 100.

Pictograms

Pictograms are something of a cross between a bar graph and a chart. Like line graphs, pictograms display numerical data, often by plotting it across an *x* and *y*

**FIGURE 14.20
An Outline
Converted to a
Tree Chart**

Shows what
items belong
together and
how they are
connected.

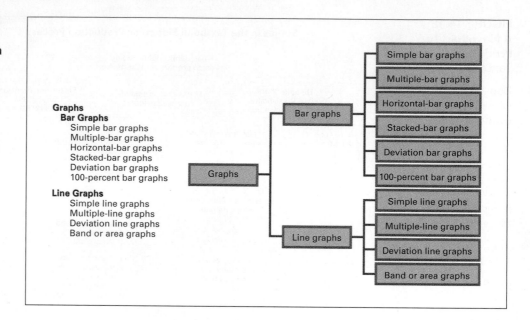

**FIGURE 14.21
A Gantt Chart**

Depicts how the
phases of a proj-
ect relate to
each other.

*Source: Chart
created in Fast
Track Schedule™.
Reprinted by per-
mission from
AEC Software.*

14.4

Learn more about
project management at
<www.ablongman.com/
lannonweb>

	Activity	Start Date	Finish Date	August	September	October	Actual Duration
1	Design			8/11 — 8/30			19.10
2	Brainstorming	8/11	8/18				7.14
3	Research	8/18	8/23				4.98
4	Marketing	8/23	8/30				6.98
5	Testing			8/17 — 9/17			31.06
6	Prototype	8/13	8/25				11.88
7	Drawings	8/25	9/2				8.28
8	Build	8/25	9/14				19.60
9	Monthly Reviews			Fri, Aug 30 — Fri, Sept 27 — Fri, Oct 25			
10	Production			9/17 — 11/7			50.96
11	Factory Prep	9/17	10/7				20.30
12	Materials Delivery	10/7	10/21				13.30
13	Production	10/21	11/2				12.18
14	Begin Shipping	11/7				Nov 7	

Monthly review meetings on last Friday of every month

axis. But like a chart, pictograms use icons, symbols, or other graphic devices rather than simple lines or bars. In Figure 14.22 stick figures illustrate population changes during a given period. Pictograms are visually appealing and can be especially useful for nontechnical or multicultural audiences. Graphics software makes it easy to create pictograms.

**FIGURE 14.22
A Pictogram**

In place of lines and bars, icons and symbols lend appeal and clarity, especially for nontechnical or multicultural audiences.
Source: U.S. Bureau of the Census.

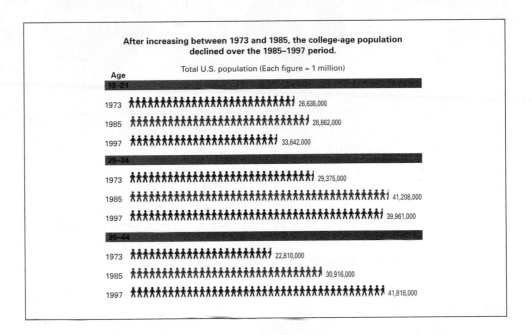

After increasing between 1973 and 1985, the college-age population declined over the 1985–1997 period.

Total U.S. population (Each figure = 1 million)

Age
19–24
1973　26,635,000
1985　28,862,000
1997　33,642,000

25–34
1973　29,375,000
1985　41,208,000
1997　39,961,000

35–44
1973　22,810,000
1985　30,916,000
1997　41,818,000

GRAPHIC ILLUSTRATIONS

An illustration is sometimes the best and only way to convey information. Illustrations can be diagrams, maps, drawings, icons, photographs, or any other visual that relies mainly on pictures rather than on data or words. For example, the diagram of a safety-belt locking mechanism in Figure 14.23 accomplishes what the verbal text alone cannot: it portrays the mechanism in operation.

Verbal text that requires a visual supplement

The safety-belt apparatus includes a tiny pendulum attached to a lever, or locking mechanism. Upon sudden deceleration, the pendulum swings forward, activating the locking device to keep passengers from pitching into the dashboard.

Illustrations are invaluable when you need to convey spatial relationships or help your audience see what something actually looks like. Drawings can be more effective than photographs because in a drawing you can simplify the view, remove any unnecessary features, and focus on what is important.

**FIGURE 14.23
A Diagram of a
Safety-Belt
Locking
Mechanism**

Shows how the
basic parts work
together.
*Source: U.S.
Department of
Transportation.*

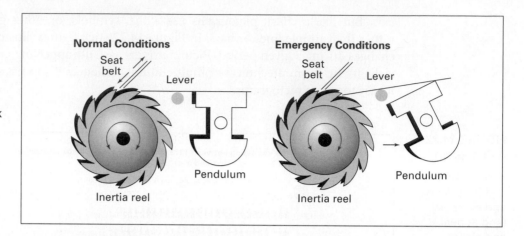

Diagrams

Diagrams are especially effective for presenting views that could not be captured by photographing or observing the object.

Exploded diagrams, like that of a brace for an adjustable basketball hoop in Figure 14.24, show how the parts of an item are assembled; they often appear in repair or maintenance manuals. Notice how parts are numbered for the user's easy reference to the written instructions.

**FIGURE 14.24
An Exploded
Diagram of a
Brace for a
Basketball Hoop**

Shows how
the parts are
assembled.
*Source: Courtesy
of Spalding.*

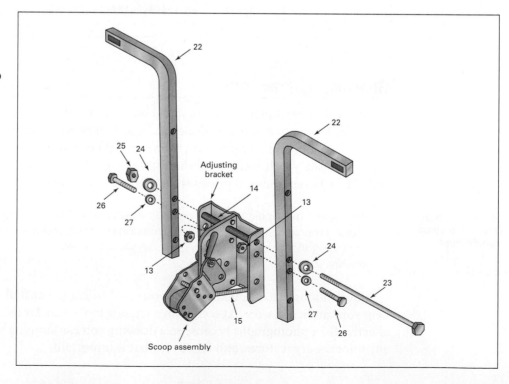

Cutaway diagrams show the item with its exterior layers removed in order to reveal interior sections, as in Figure 14.25. Unless the specific viewing perspective is immediately recognizable (as in Figure 14.25), name the angle of vision: "top view," "side view," and so on.

**FIGURE 14.25
Cutaway
Diagram
of a Surgical
Procedure**

Shows what
is inside.
Source: Trans-
sphenoidal
Approach for
Pituitary Tumor,
© *1986 by The
Ludann Co.,
Grand Rapids,
MI.*

THE OPERATION

Incision

Transsphenoidal surgery is performed with the patient under general anesthesia and positioned on his back. The head is fixed in a special headrest, and the operation is monitored on a special x-ray machine (fluoroscope).

In the approach illustrated here (not used by all surgeons), a small incision is made in one side of the nasal septum **(Fig. 2)**. Part of the septum is then removed to provide access to the sphenoid sinus cavity **(Fig. 3)**.

Figure 2
Incision into nasal septum

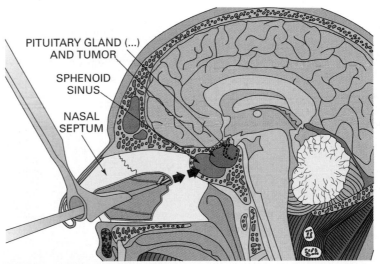

PITUITARY GLAND (...)
AND TUMOR

SPHENOID
SINUS

NASAL
SEPTUM

Figure 3
Removal of nasal septum to reach pituitary chamber

Block diagrams are simplified sketches that represent the relationship between the parts of an item, principle, system, or process. Because block diagrams are designed to illustrate *concepts* (such as current flow in a circuit), the parts are represented as symbols or shapes. The block diagram in Figure 14.26 illustrates how any process can be controlled automatically through a feedback mechanism. Figure 14.27 shows the feedback concept applied as the cruise-control mechanism on a motor vehicle.

**FIGURE 14.26
A Block
Diagram
Illustrating the
Concept of
Feedback**

Depicts a single
concept.

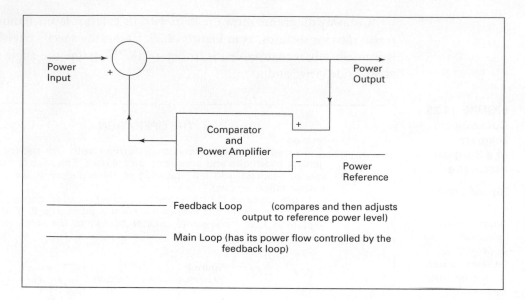

**FIGURE 14.27
A Block
Diagram
Illustrating a
Cruise-Control
Mechanism**

Depicts a spe-
cific application
of the feedback
concept.

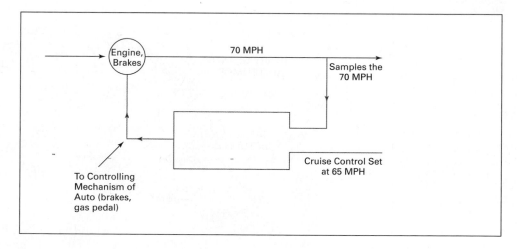

It is easy to create impressive-looking visuals by using electronic drawing programs, clip art, and image banks. However, specialized diagrams generally require
the services of graphic artists or technical illustrators. The client requesting or
commissioning the visual provides the art professional with an *art brief* (often prepared by writers and editors) that spells out the visual's purpose and specifications. The art brief is usually reinforced by a *thumbnail sketch,* a small, simple
sketch of the visual being requested. (See Chapter 15 for thumbnail sketches also
used in planning page layouts.) For example, part of the brief addressed to the
medical illustrator for Figure 14.25 might read as follows:

An art brief for
Figure 14.25

- **Purpose:** to illustrate transsphenoidal adenomectomy for laypersons
- **View:** full cutaway, sagittal
- **Range:** descending from cranial apex to a horizontal plane immediately below the upper jaw and second cervical vertebra
- **Depth:** medial cross-section
- **Structures omitted:** cranial nerves, vascular and lymphatic systems

A thumbnail
sketch of Figure
14.25

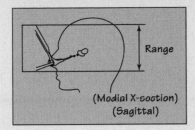

(Medial X-section)
(Sagittal)

- **Structures included:** gross anatomy of bone, cartilage, and soft tissue—delineated by color, shading, and texture
- **Structures highlighted:** nasal septum, sphenoid sinus, and sella turcica, showing the pituitary embedded in a 1.5 cm tumor invading the sphenoid sinus via an area of erosion at the base of the sella

Maps

Besides being visually engaging and easily remembered, maps are useful for showing comparisons and for helping users to *visualize* position, location, and relationships among complex data. Figure 14.28 conveys important statistical information in a format that is both accessible and understandable. Color enhances the percentage comparisons.

**FIGURE 14.28
A Map Rich
in Statistical
Significance**

Shows the
geographic
distribution
of data.
*Source: U.S.
Bureau of the
Census.*

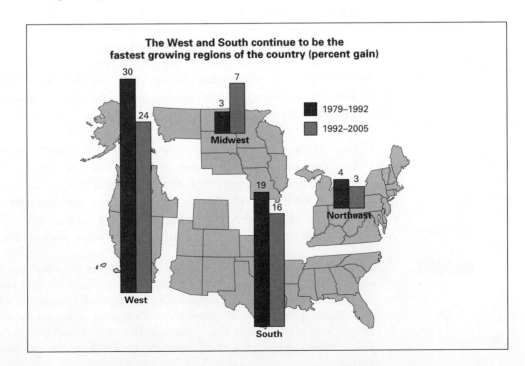

Photographs

Photographs are useful for showing exactly how something looks (Figure 14.29) or how something is done (Figure 14.30). But a photograph is hard to interpret if it includes needless details or fails to identify or emphasize the important material. One graphic design expert offers this advice for using photos in technical documents:

> To use pictures as tools for communication, pick them for their capacity to carry meaning, not just for their prettiness as photographs, . . . [but] for their inherent significance to the [document]. (White, *Great Pages* 110, 122)

Specialized photographs often require the services of a professional who knows how to use angles, lighting, and special film to achieve the desired focus and emphasis.

Whenever you include photographs in a document or presentation, follow these suggestions:

How to display a photograph

- *Try to simulate the approximate angle of vision readers would have in identifying or viewing the item or, for instruction, in doing the procedure (Figure 14.31).*
- *Trim (or crop) the photograph to eliminate needless details (Figures 14.32 and 14.33).*
- *For emphasizing selected features of a complex mechanism or procedure, consider using diagrams in place of photographs or as a supplement (Figures 14.34 and 14.35).*
- *Label all the parts readers need to identify (Figure 14.36).*
- *For an image unfamiliar to readers, provide a sense of scale by including a person, a ruler, or a familiar object (such as a hand) in the photo.*
- *If your document will be published, obtain a signed release from any person depicted in the photograph and written permission from the copyright holder. Beneath the photograph, cite the photographer and the copyright holder.*
- *In your discussion, refer to the photograph by figure number and explain what readers should look for. Or include a prose caption.*

WWW

14.5

Learn some basic photo editing techniques at <www.ablongman.com/lannonweb>

Digital imaging technology allows you to work with photographs in all sorts of ways. Also, commercial vendors such as PhotoDisc, Inc. <www.photodisc.com> offer digital libraries of royalty-free stock photographs via CD-ROM or the Web. (See page 329 for more Web sources.) For a fee, you can download photographs, edit, and alter them as needed by using a program such as *Adobe Photoshop*, and then insert these images in your own print or electronic documents.

NOTE *This capacity for altering photographic content creates potential for distortion and raises ethical questions about digital manipulation as well as legal questions about copyright and privacy infringement. On the ethical front, for example, a few key clicks can edit out, or insert, people or objects in a photograph or a video. Even live television video is subject to the same type of manipulation in real time. On the legal front, the unauthorized reproduction of a person's image could involve you in an expensive lawsuit.*

FIGURE 14.29   Shows the Appearance of Something   A Fixed-Platform Oil Rig

Source: SuperStock.

FIGURE 14.30   Shows How Something Is Done   Antibody Screening Procedure

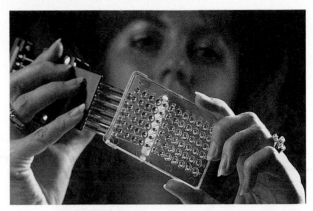

Source: SuperStock.

FIGURE 14.31   Shows a Realistic Angle of Vision   Titration in Measuring Electron-Spin Resonance

Source: SuperStock.

FIGURE 14.32   A Photograph That Needs to Be Cropped   Replacing the Microfilter Activation Unit

Source: SuperStock.

FIGURE 14.33   The Cropped Version of Figure 14.32

Source: SuperStock.

FIGURE 14.34   Shows a Complex Mechanism   Sapphire Tunable Laser

Source: SuperStock.

FIGURE 14.35   A Simplified Diagram of Figure 14.34   Major Parts of the Laser

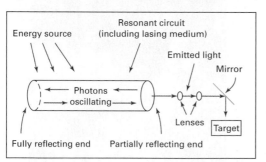

FIGURE 14.36   Shows Essential Features Labeled   Standard Flight Deck for a Long-Range Jet

Source: SuperStock.

COMPUTER GRAPHICS

Many of the tasks formerly done by graphic designers and technical illustrators now fall to people with little or no formal training. Whatever your career, you could be expected to produce high-quality graphics for conferences, presentations, and in-house publications. This text offers only a brief introduction to these matters. Your best bet is to learn all you can about computer graphics and graphic design (perhaps by taking a class).

Selecting Design Options

Here is a brief listing of design options offered by various computer programs:

A sampling of resources for electronic visual design

- Using spreadsheet software such as *Microsoft Excel,* you can create a variety of charts and graphs and update them easily whenever the data change. Or you can use the "Insert Chart" or "Insert Table" features in most word-processing programs.
- Using a drawing program such as *CorelDraw* or *Adobe Illustrator,* you can sketch, edit, and refine your diagrams and drawings on screen.
- Using project management software such as *Microsoft Project,* you can create Gantt charts, PERT charts, and other organizational and scheduling charts.
- Using presentation software such as *Microsoft PowerPoint,* you can create dynamic slides and other animated presentations.
- Using Web resources, you can import an endless array of graphics. For more on Web-based visual resources, see page 329.

Other, more specialized, programs for visual design are also available, but the ones above are fairly easy to master and useful to know.

Using Clip Art

Clip art is a generic term for collections of ready-to-use images (of computer equipment, maps, machinery, medical equipment, and so on), all stored electronically. Clip art packages allow you to import into your document countless images like the one in Figure 14.37. Using a drawing program, you can enlarge, enhance, or customize the image, as in Figure 14.38.

NOTE   *Although handy, clip art often has a generic or crude appearance that makes a document look unprofessional. Consider using clip art for icons only, for in-house documents, or for situations in which time or budget preclude using original artwork (Menz 5).*

One form of clip art especially useful in technical writing is the icon (an image with all nonessential background removed). Icons convey a specific idea visually as in Figure 14.39. Icons appear routinely in computer documentation and in other types of instructions because the images immediately signal the action desired.

**FIGURE 14.37
A Clip Art Image**

Offers a ready-made image that can be customized.

Source: Desktop Art®; Business I, © *Dynamic Graphics, Inc.*

**FIGURE 14.38
A Customized Image**

Source: Professor R. Armand Dumont.

Whenever you use an icon, be sure it is "intuitively recognizable" to multicultural users ("Using Icons" 3). Otherwise, people could misinterpret its meaning—in some cases with disastrous results.

NOTE

Certain icons have offensive connotations in certain cultures. Hand gestures, for example, are especially problematic: some Arab cultures consider the left hand unclean; a pointing index finger—on either hand—as in Figure 14.39, is a sign of rudeness in Venezuela or Sri Lanka (Bosley 5–6).

Using Color

Color often makes a presentation more interesting. Moreover, color attracts and focuses users' attention and helps them identify the various elements. In Figure 14.21, for example, color helps users sort out the key schedule elements of a Gantt chart for a major project: activities, time lines, durations, and meetings.

Color can help clarify a concept or dramatize how something works. In Figure 14.40 bright colors against a darker, duller background enable users to *visualize* the "heat mirror" concept.

Color can help clarify complex relationships. In Figure 14.41, an area map using six distinctive colors allows users to make various comparisons at a glance.

**FIGURE 14.39
Icons**

Provide a simple picture of the object, action, or concept each image represents.

Source: <4YEO.com/page elements/icons/signs/index.htm>

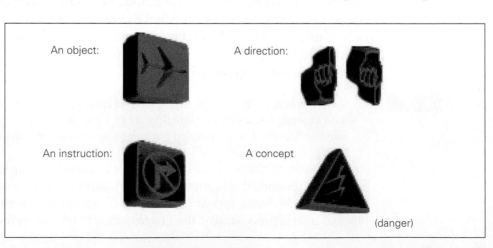

**FIGURE 14.40
Color Used as
a Visualizing
Tool**

*Source: Courtesy
of National
Audubon Society.*

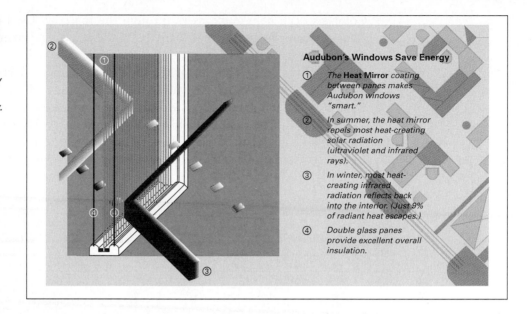

**FIGURE 14.41
Colors Used
to Show
Relationships**

*Source: National
Oceanic and
Atmospheric
Administration
(NOAA), National
Climatic Data
Center <www.
ncdc.noaa.gov>.*

14.6

Where is technical
visualization going
with 3-D graphics?
Find out more at
<www.ablongman.com/
lannonweb>

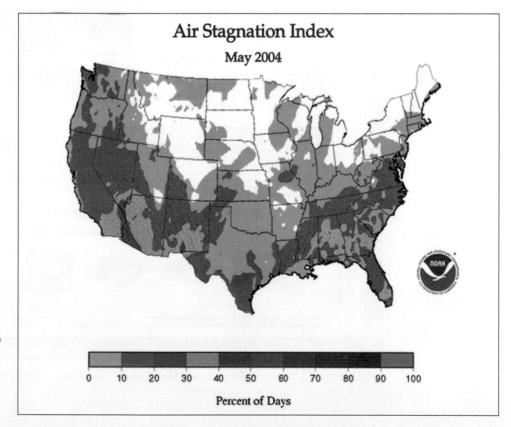

Color also can help guide users through the material. Used effectively on a printed page, color helps organize the user's understanding, provides orientation, and emphasizes important material.

On a Web page, color can mirror the site's main theme or "personality," orient the user, and provide cues for navigating the site. Figure 14.42 is a page from the National Oceanic and Atmospheric Administration. Sky-blue (or ocean-blue) as the dominant color (set against the blackness of outer space) reflects NOAA's mission in monitoring global climate. The striking image of the sun intersecting (or piercing) a depleted ozone shield helps underscore the urgency of ozone depletion

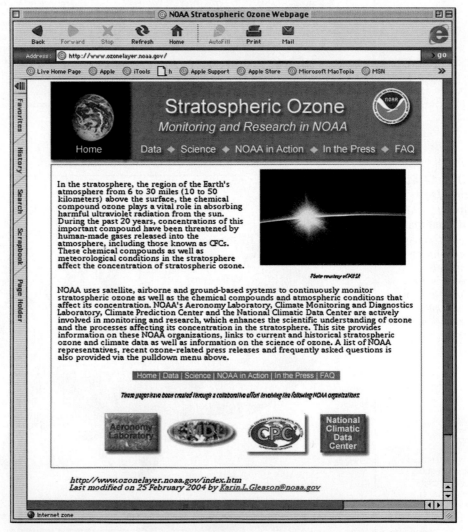

FIGURE 14.42   A Web Page That Uses Color Effectively This design embodies the advice on page 328 for incorporating color.

Source: National Oceanic and Atmospheric Administration <www.ozonelayer.noaa.gov>.

as an environmental issue. In the masthead and elsewhere, subtle links in gray reversed type evoke the subtlety of the ozone depletion itself—a gradual, subtle process, but one with potentially grave consequences.

Following are just a few possible uses of color in page design (White, *Color* 39–44; Keyes 647–49). For more on designing pages, see Chapter 15.

USE COLOR TO ORGANIZE.   Users look for ways of organizing their understanding of a document (Figure 14.43). Color can reveal structure and break material up into discrete blocks that are easier to locate, process, and digest.

**FIGURE 14.43
Color Used for
Organization**

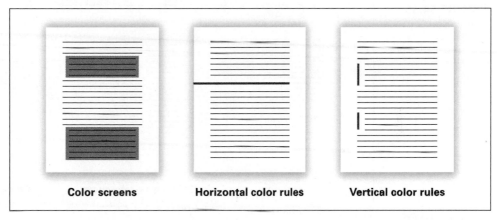

How color
reveals
organization

- A color background screen can set off like elements such as checklists, instructions, or examples.
- Horizontal rules can separate blocks of text, such as sections of a report or areas of a page.
- Vertical rules can set off examples, quotations, captions, and so on.

USE COLOR TO ORIENT.   Users look for signposts that help them find their place or find what they need (Figure 14.44).

**FIGURE 14.44
Color Used for
Orientation**

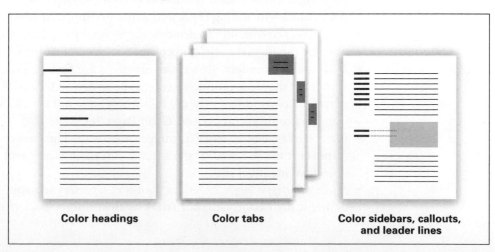

How color
provides
orientation

- Color can help headings stand out from the text and differentiate major from minor headings.
- Color tabs and boxes can serve as location markers.
- Color sidebars (for marginal comments), callouts (for labels), and leader lines (for connecting a label to its referent) can guide the eyes.

USE COLOR TO EMPHASIZE.   Users look for places to focus their attention in a document (Figure 14.45).

**FIGURE 14.45
Color Used for
Emphasis**

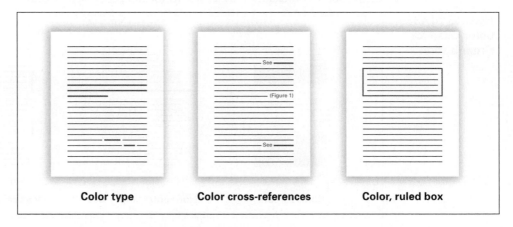

| Color type | Color cross-references | Color, ruled box |

How color
emphasizes

- Color type can highlight key words or ideas.
- Color can call attention to cross-references or to links on a Web page.
- A color ruled box can frame a warning, caution, note, or hint.

HOW TO INCORPORATE COLOR.   To use color effectively, follow these suggestions:

How to use color
for greatest effect

- *Use color sparingly.* Color gains impact when it is used selectively. It loses impact when it is overused (*Aldus Guide* 39). Use no more than three or four distinct colors—including black and white (White, *Great Pages* 76).
- *Apply color consistently to elements throughout your document.* Inconsistent use of color can distort users' perception of the relationships (Wickens 117).
- *Make color redundant.* Be sure all elements are first differentiated in black and white: by shape, location, texture, type style, or type size. Different readers perceive colors differently or, in some cases, not at all. Many readers have impaired color vision (White, *Great Pages* 76).
- *Use a darker color to make a stronger statement.* The darker the color the more important the material. Darker items can seem larger and closer than lighter objects of identical size.
- *Make color type larger or bolder than text type.* Try to avoid color for text type, or use a high-contrast color (dark against a light background). Color is less

**FIGURE 14.46
A Color Density
Chart**

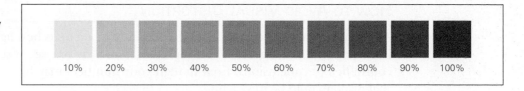

| 10% | 20% | 30% | 40% | 50% | 60% | 70% | 80% | 90% | 100% |

visible on the page than black ink on a white background. The smaller the image or the thinner the rule, the stronger or brighter the color should be (White, *Editing 229, 237*).

- *Create contrast.* For contrast in a color screen, use a very dark type against a very light background, say a 10- to 20-percent screen (Gribbons 70). The larger the screen area, the lighter the background color should be (Figure 14.46).

Colors have different meanings in different cultures. In the United States for example, red signifies danger and green traditionally signifies safety. But in Ireland, green or orange carry political connotations in certain contexts. In Muslim cultures, green is a holy color (Cotton 169).

USING WEB SITES FOR GRAPHICS SUPPORT

The World Wide Web offers a broad array of visual resources. Following is a sampling of useful Web sites and gateways:

Sources for
visuals on the
Web

- *Clip art:* For a comprehensive and updated directory to clip art sites, including numerous free sources, go to <www.clipart.com>.
- *Photographs:* Vintage photos of people, places, and products can be found at <www.classicphotos.com>. For links to all types of photography sites, go to <www.photolinks.net>, which offers a search engine as well.
- *Art images (of paintings, sculpture, and so on):* Go to <www.artresources.com> for a search engine and links to art sites worldwide.
- *Maps:* For links to countless varieties of local, global, and political maps, check out <www.nationalgeographic.com/maps/index.html>.
- *Audio and video:* For examples, instructions, and software for adding audio and video to your own Web site, go to <www.streamingmediaworld.com/>.
- *Miscellaneous resources:* For photographic images, illustrations, clip art, motion, audio, type fonts, and graphics software, go to <www.eyewire.com>.

NOTE

Be extremely cautious about downloading visuals (or any material, for that matter) from the Web and then using them. Review the copyright law (page 138). Originators of any work on the Web own the work and the copyright. Keep in mind that any photograph—including one that might be offered as "free" clip art—is protected by copyright.

How to Avoid Visual Distortion

Although you are perfectly justified in presenting data in its best light, you are ethically responsible for avoiding misrepresentation. Any one set of data can support contradictory conclusions. Even though your numbers may be accurate, your visual display could be misleading.

Present the Real Picture

Visual relationships in a graph should accurately portray the numerical relationships they represent. Begin the vertical scale at zero. Never compress the scales to reinforce your point.

Notice how visual relationships in Figure 14.47 become distorted when the value scale is compressed or fails to begin at zero. In version A, the bars accurately depict the numerical relationships measured from the value scale. In version B, item Z (400) is depicted as three times X (200). In version C, the scale is overly compressed, causing the shortened bars to understate the quantitative differences.

Deliberate distortions are unethical because they imply conclusions contradicted by the actual data.

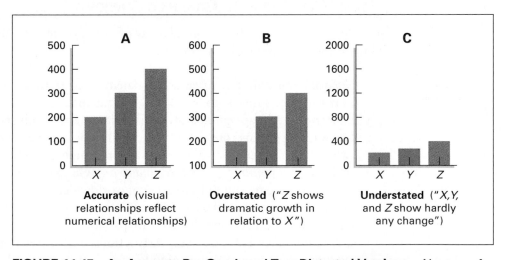

FIGURE 14.47   An Accurate Bar Graph and Two Distorted Versions   Absence of a zero baseline in B shrinks the vertical axis and exaggerates differences among the data.

In C, the excessive value range of the vertical axis dwarfs differences among the data.

Present the Complete Picture

Without getting bogged down in needless detail, an accurate visual includes all essential data. Figure 14.48 shows how distortion occurs when data that would provide a complete picture are selectively omitted. Version A accurately depicts the

**FIGURE 14.48
An Accurate
Line Graph and
a Distorted
Version**

Selective
omission of
data points in
B causes the
lines to flatten.

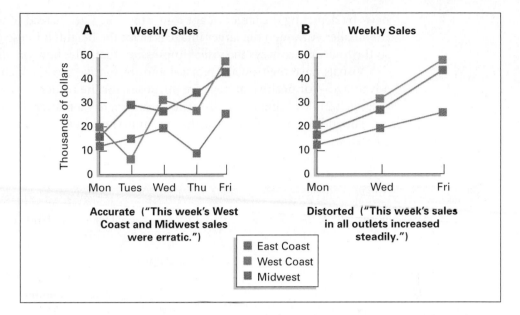

A   **Weekly Sales**   B   **Weekly Sales**

**Accurate ("This week's West
Coast and Midwest sales
were erratic.")**

**Distorted ("This week's sales
in all outlets increased
steadily.")**

East Coast
West Coast
Midwest

numerical relationships measured from the value scale. In version B, too few points are plotted. Always decide carefully what to include and what to leave out.

Don't Mistake Distortion for Emphasis

When you want to emphasize a point (a sales increase, a safety record, etc.), be sure your data support the conclusion implied by your visual. For instance, don't use inordinately large visuals to emphasize good news or small ones to downplay bad news (Williams 11). When using clip art, pictograms, or drawn images to dramatize a comparison, be sure the relative size of the images or icons reflects the quantities being compared.

A visual accurately depicting a 100-percent increase in phone sales at your company might look like version A in Figure 14.49. Version B overstates the good

**FIGURE 14.49
An Accurate
Pictogram and a
Distorted
Version**

In B, the relative
sizes of the
images are not
equivalent to the
quantities they
represent.

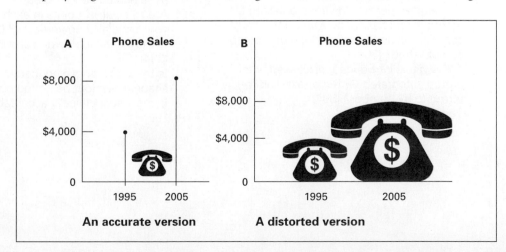

A   **Phone Sales**   B   **Phone Sales**

An accurate version   **A distorted version**

news by depicting the larger image four times the size, instead of twice the size, of the smaller. Although the larger image is twice the height, it is also twice the *width*, so the total area conveys the visual impression that sales have *quadrupled*.

Visuals have their own rhetorical and persuasive force, which you can use to advantage—for positive or negative purposes, for the reader's benefit or detriment (Van Pelt 2). Avoiding visual distortion is ultimately a matter of ethics.

For additional guidance, use the planning sheet in Figure 14.50, and the checklist below.

☑ CHECKLIST for Usability of Visuals

(Numbers in parentheses refer to the first page of discussion.)

Content

- ☐ Does the visual serve a legitimate purpose (clarification, not mere ornamentation) in the document? (292)
- ☐ Is the visual titled and numbered? (334)
- ☐ Is the level of complexity appropriate for the audience? (334)
- ☐ Are all patterns in the visual identified by label or legend? (302)
- ☐ Are all values or units of measurement specified (grams per ounce, millions of dollars)? (301)
- ☐ Are the numbers accurate and exact? (301)
- ☐ Do the visual relationships represent the numerical relationships accurately? (330)
- ☐ Are explanatory notes added as needed? (301)
- ☐ Are all data sources cited? (301)
- ☐ Has written permission been obtained for reproducing or adapting a visual from a copyrighted source in any type of work to be published? (301)
- ☐ Is the visual introduced, discussed, interpreted, integrated with the text, and referred to by number? (334)

- ☐ Can the visual itself stand alone in terms of meaning? (334)

Arrangement

- ☐ Is the visual easy to locate? (334)
- ☐ Do all design elements (title, line thickness, legends, notes, borders, white space) work to achieve balance? (334)
- ☐ Is the visual positioned on the page to achieve balance? (334)
- ☐ Is the visual set off by adequate white space or borders? (334)
- ☐ Is a broadside visual turned 90 degrees so that the left-hand side is at the bottom of the page? (300)
- ☐ Is the visual in the best location in the document? (334)

Style

- ☐ Is this the best type of visual for your purpose and audience? (295)
- ☐ Are all decimal points in each column of your table aligned vertically? (301)
- ☐ Is the visual uncrowded and uncluttered? (334)
- ☐ Is the visual engaging (patterns, colors, shapes), without being too busy? (334)
- ☐ Is the visual ethically acceptable? (330)

Focusing on Your Purpose

• What is this visual's purpose (to instruct, persuade, create interest)?_____

• What forms of information (numbers, shapes, words, pictures, symbols) will this visual depict?

  _____

• What kind of relationship(s) will the visual depict (comparison, cause-effect, connected parts, sequence of steps)? _____

• What judgment, conclusion, or interpretation is being emphasized (that profits have increased, that toxic levels are rising, that X is better than Y, that time is being wasted)?

  _____

• Is a visual needed at all? _____

Focusing on Your Audience

• Is this audience accustomed to interpreting visuals? _____

• Is the audience interested in specific numbers or an overall view? _____

• Should the audience focus on one exact value, compare two or more values, or synthesize a range of approximate values (Wickens 121)? _____

• Which type of visual will be most accurate, representative, accessible, and compatible with the type of judgment, action, or understanding expected from the audience? _____

• In place of one complicated visual, would two or more straightforward ones be preferable?

  _____

Focusing on Your Presentation

• What enhancements, if any, will increase audience interest (colors, patterns, legends, labels, varied typefaces, shadowing, enlargement or reduction of some features)? _____

  _____

• Which medium—or combination of media—will be most effective for presenting this visual (slides, transparencies, handouts, large-screen monitor, flip chart, report text)? _____

  _____

• To achieve the greatest utility and effect, where in the presentation does this visual belong?

  _____

FIGURE 14.50   A Planning Sheet for Preparing Visuals

GUIDELINES for Fitting Visuals with Printed Text

Ensure that your document's visual and verbal elements complement each other:

1. *Place the visual where it will best serve your readers.* If it is central to your discussion, place the visual as close as possible to the material it clarifies. (Achieving proximity often requires that you ignore the traditional "top or bottom" design rule for placing visuals on a page.) If the visual is peripheral to your discussion or of interest to only a few readers, place it in an appendix so that interested readers can refer to it. Tell readers when to consult the visual and where to find it.

2. *Never refer to a visual that readers cannot easily locate.* In a long document, don't be afraid to repeat a visual if you discuss it again later.

3. *Never crowd a visual into a cramped space.* Set the visual off by framing it with plenty of white space, and position it on the page for balance. To save space and to achieve proportion with the surrounding text, consider the size of each visual and the amount of space it will occupy.

4. *Number the visual and give it a clear title and labels.* Your title should tell readers what they are seeing. Label all the important material and cite the source of data or of graphics.

5. *Match the visual to your audience.* Don't make it too elementary for specialists or too complex for nonspecialists. Your intended audience should be able to interpret the visual correctly.

6. *Introduce and interpret the visual.* In your introduction, tell readers what to expect:

   INFORMATIVE   As Table 2 shows, operating costs have increased
                   7 percent annually since 1980.
   UNINFORMATIVE   See Table 2.

   Visuals alone make ambiguous statements (Girill, "Technical Communication and Art" 35); pictures need to be interpreted. Instead of leaving readers to struggle with a page of raw data, explain the relationships displayed. Follow the visual with a discussion of its important features:

   INFORMATIVE   This cost increase means that . . . .

   Always tell readers what to look for and what it means.

7. *Use prose captions to explain important points made by the visual.* Captions help readers interpret a visual. Use a smaller type size so that captions don't compete with text type (*Aldus Guide* 35).

8. *Never include excessive information in one visual.* Any visual that contains too many lines, bars, numbers, colors, or patterns will overwhelm readers. In place of one complicated visual, use two or more straightforward ones.

9. *Be sure the visual can stand alone.* Even though it repeats or augments information already in the text, the visual should contain everything users will need to interpret it correctly.

EXERCISES   For more exercises, visit
<www.ablongman.com/lannon>

1. The following statistics are based on data from three colleges in a large western city. They give the number of applicants to each college over six years.

   - In 2000, X college received 2,341 applications for admission, Y college received 3,116, and Z college 1,807.
   - In 2001, X college received 2,410 applications for admission, Y college received 3,224, and Z college 1,784.
   - In 2002, X college received 2,689 applications for admission, Y college received 2,976, and Z college 1,929.
   - In 2003, X college received 2,714 applications for admission, Y college received 2,840, and Z college 1,992.
   - In 2004, X college received 2,872 applications for admission, Y college received 2,615, and Z college 2,112.
   - In 2005, X college received 2,868 applications for admission, Y college received 2,421, and Z college 2,267.

   Display these data in a line graph, a bar graph, and a table. Which version seems most effective for a reader who (a) wants exact figures, (b) wonders how overall enrollments are changing, or (c) wants to compare enrollments at each college in a certain year? Include a caption interpreting each version.

2. Devise a flowchart for a process in your field or area of interest. Include a title and a brief discussion.

3. Devise an organization chart showing the lines of responsibility and authority in an organization where you work.

4. Devise a pie chart to depict your yearly expenses. Title and discuss the chart.

5. Obtain enrollment figures at your college for the past five years by gender, age, race, or any other pertinent category. Construct a stacked-bar graph to illustrate one of these relationships over the five years.

6. Keep track of your pulse and respiration at thirty-minute intervals over a four-hour period of changing activities. Record your findings in a line graph, noting the times and specific activities below your horizontal coordinate. Write a prose interpretation of your graph and give the graph a title.

7. In textbooks or professional journal articles, locate each of these visuals: a table, a multiple-bar graph, a multiple-line graph, a diagram, and a photograph. Evaluate each according to the revision checklist, and discuss the most effective visual in class.

8. Choose the most appropriate visual for illustrating each of these relationships. Justify each choice in a short paragraph.

   a. A comparison of three top brands of skis, according to cost, weight, durability, and edge control.
   b. A breakdown of your monthly budget.
   c. The changing cost of an average cup of coffee, as opposed to that of an average cup of tea, over the past three years.
   d. The percentage of college graduates finding desirable jobs within three months after graduation, over the last ten years.
   e. The percentage of college graduates finding desirable jobs within three months after graduation, over the last ten years—by gender.
   f. An illustration of automobile damage for an insurance claim.
   g. A breakdown of the process of radio wave transmission.
   h. A comparison of five cereals on the basis of cost and nutritive value.
   i. A comparison of the average age of students enrolled at your college in summer, day, and evening programs, over the last five years.
   j. Comparative sales figures for three items made by your company.

9. *Computer graphics:* Compose and enhance one or more visuals electronically. You might begin by looking through a recent edition of the *Statistical Abstract of the United States* (in the

government documents, reserve, or reference section of your library). From the *Abstract,* or from a source you prefer, select a body of numerical data that will interest your classmates. After completing the planning sheet in Figure 14.50 (page 333), compose one or more visuals to convey a message about your data to make a point, as in these examples:

- Consumer buying power has increased or decreased since 1990.
- Defense spending, as a percentage of the federal budget, has increased or decreased since 2000.
- Average yearly temperatures across the United States are rising or falling.

Experiment with formats and design options, and enhance your visual(s) as appropriate. Add any necessary prose explanations.

Be prepared to present your visual message in class, using either an overhead or opaque projector or a large-screen monitor.

10. Revise the layout of Table 14.5 according to the guidelines on page 334, and explain to readers the significant comparisons in the table. (*Hint:* The unit of measurement is percentage.)

11. Display each of these sets of information in the visual format most appropriate for the stipulated audience. Complete the planning sheet in Figure 14.50 for each visual. Explain why you

selected the type of visual as most effective for that audience. Include with each visual a brief prose passage interpreting and explaining the data.

a. (For general readers.) Assume that the Department of Energy breaks down energy consumption in the United States (by source) into these percentages: In 1970, coal, 18.5; natural gas, 32.8; hydro and geothermal, 3.1; nuclear, 1.2; oil, 44.4. In 1980, coal, 20.3; natural gas, 26.9; hydro and geothermal, 3.8; nuclear, 4.0; oil, 45.0. In 1990, coal, 23.5; natural gas, 23.8; hydro and geothermal, 7.3; nuclear, 4.1; oil, 41.3. In 2000, coal, 20.3; natural gas, 25.2; hydro and geothermal, 9.6; nuclear, 6.3; oil, 38.6.

b. (For experienced investors in rental property.) As an aid in estimating annual heating and air-conditioning costs, here are annual maximum and minimum temperature averages from 1911 to 2000 for five Sunbelt cities (in Fahrenheit degrees): In Jacksonville, the average maximum was 78.4; the minimum was 57.6. In Miami, the maximum was 84.2; the minimum was 69.1. In Atlanta, the maximum was 72.0; the minimum was 52.3. In Dallas, the maximum was 75.8; the minimum was 55.1. In Houston, the maximum was 79.4; the minimum was 58.2. (From U.S. National Oceanic and Atmospheric Administration.)

TABLE 14.5
An Example of a Poor Layout

Educational Attainment of Persons 25 Years Old and Over

	Highest level completed	
Year	High school (4 years or more)	College (4 years or more)
1970	52.3164	10.7431
1980	66.5432	16.2982
1982	71.0178	17.7341
1984	73.3124	19.1628
1993	80.2431	21.9316
1996	81.7498	23.6874
2000	84.1354	26.7216

Source: Adapted from Statistical Abstract of the United States: 2003 (123rd ed.). *Washington, DC: GPO. 154.*

c. (For the student senate.) Among the students who entered our school four years ago, here are the percentages of those who graduated, withdrew, or are still enrolled: In Nursing, 71 percent graduated; 27.9 percent withdrew; 1.1 percent are still enrolled. In Engineering, 62 percent graduated; 29.2 percent withdrew; 8.8 percent are still enrolled. In Business, 53.6 percent graduated; 43 percent withdrew; 3.4 percent are still enrolled. In Arts and Sciences, 27.5 percent graduated; 68 percent withdrew; 4.5 percent are still enrolled.

d. (For the student senate.) Here are the enrollment trends from 1993 to 2005 for two colleges in our university. In Engineering: 1993, 455 students enrolled; 1994, 610; 1995, 654; 1996, 758; 1997, 803; 1998, 827; 1999, 1046; 2000, 1200; 2001, 1115; 2002, 1075; 2003, 1116; 2004, 1145; 2005, 1177. In Business: 1993, 922; 1994, 1006; 1995, 1041; 1996, 1198; 1997, 1188; 1999, 1227; 1999, 1115; 2000, 1220; 2001, 1241; 1992, 1366; 2003, 1381; 2004, 1402; 2005, 1426.

12. Anywhere on campus or at work, locate at least one visual that needs revision for accuracy, clarity, appearance, or appropriateness. Look in computer manuals, lab manuals, newsletters, financial aid or admissions or placement brochures, student or faculty handbooks, newspapers, or textbooks. Use the planning sheet in Figure 14.50 and the checklist (page 334) as guides to revise and enhance the visual. Submit to your instructor a copy of the original, along with a memo explaining your improvements. Be prepared to discuss your revision in class.

13. Locate a document (news, magazine, or journal article, brief instructions, or the like) that lacks adequate or appropriate visuals. Analyze the document and identify where visuals would be helpful. In a memo to the document's editor or author, provide an art brief and a thumbnail sketch (page 318) for each visual you would recommend, specifying its exact placement in the document.

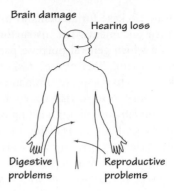

Source: U.S. Environmental Protection Agency. Protect Your Family from Lead in Your Home, 1995. 3.

Note: Be sure to provide enough detail for your audience to understand your suggestion clearly. For example, instead of merely recommending a "diagram of the toxic effects of lead on humans," stipulate a "diagram showing a frontal outline of the human body with the head turned sideways in profile view. Labels and arrows point to affected body areas to indicate brain damage, hearing problems, digestive problems, and reproductive problems."

14. Locate a Web page that uses color effectively to mirror the site's main theme or personality, to orient the user, and to provide cues for easy navigation. Download the Web page to a floppy disk and print it out using a color printer (or print the page directly from your screen if your computer has its own color printer). Using the analysis of Figure 14.42 (page 326) as a model, prepare a brief memo justifying your choice. Be prepared to discuss and illustrate the Web page's effectiveness in class.

Note: If your classroom is equipped with a computer and a large-screen monitor, consider doing your presentation electronically.

COLLABORATIVE PROJECTS

1. Assume that your technical writing instructor is planning to purchase five copies of a graphics

software package for students to use in designing their documents. The instructor has not yet decided which general-purpose package would be most useful. Your group's task is to test one package and to make a recommendation.

In small groups, visit your school's microcomputer lab and ask for a listing of the graphics packages that are available to students and faculty. Select one package and learn how to use it. Design at least four representative visuals. In a memo or presentation to your instructor and classmates, describe the package briefly and tell what it can do. Would you recommend purchasing five copies of this package for general-purpose use by writing students? Explain. Submit your report, along with the sample graphics you have composed.

Do the same assignment, comparing various clip art packages. Which package offers the best image selection for writers in your specialty?

2. Compile a list of six World Wide Web sites that offer graphics support by way of advice, image banks, design ideas, artwork catalogs, and the like. Provide the address for each site, along with a description of the resources offered and their approximate cost. Report your findings in the format stipulated by your instructor. See page 329 for URLs that will get you started.

15

Designing Pages and Documents

15.1

How does page design affect workplace dynamics? Find out more at <www.ablongman.com/lannonweb>

Page design, the layout of words and graphics, determines the look of a document. Well-designed pages invite users in, guide them through the material, and help them understand and remember it.

In this electronic age the term "page" takes on broad meanings: On the computer screen, a page can scroll on endlessly. Also, *page* might mean a page of a report, but it can also mean one panel of a brochure or part of a reference card for installing printer software. The following discussion focuses mainly on traditional paper (printed) pages. See Designing On-Screen Documents later in this chapter for a discussion of pages in electronic documents.

PAGE DESIGN IN WORKPLACE DOCUMENTS

Technical documents rarely get undivided attention

People read work-related documents only because they have to. If they have easier ways of getting the information, people will use them. In fact, busy users often only skim a document, or they refer to certain sections during a meeting or presentation. Amid frequent distractions, users want to be able to leave the document and then return and locate what they need easily

NOTE

The so-called "paperless office" is largely a myth. In fact, information technology produces more paper than ever. Also, as computers generate more and more written messages, both electronic and hard copy, any document competes for audience attention. Overwhelmed by information overload, people resist any document that looks hard to get through.

Readers are attracted by documents that appear inviting and accessible

Before actually reading the document, people usually scan it first, to get a sense of what it's about and how it's organized. An audience's first impression tends to involve a purely visual, esthetic judgment: "Does this look like something I want to read, or like too much work?" Instead of an unbroken sequence of paragraphs, users look for charts, diagrams, lists, various type sizes and fonts, different levels of headings, and other aids to navigation. Having decided at a glance whether your document is visually appealing, logically organized, and easy to navigate, users will draw conclusions about the value of your information, the quality of your work, and your overall credibility.

HOW PAGE DESIGN TRANSFORMS A DOCUMENT

To appreciate the impact of page design, consider Figures 15.1 and 15.2: Notice how the information in Figure 15.1 resists interpretation. Without design cues, we have no way of chunking this information into organized units of meaning. Figure 15.2 shows the same information after a design overhaul.

**FIGURE 15.1
Ineffective Page
Design**
This design
provides no
"road map" to
indicate how the
document is
organized or
what main ideas
it conveys.
*Source: U.S.
Department of
Energy.*

Sunspaces

Either as an addition to a home or as an integral part of a new home, sunspaces have gained considerable popularity.

A sunspace should face within 30 degrees of true south. In the winter, sunlight passes through the windows and warms the darkened surface of a concrete floor, brick wall, water-filled drums, or other storage mass. The concrete, brick, or water absorbs and stores some of the heat until after sunset, when the indoor temperature begins to cool. The heat not absorbed by the storage elements can raise the daytime air temperature inside the sunspace to as high as 100 degrees Fahrenheit. As long as the sun shines, this heat can be circulated into the house by natural air currents or drawn in by a low-horsepower fan.

To be considered a passive solar heating system, any sunspace must consist of these parts: a collector, such as a double layer of glass or plastic; an absorber, usually the darkened surface of the wall, floor, or water-filled containers inside the sunspace; a storage mass, normally concrete, brick, or water, which retains heat after it has been absorbed; a distribution system, the means of getting the heat into and around the house by fans or natural air currents; and a control system, or heat-regulating device, such as movable insulation, to prevent heat loss from the sunspace at night. Other controls include roof overhangs that block the summer sun, and thermostats that activate fans.

DESIGN SKILLS NEEDED IN TODAY'S WORKPLACE

As more and more software is developed to help people with page layout and document design, you very well may be responsible for preparing actual publications as part of your job—often without the help of clerical staff, print shops, and graphic artists. In such cases, you will need to master a variety of technologies and to observe specific guidelines.

Sunspaces

Either as an addition to a home or as an integral part of a new home, sunspaces have gained considerable popularity.

How Sunspaces Work

A sunspace should face within 30 degrees of true south. In the winter, sunlight passes through the windows and warms the darkened surface of a concrete floor, brick wall, water-filled drums, or other storage mass. The concrete, brick, or water absorbs and stores some of the heat until after sunset, when the indoor temperature begins to cool.

The heat *not* absorbed by the storage elements can raise the daytime air temperature inside the sunspace to as high as 100 degrees Fahrenheit. As long as the sun shines, this heat can be circulated into the house by natural air currents or drawn in by a low-horsepower fan.

The Parts of a Sunspace

To be considered a passive solar heating system, any sunspace must consist of these parts:

1. A *collector,* such as a double layer of glass or plastic.

2. An *absorber,* usually the darkened surface of the wall, floor, or water-filled containers inside the sunspace.

3. A *storage mass,* normally concrete, brick, or water, which retains heat after it has been absorbed.

4. A *distribution system,* the means of getting the heat into and around the house (by fans or natural air currents).

5. A *control system* (or heat-regulating device), such as movable insulation, to prevent heat loss from the sunspace at night. Other controls include roof overhangs that block the summer sun, and thermostats that activate fans.

FIGURE 15.2 Effective Page Design   Headings, spacing, color, italics, and listed items provide immediate clues as to how this document is organized, which ideas are most important, and how the ideas relate.

Desktop Publishing

Desktop publishing (DTP) systems such as *PageMaker, Adobe Framemaker,* or *Quark* combine word processing, typesetting, and graphics. Using this software along with optical scanners, and laser printers, one person, or a group working collaboratively, controls the entire production cycle: designing, illustrating, laying out, and printing the final document (Cotton 36–47):

What DTP systems can do

- Text can be typed or scanned into the program and then edited, checked for spelling and grammar, displayed in columns or other spatial arrangements, set in a variety of sizes and fonts—or sent electronically.
- Page highlights and orienting devices can be added: headings, ruled boxes, vertical or horizontal rules, colored background screens, marginal sidebars or labels, page locator tabs, shadowing, shading, and so on.
- Images can be drawn directly or imported into the program via scanners; charts, graphs, and diagrams can be drawn with graphics programs. These visuals then can be enlarged, reduced, cropped, and pasted electronically on the text pages.
- All work at all stages can be stored electronically for later use, adaptation, or updating. Documents or parts of documents used repeatedly (*boilerplate*) can be retrieved when needed, or modified or inserted in some other document.

Many of these DTP features are now contained in today's sophisticated word-processing programs. And with the enhancement of *groupware* (group authoring systems), writers from different locations can produce and distribute drafts online, incorporate reviewers' comments into their drafts, and publish documents collaboratively.

Electronic Publishing

Your work may involve electronic publishing (epublishing), in which you use programs such as *RoboHelp, Rainmaker,* or *Dreamweaver* to create documents in digital format for the Web, the company intranet, or as online help screens. You also might produce Portable Document Files, PDF versions of a document, using software such as **Build***fire.*

For projects that will be shared across different types of computer platforms, you might use markup languages. These languages use marks, or "tags," to indicate where the text should be bold, indented, italicized, and so on. Word-processing or page layout files from different programs or platforms are not always compatible, but with markup languages, once the tags are inserted, documents can be shared across many platforms.

Two examples of markup languages are standardized general markup language (SGML) and hypertext markup language (HTML). The first, SGML, is used for printed documents. The second, HTML, is used for hypertext pages—electronic documents such as online help screens or Web pages. For more on markup languages, see pages 462–63.

Using Style Sheets and Company Style Guides

Style sheets are helpful guides that ensure consistency across a single document or a set of documents. If you are working as part of a team, each writer needs to be using the same typefaces, fonts, headings, and other elements in identical fashion.

Possible style-sheet entries

- The first time you use or define a specialized term, highlight it with italics or **boldface.**
- In headings, capitalize prepositions of five or more letters ("Between," "Versus").

The more complex the document, the more specific the style sheet should be. All writers and editors should have a copy. Consider keeping the style sheet on a Web page for easy access and efficient updating.

In addition to style sheets for specific documents, some organizations produce style guides containing rules for proper use of trade names, appropriate punctuation, preferred fonts and typefaces, and so on. Style guides help ensure a consistent look across a company's various documents and publications. See Figure 15.3 for a sample style-guide page.

CREATING A USABLE DESIGN

Approach your design decisions from the top down. First, consider the overall look of your pages; next, the shape of each paragraph; and finally, the size and style of individual words and letters (Kirsh 112). Figure 15.4 depicts how design considerations move from large matters to small.

NOTE   *All design considerations are influenced by the budget for a publication. For instance, adding a single color, say, to major heads, can double the printing cost.*

If your organization prescribes no specific guidelines, the following design principles should serve in most situations.

Shaping the Page

In shaping a page, consider its look, feel, and overall layout. The following suggestions will help you shape appealing and usable pages.

USE THE RIGHT PAPER AND INK.   For routine documents (memos, letters, in-house reports) print in black ink, on $8\frac{1}{2}$-by-11-inch low-gloss, white paper. Use rag-bond paper (20 pound or heavier) with a high fiber content (25 percent minimum). Shiny paper produces glare that tires the eyes. Flimsy or waxy paper feels inferior.

**FIGURE 15.3
A Sample
Page from a
Publishing
Company's
Style Guide for
Authors**

Source: Guide
for Authors,
*published by
Addison Wesley
Longman, Inc.,
1998. Reprinted
by permission
of Pearson
Education, Inc.*

TEXT HEADINGS

Headings break up the text and help students understand the material. As a rule, they should be short phrases or single words, not sentences. Generally, three levels of headings are sufficient for a well-organized textbook. Headings should generally not occur together without text between them, and good structure dictates that there should be more than one subhead to a section to legitimize dividing it up. Try to be consistent in the frequency and use of headings from one chapter to another. For example, if you begin one chapter with a heading, do this in all other chapters. It will be most helpful to the copy editor, designer, and compositor if you type all headings of the same value in a consistent way.

{A} A-Heads

First-level heads (or A-heads) are the main chapter heads; type them on a line alone, centered on the page, in upper- and lower-case letters. Leave one line of space above and below. In addition, label the heading with its value by typing {A} preceding the heading.

Traditionally, first-level headings are numbered in math and some science texts. Double numbers, consecutive within chapters, are used. For example, 1.6 would designate the sixth A-head in Chapter 1.

{B} B-heads

Second-level heads (or B-heads) are subdivisions of the first-level head; position them on a line alone, flush left on the page, and in upper- and lower-case letters. Leave one line of space above and below. Label each B-head with a {B} preceding the heading. Since B-heads are subdivisions of A-heads, they should only be used when an A-head has previously been introduced.

{C} C-heads The third-level head is positioned as a "run-in" heading. Type it on a paragraph indent, in upper- and lower-case letters, and run into the text that follows. Leave one line of space above. Label each C-head with a {C} preceding the heading.

In some textbooks, particularly in technical disciplines, additional head levels or types of headings may be required. You should discuss the styling of such headings with your acquisitions or developmental editor. Type all headings of equal value in the same way.

**FIGURE 15.4
A Flowchart for
Decisions in
Page Design**
A top-down
design strategy
moves from
large elements
to small.

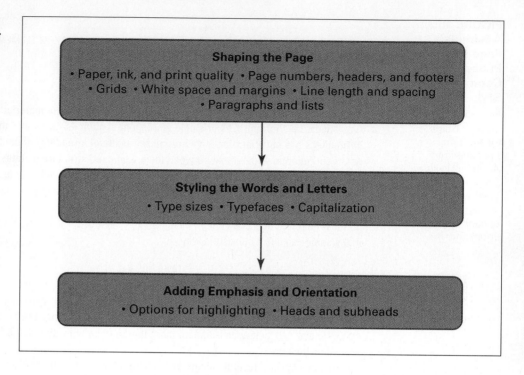

For documents that will be published (manuals, marketing literature), consider the paper's grade and quality. Paper varies in weight, grain, and finish—from low-cost newsprint, with noticeable wood fiber, to wood-free, specially coated paper with custom finishes. Choice of paper depends on the artwork to be included, the type of printing, and the intended esthetic effect: For example, you might choose specially coated, heavyweight, glossy paper for an elegant effect in an annual report (Cotton 73).

USE HIGH-QUALITY TYPE OR PRINT.   Print hard copy on an inkjet or laser printer. If your inkjet's output is blurry, consider purchasing special inkjet paper.

USE CONSISTENT PAGE NUMBERS, HEADERS, AND FOOTERS.   For a long document, count your title page as page i, without numbering it, and number all front matter pages, including the table of contents and abstract, with lowercase roman numerals (ii, iii, iv). Number the first text page and subsequent pages with arabic numerals (1, 2, 3). Along with page numbers, *headers* or *footers* appear in the top or bottom page margins, respectively. These provide chapter or article titles, authors' names, dates, or other publication information. (See, for example, the headers on the pages in this book and on page 345.)

USE A GRID Readers make sense of a page by looking for a predictable and consistent underlying structure, with the various elements located where they expect. By subdividing a page (or screen) into square and rectangular modules, a grid helps you organize your layout (Hilligoss 97).

With a view of a page's Big Picture, you can plan the size and placement of your visuals and calculate the number of lines available for written text. Most important, you can rearrange text and visuals repeatedly to achieve a balanced and consistent design (White, *Editing* 58). Here are just a few of the many possible grid patterns:

Grids provide a blueprint for page design

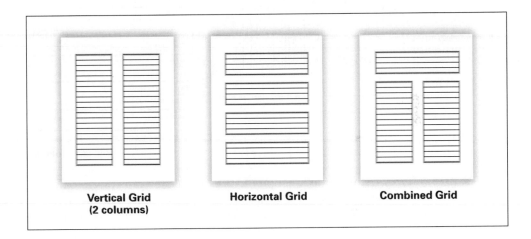

Vertical Grid (2 columns) Horizontal Grid Combined Grid

Some form of two-column grid is commonly used in manuals. See also the *Consider This* boxes and Checklists for Usability in this text. Brochures and newsletters typically use a two- or three-column grid. Web pages (see page 362) often use a combined vertical/horizontal grid. Figures 15.1 and 15.2 use a single-column grid, as do most memos, letters, and reports. (Grids are also used in storyboarding; see page 227.)

Figure 15.5 illustrates how a horizontal grid can transform the design of important medical information for consumers.

NOTE

While grid structures are especially useful in laying out newsletters and Web pages, they can be overly restrictive, allowing too much or too little space for the text that is intended. As a result, the text is forced to fit into the "mold" imposed by the grid (White, Editing 58). If this happens, be prepared to reconfigure your grid.

15.2
Find out how white space conveys attitude at <www.ablongman.com/lannonweb>

USE ADEQUATE WHITE SPACE. White space is all the space not filled by text or images. White space divides printed areas into small, digestible chunks. For instance, it separates sections in a document, headings and visuals from text, paragraphs on a page.

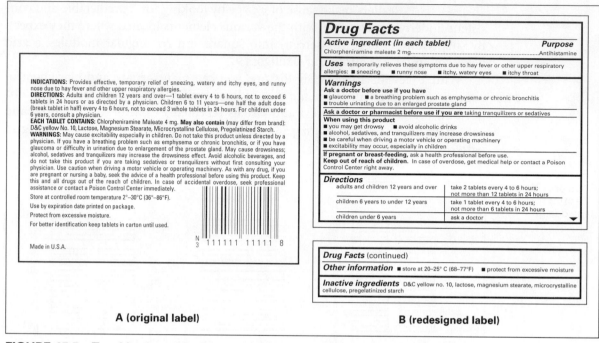

A (original label)   **B (redesigned label)**

FIGURE 15.5   Two Versions of a Consumer Label   Notice how the discrete horizontal modules in version B provide an underlying structure that is easy to navigate.
Source: Nordenberg, Tamar. "New Drug Label Spells It Out Simply." FDA Consumer (Reprint) July 1999.

Well-designed white space imparts a shape to the whole document, a shape that orients users and lends a distinctive visual form to the printed matter by:

1. keeping related elements together
2. isolating and emphasizing important elements
3. providing breathing room between blocks of information

Use white space to orient the readers

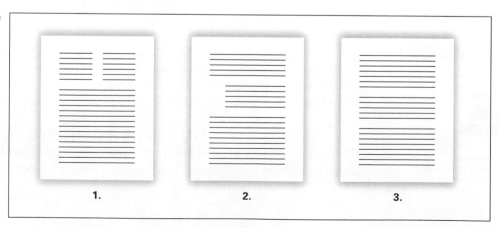

1.   2.   3.

Pages that look uncluttered, inviting, and easy to follow convey an immediate sense of user-friendliness.

PROVIDE AMPLE AND APPROPRIATE MARGINS.   Small margins crowd the page and make the material look difficult. On your $8\frac{1}{2}$-by-11-inch page, leave margins of at least 1 or $1\frac{1}{2}$ inches. If the manuscript is to be bound in some kind of cover, widen the inside margin to two inches.

Headings, lines of text, or visuals that abut the right or left margin, without indentation, are designated as *flush right* or *flush left*.

Choose between *unjustified* text (uneven or "ragged" right margins) and *justified* text (even right margins). Each arrangement creates its own "feel."

Justified lines are set flush left and right.

To make the right margin even in justified text, the spaces vary between words and letters on a line, sometimes creating channels or rivers of white space. The eyes are then forced to adjust continually to these space variations within a line or paragraph. Because each line ends at an identical vertical space, the eyes must work harder to differentiate one line from another (Felker 85). Moreover, to preserve the even margin, words at line's end are often hyphenated, and frequently hyphenated line endings can be distracting.

Unjustified lines are set flush left only.

Unjustified text, on the other hand, uses equal spacing between letters and words on a line, and an uneven right margin (as traditionally produced by a typewriter). For some readers, a ragged right margin makes reading easier. These differing line lengths can prompt the eye to move from one line to another (Pinelli 77). In contrast to justified text, an unjustified page looks less formal, less distant, and less official.

Justified text seems preferable for books, annual reports, and other formal materials. Unjustified text seems preferable for more personal forms of communication such as letters, memos, and in-house reports.

KEEP LINE LENGTH REASONABLE.   Long lines tire the eyes. The longer the line, the harder it is for the reader to return to the left margin and locate the beginning of the next line (White, *Visual Design* 25).

Notice how your eye labors to follow this apparently endless message that seems to stretch in lines that continue long after your eye was prepared to move down to the next line. After reading more than a few of these lines, you begin to feel tired and bored and annoyed, without hope of ever reaching the end.

Short lines force the eyes back and forth (Felker 79). "Too-short lines disrupt the normal horizontal rhythm of reading" (White, *Visual Design* 25).

Lines that are too
short cause your eye
to stumble from one
fragment to another
at a pace that too
soon becomes
annoying, if not
nauseating.

A reasonable line length is sixty to seventy characters (or nine to twelve words) per line for an $8\frac{1}{2}$-by-11-inch single-column page. The number of characters will depend on print size. Longer lines call for larger type and wider spacing between lines (White, *Great Pages* 70).

Line length, of course, is affected by the number of columns (vertical blocks of print) on your page. Two-column pages often appear in newsletters and brochures, but research indicates that single-column pages work best for complex, specialized information (Hartley 148).

KEEP LINE SPACING CONSISTENT. For any document likely to be read completely (letters, memos, instructions), single-space within paragraphs and double-space between. Instead of indenting the first line of single-spaced paragraphs, separate them with one line of space. For longer documents likely to be read selectively (proposals, formal reports), increase line spacing within paragraphs by one-half space. Indent these paragraphs or separate them with one extra line of space.

NOTE   *Although academic papers generally call for double spacing, most workplace documents do not.*

TAILOR EACH PARAGRAPH TO ITS PURPOSE. Users often skim a long document to find what they want. Most paragraphs, therefore, begin with a topic sentence forecasting the content.

Shape each
paragraph

Use a long paragraph (no more than fifteen lines) for clustering material that is closely related (such as history and background, or any body of information best understood in one block).

Use short paragraphs for making complex material more digestible, for giving step-by-step instructions, or for emphasizing vital information.

Instead of indenting a series of short paragraphs, separate them by inserting an extra line of space (as here).

Avoid "orphans," leaving a paragraph's opening line on the bottom of a page, or "widows," leaving a paragraph's closing line on the top of the page.

MAKE LISTS FOR EASY READING.   Users often prefer lists rather than continuous prose paragraphs (Hartley 51). Types of items you might list: advice or examples, conclusion and recommendations, criteria for evaluation, errors to avoid, materials and equipment for a procedure, parts of a mechanism, or steps or events in a sequence. Notice how the preceding information becomes easier to grasp and remember when displayed in the list below.

Types of items you might list:

- advice or examples
- conclusions and recommendations
- criteria for evaluation
- errors to avoid
- materials and equipment for a procedure
- parts of a mechanism
- steps or events in a sequence

A list of brief items usually needs no punctuation at the end of each line. A list of full sentences or questions requires appropriate punctuation after each item.

Depending on the list's contents, set off each item with some kind of visual or verbal signal. If the items require a strict sequence or chronology (say, parts of a mechanism or a set of steps), use arabic numbers (1, 2, 3) or the words *First, Second, Third,* and so on. If the items require no strict sequence (as in the list above), use dashes, asterisks, or bullets. For a checklist, use open boxes.

Use lists to help
readers organize
their understanding

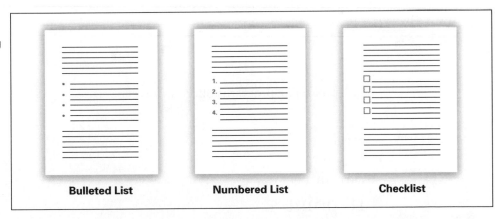

Bulleted List   Numbered List   Checklist

Introduce your list with an explanation. Phrase all listed items in parallel grammatical form. If the items suggest no strict sequence, try to impose some logical ranking (most to least important, alphabetical, or some such). Set off the list with extra white space above and below.

NOTE   *A document with too many lists appears busy, disconnected, and splintered (Felker 55). And long lists could be used by unethical writers to camouflage bad or embarrassing news (Williams 12).*

Styling the Words and Letters

In styling words and letters, we consider typographic choices that will make the text easy to read.

USE STANDARD TYPE SIZES.   To figure out the number of words that can fit on a page, designers traditionally measure the size of type and other page elements (such as visuals and line length) by *picas* and *points*. One pica equals roughly $\frac{1}{6}$ of an inch and one point equals $\frac{1}{12}$ of a pica (or $\frac{1}{72}$ of an inch).

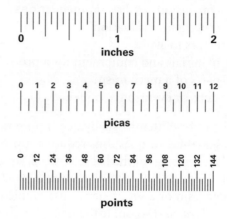

The height of a typeface, the distance from the top of the *ascender* to the base of the *descender*, is measured in points.

Word-processing programs offer various type sizes:

Select the
appropriate point
size

9 point
10 point
12 point
14 point
18 point
24 point

Standard type sizes for manuscripts run from 10 to 12 point. Use larger or smaller sizes for headings, titles, captions (brief explanation of a visual), sidebars (marginal comments), or special emphasis. Use a consistent type size for similar elements throughout the document. For overhead transparencies or computer projection in oral presentations, use 18- or 20-point type for body text and 20 or greater for headings.

SELECT APPROPRIATE FONTS.   A font, or typeface, is the style of individual letters and characters. Each font has its own *personality:* "The typefaces you select for . . . [heads], subheads, body copy, and captions affect the way readers experience your ideas" (*Aldus Guide* 24).

Particular fonts can influence reading speed by as much as 30 percent (Chauncey 26).

Word-processing programs offer a variety of fonts like the examples below, listed by name.

Select a font for
its personality

11-point New York
11-point Courier
11-point Palatino
11-point Geneva
11-point Monaco
11-point Chicago
11-point Helvetica
11-point Times

For visual unity, use different sizes and versions (**bold,** *italic,* SMALL CAPS) of the same font throughout your document—except possibly for headings, captions, sidebars, or visuals. Try to use no more than two different typeface families throughout a document.

All fonts divide into two broad categories: *serif* and *sans serif.* Serifs are the fine lines that extend horizontally from the main strokes of a letter.

Serif type makes printed body copy more readable because the horizontal lines "bind the individual letters" and thereby guide the reader's eyes from letter to letter—as in the type you are now reading (White, *Visual Design* 14).

Decide between
serif and sans
serif type

In contrast, sans serif type is purely vertical (like this). Clean looking and "businesslike," sans serif is considered ideal for technical material (numbers, equations, etc.), marginal comments, headings, examples, tables, and captions to pictures and visuals, and any other material set off from the body copy (White, *Visual Design* 16). Sans serif fonts are also more readable in *projected* environments such as overhead transparencies and slides.

Font prefer-
ences are
culturally
determined

NOTE

European readers generally prefer sans serif fonts, and other cultures have their own preferences as well. Learn all you can about the design conventions of the culture you are addressing.

Except for special emphasis, use conservative fonts; the more ornate ones are harder to read and inappropriate for most workplace documents.

AVOID SENTENCES IN FULL CAPS. Sentences or long passages in full capitals (uppercase letters) are hard to read because uppercase letters lack ascenders and descenders (page 352), and so all words in uppercase have the same visual outline (Felker 87). The longer the passage, the harder readers work to grasp your emphasis.

MY DOG HAS MANY FLEAS.

My dog has many fleas.

FULL CAPS are
good for
emphasis but
they make long
passages hard to
read.

HARD	ACCORDING TO THE NATIONAL COUNCIL ON RADIATION PROTECTION, YOUR MAXIMUM ALLOWABLE DOSE OF LOW-LEVEL RADIATION IS 500 MILLIREMS PER YEAR.
EASIER	According to the National Council on Radiation Protection, your MAXIMUM allowable dose of low-level radiation is 500 millirems per year.

Lowercase letters take up less space, and the distinctive shapes make each word easier to recognize and remember (Benson 37).

Use full caps as section headings (INTRODUCTION) or to highlight a word or phrase (WARNING: NEVER TEASE THE ALLIGATOR). As with other highlighting options discussed below, use full caps sparingly.

Highlighting for Emphasis

Effective highlighting helps users distinguish important from less important elements. Highlighting options include fonts, type sizes, white space, and other graphic devices that:

Purposes of
highlighting

- emphasize key points
- make headings prominent
- separate sections of a long document
- set off examples, warnings, and notes

On a typewriter, you can highlight with <u>underlining</u>, FULL CAPS, dashes, parentheses, and asterisks.

> You can indent to set off examples, explanations, or any material that should be distinguished from other elements in your document.

Using ruled (or typed) horizontal lines, you can separate sections in a long document:

Using ruled lines, broken lines, or ruled boxes, you can set off crucial information such as a warning or a caution:

> *Caution:* A document with too many highlights can appear confusing, disorienting, and tasteless.

See pages 327–28 for more on background screens, ruled lines, and ruled boxes.

Word processors offer highlighting options that include **boldface,** *italics,* SMALL CAPS, varying type sizes and fonts, and color. For specific highlighted items, some options are better than others:

Not all highlighting is equal

Boldface works well for emphasizing a single sentence or brief statement, and is perceived by readers as being "authoritative" (*Aldus Guide* 42).

Italics suggest a more subtle or "refined" emphasis than boldface (Aldus Guide 42). *Italics can highlight words, phrases, book titles, or anything else you would have underlined on a typewriter. But multiple lines (like these) of italic type are hard to read.*

SMALL CAPS WORK FOR HEADINGS AND SHORT PHRASES, BUT ANY LONG STATEMENT ALL IN CAPS IS HARD TO READ.

Small type sizes (usually sans serif) work well for captions and credit lines and as labels for visuals or to set off other material from the body copy.

Large type sizes and dramatic typefaces are both hard to miss and hard to digest. Be conservative—unless you really need to convey forcefulness.

Color is appropriate only in some documents, and only when used sparingly. Pages 324–29 discuss how color can influence audience perception and interpretation of a message.

Whichever highlights you select, be consistent. Make sure that all headings at one level are highlighted identically, that all warnings and cautions are set off identically, and so on. And *never* combine too many highlights.

15.3

For more on visual
"chunking" visit
<www.ablongman.com/
lannonweb>

Using Headings for Access and Orientation

Readers of a long document often look back or jump ahead to sections that interest them most. Headings announce how a document is organized, point readers to what they need, and divide the document into accessible blocks or "chunks." An informative heading can help a person decide whether a section is worth reading (Felker 17). Besides cutting down on reading and retrieval time, headings help readers remember information (Hartley 15).

DECIDE HOW TO PHRASE YOUR HEADINGS.   Depending on your purpose, you can phrase your headings in three different ways: as a topic phrase, a statement, or a question (*Writing User-Friendly Documents* 17):

Heading Type	Example	When to Use
Topic headings use a word or short phrase.	**Usable Page Design**	When you have lots of headings and want to keep them short and sweet. Or to sound somewhat formal. Frequent drawback: too vague.
Statement headings use a sentence or explicit phrase.	**How to Create a Usable Page Design**	To assert something specific about the topic. Occasional drawback: wordy and cumbersome.
Question headings pose the questions in the same way readers are likely to ask them.	**How Do I Create a Usable Page Design?**	To invite readers in and to personalize the message, making people feel directly involved. Occasional drawbacks: too "chatty" for formal reports or proposals; overuse can be annoying.

To avoid verbal clutter, brief topic headings can be useful in documents that have numerous subheads (as in a textbook or complex report)—as long as readers understand the context for each brief heading. Statement headings work well for explaining how something happens or operates (say, "How the Fulbright Scholarship Program Works"). Question headings are most useful for explaining how to do something because they address the actual questions users will have about a procedure (say, "How Do I Apply for a Fulbright Scholarship?").

Phrase your headings to summarize the content as concisely as possible (Horn 190). But keep in mind that a vague or overly general heading can be more misleading or confusing than no heading at all (Redish et al. 144). Compare, for example, a heading titled "Evaluation" versus "How the Fulbright Commission Evaluates a Scholarship Application"; the second version tells readers exactly what to expect.

MAKE HEADINGS SPECIFIC AS WELL AS COMPREHENSIVE.   Focus the heading on a specific topic. Do not preface a discussion of the effects of acid rain on lake trout with a broad heading such as "Acid Rain." Use instead "The Effects of Acid Rain on Lake Trout."

Also, provide enough headings to delineate each discussion section. If chemical, bacterial, and nuclear wastes are three *separate* discussion items, provide a heading for each. Do not simply lump them under the sweeping heading "Hazardous Wastes." If you have prepared an outline for your document, adapt major and minor headings from it.

MAKE HEADINGS GRAMMATICALLY CONSISTENT.   All major topics or all minor topics in a document share equal rank; to emphasize this equality, express topics at the same level in identical—or parallel—grammatical form.

NONPARALLEL
HEADINGS

How to Avoid Damaging Your Disks:

1.   Clean Disk Drive Heads
2.   Keep Disks Away from Magnets
3.   Writing on Disk Labels with a Felt-Tip Pen
4.   It is Crucial That Disks Be Kept Away from Heat
5.   Disks Should Be Kept Out of Direct Sunlight
6.   Keep Disks in Their Protective Jackets

In items 3, 4, and 5, the lack of verbs in the imperative mood obscures the relationship between individual steps. This next version emphasizes the equal rank of these items.

PARALLEL
HEADINGS

3.   Write on Disk Labels with a Felt-Tip Pen
4.   Keep Disks Away from Heat
5.   Keep Disks Out of Direct Sunlight

Parallelism helps make a document readable and accessible.

MAKE HEADINGS VISUALLY CONSISTENT.   "Wherever heads are of equal importance, they should be given similar visual expression, because the regularity itself becomes an understandable symbol" (White, *Visual Design* 104). Use identical type size, typeface, and indentation for all headings at a given level.

LAY OUT HEADINGS BY LEVEL.   Like a good road map, your headings should clearly announce the large and small segments in your document. (Use the logical divisions from your outline as a model for heading layout.) Think of each heading at a particular level as an "event in a sequence" (White *Visual Design* 95).

Headings vary in positioning and highlighting, depending on their level (Figure 15.6). Follow these suggestions for using headlines effectively:

- *Ordinarily, use no more than four levels of heading (section, major topic, minor topic, subtopic).* Excessive heads and subheads make a document seem cluttered or fragmented.
- *To divide logically, be sure each higher-level heading yields at least two lower-level headings.*
- *Insert one additional line of space above each heading.* For double-spaced text, triple-space before the heading, and double-space after; for single-spaced text, double-space before the heading, and single-space after.

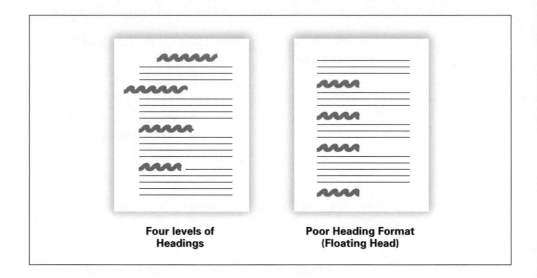

**Four levels of   Poor Heading Format
Headings   (Floating Head)**

- *Never begin the sentence right after the heading with "this," "it," or some other pronoun referring to the heading.* Make the sentence's meaning independent of the heading.
- *Never leave a heading floating as the final line of a page.* Unless two lines of text can fit below the heading, carry it over to the top of the next page.
- *Use different type sizes to reflect levels of heads.* Readers often equate large type size with importance (White, *Visual Design* 95; Keyes 641). Set major heads in a larger type size.
- *Use running heads (headers) or feet (footers) in long documents.* To help users navigate a document with multiple sections or chapters, include a chapter or section heading across the top or bottom of each page.

When headings show the relationships among all the parts, readers can grasp at a glance how a document is organized.

For two-sided pages (say, in a book), running heads or feet on each left page would align toward the page's left side.

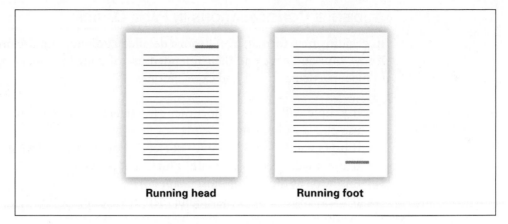

Running head   **Running foot**

Section Heading

Section headings in boldface and enlarged type are more appealing and readable than headings in full caps. Use a type size roughly 4 points larger than body copy (say, 16-point section heads for 12-point body copy). Avoid overly large heads, and use no other highlights. Set these and all lower heads one extra line space below any preceding text.

Major Topic Heading

Major topic heads abut the left margin (flush left), and each important word begins with an uppercase letter. Use boldface and a type size roughly 2 points larger than body copy, with no other highlights.

Minor Topic Heading

Minor topic heads are also set flush left. Use boldface, italics (optional), and the same type size as in the body copy, with no other highlights.

Subtopic Heading.   Instead of indenting the first line of body copy, place subtopic heads flush left on the same line as following text, and set off by a period. Use boldface and the same type size as in the body copy, with no other highlights.

FIGURE 15.6 Recommended Format for Word-Processed Headings Be sure that all headings at a given level are phrased identically (as a topic phrase, a statement, or a question).

AUDIENCE CONSIDERATIONS IN PAGE DESIGN

In deciding on a format, work from a detailed audience and use profile (Wight 11). Know your audience and their intended use of your information. Create a design to meet particular needs and expectations:

How users' needs determine page design

- If people will use your document for reference only (as in a repair manual), use plenty of headings.
- If users will follow a sequence of steps, show that sequence in a numbered list.
- If users will need to evaluate something, provide a checklist of criteria (as in this book at the end of most chapters).
- If users need a warning, highlight the warning so that it cannot possibly be overlooked.
- If users have asked for a one-page report or résumé, save space by using the 10-point type size.
- If users will be facing complex information or difficult steps, widen the margins, increase all white space, and shorten the paragraphs.

Regardless of the audience, never make the document look "too intellectually intimidating" (White, *Visual Design* 4).

Consider also your audience's cultural expectations. For instance, Arabic and Persian text is written from right to left instead of left to right (Leki 149). In other cultures, readers move up and down the page, instead of across. A particular culture might be offended by certain icons or by a typeface that seems too plain or too fancy (Weymouth 144). Ignoring a culture's design conventions can be interpreted as disrespect.

NOTE   *Even the most brilliant page design cannot redeem a document with worthless content, chaotic organization, or unreadable style. The value of any document ultimately depends on elements beneath the visual surface.*

15.4

For more on screen versus paper design visit <www.ablongman.com/lannonweb>

DESIGNING ON-SCREEN DOCUMENTS

Most of the techniques discussed so far in this chapter are appropriate for both paper and electronic documents. However, electronic documents (including Web pages, online help, and CD-ROMs) have certain special design requirements.

Web Pages

Elements of on-screen page design

Each "page" of a Web document typically stands alone as a discrete "module," or unit of meaning. Instead of a traditional introduction-body-conclusion sequence of pages, material is displayed in screen-sized chunks, each linked as hypertext. Links serve the purpose of headings. Each link takes users to a deeper level of information. Also, to be read on a computer screen, pages must accommodate small screen size, reduced resolution, and reader resistance to scrolling—among other restrictions.

In designing Web pages, follow these general guidelines:

How to design on-screen documents

- *Provide margins so that your text won't drift (or run off) the edge of the user's screen.*
- *Display the main point close to the top of each page.*
- *Keep sentences and paragraphs shorter and more concise than for hard copy.*
- *Display links, navigation bars, hot buttons, and help options on each page.*
- *As with printed text headings, make your links consistent: Use the same typeface and font for the same level heading. (Always test your links to be sure they actually work.)*
- *Don't use underlines for emphasis because these might be confused with hyperlinks.*
- *Don't mix and match too many typefaces.*
- *Use sans serif type for body text.*
- *Don't use small type: anything under 12 point is hard to read.*

Figure 15.7 incorporates these elements. See Chapter 19 for more on designing Web pages and other electronic documents.

Special authoring software such as *Adobe FrameMaker* or *RoboHelp* automatically converts hard copy document format to various on-screen formats, chunked and linked for easy navigation. However, whenever possible, work with a professional Web designer to be sure your on-screen document looks and functions the way you want it to.

NOTE

To learn about Web page design in full, you will need to take classes or read books about this topic. Many organizations have employees with job titles such as Webmaster or Web Designer who are responsible for designing Web pages.

Online Help

Like Web page design, designing online help screens is a specialty. Many organizations, especially those that produce software, hire technical communicators who know how to produce online help screens. As with all page design, paper or electronic, producing online help screens requires consistency. For more on this topic see pages 460, 534.

CD-ROMs

If you are designing information for a CD-ROM, the same concepts apply. Typefaces need to be clear and legible for the screen. Different fonts should be used consistently. Headings should also be consistent, and any links should be tested.

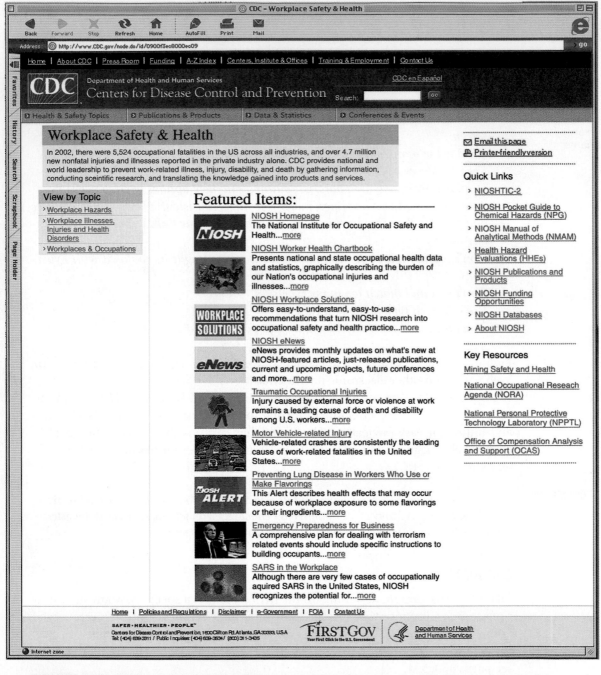

FIGURE 15.7 A User-Friendly Web Page Immediately below the headline and the main site links, a prominent heading announces the page's main topic. This is followed by a paragraph summarizing the safety and health problem and describing CDC's mission. Main topic links appear in the left column, with links to key subtopics and resources in the right column. In the center column, special features are highlighted with graphics. Links at the bottom of the page connect to additional resources and provide contact information. Despite these various layers of information, the page is easy to navigate.

Source: Workplace Safety and Health from Centers for Disease Control and Prevention, <www.cdc.gov>.

✓ CHECKLIST for Usability of Page Design

(Numbers in parentheses refer to the first page of discussion.)

Shape of the Page

☐ Is the paper white, low-gloss, rag bond, with black ink? (344)
☐ Is all type or print neat and legible? (346)
☐ Are pages numbered consistently? (346)
☐ Is the grid structure effective? (347)
☐ Does the white space adequately orient the readers? (347)
☐ Are the margins ample? (349)
☐ Is line length reasonable? (349)
☐ Is the right margin unjustified? (349)
☐ Is line spacing appropriate and consistent? (350)
☐ Does each paragraph begin with a topic sentence? (350)
☐ Does the length of each paragraph suit its subject and purpose? (350)
☐ Are all paragraphs free of "orphan" lines or "widows"? (350)
☐ Do parallel items in strict sequence appear in a numbered list? (351)
☐ Do parallel items of any kind appear in a list whenever a list is appropriate? (351)

Style of Words and Letters

☐ Is the body type size 10 to 12 points? (352)
☐ Are fonts used effectively and consistently? (353)
☐ Do full caps highlight only single words or short phrases? (354)

Emphasis and Orientation

☐ Is the highlighting consistent, tasteful, and subdued? (354)
☐ Are all patterns distinct enough so that readers will find what they need? (355)
☐ Are there enough headings for readers to know where they are in the document? (356)
☐ Are headings informative, comprehensive, specific, parallel, and visually consistent? (357)
☐ Are headings clearly differentiated according to level? (357)

Audience Considerations

☐ Does this design meet the audience's needs and expectations? (360)
☐ Does this design respect the cultural conventions of the audience? (360)

EXERCISES   For more exercises, visit <www.ablongman.com/lannon>

1. Find an example of effective page design, in a textbook or elsewhere. Photocopy a selection (two or three pages), and attach a memo explaining to your instructor and classmates why this design is effective. Be specific in your evaluation. Now do the same for an example of bad page design, making specific suggestions for improvement. Bring your examples and explanations to class, and be prepared to discuss why you chose them.

   As an alternative assignment, imagine that you are a technical communication consultant, and address each memo to the manager of the respective organization that produced each document.

2. These are headings from a set of instructions for listening. Rewrite the headings to make them parallel.

- You Must Focus on the Message
- Paying Attention to Nonverbal Communication
- Your Biases Should Be Suppressed
- Listen Critically
- Listen for Main Ideas
- Distractions Should Be Avoided
- Provide Verbal and Nonverbal Feedback
- Making Use of Silent Periods
- Are You Allowing the Speaker Time to Make His or Her Point?
- Keeping an Open Mind Is Important

3. Using the usability checklist on page design above, redesign an earlier assignment or a document you've prepared on the job. Submit to your instructor the revision and the original, along with a memo explaining your improve-

ments. Be prepared to discuss your format design in class.

4. Anywhere on campus or at work, locate a document with a design that needs revision. Candidates include career counseling handbooks, financial aid handbooks, student or faculty handbooks, software or computer manuals, medical information, newsletters, or registration procedures. Redesign the document or a two- to five-page selection from it. Submit to your instructor a copy of the original, along with a memo explaining your improvements. Be prepared to discuss your revision in class.

5. Figure 15.8 shows two different designs for the same message. Which version is most effective, and why? Prepare a list of the specific elements in the improved version, and be prepared to discuss your list in class.

COLLABORATIVE PROJECT

Working in small groups, redesign a document you select or your instructor provides. Prepare a detailed explanation of your group's revision. Appoint a group member to present your revision to the class, using an opaque or overhead projector, a large-screen monitor, or photocopies.

**FIGURE 15.8
Two Different
Designs for the
Same Message**
Source: Writing
User-Friendly
Documents.
Washington DC:
U.S. Bureau of
Land Management, 2001. 41.

§ 2653.31 Native group selections.

(a) Selections must not exceed the amount recommended by the regional corporation or 320 acres for each Native member of a group, or 7,680 acres for each Native group, whichever is less. Native groups must identify any acreage over that as alternate selections and rank their selections. Beyond the reservations in sections 2650.32 and 2650.46 of this Part, conveyances of lands in a National Wildlife Refuge are subject to the provisions of section 22(g) of ANCSA and section 2651.41 of this chapter as though they were conveyances to a village corporation.

(b) Selections must be contiguous and the total area selected must be compact except where separated by lands that are unavailable for selection. BLM will not consider the selection compact if it excludes lands available for selection within its exterior boundaries; or an isolated tract of public land of less than 640 acres remains after selection. The lands selected must be in quarter sections where they are available unless exhaustion of the group's entitlement does not allow the selection of a quarter section. The selection must include all available lands in less than quarter sections. Lands selected must conform as nearly as practicable to the United States lands survey system.

§ 2653.31 What are the selection criteria for Native group selections and what lands are available?

You may select only the amount recommended by the regional corporation or 320 acres for each Native member of a group, or 7,680 acres for each Native group, whichever is less. You must identify any acreage over 7,680 as alternate selections and rank their selection.

§ 2653.32 What are the restrictions in conveyances to Native groups?

Beyond the reservations described in this part conveyances of lands in a National Wildlife Refuge are subject to section 22(g) of ANSCA as though they were conveyances to a village.

§ 2653.33 Do Native group selections have to be contiguous?

Yes, selections must be contiguous. The total area you select must be compact except where separated by lands that are unavailable for selection. BLM will not consider your selection if:

(a) It excludes lands available for selection within its exterior boundaries; or

(b) An isolated tract of public land of less than 640 acres remains after selection.

§ 2653.34 How small a parcel can I select?

Select lands in quarter sections where they are available unless there is not enough left in your group's entitlement to allow this. Your selection must include all available lands in areas that are smaller than quarter sections. Conform your selections as much as possible to the United States land survey system.

16

Designing and Testing the Document for Usability

WHY A USABLE DESIGN IS ESSENTIAL

HOW TO ACHIEVE A USABLE DESIGN

HOW TO TEST YOUR DOCUMENT FOR USABILITY

USABILITY ISSUES IN ONLINE OR MULTIMEDIA DOCUMENTS

USABILITY TESTING IN THE CLASSROOM

GUIDELINES for Testing a Document's Usability

CHECKLIST for Usability

Usability defined

W hether online or in hard copy, a *usable* document is safe, dependable, and easy to read and navigate. Whatever their goals in using the document, people must be able to do at least three things (Coe, *Human Factors* 193; Spencer 74):

What a usable document enables readers to do

- easily locate the information they need
- understand the information immediately
- use the information successfully

To assess the usability of a manual that accompanies a lawnmower, for instance, you would ask: "How effectively do these instructions enable all users to assemble, operate, and maintain the mower safely, efficiently, and effectively?"

WHY A USABLE DESIGN IS ESSENTIAL

Companies routinely measure the usability of their products, including the communication that accompanies the product (warnings, explanations, assembly or operating instructions, and other types of "product documentation"). From lawnmowers to consumer electronics, technology creates ever more elaborate products and gadgets. Unfortunately, many of these products "seem to become continuously more complicated to learn and use" (Hughes 488). The more complicated the product or task, the greater the need for usable documentation.

Flaws that reduce a document's usability

To keep their customers—and to avoid lawsuits—companies go to great lengths to eliminate flaws in their products and documents and to anticipate all the ways a product might fail or be misused. In a document, for example, usability can be compromised by inaccurate content, poor organization, unreadable style, inadequate visuals, or bad page design. Any such flaws—ranging from too much information to too few headings or hard-to-read type—can spell frustration or even disaster for people who use the document "in ways other than those intended" (van Der Meij 219).

16.1

For more on how usability criteria are culture-specific visit <www.ablongman.com/lannonweb>

HOW TO ACHIEVE A USABLE DESIGN

A usable design meets specific criteria that we identify by asking and answering these basic questions about the tasks, the user, and the setting:

Questions for achieving usability

- *What tasks will users need to perform to achieve their goals?*
- *What do we know about the specific users' abilities and limitations?*
- *In what setting will the document be read/used?*

Figure 16.1 outlines specific ways of applying these questions.

**FIGURE 16.1
Defining a
Document's
Usability
Criteria**
The document's
final shape is
based on careful
analysis of the
tasks involved,
the users, and
the setting.

Type of Task	User Characteristics	Constraints of the Setting
• Learn facts	• Motivation/attitude	• Distractions
• Understand concepts	• Prior knowledge/skills	• Interruptions
• Follow directions	• Experience	• Other constraints
• Make judgments	• Limitations	• Ways the document will be read
• Make decisions	• Cultural background	• Things that can go wrong

Specific usability criteria for this document
• Worthwhile content
• Sensible organization
• Readable style
• Accessible design
• Ethical, legal, and cultural considerations

Outline the Main Tasks Involved

Spell out the *performance objectives,* the precise tasks that users need to accomplish successfully (Carliner, "Physical" 564; Zibell 13). In a set of instructions, for example, these tasks might involve installing a computer program, using medical equipment, or the like. But, beyond following directions, users consult documents for other kinds of tasks. To learn facts, for example, a project manager might read daily progress reports to keep track of a team's work activities. Or, to understand a concept, a potential investor might read about gene therapy before deciding whether to invest in a biomedical company. People read the document because they want to *do* something (Gurak and Lannon 31).

ON THE JOB...
Designing documents

"Sometimes I literally design documents (that is, textbooks or supplements) if a book I'm working on needs a light revision. I have a coworker who designs documents professionally though, so if I need something fancy, he designs it and I follow the specs. If an email is long and complex sometimes I try to impose some structure to it. If you are giving someone a written correspondence more than a page long, you probably should be worrying about the design of it. If you don't make things easy to follow, sometimes people panic and won't read what you've written."

—Lorraine Patsco, Director of Prepress
and Multimedia Production

For defining the tasks your document will cover, develop an outline of steps and substeps, as in this next example (adapted from Gurak and Lannon 32):

A task outline

> ## PERFORMANCE OBJECTIVES FOR USING THE MODEL 76 BOBAN LAWNMOWER
>
> 1. Assemble the lawnmower
>   a. Remove the unit from the carton.
>   b. Assemble the handle.
>   c. Assemble the cover plate.
>   d. Connect the spark plug wire.
> 2. Operate the lawnmower
>   a. Add oil and fuel.
>   b. Adjust the cutting height.
>   c. Adjust the engine control.
>   d. Start the engine.
>   e. Mow the grass.
>   f. Stop the engine.
> 3. Maintain the lawnmower
>   a. Clean the mower after each use.
>   b. Keep oil level full and key parts lubricated.
>   c. Replace the air filter as needed.
>   d. Keep the blade sharpened

Many of these substeps can be divided further: for example, "Press primer bulb" and "Hold down control handle" are part of starting the engine.

(See also the specific "general outlines" for various documents in Part V.)

Analyze the Audience and the Setting

16.2

For other methods of audience analysis visit <www.ablongman.com/lannonweb>

To identify the intended users and their goals for this document, use a version of the Audience and Use Profile Sheet in Chapter 3. Also, consider the conditions under which this document will be used. (See also *human factors* on page 376.)

Do the Research

Learn all you can about the typical users of the product being documented (age, education, and so on). For example, you might interview typical customers or survey previous purchasers of other models made by your company. Or you might observe first-time operators coping with a manual for an earlier model and then ask for their feedback. Find out exactly how most injuries from lawnmowers occur. Check company records for customer comments and complaints about safety problems. Consult retail dealers for their feedback. Ask your company's legal department about prior injury claims by customers. The more you know, the better.

Develop a Design Plan

Your design plan is the blueprint for meeting the performance objectives you spelled out earlier (Kostur, cited in Carliner, "Physical" 564). Here you incorporate your analysis of the tasks, users, and setting along with a specific proposal for the content, shape, style, and layout of your document—the usability criteria. A design plan is especially important in collaborative work so that everyone can coordinate their efforts throughout the document's production. Depending on the document's complexity, a design plan ranges in length from a multipage report to a relatively brief memo, as in this next example (adapted from Gurak and Lannon 34):

ON THE JOB...
Designing documents

"As an editor of a journal and of two major publishing projects, I take a keen interest in design, or how information is displayed on the page. I tend to divide my essays into sections with headings in order to accentuate the main movements of my argument, and I often suggest to the other writers I edit that they do the same. If I am dealing with the study of a writer's revisions (which is a major part of my research), I develop graphics that allow readers to see the different revisions more clearly."

—John Bryant, Professor of English

A design plan for the lawnmower manual

BOBAN LAWNMOWERS, INC.

7/5/02

To: Manual Design Team
From: Jessica Brown and Fred Bowen, team leaders
Subject: Usability Analysis and Design Plan for the Model 76 User Manual

Based on the following usability analysis, we offer a design plan for the Model 76 user manual, which will provide safe and accurate instructions for assembling, operating, and maintaining the lawnmower.

Audience

"Who will be using the document?"

The audience for this manual is extremely diverse, ranging from early teens to retirees. Some may be using a walk-behind power mower for the first time; others may be highly experienced. Some are mechanically minded; others are not. Some will approach the mowing task with reasonable caution; others will not.

Setting and Hazards

"Under what conditions?"

The 76 mower will be used in the broadest possible variety of settings and conditions, ranging from manicured lawns to wet grass or rough terrain littered with branches, stones, and even small tree stumps—often obscured by weeds and tall grass. Also, the operator will need to handle gasoline on a regular basis.

Our research indicates that most user injuries fall into three categories, in descending order of frequency: foreign objects thrown into the eyes, accidental contact with the rotating blade, and fire or explosion from gasoline.

Our research also suggests that most users read a lawnmower manual (with varying degrees of attention) before initially using the mower, and they consult it later only when the mower malfunctions. Therefore, our manual needs to highlight safety issues as early as the cover page.

Purpose

"Why will they be using it?"

This manual has three purposes:

1. Instruct the user in assembling, operating, and maintaining the lawnmower.
2. Provide adequate safety instructions to protect the user and to comply with legal requirements.
3. Provide a phone number, Web address, and other contact information for users who have questions or who need replacement parts.

Performance Objectives

"What tasks do we want them to accomplish?"

This manual will address three main tasks:

1. How to assemble the lawnmower. This task is fairly straightforward, involving four simple steps—but they have to be done correctly to avoid damage to the unit and to ensure smooth operation.
2. How to operate the lawnmower. This task requires accurate measurements and adjustments and constant vigilance. Because this is the part of the procedure during which the vast proportion of injuries occur, we need to stress safety at all times.
3. How to maintain the lawnmower. To keep the mower in good operating condition, users need to attend to these steps regularly and to be especially careful to wear eye protection if they sharpen their own blade.

Design Plan

"What should the document contain and how should it look?"

For a manual that is inviting, readable, and easily accessible, we recommend the following plan:

- Size, page layout, and color: $8\frac{1}{2}$ x 11 trim size; two-column pages; black ink on white paper.
- Cover page: Includes a drawing of the 76 mower and a highlighted reference to safety instructions throughout the manual.
- Safety considerations: Cautions and warnings displayed prominently before a given step.
- Visuals: Drawings and/or diagrams to accompany each step as needed.
- Inside front cover: Table of contents, and customer contact information.
- Introduction: A complete listing of all safety warnings that appear at various points in the manual.
- Section One: A numbered list of steps for assembling the mower.

- Section Two: A numbered list of steps for operating the mower.
- Section Three: A bulleted list of tasks required for maintaining the mower, with diagrams and cautions, and warnings as needed.

Production Schedule

The first units of our Model 76 are scheduled for shipment May 4, 2003, and completed manuals are to be included in that shipment. To meet this deadline, we propose the following production schedule:

January 15: First draft, including artwork, is completed and usability tested on sample users.

February 15: Manual is revised based on results of usability test.

March 15: After copyediting, proofreading, and final changes, manual goes to the compositor.

April 10: Page proofs and art proofs are reviewed and corrected.

April 17: Corrected proofs go to the printer.

April 30: Finished manual returns from printer and is packaged for May 4 shipment.

"How do we meet the document deadline?"

HOW TO TEST YOUR DOCUMENT FOR USABILITY

The purpose of usability testing is to keep what works in a product or a document and to fix what doesn't. Figure 16.2, for example, is designed to assess user needs and preferences for the *Statistical Abstract of the United States*, a widely consulted publication updated yearly.

Usability testing usually occurs at two levels (Petroski 90): (1) *alpha testing*, by the product's designers or the document's authors, and (2) *beta testing*, by the actual users of the product or document. At the beta level, two types of testing can be done: *qualitative* and *quantitative*.

Qualitative Testing

Qualitative testing shows which specific parts of the document work or don't work

To identify which parts of the document work or don't work, observe how users react or what they say or do. Qualitative testing employs either focus groups or protocol analysis ("Testing Your Documents" 1–2):

Designing documents

"We are always thinking about design. I've just hired someone to help us come up with a new look to our stuff. Formatting, layout, font, graphics, artwork, color are discussed every time we do a major piece. No one is really qualified to do this based on traditional academic training. So we have brought in a design person half-time to take on editorial and design responsibilities."

—Paul Harder, President, mid-sized consulting company

FORM S-555
(8-3-94)

STATISTICAL ABSTRACT SURVEY

U.S. DEPARTMENT OF COMMERCE
BUREAU OF THE CENSUS

Please take a few minutes to answer the questions below. Your voluntary cooperation will help us continue to serve your needs as data users. When completed, please refold, **apply tape to the open edges at the top,** and drop in the mail. Thank you.

1a. **Which sections do you refer to frequently?** Mark (x) all that apply.

☐ Population
☐ Vital statistics
☐ Health
☐ Education
☐ Law enforcement
☐ Geography
☐ Parks and recreation
☐ Elections

☐ Federal Government
☐ State and local government
☐ National defense
☐ Social insurance
☐ Labor force
☐ Income
☐ Prices
☐ Banking

☐ Business
☐ Communications
☐ Energy
☐ Science
☐ Land transportation
☐ Air and water transportation
☐ Agriculture

☐ Forests and fisheries
☐ Mining
☐ Construction and housing
☐ Manufactures
☐ Domestic trade
☐ Foreign commerce
☐ Outlying areas
☐ International statistics

b. Indicate topics for which you would like to see more coverage.

2. **To what degree do you find our current presentation of data in tables clear, meaningful, and easy to understand?** Mark (X) one.

☐ Very much   ☐ Somewhat   ☐ Not at all

3. **Which of the following do you feel currently interferes with the clarity of the tables presented?** Mark (X) all that apply.

☐ Hard to understand column headings and row indentations
☐ Too many numbers in the tables—too much data to absorb
☐ Not clear what numbers mean when they are rounded ("In thousands" or "in millions," for example)
☐ Too many notes in the tables

☐ Some concepts too difficult to understand
☐ Other — Specify

_____

_____

4. Please indicate which of these features you might like to see expanded or reduced in future editions.

	Mark (X) the appropriate column for each feature.		
	Expand	Reduce	OK as is
a. State rankings *(pp. xii–xxi)*			
b. Telephone contact list *(pp. xxii–xxiv)*			
c. Introductory text for sections			
d. Guide to Sources *(pp. 887–925)*			
e. Metropolitan Concepts and Components *(pp. 926–935)*			
f. Statistical Methodology (now called Limitations of the Data) *(pp. 936–950)*			
g. Index *(pp. 959–1011)*			
h. Charts and graphs			

5. **Indicate your level of satisfaction with the Abstract.** Mark (x) one.

☐ Very satisfied   ☐ Satisfied   ☐ Indifferent   ☐ Unsatisfied   ☐ Very unsatisfied

FIGURE 16.2 A Usability Survey   This survey yields valuable qualitative data in the form of user feedback that can be used for improving future editions of *The Statistical Abstract.*
Source: U.S. Bureau of the Census.

Two types of qualitative testing

For other methods of qualitative testing visit <www.ablongman.com/lannonweb>

- *Focus groups.* Based on a list of targeted questions about the document's content, organization, style, and design (as in the Basic Usability Survey in Figure 16.4), users discuss what information they think is missing or excessive, what they like or dislike, and what they find easy or hard to understand. They also might suggest ways of revising the document's graphics, format, or level of technical information. When group discussion is not possible, mail surveys like the one in Figure 16.2 can be effective with users who are diverse and dispersed.
- *Protocol analysis.* In a one-on-one interview, a user reads a specific section of a document and then explains what that section means. For long documents, the interviewer also observes how the person actually reads the document, for example, how often she/he flips pages or refers to the index or table of contents to find information. In another version of protocol analysis, users read the material and think out loud about what they find useful or confusing as they perform the task (Ostrander 20).

Quantitative testing shows whether the document succeeds as a whole

For other methods of quantitative testing visit <www.ablongman.com/lannonweb>

For more on the limits of usability questions visit <www.ablongman.com/lannonweb>

Quantitative Testing

Assess a document's overall effectiveness by using a *control group:* For example, you can compare success rates among people using different versions of your document or count the number of people who performed the task accurately ("Testing Your Documents" 2–4). You can also measure the time required to complete a task and the types and frequency of user errors (Hughes 489). Although it yields hard numerical data, quantitative testing is more complicated, time-consuming, and expensive than its qualitative counterpart.

When to Use Which Test

If time, budget, and available users allow, consider doing both qualitative and quantitative testing. Quantitative testing ordinarily is done last, as a final check on usability. Each test has its benefits and limitations: "Control [quantitative] testing will tell you *if* the new document is a success, but it won't tell you *why* it is or isn't a success" ("Testing Your Documents" 3). In short, to find out if the document succeeds as a whole, use quantitative testing; to find out exactly which parts of the document work or don't work, use qualitative testing.

USABILITY ISSUES IN ONLINE OR MULTIMEDIA DOCUMENTS

In contrast to printed documents, online or multimedia documents pose unique usability considerations (Holler 25; Humphreys 754–55):

- Online documents tend to focus more on "doing" than on detailed explanations. Workplace readers typically use online documents for reference or training rather than for study or memorizing.
- Users of online instructions rarely need persuading (say, to pay attention or follow instructions) because they are guided interactively through each step of the procedure.
- Visuals play a huge role in online instruction.
- Online documents are typically organized to be read interactively and selectively rather than in linear sequence. Users move from place to place, depending on their immediate needs. Organization is therefore flexible and modular, with small bits of easily accessible information that can be combined to suit a particular user's needs and interests.
- Online readers can easily lose their bearings. Unable to shuffle or flip through a stack of printed pages in linear order, readers need constant orientation (to retrieve some earlier bit of information or to compare something on one page with something on another). In the absence of page or chapter numbers, index, or table of contents, "Find," "Search," and "Help" options need to be plentiful and complete.

Figure 16.3 shows the home page for IBM's "Ease of Use" Web site, which offers a wealth of information on usability. Although the emphasis here is on Web pages, much of this material also applies to hard copy documents. See Chapter 19 for more on usability in Web pages and other electronic documents.

USABILITY TESTING IN THE CLASSROOM

Ideally, usability tests occur in a setting that simulates the real conditions, with people who will actually use the document (Redish and Schell 67; Ruhs 8). But even in a classroom setting you can use qualitative testing to assess a document on the basis of standard usability criteria.

The Checklist for Usability (page 379) identifies criteria shared by many technical documents. In addition, specific elements (visuals, page design) and document types (proposals, memos, instructions) have their own usability criteria. These are detailed in the individual checklists for usability throughout this book (for example, pages 406, 457, 478).

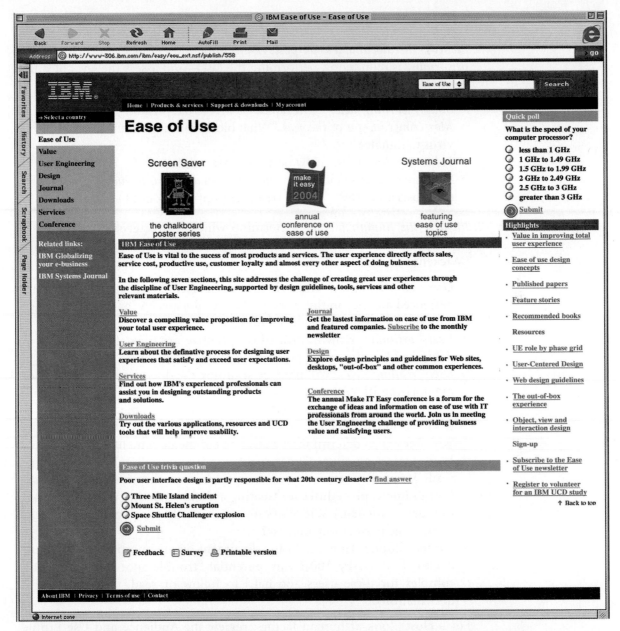

FIGURE 16.3 **IBM's "Ease of Use" Web Site** Begin with the *User-Centered Design* link on this home page for up-to-date coverage of usability issues.

Source: From <www.ibm.com/easy>. Copyright © IBM Corporation, 2004. Reproduced with permission. All rights reserved.

GUIDELINES for Testing a Document's Usability

1. *Identify the document's purpose.* Determine how much *learning* versus *performing* is required in performing the task—and assess the level of difficulty (Mirel et al. 79; Wickens 232, 243, 250):

   - *Simply learning facts.* "How rapidly is this virus spreading?"
   - *Mastering concepts or theories.* "What biochemical mechanism enables this virus to mutate?"
   - *Following directions.* "How do I inject the vaccine?"
   - *Navigating a complex activity that requires decisions or judgments.* "Is this diagnosis accurate?" "Which treatment option should I select?"

2. *Identify the human factors.* Determine which characteristics of the user and the work setting (**human factors**) enhance or limit performance (Wickens 3):

   - *User's abilities/limitations.* What do these users know already? How experienced are they in this task area? How educated? What cultural differences could create misunderstanding?
   - *Users' attitudes.* How motivated or attentive are they? How anxious or defensive? Do users need persuading to pay attention or be careful?
   - *Users' reading styles.* Will users be scanning the document, studying it, or memorizing it? Will they read it sequentially (page by page) or consult the document periodically and randomly? (For example, expert users often look only for key words and key concepts in titles, abstracts, or purpose lines, then determine what sections of the document they will read.)
   - *Workplace constraints.* Under what conditions will the document be read? What distractions or interruptions does the work setting pose? (For example, procedures for treating a choking victim, posted in a busy restaurant kitchen.) Will users always have the document in front of them while performing the task?
   - *Possible failures.* How might the document be misinterpreted or misunderstood (Boiarsky 100)? Any potential "trouble spots" (material too complex for these users, too hard to follow or read, too loaded with information)?

   For a closer look at human factors, review the Audience and Use Profile Sheet (page 65).

3. *Design the usability test.* Ask respondents to focus on specific problems (Hart 53–57; Daugherty 17–18):

   - *Content:* "Where is there too much or too little information?" "Does anything seem inaccurate?"

- *Organization:* "Does anything seem out of order, or hard to find or follow?"
- *Style:* "Is anything hard to understand?" "Are any words inexact or too complex?" "Do any expressions seem wordy?"
- *Design:* "Are there any confusing headings, or too many or too few?" "Are any paragraphs, lists, or steps too long?" "Are there any misleading or overly complex visuals?" "Could any material be clarified by a visual?" "Is anything cramped and hard to read?"

- *Ethical, legal, and cultural considerations:* "Does anything mislead?" "Could anything create potential legal liability or cross-cultural misunderstanding?"

Figure 16.4 shows a sample usability survey (Carliner, "Demonstrating Effectiveness" 258). Notice how the phrasing encourages users to respond with examples instead of just "yes" or "no."

4. *Administer the test.* Have respondents perform the task under controlled conditions. To ensure dependable responses, look for these qualities (Daugherty 19–20):

- *A range of responses.* Select evaluators at various levels of expertise with this task.
- *Independent responses.* Ask each evaluator to work alone when testing the document and recording the findings.
- *Reliability.* Have evaluators perform the test twice.
- *Group consensus.* After individual testing, arrange a meeting for evaluators to compare notes and to agree on needed revisions.
- *Thoroughness.* Once the document is revised/corrected, repeat this entire testing procedure.

5. *Revise your design plan and your draft documents based on user feedback.* If users find a technical term hard to understand, define it clearly or use a simpler word or concept. If a graphic makes no sense, find one that does. These changes are easier when made earlier rather than later, after your document has already been printed, distributed, or posted to the Web.

6. *Provide mechanisms for user feedback.* You can include ways for users to provide feedback on the document: customer comment cards, email addresses, phone numbers, Web sites. If your instructions contain a mistake on page 6, you can be sure customers will let you know, provided you give them a way to contact you.

Basic Usability Survey

1. Briefly describe why this document is used. _____

   _____

2. Evaluate the *content:*
   - Identify any irrelevant information. _____

     _____

   - Indicate any gaps in the information. _____

     _____

   - Identify any information that seems inaccurate. _____

     _____

   - List other problems with the content. _____

     _____

3. Evaluate the *organization:*
   - Identify anything that is out of order or hard to locate or follow. _____

     _____

   - List other problems with the organization. _____

     _____

4. Evaluate the *style:*
   - Identify anything you misunderstood on first reading. _____

     _____

   - Identify anything you couldn't understand at all. _____

     _____

   - Identify expressions that seem wordy, inexact, or too complex. _____

     _____

   - List other problems with the style. _____

     _____

5. Evaluate the *design:*
   - Indicate any headings that are missing, confusing, or excessive. _____

     _____

   - Indicate any material that should be designed as a list. _____

     _____

   - Give examples of material that might be clarified by a visual. _____

     _____

   - Give examples of misleading or overly complex visuals. _____

     _____

   - List other problems with design. _____

     _____

6. Identify anything that seems misleading or that could create legal problems or
   cross-cultural misunderstanding. _____

   _____

7. Please suggest other ways of making this document easier to use. _____

   _____

FIGURE 16.4 A Basic Usability Survey Versions of these questions can serve as a basis
for beta testing (by the document's users).

☑ CHECKLIST for Usability

(Numbers in parentheses refer to the first page of discussion.)

Content

- ☐ Is all material relevant to this user for this task? (38)
- ☐ Is all material technically accurate? (169)
- ☐ Is the level of technicality appropriate for this audience? (30)
- ☐ Are warnings and cautions inserted where needed? (546)
- ☐ Are claims, conclusions, and recommendations supported by evidence? (56)
- ☐ Is the material free of gaps, foggy areas, or needless details? (39)
- ☐ Are all key terms clearly defined? (481)
- ☐ Are all data sources documented? (685)

Organization

- ☐ Is the structure of the document visible at a glance? (219)
- ☐ Is there a clear line of reasoning that emphasizes what is most important? (222)
- ☐ Is material organized in the sequence users are expected to follow? (235)
- ☐ Is everything easy to locate? (219)
- ☐ Is the material "chunked" into easily digestable parts? (239)

Style

- ☐ Is each sentence understandable the first time it is read? (245)
- ☐ Is rich information expressed in the fewest words possible? (253)
- ☐ Are sentences put together with enough variety? (260)
- ☐ Are words chosen for exactness, and not for camouflage? (264)
- ☐ Is the tone appropriate for the situation and audience? (274)

Design

- ☐ Is page design inviting, accessible, and appropriate for the user's needs? (340)
- ☐ Are there adequate aids to navigation (heads, lists, type styles)? (344)
- ☐ Are adequate visuals used to clarify, emphasize, or summarize? (291)
- ☐ Do supplements accommodate the needs of a diverse audience? (643)

Ethical, Legal, and Cultural Considerations

- ☐ Does the document indicate sound ethical judgment? (74)
- ☐ Does the document comply with copyright law and other legal standards? (85)
- ☐ Does the document respect users' cultural diversity? (59)

EXERCISES

 For more exercises, visit <www.ablongman.com/lannon>.

1. As a class, identify an activity that could require instructions for a novice to complete (for example, surviving the first week of college as a commuter). Using the outline on page 368 as a model, prepare a task outline of the steps and substeps for this activity (for example, "1. Obtain essential items: campus map, ID card, parking sticker," and so on; "2. Get to know the library"; "3. Get to know your advisor"; "4. Establish a campus support network"). Exchange task outlines with another student in your class and critique each other's outlines. Revise your outline and be prepared to discuss it in class.

2. Find a set of instructions or some other technical document that is easy to use. Assume that you are Associate Director of Communications for the company that produced this document and you are doing a final review before the document is released. With the Checklist for Usability as a guide, identify those features that make the document usable and prepare a memo to your boss that justifies your decision to release the document.

Following the identical scenario, find a document that is hard to use, and identify the features that need improving. Prepare a memo to your boss that spells out the needed improvements. Submit both memos and the examples to your instructor.

3. As Communications Director for your software company, you've decided to institute a "Usability Awareness" workshop for members of your writing team. As a first step, you've checked out the following Web sites on usability and decided to write a one-paragraph summary (or bulleted list) of the material to be found on each site. Your final memo will serve as a quick reference to usability resources for the writing team. Bring your completed memo to class for discussion.

Usability Resources on the Web
<www.useit.com>
<www.best.com/~jthom/usability>
<www.3.ibm.com/ibm/easy>

<www.upassoc.org>
<www.plainlanguage.gov/howto/test.htm>

COLLABORATIVE PROJECT

Test the usability of a document prepared for this course.

a. As a basis for your *alpha test,* adapt the guidelines on page 377, the Audience and Use Profile Sheet on page 65, the general checklist for Usability (379), and whichever specific checklist applies to this particular document (as on page 561, for example).

b. As a basis for your *beta test,* adapt the Usability Survey on page 378.

c. Revise the document based on your findings. Obtain your data qualitatively, through focus group discussions and/or protocol analysis.

Appoint a group member to explain the usability testing procedure and the results to the class.

PART V

Specific Documents and Applications

Ninety Million Americans Lack Health Literacy

Technical language and arcane terminology cause nearly half of American adults to face higher risks of health problems, the Institute of Medicine has recently reported (Neergaard). Even educated readers have trouble with medical jargon, but for millions who don't read well or aren't fluent in English, the problems are compounded. Such people face what the Institute report calls "limited health literacy," and they have problems following instructions on drug labels, interpreting hospital consent forms, and understanding a doctor's diagnosis.

The American Medical Association has sponsored efforts to improve doctors' ability to communicate with patients, but most doctors would be "stunned" if they actually quizzed patients about what they understood after a visit. The Institute report makes a number of recommendations, including government-sponsored research on ways to improve health literacy and a requirement that health organizations and medical schools teach health literacy and patient communication. ■

17

Memo Reports
and Electronic Mail

INFORMATIONAL VERSUS ANALYTICAL REPORTS

In the professional world, decision makers rely on two broad types of reports: Some reports focus primarily on *information* ("what we're doing now," "what we did last month," "what our customer survey found," "what went on at the department meeting"). But beyond merely providing information, many reports also include *analysis* ("what this information means for us," "what courses of action should be considered," "what we recommend, and why").

The role of analysis in a technical report

Analysis is the heart of technical communication: It involves evaluating your information, interpreting it accurately, drawing valid conclusions, and making persuasive recommendations. Although gathering and reporting information are invaluable workplace skills, *analysis* is ultimately what professionals largely do to earn their pay. And the results of any detailed analysis almost invariably get recorded in a written report. Chapter 24 covers formal analytical reports.

FORMAL VERSUS INFORMAL REPORTS

For every long (formal) report, countless short (informal) reports lead to informed decisions on matters as diverse as the most comfortable office chairs to buy or the best recruit to hire for management training. Unlike long reports (pages 606–40), most short reports require no extended planning, are quickly prepared, contain little or no background information, and have no front or end matter (title page, table of contents, glossary, etc.). But despite their conciseness, short reports do provide the information and analysis that readers need.

Although various formats can be used, short reports often take the form of a memorandum.

PURPOSE OF MEMO REPORTS

Memos have many uses

Memos, the major form of communication in most organizations, leave a *paper trail* of directives, inquiries, instructions, requests, and recommendations, and daily reports for future reference.

Despite the explosive growth of email, paper memos continue to be used widely, especially when an email would be considered too informal or when the message is lengthy. (See page 405 for guidelines.) Different organizations have different preferences about paper memos versus email (Gurak and Lannon 189).

NOTE   *Email memos leave their own trail. Although generally less formal and more quickly written than paper documents, email messages are saved in both hard copy and online— and can inadvertently be forwarded to someone never intended to receive or read them.*

Memos have legal and ethical implications

Organizations rely on memos to trace decisions and responsibilities, track progress, and recheck data. Therefore, any memo you write can have far-reaching ethical and legal implications. Be sure your memo includes the date and your initials or signature. Also make sure that your information is specific, unambiguous, and accurate.

ELEMENTS OF A USABLE MEMO

A usable memo is easy to scan, file, and retrieve. The paper memo has a header that names the organization and identifies the sender, recipient, subject (often in caps or underlined for emphasis), and date. (Placement of these items may differ among firms.) Other memo elements are summarized in Figure 17.1.

INTERPERSONAL CONSIDERATIONS IN WRITING A MEMO

A form of "in-house" correspondence, memos circulate among colleagues, subordinates, and superiors to address questions like these:

What memo recipients want to know

- *What are we doing right, and how can we do it better?*
- *What are we doing wrong, and how can we improve?*
- *Who's doing what, and when, and where?*

Memo topics often involve evaluations or recommendations about policies, procedures, and, ultimately, the *people with whom we work*.

Because people are sensitive to criticism (even when it is merely implied) and often resistant to change, an ill-conceived or aggressive memo can spell disaster for its author. Consider, for instance, this evaluation of one company's training program for new employees:

A hostile approach

> No one tells new employees what it's *really* like to work here—how to survive politically: for example, never tell anyone what you *really* think; never observe how few women are in management positions, or how disorganized things seem to be. New employees shouldn't have to learn these things the hard way. We need to demand clearer behavioral objectives.

Instead of emphasizing deficiencies, the following version focuses on the *benefits* of change:

A more reasonable approach

> New employees would benefit from a concrete guide to the personal and professional traits expected in our company. Training sessions could be based on the attitudes, manners, and behavior appropriate in business settings.

Here are some common mistakes that can offend coworkers:

When memos go wrong

- *Griping or complaining.* Everyone has problems of their own. Never complain without suggesting a solution.
- *Being too critical or judgmental.* Making someone look bad means making an enemy.

NAME OF ORGANIZATION

MEMORANDUM

To: Name and title of recipient (the title also serves as a record for reference)
From: Your name and title (and handwritten initials)
Date: (also serves as a chronological record for future reference)
Subject: ELEMENTS OF A USABLE MEMO (or, replace "Subject" with "Re" for
          "in reference to")

Subject Line
Preview your main point, to help readers assess this memo's relevance to them personally.
An explicit title also makes filing by subject easier.

Introductory Paragraph
Unless you have reason for being indirect, emphasize the important information by stating
your main point immediately.

Topic Headings
When discussing a number of subtopics, include headings (as shown here). Headings help
you organize and they help readers locate information quickly.

Visuals
To summarize numerical data, your memo may include one or more visuals.

Paragraph Spacing
Do not indent the first line of paragraphs. Single-space within paragraphs and double-space
between them.

Second-Page Header
Keep the memo as brief as possible. If you must exceed one page, include a header on
subsequent pages (for example: J. Baxter, 6/12/2005, page 2).

Memo Verification
Do not sign your memos. Initial the "From" line, after your name.

Copy, Distribution, and Enclosure Notations
These items are illustrated under "Letters," on page 416, and used in the same way with
memos, as needed.

FIGURE 17.1   Elements of a Usable Memo

- *Being too bossy.* The imperative mood is best reserved for instructions.
- *Neglecting to provide a copy to each appropriate person.* No one appreciates being left "out of the loop."

Before releasing any memo designed to influence people's thinking, review Chapter 4 carefully.

NOTE
Busy people are justifiably impatient with any memo that seems longer than it needs to be or that contains typos or grammar and spelling errors. Use your spelling and grammar checker, but also proofread carefully. (See page 24 for proofreading guidelines.)

DIRECT VERSUS INDIRECT ORGANIZING PATTERNS

In planning a memo, you can choose from two basic organizing patterns. First is the *direct* pattern, in which you begin with your main point (your request, recommendation, or position on the issue) and then present the details or analysis supporting your case. The second pattern is *indirect,* in which you lay out the details of your case before delivering the bottom line.

"Should the details or the bottom line come first?"

Readers generally prefer the direct approach, especially for analytical reports, because they can get right to your main point. For example, if the Payroll Division of a large company has to announce to employees that weekly paychecks will be delayed because of a virus in the computer system, a direct approach would be appropriate. But when you need to convey bad news or make an unpopular request or recommendation (as in asking for a raise or announcing employee layoffs), you might consider an indirect approach so that you can present your case first. The danger of an indirect approach, however, is that readers could think you are being evasive. For more on direct versus indirect organizing patterns, see page 420.

INFORMATIONAL REPORTS IN MEMO FORM

Among the informational reports that help keep organizations running from day to day are progress reports, periodic activity reports, and meeting minutes.

Progress Reports

Progress reports serve many purposes

Many organizations depend on progress reports (also called status reports) to track activities, problems, and progress on various projects. Some professions require regular progress reports (daily, weekly, monthly), while others only use such reports as needed for a specific project or task. Daily progress reports are vital in a business that assigns crews to many projects.

Managers often rely on progress reports to decide how to allocate funds. Managers also need to know about delays that could slow a project and increase costs. Also managers need information in order to coordinate the efforts of various groups working on a project.

When, say, construction work is performed for a client, regular reports spell out for the client (and investors and loan officers) how time and money are being spent and how problems and setbacks are being addressed. Such reports therefore can help predict whether the project will be completed on schedule and within budget. Many contracts stipulate the dates and stages at which progress will be reported. Failing to report on time may incur legal penalties.

Often, a progress report is one of a series. Together, the project proposal (Chapter 23), progress reports (the number varies with the scope and length of the project), and the final report (Chapter 24) provide a record and history of the project.

To meet managers' and clients' needs, progress reports must answer these questions:

What recipients of a progress report want to know

- *How much has been accomplished since the last report?*
- *Is the project on schedule?*
- *If not, what went wrong?*
- *How was the problem corrected?*
- *How long will it take to get back on schedule?*
- *What else needs to be done?*
- *What is the next step?*
- *Have you encountered any unexpected developments?*
- *When do you anticipate completion?* Or (on a long project) *when do you anticipate completion of the next phase?*

If the report is part of a series, you might also refer to prior problems or developments.

Many organizations have forms for organizing progress reports, so no one format is best. But each report in a series should be organized identically. The following memo illustrates how one writer organized her report.

PROGRESS REPORT (ON THE JOB)

MEMORANDUM
To: P. J. Stone, Senior Vice President
From: B. Poret, Group Training Manager
Date: June 6, 20xx
Subject: *Progress Report: Training Equipment for New Operations Building*

Work Completed

Summarizes achievements to date

Our training group has met twice since our May 12 report in order to answer the questions you posed in your May 16 memo. In our first meeting, we identified the types of training we anticipate.

Types of Training Anticipated

- Divisional Surveys
- Loan Officer Work Experience

**Details the
achievements**

- Divisional Systems Training
- Divisional Clerical Training (Continuing)
- Divisional Clerical Training (New Employees)
- Divisional Management Training (Seminars)
- Special/New Equipment Training

In our second meeting, we considered various areas for the training room.

Training Room

The frequency of training necessitates having a training room available daily. The large training room in the Corporate Education area (10th floor) would be ideal. Before submitting our next report, we need your confirmation that this room can be assigned to us.

To support the training programs, we purchased this equipment:

- Audioviewer
- LCD monitor
- Videocassette recorder and monitor
- CRT
- Software for computer-assisted instruction
- Slide projector
- Tape recorder

This equipment will allow us to administer training in a variety of modes, ranging from programmed and learner-controlled instruction to group seminars and workshops.

Work Remaining

To support the training, we need to furnish the room appropriately. Because the types of training will vary, the furniture should provide a flexible environment. Outlined here are our anticipated furnishing needs.

- Tables and chairs that can be set up in many configurations. These would allow for individual or group training and large seminars.
- Portable room dividers. These would provide study space for training with programmed instruction, and allow for simultaneous training.
- Built-in storage space for audiovisual equipment and training supplies. Ideally, this storage space should be multipurpose, providing work or display surfaces.
- A flexible lighting system, important for audiovisual presentations and individualized study.
- Independent temperature control, to ensure that the training room remains comfortable regardless of group size and equipment used.

**Describes what
remains to be
done**

**Gives a rough
timetable**

The project is on schedule. As soon as we receive your approval of these specifications, we will proceed to the next step: sending out bids for room dividers, and having plans drawn for the built-in storage space.

cc.   R. S. Pike, SVP
      G. T. Bailey, SVP

As you work on a longer report or term project, your instructor might require a progress report. In this next memo, Karen Granger documents her progress on her term project: an evaluation of the Environmental Protection Agency's effectiveness in cleaning a heavily contaminated harbor.

PROGRESS REPORT ON TERM PROJECT

Progress Report

To: Dr. John Lannon
From: Karen P. Granger
Date: April 17, 20xx
Subject: **Evaluation of the EPA's *Remedial Action Master Plan***

Project Overview
As my term project, I have been evaluating the issues of politics, scheduling, and safety surrounding the EPA's published plan to remove PCB contaminants from New Bedford Harbor.

Work Completed

February 23:   Began general research on the PCB contamination of the New Bedford Harbor.

March 8:   Decided to analyze the *Remedial Action Master Plan* (RAMP) in order to determine whether residents are being "studied to death" by the EPA.

March 9–19:   Drew a map of the harbor to show areas of contamination. Obtained the RAMP from Pat Shay of the EPA.
Interviewed Representative Grimes briefly by phone; made an appointment to interview Grimes and Sharon Dean on April 13. Interviewed Patricia Chase, President of the New England Sierra Club, briefly by phone.

March 24:   Obtained *Public Comments on the New Bedford RAMP,* a collection of reactions to the plan.

April 13:   Interviewed Grimes and Dean; searched Grimes's files for information. Also searched the files of Raymond Soares, New Bedford Coordinator, EPA.

Work in Progress
Contacting by telephone the people who commented on the RAMP.

Work to Be Completed

April 25:   Finish contacting commentators on the RAMP.

April 26:   Interview an EPA representative about the complaints that the commentators raised on the RAMP.

Date for Completion: May 3, 20xx

Margin notes:

Summarizes achievements to date

Describes work remaining, with timetable

Complications

The issue of PCB contamination is complicated and emotional. The more I uncovered, the more difficult I found it to remain impartial in my research and analysis. As a New Bedford resident, I expected to find that we are indeed being studied to death; because my research seems to support my initial impression, I am not sure I have remained impartial.

Lastly, the people I talk to do not always have the time to answer all my questions. Everyone, however, has been interested and encouraging, if not always informative.

Describes problems encountered (margin note)

Periodic Activity Reports

The periodic activity report resembles the progress report in that it summarizes activities over a specified time frame. But unlike progress reports, which summarize specific accomplishments on a given *project*, periodic reports summarize general activities during a given *period*. Manufacturers requiring periodic reports often have prepared forms, because most of their tasks are quantifiable (e.g., units produced). But most white-collar jobs do not lend themselves to prepared-form reports. You may have to develop your own format, as the next writer does.

Fran DeWitt's report answers her boss's primary question: *What did you accomplish last month?* Her response has to be detailed and informative.

PERIODIC ACTIVITY REPORT

Date:	6/18/xx
To:	N. Morgan, Assistant Vice President
From:	F. C. DeWitt
Subject:	**Recent Meetings for Computer-Assisted Instruction**

Gives overview of recent activities, and their purpose (margin note)

For the past month, I've been working on a cooperative project with the Banking Administration Institute, Computron Corporation, and several banks. My purpose has been to develop training programs, specific to banking, appropriate for computer-assisted instruction (CAI).

We have focused on three major areas: Proof/Encoding Training, Productivity Skills for Management, and Banking Principles.

Gives details (margin note)

I hosted two meetings for this task force. On June 6, we discussed Proof/Encoding Training, and on June 7, Productivity Training. The objective for the Proof/Encoding meeting was to compare ideas, information, and current training packages available on this topic. We are now designing a training course.

The objective of the meeting on Productivity was to discuss skills that increase productivity in banking (specifically Banking Operations). Discussion included instances in which computer-assisted instruction is appropriate for teaching productivity skills. Computron also discussed computer applications used to teach productivity.

On June 10, I attended a meeting in Washington, D.C., to design a course in basic banking principles for high-level clerical/supervisory-level employees. We also discussed the feasibility of adapting this course to CAI. This type of training, not currently available through Corporate Education, would meet a definite supervisor/management need in the division.

Explains the benefits of these activities

My involvement in these meetings has two benefits. First, structured discussions with trainers in the banking industry provide an exchange of ideas, methods, and experiences. This involvement expedites development of our training programs because it saves me time on research. Second, automation will continue to affect future training practices. With a working knowledge of these systems and their applications, I now am able to assist my group in designing programs specific to our needs.

Employees use periodic reports to inform management of what they are doing and how well they are doing it. Therefore, accuracy, clarity, and appropriate level of detail are important—as is the persuasive (and ethical) dimension. Make sure that recipients have all the necessary facts—and that they understand these facts as clearly as you do.

Meeting Minutes

Many team or project meetings require someone to record the proceedings. Minutes are the records of such meetings. Copies of minutes usually are distributed (often via email) to all members and interested parties, to track the proceedings and to remind members of their designated responsibilities. The appointed secretary records the minutes.

When you record minutes, answer these questions:

What recipients of minutes want to know

- *What group held the meeting? When, where, and why?*
- *Who chaired the meeting? Who else was present?*
- *Were the minutes of the last meeting approved (or disapproved)?*
- *Who said what?*
- *Was anything resolved?*
- *Who made which motions and what was the vote? What discussion preceded the vote?*
- *Who was given responsibility for which tasks?*

Minutes are filed as part of an official record, and so must be precise, clear, highly informative, and free of the writer's personal commentary ("As usual, Ms. Jones disagreed with the committee") or judgmental words ("good," "poor," "irrelevant").

MEETING MINUTES

Subject: **Minutes of Managers' Meeting, October 5, 20xx**

Members Present
Harold Tweeksbury, Jeannine Boisvert, Sheila DaCruz, Ted Washington, Denise Walsh, Cora Parks, Cliff Walsh, Joyce Capizolo

Agenda
1. The meeting was called to order on Wednesday, October 5, at 10 A.M. by Cora Parks.
2. The minutes of the September meeting were approved unanimously.
3. The first order of new business was to approve the following policies for the Christmas season:

   a. Temporary employees should list their ID numbers in the upper-left corner of their receipt envelopes to help verification. Discount Clerical assistant managers will be responsible for seeing that this procedure is followed.
   b. When temporary employees turn in their envelopes, personnel from Discount Clerical should spot-check them for completeness and legibility. Incomplete or illegible envelopes should be corrected, completed, or rewritten. *Envelopes should not be sealed.*

4. Jeannine Boisvert moved that we also hold one-day training workshops for temporary employees in order to teach them our policies and procedures. The motion was seconded. Joyce Capizolo disagreed, saying that on-the-job training (OJT) was enough. The motion for the training session carried 6–3. The first workshop, which Jeannine agreed to arrange, will be held October 25.
5. Joyce Capizolo requested that temporary employees be sent a memo explaining the temporary employee discount procedure. The request was converted to a motion and seconded by Cliff Walsh. The motion passed by a 7–2 vote.
6. Cora Parks adjourned the meeting at 11:55 A.M.

Tells who attended

Summarizes discussion of each item

Tells who said what

Tells what was voted

Different organizations often have templates or special formats for recording minutes.

ANALYTICAL REPORTS IN MEMO FORM

Analytical reports not only provide information, but also present an *analysis* of that information: what the information means and what action it suggests. Common types of short analytical reports include feasibility reports, recommendation reports, and justification reports.

Feasibility Reports

"Should we or shouldn't we?"

Feasibility reports are used when decision makers need to assess whether an idea or plan or course of action is realistic and practical: "How *doable* is this idea?" Although a particular course of action might be *possible*, it might not be *practical*—because such action might be too costly or hazardous or poorly timed, among other reasons. For example, a maker of precision tools might examine the feasibility of automating several key manufacturing processes. While automating these tasks might lower costs and increase productivity for the short term, the dampening effect of layoffs on company morale could lead to reduced productivity as well as quality control problems for the long term.

Testing for usability

We create surveys, and even when we devise questions that we're convinced are unambiguous, we pretest them—we often find that questions have been misinterpreted. We rewrite them and test them again, until we get them right. The basic element in my writing process is checking and rechecking for ambiguous messages and revising as often as time allows.

—James North, project manager
for a market-research firm

A feasibility analysis provides answers to questions like these:

What recipients of a feasibility report want to know

- *Is this course of action likely to succeed?*
- *Why or why not?*
- *What are the assessment criteria (e.g., cost, safety, productivity)?*
- *Do the benefits outweigh the drawbacks or risks?*
- *What are the pros and cons?*
- *What alternatives do we have?*
- *Can we get the funding?*
- *Should we do anything at all? Should we wait?*

NOTE   *An assessment of feasibility often requires two additional types of analysis: an examination of what caused a problem or situation and a comparison of two or more alternative solutions or courses of action. For detailed discussion of these three analytical approaches, see pages 607–09.*

In the following memo, a securities analyst for a state pension fund reports to the fund's manager on the feasibility of investing in a rapidly growing computer maker. (Notice the technical language, appropriate for an audience familiar with the specialized terminology of finance.)

A FEASIBILITY ANALYSIS

State Pension Fund

MEMORANDUM

To: Mary K. White, Fund Manager
From: Martha Mooney
Date: April 1, 20xx
Subject: *The Feasibility of Investing in WBM Computers, Inc.*

Gives brief background

Our zero-coupon bonds, comprising 3.5 percent of State Pension Fund's investment portfolio, will mature on April 15. Current inflationary pressures are making fixed-income investments less attractive than equities. As you requested, I have researched and compared the feasibility of various investment alternatives based on these criteria: market share, earnings, and dividends.

Recommendation

Makes a direct recommendation

Given its established market share, solid earnings, and generous dividends, WBM Computers, Inc. is a sound and promising company. I recommend that we invest our maturing bond proceeds in WBM's Class A stock.

Market Share

Explains the criteria supporting the recommendation

Although only ten years old, WBM successfully competes with well established computer makers. WBM's market share has grown steadily for the past five years. For this past year, total services and sales ranked 367th in the industrial United States, with orders increasing from $750 million to $1.25 billion. Net income places WBM 237th in the country, and 13th on return to investors.

Earnings

WBM's margin for profit on sales is 9 percent, a roughly steady figure for the past three years. Whereas 1995 earnings were only $ .09 per share, this year's earnings are $1.36 per share. Included in these ten-year earnings is a two-for-one stock split issued on November 2, 2004. Barring a global downturn in computer sales, WBM's outlook for continued strong earnings is promising.

Dividends

Investors are offered two types of common stock, listed on the American Stock Exchange. The assigned par value of both classes is $.50 per share. Class A stock pays an additional $.25 per share dividend but restricts voting privileges to one vote for every ten shares held by the investor. Class B stock is not entitled to the extra dividend but carries full voting rights. The additional dividend from Class A shares would enhance income flow into our portfolio.

Encourages reader action

WBM shares now trade at 14 times earnings, with a current share price of $56.00, a relative bargain in my estimation. An immediate investment would add strength and diversity to our portfolio.

Feasibility analysis is an essential basis for any well-conceived recommendation. Notice how the following recommendation and justification reports include implicit considerations of feasibility. Before people will accept your recommendation, you have to persuade them that this is a good idea.

Recommendation Reports

Recommendation reports interpret data, draw conclusions, and recommend a course of action, usually in response to a specific problem. The following recommendation report is addressed to the writer's supervisor. This is just one example of a short report used to examine a problem and recommend a solution.

17.1

For more sample memos visit <www.ablongman.com/lannonweb>

A Problem-Solving Recommendation

Bruce Doakes is assistant manager of occupational health and safety for a major airline that employs over two hundred reservation and booking agents. Each agent spends eight hours daily seated at a workstation that has a computer, telephone, and other electronic equipment. Many agents have complained of chronic discomfort from their work: headache, eyestrain and irritation, blurred or double vision, backache, and stiff neck and joints.

Bruce's boss asked him to study the problem and recommend improvements in the work environment. Bruce surveyed employees and consulted ophthalmologists, chiropractors, orthopedic physicians, and the latest publications on ergonomics (tailoring work environments for employees' physical and psychological well-being). After completing his study, Bruce composed his report. ❏

Bruce's report had to be persuasive as well as informative.

RECOMMENDATION MEMO

Trans Globe Airlines

MEMORANDUM
To: R. Ames, Vice President, Personnel
From: B. Doakes, Health and Safety
Date: August 15, 20xx
Subject: **Recommendations for Reducing Agents' Discomfort**

In our July 20 staff meeting, we discussed physical discomfort among reservation and booking agents, who spend eight hours daily at automated workstations. Our agents complain of headaches, eyestrain and irritation, blurred or double vision, backaches, and stiff joints. This report outlines the apparent causes and recommends ways of reducing discomfort.

Causes of Agents' Discomfort
For the time being, I have ruled out the computer display screens as a cause of headaches and eye problems for the following reasons:

Provides immediate orientation by giving brief background and main point

Interprets findings and draws conclusions

1. Our new display screens have excellent contrast and no flicker.
2. Research findings about the effects of low-level radiation from computer screens are inconclusive.

The headaches and eye problems seem to be caused by the excessive glare on display screens from background lighting.

Other discomforts, such as backaches and stiffness, apparently result from the agents' sitting in one position for up to two hours between breaks.

Recommended Changes

Makes general recommendations

We can eliminate much discomfort by improving background lighting, workstation conditions, and work routines and habits.

Background Lighting.   To reduce the glare on display screens, these are recommended changes in background lighting:

Expands on each recommendation

1. Decrease all overhead lighting by installing lower-wattage bulbs.
2. Keep all curtains and adjustable blinds on the south and west windows at least half-drawn, to block direct sunlight.
3. Install shades to direct the overhead light straight downward, so that it is not reflected by the screens.

Workstation Conditions.   These are recommended changes in the workstations:

1. Reposition all screens so light sources are neither at front nor back.
2. Wash the surface of each screen weekly.
3. Adjust each screen so the top is slightly below the operator's eye level.
4. Adjust all keyboards so they are 27 inches from the floor.
5. Replace all fixed chairs with adjustable, armless, secretarial chairs.

Work Routines and Habits.   These are recommended changes in agents' work routines and habits:

1. Allow frequent rest periods (10 minutes each hour instead of 30 minutes twice daily).
2. Provide yearly eye exams for all terminal operators, as part of our routine healthcare program.
3. Train employees to adjust screen contrast and brightness whenever the background lighting changes.
4. Offer workshops on improving posture.

Discusses benefits of following the recommendations

These changes will give us time to consider more complex options such as installing hoods and antiglare filters on terminal screens, replacing fluorescent lighting with incandescent, covering surfaces with nonglare paint, or other disruptive procedures.

cc.   J. Bush, Medical Director
      M. White, Manager of Physical Plant

For more examples and advice on formulating, evaluating, and refining your recommendations, see pages 612–15.

Justification Reports

Many recommendation reports respond to reader requests for a solution to a problem; others originate with the writer, who has recognized a problem and developed a solution. This latter type is often called a *justification report*. As the name implies, such reports *justify* the writer's position on some issue. Justification reports therefore typically follow a direct organizing plan, beginning with the request or recommendation. Such reports answer this key question for recipients: *Why should we?*

Typically, justification reports follow a version of this arrangement:

How to organize
a justification
report

1. State the problem and your recommendations for solving it.
2. Point out the cost, savings, and benefits of your plan.
3. If needed, explain how your suggestion can be implemented.
4. Conclude by encouraging the reader to act.

The next writer uses a version of the preceding arrangement: she begins with the problem and recommended solution, spells out costs and benefits, and concludes by reemphasizing the major benefit. The tone is confident yet diplomatic—appropriate for an unsolicited recommendation to a superior.

JUSTIFICATION MEMO

Greentree Bionomics, Inc.

MEMORANDUM

To: D. Spring, Personnel Director
           Greentree Bionomics, Inc. (GBI)
From: M. Noll, Biology Division
Date: April 18, 20xx
Subject: **The Need to Hire Additional Personnel**

Introduction and Recommendation

Opens with the
problem

With twenty-six active employees, GBI has been unable to keep up with scheduled contracts. As a result, we have a contract backlog of roughly $500,000. This backlog is caused by understaffing in the biology and chemistry divisions.

Recommends a
solution

To increase production and ease the workload, I recommend that GBI hire three general laboratory assistants.

Expands on the recommendation

The lab assistants would be responsible for cleaning glassware and general equipment; feeding and monitoring fish stocks; preparing yeast, algae, and shrimp cultures; preparing stock solutions; and assisting scientists in various tests and procedures.

Costs and Benefits

Shows how benefits would offset costs

While costing $74,880 yearly (at $12.00/hour), three full-time lab assistants would have a positive effect on overall productivity:

1. Uncleaned glassware would no longer pile up, and the fish holding tanks could be cleaned daily (as they should be) instead of weekly.
2. Because other employees would no longer need to work more than forty hours weekly, morale would improve.
3. Research scientists would be freed from general maintenance work (cleaning glassware, feeding and monitoring the fish stock, etc.). With more time to perform client tests, the researchers could eliminate our backlog.
4. With our backlog eliminated, clients would no longer have cause for impatience.

Conclusion

Encourages acceptance of recommendation

Increased production at GBI is essential to maintaining good client relations. These additional personnel would allow us to continue a reputation of prompt and efficient service, thus ensuring our steady growth and development.

ELECTRONIC MAIL

17.2

How can email disrupt the workplace? Find out more at <www.ablongman.com/lannonweb>

Surveys show that email, in the words of one researcher, "is by far the most frequently used and highly prized feature of the Internet," with messages sent each day numbering in billions (Specter, 95, 101). See Figure 17.2 for elements of a typical email message.

Email Benefits

Compared to phone, fax, or conventional mail (or even face-to-face conversation, in some cases), email can offer real advantages:

Email advantages in the workplace

- *Lack of real-time constraints.* Email allows people to communicate at any time. Besides eliminating "telephone tag," email offers the choice of when to read a message and whether to respond.
- *Efficient filing, retrieval, and forwarding.* Email messages can be filed for future reference, cut and pasted into other documents, and forwarded to others in a single keystroke.
- *Attachments.* Documents or electronic files can be attached and sent for the recipient to download, usually with the original formatting intact.

```
From:    "Marcia lannon" <marcialannon@earthlink.net>
To:   ergo@valinet.com
CC:   rdumont@umassd.edu
Date:   Mon, 16 Jul 2004 10:30:39 +0000
Subject:   Elements of a Typical Email Message


  Header
  Besides the standard information from a memo (Date, To, From, Subject) the
  email header includes a copy notation ("CC"), and the exact time the message
  was sent. Depending on the email program, additional information (host server,
  message ID number, time the message was read) may be included as well.

  Message Body
  To avoid scrolling, try to limit the message to one page. For readability,
  keep the paragraphs short.

  Headings and Lists
  Use headings and lists to break up a long message.

  Enclosures
  Documents, software, images, and other files can be included as
  enclosures (or attachments) to an email message.

  Signature Block
  Your name, company, and contact information can be added automatically
  to each email. For example:

  Marcia W. Lannon
  Manager, Customer Relations
  Apex, Inc.
  Tel. 467-896-5698
  Fax. 467-896-6845


  REPLY | REPLY ALL | GET ENCLOSURE | FORWARD [As Attachment]
  Previous | Next | Delete | Done
```

FIGURE 17.2 Elements of a Typical Email Message Different email programs produce slightly different formats.

- *Democratic communication.* Email allows anyone in an organization to contact anyone else. The filing clerk could conceivably email the company president directly, whereas a conventional memo or phone call would be routed through the management chain or screened by assistants (D. Goodman 33–35). In addition, shy people may be more willing to speak out during an email conversation.

NOTE *Communication expert Stuart Selber offers this important caution: "The mere presence of email on the job does not mean that someone's work environment supports democratic uses of email. In fact, just the opposite could be true: a company can use email to monitor and [intimidate employees]."*

- *Creative thinking.* Email users generally communicate spontaneously—without worry about page design, paragraph structure, or perfect phrasing. This relatively free exchange of views can lead to new insights or ideas (Bruhn 43).
- *Collaboration and research.* Teams can keep in touch via email, and researchers can contact people for answers they need.

NOTE   *While electronically mediated collaboration increases the quality and sharing of ideas, groups who meet only via network develop less trust than groups who have some face-to-face contact (Ross-Flanigan 57, 58).*

Email Copyright Issues

Any email message you receive is copyrighted by the person who wrote it. Under current law, forwarding this message to anyone for any purpose is a violation of the owner's copyright. The same is true for reproducing an email message as part of any type of publication—unless your use of this material falls within the boundaries of "fair use." (See pages 138, 144.)

Email Privacy Issues

Gossip, personal messages, risqué jokes, or complaints about the boss or a colleague—all might reach unintended recipients. While phone companies and other private carriers are governed by laws protecting privacy, no such legal protection yet exists for Internet communication (Peyser and Rhodes 82). The Electronic Privacy Act of 1986 offers limited protection against unauthorized reading of another person's email, but employers are exempt (Extejt 63).

Employers are legally entitled to monitor employee email   In some instances it may be proper for an employer to monitor E-mail, if it has evidence of safety violations, illegal activity, racial discrimination, or sexual improprieties, for instance. Companies may also need access to business information, whether it is kept in an employee's drawer, file cabinet, or computer E-mail (Bjerklie 15).

Email privacy can be compromised in other ways as well:

Email offers no privacy
- Everyone on a group mailing list—intended recipient or not—automatically receives a copy.
- Even when "deleted" from the system, messages can live on, saved in a backup file.
- Besides infringing on copyright (see page 138), forwarding a message without the author's consent violates that person's privacy.
- Anyone with access to your network and password can read your document, alter it, use parts of it out of context, pretend to be its author, forward it, plagiarize your ideas, or even author a document or conduct illegal activity in your name. (One partial safeguard is encryption software, which scrambles

the message, and the only people who can unscramble it are those who have the code. Another security strategy is to circulate any sensitive documents as an email attachment in Adobe *Portable Document File* format. PDF format prevents a document from being altered or rewritten.)

GUIDELINES for Using Electronic Mail*

Observe Proper Etiquette

1. *Check and answer your email daily.* Like an unreturned phone call, unanswered email is annoying. If you're really busy, at least acknowledge receipt and respond later.

2. *Check your distribution list before each mailing.* Verify that your message will reach all intended recipients—but no unintended ones.

Consider the Ethical, Legal, and Interpersonal Implications

1. *Assume that your email is permanent and could be read by anyone at any time.* Don't write anything you couldn't say face-to-face. Avoid *spamming* (sending junk mail) and *flaming* (making rude remarks).

2. *Think twice before making humorous remarks.* What seems amusing to you may be offensive to others, including recipients from different cultures. Any email judged to be harassing or discriminatory brings immediate dismissal and often leads to legal action against the company as well as the guilty employee.

3. *Don't use email for confidential information.* Avoid complaining, criticizing, or evaluating people, or anything that should be kept private (say, an employee reprimand).

4. *Don't use the company email network for personal correspondence or for anything not work-related.* Employers increasingly monitor their email networks.

5. *Before you forward an incoming message, obtain permission from the sender.* Assume that anything you receive is the private property of the sender. (See page 138 for email copyright issues.)

Make the Message Usable

1. *Limit your message to a single topic.* Remain focused and concise.

2. *Limit your message to a single screen, if possible.* Don't force recipients to scroll needlessly.

3. *Use a clear subject line to identify your topic.* Instead of "Test Data" or "Data Request," announce your purpose clearly: ("Request for Beta Test Data for Project #16"). Recipients scan subject lines in deciding which new mail to read immediately.

(continues)

Guidelines (continued)

4. *Refer clearly to the message to which you are responding:* ("Here are the Project 16 Beta test data you requested on Oct. 10").

5. *Keep sentences and paragraphs short.*

6. *Use a block format.* Don't indent paragraphs.

7. *Don't write in FULL CAPS—unless you want to SCREAM at the recipient!*

8. *Where appropriate in formal emails, use graphic highlighting.* Headings, bullets, numbered lists, boldface, and italics improve readability and impart professionalism.

9. *Where appropriate, use formal salutations and closings.* Choose the level of formality that reflects your recipient and your purpose. When addressing someone you don't know or someone in a position of authority, begin with a formal salutation ("Dear Doctor Gomez") and end with formal closing ("Sincerely"). But for a familiar recipient, be less formal ("Hello," "Regards").

10. *Use smiley faces and abbreviations sparingly.* Smiley faces, made from a colon, dash, and right-hand parentheses:-) are used to signify humor. Use these and other emoticons infrequently and only in informal messages to people you know well. Also, common email abbreviations (FYI, TW, HAND—which mean "for your information," "by the way," and "have a nice day") may annoy some recipients.

11. *Close with a signature section.* Include the name of your company or department, your telephone and fax number, and any other contact information the recipient might consider relevant.

12. *Don't send huge or specially formatted attachments (or enclosures) without checking with the recipient.* Not all email browsers can handle formatted files, photos, and so on. Also, if a recipient has a slow Internet connection, downloading a long or complex attachment will take forever. Always ask beforehand about whether the recipient's browser can accept attachments and about which file types (*Simple Text, PDF File, Rich Text Format,* and so on) the equipment can handle.

13. *Proofread for spelling, punctuation, and grammar.* A mechanically correct message is always more credible than a sloppy one.

*Adapted from Bruhn 43; D. Goodman 33–35, 167; "Email Etiquette" 3; Gurak and Lannon 186; Kawasaki, "The Rules" 286; D. Munger; Nantz and Drexel 45–51; Peyser and Rhodes 82.

INSTANT MESSAGING

A faster medium than email, instant messaging (IM) allows for text-based conversation in real time: The user types a message in a pop-up box and the recipient can respond instantly. IM groupware enables multiple users to converse and collaborate from various locations. According to *Fortune* magazine, "instant messaging is rising fast in corporate America," rapidly displacing email for routine communication (Varchaver 102).

Although instant messaging has been popular among teens and college students, its more recent advent as a business tool means that few rules govern its use. Also, most current IM software does not automatically save these messages electronically. But as IM becomes more pervasive in the workplace, companies will likely monitor its use by employees and save all messages as a permanent record.

GUIDELINES for Choosing Email Versus Paper or Telephone

Email is excellent for reaching a lot of people quickly with a relatively brief, informal message. Instant messaging is good for conversation. But there are often good reasons to put something on paper—or to speak directly with the recipient instead.

1. *Don't use email when a more personal medium is preferable.* Sometimes an issue is best resolved by a phone call, or even by voice mail—whereas email might imply that the sender can't be bothered to speak with the recipient directly.

2. *Don't use email for a detailed discussion.* In contrast to a rapid-fire email message, preparing a paper document is generally more deliberate, giving you a chance to shape and clarify your thinking, to choose words carefully, and to revise. Also, a well-crafted paper document (say, a memo or a letter) is likely to be read more attentively. (For in-house recipients, consider transmitting your paper document as a file attached to a brief, introductory email.)

3. *Don't use email for most formal correspondence.* Because of the volume and causal style of email, recipients might overlook the significance of a message. Don't use email to apply for a job, request a raise, resign from a job, or respond to clients or customers unless recipients specifically request this method. (Chapter 18 discusses letters—including electronic job hunting.)

☑ CHECKLIST for Usability of Memo Reports

(Numbers in parentheses refer to the first page of discussion.)

Ethical, Legal, and Interpersonal Considerations

- ☐ Is the information specific, accurate, and unambiguous? (386)
- ☐ Is this the best report medium (paper, email, phone) for the situation? (405)
- ☐ Is the memo inoffensive to all parties? (386)
- ☐ Are all appropriate parties receiving a copy? (388)

Organization

- ☐ Is the direct or indirect pattern used appropriately to present the report's bottom line? (388)
- ☐ Is the material "chunked" into easily digestible parts? (387)

Format

- ☐ Does the memo have a complete heading? (387)
- ☐ Does the subject line forecast the memo's contents? (387)

- ☐ Are paragraphs single spaced within and double spaced between? (387)
- ☐ Do headings announce subtopics? (387)
- ☐ If more than one reader is receiving copies, does the memo include a distribution notation (cc:) to identify other readers? (416)
- ☐ Does the document's appearance create a favorable impression? (387)

Content

- ☐ Is the message short and to the point? (385)
- ☐ Are tables, charts, and other graphics used as needed? (387)
- ☐ Are recipients given enough information for an *informed* decision? (39)
- ☐ Are the conclusions and recommendations clear? (399)

Style

- ☐ Is the writing clear, concise, exact, fluent, and appropriate? (244)
- ☐ Is the tone appropriate? (274)
- ☐ Has the memo been carefully proofread? (24)

EXERCISES   For more exercises, visit <www.ablongman.com/lannon>

1. We would all like to see changes in our schools' policies or procedures, whether they are changes in our majors, school regulations, social activities, grading policies, or registration procedures. Find some area of your school that needs obvious changes, and write a justification report to the person who might initiate that change. Explain why the change is necessary and describe the benefits. Follow the format on page 399.

2. Think of an idea you would like to see implemented in your job (e.g., a way to increase productivity, improve service, increase business, or improve working conditions). Write a justification memo, persuading your audience that your idea is worthwhile.

3. Write a memo to your employer, justifying reimbursement for this course. *Note:* You might have written another version of this assignment for Exercise 2 in Chapter 1. If so, compare early and recent versions for content, organization, style, and format.

4. Identify a dangerous or inconvenient area or situation on campus or in your community (endless cafeteria lines, a poorly lit intersection, slippery stairs, a poorly adjusted traffic light). Observe the problem for several hours during a peak use period. Write a justification report to a *specifically identified* decision maker, describing the problem, listing your observations, making recommendations, and encouraging reader support or action.

5. Assume that you have received a $10,000 scholarship, $2,500 yearly. The only stipulation for re-

ceiving installments is that you send the scholarship committee a yearly progress report on your education, including courses, grades, school activities, and cumulative average. Write the report.

6. In a memo to your instructor, outline your progress on your term project. Describe your accomplishments, plans for further work, and any problems or setbacks. Conclude your memo with a specific completion date.

7. Keep accurate minutes for one class session (preferably one with debate or discussion). Submit the minutes in memo form to your instructor.

8. Conduct a brief survey (e.g., of comparative interest rates from various banks on a car loan, comparative tax and property evaluation rates in three local towns, or comparative prices among local retailers for an item). Arrange your data and report your findings to your instructor in a memo that closes with specific recommendations for the most economical choice.

9. Recommendation report (choose one)
   a. You are legal consultant to the leadership of a large autoworkers' union. Before negotiating its next contract, the union needs to anticipate effects of robotics technology on assembly-line autoworkers within ten years. Do the research and write a report recommending a course of action.

   b. You are a consulting engineer to an island community of two hundred families suffering a severe shortage of fresh water. Some islanders have raised the possibility of producing drinking water from salt water (desalination). Write a report for the town council, summarizing the process and describing instances in which desalination has been used successfully or unsuccessfully. Would desalination be economically feasible for a community this size? Recommend a course of action.

   c. You are a health officer in a town less than one mile from a massive radar installation. Citizens are disturbed about the effects of microwave radiation. Do they need to worry? Should any precautions be taken? Find the facts and write your report.

   d. You are an investment broker for a major firm. A longtime client calls to ask your opinion. She is thinking of investing in a company that is fast becoming a leader in nanotechnology. "Should I invest in this technology?" your client wants to know. Find out, and give her your recommendations in a short report.

   e. The buildings in the condominium complex you manage have been invaded by carpenter ants. Can the ants be eliminated by any insecticide *proven* nontoxic to humans or pets? (Many dwellers have small children and pets.) Find out, and write a report making recommendations to the maintenance supervisor.

   f. The "coffee generation" wants to know about the properties of caffeine and the chemicals used on coffee beans. What are the effects of these substances on the body? Write your report, making specific recommendations about precautions that coffee drinkers can take.

   g. As a consulting dietitian to the school cafeteria in Blandville, you've been asked by the school to report on the most dangerous chemical additives in foods. Parents want to be sure that foods containing these additives are eliminated from school menus, insofar as possible. Write your report, making general recommendations about modifying school menus.

   h. Dream up a scenario of your own in which information and recommendations would make a real difference. (Perhaps the question could be one you've always wanted answered.)

10. Individually or in small groups, decide whether each of the following documents would be appropriate for transmission via a company email network. Be prepared to explain your decisions.

Sarah Burnes' memo about benzene levels (page 19)

The "Rational Connection" memo (page 50)

The "better" memo to the maintenance director (page 53)

Tom Ewing's letter to a potential customer (page 63)

The medical report written for expert readers (page 30)

A memo reporting illegal or unethical activity in your company

A personal note to a colleague

A request for a raise or promotion

Minutes of a meeting

Announcement of a no-smoking policy

An evaluation or performance review of an employee

A reprimand to an employee

A notice of a meeting

Criticism of an employee or employer

A request for volunteers

A suggestion for change or improvement in company policy or practice

A gripe

A note of praise or thanks

A message you have received and have decided to forward to other recipients

COLLABORATIVE PROJECTS

1. Organize into groups of four or five and choose a topic for which group members can agree on a position. Here are some possibilities:

   - Should your college abolish core requirements?
   - Should every student in your school pass a writing proficiency exam before graduating?
   - Should courses outside your major be graded pass/fail at your request?
   - Should your school drop or institute student evaluation of teachers?
   - Should all students be required to be computer literate before graduating?
   - Should campus police carry guns?
   - Should school security be improved?
   - Should students with meal tickets be charged according to the type and amount of food they eat, instead of paying a flat fee?

   As a group, decide your position on the issue, and brainstorm collectively to justify your recommendation to a stipulated primary audience in addition to your classmates and instructor. Complete an audience and use profile (page 65), and compose a justification report. Appoint one member to present the report in class.

2. Divide into groups and respond to the following scenario:

   As a legal safeguard against discriminatory, harassing, or otherwise inappropriate email mes-

sages, a legal consultant to your company or college has proposed a plan for electronic monitoring of email use at your organization. Your employer or college dean has asked your team to study the issue and to answer this question: "Should we support this plan?" Among the many subordinate questions to consider:

   - What are the rights of the people who would be monitored?
   - What are the rights of the organization?
   - Is the plan ethical?
   - Could the plan backfire? Why?
   - How would the plan affect people's perception of the organization?
   - Should monitoring be done selectively or routinely?
   - Should the entire organization be given a voice in the decision?
   - Are there acceptable alternatives to monitoring?

   Begin by reviewing Chapters 5 and 6 and consulting Figure 24.3. Then do the research and prepare a memo that makes a persuasive case for your team's recommendation.

   Web sites that address privacy issues:

   - *Electronic Privacy Information Center,* a public interest research center at <www.epic.org>.
   - *Privacy International,* a human rights group at <www.privacy.org/pi/>.
   - *Computer Professionals for Social Responsibility,* a public interest group at <www.cpsr.org>.

3. People regularly contact your organization (your company, agency, or college department) via email, letter, or your Web site to request information. You decide to prepare a FAQ list in response to the ten most frequently asked questions about products, services, specific concentrations within the major, admission requirements, or the like. In addition to being posted on your Web site, this list can be sent as an email attachment or mailed out as hard copy, depending on the reader's preference.

   After analyzing your specific audience and purpose and doing the research, prepare your list in a short report format.

18

Letters and Employment Correspondence

W̲e often have good reason to correspond in a more formal and personal medium than a memo or email message. A well-crafted letter is appropriate in situations like these:

When to send a
letter instead of a
memo or email

- To personalize your correspondence, conveying the sense that this message is prepared exclusively for your recipient.
- To convey a dignified, professional impression.
- To act as a representative of your company or organization.
- To present a reasoned, carefully constructed case.
- To respond to clients, customers, or anyone outside your organization.
- To provide an official notice or record (as in a letter announcing legal action or confirming a verbal agreement).

Because a letter often has a persuasive purpose, proper tone is essential for connecting with the recipient. Because your signature indicates your approval—and responsibility—for the contents of the letter (which may serve as a legal document), precision is crucial.

This chapter covers three common letter types: inquiry letters, claim letters, and letters of application, along with résumés. Other types are discussed in Chapters 23 and 25.

ELEMENTS OF USABLE LETTERS

The conventional arrangement of workplace letters allows recipients to locate what they need immediately (Figure 18.1).

Basic Parts of Letters

A typical business letter has six parts:

HEADING AND DATE.   If your stationery has a company letterhead, simply include the date a few lines below the letterhead, flush against the right or left margin. On blank stationery, include your return address and the date (but not your name):

Street address
City, state, zip
Month, day, year

> 154 Sea Lane
> Harwich, MA 02163
> July 15, 20xx

Use the Postal Service's two-letter state abbreviations (MA for Massachusetts, VT for Vermont) in your heading, inside address, and on the envelope.

INSIDE ADDRESS.   Two to six line spaces below the heading, flush against the left margin is the inside address (the address to which you are sending the letter).

Heading

LEVERETT LAND & TIMBER COMPANY, INC.
18 River Rock Road,
Leverett, MA 01054

creative land use
quality building materials
architectural construction

Date

January 17, 20xx

Inside address

Mr. Thomas E. Muffin
Clearwater Drive
Amherst, MA 01022

Salutation

Dear Mr. Muffin:

Letter text

I have examined the damage to your home caused by the ruptured water pipe and consider the following repairs to be necessary and of immediate concern:

    Exterior:
      Remove plywood soffit panels beneath overhangs
      Replace damaged insulation and plumbing
      Remove all built-up ice within floor framing
      Replace plywood panels and finish as required

    Northeast Bedroom—Lower Level:
      Remove and replace all sheetrock, including closet
      Remove and replace all door casings and baseboards
      Remove and repair windowsill extensions and moldings
      Remove and reinstall electric heaters
      Respray ceilings and repaint all surfaces

This appraisal of damage repair does not include repairs and/or replacements of carpets, tile work, or vinyl flooring. Also, this appraisal assumes that the plywood subflooring on the main level has not been severely damaged.

Leverett Land & Timber Company, Inc. proposes to furnish the necessary materials and labor to perform the described damage repairs for the amount of six thousand one hundred and eighty dollars ($6,180).

Complimentary closing

Sincerely,

Signature

G.A. Jackson

Title

Gerald A. Jackson, President

Typist's initials

GAJ/ob

Enclosure notation

Encl. Itemized estimate

Phone: 410-555-9879   Fax: 410-555-6874   Email: llt@yonet.com

FIGURE 18.1 A Standard Design for a Workplace Letter

Name and
position
Company name
Street address
City, state, zip

Dr. Ann Mello, Dean of Students
Western University
30 Mogul Hill Road
Stowe, VT 51350

Whenever possible, address a specifically named recipient, and include the person's title. Using "Mr." or "Ms." before the name is optional. (See page 281 for avoiding sexist usage in titles and salutations.)

NOTE *Depending on the letter's length, adjust the horizontal placement of your return address and inside address to achieve a balanced page.*

SALUTATION. The salutation, two line spaces below the inside address, begins with *Dear* and ends with a colon (*Dear Ms. Smith:*). If you don't know the recipient's name, use the position title (*Dear Manager*) or, preferably, an attention line (page 416). Only address the recipient by first name if that is the way you would address this individual in person.

Typical
salutations

Dear Ms. Smith:
Dear Managing Editor:
Dear Professor Lexington-Trudeau:

No satisfactory guidelines exist for addressing several recipients simultaneously. *Gentlemen* or *Dear Sirs* implies bias. *Ladies and Gentlemen* sounds too much like the beginning of a speech. *Dear Sir or Madam* is old-fashioned. *To Whom It May Concern* is vague and impersonal. Your best bet is to eliminate the salutation by using an attention line.

 NOTE *For international audiences, an inappropriate salutation is highly offensive. In France or England, for example, a person's title should be used in the greeting, as in "Monsieur le Professeur Larrouse" (Sabath 164); in England, "Dear Madam" and "Dear Sir" continue to be acceptable for people not known well by the writer (Scott 55). Whenever possible, learn about your recipient's culture and preferences beforehand.*

The shape of
workplace letters

LETTER TEXT. Typically, your letter text begins two line spaces (returns) below the salutation. Workplace letters typically include (1) a brief introductory paragraph (five or fewer lines) that identifies your purpose and connects with the recipient's interest, (2) one or more body paragraphs containing details of your message, and (3) a concluding paragraph that sums up and encourages action.

Keep the paragraphs short, usually fewer than eight lines. If a paragraph goes beyond eight lines, consider using a bulleted or numbered list to make the paragraph readable. (In the letter on page 421, notice that the body section is divided into four questions for easy answering.)

COMPLIMENTARY CLOSING. The closing, two line spaces (returns) below the last line of text, should parallel the level of formality used in the salutation and should reflect your relationship to the recipient (polite but not overly intimate). These possibilities are listed in decreasing order of formality:

Complimentary closings

> Respectfully,
> Sincerely, (*most often used*)
> Cordially,
> Best wishes,
> Warmest regards,
> Regards,
> Best,

In a modified block letter (Figure 18.2), align the closing with your heading or the date. In a full block letter (Figure 18.3), position the closing flush against the left margin.

SIGNATURE. Type your full name and title on the fourth and fifth lines below and aligned with the closing. Sign in the triple space between the closing and your typed name.

The signature block

> Sincerely yours,
>
> *Martha S. Jones*
>
> Martha S. Jones
> Personnel Manager

If you are representing your company or a group that bears legal responsibility for the correspondence, type the company's name in full caps two line spaces below your complimentary closing; place your typed name and title four line spaces below the company name and sign in the triple space between.

Signature block representing the company

> Yours truly,
>
> HASBROUCK LABORATORIES
>
> *Lester Fong*
>
> L. H. Fong
> Research Associate

NOTE

To save space, sample letters in this chapter often show only the letter text, with heading, date, inside address, salutation, closing, and signature omitted. Your own letters of course would include these basic parts, as well as any needed specialized parts discussed below. See Figures 18.1 and 18.7, for example.

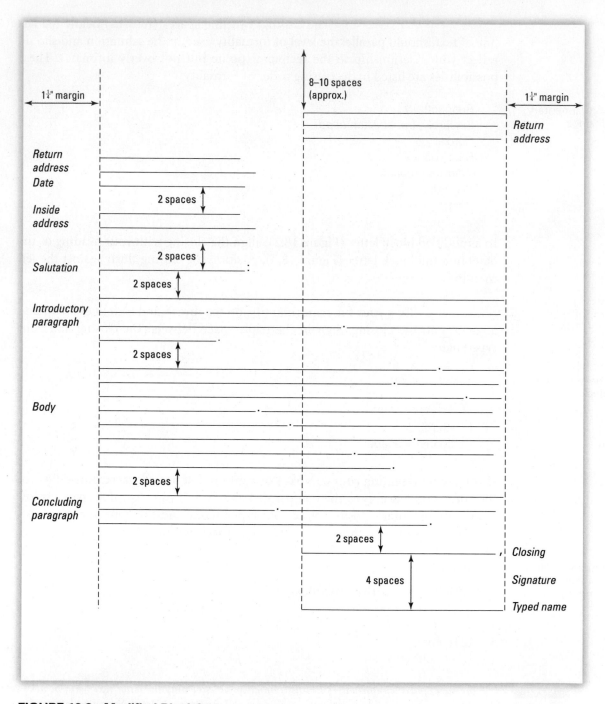

FIGURE 18.2   Modified Block Letter

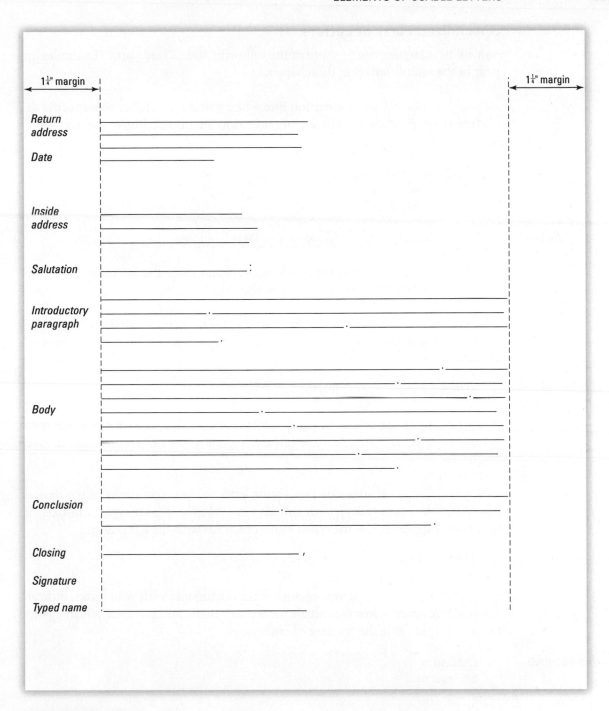

FIGURE 18.3   Block Letter

Specialized Parts of Letters

Some letters require one or more of the following specialized parts. (Examples appear in the sample letters in this chapter.)

ATTENTION LINE.   Use an attention line when you direct a letter to a specific department or position within an organization but don't know the recipient's name.

> Glaxol Industries, Inc.
> 232 Rogaline Circle
> Missoula, MT 61347
>
> ATTENTION: Director of Research and Development

An attention line can replace your salutation

Drop two line spaces below the inside address and place the attention line either flush with the left margin or centered on the page.

SUBJECT LINE.   Typically, subject lines are used with memos, but if the recipient is not expecting your letter, a subject line is a good way of catching a busy person's attention.

A subject line can attract attention

> SUBJECT: Placement of the Subject Line

Place the subject line below the inside address or attention line with one line space before and after it. You can underline or highlight the subject to make it more prominent.

TYPIST'S INITIALS.   If someone types your letter for you (common in the days of typewriters but rare today), your initials (in CAPS) and your typist's initials (in lower case) appear below the typed signature, flush with the left margin.

Typist's initials

> JJ/pl

ENCLOSURE NOTATION.   If you enclose other documents with your letter, indicate this one line space below the initials (or writer's name and position), flush against the left margin. State the number of enclosures.

Enclosure noted

> Enclosure
> Enclosures 2
> Encl. 3

If the enclosures are important documents such as legal certificates, checks, or specifications, name them in the notation.

Enclosure named | Enclosures: 2 certified checks, 1 set of KBX plans.

DISTRIBUTION NOTATION. If you distribute copies of your letter to other recipients, indicate this by inserting the notation "Copy" or "cc" one line below the previous line (such as an enclosure line). The "cc" notation once stood for "carbon copy," but no one uses carbon paper any more, so now it is said to stand for "courtesy copy."

> cc: office file
>   Melvin Blount
>
> c: S. Furlow
>   B. Smith

Most copies are distributed on an *FYI* (*For Your Information*) basis, but writers sometimes use the distribution notation to maintain a paper trail or to signal the primary recipient that this information is being shared with others (e.g., superiors, legal authorities).

POSTSCRIPT. A postscript (typed or handwritten) draws attention to a point you wish to emphasize or adds a personal note. Do not use a postscript if you forget to mention a point in the body of the letter. Rewrite the body section instead.

A postscript | P. S. Because of its terminal position in your letter, a postscript can draw attention to a point that needs reemphasizing.

Place the postscript two line spaces below any other notation, and flush against the left margin. Because readers often regard postscripts as sales letter gimmicks, use them sparingly in professional correspondence.

Design Factors

Although several formats are acceptable, and your company may have its own, the two most popular formats for workplace letters are *modified block* and *block*.

In the modified block form, the first line of a paragraph is not indented. Paragraphs are separated by a line space. The return address, complimentary closing, and signature align at page center (Figure 18.2).

In the block form, every line begins at the left margin (Figure 18.3). This form is popular because it looks businesslike and eliminates the need to tab and center.

These additional design factors make workplace letters appear inviting, accessible, and professional:

QUALITY STATIONERY. Use high-quality, 20-pound bond, $8\frac{1}{2}"$ × 11" white stationery with a minimum fiber content of 25 percent.

UNIFORM MARGINS AND SPACING.   When using stationery without a letterhead, frame your letter with $1\frac{1}{2}$-inch top and side margins and bottom margins of 1 to $1\frac{1}{2}$ inches. Use single spacing within paragraphs and double spacing between. Vary these guidelines based on the amount of space required by the letter's text, but strive for a balanced look.

HEADERS FOR SUBSEQUENT PAGES.   Head each additional page with a notation identifying the recipient, date, and page number.

Subsequent-page
header

> Adrianna Fonseca, June 25, 20xx, p. 2

Align your header with the right-hand margin. See page 64 for an example.

NOTE   *Never use an additional page solely for the closing section. Instead, reformat the letter so that the closing appears on the first page or so that at least two lines of text appear above the closing on the subsequent page.*

THE ENVELOPE.   Your envelope (usually a #10 envelope) should be of the same quality as your stationery. Place the recipient's name and address at a fairly central point on the envelope. Place your own name and address in the upper-left corner. Single-space these elements. Your word processor likely has an envelope printing function that will automatically place these elements. (See your printer's operating manual for instructions.)

INTERPERSONAL CONSIDERATIONS IN WORKPLACE LETTERS

In addition to presenting an accessible and inviting design, an effective letter enhances the relationship between sender and recipient. Interpersonal elements make a *human* connection.

FOCUS ON YOUR RECIPIENT'S INTERESTS: THE "YOU" PERSPECTIVE.   When speaking face-to-face, you unconsciously modify your statements and expression as you read the listener's signals: a smile, a frown, a raised eyebrow, a nod. In a telephone conversation, a voice provides cues that signal approval, dismay, anger, or confusion. In writing a letter, however, you might forget that a flesh-and-blood recipient will be reacting to what you are saying—or seem to be saying.

A letter conveying a "you" perspective focuses on content important to the recipient and shows respect for that person's feelings and attitudes.

To achieve a "you" perspective, ask yourself how the recipient will react to what you have written. Even a single word, carelessly chosen, can offend. In trying to correct a billing error, for example, you might feel tempted to write this:

A needlessly
offensive tone

> Our record keeping is very efficient, so this is obviously your error.

Such an accusatory tone might be appropriate after numerous failed attempts to achieve satisfaction on your part, but in your initial correspondence it would be offensive. Here is a more considerate version:

A tone that
conveys the
"you"
perspective

> If my paperwork is wrong, please let me know and I will send you a corrected version immediately.

Instead of hastily judging the recipient, this second version invites a response.

USE PLAIN ENGLISH. Workplace correspondence often suffers from *letterese*, those tired, stuffy, and overblown phrases some writers think they need to make their communications sound important.

Letterese

> Humbly thanking you in anticipation of your kind assistance, I remain Faithfully yours,

Instead you might simply write this.

Clear phrasing

> I would appreciate your help.

Here are a few of the old standards with some clearer, more direct translations.

Letterese	**Plain-English**
As per your request	As you requested
Contingent upon receipt of	As soon as we receive
I am desirous of	I want, I would like
Please be advised that I	I
This writer	I
In the immediate future	Soon
In accordance with your request	As you requested
Due to the fact that	Because

Plain-English
translations

Be natural. Write as you would speak in a classroom or office.

NOTE

Comunications expert Laura Gurak notes that "in the legal profession (and others), certain phrases such as these are known as 'terms of art' and connote a specific meaning. In these cases, you may not be able to avoid such [elaborate] phrases" (Gurak and Lannon, 2nd ed. 202).

FOCUS ON THE HUMAN CONNECTION. As you plan and write, answer these questions:

Questions for
connecting with
the recipient

- *What do I want this person to do, think, or feel?* Offer a job, give advice or information, follow my instructions, grant a favor, accept bad news?
- *What details and emphasis does this person expect?* Measurements, dates, costs, model numbers, enclosures, other details?
- *To whom am I writing?* When possible, write to a person, not a title.
- *What is my relationship to this person?* Is this a potential employer, an employee, a person doing a favor, a person whose products are disappointing, an acquaintance, an associate, a stranger?

ANTICIPATE THE RECIPIENT'S REACTION.   After you have written a draft, answer the following three questions, which pertain to the effect of your letter. Will the recipient feel inclined to respond favorably?

Questions for
anticipating the
recipient's
reaction

- *How will this person react?* With anger, hostility, pleasure, confusion, fear, guilt, resistance, satisfaction?
- *What impression of me does this letter convey?* Do you seem intelligent, courteous, friendly, articulate, pretentious, illiterate, confident?
- *Am I ready to sign my letter with confidence?* Think about it.

DECIDE ON A DIRECT OR INDIRECT ORGANIZING PATTERN.   The reaction you anticipate should determine the organizational plan of your letter: either *direct* or *indirect*.

Questions for
organizing your
message

- *Will the recipient feel pleased, angry, or neutral?*
- *Will the message cause resistance, resentment, or disappointment?*

When to be direct

The direct pattern puts the main point in the first paragraph, followed by the explanation. Be direct when you expect the recipient to react with approval or when you want to convey immediately the point of your letter (e.g., in good news, inquiry, or application letters—or other routine correspondence).

When to be
indirect

If you expect resistance or if your recipient is not from the United States, consider an indirect pattern. Give the explanation before the main point (as in requesting a pay raise or making a controversial recommendation such as increasing the number of credits required for an undergraduate degree).

Research indicates that "readers will always look for the bottom line" (*Writing User-Friendly Documents* 14). Therefore, a direct pattern, even for bad news, may be preferable—as in complaining about a faulty product. But whenever you must give bad news, don't just blurt it out:

Blunt and
impersonal

Your application for admission to the Program has been denied.

Instead, try to give information recipients can use, and do this inoffensively:

A "you"
perspective

Unfortunately we are unable to offer you admission to this year's Program. This letter will explain why we made this decision and how you can reapply.

NOTE   *Whenever you consider using an indirect pattern, think carefully about its ethical implications. Never try to deceive the recipient—and never create an impression that you have something to hide.*

For more on direct versus indirect organizing patterns, see page 388.

18.1

For more model letters visit <www.ablongman.com/lannonweb>

INQUIRY LETTERS

You may have questions about a product, service, a procedure, or some other item. Before you write, do your homework so that you can ask the right questions. A vague request ("Please send me all your data on . . .") is likely to be ignored.

A Sample Situation

You are preparing a report on the feasibility of harnessing solar energy for solar heating in northern climates. After learning that a nonprofit research group has been experimenting with solar applications, you decide to write for details.

Keep the inquiry short and to the point. Follow the direct approach and state clearly at the outset what you are requesting and why. Maintain the "you" perspective.

In the body of your letter write specific and clearly worded questions that are easy to understand and answer. If you have multiple questions put them in a numbered list, to help readers organize their answer and to increase your chances of getting all the information you want. Consider leaving space for responses below each question. (For more than five questions, consider using an attached questionnaire.) Provide multiple ways for the recipient to reach you: email, fax, phone, surface mail.

INQUIRY LETTER

States the purpose

As a student at Evergreen College, I am preparing a report (April 15 deadline) on the feasibility of solar energy as a viable source of home heating in northern climates.

While gathering data on home solar heating, I encountered references (in *Scientific American* and elsewhere) to your group's pioneering work in solar energy systems. Would you please allow me to benefit from your experience? I would appreciate answers to these questions in particular:

Makes a reasonable request

Presents a list of specific questions

1. At this stage of development, do you consider active or passive heating more practical? (Please explain briefly.)

Leaves space for response to each question

2. Do you expect to surpass the 60 percent limit of heating needs supplied by the active system? If so, at what level of efficiency and how soon?

3. What is the cost of materials for building your active system, per cubic foot of living space?

4. What metal do you use in collectors, to obtain the highest thermal conductivity at the lowest maintenance costs?

Provides complete contact information

Please write your answers in the spaces provided and drop the return envelope in the mail. If an alternative way to respond is more convenient, here is my contact information: phone—555-986-6578 (collect); fax—555-986-5432; email—agreene245@hotmail.com.

Tells how the material will be used, and offers to share findings

Your answers, along with any recent findings you can share, will enrich a learning experience I hope to put into practice after graduation by building my own solar-heated home. I would be glad to send you a copy of my report, along with the house plans I have designed. Thank you.

If your questions are too numerous or involved to be answered in print, you might request an interview (if the respondent is nearby). Karen Granger, the next writer, sought a state representative's "opinions on the EPA's progress" in cleaning up local contamination. Anticipating a complex answer, she asked for an interview.

LETTER REQUESTING AN INFORMATIVE INTERVIEW

As a technical writing student at the University of Massachusetts, I am preparing a report evaluating the EPA's progress in cleaning up PCB contamination in New Bedford Harbor.

In my research, I have encountered your name repeatedly. Your dedicated work has had a definite influence on this situation, and I am hoping to benefit from your knowledge.

I was surprised to learn that, although this contamination is considered the most extensive anywhere, the EPA still has not moved beyond conducting studies. My own study questions the need for such extensive data gathering. Your opinion, as I can ascertain from *Standard Times* articles, is that the EPA is definitely moving too slowly.

The EPA refutes that argument by asserting they simply do not yet have the information necessary to begin a clean-up operation.

As both a writer and a New Bedford resident, I am very interested in your opinions on the EPA's progress. Could you find time in your busy schedule to grant me an interview? With your permission, I will phone your office in a few days to ask about arranging an appointment.

I would deeply appreciate your assistance and I would gladly send you a copy of my completed report.

Whenever you seek a written response to an inquiry, include a stamped, self-addressed envelope for the reply.

Telephone and Email Inquiries

Why inquiries via regular mail may be preferable

Unsolicited inquiries via telephone, email, or fax can be efficient and productive but also unwelcome and intrusive. A traditional letter implies greater respect for the recipient's privacy and provides a certain distance from which that person can contemplate a response—or even decide not to respond.

Etiquette for telephone or email inquiries

One alternative is to inquire by traditional letter and to invite a response by phone (collect), fax, or email, if your recipient prefers. Or, if you must inquire via telephone or email, consider this suggestion: Establish a brief initial contact in which you apologize for any intrusion and then ask if the respondent is willing to answer your questions at a convenient future time. In this or any communication, never sacrifice goodwill in the interest of efficiency.

CLAIM LETTERS

Claim letters request adjustments for defective goods or poor services, or they complain about unfair treatment or the like.

Routine Claims

Routine claims follow a direct organizational pattern, because the claim is backed by a contract, guarantee, or the company's reputation. State the request or problem in your introductory paragraph; then explain in the body section. Close courteously, restating the action you request.

Make the tone courteous and reasonable. Your goal is not to express dissatisfaction but to achieve results: a refund, replacement, apology. Press your claim objectively yet firmly by explaining it clearly and by stipulating the *reasonable* action that will satisfy you.

Explain the problem in enough detail to clarify the basis for your claim. Explain that your new alarm clock never rings instead of merely saying it's defective. Identify the faulty item clearly, giving serial and model numbers, and date and place of purchase. Then propose a fair adjustment. Conclude by expressing goodwill and confidence in the company's integrity.

The next writer asks directly how to return his defective skis for repair. The attention line directs his claim to the appropriate department. The subject line, and its reemphasis in the first sentence, makes clear the nature of the claim.

ROUTINE CLAIM LETTER

Attention: <u>Consumer Affairs Department</u>

Subject: <u>Delaminated Skis</u>

States problem and action desired

This winter, my Tornado skis began to delaminate. I want to take advantage of your lifetime guarantee to have them relaminated.

Provides details

I bought the skis from the Ski House in Erving, Massachusetts, in November 1983. Although I no longer have the sales slip, I did register them with you. The registration number is P9965.

Explains basis for claim

I'm aware that you no longer make metal skis, but as I recall, your lifetime guarantee on the skis I bought was a major selling point. Only your company and one other were backing their skis so strongly.

Courteously states desired action

Would you please let me know how to go about returning my delaminated skis for repair?

Arguable Claims

When your request is in some way unusual or debatable, you must *persuade* the recipient to grant your claim.

Use an indirect organizing pattern for an arguable claim. People are more likely to respond favorably *after* reading your explanation. Begin with a neutral statement both parties can agree to—but that also serves as the basis for your request (e.g., "Customer goodwill is often an intangible quality, but a quality that brings tangible benefits").

Once you've established agreement, explain and support your claim. Include enough information for a fair evaluation: date and place of purchase, order or policy number, dates of previous letters or calls, and background.

Conclude by requesting a *specific action* (a credit to your account, a replacement, a rebate). Ask confidently.

The next writer employs a tactful, reasonable tone and an indirect pattern.

ARGUABLE CLAIM LETTER

Establishes early agreement

Your company has an established reputation as a reliable wholesaler of office supplies. For eight years we have counted on that reliability, but a recent episode has left us annoyed and disappointed.

Presents facts to support claim

On January 29, 2005, we ordered five cartons of 2.0 megabyte KAO diskettes (#A74–866) and thirteen cartons of Epson MX 70/80 black ribbon cartridges (#A19–556).

Offers more support

On February 5, the order arrived. But instead of the 2.0 MB KAO diskettes ordered, we received 1.4 MB 3M diskettes. And the Epson ribbons were blue, not the black we had ordered. We returned the order the same day.

Includes all relevant information

Also on the 5th, we called John Fitzsimmons at your company to explain our problem. He promised delivery of a corrected order by the 12th. Finally, on the 22nd we did receive an order—the original incorrect one—with a note claiming that the packages had been water damaged while in our possession.

Sticks to the facts—accuses no one

Our warehouse manager insists the packages were in perfect condition when he released them to the shipper. Because we had the packages only five hours, and had no rain on the 5th, we are certain the damage did not occur here.

Requests a specific adjustment

Responsibility for damages therefore rests with either the shipper or your warehouse staff. What bothers us is our outstanding bill from Hightone ($1,049.50) for the faulty shipment. We insist that the bill be canceled and that we receive a corrected statement. Until this misunderstanding, our transactions with your company were excellent. We hope they can be again.

Stipulates a reasonable response time

We would appreciate having this matter resolved before the end of this month.

18.2

Online career sites change quickly. Stay current at <www.ablongman.com/lannonweb>

Résumés and Job Applications

Attractive jobs are highly competitive. Large companies typically receive a flood of résumés for a mere handful of openings. Whether you are applying for your first professional job or changing careers, you must market your skills effectively. Your résumé and letter of application must stand out among the competition.

Employment Outlook in the Twenty-First Century

In today's workplace, the name of the game is *change* (as page 12 shows in detail). The U.S. Labor Department estimates that "a typical 32-year-old has already held 9 jobs" (Conlin 170). What this means is that the typical "employee" is becoming someone who works for various employers just long enough to complete a particular project (Bolles 141). These "free agents" will increase from 26 percent of the U.S. workforce in 2000 to an estimated 41 percent by 2010 (Conlin 170).

In addition, more and more jobs (ranging from accounting to reading x-rays) are being "outsourced" to other countries that have far lower labor costs. Other jobs (such as programming or processing) are being automated, in many cases.

A glimpse at your job future

Whatever your major, the message is clear. (a) expect multiple employers and careers; (b) expect to rely on skills that involve working well with others, being flexible and adaptable to change and life-long learning, and having a talent for creative problem solving. (For more on these twenty-first century skills, review page 12.)

NOTE   *Although email and online job listings and résumé postings have provided new tools for job seekers, today's job searches require the same basic approach and communication skills that people have relied on for decades.*

Prospecting for Jobs

Begin your employment search by studying the job market to identify careers and jobs for which you best qualify.

ASSESS YOUR SKILLS AND APTITUDES.   Beyond the portable skills mentioned above, what specific qualities can you bring to the job search?

Identify your
assets

- What skills have you acquired in school, on the job, through hobbies or other interests?
- Do you have skills in leadership or in group projects (as demonstrated in employment, social organizations, or extracurricular activities)?
- Do you speak a second language? Have musical or artistic talent?
- Do you have communication skills? Are you a good listener?
- Do you perform well under pressure?

Besides helping focus your search, answers to these questions will be handy when you write your résumé and prepare for interviews.

RESEARCH THE JOB MARKET.   Launch your search early. Don't wait for the job to come to you. Begin by scanning the Help Wanted section in major Sunday newspapers for job descriptions, salaries, and qualifications. The Web provides an endless resource for job seekers. Check out, for example, the Bureau of Labor statistics at <www.bls.gov>. (For more on Internet job sites, see pages 441–43). Also, consult specific resources such as these (increasingly available in electronic versions):

Know what is
available

- *Fortune, Business Week, Forbes,* the *Wall Street Journal,* or trade publications in your field—for the latest developments and the big picture on business, economy, or technology. Articles about specific topics and companies can be searched in the *Business Index.*
- *Occupational Outlook Handbook* and its quarterly update, *Occupational Outlook Quarterly,* published by the U.S. Department of Labor—for occupations, qualifications, employment prospects, and salaries.
- *Almanac of Jobs and Salaries, Dun's Employment Opportunities Directory*, or *Federal Career Opportunities*—for government and private-sector organizations that hire college graduates.
- *Moody's Industrial Index* or Standard and Poor's *Register of Corporations*—for company locations, major subsidiaries, products, executive officers, and assets.
- Annual reports, for a company's subsidiaries, financial health, innovations, performance, and prospects. (Many libraries collect annual reports.)

Ask a reference librarian to suggest additional handbooks, government publications, newsletters, and magazines or journals in your field.

Launch your search well before your senior year to learn whether certain courses make you more marketable. If you are changing jobs or careers, employers will be interested in what you have accomplished *since* college. Be prepared to show how your experience is relevant to the new job.

You might also register with an employment agency. A fee is payable after you are hired, but employers often pay this fee. Ask about fee arrangements *before* you sign up. Also, consider investing in a *career coach,* an expert in preparing job seekers for their quest.

LEARN TO NETWORK.   Do all the *networking* you can. Former recruiter Brad Karsh offers recent graduates this advice: "Your best chance of getting a job is through someone who knows someone. Use your college alumni network. Find out who is working in the field you want to enter, and call them" (qtd. in Fisher, "I didn't" 178). Karsh also advises that you consult family friends about whom they might know. (For more advice, visit Karsh's job-search Web site for recent grads at <www.jobbound.com>.

Here are additional suggestions for networking:

How to make contacts

- Visit your college placement service. Openings and job fair notices are posted there; interviews are scheduled; and counselors provide job-hunting advice. Sign up for interviews with recruiters who visit the campus. Go to job fairs.
- Speak with people in your field to get an inside view and practical advice.
- Seek advice from faculty who do outside consulting or who have worked in business, industry, or government.
- Look for a summer job or internship in your field; this experience may count as much as your academic credentials.
- Do related volunteer work. (Visit <www.volunteermatch.org> for organizations that seek volunteers.)
- Register with agencies that provide temporary staffing. Even the most humble and temporary job offers the chance to make contacts and could be a way of getting your foot in the door.
- Join a professional organization in your field. Student memberships usually are available at reduced fees. Such affiliations can generate excellent contacts, and they look good on your résumé. Try to attend meetings of the local chapter.

Notice that several job-search methods in Figure 18.4 include talking with people and exploring useful contacts.

Once you have a clear picture of where you fit into the job market, you must answer the big question asked by all employers: *What do you have to offer?* Your answer must be a highly polished presentation of yourself, your education, work history, interests, and skills—in short, your résumé.

**FIGURE 18.4
How People
Usually Find
Jobs**

*Source: Tips for
Finding the Right
Job. U.S. Depart-
ment of Labor.*

Most Commonly Used Job-Search Methods

Percent of Total Job-seekers Using the Method	Method	Effectiveness Rate*
66.0%	Applied directly to employer	47.7%
50.8	Asked friends about jobs where they work	22.1
41.8	Asked friends about jobs elsewhere	11.9
28.4	Asked relatives about jobs where they work	19.3
27.3	Asked relatives about jobs elsewhere	7.4
45.9	Answered local newspaper ads	23.9
21.0	Private employment agency	24.2
12.5	School placement office	21.4
15.3	Civil Service test	12.5
10.4	Asked teacher or professor	12.1
1.6	Placed ad in local newspaper	12.9
6.0	Union hiring hall	22.2

*A percentage obtained by dividing the number of jobseekers who actually found work using the method, by the total number of jobseekers who tried to use that method, whether successfully or not.

Preparing Your Résumé

The résumé is a summary of your experience and qualifications. Written before your application letter, the résumé provides background information to support your letter. This information supplies an employer with a one- or two-page reference. The letter will emphasize specific parts of your résumé and will discuss how your qualifications fit the requirements of a particular job.

Employers generally spend fifteen to forty-five seconds initially scanning a résumé. They look for an obvious and persuasive answer to this question: *What can you do for us?* Employers are impressed by a résumé that

What employers
expect in a
résumé

1. looks good (conservative, tasteful, uncluttered, on quality paper);
2. reads easily (headings, typeface, spacing, and punctuation that provide clear orientation); and
3. provides information the employer needs for deciding whether to interview the applicant.

Employers generally discard résumés that are mechanically flawed, cluttered, sketchy, or hard to follow. Make your résumé perfect.

Most résumés organize the information within these categories:

Résumé
categories

- contact information
- job and career objectives
- education

- work experience
- personal data
- interests, activities, awards, and skills
- references
- portfolio (if applicable)

Select and organize material to emphasize what you can offer. Don't just list *every-thing*. Don't abbreviate, because some people might not know the referent. Use punctuation to clarify and emphasize, not to be "artsy." Try to limit your résumé to a single page, as most employers prefer. (Of course, if you are changing jobs or careers, or if your résumé looks cramped, you might need a second page.)

Begin your résumé well before your job search. Your final version can be printed for various similar targets—but each new type of job requires a new résumé that is tailored to fit the advertised demands of that job.

NOTE *Never "invent" credentials. Your résumé should make you look as good as the facts allow. Distorting the facts is unethical and counterproductive. Companies routinely investigate claims made in a résumé, and people who have lied are fired.*

CONTACT INFORMATION. Include your full name, mailing and email address, and phone number (many interview invitations and job offers are made by phone). If you use an answering machine, be sure to record an outgoing message that sounds professional. If you anticipate an address change after a certain date, include both your current and future addresses and the date of the change.

JOB AND CAREER OBJECTIVES. Spell out the kind of job you want. Avoid vague statements such as "A position in which I can apply my education and experience." Prepare different statements to focus on the requirements of different jobs.

The key to a successful résumé is the image of *you* it projects—disciplined and purposeful, yet flexible. State your immediate and long-range goals, including any plans to continue your education. If the company has various branches, include "Willing to relocate."

Statement of career objective | Intensive-care nursing in a teaching hospital, with the eventual goal of supervising and instructing.

One hiring officer for a major computer firm offers this advice: "A statement should show that you know the type of work the company does and the type of position it needs to fill" (Beamon, qtd. in Crosby, *Résumés* 3).

NOTE *Below career objectives, you might insert a summary of qualifications (Figure 18.6). This section is vital in a computer-scannable résumé (18.8) but, even in a conventional résumé, a "Qualifications" section can highlight your strengths. Make the summary specific and* *concrete: replace "proven leadership" with "team and project management," "special-event planning," or "instructor-led training"; replace "persuasive communicator" with "fundraising" "publicity campaigns," "environmental/public-interest advocacy, and "door-to-door canvassing." In short, allow the reader to* **visualize** *your activities.*

EDUCATION. If your education is more substantial than your work experience, place it first. Begin with your most recent school and work backward, listing degrees, diplomas, and schools attended *beyond* high school (unless the high school's prestige, its program, or your achievements warrant its inclusion). List your major, minor, and selected courses that have directly prepared you for the job you seek. If your class rank is in the upper 30 percent and your grade point average is 3.0 or above, list them. Include schools or specialized training during military service. If you finance part or all of your education by working, say so, indicating the percentage of your contribution.

WORK EXPERIENCE. If your experience relates to the job you seek, list it before your education. Start with your most recent job and work backward. Provide dates and names of employers. Indicate whether the job was full-time, part-time (hours weekly), or seasonal. Describe your exact duties in each job, indicating promotions. If the job was major (and related to this one), describe it in detail, and give your reason for leaving. Include military experience. If you have no paid experience, show your potential by emphasizing your education (including internships and special projects) and by writing an enthusiastic letter.

NOTE

Complete sentences are unnecessary in résumés. They take up room best left for emphasis on your other qualifications. Use action verbs (supervised, developed, built, taught, installed, managed, trained, solved, planned, directed) *to stress your ability to produce results. If your résumé is likely to be scanned electronically, list key words as nouns* (leadership skills, software development, data processing, editing) *immediately below your contact information. (See page 446.)*

PERSONAL DATA. An employer cannot legally discriminate on the basis of sex, religion, race, age, national origin, disability, or marital status. Therefore, you aren't required to provide this information or a photograph. But if you believe that any of this information could advance your prospects, include it.

PERSONAL INTERESTS, ACTIVITIES, AWARDS, AND SKILLS. List hobbies, sports, and other pastimes only if they are *relevant* to the position; memberships in teams and organizations; offices held; and any special recognition you have received. Include dates and types of volunteer work. Employers know that people with well-rounded lifestyles are likely to take an active interest in their jobs. List work-related skills, say, in foreign language, typing, first aid, or computers. Be selective: List only those items that reflect the qualities employers seek.

REFERENCES. List three to five people *who have agreed to provide* strong, positive assessments of your qualifications and personal qualities. Some references will be asked to provide letters or email responses. Some will be asked to complete reference forms. Other requests will be made by telephone calls from employers doing reference checks. In any case, it's a good idea to keep your own file containing letters from each of your references.

How to choose references

Select references who can speak genuinely about your ability and character. Choose among former employers, professors, and respected community figures who know you well enough to speak *concretely* on your behalf. Do not choose members of your family or friends not in your field.

How to request a reference

A lukewarm letter of reference is more damaging than no letter at all. Don't simply ask, "Could you please be one of my references?" This is hard to refuse, but a person who doesn't know you well or who is unimpressed by your work might write a letter that does more harm than good. Instead, make an explicit request: "Do you know me and my work well enough to write a strong letter of reference? If so, would you be one of my references?" This second version gives your respondent the option of declining gracefully or it elicits a firm commitment to a positive recommendation.

Letters of recommendation are time-consuming to write. Few people have time to write individual letters to every prospective employer. Ask for only one letter, with no salutation. Your reference keeps a copy, you keep the original for your personal dossier (to reproduce as necessary), and a copy goes to the placement office for your placement dossier. Because the law permits you to read all material in your dossier, this arrangement provides you with your own copy of your credentials. (The dossier is discussed later in this chapter.)

NOTE

Under some circumstances you may—if you wish to—waive the right to examine your references. Some applicants, especially those applying to professional schools, as in medicine and law, waive this right in concession to a general feeling that a letter writer who is assured of confidentiality is more likely to provide a balanced, objective, and reliable assessment of a candidate. Seek the advice of your major adviser or a career counselor.

If the people you select as references live elsewhere, you might make your request by letter:

LETTER REQUESTING A REFERENCE

From September 19xx to August 20xx, I worked at Teo's Restaurant as a waiter, cashier, and then assistant manager. Because I enjoyed my work, I decided to study for a career in the hospitality field.

In three months I will graduate from San Jose City College with an A.A. degree in Hotel and Restaurant Management. Next month I begin my job search. Do you remember me and my work enough to write a strong letter of recommendation? If so, would you kindly serve as one of my references?

Please address your salutation "To Whom It May Concern." If you could send me the original letter, I will forward a copy to my college placement office.

To update you on my recent activities, I've enclosed my résumé. If I can answer any questions, please contact me by phone (collect) at 214-316-2419 or by email at jpur@valnet.com.

Thank you for your help and support.

NOTE   *Opinion is divided about whether names and addresses of references should appear in a résumé. If saving space is important, simply state, "References available on request," keeping your résumé only one page long, but if your résumé already takes up more than one page, you probably should include names and addresses of references. (A prospective employer might recognize a name, and thus notice your name among the crowd of applicants.) If you are changing careers, a full listing of references is extremely important.*

NOTE   *If you don't list references on your résumé, prepare a separate reference sheet that you can provide on request. Beneath your personal contact information, repeated from the résumé, list each reference, including the person's title, company address, and contact information (Crosby, "Resumes"6).*

PORTFOLIO.   As concrete evidence of your skills, organize copies of the relevant work you've done in a leather or leatherlike notebook. Depending on the nature of the work, the portfolio might contain sample documents you've written or edited (such as reports, articles, or manuals) or other evidence of your job-related skills (such as engineering drawings or software documentation). Portfolios are obviously more appropriate for jobs that generate actual writing or visual samples than for those that don't—more appropriate, say, for graphic artists or marketing specialists than for resort or hotel managers. If you do have a portfolio, indicate this on your résumé, followed by "Available on request." (See the suggestions for preparing a portfolio on page 448.)

Organizing Your Résumé

Emphasize your assets

Organize your résumé to convey the strongest impression of your qualifications, skills, and experience. Depending on your background, you can arrange your material in reverse chronological order, functional order, or a combination of both.

REVERSE CHRONOLOGICAL ORGANIZATION.   In a reverse chronological résumé you list your most recent experience first, moving backward through your earlier experiences. Use this arrangement to show a pattern of job experience or progress along a specific career track (as in Figure 18.5).

James David Purdy
203 Elmwood Avenue
San Jose, CA 95139
Tel.: 214-316-2419
Email: jpur@valnet.com

Objective	Customer relations for a hospitality chain, leading to management.
Education 2003–2006	*San Jose College, San Jose, CA* Associate of Arts Degree in Hotel/Restaurant Management, June 2006. Grade point average: 3.25 of a possible 4.00. All college expenses financed by scholarship and part-time job (20 hours weekly).
Employment 2003–2006	*Peek-a-Boo Lodge, San Jose, CA* Began as desk clerk and am now desk manager (part-time) of this 200-unit resort. Responsible for scheduling custodial and room service staff, convention planning, and customer relations.
2002–2003	*Teo's Restaurant, Pensacola, FL* Beginning as waiter, advanced to cashier and finally to assistant manager. Responsible for weekly payroll, banquet arrangements, and supervising dining room and lounge staff.
2001–2002	*Encyclopaedia Britannica, Inc., San Jose, CA* Sales representative (part-time). Received top bonus twice.
2000–2001	*White's Family Inn, San Luis Obispo, CA* Worked as busperson, then server (part-time).
Personal	*Awards* Captain of basketball team, 2001 Lion's Club Scholarship, 2003. *Special Skills* Speak French fluently; expert skier. *Activities* High school basketball and track teams (3 years); college student senate (2 years); Innkeepers' Club—prepared and served monthly dinners at the college (2 years). *Interests* Skiing, cooking, sailing, oil painting, and backpacking.
References	Available on request.

FIGURE 18.5   Résumé for an Entry-Level Candidate (Reverse Chronological Arrangement)

FUNCTIONAL ORGANIZATION.   In a functional résumé, you emphasize skills, abilities, and achievements that relate specifically to the job for which you are applying. Use this arrangement if you have limited job experience, gaps in your job record, or are changing careers.

18.3
For more model résumés visit <www.ablongman.com/lannonweb>

COMBINED ORGANIZATION.   Most employers prefer chronologically ordered résumés because they are easier to scan. However, electronic scanning of résumés (page 443) calls for a more functional pattern. One alternative is a modified-functional résumé, which preserves the logical progression that employers prefer but which also highlights your abilities and job skills (as in Figure 18.6).

NOTE   *The résumé is not the place to bring up the topic of salary. Wait until your interview, or later.*

Using a word processor and laser printer, you can make changes to suit various jobs and have a perfect document for each version. Make each version neat, attractive, and error-free.

NOTE   *The importance of proofreading your résumé cannot be overstated. By relying solely on your computer's spelling checker you might end up expressing pride in receiving a "plague" instead of a "plaque," in receiving a "bogus award" instead of a "bonus award" or in "ruining" your own business instead of "running" it.*

18.4
Learn about formatting electronic résumés at <www.ablongman.com/lannonweb>

A Sample Situation

Imagine that you are a twenty-four-year-old student about to graduate from a community college with an A.A. degree in Hotel and Restaurant Management. Before college, you worked at related jobs for more than three years. You now seek a junior management position while you continue your education part-time.

You have spent two weeks compiling information for your résumé and obtaining commitments from four references. Figure 18.5 shows your résumé.

Preparing Your Job Application Letter

Include a cover letter with each résumé you send. In the words of one employment expert, "Sending a résumé without a cover letter is like starting an interview without shaking hands" (Crosby, *Résumés* 12).

YOUR IMAGE.   Your application letter complements your résumé by explaining how your credentials fit this particular job and by conveying a sufficiently professional persona for the prospective employer to decide that you warrant an interview. Your letter can also highlight some specific qualifications or skills. For example, you may have "C++ programming" listed on your résumé under the category "Programming Languages." But for one particular job application, you may wish to call attention to this item in your application letter.

Karen P. Granger
82 Mountain Street
New Bedford, MA 02740
Telephone (617) 864-9318
Email: kgrang@swis.net

Objective	A summer internship documenting microcomputer software.
Qualifications	Software and hardware documentation. Editing. Desktop publishing. Usability testing. Computer science. Internet research. World Wide Web collaboration. Networking technology. Instructor-led training. DEC 20 mainframe and VAX 11/780 systems. *PageMaker*, Adobe *FrameMaker*, *RoboHelp*, Webworks *Publisher*, *PowerPoint*, *Excel*, and *Lotus Notes* software. Logo, Pascal, HTML, and C++ program languages.
Education	Attending University of Massachusetts at Dartmouth; B.A. expected January 2007. Major: English/Communications. Minor: Computer Science. Dean's List, all semesters. GPA: 3.54. Class rank 110 of 1,792.
Experience Intern Technical Writer	**Conway Communications, Inc.,** 39 Wall Street, Marlboro, MA 02864. Learned local area network (LAN) technology and Conway's product line. Wrote, designed, and tested five hardware upgrade manuals. Produced a hardware installation/maintenance manual from another writer's work. Specified and approved all illustrations. Designed a fully linked home page and online help for the company intranet. Summers 2004, 2005.
Writing Tutor	**Writing/Reading Center,** UMD. Tutored writing and word processing for individuals and groups. Edited WRC student newsletter. Trained new tutors. Cowrote and acted in a video about the WRC. Designed WRC home page for World Wide Web. Fall 2003–present.
Managing Editor	**The Torch,** UMD Weekly Newspaper. Organized staff meetings, generated story ideas, wrote articles and editorials, edited articles, and supervised page layout and paste-up. Fall 2004–present.
Achievements	Writing samples published in Dr. John M. Lannon's *Technical Communication,* 10th ed. (Longman, 2006); Massachusetts State Honors Scholarship, 2003–2006.
Activities	Student member, Society for Technical Communication and American Society for Training and Development; student representative, College Curriculum Committee; UMD Literary Society.
References and Portfolio	Available on request.

FIGURE 18.6 Résumé for a Summer Internship Candidate (Modified-Functional Arrangement) Because this applicant is seeking a position in writing and editing, she indicates that her writing portfolio is available.

The application letter can accentuate items on the résumé

> You will note on my résumé that I am experienced with C++ programming. In fact, I also tutor C++ programming students in our school's Learning Center.

TARGETS.   The letter should never be photocopied. Although you can base letters to different employers on the same model—with appropriate changes—each letter should be prepared anew.

Sometimes you will apply for positions advertised in print or by word of mouth (solicited applications). At other times you will write prospecting letters to organizations that have not advertised but might need someone like you (unsolicited applications). In either case, tailor your letter to the situation.

NOTE

Write to a specific person—not to some generic recipient such as "Director of Human Resources" or "Personnel Office." If you don't know who does the hiring, phone the company and ask for that person's name and title, and be sure you get the spelling right.

THE SOLICITED LETTER.   Imagine that you are James Purdy (Figure 18.5). In *Innkeeper's Monthly,* you read the following advertisement and decide to apply.

> ## RESORT MANAGEMENT OPENINGS
>
> Liberty International, Inc., is accepting applications for several junior management positions at our new Lake Geneva resort. Applicants must have three years of practical experience, along with formal training in all areas of hotel/restaurant management. Please apply by June 1, 20xx, to
>
> Sara Costanza
> Personnel Director
> Liberty International, Inc.
> 32 Apex Way
> Lansdowne, PA 24135

Now plan and compose your letter.

INTRODUCTION.   Begin by naming the job you're applying for and where you may have seen it advertised. In one sentence, identify yourself and your background. If possible, establish a connection by mentioning a mutual acquaintance who encouraged you to apply—but only if that person has given permission.

BODY.   Without merely repeating your résumé, focus on the qualifications you can bring to this specific job:

How to compose a persuasive application

- *Don't come across as a jack-of-all-trades.* Relate your qualifications specifically to the job for which you are applying.
- *Avoid flattery:* "I am greatly impressed by your remarkable company."

- *Be specific.* Replace "much experience," "many courses," or "increased sales" with "three years of experience," "five courses," or "a 35 percent increase in sales between June and October 2002."
- *Support all claims with evidence, to show how your qualifications will benefit this employer.* Instead of saying, "I have leadership skills," say, "I was student senate president during my senior year and captain of the lacrosse team."
- *Create a dynamic tone by using active voice and action verbs.*

WEAK	Management responsibilities were steadily given to me.
STRONG	I steadily assumed management responsibilities.

- *Trim the fat from your sentences.*

FLABBY	I have always been a person who enjoys a challenge.
LEAN	I enjoy a challenge.

- *Express self-confidence.*

UNSURE	It is my opinion that I have the potential to become a successful manager because . . . .
CONFIDENT	I will be a successful manager because . . . .

- *Never be vague.*

VAGUE	I am familiar with the 1022 interactive database management system, and RUNOFF, the text-processing system.
DEFINITE	As a lab grader for one semester, I kept grading records on the 1022 database management system, and composed lab procedures on the RUNOFF text-processing system.

- *Avoid letterese.* Write in plain English.
- *Be enthusiastic.* An enthusiastic attitude can be as important as your background, in some instances.

CONCLUSION.   Restate your interest and emphasize your willingness to retrain or relocate (if necessary). If the recipient is nearby, request an interview; otherwise, request a phone call, stating times you can be reached. Your conclusion should leave the impression that you are someone worth knowing.

REVISION.   *Never* settle for a first draft—or a second or third! This letter is your model for letters serving in various circumstances. Make it perfect and do not exceed one page.

After several revisions, James Purdy finally signed the letter shown in Figure 18.7.

203 Elmwood Avenue
San Jose, CA 10462
April 22, 2006

Sara Costanza
Personnel Director
Liberty International, Inc.
Lansdowne, PA 24153

Dear Ms. Costanza:

Writer identifies self and purpose

Establishes a connection

Please consider my application for a junior management position at your Lake Geneva resort. I will graduate from San Jose City College on May 30 with an Associate of Arts degree in Hotel/Restaurant Management. Dr. H. V. Garlid, my nutrition professor, described his experience as a consultant for Liberty International and encouraged me to apply.

Relates specific qualifications to the job opening

For two years I worked as a part-time desk clerk, and I am now the desk manager at a 200-unit resort. This experience, combined with earlier customer relations work in a variety of situations, has given me a clear and practical understanding of customers' needs and expectations.

As an amateur chef, I know of the effort, attention, and patience required to prepare fine food. Moreover, my skiing and sailing background might be assets to your resort's recreation program.

Expresses confidence and enthusiasm throughout

I have confidence in my hospitality management skills. My experience and education have prepared me to work well with others and to respond creatively to changes, crises, and added responsibilities.

Makes follow-up easy for the reader

If my background meets your needs, please phone me any weekday after 4 p.m. at 214–316–2419.

Sincerely ,

James D. Purdy

James D. Purdy

Enclosure

FIGURE 18.7   A Sample Cover Letter for a Job Application   Purdy wisely emphasizes practical experience because his background is varied and impressive. An applicant with less practical experience would emphasize education instead, discussing related courses, extracurricular activities, and aptitudes.

As an additional example, here is the letter composed by Karen Granger in her quest for a summer internship.

INTERNSHIP APPLICATION LETTER

Dear Mr. White:

Begins by stating purpose

I read in *InternWeb.com* that your company offers a summer documentation internship. Because of my education and previous technical writing employment, I am very interested in such a position.

Identifies herself and college background

In January 2007, I will graduate from the University of Massachusetts with a B.A. in English/Writing. I have prepared specifically for a computer documentation career by taking computer science, mathematics, and technical writing courses.

Expands on background

In one writing course, the Computer Documentation Seminar, I wrote three software manuals. One manual uses a tutorial to introduce beginners to the Apple Macintosh and *Microsoft Word*. The other manuals describe two IBM PC applications that arrived at the university's computer center with no documentation.

Describes work experience

The enclosed résumé describes my work as the intern technical writer with Conway Communications, Inc. for two summers. I learned local area networking (LAN) by documenting Conway's LAN hardware and software. I was responsible for several projects simultaneously and spent much of my time talking with engineers and testing procedures. If you would like samples of my writing, please let me know.

Explains interest in this job

Although Conway has invited me to return next summer and to work full-time after graduation, I would like more varied experience before committing myself to permanent employment. I know I could make a positive contribution to Birchwood Group, Inc. May I telephone you next week to arrange a meeting?

Sincerely,

THE UNSOLICITED LETTER.   Ambitious jobseekers do not limit their search to advertised openings. (Fewer than 20 percent of all job openings are advertised.) The unsolicited, or prospecting, letter is a good way to uncover possibilities beyond the Help Wanted section.

Drawbacks of unsolicited applications

Unsolicited applications do have drawbacks: (a) You can waste time writing to organizations that have no openings, and (b) you cannot tailor your letter to specific requirements. But there also are advantages: For advertised openings, you compete with legions of applicants, whereas your unsolicited letter might arrive

Advantages of
unsolicited
applications

just as an opening materializes. Even employers with no openings often welcome and file impressive applications or pass them to another employer who has an opening.

Spark reader
interest

Because an unsolicited letter arrives unexpectedly, you need to get the reader's immediate attention. Don't begin: "I am writing to inquire about the possibility of obtaining a position with your company." If you can't establish a connection through a mutual acquaintance, use a forceful opening:

Opens forcefully

> Does your hotel chain have a place for a junior manager with a college degree in hospitality management, proven commitment to quality service, and customer relations experience that extends far beyond textbooks? If so, please consider my application for a position.

Address your letter to the person most likely in charge of hiring. (Consult company Web sites or the business directories listed on page 426 for names of company officers.) Then call the company to verify the person's name and title.

NOTE

For samples of all types of job-related letters (applications, pay raise requests, and so on) see the free, online version of 200 Letters for Job-Hunters *by William S. Frank at <www.careerlab.com/letters/> (Bolles 61). While these samples give some ideas for approaching your own writing situation, never borrow them wholesale; many employers will spot a "canned" letter immediately—either because employers themselves may have read the book or because someone else has tried submitting the same borrowed sample!*

CONSIDER THIS How Applicants Are Screened for Personal Qualities

As many as 25 percent of résumés contain falsified credentials, such as a nonexistent degree or a contrived affiliation with a prestigious school (Parrish 1+). A security director for one major employer estimates that 15 to 20 percent of job applicants have something personal to hide: a conviction for drunk driving or some other felony, trouble with the IRS, bad credit, or the like (Robinson 285).

With yearly costs of employee dishonesty or bad judgment amounting to billions of dollars, companies use preemployment screening for integrity, emotional stability, and a host of other personal qualities (Hollwitz and Pawlowski 203, 209). Screening often begins with a background check of education, employment history, and references. One corporation checks up to ten references (from peers, superiors, and subordi-

nates) per candidate (Martin, "So" 78). In addition, roughly 95 percent of corporations check on the applicant's character, trustworthiness, and reputation: they may examine driving, credit, and criminal records, and interview neighbors and coworkers (Robinson 285).

The law affords some protection by requiring employers to notify the applicant before checking on character, reputation, and credit history and to provide a copy of any report that leads to a negative hiring decision (Robinson 285). Once an applicant is hired, however, the picture changes: more than 50 percent of companies provide personal information to credit agencies, banks, and landlords without informing employees, and 40 percent don't inform employees about what kinds of records are being kept on them (Karaim 72).

Beyond screening for background, employers use aptitude and personality tests to pinpoint desirable qualities. A sampling of test questions (Garner 86; Kane 56; Martin, "So" 77, 78):

- *Ability to perform under pressure:* "Do you get nervous and confused at busy intersections?"
- *Emotional stability and even temper:* "Do you honk your horn often while driving?"
- *Sense of humor:* "Tell us a joke."
- *Ability to cope with people in stressful situations:* "Do you like to argue and debate?" "Are you good at taking control in a crisis?"
- *Persuasive skills:* "Write a brief memo to a client, explaining why X [stipulated on the test] can't be done on time."
- *Presentation skills:* "Prepare and give a five-minute speech on some aspect of the industry as it relates to this company."

These tests may be given online, before an applicant is considered for an interview.

Above all, most employers look for candidates who are *likable.* One employer checks with each person an applicant speaks with during the company visit—including the receptionist. Another employer has candidates join in a company softball game (Martin, "So" 77).

ELECTRONIC JOB HUNTING

The computer and the World Wide Web continue to improve the quality and speed of contact between jobseekers and employers. Beyond ease and efficiency, electronic job hunting offers real benefits:

Benefits of online job searches

- You can search for jobs worldwide.
- You can focus your search by region, industry, or job category.
- You can research companies comprehensively from many perspectives.
- You can create your own Web site, with hyperlinks to samples of your work, employment references, or other supporting material.
- You can search "passively" and discretely (say, while employed elsewhere) by specifying preferences for salary, region, types of industry, and then receive an email message when the service provider identifies an opening that matches your "profile" (Martin, "Changing Jobs?" 206).
- Your search can be ongoing, in that your résumé remains part of an active computer file, until you delete it.

NOTE *Keep in mind that even the largest job-posting sites represent only a fraction of all jobs available; don't overlook traditional job-hunting sources listed on page 426 and in Figure 18.4.*

18.5

For more on electronic job hunting visit <www.ablongman.com/lannonweb>

Online Employment Resources

Most major companies recruit on sites such as these:

- *Career Mosaic* at <http://www.careermosaic.com> provides lists of "hot" jobs, direct access to employer Web sites, free posting of résumés, online job fairs (at

A sampling of job sites on the Web

which you might chat with an interviewer online), career advice, and domestic and international job listings.

- *CareerPath.com* at <http://www.careerpath.com> posts job listings from major newspapers nationwide, and enables jobseekers to post résumés free.
- A Web search via *Yahoo!* or *HotBot* of the linked categories "Business and Economy: Employment: Jobs" yields job listings in specific fields or locations, company profiles, employers who hire in particular specialties, sites on which résumés or onscreen application forms can be posted free, and templates for creating online résumés. One limitation of using a keyword search of major search engines or directories is the countless number of "hits" you would have to sift through. For a far more efficient alternative, consider the gateway sites discussed below.
- *College Grad Job Hunter* at <www.collegegrad.com> focuses on entry-level job hunting for students and recent graduates.
- Gateway sites for job hunters offer countless targeted resources that have been organized for easy navigation and evaluated for usefulness (Bolles 10–11). Two highly regarded gateway employment sites are *The Riley Guide* at <www.dbm.com/jobguide/> and *JobHunt: A Meta-list of Online Search Resources and Services* at <www.job-hunt.org>.

Electronic recruiting centers such as careerWEB match applicants with employers in engineering, marketing, data processing, and technical fields worldwide. Countless specialized sites include *Boston Job Bank, College Grad Job Hunter, Euro-jobs On-line, HiTechCareers, Hospitality Industry Job Exchange, Jobs in Atomic and Plasma Physics,* and *Positions in Bioscience and Medicine.*

Major companies such as Johnson & Johnson, Boeing, John Deere, and IBM—to name a few—post openings on their own Web sites. You can visit a company's site, learn about its history, products, and priorities, search its job listings, email your résumé and application letter, and arrange an interview.

NOTE *For a company's real story, look beyond its Web site (for which information has been selected to paint the rosiest possible picture). To see what industry analysts think about the company's prospects, consult an impartial research site such as* Hoover's Online *at* <http://www.hoovers.com>. *Here, in addition to relatively objective financial and management profiles, you can access recent articles about the company (Martin, "Changing Jobs?" 208). For more on researching a company, see pages 426 and 450.*

Web sites maintained by professional organizations offer additional job listings, along with career advice and industry prospects. Some sites also allow for posting of interactive, hyperlinked résumés.

Professional organizations on the Web

- Society for Technical Communication <http://www.stc-va.org>
- International Television and Video Association <http://www.itva.org>
- Institute for Electrical and Electronic Engineers <http://www.ieee.org>

For internship postings and related information, go to *InternWeb.com* at <www.internweb.com>, *RisingStar Internships* at <www.rsinternships.com>, and *InternshipPrograms.com* at <www.internships.wetfeet.com/home.sap>.

NOTE

Record each date and site at which you post an online résumé (or fill out an onscreen application) so you can keep track of responses and edit or delete your material as needed (Curry 100).

Electronic Scanning of Résumés

Countless large and midsize companies scan résumés electronically. Electronic storage of online or hard copy résumés offers an efficient way to screen applicants, to compile a database of applicants (for later openings), and to evaluate all applicants fairly.

How scanning works

An optical scanner feeds in the printed page, stores it as a file, and searches the file for keywords associated with the job opening (nouns instead of traditional "action verbs"). Those résumés containing the most keywords ("hits") make the final cut (Pender 120).

How to Prepare Content for a Scannable Résumé

Using nouns as keywords, list all your skills, credentials, and job titles. (Help Wanted ads or postings are a good source for keywords.)

Keywords to list on a résumé

- *List specialized skills:* marketing, C++ programming, database management, user documentation, Internet collaboration, software development, graphic design, hydraulics, fluid mechanics, editing, surveying, soil testing.
- *List general skills:* teamwork coordination, conflict management, oral communication, report and proposal writing, problem solving, troubleshooting, bilingual in Spanish and English.
- *List credentials:* student member, Institute of Electrical and Electronic Engineers, board-certified in Medical Technology, B.S. Electrical Engineering, top 5 percent of class.
- *List job titles:* manager, director, supervisor, intern, coordinator, project leader, technician.
- *List synonymous versions of key terms (to increase the chances of a hit):* procurement *and* purchasing multimedia *and* hypermedia. Web page design *and* HTML management *and* supervision (McNair 14). If you lack skills or experience, emphasize your personal qualifications: *analytical skills, energy, efficiency, flexibility, imagination, motivation.*
- *Indicate willingness to relocate and travel (if you are truly willing to do so):* This is especially vital in any résumé for the global marketplace.

NOTE

Don't hesitate to make your scannable résumé longer than your hard copy version (but no more than three pages total). The longer the résumé being scanned, the more hits are possible (McNair 13).

How to Design a Scannable Résumé

Making a
scannable
résumé computer
accessible

For designing a résumé that can be scanned effectively, experts offer the following suggestions:

- *Keep the print standard and simple.* Scanners work best with standard typefaces such as Optima, Courier, Futura, Helvetica, or Times. Depending on the font, choose a type size ranging from 10 to 14 points. Avoid small print. For example, you might use 12-point Helvetica for headings and 12-point Times Roman for body text (McNair 13).
- *Avoid fancy highlighting.* Use **boldface** or FULL CAPS for emphasis. Avoid fancy fonts, italics, underlines, bullets, slashes, dashes, parentheses, or ruled lines (Pender 120).
- *Arrange lines that the scanner can differentiate.* To indent, use the space bar instead of tabs. Instead of allowing lines to "wrap around," end each line by hitting the Return key (Le Vie 11).
- *Avoid a two-column format.* Multiple columns can be jumbled by scanners that read across the page.
- *Do not fold or staple your pages.* Submit the unfolded résumé in a large envelope.

You might submit two versions of your résumé: one traditionally designed and one scannable—or include a keyword section, as in Figure 18.6 on page 435. Or you might submit a version via email and bring hard copies to the interview. Figure 18.8 shows a scannable version of Figure 18.6.

Electronic Résumés

18.6

Keep up with the
evolving conventions
of electronic
résumés at
<www.ablongman.com/
lannonweb>

Online versions of hard copy résumés can be submitted as email or hypertext.

EMAIL RÉSUMÉS. You might submit your résumé directly as email text or attach a file containing a version formatted for hard copy. Check with the recipient beforehand. Not all systems can receive or decode attached files—so when in doubt, paste your résumé directly into your email message (and reformat as needed).

One way to ensure that your résumé can be read by any computer is to create an ASCII version (or "text file"). Select "Save As Text Only" from your desktop menu, and reformat your ASCII page as needed (Robart 14). When you do send an ASCII version of your résumé, career expert Martin Kimeldorf suggests that you include at the end a sentence like this: "An attractive and fully formatted hard copy version of this document is available upon request" (qtd. in Bolles 60).

NOTE

When submitting a résumé in electronic form, always include a cover letter, either as an email document or as an attractively formatted attachment. If you do send these documents as attachments, be sure they can be translated by the recipient's software. Indicate the software (and version) used in composing the résumé and cover letter, or ask the employer to specify the desired software (Robart 13). When in doubt, send your material as email text, or include an ASCII version as an additional attachment.

HYPERLINKED RÉSUMÉS.   Consider placing a hyperlinked résumé (which might include an online portfolio containing a personal statement, blueprints, writing samples, or indicators of your talents) on your own Web site or that of your school or professional society (Krause 159–60). Figure 18.9 shows a hyperlinked version of Karen Granger's résumé (Figures 18.6 and 18.8).

You then include the address for your hyperlinked résumé on your hard copy or scannable résumé. Some employers refuse to track down a résumé on a Web page, so be sure to provide other delivery options as well (Robart 13–14).

PREFERRED FORMS OF RÉSUMÉS FOR DIFFERENT PURPOSES

Purpose	Preferred Form
• *For applying via traditional mail, fax, or email attachment; and for job interviews*	• *Attractively formatted and highlighted word-processed document, or scannable version, or both*
• *For translation by any computer*	• *ASCII (text only) file*
• *For use as a Web page or as an online posting on a Web site job board or database*	• *Hyperlinked document with links to materials supporting your application, or ASCII version, or both*
• *For applying via email*	• *Direct pasting into email, or word-processed attachment, or ASCII version*

NOTE   *Be sure your hyperlinked résumé can download quickly. Keep in mind that complex graphics and multimedia download very slowly.*

Protecting Privacy and Security When You Post a Résumé Online

Indiscriminate posting of résumés as Web pages or on job boards can be hazardous to your career or your welfare. If you already have a job, your present employer could discover that you're looking elsewhere. Even more important, placing certain types of information about yourself on the Internet can jeopardize your personal safety.

Your safest bet is to limit the contact and personal information that you list on your personal home page. Job expert Richard Bolles suggests including only your email address and phone number; home and work addresses and names of past employers and references can be supplied *after* a potential employer has contacted you by phone or email (60). In a hyperlinked résumé, withholding information about employers, schools, or training programs is often impossible. But personal contact information and names of references can still be protected, as shown in Figure 18.9.

KAREN P. GRANGER
82 Mountain Street
New Bedford, MA 02720
Telephone (617) 864-9318
Email: kgrang@swis.net

OBJECTIVE
A summer internship in software documentation.

QUALIFICATIONS
Software and hardware documentation. Editing. Desktop publishing. Usability testing. Computer science. Internet research. World Wide Web collaboration. Networking technology. Instructor-led training. DEC 20 mainframe and VAX 11/780 systems. *PageMaker,* Adobe *FrameMaker, RoboHelp,* Webworks *Publisher, PowerPoint, Excel,* and *Lotus Notes* software. Logo, Pascal, HTML, and C++ program languages.

EDUCATION
UNIVERSITY OF MASSACHUSETTS AT DARTMOUTH: B.A. expected January 2007. English and Communications major. Computer Science minor. GPA: 3.54. Class rank: top 7 percent.

EXPERIENCE
CONWAY COMMUNICATIONS, INC., 39 WALL STREET, MARLBORO, MA 02864: Intern Technical Writer. LAN technology. Writing, designing, and testing hardware upgrade manuals. Desktop publishing of installation and maintenance manual. Specifying art and illustrations. Designed a fully linked home page and online help for the company intranet. Summers 2004 and 2005.

WRITING AND READING CENTER, UMD: Tutor. Individual and group instruction in writing and word processing. Training new tutors. Newsletter editing. Scriptwriting and acting in a training video. Designing home page. Fall 2003–present.

THE TORCH, UMD WEEKLY NEWSPAPER: Managing Editor. Responsible for conducting staff meetings, generating story ideas, writing editorials and articles, and for supervising page layout, paste-up, and copyediting. Fall 2004–present.

ACHIEVEMENTS AND AWARDS
Writing samples published in Dr. John M. Lannon's TECHNICAL COMMUNICATION, 10th ed., Longman, 2006. Massachusetts State Honors Scholarship, 2003–2006. Dean's list, each semester.

ACTIVITIES
Student member, Society for Technical Communication and American Society for Training and Development. Student representative, College Curriculum Committee. UMD Literary Society.

REFERENCES AND WRITING PORTFOLIO
Available on request.

FIGURE 18.8   A Computer-Scannable Résumé   Notice the standard print and the absence of fancy highlighting.

**FIGURE 18.9
The First
Page of a
Hyperlinked
Résumé**

Links connect to
various types of
information, in-
cluding several
links to Karen's
writing portfolio.
For security rea-
sons, personal
contact informa-
tion is limited to
the applicant's
phone number
and email ad-
dress.

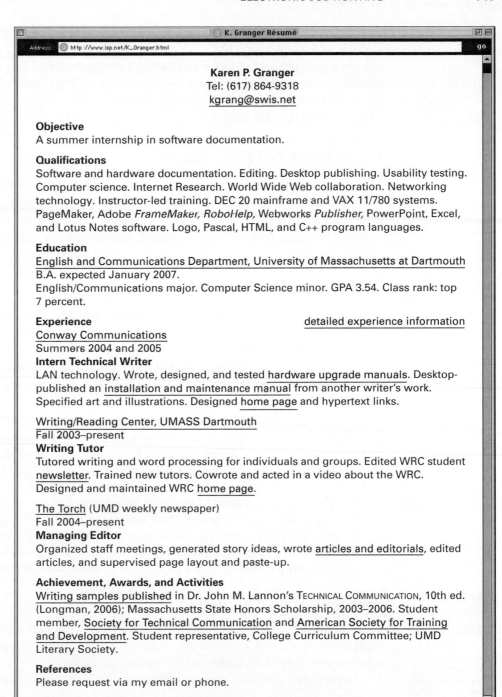

K. Granger Résumé

Address: http://www.isp.net/K_Granger.html   go

Karen P. Granger
Tel: (617) 864-9318
kgrang@swis.net

Objective
A summer internship in software documentation.

Qualifications
Software and hardware documentation. Editing. Desktop publishing. Usability testing.
Computer science. Internet Research. World Wide Web collaboration. Networking
technology. Instructor-led training. DEC 20 mainframe and VAX 11/780 systems.
PageMaker, Adobe *FrameMaker, RoboHelp,* Webworks *Publisher,* PowerPoint, Excel,
and Lotus Notes software. Logo, Pascal, HTML, and C++ program languages.

Education
English and Communications Department, University of Massachusetts at Dartmouth
B.A. expected January 2007.
English/Communications major. Computer Science minor. GPA 3.54. Class rank: top
7 percent.

Experience   detailed experience information
Conway Communications
Summers 2004 and 2005
Intern Technical Writer
LAN technology. Wrote, designed, and tested hardware upgrade manuals. Desktop-
published an installation and maintenance manual from another writer's work.
Specified art and illustrations. Designed home page and hypertext links.

Writing/Reading Center, UMASS Dartmouth
Fall 2003–present
Writing Tutor
Tutored writing and word processing for individuals and groups. Edited WRC student
newsletter. Trained new tutors. Cowrote and acted in a video about the WRC.
Designed and maintained WRC home page.

The Torch (UMD weekly newspaper)
Fall 2004–present
Managing Editor
Organized staff meetings, generated story ideas, wrote articles and editorials, edited
articles, and supervised page layout and paste-up.

Achievement, Awards, and Activities
Writing samples published in Dr. John M. Lannon's Technical Communication, 10th ed.
(Longman, 2006); Massachusetts State Honors Scholarship, 2003–2006. Student
member, Society for Technical Communication and American Society for Training
and Development. Student representative, College Curriculum Committee; UMD
Literary Society.

References
Please request via my email or phone.

Internet zone

Some résumé-posting sites are more private than others. On a site such as *E.SPAN* at <www.espan.com>, for example, you can conceal your contact information and your name. Moreover, you must give permission (via email) before the résumé can be sent to an employer who requests it (Imperato 197). Other sites offer similar options for anonymity while some offer no privacy at all. Find out which is which by checking the site's privacy statement.

SUPPORT FOR THE APPLICATION

An employer impressed by your résumé and cover letter will want answers to these three questions: How highly do other people think of you and your work? What evidence can you show as proof of your skills? How well do you communicate orally? These questions will be answered, respectively, by your dossier, portfolio, and job interview.

Your Dossier

Your dossier contains your credentials: college transcript, recommendation letters, and other items (such as a scholarship award or commendation letter) that document your achievements. An employer impressed by what you say about yourself will want to read what others think and will request your dossier. By collecting recommendations in one folder, you spare your references from writing the same letter repeatedly.

Your college placement office will keep your dossier (or placement folder) on file and send copies to employers. Always keep your own copy as well. Then, if an employer requests your dossier, you can make a photocopy and mail it, advising your recipient that the placement copy is on the way.

NOTE   *This is not needless repetition! Most employers establish a specific timetable for (1) advertising an opening, (2) reading letters and résumés, (3) requesting and reviewing dossiers, (4) holding interviews, and (5) making an offer. Timing is crucial. Too often, dossier requests may sit on someone's desk and may even get lost in a busy placement office. Weeks can pass before your dossier is mailed. In these situations the only loser is you.*

18.7

For some examples of portfolio formatting visit <www.ablongman.com/lannonweb>

How to prepare a portfolio

Your Professional Portfolio

An organized, professional-looking portfolio shows you can apply your skills and makes you stand out as a job candidate. Also, the portfolio gives you specific achievements and skills to discuss during job interviews.

To prepare a portfolio that shows you at your best, follow these suggestions:

• *Collect materials relevant to the job.* Gather documents or graphics you've prepared in school or on the job, presentations you've given, and projects or experiments you've worked on. Possible items: campus newspaper articles, reports on course projects, papers that earned an "A", examples of persuasive argument, documents from an internship, visuals you've designed for an oral presentation, and so on. Once you've gathered your samples, select those that relate specifically to the job you seek.

- *Sort your materials according to the major requirements of the job.* If requirements include desktop publishing, editing, and marketing, select two or three items for each category (a brochure you've designed, pages from a manual you've edited, slides from an oral presentation, and so on).
- *Assemble your portfolio.* Encase each page in its own clear plastic envelope and arrange your items in a portfolio-type notebook (found in office supply or art supply stores). Place your résumé first, followed by a table of contents giving the title of each item and one or two brief sentences about the item's purpose and audience. Use divider pages to group the items into categories. Add, remove, or reassemble items for various job requirements.
- *Make copies as employers request.* When an employer requests the portfolio before the interview, send a photocopy. In case you need to leave a copy after your interview, bring one along with the original, bound portfolio.

As you create your portfolio, seek advice and feedback from major professors and other people in the field. For portfolio advice, go to <www.talent-net.com> and <www.prospring.net>.

Employment Interviews

An employer impressed by your credentials will arrange an interview. The interview's purpose is to confirm the employer's impressions from your application.

Interviews come in various shapes and sizes. They can be face-to-face or via telephone or video conference. You might meet with one interviewer, a hiring committee, or several committees in succession. You might be interviewed alone or as part of a group of candidates. Interviews can last one hour or less, a full day, or several days. The interview can range from a pleasant chat to a grueling interrogation. Some interviewers may antagonize you deliberately to observe your reaction. Unprepared interviewees make mistakes like the following (Dumont and Lannon 620):

How people fail job interviews

- know little or nothing about the company or what role they would play as an employee
- have inflated ideas about their own worth
- have little idea of how their education prepares them for work
- dress inappropriately
- exhibit no self-confidence
- have only vague ideas of how they could benefit the employer
- inquire only about salary and benefits
- speak negatively of former employers or coworkers

Careful preparation is the key to a productive interview

Prepare by learning about the company in trade journals, industrial indexes, and other resources listed on page 426. Request company literature, including its annual report. Speak with people who know about the company. Visit the company's Web site and, for a more objective view, check sites such as <www.career-mosaic.com> for "insider" information about a company. Also, do a keyword search (using the company's name) of business magazine Web sites such as <Fortune.com>, <Forbes.com>, or <BusinessWeek.com> for articles about a company's financial health, working conditions, environmental record, chance of merger (which often means big layoffs), or some impending crises (an automaker's tire recall, for example), and the like. Once you've done all this ask yourself, "Does this job seem like a good fit?"

NOTE   *Taking the wrong job can be far worse than taking no job at all—especially for a recent college graduate trying to build solid working credentials.*

Once you've learned enough to decide that you actually want to work for this employer, prepare—and practice—specific answers to the obvious questions:

- *Why does this job appeal to you?*
- *What do you know about our company?*
- *What do you know about our core values (say, informal management structure, commitment to diversity or the environment)?*
- *What do you know about the expectations and demands of this job?*
- *What are the major issues affecting this industry?*
- *How would you describe yourself?*
- *What do you see as your biggest weakness? Biggest strength?*
- *Can you describe an instance in which you came up with a new and better way of doing something?*
- *What are your short-term and long-term career goals?*

Plan direct answers to questions about your background, training, experience, and salary requirements.

Prepare your own list of *well-researched* questions about the job and the organization; you will be invited to ask questions, and what you ask can be as revealing as any answers you give.

NOTE   *Being truthful during a job interview is not only ethical but also smart. Companies routinely verify an applicant's claims about education, prior employment, positions held, salary, and personal background. Say you have some past infraction such as a bad credit rating or a brush with the law or some pressing personal commitment such as caring for an elderly parent or a disabled child. Experts suggest that it's better to air these issues right up front—before the employer finds out from other sources. The employer will appreciate your honesty and you will know exactly where you stand before accepting the job (Fisher, "Truth" 292). For more on applicant screening, see page 440.*

GUIDELINES for Surviving a Job Interview

The Face-to-Face Interview

1. *Get your timing right.* Confirm the interview's exact time and location. Give yourself ample time to get there. Arrive early but no more than 10 minutes or so.

    NOTE   *If you are offered a choice of interview times, choose mid-morning over late afternoon: According to an Accountemps survey of 1,400 managers, 69 percent prefer mid-morning for doing their hiring, whereas only 5 percent prefer late afternoon (Fisher, "My Company" 184).*

2. *Don't show up empty-handed.* Have a briefcase, pen, and notepad. Have your own questions organized and written out. Bring extra copies of your résumé (unfolded) and a portfolio (if appropriate) with examples of your work.

3. *Make a positive first impression:*
    - Come dressed as if you already work for the company.
    - Learn the name of your interviewer beforehand, so you can greet this person by name—but never by first name unless invited.
    - Extend a firm handshake, smile, and look the interviewer in the eye.
    - Wait to be asked to take a chair.
    - Relax but do not slouch.
    - Keep your hands in your lap.
    - Do not fiddle with your face, hair, or other body parts.
    - Maintain eye contact much of the time, but don't stare.

4. *Don't worry about having all the answers.* When you don't know the answer to a question, say so, and relax. Interviewers typically do about 70 percent of the talking (Kane 56).

5. *Avoid abrupt yes or no answers—as well as life stories.* Keep answers short and to the point.

6. *Don't answer questions by merely repeating the material on your résumé.* Instead, explain how specific skills and types of experience could be assets to this particular employer. For evidence, refer to your portfolio whenever possible.

7. *Remember to smile often and to be friendly and attentive throughout the interview.* In the end, people hire the candidate they **LIKE** the best!

8. *Never criticize a previous employer.* Above all, interviewers like people who have positive attitudes.

9. *Be prepared to ask intelligent questions.* When asked if you have questions, focus on the nature of the job: travel involved, level of responsibility, typical job assignments, opportunities for further training, types of clients, and so on. Avoid questions that could easily have been answered by your own prior research.

(continues)

Guidelines (continued)

10. *Don't be afraid to allow silence.* An interviewer may simply stop talking, just to observe your reaction to silence. If you have nothing more to say or ask, don't feel compelled to speak. Let the interviewer make the next move.

11. *Take a hint.* When your interviewer hints that the meeting is ending (perhaps by checking a watch), restate your interest, ask when a hiring decision is likely to be made, thank the interviewer, and leave.

12. *Display etiquette and restraint.* If you are invited to lunch, don't order the most expensive dish on the menu; don't order an alcoholic beverage; don't smoke; don't salt your food before tasting it; don't eat too quickly; don't put your elbows on the table; don't speak with your mouth full; and don't order a huge dessert. In short, remember this adage: "There's no such thing as a free lunch." And try to order last. For more on dining etiquette go to <www.epicurious.com/c_play/cO2_polite/polite.html>.

13. *Send a follow-up letter.* As soon as possible, send a brief thank-you note (page 453).

The Telephone Interview

As a screening device, many employers interview candidates initially over the telephone. (The interviewer(s) will usually phone beforehand to arrange a mutually convenient time.) A phone interview gives you the chance of making a good first impression by speaking from your home turf and having notes, questions, and other backup materials organized within easy reach. A few guidelines (Crosby, *Employment* 20–21; Ford, "Phone" 19):

1. *If you have "call waiting," disable it temporarily, to avoid beeping and interruptions.* On most phones, you can disable this feature by pressing *70 as soon as the call has been connected.

2. *Arrange all your materials where you can reach them.* Have your list of questions, job description, talking points, résumé, pen, paper, and anything else you might need. Tape things on the walls, spread them on the floor, or whatever works for you.

3. *Sit in a straight-backed chair or remain standing.* These postures may help you speak more emphatically and confidently. They are also likely to keep you more attentive and businesslike than if you were lounging in a comfy armchair.

4. *Identify the interviewer clearly.* Ask for the interviewer's name (spelled) and contact information, including email, and a mailing address to which you can send a thank-you letter.

5. *As the interview ends, encourage further contact.* Restate your interest in the position and your desire to visit the organization and meet people in person.

6. *Send your thank-you letter as soon as possible.*

The Follow-Up Letter

Within a day or so after the interview, send a thank-you letter. Not only is this courteous, but it also reinforces a positive impression. Keep your letter brief, but try to personalize your connection with the reader (Crosby, *Employment* 20):

What to say in a follow-up letter

- Open by thanking the interviewer and reemphasizing your interest in the position.
- Refer to some details from the interview or some aspect of your visit that would help the interviewer reconnect in his/her mind with you specifically. (If you forgot to mention something important during the interview, include it here—briefly.)
- Close with genuine enthusiasm, and make it easy for the reader to respond.

James Purdy (page 433) sent Sara Costanza this follow-up letter:

Refresh the employer's memory

Thank you for your hospitality during my Tuesday visit to Greenwoods resort. I am very interested in the restaurant-management position, and was intrigued by our discussion about developing an eclectic regional cuisine.

Everything about my tour was enjoyable, but I was especially impressed by the friendliness and professionalism of the resort staff. People seem to love working here, and it's not hard to see why.

I'm convinced I would be a productive employee at Greenwood, and would welcome the chance to prove my abilities. If you need additional information, please call me at (214) 316-2419.

NOTE *Employment expert Olivia Crosby offers these suggestions: Instead of email, send hard copy, either in a business-letter format or as a tasteful, handwritten note. Write to each person with whom you spoke or to the person in charge of the group interview. Be sure to spell each person's name correctly and to proofread repeatedly (Employment 20).*

Letters of Acceptance or Refusal

You may receive a job offer by phone or letter. If by phone, request a written offer, and respond with a formal letter of acceptance. This letter may serve as part of your contract; spell out the terms you are accepting. Here is Purdy's letter of acceptance:

Accept an offer with enthusiasm

I am delighted to accept your offer of a position as assistant recreation supervisor at Liberty International's Lake Geneva Resort, with a starting salary of $44,500.

As you requested, I will phone Bambi Druid in your personnel office for instructions on reporting date, physical exam, and employee orientation.

I look forward to a long and satisfying career with Liberty International.

You may have to refuse offers. Even if you refuse by phone, write a prompt and cordial letter of refusal, explaining your reasons, and allowing for future possibilities. Purdy handled one refusal this way:

Decline an offer diplomatically

18.8

For more job hunting resources visit <www.ablongman.com/lannonweb>

> Although I was impressed by the challenge and efficiency of your company's operations, I must decline your offer of a position as assistant desk manager of your London hotel.
>
> I have taken a position with Liberty International because Liberty has offered me the chance to participate in its manager-trainee program. Also, Liberty will pay tuition for the courses I take in completing my B.S. degree in hospitality management.
>
> If any future openings should materialize at your Aspen resort, however, I would again appreciate your considering me as a candidate.
>
> Thank you for your interest in me and for your courtesy.

CONSIDER THIS How to Evaluate a Job Offer

Fortunately, most organizations will not expect you to accept or reject an offer on the spot. You will probably be given at least a week to make up your mind. Although there is no way to remove all risks from this career decision, you will increase your chances of making the right choice by thoroughly evaluating each offer—weighing all the advantages against all the disadvantages of taking the job.

The Organization

Background information on the organization—be it a company, government agency, or non-profit concern—can help you decide whether it is a good place for you to work.

Does the organization's business or activity match your own interests and beliefs? It will be easier to apply yourself to the work if you are enthusiastic about what the organization does.

How will the size of the organization affect you? Large firms generally offer a greater variety of training programs and career paths, more managerial levels for advancement, and better employee benefits than small firms. Large employers also have more advanced technologies in their laboratories, offices, and factories. How-

ever, jobs in large firms tend to be highly specialized—workers are assigned relatively narrow responsibilities. On the other hand, jobs in small firms may offer broader authority and responsibility, a closer working relationship with top management, and a chance to clearly see your contribution to the success of the organization.

Should you work for a fledgling organization or one that is well established? New businesses have a high failure rate, but for many people, the excitement of helping create a company and the potential for sharing in its success more than offset the risk of job loss. It may be just as exciting and rewarding, however, to work for a young firm that already has a foothold on success.

Does it make a difference if the company is private or public? A private company may be controlled by an individual or a family, which can mean that key jobs are reserved for relatives and friends. A public company is controlled by a board of directors responsible to the stockholders. Key jobs are open to anyone with talent.

Is the organization in an industry with favorable long-term prospects? The most successful firms tend to be in industries that are growing rapidly.

Where is the job located? If it is in another city, you need to consider the cost of living, the availability of housing and transportation, and the quality of educational and recreational facilities in the new location. Even if the place of work is in your area, consider the time and expense of commuting and whether you can use public transportation.

Where are the firm's headquarters and branches located? Although a move may not be required now, future opportunities could depend on your willingness to move to these places.

It is usually easy to get background information on an organization simply by telephoning its public relations office. A public company's annual report to the stockholders tells about its corporate philosophy, history, products or services, goals, and financial status. Most government agencies can furnish reports that describe their programs and missions. Press releases, company newsletters or magazines, and recruitment brochures can also be useful. Ask the organization for any other items that might interest a prospective employee.

Background information on the organization may also be available at your public or school library. If you cannot get an annual report, check the library for reference directories that provide basic facts about the company, such as earnings, products and services, and number of employees.

Stories about an organization in magazines and newspapers can tell a great deal about its successes, failures, and plans for the future. You can identify articles on a company by looking under its name in periodical or computerized indexes such as the *Business Periodicals Index, Reader's Guide to Periodical Literature, Newspaper Index, Wall Street Journal Index,* and *New York Times Index.* It will probably not be useful to look back more than two or three years.

The library may also have government publications that present projections of growth for the industry in which the organization is classified. Long-term projections of employment and output for more than two hundred industries, covering the entire economy, are developed by the Bureau of Labor Statistics and revised every other year—consult the current *Monthly Labor Review* for the most recent projections. The *U.S. Industrial Outlook,* published annually by the U.S. Department of Commerce, presents detailed analysis of growth prospects for a large number of industries. Trade magazines also have frequent articles on the trends for specific industries.

Career centers at colleges and universities often have information on employers that is not available in libraries. Ask the career center librarian how to find out about a particular organization.

The Nature of the Work

Even if everything else about the job is attractive you will be unhappy if you dislike the day-to-day work. Determining in advance whether you will like the work may be difficult. However, the more you find out about it before accepting or rejecting the job offer, the more likely you are to make the right choice. Ask yourself questions like the following.

Does the work match your interests and make good use of your skills? The duties and responsibilities of the job should be explained in enough detail to answer this question.

How important is the job in this company? An explanation of where you fit in the organization and how you are supposed to contribute to its overall objectives should give an idea of the job's importance.

Are you comfortable with the supervisor?

Do employees seem friendly and cooperative?

Does the work require travel?

Does the job call for irregular hours?

How long do most people who enter this job stay with the company? High turnover can mean dissatisfaction with the nature of the work or something else about the job.

The Opportunities

A good job offers you opportunities to grow and move up. It gives you chances to learn new skills, increase your earnings, and rise to positions of greater authority, responsibility, and prestige.

(continues)

Consider This (continued)

The company should have a training plan for you. You know what your abilities are now. What valuable new skills does the company plan to teach you?

The employer should give you some idea of promotion possibilities within the organization. What is the next step on the career ladder? If you have to wait for a job to become vacant before you can be promoted, how long does this usually take? Employers differ on their policies regarding promotion from within the organization. When opportunities for advancement do arise, will you compete with applicants from outside the company? Can you apply for jobs for which you qualify elsewhere within the organization or is mobility within the firm limited?

The Salary and Benefits

Wait for the employer to introduce these subjects. Most companies will not talk about pay until they have decided to hire you. In order to know if their offer is reasonable, you need a rough estimate of what the job should pay. You may have to go to several sources for this information. Talk to friends who were recently hired in similar jobs. Ask your instructors and the staff in the college placement office about starting pay for graduates with your qualifications. Scan the Help Wanted ads in newspapers. Check the library or your school's career center for salary surveys, such as the College Placement Council Salary Survey and Bureau of Labor Statistics occupational wage surveys. If you are considering the salary and benefits for a job in another geographic area, make allowances for differences in the cost of living, which may be signif-

icantly higher in a large metropolitan area than in a smaller city, town, or rural area. Use the research to come up with a base salary range for yourself, the top being the best you can hope to get and the bottom being the least you will take. An employer cannot be specific about the amount of pay if it includes commissions and bonuses. The way the plan works, however, should be explained. The employer also should be able to tell you what most people in the job earn.

Also take into account that the starting salary is just that, the start. Your salary should be reviewed on a regular basis—many organizations do it every twelve months. If the employer is pleased with your performance, how much can you expect to earn after one, two, three, or more years?

Don't think of your salary as the only compensation you will receive—consider benefits. Benefits can add a lot to your base pay. Health insurance and pension plans are among the most important benefits. Other common benefits include life insurance, paid vacations and holidays, and sick leave. Benefits vary widely among smaller and larger firms, among full-time and part-time workers, and between the public and private sectors. Find out exactly what the benefit package includes and how much of the costs you must bear.

Asking yourself these kinds of questions won't guarantee that you make the best career decision—only hindsight could do that—but it will probably help you make a better choice than if you act on impulse.

Source: Excerpted from U.S. Department of Labor. Tomorrow's Jobs. *Washington, DC, GPO, 2000.*

☑ CHECKLIST for Usability of Letters

(Numbers in parentheses refer to the first page of discussion.)

Content

☐ Is the letter addressed to a specifically named person? (412)
☐ Does the letter contain all the standard parts? (410)
☐ Does the letter have all needed specialized parts? (416)
☐ Have you given the recipient all necessary information? (39)
☐ Have you identified the name and position of your recipient? (412)

Arrangement

☐ Does the introduction immediately engage the reader and lead naturally to the body? (412)
☐ Are transitions between letter parts clear and logical? (772)

☐ Does the conclusion encourage the reader to act? (412)
☐ Is the format correct? (417)
☐ Is the design acceptable? (411)

Style

☐ Is the letter in conversational language (free of letterese)? (419)
☐ Does the letter reflect a "you" perspective throughout? (418)
☐ Does the tone reflect your relationship with the recipient? (274)
☐ Is the recipient likely to react favorably to this letter? (420)
☐ Is the style clear, concise, and fluent throughout? (244)
☐ Is the letter grammatical? (Appendix C)
☐ Does the letter's appearance enhance your image? (340)

EXERCISES

 For more exercises, visit
<www.ablongman.com/lannon>

1. Bring to class a copy of a business letter addressed to you or a friend. Compare letters. Choose the most and least effective.

2. Write and mail an unsolicited letter of inquiry about the topic you are investigating for an analytical report or research assignment. In your letter you might request brochures, pamphlets, or other informative literature, or you might ask specific questions. Submit a copy of your letter and the response to your instructor.

3. a. As a student in a state college, you learn that your governor and legislature have cut next year's operating budget for all state colleges by 20 percent. This cut will cause the firing of young and popular faculty members; drastically reduce admissions, financial aid, and new programs; and wreck college morale. Write a claim letter to your governor or representative, expressing your strong disapproval and justifying a major adjustment in the proposed budget.

   b. Write a claim letter to a politician about some issue affecting your school or community.

   c. Write a claim letter to an appropriate school official to recommend action on a campus problem.

4. Write a five hundred to seven hundred word personal statement applying to a college for transfer or for graduate or professional school admission. Cover two areas: (1) what you can bring to this school by way of attitude, background, and talent; and (2) what you expect to gain in personal and professional growth.

5. Write a letter applying for a part-time or summer job, in response to a specific ad. Choose an organization related to your career goal. Identify the exact hours and calendar period during which you are free to work.

6. Most of these sentences need to be overhauled before being included in a letter. Identify the weakness in each statement, and revise as needed. For example, revise the accusatory "You were not very clear" to "We did not understand your message."

a. Pursuant to your ad, I am writing to apply for the internship.
b. I need all the information you have about methane-powered engines.
c. You idiots have sent me a faulty disk drive!
d. It is imperative that you let me know of your decision by January 15.
e. You are bound to be impressed by my credentials.
f. I could do wonders for your company.
g. I humbly request your kind consideration of my application for the position of junior engineer.
h. If you are looking for a winner, your search is over!
i. I have become cognizant of your experiments and wish to ask your advice about the following procedure.
j. You will find the following instructions easy enough for an ape to follow.
k. I would love to work for your wonderful company.
l. As per your request I am sending the country map.
m. I am in hopes that you will call soon.
n. We beg to differ with your interpretation of this leasing clause.
o. I am impressed by the high salaries paid by your company.

7. Write a complaint letter about a problem you've had with goods or service. State your case clearly and objectively, and request a specific adjustment.

8. *For Class Discussion:* Under what circumstances might it be acceptable to contact a potential inquiry respondent by email? When should you just leave the person alone?

COLLABORATIVE PROJECTS

1. Form groups according to college majors. Prepare a set of instructions for entry-level jobseekers in your major, telling them how to launch their search. Base at least part of your advice on your analysis of Figure 18.4. Limit your document to one double-sided page, using an inviting and accessible design and any visuals you consider appropriate. Appoint one group member to present your final document to the class.

2. Divide into groups and prepare a listing of five Web sites that jobseekers should visit for advice about cover letters and résumés, including on-line postings. Include a one-paragraph summary of the material to be found on each site. Compare the findings of your group with others in your class. In addition to sites mentioned in this chapter, here are other sources where you might begin:

Web Resources for Résumés and Cover Letters

\<www.jobsmart.org/tools/resume\>

\<www.rileyguide.com\>

\<www.eresumes.com\>

\<www.quint.careers.com\>

\<www.damngood.com/jobseekers/tips.html.60\>

Note: Expand your search beyond these sites.

3. Divide into groups and prepare a Web site guide for entry-level jobseekers in your field, based on answers to questions like these:

"Where can I find listings for job opportunities in our state or region?"

"Where can I find listings for internships in our field?"

"What Web site focuses on jobs in our field?" (such as hi-tech)

"Where can I find listings for temporary or contract work in our field?"

Once you've identified ten likely questions, list one site that could answer each question. For example:

\<www.craiglist.com\> for jobs in a particular region

\<www.4work.com\> for internship opportunities

\<www.firsttuesday.com\> for hi-tech jobs

\<www.workflex.monster.com\> for contract or temporary jobs

Note: Expand your search beyond these sites.

Report your findings in a memo to your instructor and classmates.

19

Web Pages and Other Electronic Documents

Providing information online has become a critical communication strategy for several reasons.[1] Once text is printed it remains the same for as long as the paper or the disk it was printed on lasts. If the text needs to be updated, a new (and costly) print run has to be made. At the same time, printed copy takes up space and costs money to distribute. But information that is saved on a server and delivered fresh to each user upon request takes up negligible physical space, can be updated for only the cost of the labor, and can in fact be updated automatically, without direct human intervention. Moreover, one can be relatively confident that only the most current information is circulating.

Writing for online delivery is a complex topic and one that ceaselessly changes as the technologies used to deliver it change. Writing online requires regular training and a wide range of skills including visual rhetoric, information design, and computer–human interaction, or usability. It is primarily a collaborative activity, often involving content providers, information architects, graphic designers, computer programmers, and marketers. Nevertheless, especially in small businesses and nonprofit organizations, online writing often becomes the responsibility of the communications person.

This chapter introduces types of electronic documents essential to the workplace and the basic technologies of online writing.

ONLINE DOCUMENTATION

People who use computers in their jobs need instructions and training for operating the systems and understanding their equipment's many features. Online documentation is designed to support specific tasks and provide answers to specific questions.

Although computers come with printed manuals, the computer itself is often the preferred training medium. In computer-based training, on-screen documentation explains how the system works and how to use it. Examples of online help:

- error messages and troubleshooting advice
- reference guides to additional information or instructions
- tutorial lessons that include interactive exercises with immediate feedback
- help and review options to accommodate different learning styles
- link to software manufacturer's Web site

Instead of leafing through a printed manual, users find what they need by typing a simple command, clicking a mouse button, using a help menu, or following an electronic prompt.

---

[1] My thanks to Professor George Pullman, Georgia State University, for revising and updating this chapter.

19.1

To see examples of *RoboHelp* visit <www.ablongman.com/lannonweb>

What users prefer

Why businesses prefer online documentation

The origin of "hypertext"

A hypertext document can exist on paper as well as online

Hypertext accommodates inquiry in various directions

Special software such as *RoboHelp*™ or *Doc-to-Help*™ can convert print material into online help files that appear (a) as dialog boxes that ask the user to input a response or click on an option, or (b) as pop-up or balloon help that appears when the user clicks on an icon or points to an item on the screen for more information. (Explore, for example, the online help resources on your own computer.)

Research indicates that users—especially inexperienced users—often prefer printed manuals over their online counterparts. Also, users in general prefer having both printed and online options available (Smart, Whiting, and DeTienne 291, 301).

However, the cost of producing and distributing printed materials makes online documentation attractive to software producers. Also, because most business software is sold on a subscription basis, with a new version coming out every sixteen months or so, providing paper documentation becomes increasingly cost prohibitive. (See Chapter 22 for more on online documentation.)

HYPERTEXT

In 1965, Ted Nelson coined the term "hypertext" to describe a kind of writing that would enable multidimensional reading strategies. The word has taken on a life of its own, most notably in referring to descriptions of the World Wide Web, a result that greatly irritates Mr. Nelson <http://ted.hyperland.com/buyin.txt>.

There is, however, nothing necessarily electronic about hypertext. In a paper document, indexes, page numbers, section headings, cross-references, and other typographical conventions help readers navigate the text in multiple dimensions: in other words, they are able to jump from place to place searching for the specific thing they need. The reader's purpose often has a greater effect on how something gets read than the medium employed. For example, a dictionary from circa 1900 is multimedia and hypertextual because it has drawings as well as text and because it has cross-references. There are novels, Julio Cortazar's *Hopscotch* (1965) for instance, which are printed on paper and look like any other novel but which consist of chapters that can be read in different sequences, creating different "stories" out of the same material. Newspapers and magazines aren't meant to be read linearly, either. At the same time, it is possible to create online text that has no links and so can be read sensibly in only one direction, linearly.

Nevertheless, because multidimensional reading is typical of information-seeking readers, hypertext serves as a research and reference tool. In a hypertext system, a topic can be explored from any angle, at any level of detail. Assume, for instance, that you are using a hypertext database to research the AIDS epidemic. The database contains chunks of related topics organized in a network (or web) of files linked electronically (Horton, "Is Hypertext" 22), as in Figure 19.1.

After accessing the initial file ("The AIDS Epidemic"), you can navigate the network in any direction, choosing which file to open and where to go next, thereby customizing the direction of your search. The files themselves might be printed words, graphics, sound, video, or animation.

**FIGURE 19.1
Topics (or Files)
in a Hypertext
Network Are
Linked
Electronically**

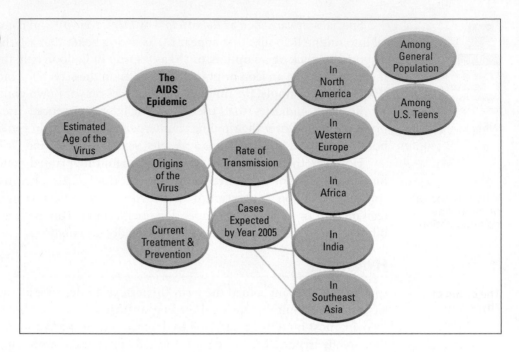

MARKUP

How markup
language works

Any document, whether typeset, served over the Web, or even handwritten, has a structure that can be labeled. With a letter, for example, the markup would consist of DATE, ADDRESSEE, RETURN ADDRESS, SALUTATION, BODY, CLOSING, SIGNATURE. At the moment, information designed to be presented online, whether on disk or over the Web, is marked up in Hypertext Markup Language (HTML). HTML consists of a set of tags used to label the structure of a document and so make it theoretically possible for a document to be read in the same format on any computer regardless of operating system or browser.

How HTML tags
work

A Web browser's default style sheet represents labels such as those in Figure 19.2 as visual cues. H1 is printed to the screen as text roughly two-and-a-half times the size of normal and three times as dark, for example. This default style sheet varies slightly from Web browser to Web browser and so what looks one way in one browser will look different in another, a fact that adds to the complexity of designing information for electronic delivery.

There are several WYSIWYG ("what you see is what you get") software packages for HTML, *DreamWeaver*™ being the current industry standard. While such packages enable a writer to do much more much more rapidly, it is a good idea to learn to recognize the basic tags so that if problems arise they can be dealt with. Learning to code by hand, using the tags directly, may also improve one's proofreading skills. For more on HTML, visit <http://www.w3schools.com>.

FIGURE 19.2
A Sampling of HTML Commands

- To denote a Web page (so browsers can identify this page as a HTML file):

    <HTML> entire HTML file </HTML> *(slash [/] denotes an ending tag)*

- To denote the main parts of a HTML file:

    <HEAD> . . . . </HEAD> *(sets off prefatory material, such as title)*
    <BODY> . . . . </BODY> *(sets off all material browsers will display)*

- To signify a break in the text*:

     (line break; begins next line at left margin)
    <P>   *(paragraph)*
    <HR>   *(displays horizontal rule, across text width of page)*

*No closing tag is needed for a mere insertion into the page, in which no specific content is modified. These are called "empty tags."

- To specify headings:

    <H1> . . . . </H1> *(for highest level head)*
    <H2> . . . . </H2> *(and so on, down to as far as sixth-level heads)*

- To align elements elsewhere than at left margin *(center, right, justify)*:

    <H2 ALIGN = right> . . . . </H2> (aligns head on right margin)
    <P ALIGN = center> . . . . </P> (centers the paragraph)
    <BLOCKQUOTE> . . . . </BLOCKQUOTE> (sets off quoted material)

- To display a list:

    <UL>   *(unnumbered list, with bullets displayed before each item)*
     <LI>   *(1st item in list)*
     <LI>   *(2nd item, and so on)*
    </UL>   *(end of list)*
    <OL>   *(numbered list, with a number displayed before each item)*
     <LI>   *(1st item in list)*
     <LI>   *(2nd item, and so on)*
    </OL>   *(end of list)*

- To specify typestyles:

    <Q> . . . . </Q>   "Show enclosed content in quotation marks."
    <B> . . . . </B>   **Boldface**
    <I> . . . . </I>   *Italics*
    <BIG> . . . . </BIG>   Larger than current font
    <SMALL> . . . . </SMALL>   Smaller than current font

- To insert a figure in the text:

    <FIG SRC = "filename or URL"> . . . </FIG>

- To create links (HREF stands for Hypertext Reference):

    <A HREF = "some URL">Go to some URL </A> *(to other Web sites)*
    <A HREF = "#price list">price list</A> *(to another part of document)*

BEYOND HTML

Over the years computer equipment has become more visually oriented, and electronic texts have reached a far wider audience than was originally imagined. Because HTML was never intended as a layout tool, it offers the designer very little control over the appearance of text. In the late 90s, Web designers solved this problem by using *Photoshop™* to create visually impressive images and then slice those images into small sections that would transfer quickly over the Web. The problem with this solution, however, was that whenever business (or fashion) required a change, the entire image had to be redone. Most Web sites today are dynamic in the sense that they are constantly receiving new content from several sources, from news-feeds as well as from contributing editors. It is far more efficient to regularly update or replace text than it is to update pictures. Cascading Style Sheets (CSS) enables text-based layouts that are nearly as visually sophisticated as graphics-based layouts but far more flexible. (For more on CSS, go to <http://www. w3schools.com/css">.)

Changes in online information systems are inevitable: prices change, products come and go, new information becomes available, and so on. Static Web "pages," that is, information whose appearance is outdated or which seem to have been posted and then abandoned, do not inspire confidence.

TEXT VERSUS IMAGES: CURRENT PREFERENCES

What users
expect in a Web
page

People have come to expect fresh content and up-to-date designs. Compare a current site from its predecessor of only a year or two ago and you will see how things have changed: In the earlier version of the EPA site (Figure 19.3), there is very little text, and images provide access to the content. Drop-down menus provide navigation. In the current version (2004), there is more text and navigation is provided by embedded links (Figure 19.4). The current version is somewhat like a newspaper, with freshness of content taking primary place over visual appeal.

The differences
three years make

The traditional advice regarding writing for online delivery is to use as few words as possible because people don't like to read from a computer screen, the premise upon which the earlier version of epa.gov seems to have been based. The prevalence today of Web sites like the current epa.gov suggests that the traditional advice may no longer apply. If the content is interesting and fresh, people will read online.

As technologies change, online writing changes. But not always in ways one would assume. Five years ago, when more people had slow dial-up connections, graphics were prevalent. Today, when more people have faster connections, graphics are less prevalent. At the same time, animations like those made possible by software packages such as *Flash™* have also become popular. These image-intensive virtual movies would be impossibly slow over a modem connection. Change is so prevalent on the Web that it is important for technical communicators to become students of

FIGURE 19.3
The 2001
Version of the EPA Site
Notice the high ratio of images to text.
Source: U.S. Environmental Protection Agency.
<www.epa.gov/enviro>.

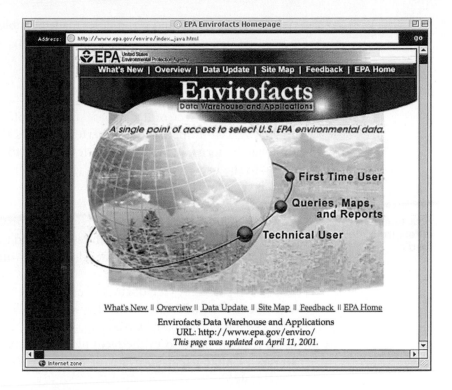

FIGURE 19.4   The 2004
Version of the EPA Site
In this version, content takes primacy over visual appeal.
Source: U.S. Environmental Protection Agency.
<www.epa.gov/enviro>.

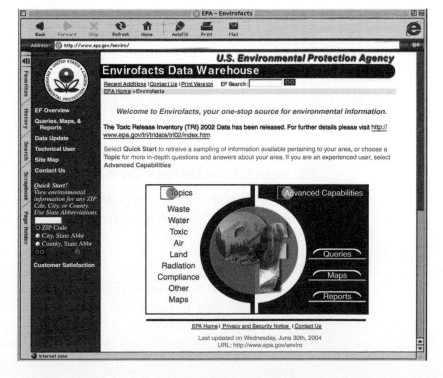

the Web, to notice trends and possibilities made possible by new technologies, to develop practices that can be adapted and abandoned on a moment's notice.

THE WEB

Like a CD-ROM or electronic database, the Web offers a collection of electronic documents and multimedia. But hypertext enables the Web to link information in nonlinear patterns, providing countless routes to be explored—worldwide—according to the user's particular needs (Hunt 377). Some unique characteristics of the Web (December 371–72):

How the Web differs from other media

- *The Web is interactive.* Each user constructs his/her own hypertextual path through the material and often can respond/add to the message.
- *The Web allows reciprocal use.* Besides getting information, users often also provide it.
- *The Web is porous.* A Web document can be entered at various points because a Web site usually offers multiple files, which are linked.
- *The Web is ever changing.* A Web page or site is "a work in progress"—not only in its content but also in the technology itself (software, hardware, modems, servers). Unlike paper, software, or nonrewritable CD-ROM, the Web has no "final state."

These features allow Web users to discover and create their own connections among an endless array of information.

NOTE   *Keep in mind that Web pages, like all online screens, take at least 25 percent longer to read than paper documents. One possible solution: high-resolution screens, as readable as paper copy, should be available and affordable within a decade (Neilsen, "Be Succinct").*

ELEMENTS OF A USABLE WEB SITE

Although more diverse than typical users of paper documents, Web users share common expectations. Following are basic usability requirements for a Web site.

Accessibility

Users expect a site to be easy to enter, navigate, and exit. Instead of reading word for word, they tend to skim, looking for key material without having to scroll through pages of text. They look for chances to interact, and they want to download material at a reasonable speed (roughly 8–12 seconds per page).

CONSIDER THIS Web Sites Enhance Workplace Transactions

The Web is a tool for advertising, learning about new products or companies, updating product information, or ordering products (Teague 236, 238). Each organization advertises its services and products via its own *home page,* a type of electronic billboard that introduces the organization and provides links to additional pages.

Specific Benefits

- *Visibility.* A Web site attracts business by establishing a presence in markets worldwide.
- *Access.* A Web site is accessible twenty-four hours a day.
- *Customer relations.* Through enhanced customer service and support and rapid response, a Web site increases customer satisfaction and enhances a company's caring persona (Hoger, Cappel, and Myerscough 41).
- *Efficiency.* Two-way, real-time communication allows sudden problems, errors, or areas of danger to be broadcast rapidly. The audience can control the viewing of messages and respond immediately. On-screen instructions (say, for installing a modem) can be enhanced with high-resolution, 3-D graphics; parts can be color coded for assembly; and material can be updated instantly (Dulude 49–60).
- *Economy.* The cost of an Internet/Web bank transaction drops from over $1 to roughly 1 cent; the cost of processing an airline ticket drops from $8 to $1 (IBM). By radically reducing printing, mailing, and distribution costs, a site can facilitate mass publishing. Also, an advertiser can embed deeper and deeper levels of product details, without using extra page space. Ultimately, as the cost of business transactions drops, so does the number of required employees.
- *Data gathering.* Tracking software provides customer data by recording who uses the Web site, how often they use it, and exactly where they go. Employees access reference materials from journal and trade magazines, as well as addresses of researchers, and the latest information about legal issues and government regulations (Ritzenthaler and Ostroff 17–18).
- *Information sharing.* Intranets and extranets (Chapter 8) increase the flow of ideas up and down, and from outside to inside the company and vice versa. Knowledge audits identify who knows what and this information is then listed in the company intranet directory ("yellow pages").
- *Collaboration.* Company sites help reduce the length of and need for face-to-face meetings. And people who do meet are well prepared because they have shared information beforehand.

Web Applications in Major Companies

- For training employees in rapidly changing job skills, companies rely on distance-learning programs offered by colleges and universities. Such programs include email correspondence with faculty, online discussion groups, assignments and examples downloaded from the school Web site, and searches of virtual libraries (for special training or MBA work, etc.).
- General Electric is saving hundreds of millions in operating costs by using the Internet for all sorts of transactions, including purchasing and marketing (Reinhardt 130).
- NASA posts requests for proposals (RFPs) on an engineering Web site, to which bidders and contractors can respond via email—thereby speeding the whole process, eliminating fax, phone, and copying time (Machlis 45).
- The Volvo Corporation is connecting all its branches, warehouses, and truck dealerships in Sweden and the United States via a global network to keep track of parts, specifications, and product updates, and to allow authorized employees worldwide access to company databases so that "all data will be available anywhere" (Hamblen 51+).

Worthwhile Content

Users expect the site to contain all the explanations they need (help screens, links, and so on). They want material that is accurate and constantly up-to-date (say, product and price updates). They expect clear error messages that spell out appropriate corrective action. They look for links to other, high-quality sites as indicators of credibility. They look for an email address and other contact information to be prominently displayed.

Sensible Arrangement

Users want to know where they are, and where they are going. They expect a recognizable design and layout, with links easily navigated forward or backward, back links to the home page, and no dead ends. They look for navigation bars and hot buttons to be explicitly labeled ("Company Information," "Ordering," "Job Openings," and so on).

Instead of a traditional introduction, discussion, and conclusion, users expect the punch line right up front. Because they hate to scroll, users often read only what is on the first screen, "above the fold."

Good Writing and Page Design

Users expect a writing style that is easy to read and error-free. They look for concise pages that are quick to scan, with short sentences and paragraphs, headings, and bulleted lists. Instead of having to wade through overstatement and exaggeration to "get at the facts," users expect restrained, impartial language (Neilsen, "Be Succinct" 2). Figure 19.5 illustrates the effect of good writing on usability.

Good Graphics and Special Effects

Some users look for images or multimedia special effects—as long as they are neither excessive nor gratuitous. Since other users often disable their browser's visual capability (to save memory and downloading time) they look for a prose equivalent of each visual (*visual/prose redundancy*). They expect to recognize each icon and screen element—hot buttons, links, help options, and the like.

Figure 19.6 is an updated version of an award-winning Web page designed for usability. Figure 19.7 shows a highly simplified but usable design.

Site Version	Sample Paragraph	Usability Improvement (relative to control condition)
Promotional writing (control condition) using the "marketese" found on many commercial Web sites	Nebraska is filled with internationally recognized attractions that draw large crowds of people every year, without fail. In 1996, some of the most popular places were Fort Robinson State Park (355,000 visitors), Scotts Bluff National Monument (132,166), Arbor Lodge State Historical Park & Museum (100,000), **Carhenge** (86,598), Stuhr Museum of the Prairie Pioneer (60,002), and Buffalo Bill Ranch State Historical Park (28,446).	0% (by definition)
Concise text with about half the word count as the control condition	In 1996, six of the best-attended attractions in Nebraska were Fort Robinson State Park, Scotts Bluff National Monument, Arbor Lodge State Historical Park & Museum, **Carhenge,** Stuhr Museum of the Prairie Pioneer, and Buffalo Bill Ranch State Historical Park.	58%
Scannable layout using the same text as the control condition	Nebraska is filled with internationally recognized attractions that draw large crowds of people every year, without fail. In 1996, some of the most popular places were: • Fort Robinson State Park (355,000 visitors) • Scotts Bluff National Monument (132,166) • Arbor Lodge State Historical Park & Museum (100,000) • **Carhenge** (86,598) • Stuhr Museum of the Prairie Pioneer (60,002) • Buffalo Bill Ranch State Historical Park (28,446)	47%
Objective language using neutral rather than subjective, boastful, or exaggerated language (otherwise the same as the control condition)	Nebraska has several attractions. In 1996, some of the most visited places were Fort Robinson State Park (355,000 visitors), Scotts Bluff National Monument (132,166), Arbor Lodge State Historical Park & Museum (100,000), **Carhenge** (86,598), Stuhr Museum of the Prairie Pioneer (60,002), and Buffalo Bill Ranch State Historical Park (28,446).	27%
Combined version using all three improvements in writing style together: concise, scannable, and objective	In 1996, six of the most visited places in Nebraska were: • Fort Robinson State Park • Scotts Bluff National Monument • Arbor Lodge State Historical Park & Museum • **Carhenge** • Stuhr Museum of the Prairie Pioneer • Buffalo Bill Ranch State Historical Park	124%

FIGURE 19.5   The Effect of Good Writing on Usability

Source: From "How Users Read on the Web" *by Jakob Nielsen, copyright © 1997 by Jakob Nielsen. All rights reserved. Reprinted with permission from Jakob Neilsen's Alertbox at <www.useit.com>.*

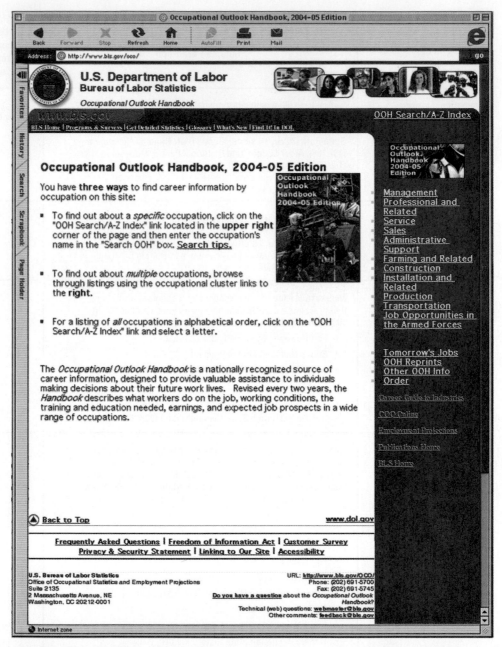

FIGURE 19.6   An Award-Winning Web Page   The previous version of this home page received an Award of Excellence from the Society for Technical Communication. Notice the prominence of the search feature and the dominance of text over graphics.

Source: U.S. Bureau of Labor Statistics, Office of Occupational Statistics and Employment Projections http://www.bls.gov/oco/.

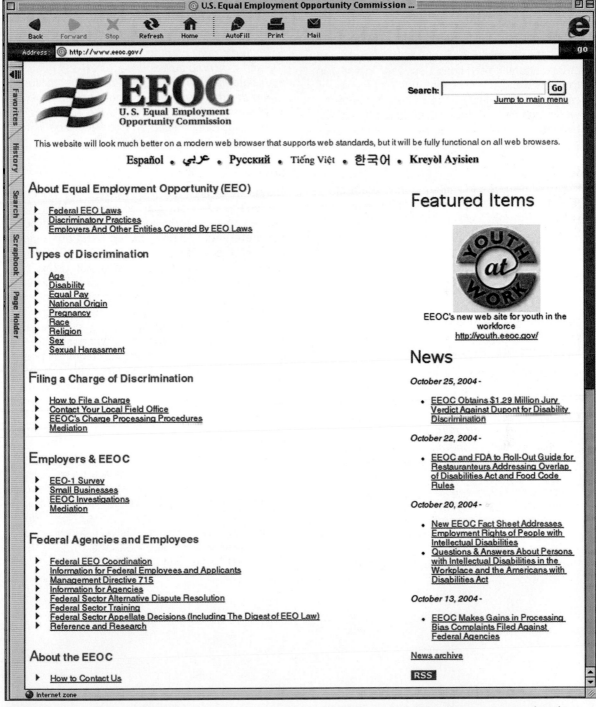

FIGURE 19.7   A Simplified Design   Typical visitors to this site are seeking direct access to legal information, without design frills. All the links, therefore, are grouped under topic headings, as one would find in the index of a book.   *Source: U.S. Equal Employment Opportunity Commission <www.eeoc/gov>.*

CONSIDER THIS Web Site Needs and Expectations Differ Across Cultures

Despite its U.S. origins, the Internet has rapidly become international and cross-cultural. Yet, countries vary greatly in their level of "Internet maturity." With the exception of Scandinavian countries, much of the world lags behind the United States. An effective international Web site therefore addresses a broad range of needs and expectations.

Cost

High phone rates in many countries affect Internet costs. In Japan, for example—whose Internet use ranks second to that in the United States—monthly cost for one hour daily online is more than double the U.S. cost for unlimited access (Nielsen, "Global"). A usable site therefore omits graphics that are slow to load.

Clarity

To facilitate access and avoid misunderstandings, international communication via the Internet incorporates measures like these:

- *Sites often provide home page versions in various languages (or links to a translation package).*
- *Time zones, currencies, and other units of measurement differ (10 A.M. in San Francisco equals 6 P.M. in London, 7 P.M. in Stockholm, or 3 A.M. in Tokyo). In arranging real-time interaction (say, an online conference), the host specifies the recipient's time as well as the home time (Nielsen, "International").*

  Because the value of a "dollar" in countries such as Australia, Canada, Singapore, or Zimbabwe differs from the value of the U.S. dollar, businesses specify "US $12.50," and

so on. Also, offering payment options in the culture's own currency helps avoid currency-exchange ambiguities (Hodges 28–29).
- *A date listed as "6/10/98" might be confusing in other cultures. It would be preferable to say "10 June 1998" or "June 10, 1998."*
- *Temperature measurements are specified as "Fahrenheit" or "Celsius."*

Privacy

Commercial U.S. Web sites routinely collect and sell to other companies personal information about a visitor's purchasing habits, product preferences, types of sites most often visited, and so on. (See page 475.) But the European Union's Data Privacy Act prohibits companies from collecting personal information without permission and gives individuals the right to easily change or remove such information. This law applies to any companies from any country who do business with a country in the European Union. Insofar as possible, a U.S. company Web site with a global audience should adjust its privacy policy to respect the laws of specific cultures (Gurak and Lannon 119).

Cultural Sensitivity

A site that is truly "international" in ambiance—and not only "American"—enables anyone anywhere to feel at home (Nielsen, "Global"). For example, it avoids sarcasm or irreverence (which some cultures consider highly offensive), and exclusive references or colloquialisms such as "bear markets," "the Wild West," and "Super Bowl."

For more on Web collaboration visit <www.ablongman.com/lannonweb>.

GUIDELINES for Creating a Web Site

> **NOTE** *Organizational Web sites are generally developed by a Web team: content developers, graphic designers, programmers, and managers. Whether or not you are an actual team member, expect a collaborative role in your organization's site development and maintenance.*

Planning Your Site

1. *Identify the site's intended audience.* Are they potential customers seeking information, people purchasing a product or service, customers seeking product support or updates or troubleshooting advice (Wilkinson 33)? Will different audiences be seeking different material?

2. *Decide on the site's purpose.* Is the purpose to publish information, sell a product, promote an idea, solicit customer feedback, advertise talents, create goodwill? Should the site convey the image of a "cool" cutting-edge company (or individual), displaying skill with the latest Web technologies (animation, interaction, fancy design)? For specific ideas, find and examine other sites that display the features you are considering.

3. *Decide on what the site will contain.* Will it display only print documents or graphics, audio, and video as well? Will links be provided and, if so, how many and to where? Will user feedback be solicited and, if so, in what form: survey questions, email comments, or the like?

4. *Decide on the level of user interaction.* Will this be a document-only site, offering no interaction beyond downloading and printing? Will it offer dynamic marketing (Dulude 69): online questions and answers, technical support, downloadable software, online catalogs? Will users be able to download documents, software, or documentation? Will an email button be included?

For more models of interactive sites visit <www.ablongman.com/lannonweb>

Laying Out Your Pages

1. *Visit other sites for design ideas.* When you find a site that looks good and navigates easily (or vice versa) analyze what works or doesn't work in terms of type style, color, layout, graphics, highlighting, and overall design (Fugate, "Wowing" 33). For a good look at award-winning Web sites of all types, go to the *Webby Awards* site at <www.webbywawards.com> (Figure 19.8).

2. *Design your pages (Chapter 15) to guide the user.* Highlight important material with headings, lists, type styles, color, and white space. Remember that too much white space causes excessive scrolling. Prefer sans serif fonts (page 353). Use storyboards (page 227) to sketch the basic elements of each page. Limit page size to 30K, to speed downloading.

(continues)

Guidelines (continued)

3. *Use graphics that download quickly.* Avoid excessive complexity and color, especially in screen backgrounds. Keep maximum image size below 30K. Create an individual file for each graphic and use thumbnail sketches on the home page, with links to the larger images, each in its own file. One expert suggests saving on download time by using tables for graphic presentation (Fugate 33).

4. *Include text-only versions of all visual information.* Roughly 20 percent of users turn off the graphics function on their browsers (Gannon 22–23).

5. *Make the content broadly accessible.* Some people may want to choose how the content of your site appears on their screen: for example, people with limited vision or people who use hand-held devices or older Web browsers. To create these options, consult the *World Wide Web Consortium's Content Accessibility Guidelines 1.0* at <www.w3.org/TR/WAI-WEB-CONTENT>.

6. *Provide orientation.* Structure the content to reflect its relative importance and frequency of use. Place the material that is most important to users right up front and create links to more detailed information. Date each page to announce the exact time of each update—or include a "What's New" head, so readers can keep track of changes.

7. *Provide navigational aids.* Keep links logical and always link back to the home page. Don't overwhelm the user with excessive choices. Label each link explicitly (for example, use "Product Updates," instead of merely "Click here"). Also, use the color blue for denoting links not yet visited and red for links already visited by that user (Fugate 34).

8. *Include an alternate, printer-based style sheet or a link to a printable version of the content.* Use a PDF (Portable Document File) if the content needs to appear exactly like the original paper document (page layout, type style, graphics, color). Also, provide a link for downloading the free Adobe *Acrobat Reader*™ that enables users to view and print the document in its original format.

9. *Define and shape the content.* Use hypertext to chunk information into subtopics, each in a digestible node, and link the nodes—but remember that hypertext takes longer to download and print (Nielsen, "Be Succinct"). Structure each hypertext node as an "inverse pyramid," in which you begin with the conclusion (Nielsen, "How Users Read"). The inverted pyramid works like a newspaper article, in which the major news/conclusion appears first (say, "The jury deliberated only two hours before returning a guilty verdict"), followed by the details (Nielsen, "Inverted Pyramids"). Last but not least, use restraint: Give users the opportunity to receive *less* information (Outing). Think hard about what users need and give them only that.

To identify academic and research sites that offer excellent content, go to the *Internet Scout Project* at <www.scout.cs.wisc.edu/about> (Figure 19.9).

10. *Sharpen the style.* Make the online text at least 50 percent shorter than its hard copy equivalent. Try to summarize (Chapter 11). Use short sentences and paragraphs. Avoid "marketese" or promotional language that exaggerates ("breakthrough," "revolutionary," "cutting edge").

Checking, Testing, and Monitoring Your Site

1. *Check your site.* Double-check the accuracy of numbers, dates, data, and such; check for broken links; and check for correct spelling, grammar, and so on.

2. *Attend to legal considerations.* Have your legal department approve all material before you post it (Wilkinson 33). Obtain written permission before linking to other Web pages or borrowing any graphic element from another site. Display a privacy notice that explains how each transaction is being recorded, collected, and used. To protect your own intellectual property, display a copyright notice on every page of the site (Evans, "Whose" 48, 50). For more on Internet legal issues (copyright, fair use, privacy, and so on) go to <www.publaw.com>.

19.4

For more on usability testing visit <www.ablongman.com/lannonweb>

3. *Test your site for usability.* Test for usability with unfamiliar users (beta testing) and keep track of their problems and questions. What do users like and dislike? Can they navigate effectively to get to what they need? Are the icons recognizable? Is the site free of needless complexity or interactivity? Test your document with various browsers to be sure it can be downloaded.

4. *Maintain your site.* Review the site regularly, update often, and redesign as needed. If the site accommodates email queries, respond within one business day (Dulude 117).

> **NOTE** *These guidelines scratch only the surface of Web site design issues. For detailed advice consult these resources:* The Yale CAIM Style Guide *at* <www.info.med.yale.edu/caim/manual> *or* IBM's Web Design Guidelines *at* <www.–3.ibm.com/easy/eou/>.

PRIVACY ISSUES IN ONLINE COMMUNICATION

Information sharing between computers makes the Internet and World Wide Web possible. For instance, when someone visits a site, the host computer needs to know what browser is being used. Also, for improved client service, a host site often tracks the links visitors follow, the files they open or download, and the pages they visit most often (Reichard 106). This user information is captured via

FIGURE 19.8
Listing of Award-Winning Sites
The Webby Awards are the equivalent of the Oscars for Web sites. This site includes archives of previous winners, offering a revealing chronicle of the evolution of Web design.
Source: The Webby Awards. <www. webbyawards.com/ main>, 2003. Reprinted with permission.

FIGURE 19.9
A Listing of Sites That Offer High-Quality Content
Based at the University of Wisconsin-Madison, the *Internet Scout Project* publishes its weekly *Scout Report,* providing reviews and links to top Web sites in Science and Engineering, Social Sciences and Humanities, and Business and Economics.

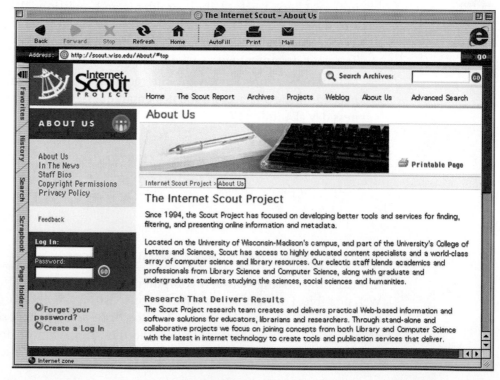

"cookies" (files the Web site sends to any computer that has connected to the host site), which record that person's usage data. Too often, however, more information gets "shared" than the user intended (James-Catalano 32). Commercial U.S. sites routinely share customer information with other companies.

Servers and sites often display privacy notices explaining how usage patterns or transactions are being recorded, collected, and used. But even the most stringent privacy policies offer only limited protection. Any Internet transaction is routed through various browsers and servers and can be intercepted anywhere along the way.

In the U.S. workplace, electronic monitoring of employees is becoming standard practice. Some types of workplace monitoring presumably have legitimate purposes. Page 402, for example, lists arguments for monitoring of workplace email. Employers claim they have valid reasons for monitoring workplace Web sites as well:

Claims in support of monitoring workplace Web sites

- *Troubleshooting.* Monitoring software (*AlertPage,*™ *Net.Medic*™) can scan a company site for broken links, and identify server glitches, software bugs, modem problems, or faulty hardware connections (Reichard 106).
- *Productivity.* Companies track intranet use for the number of queries per employee, types of questions asked, by whom, and the length of time required for employees to find what they need. These data help Webmasters decide whether the search mechanism (user interface) can be improved or whether online documents can be organized or written more clearly (Cronin, "Knowing How Employees Use the Intranet" 103). Monitoring can also reveal software bugs or recurring errors made by employees who might benefit from further training.
- *Security.* Software can track employees' visits to other Web sites, as well as files opened for recreational or personal use, email sent and received, and can even provide snapshots of an employee's computer screen (Karaim 73). Access to unauthorized Web sites can be denied and the employer can be informed about the employee's attempt. Such monitoring can be a justifiable precaution against employee theft, drug abuse, security violations (such as publishing trade secrets on the Internet)—or wasted time. For example, U.S. businesses lose an estimated 26 million worker hours yearly to computer game-playing by employees (Hutheesing 369).

Beyond these legitimate uses, monitoring also carries potential for the abuse of personal privacy.

Privacy abuses in workplace monitoring

- Employers have more freedom to violate employee privacy than the police (Karaim 72). Andre Bacard, author of *The Computer Privacy Handbook,* notes that supervisors can "tap an employee's phones, monitor her e-mail, watch her on closed-circuit TV, and search her computer files, without giving her notice" (qtd. in Karaim 72).

- Some companies notify their employees that their electronic transactions are subject to monitoring, but many do not.
- Even face-to-face transactions are subject to monitoring: An electronic "Active Badge" tracks employees as they move about their work site, recording how much time they spend in the bathroom or at the water cooler and who they talk to during the work day (Karaim 72).

☑ CHECKLIST for Usability of Web Sites

(Numbers in parentheses refer to first page of discussion.)

Accessibility
- ☐ Is the site easy to enter, navigate, and exit? (466)
- ☐ Is required scrolling kept to a minimum? (468)
- ☐ Is downloading speed reasonable? (466)
- ☐ If interaction is offered, is it useful—not superfluous? (468)
- ☐ Does the site avoid overwhelming the user with excessive choices? (474)

Content
- ☐ Are all needed explanations, error messages, and help screens provided? (468)
- ☐ Is the time of each update clearly indicated? (475)
- ☐ Is everything accurate and up-to-date? (475)
- ☐ Are links connected to high-quality sites? (468)
- ☐ Does everything belong (nothing excessive or superfluous or needlessly complex)? (474)
- ☐ Is an email button or other contact method prominently displayed? (468)
- ☐ Does the content accommodate international users? (472)

Arrangement
- ☐ Is the key part of the message on the first page? (474)
- ☐ Are navigation bars, hot buttons, help options, and links to PDF files clearly displayed and explicitly labeled? (474)
- ☐ Are links easily navigated—backward and forward—with back links to the home page? (474)

Writing and Page Design
- ☐ Is the text easy to scan, with short sentences and paragraphs, and do headings, lists, typestyles, and color highlight important material? (473)
- ☐ Is overall word count roughly one-half of the hard copy equivalent? (475)
- ☐ Is the tone reasonable and restrained—free of overstatement and "marketese"? (475)

Graphics and Special Effects
- ☐ Is each graphic easy to download? (468)
- ☐ Is each graphic backed up by a text-only version? (468)
- ☐ Is each graphic or special effect necessary? (474)

Legal Considerations
- ☐ Does the site display a privacy notice that explains how the transaction is being recorded, collected, and used? (475)
- ☐ Does each page of the site display a copyright notice? (475)
- ☐ Has written permission been obtained for each link to other sites and for each graphic element borrowed from another site? (475)
- ☐ Has all posted material received prior legal approval? (475)

INDIVIDUAL OR COLLABORATIVE PROJECTS

1. Consult the previous checklist and evaluate a Web site for usability. You might select a site at your school or place of employment. You might begin by deciding on specific information you seek (such as "internship opportunities" or "special programs" or "campus crime statistics" or "average SAT scores of admitted students") and use this as a basis for assessing the site's accessibility, content, arrangement, and so on.

   Complete your evaluation and report any problems, or suggest improvements in a memo to a designated decision maker. (Your instructor might ask different class groups to evaluate the same site and to compare their findings in class.)

2. Download and print pages from a Web site. Edit these pages to improve their layout and writing style. Submit copies to your instructor.

3. Examine Web sites from three or four competing companies (say, computer makers IBM™, Apple™, Gateway™, Dell™, and Compaq™—or automakers, and so on). Which site do you think is the most effective; the least effective; why? Report your findings in a memo to your classmates.

4. Think of a specific procedure for which you might need help as you prepare a document (say, positioning text and graphics on a page or creating a table). Compare your word-processing software's online help information on this topic with the information in the paper manual. Which version is easier to use? In which can you find the help you need more quickly? Write a short report comparing the two media. Illustrate your comparison with hard copy examples and printouts of online examples.

5. 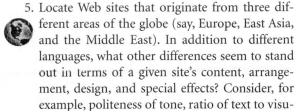Locate Web sites that originate from three different areas of the globe (say, Europe, East Asia, and the Middle East). In addition to different languages, what other differences seem to stand out in terms of a given site's content, arrangement, design, and special effects? Consider, for example, politeness of tone, ratio of text to visuals, use of colors and type styles, privacy policies, and relative ease of navigation.

   Summarize your main points, bookmark each site, and be prepared to discuss and illustrate your findings in class, preferably via interactive demonstration on the computer. If this is impossible, distribute printouts.

6. Return to Figure 19.6 and make a list of the specific features that make this an award-winning Web page. Discuss and illustrate your findings in class by using printouts or via interactive demonstration on the computer.

SERVICE-LEARNING PROJECT

Working in groups, offer to design or redesign a Web site for your school or for a community service organization.

20
Technical Definitions

PURPOSE OF TECHNICAL DEFINITIONS

LEVELS OF DETAIL IN A DEFINITION

EXPANSION METHODS

SITUATIONS REQUIRING DEFINITIONS

PLACEMENT OF DEFINITIONS

GUIDELINES for Defining Clearly and Precisely

CHECKLIST for Usability of Definitions

Definitions explain a term or concept that is specialized or unfamiliar to an audience. In many cases, a term may have more than one meaning, and a clearly written definition tells audiences exactly how the term is being used. Unless you are sure that your audience knows the exact meaning you intend, always define a term the first time you use it.

PURPOSE OF TECHNICAL DEFINITIONS

What users of a technical definition want to know

Definitions answer the question *What, exactly, are we talking about?* by spelling out the precise meaning of a term that can be interpreted in different ways; for example, a person buying a new computer needs to understand exactly what "manufacturer's guarantee" or "expandable memory" means in the context of that purchase.

Definitions can also answer the question *What, exactly, is it?* by explaining what makes an item, concept, or process unique; for example, an engineering student needs to understand the distinction between "elasticity" and "ductility." Inside or outside any field, people have to grasp precisely what "makes a thing what it is and distinguishes that thing from all other things" (Corbett 38).

Definitions have legal implications

Contracts are detailed (and legally binding) definitions of the specific terms of an agreement. If you lease an apartment or a car, for example, the printed contract will define both the *lessee's* and *lessor's* specific responsibilities. An employment contract will spell out responsibilities for both employer and employee. Many other documents, such as employee handbooks, are considered implied contracts ("Handbooks" 5). In preparing an employee handbook for your company, you would need to define such terms as "acceptable job performance" on the basis of clear objectives that each employee can understand, such as "submitting weekly progress reports, arriving on time for meetings, and so on ("Performance Appraisal" 5–6). Because you are legally responsible for any document you prepare, clear and precise definitions are essential.

Definitions have ethical implications

Definitions have ethical requirements, too. For example, on January 28, 1986, the space shuttle *Challenger* exploded 73 seconds after launch, killing all seven crew members. (Two rubber O-ring seals in a booster rocket had failed, allowing hot exhaust gases to escape and igniting the adjacent fuel tank.) Hours earlier—despite vehement objections from the engineers—management had decided that going ahead with the launch was a risk worth taking. This definition of "acceptable risk" was based not on the engineering facts but rather on bureaucratic pressure to launch on schedule. Agreeing on meaning in such cases rarely is easy, but you are ethically bound to convey an accurate interpretation of the facts as you understand them.

Definitions have societal implications

Clear and accurate definitions help the general public understand and evaluate complex technical and social issues. For example, we hear and read plenty about the debates over genetic engineering. But as a first step in understanding this debate, we would need at least this basic definition:

A general but
informative
definition

> Genetic engineering refers to [an experimental] technique through which genes can be isolated in a laboratory, manipulated, and then inserted stably into another organism. Gene insertion can be accomplished mechanically, chemically, or by using biological vectors such as viruses. (Office of Technology Assessment 20)

Of course, to follow the debate, we would need increasingly more detailed information (about specific procedures, risks, benefits, and so on). But the above definition gets us started, by enabling us to *visualize* the basic concept.

LEVELS OF DETAIL IN A DEFINITION

How much detail will your audience need to grasp your exact meaning? Can you define the term by using a synonym or will you need a full sentence, an entire paragraph, or even several pages?

Parenthetical Definition

Often, you can clarify the meaning of a word by using a more familiar synonym or a clarifying phrase:

Parenthetical
definitions

> The *leaching field* (sievelike drainage area) requires crushed stone.
>
> The trees on the site are mostly *deciduous* (shedding foliage at season's end).

NOTE   *Be sure that the synonym or explanatory phrase will clarify your meaning, not obscure it. Don't say:*

> A tumor is a neoplasm.
>
> A solenoid is an inductance coil that serves as a tractive electromagnet.

(The solenoid definition would be appropriate for an engineering manual, but too specialized for general readers.) Do say:

> A tumor is a growth of cells that occurs independently of surrounding tissue and serves no useful function.
>
> A solenoid is a coil that converts electrical energy to magnetic energy capable of performing mechanical functions.

In an electronic document, such as a Web page or online help system, these types of short definitions can easily be linked to the main word or phrase. The user who clicks on "leaching field," say, would go to a window that contains the definition and other important information.

Sentence Definition

More complex terms may require a *sentence definition* (which may be stated in more than one sentence). These definitions follow a fixed pattern: (1) the name of the item to be defined, (2) the class to which the item belongs, and (3) the features that differentiate the item from all others in its class.

Elements of sentence definitions

ITEM	CLASS	DISTINGUISHING FEATURES
carburetor	a mixing device	in gasoline engines that blends air and fuel into a vapor for combustion within the cylinders
diabetes	a metabolic disease	caused by a disorder of the pituitary gland or pancreas and characterized by excessive urination, persistent thirst, and inability to metabolize sugar
brief	a legal document	containing all the facts and points of law pertinent to a specific case, and filed by an attorney before the case is argued in court
stress	an applied force	that strains or deforms a body
fiber optics	a technology	that uses light energy to transmit voices, video images, and data via hair-thin glass fibers

These elements are combined into one or more complete sentences:

A complete sentence definition

Diabetes is a metabolic disease caused by a disorder of the pituitary gland or pancreas. This disease is characterized by excessive urination, persistent thirst, and inability to metabolize sugar.

Sentence definition is especially useful if you need to stipulate your precise working meaning of a term that has several possible meanings. State your working definitions at the beginning of your document:

A working definition

Throughout this report, the term "disadvantaged student" means . . . .

Brief definitions are fine when the audience requires only a general understanding. For example, the sentence definition on page 482 about the leaching field might be adequate in a progress report to a client whose house you are building. But a report for the public health department on groundwater contamination from leaching fields would call for an expanded definition.

Expanded Definition

The sentence definition of "solenoid" on page 482 is good for a layperson who simply needs to know what a solenoid is. An instruction manual for mechanics, however, would define solenoid in much greater detail (page 490); mechanics need to know how a solenoid works and how to use and repair it.

The problem with defining an abstract and general word, such as "condominium" or "loan," is different. "Condominium" is a vaguer term than leaching field (leaching field A is pretty much like leaching field B) because the former refers to many types of ownership agreements, and so requires expanded definition for almost any audience.

Depending on audience and purpose, an expanded definition may be a short paragraph or may extend to several pages. For example, Figure 20.1, aimed at a general audience, employs only two paragraphs (aided by visuals) to define the differences between two weapons of mass destruction. However, if a complex device, such as a digital dosimeter (used for measuring radiation exposure), is being introduced to an audience who needs to understand how this instrument works, your definition would require at least several paragraphs, if not pages.

FIGURE 20.1 An Expanded Definition.
In this example, two items are defined, to clarify an important distinction for the general public.
Source: From "Weapons of Mass Disruption," text by Michael A. Levi and Henry C. Kelly, illustrations by Sara Chen. Published in Scientific American, November 2002. Reprinted with permission.

DIRTY VERSUS NUCLEAR BOMBS

People sometimes confuse radiological with nuclear weapons.

A DIRTY BOMB is likely to be a primitive device in which TNT or fuel oil and fertilizer explosives are combined with highly radioactive materials. The detonated bomb vaporizes or aerosolizes the toxic isotopes, propelling them into the air.

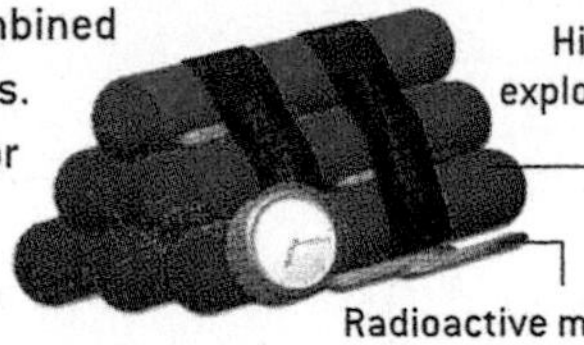

High explosives

Radioactive materials

A FISSION BOMB is a more sophisticated mechanism that relies on creating a runaway nuclear chain reaction in uranium 235 or plutonium 239. One type features tall, inward-pointing pyramids of plutonium surrounded by a shell of high explosives. When the bomb goes off, the explosives produce an imploding shock wave that drives the plutonium pieces together into a sphere containing a pellet of beryllium/polonium at the center, creating a critical mass. The resulting fission reaction causes the bomb to explode with tremendous force, sending high-energy electromagnetic waves and fallout into the air.

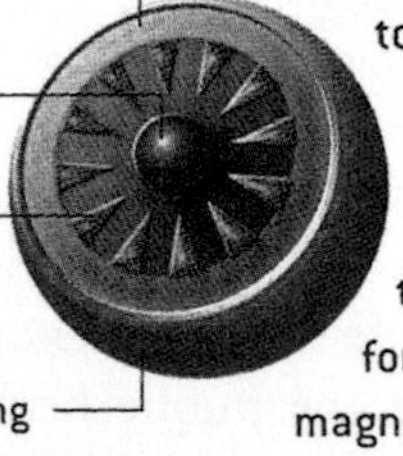

High explosives
Beryllium/polonium core
Plutonium pieces
Heavy casing

EXPANSION METHODS

How you expand a definition depends on the audience questions you can anticipate (Figure 20.2). Begin with a sentence definition, and then select from the following expansion strategies.

**FIGURE 20.2
Directions in
Which a
Definition Can
Be Expanded**

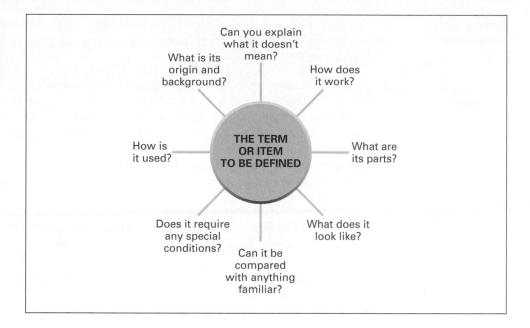

Etymology

Sometimes, a word's origin (its development and changing meanings) can help users understand its meaning. "Biological control" of insects, for example, is derived from the Greek "bio," meaning "life" or "living organism" and the Latin "contra," meaning "against" or "opposite." Biological control, then, is the use of living organisms against insects. College dictionaries contain etymological information, but your best bets are *The Oxford English Dictionary* (or its Web site) and dictionaries of science, technology, and business.

Some technical terms are acronyms, derived from the first letters or parts of several words. *Laser* is an acronym for *light amplification by stimulated emission of radiation.*

Other terms developed as jargon. For instance, "bug" (jargon for "programming error") comes from an early computer malfunction at Harvard caused by a dead bug blocking the contacts of an electrical relay. Because programmers typically hated to admit mistakes, the term became a euphemism for "error." And "debugging," of course, means to eliminate errors in a program.

History and Background

The meaning of specialized terms such as "radar," "bacteriophage," "silicon chips," or "x-ray" can often be clarified through a background discussion: discovery or history of the concept, development, method of production, applications, and so on. Specialized encyclopedias are a good background source.

"Where did it come from?"

> The idea of lasers . . . dates back as far as 212 B.C., when Archimedes used a [magnifying] glass to set fire to Roman ships during the siege of Syracuse. (Gartaganis 22)

"How was it perfected?"

> The early researchers in fiber optic communications were hampered by two principal difficulties—the lack of a sufficiently intense source of light and the absence of a medium which could transmit this light free from interference and with a minimum signal loss. Lasers emit a narrow beam of intense light, so their invention in 1960 solved the first problem. The development of a means to convey this signal was longer in coming, but scientists succeeded in developing the first communications-grade optical fiber of almost pure silica glass in 1970. (Stanton 28)

For students and researchers who want in-depth information, history and background is appropriate. However, for users trying to perform a task, history and background can be cumbersome and unnecessary. If you wanted to install a new modem, for example, you might be interested in a quick sentence explaining that "modem" stands for "modulator-demodulator." But you would not really care about the history of how modems were developed.

Negation

Some definitions can be clarified by an explanation of what the term *does not* mean:

"What does this term not mean?"

> Raw data is not "information"; data only becomes information after it has been evaluated, interpreted, and applied.

Operating Principle

Anyone who wants to use a product correctly will need to know how it operates:

"How does it work?"

> A clinical thermometer works on the principle of heat expansion: As the temperature of the bulb increases, the mercury inside expands, forcing a mercury thread up into the hollow stem.

> Air-to-air solar heating involves circulating cool air, from inside the home, across a collector plate (heated by sunlight) on the roof. This warmed air is then circulated back into the home.

> Basically, a laser [uses electrical energy to produce] coherent light, light in which all the waves are in phase with each other, making the light hotter and more intense. (Gartaganis 23)

Even abstract concepts or processes can be explained on the basis of their operating principle:

> Economic inflation is governed by the principle of supply and demand: If an item or service is in short supply, its price increases in proportion to its demand.

Analysis of Parts

When users need to understand a complex item or concept, be sure to explain each part or element:

"What are its parts?"

> The standard frame of a pitched-roof wooden dwelling consists of floor joists, wall studs, roof rafters, and collar ties.

> Psychoanalysis is an analytic and therapeutic technique consisting of four parts: (1) free association, (2) dream interpretation, (3) analysis of repression and resistance, and (4) analysis of transference.

In discussing each part, of course, you would further define specialized terms such as "floor joists" and "repression."

Analysis of parts is particularly useful for helping laypersons understand a technical subject. This next analysis helps explain the physics of lasing by dividing the process into three discrete parts:

> 1. [Lasers require] a source of energy, [such as] electric currents or even other lasers.
> 2. A resonant circuit . . . contains the lasing medium and has one fully reflecting end and one partially reflecting end. The medium—which can be a solid, liquid, or gas—absorbs the energy and releases it as a stream of photons [electromagnetic particles that emit light]. The photons . . . vibrate between the fully and partially reflecting ends of the resonant circuit, constantly accumulating energy—that is, they are amplified. After attaining a prescribed level of energy, the photons can pass through the partially reflecting surface as a beam of coherent light and encounter the optical elements.
> 3. Optical elements—lenses, prisms, and mirrors—modify size, shape, and other characteristics of the laser beam and direct it to its target. (Gartaganis 23)
>
> Figure 1 shows the three parts of a laser.

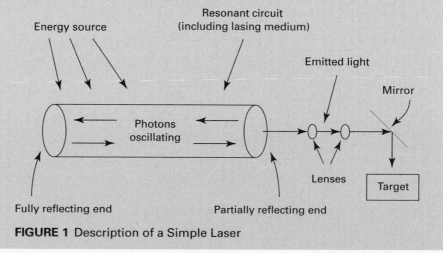

FIGURE 1 Description of a Simple Laser

Visuals

Well-labeled visuals (such as the laser drawing) help clarify definitions. Always introduce your visual and explain it. If your visual is borrowed, credit the source. Unless the visual takes up one whole page or more, do not place it on a separate page. Include the visual near its discussion.

Comparison and Contrast

By comparing or contrasting new information to information your audience already understands, you help build a bridge between their current knowledge and the new ideas. For example, for a group of nonexperts, you could explain how earthquakes start by using this *analogy* (a type of comparison, discussed on page 273):

"Does it resemble anything familiar?"

> Imagine an enormous block of gelatin with a vertical knife slit through the middle of its lower half. Gigantic hands are slowly pushing the right side forward and pulling the left side back along the slit, creating a strain on the upper half of the block that eventually splits it. When the split reaches the upper surface, the two halves of the block spring apart and jiggle back and forth before settling into a new alignment. ("Earthquake Hazard Analysis" 8)

> The average diameter of an optical cable is around two-thousandths of an inch, making it about as fine as a hair on a baby's head (Stanton 29–30).

Here is a contrast between optical fiber and conventional copper cable:

"How does it differ from comparable things?"

> Beams of laser light coursing through optical fibers of the purest glass can transmit many times more information than [conventional] communications systems. . . . A pair of optical fibers has the capacity to carry more than 10,000 times as many signals as conventional copper cable. A $\frac{1}{2}$-inch optical cable can carry as much information as a copper cable as thick as a person's arm. . . .
>
> Not only does fiber optics produce a better signal, [but] the signal travels farther as well. All communications signals experience a loss of power, or attenuation, as they move along a cable. This power loss necessitates placement of repeaters at one- or two-mile intervals of copper cable in order to regenerate the signal. With fiber, repeaters are necessary about every thirty or forty miles, and this distance is increasing with every generation of fiber. (Stanton 27–28)

Here is a combined comparison and contrast:

"How is it both similar and different?"

> Fiber optics technology results from the superior capacity of light waves to carry a communications signal. Sound waves, radio waves, and light waves can all carry signals; their capacity increases with their frequency. Voice frequencies carried by telephone operate at 1000 cycles per second, or hertz. Television signals transmit at about 50 million hertz. Light waves, however, operate at frequencies in the hundreds of trillions of hertz. (Stanton 28)

Required Materials or Conditions

Some items or processes need special materials and handling, or they may have other requirements or restrictions. An expanded definition should include this important information.

"What is needed to make it work (or occur)?"

> Besides training in engineering, physics, or chemistry, careers in laser technology require a strong background in optics (study of the generation, transmission, and manipulation of light).

Abstract concepts might also be defined in terms of special conditions:

> To be held guilty of libel, a person must have defamed someone's character through written or pictorial statements.

Example

Familiar examples showing types or uses of an item can help clarify your definition. This example shows how laser light is used as a heat-generating device:

"How is it used or applied?"

> Lasers are increasingly used to treat health problems. Thousands of eye operations involving cataracts and detached retinas are performed every year by ophthalmologists. . . . Dermatologists treat skin problems. . . . Gynecologists treat problems of the reproductive system, and neurosurgeons even perform brain surgery—all using lasers transmitted through optical fibers. (Gartaganis 24–25)

The next example shows how laser light is used to carry information:

> The use of lasers in the calculating and memory units of computers, for example, permits storage and rapid manipulation of large amounts of data. And audiodisc players use lasers to improve the quality of the sound they reproduce. The use of optical cable to transmit data also relies on lasers. (Gartaganis 25)

And this final example shows how optical fiber can relay a video signal:

> Acting, in essence, as tiny cameras, optical fibers can be inserted into the body and relay an image to an outside screen. (Stanton 28)

Examples are a powerful communication tool—as long as you tailor the examples to your audience's level of understanding.

Whichever expansion strategies you use, be sure to document your information sources.

NOTE

An increasingly familiar (and user-friendly) format for expanded definition, especially for Web users, is a listing of Frequently Asked Questions (FAQ), which organizes chunks of information as responses to questions users are likely to ask. This question-and-answer format creates a conversational style and conveys to users the sense that their particular concerns are being addressed. Consider using a FAQ list whenever you want to increase user interest and decrease resistance. For a hard copy example, see page 144. For a Web-based example, see Figure 20.4, page 495.

SITUATIONS REQUIRING DEFINITIONS

The following definitions employ expansion strategies appropriate to their audiences' needs (and labeled in the margin). Each definition, like a good essay, is unified and coherent: Each paragraph is developed around one main idea and logically connected to other paragraphs. Visuals are incorporated. Transitions emphasize the connection between ideas. Each definition is at a level of technicality that connects with the intended audience.

This example is preceded by an audience and use profile based on the worksheet on page 36.

An Expanded Definition for Semitechnical Readers

Audience and Use Profile.   The intended users of this material are beginning student mechanics. Before they can repair a solenoid, they will need to know where the term *solenoid* comes from, what a solenoid looks like, how it works, how its parts operate, and how it is used. This definition is designed as an *introduction,* so it offers only a general (but comprehensive) view of the mechanism.

Because the users are not engineering students, they do *not* need details about electromagnetic or mechanical theory (e.g., equations or graphs illustrating voltage magnitudes, joules, lines of force). ❑

EXPANDED DEFINITION: SOLENOID

Formal sentence definition

Etymology

A solenoid is an electrically energized coil that forms an electromagnet capable of performing mechanical functions. The term "solenoid" is derived from the word "sole," which in reference to electrical equipment means "a part of," or "contained inside, or with, other electrical equipment." The Greek word *solenoides* means "channel," or "shaped like a pipe."

Description and analysis of parts

A simple plunger-type solenoid consists of a coil of wire attached to an electrical source, and an iron rod, or plunger, that passes in and out of the coil along the axis of the spiral. A return spring holds the rod outside the coil when the current is deenergized, as shown in Figure 1.

Special conditions and operating principle

When the coil receives electric current, it becomes a magnet and thus draws the iron rod inside, along the length of its cylindrical center. With a lever attached to its end, the rod can transform electrical energy into mechanical force. The

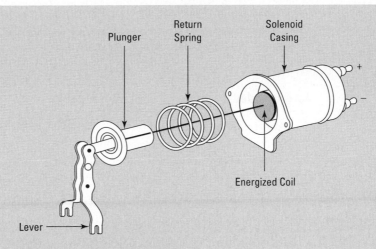

FIGURE 1 Exploded View of a Plunger-Type Solenoid

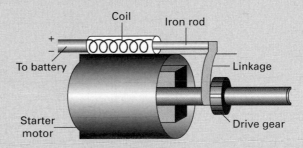

FIGURE 2 Side View of Solenoid and Starter Motor

amount of mechanical force produced is the product of the number of turns in the coil, the strength of the current, and the magnetic conductivity of the rod.

Example and analysis of parts

The plunger-type solenoid in Figure 1 is commonly used in the starter-motor of an automobile engine. This type is $4\frac{1}{2}$ inches long and 2 inches in diameter, with a steel casing attached to the casing of the starter-motor. A linkage (pivoting lever) is attached at one end to the iron rod of the solenoid, and at the other end to the drive gear of the starter, as shown in Figure 2. When the

Explanation of visual

ignition key is turned, current from the battery is supplied to the solenoid coil, and the iron rod is drawn inside the coil, thereby shifting the attached linkage. The linkage, in turn, engages the drive gear, activated by the starter-motor, with the flywheel (the main rotating gear of the engine).

Comparison of sizes and applications

Because of the solenoid's many uses, its size varies according to the work it must do. A small solenoid will have a small wire coil, hence a weak magnetic field. The larger the coil, the stronger the magnetic field; in this case, the rod in the solenoid can do harder work. An electronic lock for a standard door would, for instance, require a much smaller solenoid than one for a bank vault.

The audience for the following definition (an entire community) is too diverse to define precisely, so the writer wisely addresses the lowest level of technicality— to ensure that all readers will understand.

An Expanded Definition for Nontechnical Readers

Audience and Use Profile.   The following definition is written for members of a community whose water supply (all obtained from wells, because the town has no reservoir) is doubly threatened: (1) by chemical seepage from a recently discovered toxic dump site, and (2) by a two-year drought that has severely depleted the water table. This definition forms part of the introduction to a report that analyzes the severity of the problems and explores possible solutions.

To understand the problems, these users first need to know what a water table is, how it is formed, what conditions affect its level and quality, and how it figures into town planning decisions. The concepts of *recharge* and *permeability* are vital to understanding the problem here, so these terms are defined parenthetically. This audience has no interest in geological or hydrological (study of water resources) theory. They simply need the broadest possible picture. ❑

Formal sentence definition

Example

EXPANDED DEFINITION: WATER TABLE

The water table is the level below the earth's surface at which the ground is saturated with water. Figure 1 shows a typical water table that might be found in the East. Wells driven into such a formation will have a water level identical to that of the water table.

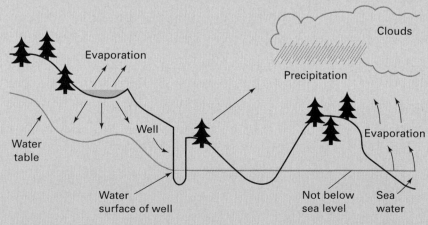

FIGURE 1 A Typical Water Table (Eastern United States)

Operating
principle

> The world's freshwater supply comes almost entirely as precipitation that originates with the evaporation of sea and lake water. This precipitation falls to earth and follows one of three courses: it may fall directly onto bodies of water, such as rivers or lakes, where it is directly used by humans; it may fall onto land, and either evaporate or run over the ground to the rivers or other bodies of water; or it may fall onto land, be contained, and seep into the earth. The latter precipitation makes up the water table.

Comparison

> Similar in contour to the earth's surface above it, the water table generally has a level that reflects such features as hills and valleys. Where the water table intersects the ground surface, a stream or pond results.

Operating
principle

> A water table's level, however, will vary, depending on the rate of recharge (replacement of water). The recharge rate is affected by rainfall or soil permeability (the ease with which water flows through the soil). A water table therefore is never static; rather, it is the surface of a body of water striving to maintain a balance between the forces which deplete it and those which

Example

> replenish it. In areas of Florida and some western states where the water table is depleted, the earth caves in, leaving sinkholes.

Special
conditions and
examples

> The water table's depth below ground is vital in water resources engineering and planning. It determines an area's suitability for wastewater disposal, or a building lot's ability to handle sewage. A high water table could become contaminated by a septic system. Also, bacteria and chemicals seeping into a water table can pollute an entire town's water supply. Another consideration in water table depth is the cost of drilling wells. These conditions obviously affect an industry's or homeowner's decision on where to locate.

Special
conditions

> The rising and falling of the water table give an indication of the pumping rate's effect on a water supply (drawn from wells) and of the sufficiency of the recharge rate in meeting demand. This kind of information helps water resources planners decide when new sources of water must be made available.

PLACEMENT OF DEFINITIONS

Poorly placed definitions interrupt the information flow. Each time an audience encounters an unfamiliar term or concept, it should be defined in the same area on the page or screen. In a printed text, do this by placing a brief, parenthetical definition immediately after the term or in the right margin. On a Web page, use a hypertext link.

More than three or four definitions on one page or screen will be disruptive. For a paper document, rewrite them as sentence definitions and place them in a "Definitions" section of your introduction or in a glossary. For a Web page, provide a link to a separate glossary page, as in Figure 20.3. Depending on its role in the document, place an expanded definition in one of these locations:

Where to place
an expanded
definition

- If the definition is essential to understanding the whole document, place it in the introduction as in Figure 1.1 (page 4).

- When the definition clarifies a major part of your discussion, place it in that section of your document but avoid doing this too often in a paper document. In a hyperlinked document, such as Figure 20.4, users can click on the item, read about it (and possibly explore deeper links), and then return to their place on the original page.
- If the definition serves only as a reference, place it in an appendix.

In online documentation, one option for making definitions available when they are needed is the "pop-up note": The term to be defined is highlighted in the text, to indicate that its definition can be called up and displayed in a small window on the actual text screen (Horton, "Is Hypertext" 25).

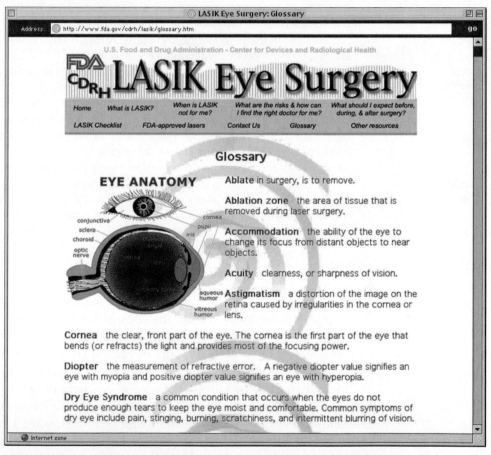

FIGURE 20.3   A Hyperlinked Glossary Page   Links clearly displayed below the site's masthead on each page enable users to access the glossary and then return to any of the main pages in an instant.

Source: U.S. Food and Drug Administration <www.fda.gov/cdrh/lasik/glossary/htm>.

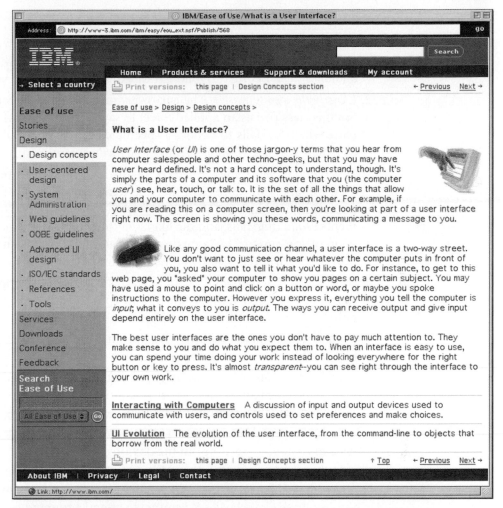

FIGURE 20.4   A Hyperlinked Expanded Definition   Embedded within a hyper-linked network, this one-page definition provides forward links to deeper levels (such as *UI Evolution*) as well as backward links to main pages—all without disrupting the discussion of "Design Concepts."

Source: From <www.ibm.com/easy/eou_ext.nsf/Publish/568>, © IBM Corporation, 1999. Reprinted with permission. All rights reserved.

GUIDELINES for Defining Clearly and Precisely

1. *Decide on the level of detail.* Definitions vary greatly in length and detail; from a few words in parentheses to a multipage document. How much does this audience need in order to follow your explanation or grasp your point?

2. *Classify the item precisely.* The narrower your class, the clearer your meaning. *Stress* is classified as an applied force; to say that stress "is what . . ." or "takes place when . . ." fails to denote a specific classification. Diabetes is precisely classified as a *metabolic disease,* not as a *medical term.*

3. *Differentiate the item accurately.* If the distinguishing features are too broad, they will apply to more than this one item. A definition of *brief* as a "legal document used in court" fails to differentiate brief from all other legal documents (*wills, affidavits,* and the like). On the other hand, a narrow differentiation of "carburetor" as "a mixing device used in boat engines" would be ignoring the carburetor's use in other gasoline engines.

4. *Avoid circular definitions.* Do not repeat, as part of the distinguishing feature, the word you are defining. "Stress is an applied force that places stress on a body" is a circular definition.

5. *Expand your definition selectively.* Begin with a sentence definition and select the best combination of development strategies for your audience and purpose.

6. *Use visuals to clarify your meaning.* No matter how clearly you explain, as the saying goes, a picture can be worth a thousand words—even more so when used with readable, accurate writing.

7. *Know "how much is enough."* Don't insult people's intelligence by giving needless details or spelling out the obvious.

8. *Consider the legal implications of your definition.* What does an "unsatisfactory job performance" mean in an evaluation of a company employee: that the employee could be fired, required to attend a training program, given one or more chances to improve, or what ("Performance Appraisal 3–4)? Failure to spell out your meaning invites a lawsuit.

9. *Consider the ethical implications of your definition.* Be sure your definition of a fuzzy or ambiguous term such as "safe levels of exposure" or "conservative investment" or "acceptable risk" is based on technical fact and not on social pressure. Consider, for example, a recent U.S. cigarette company's claim that cigarette smoking in the Czech Republic promoted "fiscal benefits," defined, in this case, by the fact that smokers die young, thus eliminating pension and health care costs for the elderly!

10. *Place your definition in an appropriate location.* Allow users to access the definition and then return to the main text with as little disruption as possible.

☑ CHECKLIST for Usability of Definitions

(Numbers in parentheses refer to the first page of discussion.)

Content

- ☐ Is the type of definition (parenthetical, sentence, expanded), suited to its purpose and user's needs? (482)
- ☐ Does the definition adequately classify the item? (496)
- ☐ Does the definition clarify, rather than obscure, the meaning? (482)
- ☐ Is the expanded definition adequately developed? (484)
- ☐ Are all data sources documented? (489)
- ☐ Are visuals employed adequately and appropriately? (488)
- ☐ Does the sentence definition describe features that distinguish the item from all other items in the same class? (482)

Arrangement

- ☐ Is the expanded definition unified and coherent (like an essay)? (490)
- ☐ Are transitions adequate? (490)
- ☐ Does the definition appear in the appropriate location? (493)

Style and Page Design

- ☐ Is the definition in plain English? (265)
- ☐ Will the level of technicality connect with the audience? (30)
- ☐ Are sentences clear, concise, and fluent? (244)
- ☐ Is word choice precise? (264)
- ☐ Is the definition grammatical? (Appendix C)
- ☐ Is the definition ethically and legally acceptable? (481)
- ☐ Is page design inviting and accessible? (366)

EXERCISES

 For more exercises, visit <www.ablongman.com/lannon>

1. Sentence definitions require precise classification and differentiation. Is each of these definitions adequate for a layperson? Rewrite those that seem inadequate. Consult dictionaries and encyclopedias as needed.

   a. A bicycle is a vehicle with two wheels.
   b. A transistor is a device used in transistorized electronic equipment.
   c. Surfing is when one rides a wave to shore while standing on a board specifically designed for buoyancy and balance.
   d. Bubonic plague is caused by an organism known as *Pasteurella pestis*.
   e. Mace is a chemical aerosol spray used by the police.
   f. A Geiger counter measures radioactivity.
   g. A cactus is a succulent.
   h. In law, an indictment is a criminal charge against a defendant.
   i. A prune is a kind of plum.
   j. Friction is a force between two bodies.
   k. Luffing is what happens when one sails into the wind.
   l. A frame is an important part of a bicycle.
   m. Hypoglycemia is a medical term.
   n. An hourglass is a device used for measuring intervals of time.
   o. A computer is a machine that handles information with amazing speed.
   p. A Ferrari is the best car in the world.
   q. To meditate is to exercise mental faculties in thought.

2. Standard dictionaries define for the layperson, whereas specialized reference books define for the specialist. Choose an item in your field and copy the definition (1) from a standard dictionary and (2) from a technical reference book. For the technical definition, label each expansion strategy. Rewrite the specialized definition for a layperson.

3. Using reference books as necessary, write sentence definitions for these terms or for terms from your field.

generator
dewpoint
microprocessor
capitalism
local area network
marsh
artificial intelligence
economic inflation
anorexia nervosa
low-impact camping
hemodialysis
modem

gyroscope
coronary bypass
oil shale
chemotherapy
estuary
Boolean logic
classical conditioning
hypothermia
thermistor
aquaculture
nuclear fission

4. Select an item from the list in Exercise 3 or from an area of interest. Identify an audience and purpose. Complete an audience and use profile sheet (page 36). Begin with a sentence definition of the term. Then write an expanded definition for a first-year student in that field. Next, write the same definition for a layperson (client, patient, or other interested party). Leave a margin at the left side of your page to list expansion strategies. Use at least four expansion strategies in each version, including at least one visual or an *art brief* (page 318) and a rough diagram. In preparing each version, consult no fewer than four outside references. Cite and document each source, using one of the documentation styles discussed in Appendix A. Submit, with your two versions, an explanation of your changes from the first version to the second.

5. Figure 20.5 shows a page from a brochure titled *Cogeneration*. The brochure provides an expanded definition for potential users of fuel conservation systems engineered and packaged by Ewing Power Systems. The intended users are plant engineers and other technical experts unfamiliar with cogeneration.

Another page of the brochure is designed in a question-and-answer (FAQ list) format. Figure 20.6 shows parts of that page.

Identify each expansion strategy in Figures 20.5 and 20.6. Is the definition appropriate for a technical audience? Why, or why not? Be prepared to discuss your analysis and evaluation in class.

COLLABORATIVE PROJECTS

1. Divide into small groups by majors or interests. Appoint one person as group manager. Decide on an item, concept, or process that would require an expanded definition for laypersons.

Examples

From computer science: an algorithm, an applications program, artificial intelligence, binary coding, top-down procedural thinking, or systems analysis.

From nursing: a pacemaker, coronary bypass surgery, or natural childbirth.

Complete an audience and use profile (page 36).

Once your group has decided on the appropriate expansion strategies (etymology, negation, etc.), the group manager will assign each member to work on one or two specific strategies as part of the definition. As a group, edit and incorporate the collected material into an expanded definition, revising as often as needed.

Your instructor may stipulate a brochure format for your definition, as in Figure 20.7 (For more on this format, refer to "brochures" in this book's Index.)

The group manager will assign one member to present the definition in class, using either opaque or overhead projection, a large-screen monitor, or photocopies.

SERVICE-LEARNING PROJECT

Revise the flyer/fact sheet you prepared for Chapter 3 page 42 to publicize your public service organization. Use all appropriate expansion strategies to show your readers "Who we are" and "What we do."

Hint: To decrease reader resistance and to help people identify with the issue, consider presenting your definition in the form of a FAQ list.

TECHNICAL CONSIDERATIONS

Turbine generator sets make electricity by converting a steam pressure drop into mechanical power to spin the generator. Conceptually, steam turbines work much the same way as water turbines. Just as water turbines take the energy from water as it flows from a high elevation to a lower elevation, steam turbines take the energy from steam as it flows from high pressure to low pressure. The amount of energy that can be converted to electricity is determined by the difference between the inlet pressure and the exhaust pressure (pressure drop) and the volume of steam flowing through the turbine.

Steam turbines have been used in industry in a variety of applications for decades and are the most common way utilities generate electricity. Exactly how a steam turbine generator can be used in your plant depends upon your circumstances.

IF YOU USE WASTE AS A BOILER FUEL

If you use wood waste or incinerator waste as a boiler fuel you can afford to condense turbine exhaust steam in a condenser. This allows you to convert waste fuel into electricity.

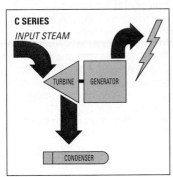

The simplest form of a condensing turbine generator set is the Ewing Power Systems C Series. All surplus steam enters the turbine at high pressure and exhausts to a condenser at a very low pressure, usually a vacuum. Because of the very low exhaust pressure, the pressure drop through the turbine is greater and more energy is extracted from each pound of steam. This is the same basic design as utilities use to produce power. The condenser can be either air or water cooled. In water cooled systems the "cooling" water can be hot enough for use as process hot water or for space heat.

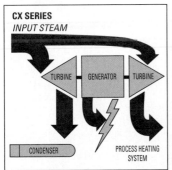

In situations where there is surplus fuel and also a need for low pressure process steam, the Ewing Power Systems CX Series is the system of choice. This arrangement includes a back pressure turbine and a condensing turbine connected to a common generator. Low pressure process or space heating loads are met with the back pressure turbine while surplus steam is directed to the condensing turbine to maximize power production.

IF YOU PURCHASE BOILER FUEL SUCH AS OIL OR GAS

If oil or gas is used as boiler fuel the best use of a turbine is as a replacement for a steam pressure reducing valve. Many plants produce steam at high pressure and then use some or all of the steam at low pressure after passing it through a pressure reducing valve. Other plants have high pressure boilers but run them at low pressure because they do not need high pressure steam for their process. In either case, a turbine generator can turn the pressure drop energy potential into electricity.

The Ewing Power Systems BP Series turbine generator sets are designed for pressure reducing (back pressure) applications. Very little energy is consumed by the turbine, so most of the inlet steam is available for process. The turbine generator uses about 3631 BTU's per hour for each kilowatt-hour produced. At 40 cents per gallon for

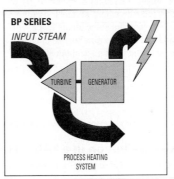

No. 6 fuel oil and 85% boiler efficiency, it will cost about 1.1 cents per kilowatt-hour to generate your own power with a BP Series turbine. Generating costs for gas-fired boilers are similar.

To generate power at this very low cost, all the exhaust steam must be used productively. Generator output is therefore completely governed by process steam demand. For example, if steam is used for space heating you will make more electricity on cold days than on warmer days because more steam will flow through the turbine.

ELECTRICAL CONSIDERATIONS

In most cases the generator will be connected to your plant electrical system and to your utility. This means that you will not give up the security of utility power. It also means that you do not have to generate *all* your own power; most cogeneration systems provide only part of the plant load. The more power the generator is making, the less you buy from the utility. If you make more power than you use, you will be able to sell the excess to the utility. If your generator is off-line for any reason, you will be able to buy power from the utility, just as you do now.

There are two primary generator designs: induction and synchronous. Induction generators are similar to induction motors and are much simpler than synchronous generators. Synchronous generators require more elaborate controls and are usually more expensive but offer the advantage of stand-alone capability. Whereas induction generators cannot operate unless they are connected to a utility grid, synchronous sets can be operated in isolation as emergency units or when it is economically advantageous to avoid interconnection with the utility.

All turbine generator sets from Ewing Power Systems include a complete electrical control panel. Our standard panels meet most utility interconnection requirements and we will customize the panel to meet unusual requirements. Synchronous panels can be built for full utility paralleling, stand-alone capability, or both.

FIGURE 20.5   Expanded Definition in a Technical Brochure
Source: Courtesy of Ewing Power Systems, So. Deerfield, MA 01373.

Q. What is cogeneration?

A. It is the simultaneous production of electricity and useful thermal energy. This means that you can generate electricity with the same steam you are now using for heating or process. *You can use the same steam twice.* In modern usage cogeneration has also come to mean using waste fuel for in-plant electricity generation.

Q. How does it save money?

A. Cogeneration saves money by allowing you to produce your own electricity for a fraction of the cost of utility power. Cogenerated power is cheaper because cogeneration systems are much more efficient than central utility plants. By using the same steam twice, cogeneration systems can achieve efficiencies of up to 80%, whereas the best utilities can do is about 40%.

Q. Are there other benefits to cogeneration?

A. Yes. For companies using waste fuel, elimination of waste disposal can be a very important benefit. Depending upon design, a cogeneration system can provide emergency standby power and can smooth out boiler load swings.

Q. Is cogeneration new?

A. No. It's been done ever since the beginning of the electrification of industrial America. Originally, most electric power was cogenerated by individual manufacturers, not the utilities. In the 1920s and '30s, as cheaper utility electricity became available, cogeneration waned. With cheap oil available, power rates continued to decline through the 1950s and '60s. Then came the 1973–74 Arab Oil Embargo. Everything changed abruptly. Since then electricity prices have risen. Further upward pressure was produced by some utilities' nuclear power plant building programs. Now many companies are getting back to their original source of power: cogeneration.

FIGURE 20.6   Expanded Definition in a FAQ List Format
Source: Courtesy of Ewing Power Systems, So. Deerfield, MA 01373.

What is a generic drug?

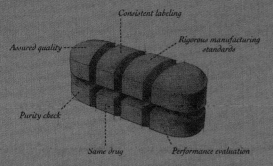

Consistent labeling

Rigorous manufacturing standards

Assured quality

Purity check

Same drug

Performance evaluation

When a brand-name drug's patent protection expires, generic versions of the drug can be approved for sale. The generic version works like the brand-name drug in dosage, strength, performance and use, and must meet the same quality and safety standards. All generic drugs must be reviewed and approved by FDA.

How does FDA ensure that my generic drug is as safe and effective as the brand-name drug?

All generic drugs are put through a rigorous, multi-step review process that includes a review of scientific data on the generic drug's ingredients and performance. FDA also conducts periodic inspections of the manufacturing plant, and monitors drug quality—even after the generic drug has been approved.

If generic drugs and brand-name drugs have the same active ingredients, why do they look different?

Generic drugs look different because certain inactive ingredients, such as colors and flavorings, may be different. These ingredients do not affect the performance, safety or effectiveness of the generic drug. They look different because trademark laws in the U.S. do not allow a generic drug to look exactly like other drugs already on the market.

Is my generic drug made by the same company that makes the brand-name drug?

It is possible. Brand-name firms are responsible for manufacturing approximately 50 percent of generic drugs.

Are generic drugs always made in the same kind of facilities as brand-name drugs?

Yes. All generic drug manufacturing facilities must meet FDA's standards of good manufacturing practices. FDA will not permit drugs to be made in substandard facilities. FDA conducts about 3,500 inspections a year to ensure standards are met.

FDA makes it tough to become a generic drug in America so you can feel confident about taking your generic drugs. If you still want to learn more, talk with your doctor, pharmacist or other health care professional. Or call **1-888-INFO-FDA** or visit **www.fda.gov/cder** today.

U.S. Food and Drug Administration

Generic Drugs: Safe. Effective. FDA Approved.

FIGURE 20.7   A Definition for Laypersons, Designed as a Two-Column Brochure

Source: U.S. Department of Health and Human Services. Food and Drug Administration.

21

Technical Descriptions and Specifications

Technical description combines words and visuals to create a picture—a clear mental image—of a product or a process. Any subject can be visualized from countless perspectives. Therefore, how you describe something—your perspective—depends on your purpose and the user's needs.

PURPOSES AND TYPES OF TECHNICAL DESCRIPTION

What users
of a technical
description want
to know

Description, like definition (Chapter 20), answers the question *What is it?* But to help users "visualize," description answers additional questions that include *What does it do? What does it look like? What is it made of? How does it work? How does it happen?* Definition and description depend on each other and provide the foundation for virtually any type of technical explanation.

Product versus
process
descriptions

Technical descriptions divide into two basic types: product descriptions and process descriptions. Anyone learning to use a particular device (say, a stethoscope) relies on product description. Anyone wanting to understand the steps or stages in a complex event (say, how lightning is produced) relies on process description.

The product description in Figure 21.1, part of an installation and operation manual, gives do-it-yourself homeowners a clear mental image of the overall device and its parts. The accompanying process description in Figure 21.2 shows the device in action.

ELEMENTS OF A USABLE DESCRIPTION

Clear and Limiting Title

Give an
immediate
forecast

The title "A Description of a Velo Ten-Speed Racing Bicycle" promises a comprehensive description. If you intend to describe the braking mechanism only, be sure your title indicates this: "A Description of the Velo's Center-Pull Caliper Braking Mechanism."

21.1
How is technicality
culturally situated?
Find out more at
<www.ablongman.com/
lannonweb>

Appropriate Level of Detail and Technicality

Give users
exactly and only
what they need

Give enough detail to create a clear picture, but omit unnecessary information. Identify your audience and its reasons for using your description.

Assume that you want to describe a specific bicycle model. The picture you create will depend on the details you select. How will your audience use this description? Is this a customer interested in the bike's appearance—its flashy looks and racy style? Is it a repair technician who needs to know how parts operate? Is it a helper in your bicycle shop who needs to know how to assemble this bike? Or is it the bike's manufacturer who needs precise technical specifications?

The descriptions of the hot water maker in Figures 21.1 and 21.2 focus on *what this model looks like, what it's made of,* and *how it works.* Its intended audience of do-it-yourselfers already knows what a hot water maker is and what it does. That audience needs no background. (A description of *how it was put together* appears with the installation and maintenance instructions later in the manual.)

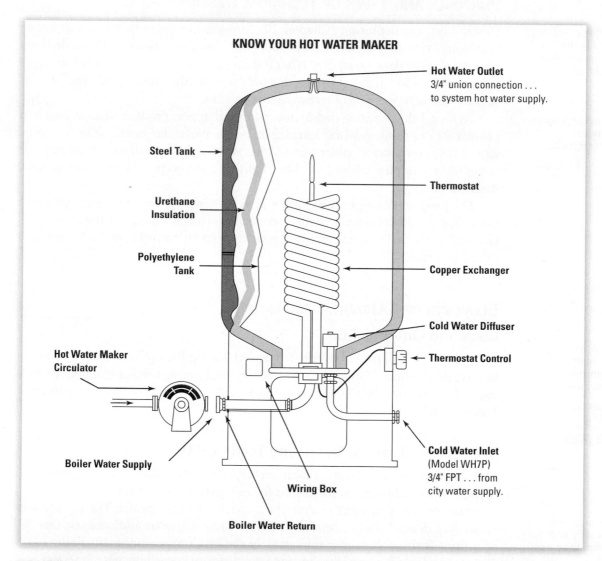

KNOW YOUR HOT WATER MAKER

Hot Water Outlet
3/4" union connection . . .
to system hot water supply.

Steel Tank

Urethane Insulation

Polyethylene Tank

Thermostat

Copper Exchanger

Cold Water Diffuser

Thermostat Control

Hot Water Maker Circulator

Boiler Water Supply

Wiring Box

Boiler Water Return

Cold Water Inlet
(Model WH7P)
3/4" FPT . . . from
city water supply.

FIGURE 21.1 A Product Description   This description allows users to visualize the basic parts of the hot water maker and their relationships.
Source: Courtesy of AMTROL Inc.

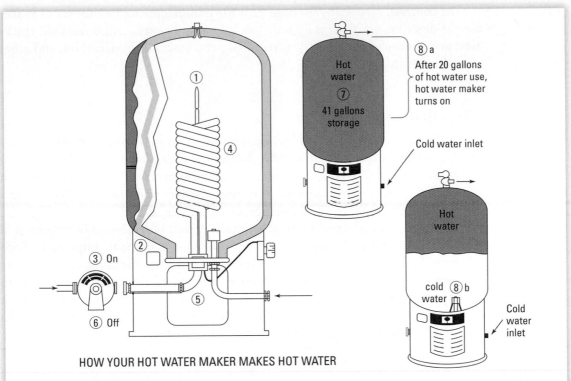

HOW YOUR HOT WATER MAKER MAKES HOT WATER

1. The thermostat calls for energy to make hot water in your Hot Water Maker.

2. The built-in relay signals your boiler/burner to generate energy by heating boiler water.

3. The Hot Water Maker circulator comes on and circulates hot boiler water through the inside of the Hot Water Maker heat exchanger.

4. Heat energy is transferred, or "exchanged" from the boiler water inside the exchanger to the water surrounding it in the Hot Water Maker.

5. The boiler water, after the maximum of heat energy is taken out of it, is returned to the boiler so it can be reheated.

6. When enough heat has been exchanged to raise the temperature in your Hot Water Maker to the desired temperature, the thermostat will de-energize the relay and turn off the Hot Water Maker circulator and your boiler/burner. This will take approximately 23 minutes—when you first start up the Hot Water Maker.

During use, reheating will be approximately 9–12 minutes.

7. You now have 41 gallons of hot water in storage . . . ready for use in washing machines, showers, sinks, etc. This 41 gallons of hot water will stay hot up to 10 hours, if you don't use it, without causing your boiler/burner to come on. (Unless, of course, you need it for heating your home in the winter.)

8. When you do use hot water, you will be able to use approximately 20 gallons, before the Hot Water Maker turns on. Then you will still have 21 gallons of hot water left for use, as your Hot Water Maker "recoups" 20 gallons of cold water. This means, during normal use (3 1/2 GPM Flow), you will never run out of hot water.

You can expect substantial energy savings with your Hot Water Maker, as its ability to store hot water and efficiently transfer energy to make more hot water will keep your boiler off for longer periods of time.

FIGURE 21.2 A Process Description   This description allows users to visualize the sequence of events in producing the hot water.

In contrast, *specifications* (page 520) for manufacturing the hot water maker would describe each part in exacting detail (e.g., the steel tank's required thickness and pressure rating as well as required percentages of iron, carbon, and other constituents in the steel alloy).

Objectivity

Objective description filters out personal impressions to whatever extent is appropriate, focusing instead on observable details ("The drilling crew worked in freezing rain and gale-force winds"). Subjective description colors objective details with personal impressions and metaphors. It usually strives to create a mood or share a feeling about the subject ("The weather was miserable").

Description has ethical implications

Except for promotional writing, technical descriptions should be impartial, if they are to be ethical. But pure objectivity is, of course, humanly impossible. Each observer has a unique perspective on the facts and their meaning, and therefore chooses what to put in and what to leave out. Nonetheless, we are expected to communicate the facts as we know and understand them. Even positive claims made in promotional writing (for example, "reliable," "rugged," and so on in Figure 21.9) should be based on objective and verifiable evidence.

NOTE *Maintaining objectivity does not mean forsaking personal evaluation in cases in which a product may be unsafe or unsound. An ethical communicator "is obligated to express her or his opinions of products, as long as these opinions are based on objective and responsible research and observation" (MacKenzie 3).*

The following suggestions will help you achieve objectivity in writing technical descriptions.

SELECT DETAILS THAT ARE CONCRETE AND SPECIFIC ENOUGH TO CONVEY AN UNMISTAKABLE PICTURE. Focus on details any observer could recognize, details a camera would record. Avoid details that provide no distinct visual image, as in the italicized words below:

Subjective

His office has an *awful* view, *terrible* furniture, and a *depressing* atmosphere.

This next version provides a clear and exact picture:

Objective

His office has broken windows looking out on a brick wall, a rug with a six-inch hole in the center, missing floorboards, broken chairs, and a ceiling with chunks of plaster missing.

Use Precise and Informative Language.   Instead of "large," "long," and "near," give exact measurements, weights, dimensions, and ingredients. Specify location and spatial relationships: "above," "oblique," "behind," "tangential," "adjacent," "interlocking," "abutting," and "overlapping." Specify position: "horizontal," "vertical," "lateral," "longitudinal," "in cross-section," "parallel." Avoid judgmental words ("impressive," "poor"), unless your judgment is requested and can be supported by facts.

Notice how the following precise words help users to *visualize:*

INDEFINITE	PRECISE
a late-model car	a 2006 Lexus ES 300
an inside view	a cross-sectional, cutaway, or exploded view
next to the foundation	adjacent to the right side
a small red thing	a red activator button with a 1-inch diameter and a concave surface

NOTE   *Never confuse precision with overly complicated technical terms or needless jargon. Don't say "phlebotomy specimen" instead of "blood," or "thermal attenuation" instead of "insulation," or "proactive neutralization" instead of "damage control." The clearest writing aimed at laypersons uses precise but plain language—as long as the simpler words do the job.*

Visuals

Let the visual repeat, restate, or reinforce the prose

Use drawings, diagrams, or photographs generously—with captions and labels that help users interpret what they are seeing. Notice how the diagram in Figure 21.3 combines words and images to provide a simple but dynamic picture of a process in action.

NOTE   *Economy in a visual often equals clarity. Avoid verbal and visual clutter. The experiment described in the caption to Figure 21.3, for example, demonstrates why it's important to "minimize distracting detail" (Rowan 214).*

Visuals generated by computers (for example, 3-D drawing or architectural drafting programs) are particularly appropriate for technical descriptions. Other sources for descriptive graphics include clip art, electronic scanning, and downloading from the Internet. (See page 329 for a sampling of useful Web sites and discussion of legal issues regarding the use of computer graphics.)

Clearest Descriptive Sequence

Organize for the user's understanding

Any item or process usually has its own logic of organization, based on (1) the way it appears as a static object, (2) the way its parts operate in order, or (3) the way its

The Process of Lightning

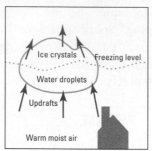

1. Warm moist air rises, water vapor condenses and forms cloud.

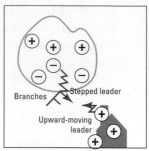

4. Two leaders meet; negatively charged particles rush from cloud to ground.

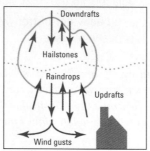

2. Raindrops and ice crystals drag air downward.

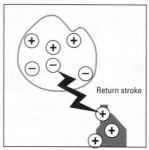

5. Positively charged particles from the ground rush upward along the same path.

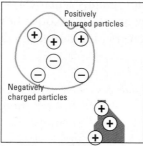

3. Negatively charged particles fall to bottom of cloud.

FIGURE 21.3   A Process Visual That Minimizes Distracting Detail   In an experiment, students who were given this visual alone were better able to understand the process of lightning than students who were given 550 words of text along with the visual.

Source: From Journal of Educational Psychology, *88, by Richard E. Mayer, et al. Copyright © 1996 by the American Psychological Association. Reprinted with permission.*

parts are assembled. These relationships—spatial, functional, and chronological—are discussed below.

A spatial sequence parallels the user's angle of vision in viewing the item

SPATIAL SEQUENCE.   Part of all physical descriptions, a spatial sequence answers these questions: *What is it? What does it do? What does it look like? What parts and material is it made of?* Use this sequence when you want users to visualize a static item or a mechanism at rest (an office interior, a document, the Statue of Liberty, a plot of land, a chainsaw, or a computer keyboard). Can this item best be visualized from front to rear, left to right, top to bottom? (What logical path do the parts create?) A retractable pen would logically be viewed from outside to inside. The specifications in Figure 21.4 (page 522) proceed from the ground upward.

A functional sequence parallels the order in which parts operate

FUNCTIONAL SEQUENCE.   The functional sequence answers: *How does it work?* It is best used in describing a mechanism in action, such as a 35-millimeter camera, a nuclear warhead, a smoke detector, or a car's cruise-control system. The logic of the item is reflected by the order in which its parts function. Like the hot water maker in Figure 21.2, a mechanism usually has only one functional sequence. The stethoscope description on page 510 follows the sequence of parts through which sound travels.

When describing a solar home-heating system, you would begin with the heat collectors on the roof, moving through the pipes, pumping system, and tanks for the heated water, to the heating vents in the floors and walls—from source to outlet. After this functional sequence of operating parts, you could describe each part in a spatial sequence.

A chronological sequence parallels the order in which parts are assembled or stages occur

CHRONOLOGICAL SEQUENCE.   A chronological sequence answers: *How is it put together? How does it work? How does it happen?* Use the chronological sequence for an item that is best visualized in terms of its order of assembly (such as a piece of furniture, an umbrella tent, or a prehung window or door unit). Architects might find a spatial sequence best for describing a proposed beach house to clients; however, they would use a chronological sequence (of blueprints) for specifying for the builder the prescribed dimensions, materials, and construction methods at each stage of the process.

NOTE   *You can combine these sequences as needed. For example, in describing an automobile jack (for a car owner's manual) you would employ a spatial sequence to help users recognize this item, a functional sequence to show them how it works, and a chronological sequence to help them assemble and use it correctly.*

AN OUTLINE AND MODEL FOR PRODUCT DESCRIPTION

Description of a complex mechanism almost invariably calls for an outline. This model is adaptable to any description.

> **I. Introduction: General Description[1]**
>   A. Definition, Function, and Background of the Item
>   B. Purpose (and Audience—for classroom only)
>   C. Overall Description (with general visuals, if applicable)
>   D. Principle of Operation (if applicable)
>   E. Preview of Major Parts
> **II. Description and Function of Parts**
>   A. Part One in Your Descriptive Sequence
>   1. Definition
>   2. Shape, dimensions, material (with specific visuals)
>   3. Subparts (if applicable)
>   4. Function
>   5. Relation to adjoining parts
>   6. Mode of attachment (if applicable)
>   B. Part Two in Your Descriptive Sequence (and so on)
> **III. Conclusion and Operating Description**
>   A. Summary (used only in a long, complex description)
>   B. Interrelation of Parts
>   C. One Complete Operating Cycle

You might modify, delete, or combine certain parts of this outline to suit your subject, purpose, and audience.

Introduction: General Description

Give users only as much background as they need to understand the product.

> ## A DESCRIPTION OF THE STANDARD STETHOSCOPE
>
> **Introduction**
> The stethoscope is a listening device that amplifies and transmits body sounds to aid in detecting physical abnormalities.
>
> This instrument has evolved from the original wooden, funnel-shaped instrument invented by a French physician, R. T. Laënnec, in 1819. Because of his female patients' modesty, he found it necessary to develop a device, other than his ear, for auscultation (listening to body sounds).

Definition and function

History and background

---

[1]In most descriptions, the subdivisions in the introduction can be combined and need not appear as individual headings in the document.

Purpose and
audience

Overall view and
operating
principle

Preview of major
parts

Visual reinforces
the prose

> This description explains to the beginning paramedical or nursing student the structure, assembly, and operating principle of the stethoscope.
>
> The standard stethoscope, roughly 24 inches long and weighing about 5 ounces, consists of a sensitive sound-detecting and amplifying device whose flat surface is pressed against a bodily area. This amplifying device is attached to rubber and metal tubing that transmits the body sound to a listening device inserted in the ear.
>
> The stethoscope's Y-shaped structure contains seven interlocking pieces: (1) diaphragm contact piece, (2) lower tubing, (3) Y-shaped metal piece, (4) upper tubing, (5) U-shaped metal strip, (6) curved metal tubing, and (7) hollow ear plugs. These parts form a continuous unit (Figure 1).

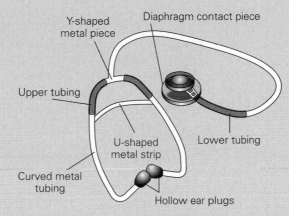

FIGURE 1 Stethoscope with Diaphragm Contact Piece (Front View)

Description and Function of Parts

The body of your text describes each major part. After arranging the parts in sequence, follow the logic of each part. Provide only as much detail as users need.

Readers of this description will use a stethoscope daily, so they need to know how it works, how to take it apart for cleaning, and how to replace worn or broken parts. (Specifications for the manufacturer would require many more technical details—dimensions, alloys, curvatures, tolerances, and so on.)

Definition, size,
shape, and
material

Subparts

> **Description and Function of Parts**
>
> **DIAPHRAGM CONTACT PIECE.** The diaphragm contact piece is a shallow metal bowl, about the size of a silver dollar (and twice its thickness). Various body sounds cause it to vibrate.
>
> Three separate parts make up the piece: hollow steel bowl, plastic diaphragm, and metal frame (Figure 2).

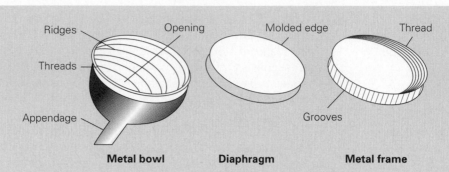

FIGURE 2 Exploded View of a Diaphragm Contact Piece

The stainless steel metal bowl has a concave inner surface, with concentric ridges that funnel sound toward an opening in the tapered base, then out through the hollow appendage. Lateral threads ring the outer circumference of the bowl to accommodate the interlocking metal frame. A fitted diaphragm covers the bowl's upper opening.

The diaphragm is a plastic disk, 2 millimeters thick, 4 inches in circumference, with a molded lip around the edge. It fits flush over the metal bowl and vibrates sound toward the ridges. A metal frame that screws onto the bowl holds the diaphragm in place.

A stainless steel frame fits over the disk and metal bowl. A $\frac{1}{4}$-inch ridge between the inner and outer edge accommodates threads for screwing the frame to the bowl. The frame's outside circumference is notched with equally spaced, perpendicular grooves—like those on the edge of a dime—to provide a gripping surface.

The diaphragm contact piece receives, amplifies, and transmits sound through the system of attached tubing. The piece attaches to the lower tubing by an appendage on its apex (narrow end), which fits inside the tubing.

Function and relation to adjoining parts

Mode of attachment

Each part of the stethoscope, in turn, is described according to its own logic of organization.

Summary and Operating Description

Conclude by explaining how the parts work together to make the whole item function.

Summary and Operating Description

The seven major parts of the stethoscope provide support for the instrument, flexibility of movement for the operator, and ease in use.

In an operating cycle, the diaphragm contact piece, placed against the skin, picks up sound impulses from the body's surface. These impulses cause the plastic diaphragm to vibrate. The amplified vibrations, in turn, are carried through a tube to a dividing point. From here, the amplified sound is carried through two separate but identical series of tubes to hollow ear plugs.

How parts interrelate

One complete operating cycle

A SITUATION REQUIRING PRODUCT DESCRIPTION

The following description of a solar collector, aimed toward a general audience, adapts the previous outline model. The author, Roxanne Payton, is a mechanical engineer specializing in nonpolluting energy technologies. Roxanne prepared this description as part of an informational booklet on solar energy systems distributed by her company, Eco-Solutions.

A Mechanism Description for a Nontechnical Audience

AUDIENCE AND USE PROFILE. The audience here will be homeowners or potential homeowners interested in incorporating solar flat-plate collectors as a heating source. Although many of these people probably lack technical expertise, they presumably have some general knowledge about active solar heating systems and principles. Therefore, this description will focus on the collectors rather than on the entire system, while omitting specific technical data (say, the heat conducting and corrosive properties of copper versus aluminum in the absorber plates). The engineer who designed the collector would include such data in research-and-development reports for the manufacturer. Informed laypersons, however, need only the information that will help them visualize and understand how a basic collector operates. ❑

DESCRIPTION OF A STANDARD FLAT-PLATE SOLAR COLLECTOR

Introduction—General Description

Definition and function

A flat-plate solar collector is an energy gathering device that absorbs sunlight and converts it into heat. Depending on a site's geographical location, a flat-plate collection system can provide between 30 and 80 percent of a home's hot water and space heating.

Background

The flat-plate solar collector has found the widest application in the solar energy industry because it is inexpensive to fabricate, install, and maintain as compared with higher temperature heat collection plates. Flat-plate collectors can easily be incorporated into traditional or modern building design, provided that the tilt and orientation are properly calculated. Collectors work best if they face the sun directly, just a few degrees west of due south, and are tilted up at an angle that equals the latitude of the site plus 10 degrees. By using direct as well as diffuse solar radiation, flat-plate collectors can attain 250 degrees Fahrenheit—well above the moderate temperatures needed for space heating and domestic hot water.

Overall view and operating principle

A standard collection unit is rectangular, nine feet long by four feet wide by four inches high. The collector operates on a heat-transfer principle: the sun's rays strike an absorber plate, which in turn transfers its heat to fluid circulating through adjacent tubes.

List of major parts (spatial sequence)

Five main parts make up the flat-plate collector: the enclosure, the glazing (and frame), the absorber plate, the flow tubes holding the transfer fluid, and the insulation (Figure 1).

Flat Plate Collector

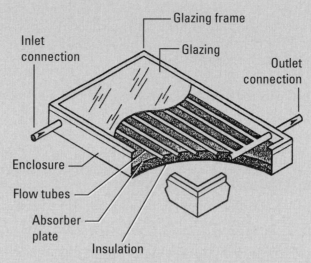

FIGURE 1   A Flat-Plate Collector (Cutaway View)
Source: Solar Water Heating. U.S. Department of Energy, March 1996.

First major part
(definition,
shape, and
material)

Second major
part, etc.

Description of Parts and Their Function

THE ENCLOSURE.   The enclosure is a rectangular metal or plastic tray that serves as a container for the remaining (four) main parts of the collector. It is mounted on the roof of a home at a precise angle for absorbing solar rays.

THE GLAZING (AND FRAME).   The glazing consists of one or more layers of transparent plastic or glass that allow the sun's rays to shine on the absorber plate. This part also provides a cover for the enclosure and serves as insulation by trapping the heat that has been absorbed. An insulated frame secures the glazing sheet to the enclosure.

THE ABSORBER PLATE.   The metallic absorber plate, coated in black for maximum efficiency, absorbs solar radiation and converts it into heat energy. This plate provides the heat source for the transfer fluid contained in the adjacent tubing.

THE FLOW TUBES AND TRANSFER FLUID.   The captured solar heat is removed from the absorber by means of a transfer medium, generally treated water. The transfer medium is heated as it passes through flow tubes attached to the absorbing plate and then transported to points of use in the home or to storage, depending on energy demand.

THE INSULATION.   Fiberglass insulation surrounds the bottom, edges, and sides of the collector. Its purpose is to retain the absorbed energy within the collector and to limit the amount of heat loss.

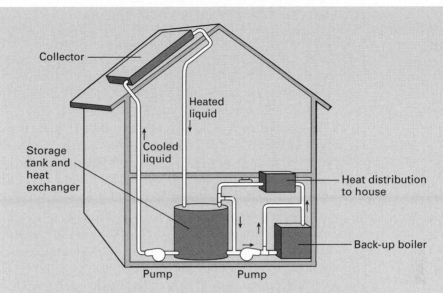

FIGURE 2   How Solar Energy is Captured and Distributed Throughout a Home
Source: Adapted from *Converting a Home to Solar Heat.* U.S. Department of Energy, December 1995.

Operating Description and Conclusion

A conclusion emphasizing the collector's efficiency

In one complete operating cycle, solar rays penetrate the transparent cover and heat the black absorber plate (Figure 2). Insulation along the enclosure's edges, sides, and bottom helps retain the heat. As the absorber plate becomes hot, it heats a liquid circulating through attached flow tubes, which is then pumped to a heat exchanger. The heat exchanger transfers the heat to the water in a storage tank. The cooled liquid is then pumped back to the collector to be re-heated while the hot water in the storage tank is pumped to various uses in the home.

One complete operating cycle (functional sequence)

The solar energy annually striking the roof of a typical house is ten times as great as its annual heat demand. Therefore, properly designed and installed, a flat-plate solar heating system can provide a large percentage of a house's space heating and domestic hot water requirements.

21.2

For more models
of process
description visit
<www.ablongman.com/
lannonweb>

AN OUTLINE FOR PROCESS DESCRIPTION

A description of how things work or happen divides the process into its parts or principles. Colleagues and clients need to know how stock and bond prices are governed, how your bank reviews a mortgage application, how an optical fiber conducts an impulse, and so on. A process description must be detailed enough to allow users to follow the process step by step.

Much of your college writing explains how things happen. Your audience is the professor, who will evaluate what you have learned. Because this person knows *more* than you do about the subject, you often discuss only the main points, omitting details.

But your real challenge comes in describing a process for audiences who know *less* than you, and who are neither willing nor able to fill in the blank spots; you then become the teacher, and the users become your students.

Introduce your description by telling what the process is, and why, when, and where it happens. In the body, tell how it happens, analyzing each stage in sequence. In the conclusion, summarize the stages, and describe one full cycle of the process.

Sections from the following general outline can be adapted to any process description:

I. Introduction
    A. Definition, Background, and Purpose of the Process
    B. Intended Audience (usually omitted for workplace audiences)
    C. Prior Knowledge Needed to Understand the Process
    D. Brief Description of the Process
    E. Principle of Operation
    F. Special Conditions Needed for the Process to Occur
    G. Definitions of Special Terms
    H. Preview of Stages
II. Stages in the Process
    A. First Major Stage
        1. Definition and purpose
        2. Special conditions needed for the specific stage
        3. Substages (if applicable)
            a.
            b.
    B. Second Stage (and so on)
III. Conclusion
    A. Summary of Major Stages
    B. One Complete Process Cycle

Adapt this outline to the structure of the process you are describing.

A SITUATION REQUIRING PROCESS DESCRIPTION

The following document is patterned after the sample outline. The author, Bill Kelly, belongs to an environmental group studying the problem of acid rain in its Massachusetts community. (Massachusetts is among the states most affected by acid rain.) To gain community support, the environmentalists must educate citizens about the problem. Bill's group is publishing and mailing a series of brochures. The first brochure provides an overview of the acid rain process.

A Process Description for a Nontechnical Audience

Audience and Use Profile.   Some will already be interested in the problem; others will have no awareness (or interest). Therefore, explanation is given at the lowest level of technicality (no chemical formulas, equations). But the explanation needs to be vivid enough to appeal to less aware or less interested readers. Visuals create interest and illustrate simply. To give an explanation thorough enough for broad understanding, the process is divided into three chronological steps: how acid rain develops, spreads, and destroys. ❏

HOW ACID RAIN DEVELOPS, SPREADS, AND DESTROYS

Introduction

Definition

Acid rain is environmentally damaging rainfall that occurs after fossil fuels burn, releasing nitrogen and sulfur oxides into the atmosphere. Acid rain, simply stated, increases the acidity level of waterways because these nitrogen and sulfur oxides combine with the air's normal moisture. The resulting rainfall is far more acidic than normal rainfall. Acid rain is a silent threat because its effects, although slow, are cumulative. This description explains the cause, the distribution cycle, and the effects of acid rain.

Purpose

Brief description of the process

Most research shows that power plants burning oil or coal are the primary causes of acid rain. The burnt fuel is not completely expended, and some residue enters the atmosphere. Although this residue contains several potentially toxic elements, sulfur oxide and, to a lesser extent, nitrogen oxide are the major problems, because they are transformed when they combine with moisture. This chemical reaction forms sulfur dioxide and nitric acid, which then rain down to earth.

Preview of stages

The major steps explained here are (1) how acid rain develops, (2) how acid rain spreads, and (3) how acid rain destroys.

The Process

First stage

HOW ACID RAIN DEVELOPS.   Once fossil fuels have been burned, their usefulness ends. Unfortunately, it is here that the acid rain problem begins.

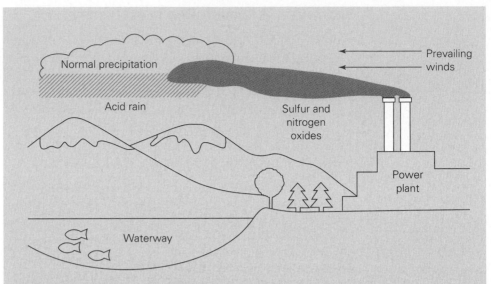

FIGURE 1 How Acid Rain Develops

Fossil fuels release various elements during combustion. Two of these, sulfur oxide and nitrogen oxide, combine with normal moisture to produce sulfuric acid and nitric acid. (Figure 1 illustrates how acid rain develops.) The released gases combine with atmospheric ozone and water vapor to produce rain or snowfall that is more acidic than normal precipitation.

Definition

Acid level is measured by pH readings. The pH scale runs from 0 through 14—a pH of 7 is considered neutral. (Distilled water has a pH of 7.) Numbers above 7 indicate increasing degrees of alkalinity. (Household ammonia has a pH of 11.) Numbers below 7 indicate increasing acidity. Movement in either direction on the pH scale, however, means multiplying by 10. Lemon juice, which has a pH value of 2, is 10 times more acidic than apples, which have a pH of 3, and is 1,000 times more acidic than carrots, which have a pH of 5.

Because of carbon dioxide (an acid substance) normally present in air, unaffected rainfall has a pH of 5.6. At this time, the pH of precipitation in the northeastern United States and Canada is between 4.5 and 4. In Massachusetts, rain and snowfall have an average pH reading of 4.1. A pH reading below 5 is considered abnormally acidic, and therefore a threat to aquatic populations.

Second stage

HOW ACID RAIN SPREADS.   Although we might expect areas containing power plants to be most severely affected, acid rain can in fact travel thousands of miles from its source. Stack gases escape and drift with the wind currents. The sulfur and nitrogen oxides are thus able to travel great distances before they return to earth as acid rain.

For an average of two to five days after emission, the gases follow the prevailing winds far from the point of origin. Estimates show that about 50 percent of the acid rain that affects Canada originates in the United States; at the same time, 15 to 25 percent of U.S. acid rain originates in Canada.

The tendency of stack gases to drift makes acid rain a widespread menace. More than 200 lakes in the Adirondacks, hundreds of miles from any industrial center, are unable to support life because their water has become so acidic.

Third stage

Substage

Substage

HOW ACID RAIN DESTROYS.   Acid rain causes damage wherever it falls. It erodes various types of building rock such as limestone, marble, and mortar. Damage to buildings, houses, monuments, statues, and cars is widespread. Some priceless monuments and carvings have already been destroyed, and even trees of some varieties are dying in large numbers.

More important is acid rain damage to waterways in affected areas. (Figure 2 illustrates how a typical waterway is infiltrated.) Acid rain dramatically lowers the pH in lakes and streams. Although its effect is not immediate, acid rain can eventually make a waterway so acidic that it dies. In areas with natural acid-buffering elements such as limestone, the dilute acid has less effect. The northeastern United States and Canada, however, lack this natural protection, and so are continually vulnerable.

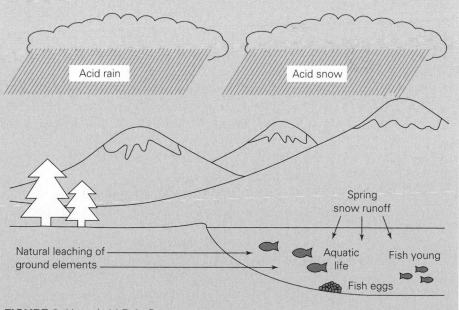

FIGURE 2 How Acid Rain Destroys

The pH level in an affected waterway drops so low that some species cease to reproduce. A pH of 5.1 to 5.4 means that entire fisheries are threatened: once a waterway reaches a pH of 4.5, fish reproduction ceases. Because each creature is part of the overall food chain, loss of one element in the chain disrupts the whole cycle.

In the northeastern United States and Canada, the acidity problem is compounded by the runoff from acid snow. During winter, acid snow sits with little melting, so that by spring thaw, the acid released is greatly concentrated. Aluminum and other heavy metals normally present in soil are also released by acid rain and runoff. These toxic substances leach into waterways in heavy concentrations, affecting fish in all stages of development.

One complete cycle

Summary

Acid rain develops from nitrogen and sulfur oxides emitted by the burning of fossil fuels. In the atmosphere, these oxides combine with ozone and water to form precipitation with a lower-than-average pH. This acid precipitation returns to earth many miles from its source, severely damaging waterways that lack natural buffering agents. The northeastern United States and Canada are the most severely affected areas in North America.

SPECIFICATIONS

21.3

For model specifications visit <www.ablongman.com/lannonweb>

Airplanes, bridges, smoke detectors, and countless other technologies are produced according to certain *specifications*. A particularly exacting type of description, specifications (or "specs") prescribe standards for performance, safety, and quality. For countless products and processes, specifications describe these features:

Specifications describe products and processes

- the methods for manufacturing, building, or installing a product
- the materials and equipment to be used
- the size, shape, and weight of the product
- specific testing, maintenance, and inspection procedures

Specifications are often used to ensure compliance with a particular safety code, engineering standard, or government or legal ruling.

Specifications have ethical and legal implications

Because specifications define an "acceptable" level of quality, any product that fails to meet these specs may provide grounds for a lawsuit. When injury or death results (as in a bridge collapse or an airline accident), the device is usually checked to make sure it was built and maintained according to the appropriate specifications. If not, the contractor, manufacturer, or supplier may be liable.

Federal and state regulatory agencies routinely issue specifications to ensure safety. The Consumer Product Safety Commission specifies that power lawn mowers be equipped with a "kill switch" on the handle, a blade guard to prevent foot injuries, and a grass thrower that aims downward to prevent eye and facial injury. This same agency issues specifications for baby products, as in governing the fire retardancy of pajama fabric. Passenger airline specifications for aisle width, seat belt configurations, and emergency equipment are issued by the Federal Aviation Administration. State and local agencies issue specifications in the form of building codes, fire codes, and other standards for safety and reliability.

Government departments (Defense, Interior, etc.) issue specifications for all types of military hardware and other equipment. A set of NASA specifications for spacecraft parts can be hundreds of pages long, prescribing the standards for even the smallest nuts and bolts, down to screw thread depth and width in millimeters.

The private sector issues specifications for countless products or projects, to help ensure that customers get exactly what they want. Figure 21.4 shows partial specifications drawn up by an architect for a medical clinic building. This section of the specs covers only the structure's "shell." Other sections detail the requirements for plumbing, wiring, and interior finish work.

Specifications like those in Figure 21.4 must be clear enough for identical interpretation by a broad audience (Glidden 258–59).

- *the customer,* who has the big picture of what is needed and who wants the best product at the best price
- *the designer* (architect, engineer, computer scientist, etc.), who must translate the customer's wishes into the actual specification
- *the contractor or manufacturer,* who won the job by making the lowest bid, and so must preserve profit by doing only what is prescribed
- *the supplier,* who must provide the exact materials and equipment
- *the workforce,* who will do the actual assembly, construction, or installation (managers, supervisors, subcontractors, and workers—some working on only one part of the product, such as plumbing or electrical)
- *the inspectors* (such as building, plumbing, or electrical inspectors), who evaluate how well the product conforms to the specifications

Each of these parties needs to understand and agree on exactly *what* is to be done and *how* it is to be done. In the event of a lawsuit over failure to meet specifications, the readership broadens to include judges, lawyers, and jury. Figure 21.5 depicts how a set of clear specifications unifies all users (their various viewpoints, motives, and levels of expertise) in a shared understanding.

In addition to guiding a product's design and construction, specifications can facilitate the product's use and maintenance. For instance, specifications in a computer manual include the product's performance limits, or *ratings:* its power requirements; its work, processing, or storage capacity; operating environment requirements; the

Ruger, Filstone, and Grant
Architects

MATERIAL SPECIFICATIONS FOR THE POWNAL CLINIC BUILDING

Foundation
    footings: 8" x 16" concrete (load-bearing capacity: 3,000 lbs. per sq. in.)
    frost walls: 8" x 4' @ 3,000 psi
    slab: 4" @ 3,000 psi, reinforced with wire mesh over vapor barrier

Exterior Walls
    frame: eastern pine #2 timber frame with exterior partitions set inside posts
    exterior partitions: 2" x 4" kiln-dried spruce set at 16" on center
    sheathing: 1/4" exterior-grade plywood
    siding: #1 red cedar with a 1/2" x 6" bevel
    trim: finished pine boards ranging from 1" x 4" to 1" x 10"
    painting: 2 coats of Clear Wood Finish on siding; trim primed and finished with one
                coat of bone white, oil base paint

Roof System
    framing: 2" x 12" kiln-dried spruce set at 24" on center
    sheathing: 5/8" exterior-grade plywood
    finish: 240 Celotex 20-year fiberglass shingles over #15 impregnated felt roofing paper
    flashing: copper

Windows
    Anderson casement and fixed-over-awning models, with white exterior cladding,
    insulating glass and screens, and wood interior frames

Landscape
    driveway: gravel base, with 3" traprock surface
    walks: timber defined, with traprock surface
    cleared areas: to be rough graded and covered with wood chips
    plantings: 10 assorted lawn plants along the road side of the building

FIGURE 21.4   Specifications for a Building Project (Partial)   These specifications ensure that all parties agree on the specific materials to be used.

makeup of key parts; and so on. Product support literature for appliances, power tools, and other items routinely contains ratings to help customers select a good operating environment or replace worn or defective parts (Riney 186). The ratings in Figure 21.6 are taken from the owner's manual for an ink jet printer.

FIGURE 21.5
Users and
Potential
Users of
Specifications

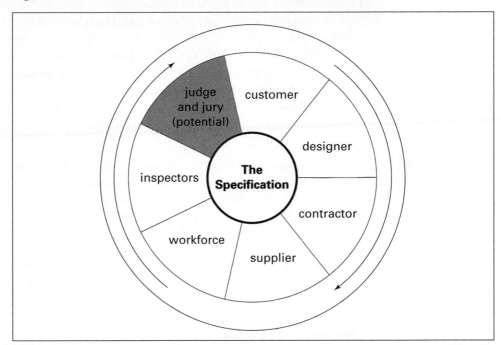

FIGURE 21.6
Specifications
for the Color
StyleWriter™
2400

Source:
Reprinted by per-
mission of Apple
Computer, Inc.

General specifications

Marking engine

- Thermal ink jet engine

Resolution

- 360 dots per inch (dpi) for text and graphics (180 dpi for draft quality)

Engine speed

- Printing speed depends on the images printed and on the Macintosh computer used.

Connector cable

- Apple System/Peripheral-8 cable

Interface

- High-speed serial (RS-422)
- Optional LocalTalk

Paper feed in pounds (lb.) and grams/meter^2 (g/m^2)

- Sheet feeder holds up to 100 sheets of 20-lb. (75-g/m^2) paper or 15 envelopes.

TECHNICAL MARKETING LITERATURE

Technical marketing literature is designed to sell a technical or scientific product or service to audiences that range from novice to highly informed. Descriptions and specifications are essential marketing tools because they help potential customers to visualize the product or service and to recognize how its special features can fit their exact needs.

Technical marketing audiences expect a factual presentation

Even though technical marketing has persuasion as its main goal, readers dislike a "hard-sell approach." They expect upbeat promotional claims such as "durable finish" and "performance of a lifetime" to be backed up by solid evidence: for example, results of objective product testing, performance ratings, and data that indicate how the product meets or exceeds industry specifications.

NOTE
Unlike proposals (Chapter 23), which are also used to sell a product or service, technical marketing materials tend to be less formal and more dynamic, colorful, and varied. A typical proposal is tailored to one specific client's specific needs and follows a fairly standard format; marketing literature, on the other hand, uses a wide variety of formats to present the product in its best light for a broad array of audiences and needs.

Marketing documents run the gamut from simple "fact sheets" to glossy booklets with colorful photographs and other visuals. Here are some common types of documents.

Common formats for technical marketing documents

- *Brochures* are a popular marketing medium. The six-panel foldout brochure in Figure 21.7 is designed to promote a brand of windows and doors by emphasizing their energy efficiency as well as their beauty; product appeal is enhanced by a visual background of striking product designs and various weather conditions and settings. Notice how the brochure's panels provide a logical sequence, with each panel offering its own discrete "chunk" (page 239) of information about the product (Hilligoss 71). The balance of technical and aesthetic details create a persuasive message something like this: "Marvin products not only look great but also work great."
- *Web pages* are especially effective for technical marketing. Depending on a person's specific interests and expertise, she or he can explore the links of particular interest, read or download the material, and easily return to the site's home page. As a marketing tool, Web pages offer these advantages: product information can easily be updated; customers can interact to ask questions, give feedback, and place orders directly online; and through the use of animation, the product can actually be depicted in operation (Gurak and Lannon 2nd ed., 141).
- *Fact sheets* offer basic data about the product or service in a straightforward, unadorned format, usually on a single $8\frac{1}{2} \times 11$-inch page, sometimes using both sides of the page. Figure 21.9 displays a double-sided fact sheet that

Outside panels

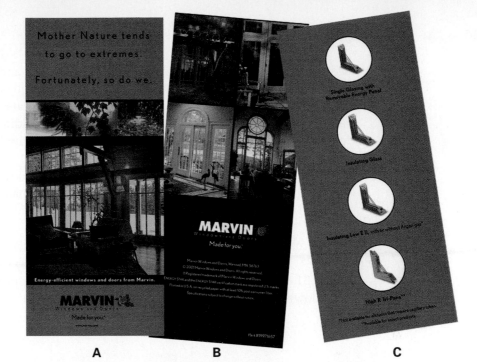

A   B   C

Inside panels

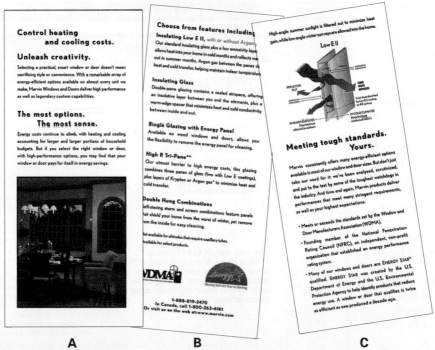

A   B   C

FIGURE 21.7 A Technical Marketing Brochure for a Broad Audience First, the upbeat message and engaging photos in outside Panel A immediately place the product in a favorable aesthetic light. Next, inside Panel A focuses on product benefits (energy savings, beauty, and versatility) followed by another attractive photo. Next, inside Panel B—at the brochure's very center—focuses on technical features and options, with contact information prominently displayed. Following a graphic depicting insulating glass in action, inside Panel C describes the rigorous standards that Marvin products meet. Next, outside Panel C shows cross-sections of various insulating options. Finally, the striking photos in outside Panel B echo the aesthetic focus of the opening panel. *Source: Reprinted courtesy of Marvin Windows and Doors. Copyright 2001.*

accompanies the expanded definition in Figures 20.5 and 20.6 (pages 499, 500). Even though this sheet contains a great deal of highly technical information, it is designed to be inviting and navigable: visuals are engaging and easy to interpret; paragraphs are concise and readable; headings provide a clear forecast of each section; and the most complex data is chunked into clearly labeled lists (Hilligoss 63).

- *Business letters* are the most personal type of marketing document. See, for example, how Tom Ewing's letter on page 63 creates a human connection with a potential customer.

(For an intimate look at technical marketing, see Richard Larkin's report in Chapter 24.)

Figure 21.8 shows a web-based description of one of the products (the double hung window) mentioned in Figure 21.9.

**FIGURE 21.8
A Technical
Marketing Web
Page**

This one-page description provides links to deeper levels about the product (such as "Specs"), background on the company, or contact and ordering information. Clicking on "View Tilt Lever Action" provides an animated view of this mechanism.

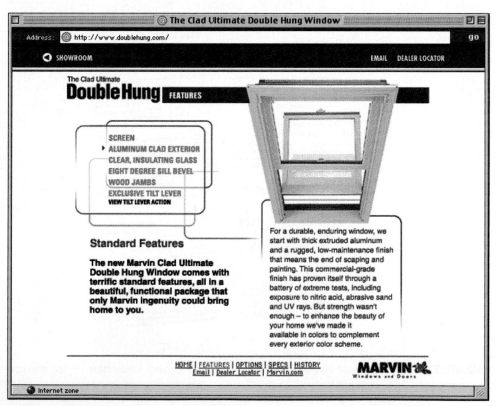

*Source: Reprinted courtesy of Marvin Windows and Doors
<www.doublehung.com/features.asp>.*

☑ **CHECKLIST** for Usability of Technical Descriptions

(Numbers in parentheses refer to first page of discussion.)

Content

☐ Does the title promise exactly what the description delivers? (503)

☐ Are the item's overall features described, as well as each part? (510)

☐ Is each part defined before it is discussed? (510)

☐ Is the function of each part explained? (511)

☐ Do visuals appear whenever they can provide clarification? (507)

☐ Will users be able to visualize the item? (506)

☐ Are any details missing, needless, or confusing for this audience? (503)

☐ Is the description ethically acceptable? (506)

Arrangement

☐ Does the description follow the clearest possible sequence? (507)

☐ Are relationships among the parts clearly explained? (510)

Style and Page Design

☐ Is the description sufficiently impartial? (506)

☐ Is the language informative and precise? (507)

☐ Is the level of technicality appropriate for the audience? (503)

☐ Is the description written in plain English? (507)

☐ Is each sentence clear, concise, and fluent? (244)

☐ Is the description grammatical? (Appendix C)

☐ Is the page design inviting and accessible? (340)

EXERCISES

 For more exercises, visit <www.ablongman.com/lannon>

1. Select an item from this list or a device used in your major field. Using the general outline as a model, develop an objective description. Include (a) all necessary visuals or (b) an "art brief" (page 318) and a rough diagram for each visual or (c) a "reference visual" (a copy of a visual published elsewhere) with instructions for adapting your visual from that one. (If you borrow visuals from other sources, provide full documentation.) Write for a specific use by a specified audience. Attach your written audience and use profile (based on the worksheet, page 36) to your document.

   soda-acid fire extinguisher
   breathalyzer
   sphygmomanometer
   transit
   Skinner box
   distilling apparatus
   saber saw
   hazardous waste site
   brand of woodstove
   photovoltaic panel
   catalytic converter
   radio

   Remember, you are simply describing the item, its parts, and its function: *do not* provide directions for its assembly or operation.

   As an optional assignment, describe a place you know well. You are trying to convey a visual image, not a mood; therefore, your description should be impartial, discussing only observable details.

2. The flat-plate collector description in this chapter is aimed toward a general audience. Evaluate it using the revision checklist. In one or two paragraphs, discuss your evaluation, and suggest revisions.

3. Figure 21.9 is designed to promote as well as describe a technical product. Answer the following questions about the document:

- When read in conjunction with Figures 20.5 and 20.6, is Figure 21.9 an effective introduction to this particular product for its intended audience of engineers and other technical experts? Why or why not?
- Is the overall page design effective? Why or why not? Be specific. (See Chapter 15 and "Using Color," pages 324–29.)
- Are the visuals adequate and appropriate? Why or why not? (See Chapter 14.) Why aren't the diagrams on page one labeled more extensively?
- Is this a sufficiently impartial description? Why or why not? Given its purpose as a marketing document, is the description ethically appropriate? (See Chapter 5.)

Be prepared to discuss your analysis and evaluation in class.

4. Locate a description and specifications for a particular brand of automobile or some other consumer product. Evaluate this material for promotional and descriptive value and ethical appropriateness.

5. Select a specialized process that you understand well and that has several distinct steps. Using the process description on pages 517–20 as a model, explain this process to classmates who are unfamiliar with it. Begin by completing an audience and use profile (page 517). Some possible topics: how the body metabolizes alcohol, how economic inflation occurs, how the federal deficit affects our future, how a lake or pond becomes a swamp, how a volcanic eruption occurs, or how the greenhouse effect may contribute to global warming.

COLLABORATIVE PROJECTS

1. Divide into small groups. Assume that your group works in the product development division of a large and diversified manufacturing company.

Your division has just thought of an idea for an inexpensive consumer item with a potentially vast market. (Choose one from the following list.)

flashlight
nail clippers
retractable ballpoint pen
scissors
stapler
any simple mechanism

Your group's assignment is to prepare three descriptions of this invention:

a. for company executives who will decide whether to produce and market the item
b. for the engineers, machinists, and so on who will design and manufacture the item
c. for the customers who might purchase and use the item

Before writing for each audience, be sure to collectively complete audience and use profiles (page 517).

Appoint a group manager, who will assign tasks (visuals, typing, etc.) to members. When the descriptions are fully prepared, the group manager will appoint one member to present them in class. The presentation should include explanations of how the descriptions differ for the different audiences.

2. Assume that your group is an architectural firm designing buildings at your college. Develop a set of specifications for duplicating the interior of the classroom in which this course is held. Focus only on materials, dimensions, and equipment (whiteboard, desk, etc.) and use visuals as appropriate. Your audience includes the firm that will construct the classroom, teachers, and school administrators. Use the same format as in Figure 21.4, or design a better one. Appoint one member to present the completed specifications in class. Compare versions from each group for accuracy and clarity.

3. As a group, select a particular product (sound system, laptop computer, video game, or the like) for which descriptions and specifications

C Series System Description

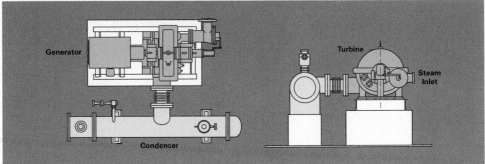

The Ewing Power Systems C Series is a complete single-stage condensing turbine generator package designed for use where maximum electricity production is desired and surplus steam is available. High-pressure steam passes through the turbine and exits at a vacuum pressure to a close-coupled condenser. The condenser may be either water cooled or air cooled. In water-cooled systems the cooling water can be used for heating or the heat can be dissipated in a cooling tower. This series is ideal for converting waste fuel into valuable electricity. It is generally not suited for applications where oil or gas is the primary boiler fuel unless the condenser cooling water will be used for heating.

Features and Specifications

Turbine Features
- Coppus RLHA turbine, fully proven in worldwide applications
- Integrated steam control system including all transmitters, actuators, and controllers
- Dual electronic and mechanical overspeed trip mechanisms
- Hand valves for maximum operating efficiency
- Very low maintenance
- Rugged, reliable design, 20 year minimum service life
- Meets all applicable NEMA and API specifications

Steam Specifications
- Recommended Inlet Pressure
  - Maximum: 700 psig
  - Minimum: 14 psig
- Recommended Exhaust Pressure
  - Maximum: 0 psig (14.7 psia)
  - Minimum: −10 psig (4.7 psia)
- Steam Flow
  2,500 pounds of steam per hour (75 boiler horsepower) or greater

Generator Features
- Louis Allis induction generator, renowned for high efficiency and dependability*
- Models from 55 kW to 800 kW continuous duty at 480 volts
- Models to 2,000 kW at higher voltages
- Extra high efficiency design is standard

*Synchronous generators also available

Standard Prewired Electrical Controls
- Shunt trip, 3-pole, motor-operated circuit breaker with stored energy trip mechanism
- Utility Grade Protective Relays
  - Over/under Voltage
  - Over/under Frequency
  - Ground overcurrent
- Stator thermostats
- Time delay relay to disconnect on motoring
- Pilot lights for operating and trip status
- Ammeter and voltmeter
- Digital tachometer
- Kilowatt meter
- Synchronous panels available

NOTE: *We will customize our control panels to meet the interconnection requirements of any utility.*

Condenser Features
- Air-cooled models
  - Specially designed to minimize power consumption
  - Freeze protection system
  - Standard design is for 97°F ambient, higher temperatures available
- Water-cooled models
  - Includes cooling tower
  - Steel shell and tubesheet and admiralty brass tubes
  - Integral hot-well
- Both models include:
  - Steam ejector to remove non-condensables
  - Condensate pump

EWING POWER SYSTEMS

FIGURE 21.9   A Technical Marketing Fact Sheet   After a product diagram and a brief introduction to the C Series Cogeneration System, the description focuses on the product's major components and specifications.

Source: Courtesy of Ewing Power Systems, So. Deerfield, MA 01073.

Typical System Schematic

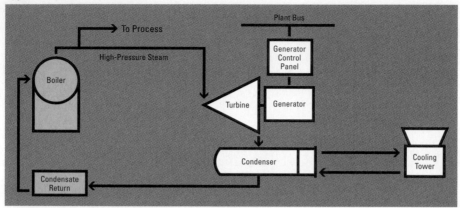

C Series with water-cooled condenser and cooling tower.

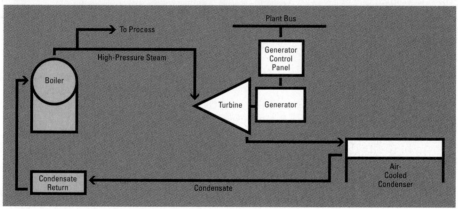

C Series with air-cooled condenser.

Services

In addition to supplying the finest available equipment, Ewing Power Systems also provides complete engineering services. We recognize that you may not be familiar with many of the engineering details of cogeneration so we go to extra lengths to assure our systems will meet your specific requirements for steam and electricity and will provide detailed prints and installation supervision. We will also work with your local utility to assure that your system is designed to take maximum advantage of their rate structure and will meet all their technical requirements. These engineering services, together with our complete economic analysis, assure that your cogeneration system will provide the shortest possible payback.

(EWING POWER SYSTEMS)

5 North Street • South Deerfield, MA 01373-0566

FIGURE 21.9   A Technical Marketing Fact Sheet *(continued)*   Representational diagrams depict the sequence of events in the cogeneration process and a brief closing paragraph describes the support services offered by the vendor.

are available (in product manuals, brochures, and so on). Using Figure 21.9 as a model, design a marketing document that describes and promotes this product in a one-page, double-sided format. Include (a) all necessary visuals or (b) an "art brief" (page 318) and a rough diagram for each visual or (c) a "reference visual" with instructions for adapting your visual from that one. (If you borrow visuals from other sources, provide full documentation.)

4. Select a specialized process you understand well (how gum disease develops, how an earthquake occurs, how steel is made, how a computer com-

piles and executes a program). Write a brief description of the process. Include visuals, as stipulated in Exercise 3 above. Exchange your description with a classmate in another major. Study your classmate's explanation for fifteen minutes and then write the same explanation in your own words, referring to your classmate's paper as needed. Now, evaluate your classmate's version of your original explanation for accuracy. Does it show that your explanation was understood? If not, why not? Discuss your conclusions in a memo to your instructor, submitted with all samples.

22

Instructions and Procedures

PURPOSE OF INSTRUCTIONAL DOCUMENTS

FORMATS FOR INSTRUCTIONAL DOCUMENTS

FAULTY INSTRUCTIONS AND LEGAL LIABILITY

ELEMENTS OF USABLE INSTRUCTION

AN OUTLINE AND MODEL FOR INSTRUCTIONS

A SITUATION REQUIRING INSTRUCTIONS

PROCEDURES

CHECKLIST for Usability of Instructions

Instructions spell out the steps required for completing a task or series of tasks (say, installing printer software on your hard drive or operating an electron microscope). The audience for a set of instructions might be someone who doesn't know how to perform the task or someone who wants to perform it better. In either case, effective instructions enable users to get the job done safely and efficiently.

Procedures, a special type of instruction, serve as official guidelines for people who typically are already familiar with a given task (say, evacuating a high-rise building). Procedures ensure that all members of a group (say, employers or employees) coordinate their activities in performing the task.

PURPOSE OF INSTRUCTIONAL DOCUMENTS

The role of instructions on the job

In this technological age, almost anyone with a responsible job writes and reads instructions. For example, you might instruct a new employee in activating his or her voice mail system or a customer in shipping radioactive waste. The employee going on vacation writes instructions for the person filling in. Computer users routinely consult hard copy or online manuals (or *documentation*) for all sorts of tasks.

What users expect to learn from a set of instructions

- *Why am I doing this?*
- *How do I do it?*
- *What materials and equipment will I need?*
- *Where do I begin?*
- *What do I do next?*
- *What could go wrong?*

Because they focus squarely on the *user*—the person who will "read" and then "do"—instructions must meet the highest standards of excellence.

FORMATS FOR INSTRUCTIONAL DOCUMENTS

Instructional documents take various formats, in hard copy or electronic versions. Here are some of the most commonly used:

Common formats for instructional documents

- *Brief reference cards* (Figure 22.1) typically fit on a single page or less. The instructions usually focus on the basic steps for users who want only enough information to start on a task and to keep moving through it.
- *Instructional brochures* (Figure 22.2) can be displayed, handed out, mailed, or otherwise distributed to a broad audience. They are especially useful for advocating procedures that increase health and safety.
- *Manuals* (Figure 22.3) contain instructions for all sorts of tasks. A manual also may contain descriptions and specifications for the product, warnings, maintenance and troubleshooting advice, and any other information the user is likely to need. For complex products (say, a word-processing program) or procedures (say, cleaning up a hazardous-waste site), the manual can be a sizable

book. In a recent trend, briefer manuals contain the basic operating tips and the more lengthy information is provided as online help.

• *Online documentation* (Figure 22.4) provides the entire contents of a hard copy manual at the click of a key or mouse button. Whereas less experienced users tend to prefer paper documentation, online help is especially popular among more experienced users.

• *Hyperlinked instructions* (Figure 22.5) enable users to explore various levels and layers of information and to choose the layer that matches their needs.

Regardless of its format, any set of instructions must meet the strict legal and usability requirements discussed on the following pages.

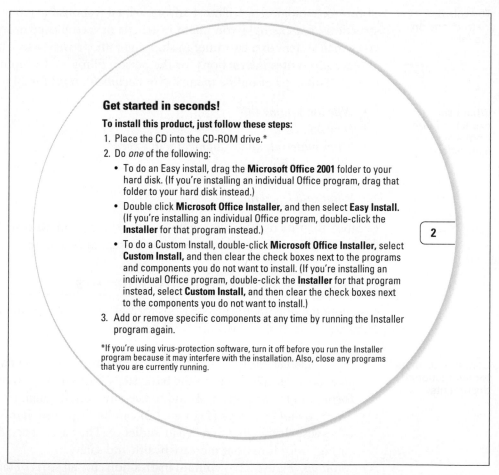

Get started in seconds!

To install this product, just follow these steps:

1. Place the CD into the CD-ROM drive.*

2. Do *one* of the following:

   • To do an Easy install, drag the **Microsoft Office 2001** folder to your hard disk. (If you're installing an individual Office program, drag that folder to your hard disk instead.)

   • Double click **Microsoft Office Installer,** and then select **Easy Install.** (If you're installing an individual Office program, double-click the **Installer** for that program instead.)

   • To do a Custom Install, double-click **Microsoft Office Installer,** select **Custom Install,** and then clear the check boxes next to the programs and components you do not want to install. (If you're installing an individual Office program, double-click the **Installer** for that program instead, select **Custom Install,** and then clear the check boxes next to the components you do not want to install.)

3. Add or remove specific components at any time by running the Installer program again.

*If you're using virus-protection software, turn it off before you run the Installer program because it may interfere with the installation. Also, close any programs that you are currently running.

FIGURE 22.1   A Brief Reference Card   This card lists the basic steps required to install *Microsoft Office 2001*™ on a Macintosh computer.

Source: From "Installation Documentation of Microsoft® Office 2001™ for Mac," *trademark owned by Microsoft Corporation. Reprinted with permission from Microsoft Corporation.*

Right now, there may be an invisible enemy ready to strike. He's called BAC (bacteria) and he can make you and those you care about sick. In fact, even though you can't see BAC — or smell him, or feel him — he and millions more like him may have already invaded the food you eat.

But you have the power to "Fight BAC!"™ and to keep your food safe from harmful bacteria. It's as easy as following these four simple steps:

Clean: Wash hands and surfaces often

Bacteria can spread throughout the kitchen and get onto cutting boards, utensils, sponges and counter tops. Here's how to "Fight BAC!"™:

- Wash your hands with hot soapy water before handling food and after using the bathroom, changing diapers and handling pets.

- Wash your cutting boards, dishes, utensils and counter tops with hot soapy water after preparing each food item and before you go on to the next food.

- Use plastic or other non-porous cutting boards. These boards should be run through the dishwasher — or washed in hot soapy water — after use.

- Consider using paper towels to clean up kitchen surfaces. If you use cloth towels, wash them often in the hot cycle of your washing machine.

Separate: Don't cross-contaminate

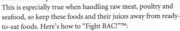

Cross-contamination is the scientific word for how bacteria can be spread from one food product to another. This is especially true when handling raw meat, poultry and seafood, so keep these foods and their juices away from ready-to-eat foods. Here's how to "Fight BAC!"™:

- Separate raw meat, poultry and seafood from other foods in your grocery shopping cart and in your refrigerator.

- If possible, use a different cutting board for raw meat products.

- Always wash hands, cutting boards, dishes and utensils with hot soapy water after they come in contact with raw meat, poultry and seafood.

- Never place cooked food on a plate which previously held raw meat, poultry and seafood.

Cook: Cook to proper temperatures

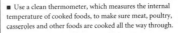

Food safety experts agree that foods are properly cooked when they are heated for a long enough time and at a high enough temperature to kill the harmful bacteria that cause foodborne illness. The best way to "Fight BAC!"™ is to:

- Use a clean thermometer, which measures the internal temperature of cooked foods, to make sure meat, poultry, casseroles and other foods are cooked all the way through.

- Cook roasts and steaks to at least 145°F. Whole poultry should be cooked to 180°F for doneness.

- Cook ground beef, where bacteria can spread during processing, to at least 160°F. Information from the Centers for Disease Control and Prevention (CDC) links eating undercooked, pink ground beef with a higher risk of illness. If a thermometer is not available, do not eat ground beef that is still pink inside.

- Cook eggs until the yolk and white are firm. Don't use recipes in which eggs remain raw or only partially cooked.

- Fish should be opaque and flake easily with a fork.

- When cooking in a microwave oven, make sure there are no cold spots in food where bacteria can survive. For best results, cover food, stir and rotate for even cooking. If there is no turntable, rotate the dish by hand once or twice during cooking.

- Bring sauces, soups, and gravy to a boil when reheating. Heat other leftovers thoroughly to at least 165°F.

Chill: Refrigerate promptly

Refrigerate foods quickly because cold temperatures keep harmful bacteria from growing and multiplying. So, set your refrigerator no higher than 40°F and the freezer unit at 0°F. Check these temperatures occasionally with an appliance thermometer. Then, "Fight BAC!"™ by following these steps:

- Refrigerate or freeze perishables, prepared foods and leftovers within two hours or sooner.

- Never defrost food at room temperature. Thaw food in the refrigerator, under cold running water or in the microwave. Marinate foods in the refrigerator.

- Divide large amounts of leftovers into small, shallow containers for quick cooling in the refrigerator.

- Don't pack the refrigerator. Cool air must circulate to keep food safe.

FIGURE 22.2 A Foldout Instructional Brochure The three inside panels of this *Fight BAC!* brochure offer "Four Simple Steps to Food Safety."
Source: Reprinted courtesy of Partnership for Food Safety Education <www.fightbac.org>.

22.1

For more on liability and public relations visit <www.ablongman.com/lannonweb>

FAULTY INSTRUCTIONS AND LEGAL LIABILITY

As many as 10 percent of workers are injured each year on the job (Clement 149). Certain medications produce depression that can lead to suicide (Caher 5). Countless injuries also result from misuse of consumer products such as power tools, car jacks, or household cleaners—misuse often caused by defective instructions.

A user injured because of unclear, inaccurate, or incomplete instructions can sue the writer. Courts have ruled that a defect in product support literature carries the same type of liability as a defect in the product itself (Girill, "Technical Communication and Law" 37).

FIGURE 22.3 Table of Contents from the *Sharp Compact Copier Operation Manual*

Source: Reprinted courtesy of Sharp Electronics Corporation. Reproduced by permission.

INTRODUCTION

Welcome to the world of compact copying on the Z-85II Copier. The Sharp Z-85II has been designed for greater copying versatility while occupying a minimum amount of space and featuring intuitive operating ease. Special features include:

- One enlargement and two reduction ratios
- Stationary Platen
- Auto start mode
- Two-way paper feed
- Automatic exposure control
- Color copying

To get full use of all Copier features, be sure to familiarize yourself with this Manual and the Copier. For quick reference during Copier use, keep the Manual in a handy location.

CONTENTS

	Page
• UNPACKING	2
• A WORD ON COPIER INSTALLATION	3
• CAUTIONS	3
• SET-UP	4
• MAKING COPIES	6
• SPECIAL PAPERS (MANUAL BY-PASS FEED)	8
• TWO-SIDED COPYING	9
• COLOR COPYING	10
• LOADING COPY PAPER	11
• MISFEED REMOVAL	12
• TD CARTRIDGE REPLACEMENT	13
• DRUM CARTRIDGE REPLACEMENT	14
• USER MAINTENANCE	15
• COPIER TROUBLE?	16
• MOVING INSTRUCTIONS	18
• SUPPLIES AND ACCESSORIES PART NUMBERS	18
• SPECIFICATIONS	19

1

Those who prepare instructions are potentially liable for damage or injury resulting from omissions such as these (Caher 5–7; Manning 13; Nordenberg 7):

Examples of faulty instructions that create legal liability

- *Failure to instruct users in the proper use of a product:* for example, a medication's proper dosage or possible interaction with other drugs.
- *Failure to warn against hazards from proper use of a product:* for example, the risk of repetitive stress injury resulting from extended use of a computer keyboard.
- *Failure to warn against the possible misuses of a product:* for example, the childhood danger of suffocation posed by plastic bags.
- *Failure to explain a product's benefits and risks in language that average consumers can understand.*
- *Failure to convey the extent of risk with forceful language.*
- *Failure to display warnings prominently.*

Some legal experts argue that defects in the instructions carry even greater liability than defects in the product because they are more easily demonstrated to a nontechnical jury (Bedford and Stearns 128).

Among all technical documents, instructions have the strictest requirements for giving users precisely what they need precisely when they need it. To design usable instructions, you must have a clear sense of (a) the specific tasks you want users to accomplish, (b) the users' abilities and limitations, and (c) the setting/circumstances in which users will be referring to this document. For advice on analyzing the tasks, users, and setting, see pages 366–68.

FIGURE 22.4 An Online Help Screen

This electronic index offers instant access to any of the topics in the entire online manual.

Source: Reprinted courtesy of Microsoft Corporation. Reprinted with permission.

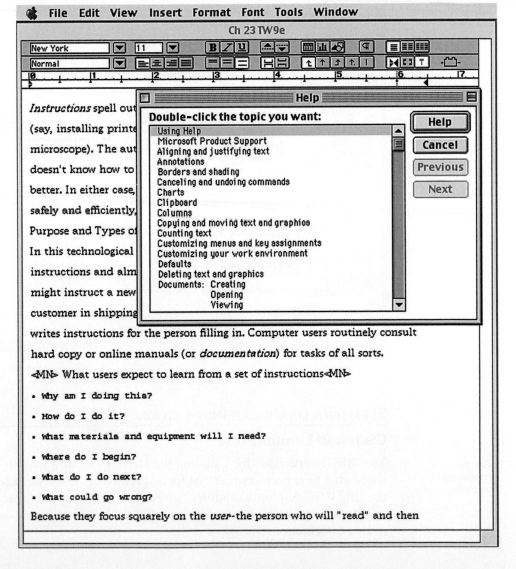

**FIGURE 22.5
A Set of
Web-Based
Instructions**

These instruc-
tions provide
links to more
specific parts of
the procedure.
*Source: U.S.
Environmental
Protection
Agency<http://
yosemite1.epa.gov/
estar/business.nsf/
content/pm>.*

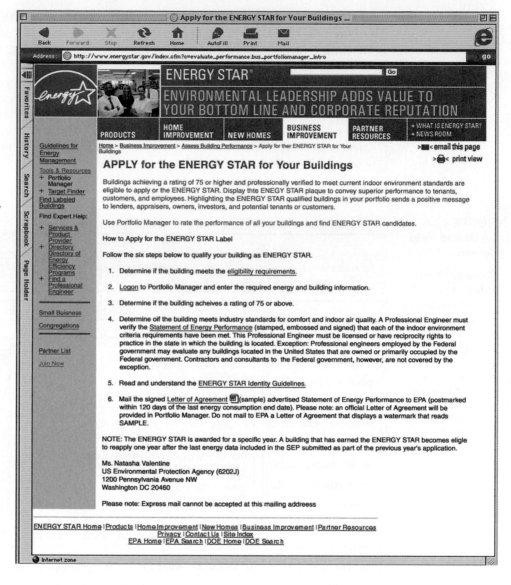

ELEMENTS OF USABLE INSTRUCTION

Clear and Limiting Title

Give an
immediate
forecast

The title "Instructions for Cleaning the Drive Head of a Laptop Computer" tells
users what to expect: instructions for a specific task involving a selected part. But
the title "The Laptop Computer" gives no forecast; a document with this title
might contain a history of the laptop, a description of each part, or a wide range of
related information.

Informed Content

Know the
procedure

Make sure you know exactly what you're talking about. Ignorance on your part makes you no less liable for faulty or inaccurate instructions:

Ignorance
provides no
legal excuse

> If the author of [a car repair] manual had no experience with cars, yet provided faulty instructions on the repair of the car's brakes, the home mechanic who was injured when the brakes failed may recover [damages] from the author. (Walter and Marsteller 165)

Unless you have performed the task often, do not try to write instructions for it.

Visuals

Instructions often include a persuasive dimension: to promote interest, commitment, or action. In addition to showing what to do, visuals attract the user's attention and help keep words to a minimum.

Types of visuals especially suited to instructions include icons, representational and schematic diagrams, flowcharts, photographs, and prose tables.

Visuals generated by computers are especially useful for instructions. Other sources for instructional graphics include clip art, electronic scanning, and downloading from the Internet. (Page 329 describes useful Web sites and discusses legal issues in the use of computer graphics.)

To use visuals effectively, consider these suggestions:

How to use
instructional
visuals

- Illustrate any step that might be hard for users to visualize. The less specialized your users, the more visuals they are likely to need.
- Parallel the user's angle of vision in performing the activity or operating the equipment. Name the angle (side view, top view) if you think people will have trouble figuring it out for themselves.
- Avoid illustrating any action simple enough for users to visualize on their own, such as "PRESS RETURN" for anyone familiar with a keyboard.

Figure 22.6 presents an array of visuals and their specific instructional functions. Each of these visuals is easily constructed and some could be further enhanced, depending on your production budget and graphics capability. Writers and editors often provide an *art brief* (page 318) and a rough sketch describing the visual and its purpose for the graphic designer or art department.

22.2
What level of
technicality is
culturally appropriate?
Learn more at
<www.ablongman.com/
lannonweb>

Appropriate Level of Detail and Technicality

Unless you know your users have the relevant background and skills, write for laypersons, and do three things:

Provide exactly
and only what
users need

1. Give them enough background to understand why they need your instructions.
2. Give them enough detail to understand what to do.
3. Give them enough examples to visualize the steps clearly.

how to locate something

Installing a communication card

1 If your communication card has ports for connecting equipment, remove the plastic access cover from the vertical plate.

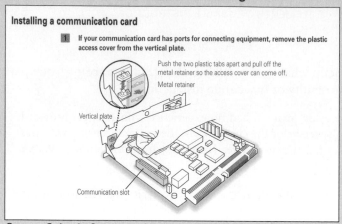

Push the two plastic tabs apart and pull off the metal retainer so the access cover can come off.

Metal retainer

Vertical plate

Communication slot

Source: © Apple Computer, Inc.*

how to operate something

Source: Superstock

how to handle something

Handling floppy disks

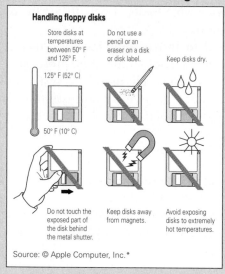

Store disks at temperatures between 50° F and 125° F.

125° F (52° C)

50° F (10° C)

Do not use a pencil or an eraser on a disk or disk label.

Keep disks dry.

Do not touch the exposed part of the disk behind the metal shutter.

Keep disks away from magnets.

Avoid exposing disks to extremely hot temperatures.

Source: © Apple Computer, Inc.*

how to assemble something

Extension Cord Retainer

1. Look into the end of the Switch Handle and you will see 2 slots. The WIDER end of the Retainer goes into the TOP slot (Figure 8).
2. Plug extension cord into Switch Handle and weave cord into Retainer, leaving a little slack (Figure 9).

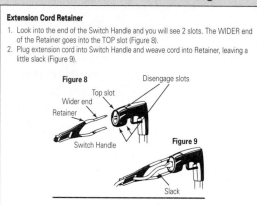

Figure 8

Disengage slots

Top slot

Wider end

Retainer

Switch Handle

Figure 9

Slack

Source: Courtesy of Black & Decker® (U.S.), Inc.

how to position something

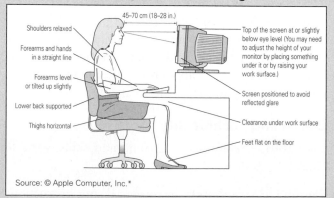

Shoulders relaxed

Forearms and hands in a straight line

Forearms level or tilted up slightly

Lower back supported

Thighs horizontal

45–70 cm (18–28 in.)

Top of the screen at or slightly below eye level (You may need to adjust the height of your monitor by placing something under it or by raising your work surface.)

Screen positioned to avoid reflected glare

Clearance under work surface

Feet flat on the floor

Source: © Apple Computer, Inc.*

how to avoid damage or injury

△ **Important:** The fixing assembly in the printer operates at very high temperatures. When you need to open the printer, be careful not to touch the fixing assembly.△

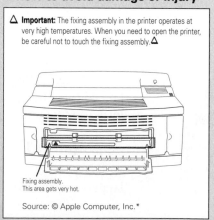

Fixing assembly. This area gets very hot.

Source: © Apple Computer, Inc.*

FIGURE 22.6   Common Types of Instructional Visuals and Their Functions

how to diagnose and solve problems

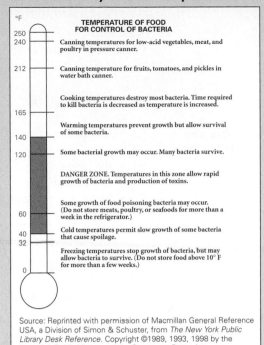

GENERAL TROUBLESHOOTING CHART

If the amplifier is otherwise operating satisfactorily the more common causes of trouble may generally be attributed to the following:
1. **Incorrect connections or loose terminal contacts.** Check the speakers, record player, tape deck, antenna and line cord.
2. **Improper operation.** Before operating any audio component, be sure to read the instructions.

3. **Improper location of audio components.** The proper positioning of components, such as speakers and turntable, is vital to stereo.
4. **Defective audio components.**

Following are some other common causes of malfunction and what to do about them.

PROGRAM	SYMPTOM	PROBABLE CAUSE	WHAT TO DO
AM, FM or MPX reception	a. Constant or intermittent noise heard at certain times or in a certain area	* Discharge or oscillation caused by electrical appliances, such as fluorescent lamps, TV sets, D.C. motors, rectifier and oscillator * Natural phenomena, such as atmospherics, static, and thunderbolt * Insufficient antenna input due to reinforced concrete walls or long distance from the station * Wave interference from other electrical appliances	* Attach a noise limiter to the electrical appliance that causes the noise, or attach it to the power source of the amplifier. * Install an outdoor antenna and ground the amplifier to raise the signal-to-noise ratio. * Reverse the power cord plug-receptacle connections. * If the noise occurs at a certain frequency. attach a wave trap to the ANT. input. * Place the set away from other electrical appliances.

Source: Courtesy of Sansui Electronic Co. Ltd.

how to proceed systematically

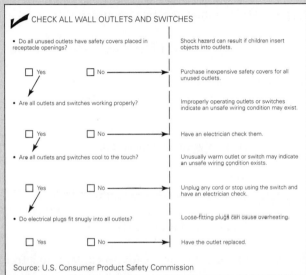

✔ CHECK ALL WALL OUTLETS AND SWITCHES

• Do all unused outlets have safety covers placed in receptacle openings?

☐ Yes   ☐ No ──→ Shock hazard can result if children insert objects into outlets.

Purchase inexpensive safety covers for all unused outlets.

• Are all outlets and switches working properly?

☐ Yes   ☐ No ──→ Improperly operating outlets or switches indicate an unsafe wiring condition may exist.

Have an electrician check them.

• Are all outlets and switches cool to the touch?

☐ Yes   ☐ No ──→ Unusually warm outlet or switch may indicate an unsafe wiring condition exists.

Unplug any cord or stop using the switch and have an electrician check.

• Do electrical plugs fit snugly into all outlets?

☐ Yes   ☐ No ──→ Loose-fitting plugs can cause overheating.

Have the outlet replaced.

Source: U.S. Consumer Product Safety Commission

how to make the right decisions

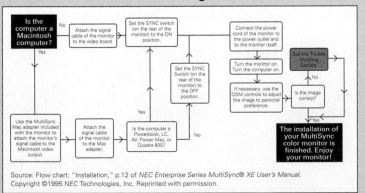

Source: Flow chart: "Installation," p.12 of *NEC Enterprise Series MultiSync® XE User's Manual.* Copyright ©1995 NEC Technologies, Inc. Reprinted with permission.

how to identify safe or acceptable limits

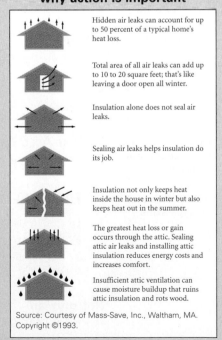

Source: Reprinted with permission of Macmillan General Reference USA, a Division of Simon & Schuster, from *The New York Public Library Desk Reference.* Copyright ©1989, 1993, 1998 by the New York Public Library and the Stonesong Press, Inc.

why action is important

Hidden air leaks can account for up to 50 percent of a typical home's heat loss.

Total area of all air leaks can add up to 10 to 20 square feet; that's like leaving a door open all winter.

Insulation alone does not seal air leaks.

Sealing air leaks helps insulation do its job.

Insulation not only keeps heat inside the house in winter but also keeps heat out in the summer.

The greatest heat loss or gain occurs through the attic. Sealing attic air leaks and installing attic insulation reduces energy costs and increases comfort.

Insufficient attic ventilation can cause moisture buildup that ruins attic insulation and rots wood.

Source: Courtesy of Mass-Save, Inc., Waltham, MA. Copyright ©1993.

FIGURE 22.6   Common Types of Instructional Visuals and Their Functions *(continued)*

These next examples show how you might adapt instructions titled "How to Create a Floppy Disk Backup to a Hard Disk" for novice Macintosh users.

PROVIDE BACKGROUND.   Begin by explaining the purpose of the task.

Tell users why they are doing this

> You might easily lose information stored on a hard disk if:
>
> • the disk is damaged by repeated use, jarring, moisture, or extreme temperature;
> • the disk is erased by a power surge, a computer malfunction, or a user error; or
> • the stored information is scrambled by a nearby magnet (telephone, computer terminal, or the like).
>
> Always make a backup copy of any disk that contains important material.

Also, state your assumptions about the user's level of technical understanding.

Spell out what users should already know

> To follow these instructions, you should be able to identify these parts of a Macintosh system: computer, monitor, keyboard, mouse, floppy disk drive, and 3.5-inch floppy disk.

Define any specialized terms that appear in your instructions.

Tell users what each key term means

> *Initialize:* Before you can store or retrieve information on a disk, you must initialize the blank disk. Initializing creates a format the computer can understand—a directory of specific memory spaces (like post office boxes) on the disk where you can store information and retrieve it as needed.

When the user understands *what* and *why*, you are ready to explain *how* to carry out the task.

Make instructions complete but not excessive

PROVIDE ADEQUATE DETAIL.   Include enough detail for users to understand and perform the task successfully, but omit general information that users probably know.

Inadequate detail for laypersons

> **FIRST AID FOR ELECTRICAL SHOCK**
>
> 1. Check vital signs.
> 2. Establish an airway.
> 3. Administer CPR as needed.
> 4. Treat for shock.

Not only are the above details inadequate, but terms such as "vital signs" and "CPR" are too technical for laypersons. Such instructions posted for workers in a high-voltage area would be useless. Illustrations and explanations are needed, as in the instructions on the following page for item 3 above, administering CPR.

Adequate detail for laypersons

METHODS OF CARDIOPULMONARY RESUSCITATION (CPR)

Mouth-to-Mouth Breathing

Step 1: If there are no signs of breathing or there is no significant pulse, place one hand under the victim's neck and gently lift. At the same time, push with the other hand on the victim's forehead. This will move the tongue away from the back of the throat to open the airway. If available, a plastic "stoma," or oropharyngeal airway device, should be inserted now.

Step 2: While maintaining the backward head tilt position, place your cheek and ear close to the victim's mouth and nose. Look for the chest to rise and fall while you listen and feel for breathing. Check for about 5 seconds.

Step 3: Next, while maintaining the backward head tilt, pinch the victim's nose with the hand that is on the victim's forehead to prevent leakage of air, open your mouth wide, take a deep breath, seal your mouth around the victim's mouth, and blow into the victim's mouth with four quick but full breaths. For an infant, give gentle puffs and blow through the mouth and nose *and* do not tilt the head back as far as for an adult.

If you do not get an air exchange when you blow, it may help to reposition the head and try again.

If there is still no breathing, give one breath every 5 seconds for an adult and one gentle puff every 3 seconds for an infant until breathing resumes.

If the victim's chest fails to expand, the problem may be an airway obstruction. Mouth-to-mouth respiration should be interrupted briefly to apply first aid for choking.

Step 1   Step 2   Step 3

Source: Reprinted with permission from New York Public Library Desk Reference, *3rd ed., copyright © 1998, 1993, 1989 by The New York Public Library and the Stonesong Press, Inc.*

Don't assume that people know more than they really do, especially when you can perform the task almost automatically. (Think about when someone taught you to drive a car—or perhaps you have tried to teach someone else.) Always assume that the user knows less than you. A colleague will know at least a little less; a layperson will know a good deal less—maybe nothing—about this procedure.

Exactly how much information is enough? Consider these suggestions:

How to provide
the right amount
of detail

- Give everything users need, so the instructions can stand alone.
- Give only what users need. Don't tell them how to build a computer when they only need to know how to copy a disk.
- Instead of focusing on the *product* ("How does it work?"), focus on the *task* ("How do I use it?" or "How do I do it?") (Grice, "Focus" 132).
- Omit steps (*Seat yourself at the computer*) that are obvious to users.
- Divide the task into simple steps and substeps. Allow users to focus on one step at a time.
- Adjust the *information rate* ("the amount of information presented in a given page," Meyer 17) to the user's background and the difficulty of the task. For complex or sensitive steps, slow the information rate. Don't make users do too much too fast.
- Reinforce the prose with visuals. Don't be afraid to repeat information if it saves users from flipping pages.
- When writing instructions for consumer products, assume "a barely literate reader" (Clement 151). Simplify.
- Recognize the persuasive dimension of the instructions. Users may need persuading that this procedure is necessary or beneficial, or that they can complete this procedure easily and competently.

OFFER EXAMPLES.   Instructions require specific examples (how to load a program, how to order a part):

> To load your program, type this command:
>
> > Load "Style Editor"
>
> Then press RETURN.

Like visuals, examples *show* users what to do.

INCLUDE TROUBLESHOOTING ADVICE.   Anticipate things that commonly go wrong when this task is performed—the paper jams in the printer, the tray of the CD-ROM drive won't open, or some other malfunction. Explain the probable cause(s) and offer solutions.

Explain what to
do when things
go wrong

> *Note:* IF *X* doesn't work, first check *Y* and then do *Z*.

In the instructions that follow, careful use of background, detailed explanation, visual examples, and troubleshooting advice create a user-friendly level of technicality for computer novices.

FIRST STEP: HOW TO INITIALIZE YOUR FLOPPY DISK

Background

Before you can copy or store information on a blank disk, you must initialize the disk. Follow this procedure:

Instructional details

1. Switch on the computer.
2. Insert your floppy disk in the floppy disk drive. Unable to recognize this new disk, the computer will ask if you wish to initialize the disk (Figure 1).

Visual example reinforces the verbal message

FIGURE 1 The "Initialize" Message

3. Using your mouse, place the on-screen pointer (small arrow) inside the "Initialize" box.
4. Press and quickly release the mouse button. Within 15–20 seconds, the initializing will be completed, and a message will appear, asking you to name your disk (Figure 2).

Visual appears close to the related step

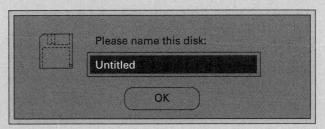

FIGURE 2 The "Disk Naming" Message

Troubleshooting tips

NOTE: If your disk is rejected, it might be improperly seated in the drive or damaged, or the drive itself might be damaged.

　　a. *Eject and reinsert the disk, and repeat steps 2–4.*
　　b. *If (a) fails, insert a different disk. If this doesn't work, have your disk drive checked.*

Make verbal and visual information redundant

In the previous sample, visuals and prose are *redundant* (Weiss 100). Chapter 13 warns against *style redundancy* (extra words that give no extra information). Effective instructions, however, often exhibit *content redundancy*, giving the same information in verbal and visual form. When you can't be sure how much is enough, risk overexplaining rather than underexplaining.

Logically Ordered Steps

Organize for
the user's
understanding

Instructions are almost always arranged in chronological order, with warnings and precautions inserted for specific steps.

Show how the
steps are
connected

> You can't splice two wires to make an electrical connection until you have removed the insulation. To remove the insulation, you will need . . . .

Notes and Hazard Notices

Alert users
to special
considerations
and hazards

Here are the only items that should interrupt the steps in a set of instructions (Van Pelt 3):

- A *note* clarifies a point, emphasizes vital information, or describes options or alternatives.

> NOTE: If you don't name a newly initialized disk, the computer automatically names it "Untitled."

While a note is designed to enhance performance and prevent error, the following hazard notices—ranked in order of severity—are designed to prevent damage, injury, or death.

- A *caution* prevents possible mistakes that could result in injury or equipment damage:

The least forceful
notice

> CAUTION: A momentary electrical surge or power failure will erase the contents of internal memory. To avoid losing your work, every few minutes save on disk what you have just typed into the computer.

- A *warning* alerts users to potential hazards to life or limb:

A moderately
forceful notice

> WARNING: To prevent electrical shock, always disconnect your printer from its power source before cleaning internal parts.

- A *danger* notice identifies an immediate hazard to life or limb:

The most forceful
notice

> DANGER: The red canister contains DEADLY radioactive material. **Do not break the safety seal** under any circumstances.

Content
requirements for
hazard notices

Inadequate notices of warning, caution, or danger are a common cause of lawsuits (page 536). Each hazard notice is legally required to (1) describe the specific hazard, (2) spell out the consequences of ignoring the hazard, and (3) offer instruction for avoiding the hazard (Manning 15).

Visual
requirements for
hazard notices

Even the most emphatic verbal notice might be overlooked by an impatient or inattentive user. Direct the user's attention with symbols, or icons, as a visual signal (Bedford and Stearns 128):

Use hazard
symbols

Warning **Do not enter** **Radioactivity** **Fire danger**

Visibility
requirements for
hazard notices

Keep the hazards prominent in the user's awareness: Preview the hazards in your introduction and place each notice, clearly highlighted (by a ruled box, a distinct typeface, larger typesize, or color), immediately before the related step.

 NOTE

Use hazard notices only when needed; overuse will dull their effect, and readers may overlook their importance.

Readability

Make instructions
immediately
readable

Instructions must be understood on the first reading because users usually take *immediate* action.

Like descriptions (page 507), instructions name parts, use location and position words, and state exact measurements, weights, and dimensions. Instructions also require strict attention to phrasing, sentence structure, and paragraph structure.

USE DIRECT ADDRESS, ACTIVE VOICE, AND IMPERATIVE MOOD.   To emphasize the user's role, write instructions in the second person, as direct address.

In general, begin all steps and substeps with action verbs, using the *active voice* and *imperative mood* ("Insert the disk" instead of "The disk should be inserted" or "You should insert the disk").

Indirect or
confusing

- The user types in his or her access code.
- You should type in your access code.
- It is important to type in the access code.
- The access code is typed in.

In this next version, the opening verb announces the specific action required.

Clear and direct

Type in your access code.

In certain cases, you may want to provide a clarifying word or phrase that precedes the verb (*Read Me* 130):

Information that might precede the verb

- [To log on,] **type in** your access code.
- [If your screen displays an error message,] **restart** the computer.
- [Slowly] **scan** the seal for gamma ray leakage.
- [In the Edit menu,] **click** on Paste.

 NOTE

Certain cultures consider the direct imperative bossy and offensive. For cross-cultural audiences, you might rephrase an instruction as a declarative statement: from "Type in your access code" to "The access code should be typed in." Or you might use an indirect imperative such as "Be sure to type in your access code" (Coe, "Writing" 18).

USE SHORT AND LOGICALLY SHAPED SENTENCES.　Use shorter sentences than usual, but don't "telegraph" your message by omitting articles (*a, an, the*), as on page 246. Use one sentence for one step, so users can perform one step at a time.

If a single step covers two related actions, describe these actions in their required sequence:

Confusing

Before switching on the computer, insert the disk in the drive.

Logical

Insert the disk in the drive; then switch on the computer.

Simplify explanations by using a familiar-to-unfamiliar sequence:

Hard

You must initialize a blank disk before you can store information on it.

Easier

Before you can store information on a blank disk, you must initialize the disk.

USE PARALLEL PHRASING.　Parallelism is important in all writing but especially in instructions, because repeating grammatical forms emphasizes the step-by-step organization.

Not parallel

To log on to the VAX 20, follow these steps:

1. Switch the terminal to "on."
2. The CONTROL key and C key are pressed simultaneously.
3. Typing LOGON, and pressing the ESCAPE key.
4. Type your user number, and then press the ESCAPE key.

All steps should be in identical grammatical form:

Parallel

To log on to the VAX 20, follow these steps:

1. Switch the terminal to "on."
2. Press the CONTROL key and C key simultaneously.
3. Type LOGON, and then press the ESCAPE key.
4. Type your user number, and then press the ESCAPE key.

PHRASE INSTRUCTIONS AFFIRMATIVELY.   Research shows that users respond more quickly and efficiently to instructions phrased affirmatively rather than negatively (Spyridakis and Wenger 205).

Negative

Affirmative

> Verify that your disk is not contaminated with dust.
>
> Examine your disk for dust contamination.

USE TRANSITIONS TO MARK TIME AND SEQUENCE.   Transitional expressions bridge related ideas. Some transitions ("in addition," "next," "meanwhile," "finally," "ten minutes later," "the next day," "before") mark time and sequence. They help users understand the step-by-step process:

Transitions
enhance
continuity

> ## PREPARING THE GROUND FOR A TENT
>
> Begin by clearing and smoothing the area that will be under the tent. This step will prevent damage to the tent floor and eliminate the discomfort of sleeping on uneven ground. **First,** remove all large stones, branches, or other debris within a level 10 × 13-foot area. Use your camping shovel to remove half-buried rocks that cannot easily be moved by hand. **Next,** fill in any large holes with soil or leaves. **Finally,** make several light surface passes with the shovel or a large, leafy branch to smooth the area.

Effective Design

An effective instructional design conveys the sense that the task is within a qualified user's range of abilities.

To help users find, recognize, and remember what they need, follow these suggestions:

How to design
instructions

- *Provide informative headings.* A heading such as "How to Initialize Your Blank Disk" is more informative than "Disk Initializing."
- *Arrange steps in a numbered list.* Unless the procedure consists of simple steps (as in "Preparing the Ground for a Tent," above), list and number each step. Numbered steps not only announce the sequence of steps, but also help users remember where they left off.
- *Separate each step visually.* Single-space within steps and double-space between.
- *Make warning, caution, and danger notices highly visible.* Use ruled boxes or highlighting, and plenty of white space.
- *Keep the visual and the step close together.* If room allows, place the visual right beside the step; if not, right after it. Set off the visual with plenty of white space.
- *Consider a multicolumn design.* If steps are brief and straightforward and require back-and-forth reference from prose to visuals, consider multiple

columns. Figure 22.7 shows how to connect peripheral devices (scanners, extra drives, and so on) to the computer using SCSI (Small Computer System Interface) cables.

- *Keep it simple.* Users can be overwhelmed by a page with excessive or inconsistent designs.
- *For lengthy instructions, consider a layered approach.* In preparing a complex manual, for instance, you might add a brief reference card or a guide for getting started or for easy reference, as in Figure 22.8.

Consult Chapter 15 for additional design considerations.

NOTE   *Like any material displayed on a computer screen, online instructions have their own design requirements, which are discussed in Chapters 15 and 19. Also, despite online documentation's increasing popularity, many users continue to find printed manuals more convenient and easier to navigate (Foster 10).*

Connecting cables

▲ **Warning:** When making SCSI connections, always turn off power to all devices in the chain. Failure to do so can cause the loss of information and damage to your equipment. ▲

1. **Shut down your PowerBook and all SCSI devices in the chain.**

2. **To connect the first device, use an Apple HDI-30 SCSI System Cable.**

   Attach the smaller end of the cable to your computer's SCSI port (marked with the icon ◈) and the larger end of the cable to either SCSI port on the device.

3. **To connect the next device, use a SCSI peripheral interface cable.**

   Both cable connectors are the same. Attach one connector to the available SCSI port on the first device, and the other connector to either SCSI port on the next device.

4. **Repeat step 3 for each additional device you want to connect.**

The illustration shows where to add cable terminators.

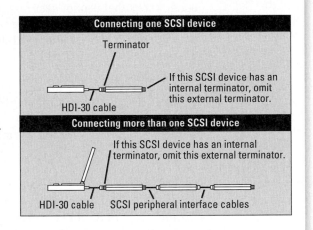

△ **Important:** The total length of an SCSI chain should not exceed 20 feet (6 meters). Apple SCSI cables are designed to meet this restriction. If you are using SCSI cables from another vendor, check the length of the chain. △

Once your SCSI devices are connected, always turn them on before turning on your PowerBook. If you turn the computer on first, it may not be able to start up, or it may not recognize the SCSI devices.

FIGURE 22.7   A Multicolumn Design
Source: Reprinted by permission of Apple Computer, Inc.

Quick Use Guide

This guide should only be used once you are familiar with the operations inside this manual. If you have any problems or need more information, please refer to the page listed.

Clock Reset (page 13)

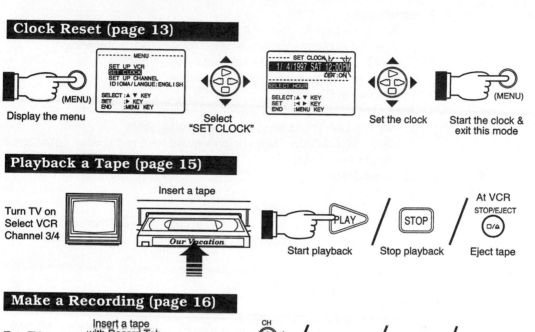

Playback a Tape (page 15)

Make a Recording (page 16)

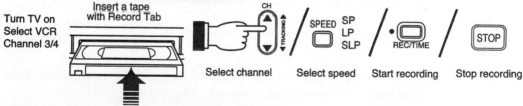

Set a Timer Recording (page 18)

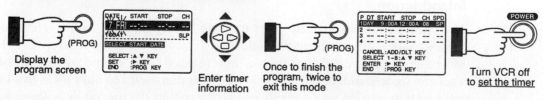

FIGURE 22.8   Layered Instructions.   Notice how this Quick Use Guide for a videocassette recorder presents concise, user-friendly instructions and cross-references to pages containing more detailed and technical information.

Source: Panasonic Omnivision VHS Operating Instructions. Panasonic Consumer Electronics. Used by permission.

22.3

For more sample
instructions visit
<www.ablongman.com/
lannonweb>

AN OUTLINE AND MODEL FOR INSTRUCTIONS

You can adapt the following outline to any instructions. Here are the possible sections to include:

I. Introduction
   A. Definition, Benefits, and Purpose of the Procedure
   B. Intended Audience (usually omitted for workplace audiences)
   C. Prior Knowledge and Skills Needed by the Audience
   D. Brief Overall Description of the Procedure
   E. Principle of Operation
   F. Materials, Equipment (in order of use), and Special Conditions
   G. Working Definitions (always in the introduction)
   H. Warnings, Cautions, Dangers (previewed here and spelled out at steps)
   I. List of Major Steps

II. Required Steps
   A. First Major Step
      1. Definition and purpose
      2. Materials, equipment, and special conditions for this step
      3. Substeps (if applicable)
         a.
         b.
   B. Second Major Step (and so on)

III. Conclusion
   A. Review of Major Steps (for a complex procedure only)
   B. Interrelation of Steps
   C. Troubleshooting or Follow-up Advice (as needed)

This outline is only tentative; you might modify, delete, or combine some elements, depending on your subject, purpose, and audience.

Introduction

The introduction should help users to begin "doing" as soon as they are able to proceed safely, effectively, and confidently (van der Meij and Carroll 245–46). Most users are interested primarily in "how to use it or fix it," and will require only a general understanding of "how it works." You don't want to bury users in a long introduction, nor do you want to set them loose on the procedure without adequate preparation. Know your audience—what they need and don't need.

Following is an introduction from instructions for people using a college library. Some users will be computer experts; some will be novices—but all will require a detailed introduction to computerized literature searches before they conduct a search on their own.

Clear and limiting title	# HOW TO USE THE OCLC TERMINAL TO SEARCH FOR A BOOK
	## Introduction
Definition of the procedure	Our library's OCLC (Online Computer Library Center) terminal offers an efficient way to search for books, journals, government publications, and other printed materials. This terminal is connected to the OCLC database in Columbus, Ohio.
Definition (continued)	The Ohio database contains more than 8 million records of books and other published materials in libraries nationwide. Each record lists the information found on a catalog card, and identifies the libraries holding the work.
Purpose	These instructions will enable you to determine whether a book you seek can be found in our library or in other libraries throughout the country.
Audience and required skills	Any library patron can use the OCLC terminal to search for a book. To operate the terminal, you only need to be familiar with the keyboard (Figure 1) and to have an operator's manual handy for general reference.
General description of the procedure	After logging on the system, you can search for a book by TITLE, AUTHOR, or AUTHOR and TITLE. Once you have viewed the catalog entry for the book you seek, you can use the terminal to determine the libraries from which you might borrow the book. These instructions show a search by TITLE only.
Materials and equipment	The only additional equipment you need is a copy of the *Manual of OCLC Participating Institutions*, to find a listing of library names according to their OCLC symbol displayed on your terminal screen.
Working definitions	Specialized terms that appear in these instructions are defined here:
	Cursor: a small, blinking rectangle that indicates the screen position of the next character you will type.
	HOME Position: the cursor location at the extreme top left of your screen.
	HOME position is where you will type most of your messages to the computer, and where the computer will display its messages to you.

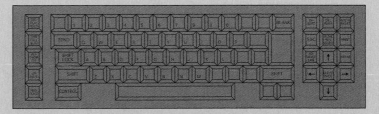

FIGURE 1   The OCLC Terminal Keyboard

Cautions and notes	Pay close attention to the NOTES that accompany the logging-on step and to the CAUTION preceding the logging-off step.
List of major steps	The major steps in using OCLC to search for a book by its TITLE are (1) logging on the system, (2) initiating the search, (3) viewing the information on your title, (4) locating the book, and (5) logging off.

Body: Required Steps

In the body section (labeled Required Steps), give each step and substep in order. Insert warnings, cautions, and notes as needed. Begin each step with its definition or purpose or both. Users who understand the reasons for a step will do a better job. A numbered list (like the following) is an excellent way to segment the steps.

Heading previews the step

Capitals and spacing set off note

Headings 1–5 offer an overview of the process

Example shows what to do

Visual reinforces the prose

Required Steps

1. *Logging on the OCLC System*
   To activate the system, you must first log on.

   a. Flick the red power switch (right edge of keyboard).

      NOTE: *Press HOME before typing, to place the cursor at HOME position.*

   b. Type the authorization number (07–34–6991).

      NOTE: *If you make a typing error, place the cursor over the error and retype. Move the cursor by using the arrow keys.*

   c. Press SEND and wait for the $\boxed{\text{HELLO}}$ response.

2. *Initiating the Search*
   To initiate a computer search, enter your title's exact code name.

   a. Type in the code for your desired title, excluding articles (*a, an, the*) when they are the first word of the title. Type your entry exactly like this:

      Suppose your title is *The Logic of Failure;* you would type **log, of, fai**

   b. Press DISPLAY/RECD and then SEND.
      The computer will display a title that matches the code you have entered (Figure 2).

```
The Logic of Failure: recognizing and avoiding error in
complex situations/Dorner, Dietrich. Metropolitan Books.
Distributed in the U.S. and Canada by Addison-Wesley,
c 1996.
```

FIGURE 2   The Title That Matches the Code You Typed

3. *Viewing the Information on Your Title*
   [Steps for viewing detailed information on the title, locating the book, and logging off continue.]

Conclusion

The conclusion of a set of instructions has several possible functions:

- You might summarize the major steps in a long and complex procedure, to help users review their performance.

- You might describe the results of the procedure.
- You might offer follow-up advice about what could be done next or refer the user to further sources of documentation.
- You might give troubleshooting advice about what to do if anything goes wrong.

You might do all these things—or none of them. If your procedural section has given users all they need, omit the conclusion altogether.

In the case of the OCLC instructions, these concluding remarks are useful and appropriate:

Follow-up advice

Troubleshooting advice

Conclusion

Once you locate the title you seek, ask a reference librarian for help requesting the book via Interlibrary Loan. (Books usually arrive within two days.)

In case the first library is unable to supply the book requested, you should initially choose several libraries from your printout. The Interlibrary Loan system will automatically forward your request to each lender you have specified, until one lender indicates it can supply the book.

A SITUATION REQUIRING INSTRUCTIONS

The instructions for felling a tree in Figure 22.9 are patterned after the general outline, shown earlier. They will appear as part of a manual for forestry students who are about to begin summer jobs with the Idaho Forestry Service.

Instructions for a Semitechnical Audience

Audience and Use Profile.   These instructions are aimed at a partially informed audience who knows how to use chainsaws, axes, and wedges but who is approaching this dangerous procedure for the first time. Therefore, no visuals of cutting equipment (chainsaws and so on) are included because the audience knows what these items look like. Basic information (such as what happens when a tree binds a chainsaw) is omitted because the audience has this knowledge also. Likewise, these users need no definition of general forestry terms such as *culling* and *thinning*, but they *do* need definitions of terms that relate specifically to tree felling (*undercut, holding wood,* and so on).

For clarity, visuals illustrate the final three steps. The conclusion, for these users, will be short and to the point—a simple summary of major steps with emphasis on safety. ❏

INSTRUCTIONS FOR FELLING A TREE   1

INTRODUCTION
Forestry Service personnel fell (cut down) trees to cull or thin a forested plot, to eliminate the hazard of dead trees standing near power lines, to clear an area for construction, and the like.

These instructions explain how to remove sizable trees for personnel who know how to use a chainsaw, axe, and wedge safely.

When you set out to fell a tree, expect to spend most of your time planning the operation and preparing the area around the tree. To fell the tree, you will make two chainsaw cuts, severing the stem from the stump. Depending on the direction of the cuts, the weather, and the terrain, the severed tree will fall into a predetermined clearing.

> **WARNING:** Although these instructions cover the basic procedure, felling is very dangerous. Trees, felling equipment, and terrain vary greatly. Even professionals are sometimes killed or injured because of judgment errors or misuse of tools.
>
> Your main concern is safety. Be sure to have an expert demonstrate this procedure before you try it. Also, pay attention to warnings in steps 1 and 4.

To fell sizable trees, you will need this equipment:
  —a 3- to 5-horsepower chainsaw with a 20-inch blade
  —a single-blade splitting axe
  —two or more 12-inch steel wedges

The major steps in felling a tree are (1) choosing the lay, (2) providing an escape path, (3) making the undercut, and (4) making the backcut.

REQUIRED STEPS
1. *How to Choose the Lay*
   The "lay" is where you want the tree to fall. On level ground in an open field, which way you direct the fall makes little difference. But such ideal conditions are rare.

   Consider ground obstacles and topography, surrounding trees, and the condition of the tree to be felled. Plan your escape path and the location of your cuts depending on surrounding houses, electrical wires, and trees. Then follow these steps:

   a. Make sure the tree is still alive.

FIGURE 22.9   A Set of Instructions

2

b. Determine the direction and amount of lean.

> **WARNING:** If the tree is dead and leaning substantially, do not try to fell it without professional help. Many dead, leaning trees have a tendency to split along their length, causing a massive slab to fall spontaneously.

c. Find an opening into which the tree can fall in the direction of lean or as close to the direction of lean as possible.

Because most trees lean downhill, try to direct the fall downhill. If the tree leans slightly away from your desired direction, use wedges to direct the fall.

2. *How to Provide an Escape Path*
Falling trees are unpredictable. Avoid injury by planning a definite escape path. Follow these steps:

a. Locate the path in the direction opposite of the fall (Figure 1).

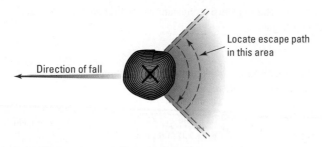

Direction of fall

Locate escape path
in this area

Figure 1 Escape Path Location

b. Clear a path 2 feet wide extending beyond where the top of the tree could land.

3. *How to Make the Undercut*
The undercut is a triangular slab of wood cut from the trunk on the side toward which you want the tree to fall. Follow these steps:

a. Start the chainsaw.

b. Holding the saw with blade parallel to the ground, make a first cut 2 to 3 feet above ground. Cut horizontally, to no more than 1/3 the tree's diameter (Figure 2).

FIGURE 22.9   A Set of Instructions (*continued*)

3

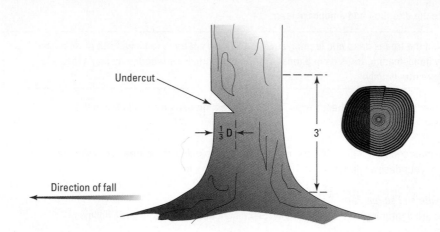

Figure 2 Making the Undercut

   c. Make a downward-sloping cut, starting 4 to 6 inches above the first so that the cuts intersect at 1/3 the diameter (Figure 2).

4. *How to Make the Backcut*
After completing the undercut, you make the backcut to sever the stem from the stump. This step requires good reflexes and absolute concentration.

> **WARNING:** Observe tree movement closely during the backcut. If the tree shows any sign of falling in your direction, drop everything and move out of its way.
>
>   Also, do not cut completely through to the undercut; instead, leave a narrow strip of "holding wood" as a hinge, to help prevent the butt end of the falling tree from jumping back at you.

To make your backcut, follow these steps:

   a. Holding the saw with the blade parallel to the ground, start your cut about 3 inches above the undercut, on the opposite side of the trunk. Leave a narrow strip of "holding wood" (Figure 3).

FIGURE 22.9   A Set of Instructions *(continued)*

4

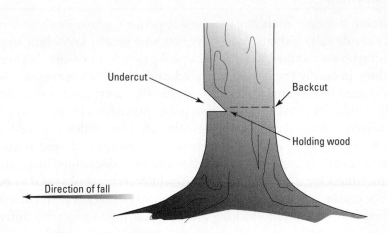

Figure 3 Making the Backcut

NOTE: If the tree begins to bind the chainsaw, hammer a wedge into the backcut with the blunt side of the axe head. Then continue cutting.

b. As soon as the tree begins to fall, turn off the chainsaw, withdraw it, and step back immediately—the butt end of the tree could jump back toward you.

c. Move rapidly down the escape path.

CONCLUSION

Felling is a complex and dangerous procedure. Choosing the lay, providing an escape path, making the undercut, and making the backcut are the basic steps—but trees, terrain, and other circumstances vary greatly.

For the safest operation, seek professional advice and help whenever you foresee *any* complications whatsoever.

FIGURE 22.9 A Set of Instructions *(continued)*

PROCEDURES

The difference between instructions and procedures

Instructions show an uninitiated user how to perform a task. *Procedures,* on the other hand, provide rules and guidance for people who usually know how to perform the task but who are required to follow accepted practice. To ensure that everyone does something in exactly the same way, procedures typically are aimed at groups of people who need to coordinate their activities so that everyone's performance meets a certain standard. Consider, for example, police procedures for properly gathering evidence from a crime scene: strict rules stipulate how evidence should be collected and labeled and how it should be preserved, transported, and stored. Evidence shown to have been improperly handled is routinely discredited in a courtroom.

Procedures help ensure safety

Procedures are useful in situations in which certain tasks need to be standardized. For example, if different people in your organization perform the same task at different times (say, monitoring groundwater pollution) with different equipment, or under different circumstances, this procedure may need to be standardized

---

V. SOIL AND GROUNDWATER ASSESSMENT

H. DECONTAMINATION PROCEDURES

The equipment decontamination procedures used during the fieldwork must be described in the site assessment report. The following procedures, at a minimum, must be used for both soil and groundwater sampling equipment:

1. **Drilling or Other Equipment**
   The drilling bits and augers must be stream cleaned between each boring and after each use.
2. **Sampling Equipment**
   a. Reusable bailers must be steam cleaned or one-time-use disposable bailers must be used.
   b. The cord used with the bailers must be discarded after each use.
   c. Sampling equipment that is not steam cleaned must be initially washed with a non-phosphate detergent, rinsed twice with tap water, and final rinsed with deionized or distilled water.
2. **Rinseate**
   The soil and water from washing, rinsing, and steam cleaning must be properly containerized and labeled for disposal.

FIGURE 22.10 A Standard Operating Procedure. Part of a manual for dealing with leaking underground fuel tanks, this SOP is aimed at technicians already familiar with techniques such as "steam cleaning" and "containerizing." However, to prevent contamination of testing equipment, each technician needs to follow this strict sequence of steps.
Source: Ventura County LUFT Guidance Manual. *Ventura, CA. April 2001.*

to ensure that all work is done with the same accuracy and precision. A document known as a *Standard Operating Procedure (SOP)* becomes the official guideline for that task (Gurak and Lannon, 2nd ed. 229), as shown in Figure 22.10.

Procedures help keep everyone "on the same page."

Organizations also need to follow strict safety procedures, say, as defined by the U.S. Occupational Safety and Health Administration (OSHA). As laws and policies change, such procedures are often updated. The written procedures must be posted for employees to read (229). Figure 22.11 shows one page outlining OSHA regulations for evacuating high-rise buildings.

The steps in a procedure may or may not be numbered, depending on whether they must be performed in strict sequence, as in Figure 22.10 versus Figure 22.11.

☑ CHECKLIST for Usability of Instructions

Use this checklist to evaluate the usability of instructions. (Numbers in parentheses refer to the first page of discussion.)

Content

☐ Does the title promise exactly what the instructions deliver? (538)
☐ Is the background adequate for the intended audience? (542)
☐ Do explanations enable users to understand what to do? (542)
☐ Do examples enable users to see how to do it correctly? (544)
☐ Are the definition and purpose of each step given as needed? (542)
☐ Is all needless information omitted? (544)
☐ Are all obvious steps omitted? (544)
☐ Do notes, cautions, or warnings appear whenever needed, before or with the step? (546)
☐ Is the information rate appropriate for the user's abilities and the difficulty of this procedure? (544)
☐ Are visuals adequate for clarifying the steps? (539)
☐ Do visuals repeat prose information whenever necessary? (545)
☐ Is everything accurate? (539)

Organization

☐ Is the introduction adequate without being excessive? (552)
☐ Do the instructions follow the exact sequence of steps? (546)
☐ Is each step numbered, if appropriate? (549)

☐ Is all the information for a particular step close together? (549)
☐ For a complex step, does each sentence begin on a new line? (544)
☐ For lengthy instructions, is a layered approach, with a brief reference card, more appropriate? (550)
☐ Is the conclusion necessary and, if necessary, adequate? (544)

Style

☐ Do introductory sentences have enough variety to maintain interest? (548)
☐ Does the familiar material appear *first* in each sentence? (548)
☐ Do steps generally have short sentences? (548)
☐ Does each step begin with an action verb? (547)
☐ Are all steps in the active voice and imperative mood? (547)
☐ Do all steps have parallel phrasing? (548)
☐ Are transitions adequate for marking time and sequence? (549)

Page Design

☐ Does each heading clearly tell users what to expect? (549)
☐ On a typed page, are steps single-spaced within, and double-spaced between? (549)
☐ Do white space and highlights set off discussion from steps? (549)
☐ Are notes, cautions, or warnings set off or highlighted? (549)
☐ Are visuals beside or near the step, and set off by white space? (549)

OSHA **FACT** *Sheet*

The National Fire Protection Association defines "high-rise building" as a building greater than 75 feet (25 m) in height where the building height is measured from the lowest level of fire department vehicle access to the floor of the highest occupiable story. Appropriate exits, alarms, emergency lighting, communication systems, and sprinkler systems are critical for employee safety. When designing and maintaining exits, it is essential to ensure that routes leading to the exits, as well as the areas beyond the exits, are accessible and free from materials or items that would impede individuals from easily and effectively evacuating. State and local building code officials can help employers ensure that the design and safety systems are adequate.

When there is an emergency, getting workers out of high-rise buildings poses special challenges. Preparing in advance to safely evacuate the building is critical to the safety of employees who work there.

What actions should employers take to help ensure safe evacuations of high-rise buildings?

Don't lock fire exits or block doorways, halls, or stairways.

- Test regularly all back-up systems and safety systems, such as emergency lighting and communication systems, and repair them as needed.
- Develop a workplace evacuation plan, post it prominently on each floor, and review it periodically to ensure its effectiveness.
- Identify and train floor wardens, including back-up personnel, who will be responsible for sounding alarms and helping to evacuate employees.
- Conduct emergency evacuation drills periodically.
- Establish designated meeting locations outside the building for workers to gather following an evacuation. The locations should be a safe distance from the building and in an area where people can assemble safely without interfering with emergency response teams.
- Identify personnel with special needs or disabilities who may need help evacuating

and assign one or more people, including back-up personnel, to help them.
- Ensure that during off-hour periods, systems are in place to notify, evacuate, and account for off-hour building occupants.
- Post emergency numbers near telephones.

What should workers know before an emergency occurs?

- Be familiar with the worksite's emergency evacuation plan;
- Know the pathway to at least two alternative exits from every room/area at the workplace;
- Recognize the sound/signaling method of the fire/evacuation alarms;
- Know who to contact in an emergency and how to contact them;
- Know how many desks or cubicles are between your workstation and two of the nearest exits so you can escape in the dark if necessary;
- Know where the fire/evacuation alarms are located and how to use them; and
- Report damaged or malfunctioning safety systems and back-up systems

What should employers do when an emergency occurs?

- Sound appropriate alarms and instruct employees to leave building.
- Notify police, firefighters, or other appropriate emergency personnel.
- Take a head count of employees at designated meeting locations, and notify emergency personnel of any missing workers.

What should employees do when an emergency occurs?

- Leave the area quickly but in an orderly manner, following the worksite's emergency evacuation plan. Go directly to the nearest fire-free and smoke-free stairwell recognizing that in some circumstances the only available exit route may contain limited amounts of smoke or fire.

Evacuating High-Rise Buildings

FIGURE 22.11   A Safety Procedure.   This page defines general safety and evacuation procedures to be followed by employers and employees. Each building in turn is required to have its own, specific procedures, based on such variables as location, design, and state law. *Source: U.S. Occupational Safety and Health Administration, 2003 <www.osha.gov>.*

EXERCISES

 For more exercises, visit <www.ablongman.com/lannon.>

1. Improve readability by revising the diction, voice, and design of these instructions.

   ### What to Do Before Jacking Up Your Car
   Whenever the misfortune of a flat tire occurs, some basic procedures should be followed before the car is jacked up. If possible, your car should be positioned on as firm and level a surface as is available. The engine has to be turned off; the parking brake should be set; and the automatic transmission shift lever must be placed in "park" or the manual transmission lever in "reverse." The wheel diagonally opposite the one to be removed should have a piece of wood placed beneath it to prevent the wheel from rolling. The spare wheel, jack, and lug wrench should be removed from the luggage compartment.

2. Select part of a technical manual in your field or instructions for a general audience and make a copy of the material. Using the checklist on page 561, evaluate the sample's usability. In a memo to your instructor, discuss the strong and weak points of the instructions. Or be prepared to explain in class why the sample is effective or ineffective.

3. Assume that colleagues or classmates will be serving six months as volunteers in agriculture, education, or a similar capacity in a developing country. Do the research and create a set of procedures that will prepare users for avoiding diseases and dealing with medical issues in that specific country. Topics might include safe food and water, insect protection, vaccinations, medical emergencies, and the like. Be sure to provide background on the specific health risks travelers will face. Design your instructions as a two-sided brief reference card, as a chapter to be included in a longer manual, or in some other format suggested by your instructor.

   Hint: Begin your research for this project by checking out the National Center for Disease Control's Web site at <www.cdc.gov/travel/>.

4. Choose a topic from this list, your major, or an area of interest. Using the general outline in this chapter as a model, outline instructions for a task that requires at least three major steps. Address a general audience, and begin by completing an audience and use profile. Include (a) all necessary visuals or (b) an "art brief" (page 318) and a rough diagram for each visual or (c) a "reference visual" (a copy of a visual published elsewhere) with instructions for adapting your visual from that one. (If you borrow visuals from other sources, provide full documentation.)

planting a tree	hitting a golf ball
hot-waxing skis	removing the rear
hanging wallpaper	wheel of a bicycle
filleting a fish	avoiding hypothermia

5. Assume that you are assistant to the communications manager for a manufacturer of outdoor products. Among the company's best-selling items are its various models of gas grills. Because of fire and explosion hazards, all grills must be accompanied by detailed instructions for safe assembly, use, and maintenance.

   One of the first procedures in the manual is the "leak test," to ensure that the gas supply-and-transport apparatus is leak free. One of the engineers has prepared the instructions in Figure 22.12. Before being published in the manual, they must be approved by communications management. Your boss directs you to evaluate the instructions for accuracy, completeness, clarity, and appropriateness, and to report your findings in a memo. Because of the legal implications, your evaluation must spell out all positive and negative details of content, organization, style, and design. (Use the checklist on page 561 as a guide.) The boss is busy and impatient, and expects your report to be no longer than two pages. Do the evaluation and write the memo.

6. Select any one of the instructional visuals in Figure 22.6 and write a prose version of those instructions—without using visual illustrations or special page design. Bring your version to

Tank must be filled prior to this step (See "Propane Tank" section for details about filling this tank)

- Hose
- Hose connections
- Valve
- Regulator fitting
- Tank valve
- Regulator
- Valve
- Valve to tank connection

Propane tank

Leak Test Check List
- Tank valve (all over) including area that screws into the tank
- Regulator fitting
- Hose connections (3 places)
- Valves (4 places)

SAFETY! **All models must be Leak Tested! Take the hose, valve, and regulator assembly outdoors in a well ventilated area. Keep away from open flames or sparks. Do Not smoke during this test. Use only a soap and water solution to test for leaks.**

1. Have propane tank filled with propane gas only by a reputable propane gas dealer.
2. Attach regulator fitting to tank valve.
   - Regulator fitting has left-hand threads. Turn counterclockwise to attach.
3. Tighten regulator fitting securely with wrench.
4. Place the two matching control knobs onto the valve stems.
5. Turn the control knobs to the right, clockwise. This is the **"Off"** position.
6. Make a solution of half liquid detergent and half water.

7. Turn gas supply "ON" at the tank valve (counterclockwise).
8. Brush soapy mixture on all connections listed in the **Leak Test Check List**.
9. Observe each place for bubbles caused by leaks.
10. Tighten any leaking connections, if possible.
    - If leak cannot be stopped. **Do Not Use Those Parts!** Order new parts.
11. Turn gas supply **"Off"** at the tank valve.
12. Push in and turn control knobs to the left ("HI" position) to release pressure in hose.
13. Disconnect the regulator fitting from the tank valve.
    - Regulator fitting has left-hand threads. Turn fitting clockwise to disconnect.
14. Leave tank outdoors and return to the grill assembly with valve, hose, and regulator.

FIGURE 22.12 Instructions for Leak Testing a Grill
Source: Instructions reprinted by permission of Thermos® Division.

class and be prepared to discuss the conclusions you've derived from this exercise.

7. Locate examples of five or more visuals from the following list.

   A visual that shows:
   - how to locate something
   - how to operate something
   - how to handle something
   - how to assemble something
   - how to position something
   - how to avoid damage or injury
   - how to diagnose and solve a problem
   - how to identify safe or acceptable limits
   - how to proceed systematically
   - how to make the right decision
   - why an action or procedure is important

   Bring your examples to class for discussion, evaluation, and comparison.

8. Find a manual and create a one-page set of layered instructions for a new user, using Figure 22.8 as a guide.

COLLABORATIVE PROJECTS

1. Draw a map of the route from your classroom to your dorm, apartment, or home—whichever is closest. Be sure to include identifying landmarks. When your map is completed, write instructions for a classmate who will try to duplicate your map from the information given in your written instructions. Be sure your classmate does not see your map! Exchange your instructions and try to duplicate your classmate's map. Compare your results with the original map. Discuss your conclusions about the usability of these instructions.

2. Divide into small groups and visit your computer center, library, or any place on campus or at work where you can find operating manuals for computers, lab or office equipment, or the like. (Or look through the documentation for your own computer hardware or software.) Locate fairly brief instructions that could use revision for improved content, organization, style, or format. Choose instructions for a procedure you are able to carry out. Make a copy of the instructions, test them for usability, and revise as needed. Submit all materials to your instructor, along with a memo explaining the improvements. Or be prepared to discuss your revision in class.

3. Visit your computer center and ask to borrow a software package that arrived without documentation or with very limited instructions. (Or perhaps you've received such programs as a member of a software club.) Run the program, and then prepare instructions for the next users. Test the usability of your instructions by having a classmate use them to run the program. Revise as needed. Remember, you are writing for a user with no experience.

4. Using word-processing or desktop publishing software, design a form to be used for advisee course scheduling, course evaluations, or some other school function. Conduct a usability study for this document and redesign it as needed. Then write a report analyzing and evaluating the form before and after the usability study.

5. Working in small groups, revise Figure 22.13 for improved usability. Appoint one member to present your group's version to the class, explaining the specific criteria used for revision.

SERVICE-LEARNING PROJECT

Do the research and prepare a set of instructions that will show general readers how to become more environmentally informed consumers and how to find, identify, evaluate, and compare environmentally friendly consumer goods such as appliances, building materials, household products, and the like. Design your instructions as a foldout brochure or a one-page (double-sided) handout, or in some other format requested by your instructor. *Hint:* Begin your research for this project by checking out the following Web sites:

- *The Gallery of Environmentally Preferable Goods* at <http://tbe.mit.edu/gallery/>
- The U.S. Environmental Protection Agency's *Energy Star* site at <www.epa.gov/energystar/>
- *The Ethical Shopper* site at <www.ethicalshopper.com>
- *Green Marketplace* at <www.greenmarketplace.com>

Proper Care Gives Safer Wear

- Follow, and save, the directions that come with your lenses. If you didn't get a patient information booklet about your lenses, request it from your eye-care practitioner.

- Use only the types of lens-care enzyme cleaners and saline solutions your practitioner okays.

- Be exact in following the directions that come with each lens-care product. If you have questions, ask your practitioner or pharmacist.

- Wash and rinse your hands before handling lenses. Fragrance-free soap is best.

- Clean, rinse, and disinfect reusable lenses each time they're removed, even if this is several times a day.

- Clean, rinse, and disinfect again if storage lasts longer than allowed by your disinfecting solution.

- Clean, rinse, and air-dry the lens case each time you remove the lenses. Then put in fresh solution. Replace the case every six months.

- Get your practitioner's okay before taking medicines or using topical eye products, even those you buy without a prescription.

- Remove your lenses and call your practitioner right away if you have vision changes, redness of the eye, eye discomfort or pain, or excessive tearing.

- Visit your practitioner every six months (more often if needed) to catch possible problems early.

Caring for Contact Lenses

Clean	Rinse	Enzyme
Remove surface dirt	Rinse away dirt	Remove deep deposits

Disinfect/Store	Rinse	Wet
Eliminate bacteria	Rinse dirt away	Prepare lens surface

Insert	Lubricate/Rewet	Clean Lens Case

Watch Out:

- Never use saliva to wet your lenses.

- Never use tap water, distilled water, or saline solution made at home with salt tablets for any part of your lens care. Use only commercial sterile saline solution.

- Never mix different brands of cleaner or solution.

- Never change your lens-care regimen or products without your practitioner's okay.

- Never let cosmetic lotions, creams, or sprays touch your lenses.

- Never wear lenses when swimming or in a hot tub.

- Never wear daily-wear lenses during sleep, not even a nap.

- Never wear your lenses longer than prescribed by your eye-care practitioner. ■

FIGURE 22.13   Procedure for Caring for Contact Lenses

Sources: Farley, Dixie. "Keeping an Eye on Contact Lenses." FDA Consumer Mar.-Apr. 1998: 17–21.

23

Proposals

Proposals attempt to *persuade* an audience to take some form of action: to authorize a project, accept a service or product, or otherwise support a specific plan for solving a problem or improving a situation.

Your own proposal might consist of a letter to your school board to suggest changes in the English curriculum; it may be a memo to your firm's vice president to request funding for a training program for new employees; or it may be an extensive document to the Defense Department to bid on a guided-missile contract (competing with proposals from other firms). As a student or as an intern at a nonprofit agency, you might submit a *grant proposal,* requesting financial support for a research or community project.

You might work alone or collaboratively, as part of a team. Developing and writing the proposal might take hours or months. If your job depends on funding from outside sources, proposals might be the most important documents you produce.

HOW PROPOSALS AND REPORTS DIFFER IN PURPOSE

While they may contain many of the same basic elements as a report, proposals have a primarily *persuasive* purpose: to move people to say "Yes. Let's move ahead on this." Of course, reports can also contain persuasive elements, as in recommending a specific course of action or justifying an equipment purchase. But reports typically serve a variety of *informative* purposes as well—such as keeping track of progress, explaining why something happened, or predicting an outcome.

A report often precedes a proposal: for example, a report on high levels of chemical pollution in a major waterway typically leads to various proposals for cleaning up that waterway. In short, once the report has *explored* a particular need, a proposal will be developed to *sell* the idea for meeting that need.

THE PROPOSAL AUDIENCE

In science, business, government, or education, proposals are written for decision makers: managers, executives, directors, clients, board members, or community leaders. Inside or outside your organization, these people review various proposals and then decide whether a specific plan is worthwhile, whether the project will materialize, or whether the service or product is useful.

Before accepting a particular proposal, reviewers look for persuasive answers to these basic questions:

What proposal reviewers want to know

- *What exactly is the problem or need, and why is this such a big deal?*
- *Why should we spend time, money, and effort on this?*
- *What exactly is your plan, and how do we know this will work?*
- *Why should we accept the following things that seem wrong or costly about your plan?*
- *What action are we supposed to take?*

Connect with your audience by addressing these questions early and systematically (as previewed on page 46 and restated here):

A proposal involves these basic persuasive tasks

1. *Spell out the problem (and its causes) clearly and convincingly.* Give enough detail for your audience to appreciate the problem's importance.
2. *Point out the benefits of solving the problem.* Explain specifically to your readers what they stand to gain.
3. *Offer a realistic, cost-effective solution.* Stick to claims or assertions you can support.
4. *Address anticipated objections to your solution.* Consider carefully your audience's level of skepticism about this issue.
5. *Induce your audience to act.* Decide exactly what you want your readers to do and give reasons why they should be the ones to take action.

Pages 578–84 offer examples and strategies for completing each of these tasks.

THE PROPOSAL PROCESS

Proposals in the commercial sector

The basic proposal process can be summarized like this: someone offers a plan for something that needs to be done. In business and government, this process has three stages:

Stages in the proposal process

1. Client *X* needs a service or product.
2. Firms *A*, *B*, and *C* propose a plan for meeting the need.
3. Client *X* awards the job to the firm offering the best proposal.

The complexity of each phase will, of course, depend on the situation. Here is a typical scenario:

Submitting a Competitive Proposal

You manage a mining engineering firm in Tulsa, Oklahoma. You regularly read the *Commerce Business Daily*, an essential online reference tool for anyone whose firm seeks government contracts. This publication lists the government's latest needs for services (salvage, engineering, maintenance) and for *supplies, equipment,* and *materials* (guided missiles, engine parts, and so on). On Wednesday, February 19, you spot this announcement:

Development of Alternative Solutions to Acid Mine Water Contamination from Abandoned Lead and Zinc Mines near Tar Creek, Neosho River, Ground Lake, and the Boone and Roubidoux aquifers in northeastern Oklahoma. This will include assessment of environmental effects of mine drainage followed by development and evaluation of alternate solutions to alleviate acid mine drainage in receiving streams. An optional portion of the contract to be bid on as an add-on and awarded at the discretion of the OWRB will be to prepare an Environmental Impact Assessment for each of three alternative solutions as selected by the OWRB. The project is expected to take six months to accomplish, with an anticipated completion date of September 30, 20XX. The projected effort for the required task is thirty person-months. The requests for proposal is available at www.owrb.org. Proposals are due March 1.

Oklahoma Water Resources Board
P.O. Box 53585
1000 Northeast 10th Street
Oklahoma City, OK 73151
(405)555–2541

Your firm has the personnel, experience, and time to do the job, so you decide to compete for the contract. Because the March 1 deadline is fast approaching, you immediately download the request for proposal (RFP). The RFP will give you the guidelines for developing and submitting the proposal—guidelines for spelling out your plan to solve the problem (methods, timetables, costs).

You then get right to work with the two staff engineers you have appointed to your proposal team. Because the credentials of your staff could affect the client's acceptance of the proposal, you ask team members to update their résumés for inclusion in an appendix to the proposal. ❑

In scenarios like this, the client will award the contract to the firm submitting the best proposal, based on the following criteria (and perhaps others):

Criteria by which reviewers evaluate proposals

- understanding of the client's needs, as described in the RFP
- clarity and soundness of the plan being offered
- quality of the project's organization and management
- ability to complete the job by deadline
- ability to control costs
- firm's experience on similar projects
- qualifications of staff to be assigned to the project
- firm's record for similar projects

A client's specific evaluation criteria are often listed (in order of importance or on a point scale) in the RFP. Although these criteria may vary, every client expects a proposal that is *clear, informative,* and *realistic.*

Proposals in the nonprofit sector

In contrast to proposals prepared for commercial purposes, museums, community service groups, and other nonprofit organizations prepare *grant propos-*

als that request financial support for worthwhile causes. Government and charitable granting agencies such as the Department of Health and Human Services, the Department of Agriculture, or the Pugh Charitable Trust solicit proposals for funding in areas such as medical research, educational TV programming, and rural development. Submission and review of grant proposals follow the same basic process used for commercial proposals. Figure 23.1 shows part of a request for funding proposals. This RFP was issued by a U.S. government health organization.

Submitting paperless proposals

23.1

For more examples of electronic proposals visit <www.ablongman.com/lannonweb>

In both the commercial and nonprofit sector, the proposal process increasingly occurs online. The National Science Foundation's *Fastlane* Web site <www.fastlane.nsf.gov>, for example, allows grant applicants to submit proposals in electronic format. This enables applicants to include sophisticated graphics, to revise budget estimates, to update other aspects of the plan as needed, and to maintain real-time contact with the granting agency while the proposal is being reviewed.

RFP No. NIH-NHLBI-HR-01-01

"Clinical Centers for the Clinical Network for the Treatment of the Adult Respiratory Distress Syndrome (ARDS)"

The National Heart, Lung, and Blood Institute (NHLBI) is soliciting proposals for clinical centers to participate in the "Clinical Centers for the Clinical Network for the Treatment of the Adult Respiratory Distress Syndrome (ARDS)". The objective of ARDSnet is to test novel therapies for the prevention and treatment of adult respiratory distress syndrome and acute lung injury (ARDS/ALI). The ARDSnet has developed a protocol to test the role of pulmonary artery catheter (PAC) in the clinical management of ARDS/ALI. The ARDSnet investigators and NHLBI staff have determined that enrollment in the protocol will be challenging and that additional sites are required to complete this protocol in a reasonable time frame. It is anticipated that this solicitation will award up to ten (10) additional Critical Care Treatment Groups (CCTG) to participate in the *PAC protocol*. Each new CCTG will enroll approximately 30 patients/year during Phase II (24 months).

FIGURE 23.1   Request for Proposal.   The complete, 44-page RFP includes a project history and background. It also stipulates the types, frequency, and format of reports required, completion dates for various phases of the project, guidelines for submitting the proposal, and additional evaluation factors (such as the diversity of people in the study population and the proposal offeror's record of performance on past projects).

Source: Adapted from the National Institutes of Health Archive copy <www.nhlbi.nih.gov/funding/inits.archive>

STATEMENT OF WORK

Independently, and not as an agent of the Government, the contractor shall furnish all the necessary services, qualified personnel, material, equipment, and facilities, not otherwise provided by the Government as needed to perform the statement of work below. Specifically, the contractor shall:

Train a staff to conduct the Pulmonary Artery Catheter (PAC) study as outlined in the protocol and manual of operations.

Participate with other study investigators in a clinical study of the PAC in the treatment of Acute Respiratory Distress Syndrome (ARDS) and Acute Lung Injury (ALI) according to the protocol and manual of operations. The protocol, manual of operations and any amendments thereto are incorporated herein, by reference, as part of the contract.

Enroll and treat a minimum of 30 patients, 13 years of age or older, with ARDS or ALI according to the PAC protocol. The patients will have a gender and racial composition similar to the population of patients that are available for study.

Perform follow-up assessment on the subjects in the manner specified in the manual of operations.

Collect the subject data as specified by the protocol and forward the data to the Clinical Coordinating Center (CCC) in accordance with procedures in the manual of operations.

Participate in the Steering Committee to monitor progress on the study.

Interact with the CCC to provide data and information necessary for data analysis work with other study investigators in the preparation and writing of reports and manuscripts for publication.

Work with other study investigators in the preparation and writing of reports and manuscripts for publication.

FIGURE 23.1   Request for Proposal. *(continued)*

23.2
What is an RFP and how might you respond to it? Learn more at
<www.ablongman.com/lannonweb>

PROPOSAL TYPES

Proposals are classified according to *origin, audience,* and *purpose.* Based on its origin, a proposal is either *solicited* or *unsolicited.* Solicited proposals are those requested by a potential client or your employer, as in the engineering firm example on page 569. Unsolicited proposals are not specifically requested. For example, if you are creating a new Web site development service in your town, you might send

TECHNICAL EVALUATION CRITERIA

Proposals submitted in response to this solicitation will be reviewed by a peer group of scientists under the auspices of the Review Branch, Division of Extramural Affairs, NHLBI, and subsequently by a review group within NHLBI. The evaluation criteria are used by the Technical Evaluation Committee when reviewing the technical proposals. The criteria below are listed in the order of relative importance with weights assigned for evaluation purposes.

Weight	Criterion
40%	Ability to enroll 30 patients with ARDS/ALI per year. Offerors shall describe their previous experience enrolling patients with ARDS or patients at risk of developing ARDS into multi-center treatment trials. If previous enrollment has been less than 30 patients a year, practical procedures must be proposed to identify and screen adequate numbers of prospective subjects.
30%	Adequacy of experience and competence of the professional and technical staff pertinent to the study, including experience in clinical studies of ARDS. In particular, the PI is expected to be expert in the treatment of critically ill patients with ARDS and also to have previously participated in multi-center treatment trials.
30%	Adequacy of laboratory/clinical facilities available, including all participating institutions. Description of the facilities and means of assuring quality control of patient care. Adequacy of plans for study coordination, including quality control of data entry.

FIGURE 23.1   Request for Proposal. *(continued)*

out short proposals to area businesses, to suggest methods for online advertising and sales.

Based on its audience, a proposal may be *internal* or *external*—written for members of your organization or for clients and funding agencies. (The situation on page 569 calls for an external proposal.)

Based on its purpose, a proposal may be a *planning, research,* or *sales* proposal. Some proposals fall within all three categories.

Planning Proposal

A planning proposal offers solutions to a problem or suggestions for improvement. It might be a request for funding to expand the campus newspaper, an architectural plan for new facilities at a ski area, or a plan to develop energy alternatives to fossil fuels.

The short planning proposal that follows is external and solicited. The XYZ Corporation is about to contract a team of communication consultants to design in-house writing workshops, and the consultants must persuade the client (the company's education officer) that their methods will succeed. After briefly introducing the problem, the authors develop their proposal under two headings and several subheadings, making the document easy to read and to the point. Because this proposal is external, it takes the form of a letter.

PLANNING PROPOSAL

States purpose

Dear Mary:

Thanks for sending the writing samples from your technical support staff. Here is what we're doing to design a realistic approach.

Needs Assessment

Identifies problem

After conferring with technicians in both Jack's and Terry's groups and analyzing their writing samples, we identified this hierarchy of needs:

- improving readability
- achieving precise diction
- summarizing information
- organizing a set of procedures
- formulating various memo reports
- analyzing audiences for upward communication
- writing persuasive bids for transfer or promotion
- writing persuasive suggestions

Proposed Plan

Proposes solution

Based on the needs listed above, we have limited our instruction package to eight carefully selected and readily achievable goals.

Course Outline.   Our eight, two-hour sessions are structured as follows:

Details what will be done

1. achieving sentence clarity
2. achieving sentence conciseness
3. achieving fluency and precise diction
4. writing summaries and abstracts
5. outlining manuals and procedures
6. editing manuals and procedures
7. designing various reports for various purposes
8. analyzing the audience and writing persuasively

Details how it will
be done

> **Classroom Format.**   The first three meetings will be lecture-intensive with weekly exercises to be done at home and edited collectively in class. The remaining five weeks will combine lecture and exercises with group editing of work-related documents. We plan to remain flexible so we can respond to needs that arise.
>
> **Limitations**
> Given our limited contact time, we cannot realistically expect to turn out a batch of polished communicators. By the end of the course, however, our students will have begun to appreciate writing as a deliberate process.

Sets realistic
expectations

Encourages
reader response

> If you have any suggestions for refining this plan, please let us know.

Notice that the word choice ("thanks," "what we're doing," "Jack and Terry") creates an informal, familiar tone—appropriate in this external document only because the consultants and client have spent many hours in conferences, luncheons, and phone conversations. Notice also that the "Limitations" section indicates that these authors are careful to promise no more than they can deliver.

Research Proposal

Research (or grant) proposals request approval (and often funding) for some type of study. A university chemist might address a research proposal to the Environmental Protection Agency for funds to identify toxic contaminants in local groundwater. Research proposals are solicited by many government and private agencies: National Science Foundation, National Institutes of Health, and others. Each granting agency has its own requirements and guidelines for proposal format and content. Successful research proposals follow those guidelines and carefully articulate the goals of the project.

Other research proposals might be submitted by students requesting funds or approval for independent study, field study, or a thesis project. A technical writing student usually submits a relatively brief research proposal that will lead to the term project (such as the long proposal that begins on page 585 or the analytical report that begins on page 617).

In the following research proposal, Tom Dewoody requests his instructor's authorization to do a feasibility study (Chapter 24) that will produce an analytical report for potential investors. Dewoody's proposal clearly answers the questions about *what, why, how, when,* and *where.* Because this proposal is internal, it is cast informally as a memo.

RESEARCH PROPOSAL

To: Dr. John Lannon
From: T. Sorrells Dewoody
Date: March 16, 20XX
Subject: *Proposal for Determining the Feasibility of Marketing Dead Western White Pine*

Introduction

Opens with background and causes of the problem

Over the past four decades, huge losses of western white pine have occurred in the northern Rockies, primarily attributable to white pine blister rust and the attack of the mountain pine beetle. Estimated annual mortality is 318 million board feet. Because of the low natural resistance of white pine to blister rust, this high mortality rate is expected to continue indefinitely.

If white pine is not harvested while the tree is dying or soon after death, the wood begins to dry and check (warp and crack). The sapwood is discolored by blue stain, a fungus carried by the mountain pine beetle. If the white pine continues to stand after death, heart cracks develop. These factors work together to cause degradation of the lumber and consequent loss in value.

Statement of Problem

Describes problem

White pine mortality reduces the value of white pine stumpage because the commercial lumber market will not accept dead wood. The major implications of this problem are two: first, in the face of rising demand for wood, vast amounts of timber lie unused; second, dead trees are left to accumulate in the woods, where they are rapidly becoming a major fire hazard here in northern Idaho and elsewhere.

Proposed Solution

Describes one possible solution

One possible solution to the problem of white pine mortality and waste is to search for markets other than the conventional lumber market. The last few years have seen a burst of popularity and growing demand for weathered barn boards and wormy pine for interior paneling. Some firms around the country are marketing defective wood as specialty products. (These firms call the wood from which their products come "distressed," a term I will use hereafter to refer to dead and defective white pine.) Distressed white pine quite possibly will find a place in such a market.

Scope

Defines scope of the proposed study

To assess the feasibility of developing a market for distressed white pine, I plan to pursue six areas of inquiry.

1. What products presently are being produced from dead wood, and what are the approximate costs of production?
2. How large is the demand for distressed-wood products?
3. Can distressed white pine meet this demand as well as other species meet it?

4. Does the market contain room for distressed white pine?
5. What are the costs of retrieving and milling distressed white pine?
6. What prices for the products can the market bear?

Methods

Describes how study will be done

My primary data sources will include consultations with Dr. James Hill, Professor of Wood Utilization, and Dr. Sven Bergman, Forest Economist—both members of the College of Forestry, Wildlife, and Range. I will also inspect decks of dead white pine at several locations and visit a processing mill to evaluate it as a possible base of operations. I will round out my primary research with a letter and telephone survey of processors and wholesalers of distressed material.

Mentions literature review

Secondary sources will include publications on the uses of dead timber, and a review of a study by Dr. Hill on the uses of dead white pine.

My Qualifications

Cites a major reference and gives the writer's qualifications for this project

I have been following Dr. Hill's study on dead white pine for two years. In June of this year I will receive my B.S. in forest management. I am familiar with wood milling processes and have firsthand experience at logging. My association with Drs. Hill and Bergman gives me the opportunity for an in-depth feasibility study.

Conclusion

Encourages reader acceptance

Clearly, action is needed to reduce the vast accumulations of dead white pine in our forests. The land on which they stand is among the most productive forests in northern Idaho. By addressing the six areas of inquiry mentioned earlier, I can determine the feasibility of directing capital and labor to the production of distressed white pine products. With your approval I will begin research at once.

Sales Proposal

The sales proposal is a marketing tool that offers a service or product. The offer may be solicited or unsolicited. If the proposal is solicited, several firms may be competing for the contract, so your proposal may be ranked against others by a committee. Because sales proposals are addressed to readers outside your organization, they are cast as letters if they are brief (as on page 411). Long sales proposals, like long reports, are formal documents with supplements (cover letter, title page, table of contents).

A successful sales proposal persuades customers that your product or service surpasses those offered by competitors. In the following solicited proposal, the writer explains why his machinery is best for the job, how the job can be done efficiently, what qualifications his company can offer, and what costs are involved. To protect himself, he points out possible causes of increased costs.

SALES PROPOSAL

Describes the subject and purpose

Gives the writer's qualifications

Explains how the job will be done

Maintains a confident tone throughout

Gives a qualified cost estimate

Encourages reader acceptance by emphasizing economy and efficiency

Subject: *Proposal to Dig a Trench and Move Boulders at Bliss Site*

Dear Mr. Haver:

I've inspected your property and would be happy to undertake the landscaping project necessary for the development of your farm.

The backhoe I use cuts a span 3 feet wide and can dig as deep as 18 feet—more than an adequate depth for the mainline pipe you wish to lay. Because this backhoe is on tracks rather than tires, and is hydraulically operated, it is particularly efficient in moving rocks. I have more than twelve years of experience with backhoe work and have completed many jobs similar to this one.

After examining the huge boulders that block access to your property, I am convinced they can be moved only if I dig out underneath and exert upward pressure with the hydraulic ram while you push forward on the boulders with your D-9 Caterpillar. With this method, we can move enough rock to enable you to farm that now inaccessible tract. Because of its power, my larger backhoe will save you both time and money in the long run.

This job should take 12 to 15 hours, unless we encounter subsurface ledge formations. My fee is $200 per hour. The fact that I provide my own dynamiting crew at no extra charge should be an advantage to you because you have so much rock to be moved.

Please phone me anytime for more information. I'm sure we can do the job economically and efficiently.

NOTE *Never underestimate costs by failing to account for all variables—a sure way to lose money or clients.*

What you include in a sales proposal will depend on the guidelines from the client or on your thorough analysis of the kinds of information your audience needs.

NOTE *The proposal categories (planning, research, and sales) discussed in this section are neither exhaustive nor mutually exclusive. A research proposal, for example, may request funds for a study that will lead to a planning proposal. The Vista proposal partially shown below combines planning and sales features; if clients accept the preliminary plan, they will hire the firm to install the automated system.*

ELEMENTS OF A PERSUASIVE PROPOSAL

Reviewers will evaluate your proposal on the basis of the following quality indicators. (See also the criteria listed on page 570.)

A Forecasting Title

Overworked reviewers facing a stack of proposals may well decide to focus on those "with the most intriguing titles" (Friedland and Folt 53). In any case, your title should announce the proposal's purpose and content. Don't be vague.

Unclear

Revised

> Proposed Office Procedures for Vista Freight, Inc.
>
> A Proposal for Automating Vista's Freight Billing System

Don't write "Recommended Improvements" when you mean "Recommended Wastewater Treatment."

Clear Understanding of the Audience's Needs

Focus on the problem and the objective

The proposal audience wants specific suggestions for meeting their specific needs. Their biggest question is "What will this plan do for me?" Show that you clearly understand your clients' problems and their expectations, and then offer an appropriate solution.

In the following proposal for automating office procedures at Vista, Inc., Gerald Beaulieu begins with a clear assessment of needs and then moves quickly into a proposed plan of action.

Gives background

Statement of the Problem

Vista provides two services. (1) It locates freight carriers for its clients. The carriers, in turn, pay Vista a 6 percent commission for each referral. (2) Vista handles all shipping paperwork for its clients. For this auditing service, clients pay Vista a monthly retainer.

Describes problem and its effects

Although Vista's business has increased steadily for the past three years, record keeping, accounting, and other paperwork are still done manually. These inefficient procedures have caused a number of problems, including late billings, lost commissions, and poor account maintenance. Updated office procedures seem crucial to competitiveness and continued growth.

Enables readers to visualize results

Objective

This proposal offers a realistic and effective plan for streamlining Vista's office procedures. We first identify the burden imposed on your staff by the current system, and then we show how to reduce inefficiency, eliminate client complaints, and improve your cash flow by automating most office procedures.

A Clear Focus on Benefits

Spell out the benefits

Do a detailed audience analysis to identify readers' major concerns and to anticipate likely questions and objections. Show your audience that you understand what they (or their organization) will gain by adopting your plan. The following bulleted list spells out exactly what tasks Vista employees will be able to accomplish once the proposed plan is implemented.

Relates benefits directly to client's needs

> Once your automated system is operational, you will be able to
>
> - identify cost-effective carriers
> - coordinate shipments (which will ensure substantial client discounts)
> - print commission bills
> - track shipments by weight, miles, fuel costs, and destination
> - send clients weekly audit reports on their shipments
> - bill clients on a 25-day cycle
> - produce weekly or monthly reports
>
> Additional benefits include eliminating repetitive tasks, improving cash flow, and increasing productivity.

(Each of these benefits will be described at length later in the "Plan" section.)

Honest and Supportable Claims

Promise only what you can deliver

23.3

For more on ethics in proposals visit <www.ablongman.com/ lannonweb>

Because they typically involve large sums of money as well as contractual obligations, proposals require a solid ethical and legal foundation. Clients in these situations often have doubts or objections about time and financial costs and a host of other risks involved whenever any important project is undertaken. Your proposal needs to address these issues openly and honestly. For example, if you are proposing to install customized virus-protection software, be clear about what this software cannot accomplish under certain circumstances. False or exaggerated promises not only damage a writer's or a company's reputation, but also invite lawsuits. (For more on supporting your claims, see page 56.)

Here is how the Vista proposal qualifies its promises:

Anticipates a major objection and offers a realistic approach

> As countless firms have learned, imposing automated procedures on employees can create severe morale problems—particularly among senior staff who feel coerced and often marginalized. To diminish employee resistance, we suggest that your entire staff be invited to comment on this proposal. To help avoid hardware and software problems once the system is operational, we have included recommendations and a budget for staff training. (Adequate training is essential to the automation process.)

If the best available solutions have limitations, say so. Notice how the above solutions are qualified ("diminish" and "help avoid" instead of "eliminate") so as not to promise more than the plan can achieve.

A proposal can be judged fraudulent if it misleads potential clients by

Major ethical and
legal violations in
a proposal

- making unsupported claims,
- ignoring anticipated technical problems, or
- knowingly underestimating costs or time requirements.

For a project involving complex tasks or phases, provide a realistic timetable (perhaps using a Gantt chart, page 312) to show when each major phase will begin and end. Also provide a realistic, accurate budget, with a detailed cost breakdown (for supplies and equipment, travel, research costs, outside contractors, or the like) to show clients exactly how the money is being spent.

NOTE   *Be absolutely certain that you spend every dollar according to the allocations that have been stipulated. For example, if a grant award allocates a certain amount for "a research assistant," be sure to spend that exact amount for that exact purpose—unless you receive written permission from the granting agency to shift funds for other purposes. Keep strict accounting of all the money you spend. Proposal experts Friedland and Folt remind us that "Financial misconduct is never tolerated, regardless of intent" (161). Even an innocent mistake or accounting lapse on your part can lead to charges of fraud.*

Appropriate Detail

Provide adequate
but not excessive
detail

Vagueness in a proposal is fatal. Spell everything out. Instead of writing, "We will install state-of-the-art equipment," enumerate the products or services to be provided.

Spells out what
will be provided

> To meet your automation requirements, we will install twelve Power Macintosh G4 computers with 60-Gigabyte hard drives. The system will be networked for rapid file transfer between offices. The plan also includes interconnection with four Hewlett-Packard 5 MP printers, and one HP Desk Jet 1600 CM color printer.

To avoid misunderstandings that could produce legal complications, a proposal must elicit *one* interpretation only.

Place support material (maps, blueprints, specifications, calculations) in an appendix so as not to interrupt the discussion.

NOTE   *While concrete and specific detail is vital, never overburden reviewers with needless material. A precise audience analysis (Chapter 3) can pinpoint specific information needs.*

Readability

Make the
proposal inviting
and easy to
understand

A readable proposal is straightforward, easy to follow, and understandable. Avoid language that is overblown or too technical for your audience. Review Chapter 13 for style strategies.

Convincing Language

Sell your ideas

Your proposal should move people to action. Review Chapter 4 for persuasion guidelines. Keep your tone confident and encouraging, not bossy and critical. For more on tone, see pages 274–82.

Visuals

Emphasize key points in your proposal with relevant tables, flowcharts, and other visuals (Chapter 14), properly introduced and discussed.

Gives a framework for interpreting the visual

Visual repeats, restates, or reinforces the prose

As the flowchart (Figure 1) illustrates, Vista's routing and billing system creates redundant work for your staff. The routing sheet alone is handled at least six times. Such extensive handling leads to errors, misplaced paperwork, and late billing.

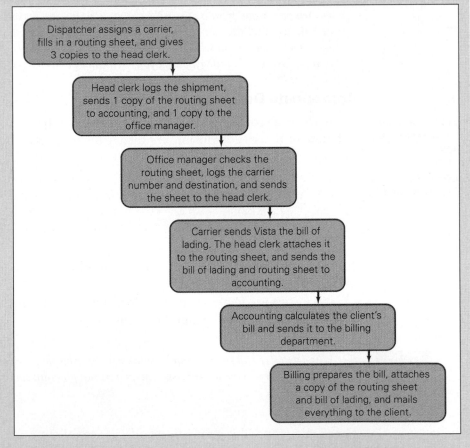

FIGURE 1 Flowchart of Vista's Manual Routing and Billing System

Accessible Page Design

Make the
audience's job
easy

Yours might be one of several proposals being reviewed. Help the audience to get in quickly, find what they need, and get out. Review Chapter 15 for design strategies.

Supplements Tailored for a Diverse Audience

Analyze the
specific needs
and interests of
each major
reviewer

A single proposal often addresses a diverse audience: executives, managers, technical experts, attorneys, politicians, and so on. Various reviewers are interested in various parts of your proposal. Experts look for the technical details. Others might be interested in the recommendations, costs, timetable, or expected results, but they will need an explanation of technical details as well.

How to give each
major reviewer
what he or she
expects

If the primary audience is expert or informed, keep the proposal text itself technical. For uninformed secondary reviewers (if any), provide an informative abstract, a glossary, and appendixes explaining specialized information. If the primary audience has no expertise and the secondary audience does, write the proposal itself for laypersons, and provide appendices with the technical details (formulas, specifications, calculations) that experts will use to evaluate your plan. See Chapter 25 for specific supplements.

If you are unsure which supplements to include in an internal proposal, ask the intended audience or study other proposals. For a solicited proposal (to an outside agency), follow the agency's instructions exactly.

Proper Citation of Sources and Contributors

Proposals rarely emerge from thin air. Whenever appropriate, especially for topics that involve ongoing research, you need to credit key information sources and contributors. Proposal experts Friedland and Folt offer these suggestions:

How to cite
sources and
contributors

- *Review the literature on this topic.* Limit your focus to "the few most important or influential" background studies (135).
- *Don't cite sources of "common knowledge" about this topic (136).* Information available in multiple sources or readily known in your discipline usually qualifies as common knowledge. For more, see page 685.
- *Provide adequate support for your plan.* "In general, cite all papers that are essential to establish credibility and feasibility" (135).
- *Provide up-to-date principal references.* Although references to earlier, groundbreaking studies are important, recent studies can be most essential (134).
- *Present a balanced, unbiased view.* Acknowledge sources that differ from or oppose your point of view; explain the key differences among the various viewpoints before making your case (134).
- *Give credit to all contributors.* Recognize everyone who has worked on or helped with this proposal: for example, coauthors, editors, data gatherers, and people who contributed various ideas (22).

Proper citation is not only an ethical requirement, but also an indicator of your proposal's credibility. See Appendix A for more on citation techniques.

23.4

For examples of oral proposals in *PowerPoint* visit <www.ablongman.com/lannonweb>

AN OUTLINE AND MODEL FOR PROPOSALS

Depending on a proposal's complexity, each section contains some or all of the subsections listed in the following general outline:

I. Introduction
   A. Statement of Problem and Objective/Project Overview
   B. Background and Review of the Literature (as needed)
   C. Need
   D. Benefits
   E. Qualifications of Personnel
   F. Data Sources
   G. Limitations and Contingencies
   H. Scope
II. Plan
   A. Objectives and Methods
   B. Timetable
   C. Materials and Equipment
   D. Personnel
   E. Available Facilities
   F. Needed Facilities
   G. Cost
   H. Expected Results
   I. Feasibility
III. Conclusion
   A. Summary of Key Points
   B. Request for Action
IV. Works Cited

These subsections can be rearranged, combined, divided, or deleted as needed. Not every proposal will contain all subsections; however, each major section must persuasively address specific information needs as illustrated in the sample proposal that begins below.

Introduction

From the beginning, your goal is *to sell your idea*—to demonstrate the need for the project, your qualifications for tackling the project, and your clear understanding of how to proceed. Readers quickly lose interest in a wordy, evasive, or vague introduction.

Following is the introduction for a planning proposal titled "Proposal for Solving the Noise Problem in the University Library." Jill Sanders, a library work-study student, addresses her proposal to the chief librarian and the administrative staff. Because this proposal is unsolicited, it must first make the problem vivid through details that arouse concern and interest. This introduction is longer than it would be in a solicited proposal, whose audience would already agree on the severity of the problem.

NOTE   *Title page, informative abstract, table of contents, and other supplements that ordinarily accompany long proposals of this type are omitted here to save space. See Chapter 25 for discussion and examples of each type of document supplement.*

INTRODUCTION

Statement of Problem
During the October 20XX Convocation at Margate University, students and faculty members complained about noise in the library. Soon afterward, areas were designated for "quiet study," but complaints about noise continue. To create a scholarly atmosphere, the library should take immediate action to decrease noise.

Objective
This proposal examines the noise problem from the viewpoint of students, faculty, and library staff. It then offers a plan to make areas of the library quiet enough for serious study and research.

Sources
My data come from a university-wide questionnaire; interviews with students, faculty, and library staff; inquiry letters to other college libraries; and my own observations for three years on the library staff.

Details of the Problem
This subsection examines the severity and causes of the noise.

Severity.   Since the 20XX Convocation, the library's fourth and fifth floors have been reserved for quiet study, but students hold group study sessions at the large tables and disturb others working alone. The constant use of computer terminals on both floors adds to the noise, especially when students converse. Moreover, people often chat as they enter or leave study areas.

On the second and third floors, designed for reference, staff help patrons locate materials, causing constant shuffling of people and books, as well as loud conversation. At the computer service desk on the third floor, conferences between students and instructors create more noise.

The most frequently voiced complaint from the faculty members interviewed was about the second floor, where people using the Reference and Government Documents services converse loudly. Students complain

Concise descriptions of problem and objective immediately alert the readers

This section comes early because it is referred to in the next section

Details help readers to understand the problem

Shows how campus feels about problem

about the lack of a quiet spot to study, especially in the evening, when even the "quiet" floors are as noisy as the dorms.

More than 80 percent of respondents (530 undergraduates, 30 faculty, 22 graduate students) to a university-wide questionnaire (Appendix A) insisted that excessive noise discourages them from using the library as often as they would prefer. Of the student respondents, 430 cited quiet study as their primary reason for wishing to use the library.

The library staff recognizes the problem but has insufficient personnel. Because all staff members have assigned tasks, they have no time to monitor noise in their sections.

Causes.   Respondents complained specifically about these causes of noise (in descending order of frequency):

1. Loud study groups that often lapse into social discussions.
2. General disrespect for the library, with some students' attitudes characterized as "rude," "inconsiderate," or "immature."
3. The constant clicking of computer terminals on all five floors, and of laptops on the first three.
4. Vacuuming by the evening custodians.

All complaints converged on lack of enforcement by library staff.

Because the day staff works on the first three floors, quiet-study rules are not enforced on the fourth and fifth floors. Work-study students on these floors have no authority to enforce rules not enforced by the regular staff. Small, black-and-white "Quiet Please" signs posted on all floors go unnoticed, and the evening security guard provides no deterrent.

Needs
Excessive noise in the library is keeping patrons away. By addressing this problem immediately, we can help restore the library's credibility and utility as a campus resource. We must reduce noise on the lower floors and eliminate it from the quiet-study floors.

Scope
The proposed plan includes a detailed assessment of methods, costs and materials, personnel requirements, feasibility, and expected results.

Margin notes:

Shows concern is widespread and pervasive

Identifies specific causes

This statement of need evolves logically and persuasively from earlier evidence

Previews the plan

Body

The body (or plan section) of your proposal will receive the most audience attention. The main goal of this section is to prove your plan will work. Here you spell out your plan in enough detail for the audience to evaluate its soundness. If this section is vague, your proposal stands no chance of being accepted. Be sure your plan is realistic and promise no more than you can deliver.

PROPOSED PLAN

This plan takes into account the needs and wishes of our campus community, as well as the available facilities in our library.

Phases of the Plan

Noise in the library can be reduced in three complementary phases: (1) improving publicity, (2) shutting down and modifying our facilities, and (3) enforcing the quiet rules.

Improving Publicity.   First, the library must publicize the noise problem. This assertive move will demonstrate the staff's interest. Publicity could include articles by staff members in the campus newspaper, leaflets distributed on campus, and a freshman library orientation acknowledging the noise problem and asking cooperation from new students. All forms of publicity should detail the steps being taken by the library to solve the problem.

Shutting Down and Modifying Facilities.   After notifying campus and local newspapers, you should close the library for one week. To minimize disruption, the shutdown should occur between the end of summer school and the beginning of the fall term.

During this period, you can convert the fixed tables on the fourth and fifth floors to cubicles with temporary partitions (six cubicles per table). You could later convert the cubicles to shelves as the need increases.

Then you can take all unfixed tables from the upper floors to the first floor, and set up a space for group study. Plans are already under way for removing the computer terminals from the fourth and fifth floors.

Enforcing the Quiet Rules.   Enforcement is the essential long-term element in this plan. No one of any age is likely to follow all the rules all the time—unless the rules are enforced.

First, you can make new "Quiet" posters to replace the present, innocuous notices. A visual-design student can be hired to draw up large, colorful posters that attract attention. Either the design student or the university print shop can take charge of poster production.

Next, through publicity, library patrons can be encouraged to demand quiet from noisy people. To support such patron demands, the library staff can begin monitoring the fourth and fifth floors, asking study groups to move to the first floor, and revoking library privileges of those who refuse. Patrons on the second and third floors can be asked to speak in whispers. Staff members should set an example by regulating their own voices.

Costs and Materials

• The major cost would be for salaries of new staff members who would help monitor. Next year's library budget, however, will include an allocation for four new staff members.

Margin notes (left column):

Tells how plan will be implemented

Describes first phase

Describes second phase

Describes third phase

Estimates costs and materials needed

- A design student has offered to make up four different posters for $200. The university printing office can reproduce as many posters as needed at no additional cost.
- Prefabricated cubicles for 26 tables sell for $150 apiece, for a total cost of $3,900.
- Rearrangement on various floors can be handled by the library's custodians.

The Student Fee Allocations Committee and the Student Senate routinely reserve funds for improving student facilities. A request to these organizations would presumably yield at least partial funding for the plan.

Personnel

The success of this plan ultimately depends on the willingness of the library administration to implement it. You can run the program itself by committees made up of students, staff, and faculty. This is yet another area where publicity is essential to persuade people that the problem is severe and that you need their help. To recruit committee members from among students, you can offer Contract Learning credits.

The proposed committees include an Antinoise Committee overseeing the program, a Public Relations Committee, a Poster Committee, and an Enforcement Committee.

Feasibility

On March 15, 20XX, I mailed survey letters to twenty-five New England colleges, inquiring about their methods for coping with noise in the library. Among the respondents, sixteen stated that publicity and the administration's attitude toward enforcement were main elements in their success.

Improved publicity and enforcement could work for us as well. And slight modifications in our facilities, to concentrate group study on the busiest floors, would automatically lighten the burden of enforcement.

Benefits

Publicity will improve communication between the library and the campus. An assertive approach will show that the library is aware of its patrons' needs and is willing to meet those needs. Offering the program for public inspection will draw the entire community into improvement efforts. Publicity, begun now, will pave the way for the formation of committees.

The library shutdown will have a dual effect: it will dramatize the problem to the community, and it will provide time for the physical changes. (An antinoise program begun with carpentry noise in the quiet areas would hardly be effective.) The shutdown will be both a symbolic and a concrete measure, leading to reopening of the library with a new philosophy and a new image.

Continued strict enforcement will be the backbone of the program. It will prove that staff members care enough about the atmosphere to jeopardize their friendly image in the eyes of some users, and that the library is not afraid to enforce its rules.

Margin notes:

Describes personnel needed

Assesses probability of success

Offers a realistic and persuasive forecast of benefits

Conclusion

The conclusion reaffirms the need for the project and induces the audience to act. End on a strong note, with a conclusion that is assertive, confident, and encouraging—and keep it short.

Reemphasizes
need and
feasibility and
encourages
action

> ### CONCLUSION AND RECOMMENDATION
>
> The noise in Margate University Library has become embarrassing and annoying to the whole campus. Forceful steps are needed to restore the academic atmosphere.
>
> Aside from the intangible question of image, close inspection of the proposed plan will show that it will work if the recommended steps are taken and—most important—if daily enforcement of quiet rules becomes a part of the library's services.

In long proposals, especially those beginning with a comprehensive abstract, the conclusion can be omitted.

SITUATION REQUIRING A PROPOSAL

The proposal in Figure 23.2 can be considered a form of grant proposal, since it requests funding for a nonprofit enterprise. As in any funding proposal, a precise plan and an itemized budget provide the essential justification for a request.

The Situation

Southeastern Massachusetts University's newspaper, the SMU *Torch*, is struggling to meet rising costs. The paper's yearly budget is funded by the Student Fee Allocation Committee which disburses money to various campus organizations. Drastic budget cuts have resulted in reduced finding for all state schools. As a result, the newspaper has received no funding increase for the last three years. Meanwhile production costs keep rising.

Bill Trippe is the *Torch*'s business manager. His task is to justify a requested increase of 20.6 percent for the coming year's budget. Before drafting his proposal, Bill constructs a detailed profile of his audience (based on the worksheet, page 65).

Audience and Use Profile for a Formal Proposal

Audience Identity and Needs.   My primary audience includes all members of the Student Fee Allocation Committee. My secondary audience is the newspaper staff, who will implement the proposed plan—if it is approved by the allocations committee.

The primary audience will use my document as perhaps the sole basis for deciding whether to grant the additional funds. Most of these readers have overseen

the newspaper budget for years, and so they already know quite a bit about our overall operation. But they still need an item-by-item explanation of the conditions created by our problems with funding and ever-increasing costs. Probable questions I can anticipate:

- *Why should the paper receive priority over other campus organizations?*
- *Just how crucial is the problem?*
- *Are present funds being used efficiently?*
- *Can any expenses be reduced?*
- *How would additional funds be spent?*
- *How much will this increase cost?*
- *Will the benefits justify the cost?*

Attitude and Personality.   My primary audience often has expressed interest in this topic. But they are likely to object to any request for more money by arguing that everyone has to economize in these difficult times. I guess I could characterize their attitude as both receptive and hesitant. (Almost every campus organization is trying to make a case for additional funds.)

I do know most committee members pretty well, and they seem to respect my management skills. But I still need to spell out the problem and propose a realistic plan, showing that the newspaper staff is sincere in its intention to eliminate nonessential operating costs. At a time when everyone is expected to make do with less, I need to make an especially strong case for salary increases (to attract talented personnel).

Expectations about the Document.   My audience has requested (solicited) this proposal, and so I know it will be carefully read—but also scrutinized and evaluated for its soundness! Especially in a budget request, my audience expects no shortcuts; I'll have to itemize every expense. The Costs sections then could be the longest part of the proposal.

And to further justify the requested budget; I can demonstrate just how well the newspaper manages its present funds. In the Feasibility section, I'll give a detailed comparison of funding, expenditures, and the size of the *Torch* in relation to the newspapers of the four other local colleges. This section should be the "clincher" because these facts are most likely to persuade the committee that my plan is cost-effective. To avoid clutter, I'll add an appendix with a table of figures for the comparison above.

To organize my document, I will (1) identify the problem, (2) establish need, (3) purpose a solution, (4) show that the plan is cost-effective, and (5) conclude with a request for action. My audience here expects a confident and businesslike—but not stuffy—tone. I want to be sure that everything in this proposal encourages readers to support our budget request. ❏

A Funding Proposal
for
The SMU *Torch*
(20XX–XX)

Prepared for
The Student Fee Allocation Committee
Southeastern Massachusetts University
North Dartmouth, Massachusetts

by
William Trippe
Torch Business Manager

May 1, 20XX

FIGURE 23.2   A Funding Proposal

The SMU *Torch*

Old Westport Road
North Dartmouth, Massachusetts 02747

May 1, 20XX

Charles Marcus, Chair
Student Fee Allocation Committee
Southeastern Massachusetts University
North Dartmouth, MA 02747

Dear Dean Marcus:

No one needs to be reminded about the effects of increased costs on our campus community. We are all faced with having to make do with less.

Accordingly, we at the *Torch* have spent long hours devising a plan to cope with increased production costs—without compromising the newspaper's tradition of quality service. I think you and your colleagues will agree that our plan is realistic and feasible. Even the "bare-bones" operation that will result from our proposed spending cuts, however, will call for a $7,710.27 increase in next year's budget.

We have received no funding increase in three years. Our present need is absolute. Without additional funds, the *Torch* simply cannot continue to function as a professional newspaper. I therefore submit the following budget proposal for your consideration.

Respectfully,

William Trippe

William Trippe
Business Manager, SMU *Torch*

FIGURE 23.2   A Funding Proposal *(continued)*

TABLE OF CONTENTS

FIGURE 23.2   A Funding Proposal *(continued)*

INFORMATIVE ABSTRACT

The SMU *Torch*, the student newspaper at Southeastern Massachusetts University is crippled by inadequate funding, having received no budget increase in three years. Increased costs and inadequate funding are the major problems facing the *Torch*. Increases in costs of technology upgrades and in printing have called for cutbacks in production. Moreover, our low staff salaries are inadequate to attract and retain qualified personnel. A nominal pay increase would make salaries more competitive.

Our staff plans to cut costs by reducing page count and by hiring a new press for the *Torch*'s printing work. The only proposed cost increase (for staff salaries) is essential.

A detailed breakdown of projected costs establishes the need for a $7,710.27 budget increase to keep the paper a weekly publication with adequate page count to serve our campus.

Compared with similar newspapers at other colleges, the *Torch* makes much better use of its money. The comparison figures in the Appendix illustrate the cost-effectiveness of our proposal.

FIGURE 23.2   A Funding Proposal *(continued)*

INTRODUCTION

Overview

Our campus newspaper faces the contradictory challenge of surviving ever-increasing production costs while maintaining its reputation for quality. The following proposal offers a realistic plan for meeting the crisis. The plan's ultimate success, however, depends on the Student Fee Allocation Committee's willingness to approve a long-overdue increase in the *Torchs'* upcoming yearly budget.

Background

In ten years, the *Torch* has grown in size, scope, and quality. Roughly 6,000 copies (24 pages/issue) are printed weekly for each fourteen-week semester. Each week, the *Torch* prints national and local press releases, features, editorials, sports articles, announcements, notices, classified ads, a calendar column, and letters to the editor. A vital part of university life, our newspaper provides a forum for information, ideas, and opinions—all with the highest professionalism. This year we published a Web-based version as well.

Statement of Problem

With much of its staff about to graduate, the *Torch* faces next year with rising costs in every phase of production, and the need to replace outdated and worn equipment.

Our newspaper also suffers from a lack of student involvement: Despite gaining valuable experience and potential career credentials, few students can be expected to work without some kind of remuneration. Most staff members do receive minimal weekly salaries: from $20 for the distributor to $90 for the Editor-in-Chief. But salaries averaging barely $3 per hour cannot possibly compete with the minimum wage. Since more and more SMU students must work part-time, the *Torch* will have to make its salaries more competitive.

The newspaper's operating expenses can be divided into four categories: hardware and software upgrades, salaries, printing costs, and miscellaneous (office supplies, mail, and so on). The first three categories account for nearly 90 percent of the budget. Over the past year, costs in all categories have increased: from as little as 2 percent for miscellaneous expenses to as much as 19 percent for technology upgrades. Printing costs (roughly one-third of our total budget) rose 9 percent in the past year, and another price hike of 10 percent has just been announced.

Need

Despite the steady increase in production costs, the *Torch* has received no increase in its yearly budget allocation ($37,400) in three years. Inadequate funding is virtually crippling our newspaper.

FIGURE 23.2 A Funding Proposal *(continued)*

Scope

The following plan includes

1. Methods for reducing production costs while maintaining the quality of our staff
2. Projected costs for technology upgrades, salaries, and services during the upcoming year
3. A demonstration of feasibility, showing our cost-effectiveness
4. A summary of attitudes shared by our personnel

PROPOSED PLAN

The following plan is designed to trim operating costs without compromising quality.

Methods

We can overcome our budget and staffing crisis by taking these steps:

Reducing Page Count. By condensing free notices for campus organizations, abolishing "personal" notices, and limiting press releases to one page, we can reduce page count per issue from 24 to 20, saving nearly 17 percent in production costs. (The items deleted from hard copy could be included as linked add-ons in the *Torch*'s Web-based version.)

Reducing Hard-Copy Circulation. Reducing circulation from 6,000 to 5,000 copies barely will cover the number of full-time SMU students, but will save nearly 17 percent in printing costs. The steadily increasing number of hits on our Web site suggests that more and more readers are using the electronic medium. (We are designing a fall survey that will help us determine how many readers rely on the Web-based version.)

Hiring a New Press. We can save money by hiring Arrow Press for printing. Other presses (including our present printer) bid at least 25 percent higher than Arrow. With its state-of-the art production equipment, Arrow will import our "camera-ready" digital files to produce the hard-copy version. Moreover, no other company offers the rapid turnover time (from submission to finished product) that Arrow promises.

Upgrading Our Desktop Publishing Technology. In order to be compatible with Arrow's specifications for submitting digital files, we will need to upgrade our desktop publishing software and supporting hardware. Upgrade costs will be largely offset during the first year by our reduced printing costs. Also, this equipment will increase efficiency and reduce labor costs, resulting in substantial payback on our technology investment.

FIGURE 23.2   A Funding Proposal (continued)

Increasing Staff Salaries. Although we seek talented students who expect little money and much experience, salaries for all positions must increase by an average of 25 percent. Otherwise, any of our staff could make as much money elsewhere by working only a little more than half the time. In fact, many students could exceed the minimum wage by working for local newspapers. To illustrate: The *Standard Beacon* pays $60 to $90 for a news article and $30 for a photo, while the *Torch* pays nothing for articles and $6 for a photo.

A striking example of low salaries is the $4.75 per hour we pay our desktop publishing staff. Our present desktop publishing cost of $3,038 could be as much as $7,000 or even higher if we had this service done by an outside firm, as many colleges do.

Without this nominal salary increase, we cannot possibly attract qualified personnel.

Costs

Our proposed budget is itemized in Table 1, but the main point is clear: If the *Torch* is to remain viable, increased funding is essential for meeting projected costs.

Table 1 Projected Costs and Requested Funding for Next Year's *Torch* Budget

PROJECTED COSTS

Hardware/Software Upgrades

Macintosh G5 (1.8 GHz) Processor	$2,494.98
23-inch Apple™ HD Display (w/rebates)	1,499.98
LaCie™ 160 GB 7200 RPM USB Hard Drive	219.99
Olympus™ Stylus 410 4MP Digital Camera	349.99
Mikrotek™ ScanMaker 6100 Pro	299.99
Extensis™ Photo Imaging Suite	499.98
Microsoft™ Office 2004 Professional-Upgrade	289.97
QuarkEXPress™ 6 software (discounted)	729.97
Adobe™ Photoshop™ CS-Upgrade	169.97
Adobe™ Illustrator™ CS-Upgrade	169.99
Macromedia™ Dreamweaver™ MX2004-Upgrade	199.98
Subtotal	**$6,924.79**

FIGURE 23.2   A Funding Proposal *(continued)*

Wages and Salaries

Desktop-publishing staff (35 hr/wk at $6.00/hr x 28 wk)	$5,880.00
Editor-in-Chief	3,150.00
News Editor	1,890.00
Features Editor	1,890.00
Advertising Manager	2,350.00
Advertising Designer	1,575.00
Webmaster	2,520.00
Layout Editor	1,890.00
Art Director	1,260.00
Photo Editor	1,890.00
Business Manager	1,890.00
Distributor	560.00
Subtotal	**$26,745.00**

Miscellaneous Costs

Graphics by SMU art students (3/wk @ $10 each)	$ 840.00
Mailing	1,100.00
Telephone	1,000.00
Campus print shop services	400.00
Copier fees	100.00
Subtotal	**$3,440.00**

Fixed Printing Costs (5,000 copies/wk x 28/wk)	**$24,799.60**

TOTAL YEARLY COSTS	**$61,909.39**
Expected Advertising Revenue ($600/wk x 28/wk)	**($16,800.00)**
Total Costs Minus Advertising Revenue	**$45,109.39**

TOTAL FUNDING REQUEST	**$45,109.39**

FIGURE 23.2   A Funding Proposal *(continued)*

Feasibility

Beyond exhibiting our need, we feel that the feasibility of this proposal can be measured through an objective evaluation of our cost-effectiveness: Compared with newspapers at similar schools, how well does the *Torch* use its funding?

In a survey of the four area college newspapers, we found that the *Torch*—by a sometimes huge margin—makes the best use of its money per page. Table 1A in the Appendix shows, that of the five newspapers, the *Torch* costs students the least, runs the most pages weekly, and spends the least money per page, *despite a circulation two to three times the size of the other papers*.

The most striking comparison is between the *Torch* and the newspaper at Fallow State College. Each student at FSC pays $24.66 yearly for a paper averaging 12 pages per issue. Here at SMU, each student pays $8.12 yearly for a paper averaging 24 pages per issue. Thus, for 33 percent of FSC's costs, SMU students are receiving twice the amount of coverage on a variety of topics.

The *Torch* has the lowest yearly cost of all five newspapers, despite having the largest circulation. With the requested budget increase, the cost would rise by only $0.88, for a yearly cost of $9.00 to each student. Although Alden College's newspaper costs each student $8.58, it is published only every third week, averages 12 pages per issue, and costs more than $71.00 yearly per page to print—in contrast to our yearly printing cost of $55.65 per page.

As the figures in the Appendix demonstrate, our cost management is responsible and effective.

Personnel

Students on the *Torch* staff are unanimous in their determination to maintain the highest professionalism. Many are planning careers in journalism, writing, editing, advertising, photography, Web design, or public relations. In any *Torch* issue, the balanced, enlightened coverage is evidence of our judicious selection and treatment of articles and our shared concern for quality.

CONCLUSION

As a broad forum for ideas and opinions, the *Torch* continues to reflect a seriousness of purpose and a commitment to free and responsible expression. Its place in the campus community is more vital than ever during these troubled times.

Every year, there are increases and decreases in allocations to student organizations. Last year, for example, eight allocations were increased by an average of $4,332. The *Torch* has received no increase in three years.

Presumably, increases materialize as priorities change and as special circumstances arise. For the *Torch*, these circumstances have been created by increasing production costs and by the need to remain abreast of the revolution in digital communication. We respectfully urge the Allocation Committee to respond to the *Torch*'s legitimate and proven needs by increasing our upcoming year's allocation to $45,109.39.

FIGURE 23.2   A Funding Proposal *(continued)*

APPENDIX (Comparative Performance)

Table 1A Allocations and Performance of Five Massachusetts College Newspapers

	Stonehorse College	Alden College	Simms University	Fallow State	SMU
Enrollment	1,600	1,400	3,000	3,000	5,000
Fee paid (per year)	$65.00	$85.00	$35.00	$50.00	$65.00
Total fee budget	$104,000	$119,000	$105,000	$150,000	$325,000
Newspaper budget	$18,300	$8,580	$36,179	$52,910	$37,392
					$45,109[a]
Yearly cost per student	$12.50	$8.58	$16.86	$24.66	$8.12
					$9.00[a]
Publication rate	Weekly	Every third week	Weekly	Weekly	Weekly
Average no. of pages	8	12	18	12	24
Average total pages	224	120	504	336	672
					560[a]
Yearly cost per page	$81.60	$71.50	$71.78	$157.47	$55.56
					$67.12[a]

[a]These figures are next year's costs for the SMU *Torch*.

Source: Figures were quoted by newspaper business managers in April 20XX.

FIGURE 23.2   A Funding Proposal *(continued)*

☑ CHECKLIST for Usability of Proposals

(Numbers in parentheses refer to the first page of discussion.)

Format

☐ Is the short internal proposal in memo form and the short external proposal in letter form? (573)

☐ Does the long proposal have adequate supplements to serve the needs of different readers? (583)

☐ Is the format professional in appearance? (340)

☐ Are headings logical and adequate? (356)

☐ Does the title forecast the proposal's subject and purpose? (579)

Content

☐ Is the problem clearly identified? (579)

☐ Is the objective clearly identified? (579)

☐ Does each key element in the proposal support its objective? (581)

☐ Are ideas and claims supported with facts or specific discussion? (580)

☐ Is the proposed plan, service, or product beneficial? (568)

☐ Are the proposed methods practical and realistic? (580)

☐ Are all foreseeable limitations and contingencies identified? (580)

☐ Is the proposal free of overstatement? (581)

☐ Are visuals used effectively and whenever appropriate? (582)

☐ Is each source and contribution properly cited? (583)

☐ Is the proposal ethically acceptable? (580)

Arrangement

☐ Is there a recognizable introduction, body, and conclusion? (584)

☐ Are all *relevant* headings from the general outline included? (584)

☐ Does the introduction provide sufficient orientation to the problem and the plan? (584)

☐ Does the body explain *how, where,* and *how much?* (586)

☐ Does the conclusion encourage acceptance of the proposal? (589)

☐ Are there clear transitions between related ideas? (Appendix C)

Style and Page Design

☐ Is the level of technicality appropriate for primary readers? (581)

☐ Do supplements follow the appropriate style guidelines? (583)

☐ Will the informative abstract be understood by laypersons? (648)

☐ Is the tone appropriate? (271)

☐ Is the writing style clear, concise, and fluent throughout? (244)

☐ Is the language convincing and precise? (582)

☐ Is the proposal written in grammatical English? (Appendix C)

☐ Is the page design inviting and accessible? (583)

EXERCISES   For more exercises, visit
<www.ablongman.com/lannon>

1. Assume that the head of your high school English department has asked you, as a recent graduate, for suggestions about revising the English curriculum to prepare students for writing. Write a proposal, based on your experience since high school. (Primary audience: the English department head and faculty; secondary audience: the school committee.) Review the outline on page 584 before selecting specific headings.

2. After identifying your primary and secondary audience, compose a short planning proposal for improving an unsatisfactory situation in the classroom, on the job, or in your dorm or apartment (e.g., poor lighting, drab atmosphere, health hazards, poor seating arrangements). Choose a problem or situation whose resolution is more a matter of common sense and lucid observation than of intensive research. Be sure to (a) identify the problem clearly, give a brief background, and stimulate interest; (b) clearly state the methods proposed to solve the problem; and (c) conclude with a statement designed to gain audience support for your proposal.

3. Write a research proposal to your instructor (or an interested third party) requesting approval for the final term project (an analytical report or formal proposal). Verify that adequate primary and secondary sources are available. Convince your audience of the soundness and usefulness of the project.

4. As an alternate term project to the formal analytical report (Chapter 24), develop a long proposal for solving a problem, improving a situation, or satisfying a need in your school, community, or job. Choose a subject sufficiently complex to justify a formal proposal, a topic requiring research (mostly primary). Identify an audience (other than your instructor) who will use your proposal for a specific purpose. Compose an audience and use profile, using the sample on page 589 as a model. Here are possible subjects for your proposal:
   - improving living conditions in your dorm or fraternity/sorority
   - creating a student-oriented advertising agency on campus
   - creating a daycare center on campus
   - creating a new business or expanding a business
   - saving labor, materials, or money on the job
   - improving working conditions
   - improving campus facilities for the disabled
   - supplying a product or service to clients or customers
   - increasing tourism in your town
   - eliminating traffic hazards in your neighborhood or on campus
   - reducing energy expenditures on the job
   - improving security in dorms or in the college library
   - improving in-house training or job orientation programs
   - creating a one-credit course in job hunting or stress management for students
   - improving tutoring in the learning center
   - making the course content in your major more relevant to student needs
   - creating a new student government organization
   - finding ways for an organization to raise money
   - improving faculty advising for students
   - purchasing new equipment
   - improving food service on campus
   - easing first-year students through the transition to college
   - changing the grading system at your school
   - establishing more equitable computer terminal use
   - designing a Web site for your employer or an organization to which you belong

SERVICE-LEARNING PROJECT[1]

In your nonprofit organization or your school, identify a particular need or project that requires outside

---

[1]This exercise was inspired by Roger H. Munger's article listed in Works Cited.

funding. Search the Web to locate an appropriate funding program. Start by using words such as *proposal, grant,* or *funding* along with keywords that describe your project, such as *remedial programs, adolescent drug treatment,* and so on.

For a general listing of foundations that provide grants, go to the *Foundation Center* at <www. fdncenter.org>. For sources of education funding, go to *efundingsolutions* at <www.efunding.com>. For sources of science funding, go to the *National Science Foundation* at <www.nsf.gov>. For grant sources for nonprofit or public service organiza-

tions, go to <www. sils.umich.edu/~nesbitt/nonprofits/nonprofits.html> or *"The Grant Getting Page"* shown in Figure 23.3.

Prepare a short report for your agency that describes the types of projects funded by your chosen source, the average amount of a grant, the number of proposals submitted in a given year, the number of grants awarded, and the specific criteria this funding program uses in evaluating different proposals. Persuade your fellow members that your organization could qualify for a grant from this source. Attach copies of relevant Web pages to your document.

**FIGURE 23.3
"The Grant
Getting Page"**

This Web site
offers a good
starting point
for applicants
seeking funding
from federal and
nonprofit grant-
ing agencies.

*Source: Reprinted
by permission of
The University of
Illinois at Chicago
<www.uic.edu/
depts/ovcr/ors/>.*

24

Formal Analytical Reports

The formal analytical report, like the shorter versions discussed in Chapter 17, usually leads to recommendations. The formal report replaces the memo when the topic requires lengthy discussion (roughly 10 or more pages). Formal reports generally include a title page, table of contents, a system of headings, a list of references or works cited, and other items of front-matter and end-matter supplements discussed in Chapter 25.

An essential component of workplace problem solving, analytical reports are designed to answer these questions:

What readers of an analytical report want to know

- *Based on the information gathered about this issue, what do we know?*
- *What conclusions can we draw?*
- *What should we do or not do?*

Assume, for example, that you receive this assignment from your supervisor:

A typical analytical problem

> Recommend the best method for removing the heavy-metal contamination from our company dump site.

To recommend the most feasible method, you will have to learn everything you can about the nature and extent of the problem. Then you will compare the advantages and disadvantages of various options based on the criteria you are using to assess feasibility: say, cost-effectiveness, time required versus time available for completion, potential risk to the public and the environment.

Recommendations have legal and ethical implications

For example, the cheapest option might also pose the greatest environmental risk and could result in heavy fines or criminal charges. But the safest option might simply be too expensive for this struggling company to afford. Or perhaps the Environmental Protection Agency has imposed a legal deadline for the cleanup. In making your recommendation, you will need to weigh all the criteria (cost, safety, time) very carefully, or you could land in jail.

The above situation calls for skills in critical thinking (Chapter 2) and in research (Chapters 7–10). Besides interviewing legal and environmental experts, you might search the literature and the Web. From these sources you can discover whether anyone has been able to solve a problem like yours, or learn about the newest technologies for toxic waste cleanup. Then you will have to decide how much, if any, of what others have done applies to your situation. (For more on analytical reasoning, see pages 607, 612.)

PURPOSE OF ANALYSIS

Using analysis on the job

You may be assigned to evaluate a new assembly technique on the production line, or to locate and purchase the best equipment at the best price. You might have to identify the cause behind a monthly drop in sales, the reasons for low employee morale, the causes of an accident, or the reasons for equipment failure. You might need to assess the feasibility of a proposal for a company's expansion or investment.

The list is endless, but the procedure is always the same: (1) asking the right questions, (2) searching the best sources, (3) evaluating and interpreting your findings, and (4) drawing conclusions and making recommendations.

TYPICAL ANALYTICAL PROBLEMS

Far more than an encyclopedia presentation of information, the analytical report traces your inquiry, your evidence, and your reasoning to show exactly how you arrived at your conclusions and recommendations.

Workplace problem solving calls for skills in three broad categories: *causal analysis, comparative analysis,* and *feasibility analysis.* Each approach relies on its own type of reasoning.

Causal Analysis: "Why Does *X* Happen?"

Designed to attack a problem at its source, the causal analysis answers questions like this: *Why do so many apparently healthy people have sudden heart attacks?* Here is how causal reasoning proceeds:

Identify the problem

Examine possible causes

Recommend solutions

> Medical researchers at the world-renowned Hanford Health Institute recently found that 20 to 30 percent of deaths from sudden heart attacks occur in people who have none of the established risk factors (weight gain, smoking, diabetes, lack of exercise, high blood pressure, or family history).
>
> To better identify people at risk, researchers are now tracking down new and powerful risk factors such as bacteria, viruses, genes, stress, anger, and depression.
>
> Once researchers identify these factors and their mechanisms, they can recommend preventive steps such as careful monitoring, lifestyle and diet changes, drug treatment, or psychotherapy (H. Lewis 39–43). ❑

A different version of causal analysis employs reasoning from effect to cause, to answer questions like this: *What are the health effects of exposure to electromagnetic radiation?*

NOTE *Keep in mind that faulty causal reasoning is extremely common, especially when we ignore other possible causes or we confuse correlation with causation (page 182).*

Comparative Analysis: "Is *X* or *Y* Better for Our Purpose?"

Designed to rate competing items on the basis of specific criteria, the comparative analysis answers questions like this: *Which type of security (firewall/encryption) program should we install on our company's computer system?* Here is how reasoning proceeds in a comparative analysis:

Identify the criteria

XYZ Corporation needs to identify exactly what information (personnel files, financial records) or functions (in-house communication, file transfer) it wants to protect from whom. Does it need both virus and tamper protection? Does it wish to restrict network access or encrypt (scramble) email and computer files so they become unreadable to unauthorized persons? Does it wish to restrict access to or from the Web? In addition to the level of protection, how important are ease of maintenance and user-friendliness?

Rank the criteria

After identifying their specific criteria, XYZ decision makers need to rank them in order of importance (for example, 1. tamper protection, 2. user-friendliness, 3. secure financial records, and so on).

Compare items according to the criteria, and recommend the best one

On the basis of these ranked criteria, XYZ will assess relative strengths and weaknesses of competing security programs and recommend the best one (Schafer 93–94). ❑

Feasibility Analysis: "Is This a Good Idea?"

Designed to assess the practicality of an idea or plan, the feasibility analysis answers questions like this: *Will increased business justify the cost of an interactive, multimedia Web site?* Here is the reasoning process in a feasibility analysis:

24.1

For model feasibility studies visit <www.ablongman.com/lannonweb>

Bigbyte, Inc., a retailer of discounted computer hardware and software, relies on catalog orders by telephone for its high-volume business. The company is now deciding whether to reallocate the bulk of its marketing resources to establish a high-profile presence on the Web.

Consider the strength of supporting reasons

Research uncovers numerous supporting reasons, including these: A Web site is an excellent medium for enhancing a company's name recognition and for announcing new products and catalog updates. Site visits can be tailored to individual customer preferences. "Cybershoppers" value the convenience of researching and comparing products from home, the absence of pressure from telephone salespersons, and the ability to customize their visits to a retailer's Web site.

Consider the strength of opposing reasons

Obstacles to Web-based sales include consumer reluctance to reveal credit card numbers and personal information online or to change their traditional shopping patterns, along with frustration with slow downloading. Also, creating and maintaining an attractive, navigable, and up-to-date site is costly, and few retail sites—even the most heavily visited—have shown actual profit.

Weigh the pros and cons and recommend a course of action

After assessing the benefits and drawbacks of Web-based marketing in *this* situation, Bigbyte decision makers can make the appropriate recommendations (Hodges 22+). ❑

For more on feasibility reports, see page 395.

Combining Types of Analysis

Analytical categories overlap considerably. Any one study may in fact require answers to two or more of the previous questions. The sample report on pages 631–38 is both a feasibility analysis and a comparative analysis. It is designed to answer these questions: *Is technical marketing the right career for me? If so, which is my best option for entering the field?*

ELEMENTS OF A USABLE ANALYSIS

The formal analytical report incorporates many elements from assignments in earlier chapters, along with the suggestions that follow.

Clearly Identified Problem or Goal

Define your goal

To solve any problem or achieve any goal, you must first identify it precisely. Always begin by defining the main questions and thinking through any subordinate questions they may imply. Only then can you determine what to look for, where to look, and how much information you will need.

Your employer, for example, might pose this question: *Will a low-impact aerobics program significantly reduce stress among my employees?* The aerobics question obviously requires answers to three other questions: *What are the therapeutic claims for aerobics? Are they valid? Will aerobics work in this situation?* With the main questions identified, you can formulate a goal (or purpose) statement:

Goal statement

> My goal is to examine and evaluate claims about the therapeutic benefits of low-impact aerobic exercise.

Words such as *examine* and *evaluate* (or *compare, identify, determine, measure, describe,* and so on) help readers understand the specific analytical activity that forms the subject of the report. For more on asking the right questions, see pages 120–21.

Adequate but Not Excessive Data

Decide how much is enough

A superficial analysis is basically worthless. Worthwhile analysis, in contrast, examines an issue in depth (as discussed on pages 122–23). In reporting on your analysis, however, you filter that material for the audience's understanding, deciding what to include and what to leave out. Do decision makers in this situation need a closer look or are we presenting excessive detail when only general information is needed? Is it possible to have too much information? In some cases, yes—as behavioral expert Dietrich Dorner explains:

Excessive
information
hampers decision
making

The more we know, the more clearly we realize what we don't know. This probably explains why . . . organizations tend to [separate] their information-gathering and decision-making branches. A business executive has an office manager; presidents have . . . advisers; military commanders have chiefs of staff. The point of this separation may well be to provide decision makers with only the bare outlines of all the available information so they will not be hobbled by excessive detail when they are obliged to render decisions. Anyone who is fully informed will see much more than the bare outlines and will therefore find it extremely difficult to reach a clear decision. (99)

Confusing the issue with excessive information is no better than recommending hasty action on the basis of inadequate information (Dorner 104).

When you might
consult an
abstract or
summary instead
of the complete
work

As you research the issue you may want to filter material for your own understanding as well. Whether to rely on the abstract or summary or to read the complete text of a specialized article or report all depends on the question you're trying to answer and the level of technical detail your readers expect. If you are expert in the field, writing for other experts, you probably want to read the entire document in order to assess the methods and reasoning behind a given study. But if you are less than expert, a summary or abstract of this study's findings might suffice. The fact sheet in Figure 24.1, for example, summarizes a detailed feasibility study for a general reading audience. Readers seeking more details, including nonclassified elements of the complete report, could visit the Transportation Security Administrations's Web site at <www.tsa.gov>.

NOTE   *If you have relied merely on the abstract or summary instead of the full article, be sure to indicate this ("Abstract," "Press Release" or the like) when you document the source in your report (as shown on page 691).*

Accurate and Balanced Data

Give readers all
they need to
make an
informed
judgment

Avoid stacking the evidence to support a preconceived point of view. Assume, for example, that you are asked to recommend the best chainsaw brand for a logging company. Reviewing test reports, you come across this information:

> Of all six brands tested, the Bomarc chainsaw proved easiest to operate. However, this brand also offers the fewest safety features.

In citing these equivocal findings, you need to present both of them accurately, and not simply the first—even though the Bomarc brand may be your favorite. Then argue for the feature (ease of use or safety) you think should receive priority. (Refer to pages 121 and 172 for exploring and presenting balanced and reasonable evidence.)

FACT SHEET: Train and Rail Inspection Pilot, Phase I

U.S. DEPARTMENT OF HOMELAND SECURITY
Transportation Security Administration
FOR IMMEDIATE RELEASE – June 7, 2004
TSA Press Office: (571) 227-2829

Objective:
Implement a pilot program to determine the feasibility of screening passengers, luggage and carry-on bags for explosives in the rail environment.

TRIP I Background:
- Secretary Ridge announced TRIP on March 22, 2004, to test new technologies and screening concepts.
- The program is conducted in partnership with the Department of Transportation, Amtrak, Maryland Rail Commuter and Washington D.C.'s Metro.
- The New Carrollton, Md. station was selected because it serves multiple types of rail operations and is located close to Washington, D.C.

TRIP I Facts:
- Screening for Phase I of TRIP began on May 4 and was completed on May 26, 2004.
- A total of 8,835 passengers and 9,875 pieces of baggage were screened during the test.
- The average time to wait in line and move through the screening process was less than 2 minutes.
- Customer Feedback cards reflect a 93 percent satisfaction rate with both the screening process and the professional demeanor of TSA personnel.

Lessons Learned:
- Overall results indicate efficient checkpoint throughput with minimal customer inconvenience.
- Passengers were overwhelmingly receptive to the screening process.
- Providing a customer service representative on-site during all screening operations helped Amtrak ensure passengers received outstanding customer service.
- Skilled TSA screeners from the agency's National Screening Force were able to quickly transition to screening in the rail environment.
- Most importantly, Phase I showed that currently available technology could be utilized to screen for explosives in the rail environment.

FIGURE 24.1 A Summary Description of a Feasibility Study. Notice that the criteria for assessing feasibility include passenger wait times, passenger receptiveness to screening, and—most important—effectiveness of screening equipment in this environment.
Source: Transportation Security Administration. Press Release.

Fully Interpreted Data

Explain the
significance of
your data

Interpretation shows the audience "what is important and what unimportant, what belongs together and what does not" (Dorner 44). For example, you might interpret the chainsaw data from page 610 in this way:

Explain the
meaning of your
evidence

> Our logging crews often work suspended by harness, high above the ground. Also, much work is in remote areas. Safety features therefore should be our first requirement in a chainsaw. Despite its ease of operation, the Bomarc saw does not meet our safety needs.

By saying "therefore" you engage in analysis—not just information sharing. Don't merely list your findings, explain what they mean. (Refer to pages 174–77.)

Subordination of Personal Bias

Evaluate and
interpret
evidence
impartially

To arrive at the truth of the matter (page 174), you need to see clearly. Don't let your biases fog up the real picture. Each stage of analysis requires decisions about what to record, what to exclude, and where to go next. You must evaluate your data (*Is this reliable and important?*), interpret your evidence (*What does it mean?*), and make recommendations (*What action is needed?*). An ethically sound analysis presents a balanced and reasonable assessment of the evidence. Do not force viewpoints that are not supported by dependable evidence. (Refer to page 124.)

Appropriate Visuals

Use visuals
generously

Graphs are especially useful in an analysis of trends (rising or falling sales, radiation levels). Tables, charts, photographs, and diagrams work well in comparative analyses. Be sure to accompany each visual with a fully interpreted "story."

NOTE　　*As the simplicity of Figure 24.2 and its brief caption illustrate, a powerful visual does not need to be complex and fancy, nor its accompanying story long and involved. Sometimes, less can be more.*

Valid Conclusions and Recommendations

Be clear about
what the
audience should
think and do

Along with the informative abstract (page 648), conclusions and recommendations are the sections of a long report that receive the most audience attention. The goal of analysis is to reach a valid conclusion—an overall judgment about what all the material means (that *X* is better than *Y*, that *B* failed because of *C*, that *A* is a good plan of action). Here is the conclusion of a report on the feasibility of installing an active solar heating system in a large building:

Offer a final judgment

1. Active solar space heating for our new research building is technically feasible because the site orientation will allow for a sloping roof facing due south, with plenty of unshaded space.
2. It is legally feasible because we are able to obtain an access easement on the adjoining property, to ensure that no buildings or trees will be permitted to shade the solar collectors once they are installed.
3. It is economically feasible because our sunny, cold climate means high fuel savings and faster payback (fifteen years maximum) with solar heating. The long-term fuel savings justify our short-term installation costs (already minimal because the solar system can be incorporated during the building's construction—without renovations).

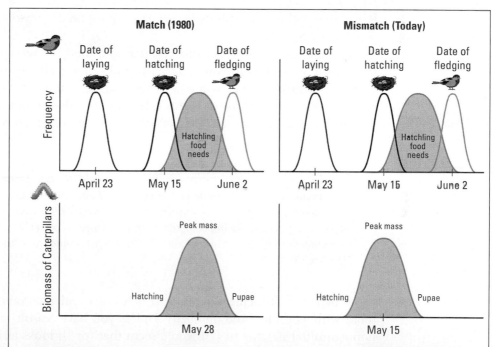

INTERDEPENDENT SPECIES can be "uncoupled" under stress from global warming. In De Hoge Veluwe National Park in the Netherlands, changes in weather patterns have caused oak buds to burst into leaf sooner. As a result, winter moth caterpillars—an important food that great tit chick hatchlings need to reach fledging size—peak in total biomass earlier today (*right*) as compared with two decades ago (*left*). Egg-laying time has not shifted.

FIGURE 24.2   A Simple but Richly Informative Visual.   The visual and caption alone capsulize the troubling implications revealed by the lengthy analysis described in the article.

Source: Grossman, Daniel, "Spring Forward." Scientific American Jan. 2004: 85–91.

Conclusions are valid when they are logically derived from accurate interpretations.

Having explained *what it all means,* you then recommend *what should be done.* Taking all possible alternatives into account, your recommendations urge the most feasible option (to invest in *A* instead of *B,* to replace *C* immediately, to follow plan *A,* or the like). Here are the recommendations based on the previous interpretations:

Tell what should be done

1. I recommend that we install an active solar heating system in our new research building.
2. We should arrange an immediate meeting with our architect, building contractor, and solar heating contractor. In this way, we can make all necessary design changes before construction begins in two weeks.
3. We should instruct our legal department to obtain the appropriate permits and easements immediately.

Recommendations are valid when they propose an appropriate response to the problem or question.

Because they culminate your research and analysis, recommendations challenge your imagination, your creativity, and—above all—your critical thinking skills. What strikes one person as a brilliant suggestion might be seen by others as irresponsible, offensive, or dangerous. (Figure 24.3 depicts the kinds of decisions writers encounter in formulating, evaluating, and refining their recommendations.)

NOTE   *Keep in mind that solving one problem might create new and worse problems—or unintended consequences. For example, to prevent crop damage by rodents, an agriculture specialist might recommend trapping and poisoning. While rodent eradication may increase crop yield temporarily, it also increases the insects these rodents feed on—leading eventually to even greater crop damage. In short, before settling on any recommendation, try to anticipate its "side effects and long-term repercussions" (Dorner 15).*

When you do achieve definite conclusions and recommendations, express them with assurance and authority. Unless you have reason to be unsure, avoid noncommittal statements ("It would seem that" or "It looks as if"). Be direct and assertive ("The earthquake danger at the reactor site is acute," or "I recommend an immediate investment"). Announce where you stand.

If, however, your analysis yields nothing definite, do not force a simplistic conclusion on your material (pages 124–26). Instead, explain your position ("The contradictory responses to our consumer survey prevent me from reaching a definite conclusion. Before we make any decision about this product, I recommend a full-scale market analysis"). The wrong recommendation is far worse than no recommendation at all. (Refer to Chapter 10, pages 192–93 for helpful guidelines.)

Consider All the Details

- What exactly should be done—if anything at all?
- How exactly should it be done?
- When should it begin and be completed?
- Who will do it, and how willing are they?
- What equipment, material, or resources are needed?
- Are any special conditions required?
- What will this cost, and where will the money come from?
- What consequences are possible?
- Whom do I have to persuade?
- How should I order my list (priority, urgency, etc.)?

Locate the Weak Spots

- Is anything unclear or hard to follow?
- Is this course of action unrealistic?
- Is it risky or dangerous?
- Is it too complicated or confusing?
- Is anything about it illegal or unethical?
- Will it cost too much?
- Will it take too long?
- Could anything go wrong?
- Who might object or be offended?
- What objections might be raised?

Make Improvements

- Can I rephrase anything?
- Can I change anything?
- Should I consider alternatives?
- Should I reorder my list?
- Can I overcome objections?
- Should I get advice or feedback before I submit this?

FIGURE 24.3 How to Think Critically About Your Recommendations

Source: Adapted from The Art of Thinking, *5th ed. Vincent R. Ruggiero, copyright © 1998. Reprinted by permission of Pearson Education, Inc.*

Self-Assessment

Assess your analysis continuously

The more we are involved in a project, the larger our stake in its outcome—making self-criticism less likely just when it is needed most! For example, it is hard to admit that we might need to backtrack, or even start over, in instances like these (Dorner 46):

Things that might go wrong with your analysis

- During research you find that your goal isn't clear enough to indicate exactly what information you need.
- As you review your findings, you discover that the information you have is not the information you need.

It is even harder to test our recommendations and admit they are not working:

- After making a recommendation, you discover that what seemed like the right course of action turns out to be the wrong one.

If you meet such obstacles, acknowledge them immediately, and revise your approach as needed.

AN OUTLINE AND MODEL FOR ANALYTICAL REPORTS

24.2

For more model reports visit <www.ablongman.com/lannonweb>

Whether you outline earlier or later, the finished report depends on a good outline. This model outline can be adapted to most analytical reports.

I. Introduction
  A. Definition, Description, and Background
  B. Purpose of the Report, and Intended Audience
  C. Method of Inquiry
  D. Limitations of the Study
  E. Working Definitions (here or in a glossary)
  F. Scope of the Inquiry (topics listed in logical order)
  G. Conclusion(s) of the Inquiry (briefly stated)
II. Collected Data
  A. First Topic for Investigation
    1. Definition
    2. Findings
    3. Interpretation of findings
  B. Second Topic for Investigation
    1. First subtopic
      a. Definition
      b. Findings
      c. Interpretation of findings
    2. Second subtopic (and so on)
III. Conclusion
  A. Summary of Findings
  B. Overall Interpretation of Findings (as needed)
  C. Recommendations (as needed and feasible)

(This outline is only tentative. Modify as necessary.)

Two sample reports in this chapter follow the model outline. The first one, "Children Exposed to Electromagnetic Radiation: A Risk Assessment" (minus the document supplements that ordinarily accompany a long report), begins on this page. The second report, "The Feasibility of a Technical Marketing Career," appears in Figure 24.4. Supplements (front and end matter) for this report appear in Chapter 25 and on page 713.

Each report responds to slightly different questions. The first tackles these questions: *What are the effects of* X *and what should we do about them?* The second tackles two questions: *Is* X *feasible, and which version of* X *is better for my purposes?* At least one of these reports should serve as a model for your own analysis.

Introduction

The introduction engages and orients the audience and provides background as briefly as possible for the given situation. Often, writers are tempted to write long introductions because they have a lot of background knowledge about the issue. But readers generally don't need long history lessons on the subject.

In your introduction, identify your topic's origin and significance, define or describe the problem or issue, and explain the report's purpose. (Stipulate your audience only in the version your instructor will read and only if you don't attach an audience and use profile.) Briefly identify your research methods (interviews, literature searches, and so on) and explain any data omissions (person unavailable for interview, research still in progress, and so on). List working definitions, but if you have more than two or three, place definitions in a glossary. List the topics you have researched. Finally, briefly preview your conclusion; don't make readers wade through the entire report to find out what you are recommending or advising.

NOTE   *Not all reports require every element. Give readers only what they need and expect.*

As you read the following introduction, think about the elements designed to engage and orient the audience (i.e., local citizens), and evaluate their effectiveness. (Review pages 120–24 for the situation that gave rise to this report.)

Children Exposed to Electromagnetic Radiation: A Risk Assessment

Laurie A. Simoneau

INTRODUCTION

Definition and background of the problem

Wherever electricity flows—through the largest transmission line or the smallest appliance—it emits varying intensities of charged waves: an *electromagnetic field* (EMF). Some medical studies have linked human exposure to EMFs with definite physiologic changes and possible illness including cancer, miscarriage, and depression.

Experts disagree over the health risk, if any, from EMFs. Some question whether EMF risk is greater from high-voltage transmission lines, the smaller distribution lines strung on utility poles, or household appliances. Conclusive research may take years; meanwhile, concerned citizens worry about avoiding potential risks.

Description of the problem

In Bocaville, four sets of transmission lines—two at 115 Kilovolts (kV) and two at 500 kV—cross residential neighborhoods and public property. The Adams elementary school is less than 100 feet from this power line corridor. EMF risks—whatever they may be—are thought to increase with proximity.

Purpose and methods of this inquiry

Based on examination of recent research and interviews with local authorities, this report assesses whether potential health risks from EMFs seem significant enough for Bocaville to (a) increase public awareness, (b) divert the transmission lines that run adjacent to the elementary school, and (c) implement widespread precautions in the transmission and distribution of electrical power throughout Bocaville.

Scope of this inquiry

This report covers five major topics: what we know about various EMF sources, what research indicates about physiologic and health effects, how experts differ in evaluating the research, what the power industry and the public have to say, and what actions are being taken locally and nationwide to avoid risk.

Conclusions of the inquiry (briefly stated)

The report concludes by acknowledging the ongoing conflict among EMF research findings and by recommending immediate and inexpensive precautionary steps for our community.

Body

The body section describes and explains your findings. Present a clear and detailed picture of the evidence, interpretations, and reasoning on which you will base your conclusion. Divide topics into subtopics, and use informative headings as aids to navigation.

NOTE

Remember your ethical responsibility for presenting a fair *and* balanced *treatment of the material, instead of "loading" the report with only those findings that support your viewpoint. Also, keep in mind the body section can have many variations, depending on the audience, topic, purpose, and situation.*

As you read the following section, evaluate how effectively it informs readers, keeps them on track, reveals a clear line of reasoning, and presents an impartial analysis.

DATA SECTION

First topic

Sources of EMF Exposure

Definition

Electromagnetic intensity is measured in *milligauss* (mG), a unit of electrical measurement. The higher the mG reading, the stronger the field. Studies suggest that consistent exposure above 1–2 mG may increase cancer risk significantly, but no scientific evidence concludes that exposure even below 2.5 mG is safe.

Findings

Table 1 gives the EMF intensities from electric power lines at varying distances during average and peak usage.

Table 1 EMF Emissions from Power Lines (in milligauss)

Types of Transmission Lines	Maximum on Right-of-Way	Distance from lines			
		50'	100'	200'	300'
115 Kilovolts (kV)					
Average usage	30	7	2	0.4	0.2
Peak usage	63	14	4	0.9	0.4
230 Kilovolts (kV)					
Average usage	58	20	7	1.8	0.8
Peak usage	118	40	15	3.6	1.6
500 Kilovolts (kV)					
Average usage	87	29	13	3.2	1.4
Peak usage	183	62	27	6.7	3.0

Source: United States Environmental Protection Agency. EMF in Your Environment. Washington: GPO, 1992. Data from Bonneville Power Administration.

Interpretation

As Table 1 indicates, EMF intensity drops substantially as distance from the power lines increases.

Although the EMF controversy has focused on 2 million miles of power lines criss-crossing the country, potentially harmful waves are also emitted by household wiring, appliances, computer terminals—and even from the earth's natural magnetic field. The background magnetic field (at a safe distance from any electrical appliance) in the average American home varies from 0.5 to 4.0 mG (United States Environmental 10). Table 2 compares intensities of various sources.

Interpretation

EMF intensity from certain appliances tends to be higher than from transmission lines because of the amount of current involved.

Definitions

Voltage measures the speed and pressure of electricity in wires, but *current* measures the volume of electricity passing through wires. Current (measured in *amperage*) is what produces electromagnetic fields. The current

Finding

flowing through a transmission line typically ranges from 200 to 400 amps.

Table 2 EMF Emissions from Selected Sources (in milligauss)

Source	Range[a,b]
Earth's magnetic field	0.1–2.5
Blowdryer	60–1400
Four in. from TV screen	40–100
Four ft from TV screen	0.7–9
Fluorescent lights	10–12
Electric razor	1200–1600
Electric blanket	2–25
Computer terminal (12 inches away)	3–15
Toaster	10–60

[a]Data from Brodeur, Paul. "Annals of Radiation: The Cancer at Slater School." *The New Yorker* 7 Dec. 1992: 88; Miltane, John. Interview 5 Apr. 2004; National Institute of Environmental Health. *Questions and Answers about EMF.* Washington: GPO, 1995:3.

[b]Readings are made with a gaussmeter, and vary with technique, proximity of gaussmeter to source, its direction of aim, and other random factors.

Most homes have a 200-amp service. This means that if every electrical item in the house were turned on at the same time, the house could run about 200 amps—almost as high as the transmission line. Consumers then have the ability to put 200 amps of current-flow into their homes, while transmission lines carrying 200 to 400 amps are at least 50 feet away (Miltane).

Proximity and duration of exposure, however, are other risk factors. People are exposed to EMFs from home appliances at close proximity, but appliances run only periodically: exposure is therefore sporadic, and intensity diminishes sharply within a few feet (Figure 1).

As Figure 1 indicates, EMF intensity drops dramatically over very short distances from the typical appliance.

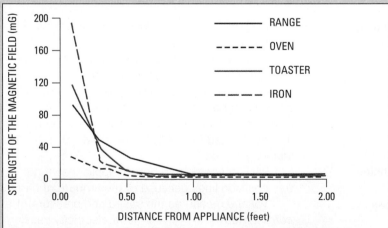

Figure 1 EMF Strengths of Typical Electric Appliances *Source: United States Environmental Protection Agency. EMF In Your Environment. Washington: GPO, 1992.*

Finding

Power line exposure, on the other hand, is at a greater distance (usually 50 feet or more), but it is constant. Moreover, its intensity can remain strong well beyond 100 feet (Miltane).

Interpretation

Research has yet to determine which type of exposure might be more harmful: briefly, to higher intensities or constantly, to lower intensities. In either case, proximity seems most significant because EMF intensity drops rapidly with distance.

Second topic

Physiologic Effects and Health Risks from EMF Exposure

Research on EMF exposure falls into two categories: epidemiologic studies and laboratory studies. The findings are sometimes controversial and inconclusive, but also disturbing.

First subtopic
Definition
General findings

Epidemiologic Studies.   Epidemiologic studies look for statistical correlations between EMF exposure and human illness or disorders. Of 77 such studies in recent decades, over 70 percent suggest that EMF exposure increases the incidence of the following conditions (Pinsky 155–215):

- cancer, especially leukemia and brain tumors
- miscarriage
- stress and depression
- learning disabilities
- heart attacks

For example, a 38-year study of nearly 140,000 electrical employees in the United States indicates that those who routinely worked in high-EMF environments were about three times more likely to die from heart attacks than coworkers in low-EMF environments (Raloff, "Electromagnetic . . . Hearts" 70).

Following are summaries of four noted epidemiologic studies implicating EMFs in cancer.

Detailed findings

A Landmark Study of the EMF/Cancer Connection.   A 1979 Denver study by Wertheimer and Leeper was the first to implicate EMFs as a cause of cancer. Researchers compared hundreds of homes in which children had developed cancer with similar homes in which children were cancer free. Victims were two to three times as likely to live in "high-current homes" (within 130 feet of a transmission line or 50 feet of a distribution line).

Critique of
findings

This study has been criticized because (1) it was not "blind" (researchers knew which homes cancer victims were living in), and (2) researchers never took gaussmeter readings to verify their designation of "high-current" homes (Pinsky 160–62; Taubes 96).

Detailed findings

Follow-up Studies.   Several major studies of the EMF/cancer connection have confirmed Wertheimer's findings:

- In 1988, Savitz studied hundreds of Denver houses and found that children with cancer were 1.7 times as likely to live in high-current homes. Unlike his predecessors, Savitz did not know whether a cancer victim lived in the home being measured, and he took gaussmeter readings to verify that houses could be designated "high-current" (Pinsky 162–63).

- In 1990, London and Peters found that Los Angeles children had 2.5 times more risk of leukemia if they lived near power lines (Brodeur 115).
- In 1992, a massive Swedish study found that children in houses with average intensities greater than 1 mG had twice the normal leukemia risk; at greater than 2 mG, the risk nearly tripled; at greater than 3 mG, it nearly quadrupled (Brodeur 115).
- Most recently, in 2002, British researchers evaluated findings from 34 studies of power line EMF effects (*a meta-analysis*). This study found "a degree of consistency in the evidence suggesting adverse health effects of living near high voltage powerlines" (Henshaw et al. 1).

Detailed findings

Workplace Studies.   More than 80 percent of 51 studies from 1981 to 1994— most notably a 1992 Swedish study—concluded that electricians, electrical engineers, and power line workers constantly exposed to an average of 1.5 to 4.0 mG had a significantly elevated cancer risk (Brodeur 115; Pinsky 177–209).

Notable Recent Studies.   Three recent workplace studies seem to support or even amplify the above findings.

- A 1994 University of Southern California study indicates that high workplace exposure to EMFs triples the risk of Alzheimer's disease (Des Marteau 38).
- A 1995 University of North Carolina study of 138,905 electric utility workers concluded that occupational EMF exposure roughly doubles brain cancer risk. This study, however, found no increased leukemia risk (Cavanaugh 8, Moore 16).
- A Canadian study of electrical-power employees published in 2000 indicates that those who had worked in strong electric fields for more than 20 years had "an eight- to tenfold increase in the risk of leukemia," along with a significantly elevated risk of lymphoma ("Strong Electric Fields" 1–2).

Interpretation

Although none of the above studies can be said to "prove" a direct cause-effect relationship, their strikingly similar results suggest a conceivable link between prolonged EMF exposure and illness.

Second subtopic

Laboratory Studies.   Laboratory studies assess cellular, metabolic, and behavioral effects of EMFs on humans and animals. EMFs directly cause the following physiologic changes (Brodeur 88; Pinsky 24–29; Raloff, "EMFs'" 30):

General findings

- reduced heart rate
- altered brain waves
- impaired immune system
- interference with the synthesis of genetic material
- disrupted regulation of cell growth

- interaction with the biochemistry of cancer cells
- altered hormonal activity
- disrupted sleep patterns

These changes are documented in the following summaries of several significant laboratory studies.

EMF Effects on Cell Chemistry.   Recent studies have demonstrated previously unrecognized effects on cell growth and division.

Detailed findings

- A 1995 University of Wisconsin study showed that cell metabolism is influenced by electromagnetic fields—the extent of effect depending on a cell's age and health. While this type of cellular stress does not appear to initiate cancer, it might help promote growth of an existing tumor (Goodman, Greenebaum, and Marron 279–338).
- A 2000 study by Michigan State University found that EMFs equal to the intensity that occurs "within a few feet" of outdoor power lines caused cells with cancer-related genetic mutations to multiply rapidly (Sivitz 196).

EMF Effects on Hormones.   Several studies have found that EMF exposure (say, from an electric blanket) inhibits production of melatonin, a hormone that fights cancer and depression, stimulates the immune system, and regulates bodily rhythms.

Detailed findings

- A 1997 study at the Lawrence National Laboratory found that EMF exposure can suppress both melatonin and the hormonelike, anticancer drug Tamoxifen (Raloff, "EMFs'" 30). A 2004 British study of Oxford University, however, found no association between decreased melatonin levels and breast cancer (Travis).
- In 1996, physiologist Charles Graham found that EMFs elevate female estrogen levels and depress male testosterone levels—hormone alterations associated respectively, with risk of breast or testicular cancer (Raloff, "EMFs'" 30)

Detailed findings

EMF Effects on Life Expectancy.   A 1994 South African study at the University of the Orange Free State measured the life span of mice exposed to EMFs. Both first and second generations of exposed mice showed significantly shortened life expectancy (de Jager and de Bruyn 221–24).

Detailed findings

EMF Effects on Behavior.   Recent studies indicate that people who live adjacent to power lines have roughly twice the normal rate of depression (Pinsky 31–32). Also, rats exposed to EMFs exhibit slower rates of learning (Beardsley 20).

Interpretation

Although laboratory studies seem more conclusive than the epidemiologic studies, what these findings *mean* is debatable.

Third topic

Debate over Quality, Cost, and Status of EMF Research
Experts differ over the meaning of EMF research findings largely because of the following limitations attributed to various studies.

First subtopic

Critiques of population studies

Limitations of Various EMF Studies
- Epidemiologic studies are criticized for overstating the evidence. For example, after reviewing the research, the American Physics Society reported "no consistent link between cancer and power line fields" (Broad A19). Some critics claim that so-called EMF-cancer links are produced by "data dredging" (making countless comparisons between types of cancers and EMF sources until certain random correlations appear) (Taubes 99). Other critics argue that news media distort the issue by publicizing studies with positive findings while often ignoring studies with negative or ambiguous findings (N. Goodman). Some studies are also accused of mistaking *coincidence* for *correlation,* without exploring "confounding factors" (e.g., exposure to toxic substances or to other adverse conditions—including the earth's natural magnetic field) (Moore 16).

Responses to critiques

  Supporters of EMF research respond that the sheer volume of epidemiologic evidence seems overwhelming (Kirkpatrick 81, 83). Moreover, the Swedish studies cited earlier seem to invalidate the above criticisms (Brodeur 115).

Critiques of lab studies

- Laboratory studies are criticized—even by scientists who conduct them—because effects on an isolated culture of cells are not always equal to effects on the total human or animal organism. Also, effects in experimental animals are not always equivalent to effects in humans (Jauchem 190–94).
- Until recently critics argued that no scientist had offered a reasonable hypothesis to explain the possible health effects (Palfreman 26). However, a 1997 study at Minnesota's Hughes Institute identified the precise biological mechanism by which EMF exposure activates tyrosine kinase enzymes that produce DNA damage (Raloff, "Electromagnetic . . . Enzymes" 119).

Responses to critiques

- Also, a 2004 University of Washington study showed that a weak electromagnetic field can break DNA strands and lead to cell death in brain cells of rats. Researchers hypothesize that EMF exposure promotes increased formation of cell-damaging agents known as free radicals (Lal and Singh).

Second subtopic

Cost objections

Costs of EMF Research.   Critics claim that research and publicity about EMFs are becoming a profit venture, spawning "a new growth industry among researchers, as well as marketers of EMF monitors" ("Electrophobia" I). Environmental expert Keith Florig identifies adverse economic effects of the EMF debate that include decreased property values, frivolous lawsuits, expensive but needless "low field" consumer appliances, and costly modifications to schools and public buildings (Monmonier 190).

Third subtopic

Present Status of EMF Research.   In July 1998, an editor at the *New England Journal of Medicine* called for ending EMF/cancer research. He cited studies

Conflicting
scientific
opinions

from the National Cancer Institute and other respected sources that showed "little evidence" of any causal connection. In a parallel development, federal and industry funding for EMF research has been reduced drastically (Stix 33). But, in August 1998, experts from the Energy Department and the National Institute of Environmental Health Sciences (NIEHS) announced that EMFs should be officially designated a "possible human carcinogen" (Gross 30).

However, one year later, in a report based on its seven-year review of EMF research, NIEHS concluded that "the scientific evidence suggesting that . . . EMF exposures pose any health risk is weak." But the report also conceded that such exposure "cannot be recognized at this time as entirely safe" (1–2).

Interpretation

In short, after more than twenty years of study, the EMF/illness debate continues, even among respected experts. While most scientists agree that EMFs exert measurable effects on the human body, they disagree about whether a real hazard exists. Given the drastic cuts in research funding, definite answers are unlikely to appear anytime soon.

Fourth topic

Views from the Power Industry and the Public

While the experts continue their debate, other viewpoints are worth considering as well.

Findings

The Power Industry's Views.   The Electrical Power Research Institute (EPRI), the research arm of the nation's electric utilities, claims that recent EMF studies have provided valuable but inconclusive data that warrant further study (Moore 17).

What does our local power company think about the alleged EMF risk? Marianne Halloran-Barney, Energy Service Advisor for County Electric, expressed this view in an email correspondence:

Findings

> There are definitely some links, but we don't know, really, what the effects are or what to do about them. . . . There are so many variables in EMF research that it's a question of whether the studies were even done correctly . . . . Maybe in a few years there will be really definite answers.

Echoing Halloran-Barney's views, John Miltane, Chief Engineer for County Electric, added this political insight:

> The public needs and demands electricity, but in regard to the negative effects of generation and transmission, the pervasive attitude seems to be "not in my back yard!" Utilities in general are scared to death of the EMF issue, but at County Electric we're trying to do the best job we can while providing reliable electricity to 24,000 customers.

Interpretation

Public Perception.   Industry views seem to parallel the national perspective among the broader population: Informed people are genuinely concerned, but remain unsure about what level of anxiety is warranted or what exactly should be done.

Finding

A 1998 survey by the Edison Electric Institute did reveal that EMFs are considered a serious health threat by 33 percent of the American public (Stix 33).

Fifth topic

Risk-Avoidance Measures Being Taken
Although conclusive answers may require decades of research, concerned citizens are already taking action against potential EMF hazards.

First subtopic

Risk Avoidance Nationwide.   Here are steps taken by various communities to protect schoolchildren from EMF exposure:

Findings

- Hundreds of individuals and community groups have taken legal action to block proposed construction of new power lines. A single Washington law firm has defended roughly 140 utilities in cases related to EMFs (Dana and Turner 32).
- Houston schools "forced a utility company to remove a transmission line that ran within 300 feet of three schools. Cost: $8 million" (Kirkpatrick 85).
- California parents and teachers are pressuring reluctant school and public health officials to investigate cancer rates in the roughly 1,000 schools located within 300 feet of transmission lines, and to close at least one school (within 100 feet) in which cancer rates far exceed normal (Brodeur 118).

Given the expense of modifying power lines to reduce EMFs (an estimated $1 billion to $3 billion yearly), critics argue that the questionable risks fail to justify the costs of measures like those above (Broad A19). Nonetheless, widespread expressions of concern about EMF exposure continue to grow.

Second subtopic
Findings

Risk Avoidance Locally.   Local awareness of the EMF issue seems low. The main public concern seems to be with property values. According to Halloran-Barney, County Electric receives one or two calls monthly from concerned customers, including people buying homes near power lines. The lack of public awareness adds another dimension to the EMF problem: People can't avoid a health threat that they don't know exists.

John Miltane stresses that County Electric takes the EMF issue very seriously: Whenever possible, new distribution lines are run underground and configured to diminish EMF intensity:

Although EMFs are impossible to eliminate altogether, we design anything we build to emit minimal intensities . . . . Also, we are considering underground cable (at $1,200 per ft.) to replace 8,000 feet of transmission lines, but the bill would ring up to nearly $10 million, for the cable alone, without labor: You don't get environmental stuff for free, which is one of the problems.

Interpretation

Before risk avoidance can be considered on a broader community level, the public must first be informed about EMFs and the associated risks of exposure.

Conclusion

The conclusion is likely to interest readers most because it answers the questions that originally sparked the analysis.

NOTE

Many workplace reports are submitted with the conclusion preceding the introduction and body sections.

In the conclusion, you summarize, interpret, and recommend. Although you have interpreted evidence at each stage of your analysis, your conclusion presents a broad interpretation and suggests a course of action, where appropriate. The summary and interpretations should lead logically to your recommendations.

Elements of a
logical conclusion

- The summary accurately reflects the body of the report.
- The overall interpretation is consistent with the findings in the summary.
- The recommendations are consistent with the purpose of the report, the evidence presented, and the interpretations given.

NOTE

Don't introduce any new facts, ideas, or statistics in the conclusion.

As you read the following conclusion, evaluate how effectively it provides a clear and consistent perspective on the whole document.

CONCLUSION
Summary and Overall Interpretation of Findings

Review of major
findings

Electromagnetic fields exist wherever electricity flows; the stronger the current, the higher the EMF intensity. While no "safe" EMF level has been identified, long-term exposure to intensities greater than 2.5 milligauss is considered dangerous. Although home appliances can generate high EMFs during use, power lines can generate constant EMFs, typically at 2 to 3 milligauss in buildings within 150 feet. Our elementary school is less than 100 feet from a high-voltage power line corridor.

Notable epidemiologic studies implicate EMFs in increased rates of medical disorders such as cancer, miscarriage, stress, depression, and learning disabilities—all directly related to intensity and duration of exposure. Laboratory studies show that EMFs cause the kinds of cellular, metabolic, and behavioral changes that could produce these disorders.

An overall
judgment about
what the findings
mean

Though still controversial and inconclusive, most of the various findings are strikingly similar and they underscore the need for more research and for risk avoidance, especially as far as children are concerned. Especially striking are the 1997 and 2004 discoveries of a direct biological link between EMF exposure and DNA damage.

Concerned citizens nationwide are beginning to prevail over resistant school and health officials and utility companies in reducing EMF risk to schoolchildren. And even though our local power company is taking reasonable risk-avoidance steps, our community can do more to learn about the issues and diminish potential risk.

Recommendations

The National Institute of Environmental Health Sciences cautions that the health evidence against EMF exposure "is insufficient to warrant aggressive regulatory concern" (NIEHS 2). In light of this "official" position, any type of government regulation anytime soon seems highly unlikely. Also, considering the limitations of what we know, drastic and enormously expensive actions (such as burying all the town's power lines or increasing the height of utility towers) seem inadvisable. In fact, these might turn out to be the wrong actions.

Despite this climate of uncertainty, however, our community still can take some immediate and inexpensive steps to address possible EMF risk:

A feasible and realistic course of action

- A version of this report should be distributed to all Bocaville residents.
- Our school board should hire a qualified contractor to take milligauss readings throughout the elementary school, to determine the extent of the problem, and to suggest reasonable corrective measures.
- Our Town Council should meet with County Electric Company representatives to explore options and costs for rerouting or burying the segment of the power lines near the school.
- A town meeting should then be held to answer citizens' questions and to solicit opinions.
- A committee (consisting of at least one physician, one engineer, and other experts) should be appointed to review emerging research as it relates to our school and town.

A closing call to action

As we await conclusive answers, we need to learn all we can about the EMF issue, and to do all we can to diminish this potentially significant health issue.

NOTE *The Works Cited section for the preceding report appears on pages 700 and 701. Simoneau uses MLA documentation style.*

Supplements

Submit your completed report with these supporting documents, in this order:

Front matter precedes the report

- title page
- letter of transmittal
- table of contents
- list of tables and figures
- abstract
- report text (introduction, body, conclusion)

End matter follows the report

- glossary (as needed)
- appendices (as needed)
- Works Cited page (or alphabetical or numbered list of references)

For discussion and examples of the above items, see Chapter 25.

A SITUATION REQUIRING AN ANALYTICAL REPORT

The report in Figure 24.4, patterned after the model outline, combines a feasibility analysis with a comparative analysis.

The Situation

Richard Larkin, author of the following report, has a work-study job fifteen hours weekly in his school's placement office. His boss, John Fitton (placement director), likes to keep up with trends in various fields. Larkin, an engineering major, has developed an interest in technical marketing and sales. Needing a report topic for his writing course, Larkin offers to analyze the feasibility of a technical marketing and sales career, both for his own decision making and for technical and science graduates in general. Fitton accepts Larkin's offer, looking forward to having the final report in his reference file for use by students choosing careers. Larkin wants his report to be useful in three ways: (1) to satisfy a course requirement, (2) to help him in choosing his own career, and (3) to help other students with their career choices.

With his topic approved, Larkin begins gathering his primary data, using interviews, letters of inquiry, telephone inquiries, and lecture notes. He supplements these primary sources with articles in recent publications. He will document his findings in APA (author-date) style.

As a guide for designing his final report, Larkin completes the following audience and use profile (based on the worksheet on page 36).

Audience and Use Profile for a Formal Report

Audience Identity and Needs

My primary audience consists of John Fitton, Placement Director, and the students who will refer to my report. The secondary audience is my writing instructor.

Because he is familiar with the marketing field, Fitton will need very little background to understand my report. Many student readers, however, will have questions like these:

- What, exactly, is technical marketing and sales?
- What are the requirements for this career?
- What are the pros and cons of this career?
- Could this be the right career for me?
- How do I enter the field?

Attitude and Personality

Readers likely to be affected by this document are students making career choices. I expect readers' attitudes will vary:

- Some readers should have a good deal of interest, especially those seeking a people-oriented career.
- Others might be only casually interested as they investigate a range of possible careers.
- Some readers might be skeptical about something written by a fellow student instead of by some expert. To connect with all these people, I need to persuade them that my conclusions are based on reliable information and careful reasoning.

Expectations about the Document

All readers expect me to spell things out, but to be concise. Visuals will help compress and emphasize material throughout.

Essential information will include an expanded definition of technical marketing and sales, the skills and attitudes needed for success, the career's advantages and drawbacks, and a description of various paths for entering the career.

This report combines feasibility and comparative analysis, so I'll want to structure the report to reveal a clear line of reasoning; in the feasibility section, reasons for and reasons against; in the comparison section, a block structure and a table that compares the four entry paths point by point. The report will close with recommendations based solidly on my conclusions.

For various readers who might not wish to read the entire report, I will include an informative abstract. ❏

NOTE   *This report's front matter (title page, information abstract, and so on) and end matter are shown and discussed in Chapter 25 and on page 713.*

☑ CHECKLIST for Usability of Analytical Reports

For evaluating your research methods and reasoning, refer also to the checklist on page 194. (Numbers in parentheses refer to the first page of discussion.)

Content

- ☐ Does the report grow from a clear statement of purpose? (609)
- ☐ Is the report's length adequate and appropriate for the subject? (609)
- ☐ Are all limitations of the analysis clearly acknowledged? (614)
- ☐ Are visuals used whenever possible to aid communication? (612)
- ☐ Are all data accurate? (610)
- ☐ Are all data unbiased? (612)
- ☐ Are all data complete? (610)
- ☐ Are all data fully interpreted? (612)
- ☐ Is the documentation adequate, correct, and consistent? (685)
- ☐ Are the conclusions logically derived from accurate interpretation? (614)
- ☐ Do the recommendations constitute an appropriate response to the question or problem? (614)

Arrangement

- ☐ Is there a distinct introduction, body, and conclusion? (616)
- ☐ Are headings appropriate and adequate? (356)
- ☐ Are there enough transitions between related ideas? (Appendix C)
- ☐ Is the report accompanied by all needed front matter? (628)
- ☐ Is the report accompanied by all needed end matter? (628)

Style and Page Design

- ☐ Is the level of technicality appropriate for the stated audience? (30)
- ☐ Is the writing style throughout clear, concise, and fluent? (244)
- ☐ Is the language convincing and precise? (264)
- ☐ Is the report written in grammatical English? (Appendix C)
- ☐ Is the page design inviting and accessible? (340)

Feasibility Analysis of a Career in Technical Marketing

INTRODUCTION

Career opportunities have narrowed for many of today's science and engineering graduates. Except for biomedical, environmental, and computer hardware engineering, employment in all engineering fields will grow at rates ranging from average to near static through 2012. Some specialities, such as mining and petroleum engineering actually will lose jobs (*The 2002–12 job outlook*, 2004, p.14). To complicate this picture, the "offshoring" of high-tech jobs to low-wage countries has meant that electrical engineers, in the words of one expert, "face some of the highest unemployment rates of any job sector in the U.S." (Lok, 2004, p. 74).

Given such bleak employment prospects, recent graduates might consider alternative careers in which they could apply their technical training. One especially attractive field that combines science and engineering expertise with "people" skills is that of *sales engineer,* a specially trained professional who markets and sells highly technical products and services.

In the growing field of technical marketing, sales engineers are in high demand (Nelson, 2001), with steady increases predicted through 2012 by the U.S. Department of Labor (Average annual job openings, 2004). Sales engineers are especially essential at this time, as customers increasingly "demand a greater role in designing their own products and a positive experience in using them" ("The answer to outsourcing," 2004, p. 136).

What specific type of work do technical marketers perform? *The Occupational Outlook Handbook* offers this job description:

> They usually sell products whose installation and optimal use requires a great deal of technical expertise and support. . . . Additionally, they provide information on their firm's products, help prospective and current buyers with technical problems, recommend improved materials and machinery . . . , design plans of proposed machinery layouts, estimate cost savings, and suggest training schedules for employees. (Manufacturers' and wholesale sales representatives, 2000)

(For a more detailed job description, refer to "The Technical Marketing Process," on page 2.)

Undergraduates interested in technical marketing need answers to basic questions:

- *Is this the right career for me?*
- *If so, how do I enter the field?*

To help answer these questions, this report analyzes information from professionals as well as from the literature.

After defining *technical marketing,* the following analysis examines employment outlook, required skills and personal qualities, career benefits and drawbacks, and various entry options.

FIGURE 24.4 An Analytical Report

COLLECTED DATA

Key Factors in a Technical Marketing Career

Anyone considering technical marketing needs to assess whether this career fits his or her interests, abilities, and aspirations.

THE TECHNICAL MARKETING PROCESS. Although *marketing* and *sales* are terms often used interchangeably, technical marketing involves far more than sales work. The process itself (identifying, reaching, and selling to customers) entails six key activities (Cornelius & Lewis, 1983, p. 44):

1. *Market research:* gathering information about the size and character of the target market for a product or service.
2. *Product development and management:* producing the goods to fill a market need.
3. *Cost determination and pricing:* measuring every expense in the production, distribution, advertising, and sales of the product, to determine its price.
4. *Advertising and promotion:* developing and implementing all strategies for reaching customers.
5. *Product distribution:* coordinating all elements of a technical product or service, from its conception through its final delivery to the customer.
6. *Sales and technical support:* creating and maintaining customer accounts, and servicing and upgrading products.

Fully engaged in all these activities, the technical marketing professional gains a detailed understanding of the industry, the product, and the customer's needs (Figure 1).

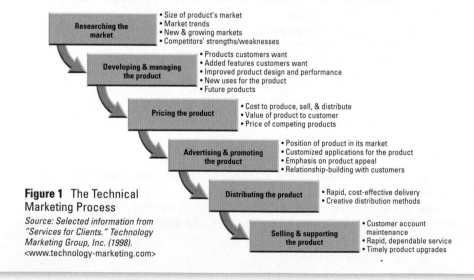

Figure 1 The Technical Marketing Process

Source: Selected information from "Services for Clients." Technology Marketing Group, Inc. (1998). <www.technology-marketing.com>

FIGURE 24.4   An Analytical Report *(continued)*

Feasibility Analysis   3

EMPLOYMENT OUTLOOK. For graduates with the right combination of technical and personal qualifications, the employment outlook for technical marketing appears excellent. While engineering jobs will increase at barely one half the average growth rate for jobs requiring a Bachelor's degree, marketing jobs will exceed the average rate (Figure 2).

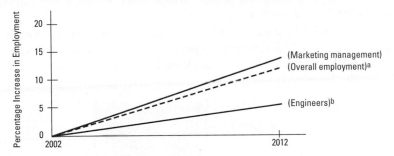

Figure 2 The Employment Outlook for Technical Marketing
[a]Jobs requiring a Bachelor's degree.
[b]Excluding outlying rates for specialties at extreme ends of the spectrum (environmental engineers: +38%; petroleum engineers: –10%).
Source: Data from U.S. Department of Labor. Bureau of Labor Statistics. Retrieved April 25, 2004, from http://www.bls.gov/EMP

Although highly competitive, these marketing positions call for the very kinds of technical, analytical, and problem-solving skills that engineers can offer—especially in an automated environment.

TECHNICAL SKILLS REQUIRED. Computer networks, interactive media, and multimedia will increasingly influence the way products are advertised and sold. Also, marketing representatives increasingly work from a "virtual" office. Using laptop computers, fax networks, and personal digital assistants, representatives in the field have real-time access to electronic catalogs of product lines, multimedia presentations, pricing for customized products, inventory data, product distribution, and customized sales contacts (Tolland, 2004).

   With their rich background in computer, technical, and problem-solving skills, engineering graduates are ideally suited for (a) working in automated environments, and (b) implementing and troubleshooting these complex and often sensitive electronic systems.

FIGURE 24.4   An Analytical Report *(continued)*

OTHER SKILLS AND QUALITIES REQUIRED. *BusinessWeek's* Peter Coy offers this distinction between routine versus non routine work:

> The jobs that will pay well in the future will be ones that are hard to reduce to a recipe. These attractive jobs—from factory floor management to sales to teaching to the professions—require flexibility, creativity, and lifelong learning. They generally also require subtle and frequent interactions with other people, often face to face. (2004, p. 50)

Technical marketing is just such a job: it involves few "cookbook-type" tasks and it requires a firm grasp of "people skills." Besides a strong technical background, success in this field calls for a generous blend of those traits summarized in Figure 3.

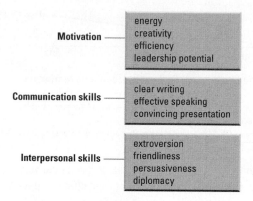

Motivation	energy creativity efficiency leadership potential
Communication skills	clear writing effective speaking convincing presentation
Interpersonal skills	extroversion friendliness persuasiveness diplomacy

Figure 3 Requirements for a Technical Marketing Career

Motivation is essential for marketing work. Professionals must be energetic and able to function with minimal supervision. Career counselor Phil Hawkins describes the ideal candidates as people who can plan and program their own tasks, who can manage their time, and who have no fear of hard work (personal interview, February 11, 2004). Leadership potential, as demonstrated by extracurricular activities, is an asset.

Motivation alone provides no guarantee of success. Marketing professionals are paid to communicate the virtues of their products or services. This career therefore requires skill in communication, both written and oral. Documents for readers outside the organization include advertising copy, product descriptions, sales proposals, sales letters, and user manuals and online help. In-house writing includes recommendation reports, feasibility studies, progress reports, memos, and email correspondence (U.S. Department of Labor, 1997, p. 8).

FIGURE 24.4   An Analytical Report *(continued)*

Skilled oral presentation is vital to any sales effort, as Phil Hawkins points out. Technical marketing professionals need to speak confidently and persuasively—to represent their products and services in the best possible light (personal interview, February 11, 2004). Sales presentations often involve public speaking at conventions, trade shows, and other similar forums.

Beyond motivation and communication skills, interpersonal skills are the ultimate requirement for success in marketing (Nelson, 2001). Consumers are more likely to buy a product or service when they like the person selling it. Marketing professionals are extroverted, friendly, and diplomatic; they can motivate people without alienating them.

ADVANTAGES OF THE CAREER. As shown in Figure 1, technical marketing offers diverse experience in every phase of a company's operation, from a product's design to its sales and service. Such broad exposure provides excellent preparation for countless upper-management positions.

In fact, sales engineers with solid experience often open their own businesses as freelance "manufacturers' agents" representing a variety of companies. These agents represent products for companies who have no marketing staff of their own. In effect their own bosses, manufacturers' agents are free to choose, from among many offers, the products they wish to represent (Tolland, 2004).

Another career benefit is the attractive salary. In addition to typically receiving a base pay plus commissions from their employer, marketing professionals are reimbursed for all business expenses. Other employee benefits often include health insurance, a pension plan, and a company car. In 2002, the median annual earnings for sales engineers ranged from $53,170 for professional and commercial equipment sales to $77,100 for computer systems design services. The highest 10 percent earned more than $108,080 annually (Bureau of Labor Statistics, 2004).

In addition, the types of interpersonal and communication skills that marketing professionals develop are highly portable. This is especially important in our current, rapidly shifting economy, in which job security is disappearing in the face of more and more temporary positions (Tolland, 2004).

DRAWBACKS OF THE CAREER. Technical marketing is by no means a career for every engineer. Sales engineer Roger Cayer cautions that personnel might spend most of their time traveling to meet potential customers. Success requires hard work over long hours, evenings, and occasional weekends. Above all, the job is stressful because of constant pressure to meet sales quotas (phone interview, February 8, 2004). Anyone considering this career should be able to work and thrive in a highly competitive environment.

The Bureau of Labor and Statistics (2004) adds that the expanding global economy means that "international travel, to secure contracts with foreign customers, is becoming more important"—typically placing more pressure on an already hectic schedule.

FIGURE 24.4 An Analytical Report *(continued)*

A Comparison of Entry Options

Engineers and other technical graduates enter technical marketing through one of four options. Some join small companies and learn their trade directly on the job. Others join companies that offer formal training programs. Some begin by getting experience in their technical specialty. Others earn a graduate degree beforehand. These options are compared below.

OPTION 1: ENTRY-LEVEL MARKETING WITH ON-THE-JOB TRAINING. Smaller manufacturers offer marketing positions in which people learn on the job. Elaine Carto, president of ABCO Electronics, believes small companies offer a unique opportunity; entry-level salespersons learn about all facets of an organization, and have a good possibility for rapid advancement (personal interview, February 10, 2004). Career counselor Phil Hawkins says, "It's all a matter of whether you prefer to be a big fish in a small pond or a small fish in a big pond" (personal interview, February 11, 2004).

Entry-level marketing offers immediate income and a chance for early promotion. A disadvantage, however, might be the loss of any technical edge one might have acquired in college.

OPTION 2: A MARKETING AND SALES TRAINING PROGRAM. Formal training programs offer the most popular entry into sales and marketing. Large to mid-size companies typically offer two formats: (a) a product-specific program, focused on a particular product or product line, or (b) a rotational program, in which trainees learn about an array of products and work in the various positions outlined in Figure 2. Programs last from weeks to months.

Former trainees Roger Cayer, of Allied Products, and Bill Collins, of Intrex, speak of the diversity and satisfaction such programs offer: specifically, solid preparation in all phases of marketing, diverse interaction with company personnel, and broad knowledge of various product lines (phone interviews, February 8, 2004).

Like direct entry, this option offers the advantage of immediate income and early promotion. With no chance to practice in their technical specialty, however, trainees might eventually find their technical expertise compromised.

OPTION 3: PRIOR EXPERIENCE IN ONE'S TECHNICAL SPECIALTY. Instead of directly entering marketing, some candidates first gain experience in their specialty. This option combines direct exposure to the workplace with the chance to sharpen technical skills in practical applications. In addition, some companies, such as Roger Cayer's, will offer marketing and sales positions to outstanding staff engineers, as a step toward upper management (phone interview, February 8, 2004).

FIGURE 24.4   An Analytical Report *(continued)*

Although this option delays a candidate's entry into technical marketing, industry experts consider direct workplace and technical experience key assets for career growth in any field. Also, work experience becomes an asset for applicants to top MBA programs (Shelley, 1997, pp. 30–31).

OPTION 4: GRADUATE PROGRAM. Instead of direct entry, some people choose to pursue an MS degree in their specialty or an MBA. According to engineering professor Mary McClane, MS degrees are usually unnecessary for technical marketing unless the particular products are highly complex (personal interview, April 2, 2004).

In general, jobseekers with an MBA have a distinct competitive advantage. More significantly, new MBAs with a technical bachelor's degree and one to two years of experience command salaries from 10 to 30 percent higher than MBAs who lack work experience and a technical bachelor's degree. In fact, no more than 3 percent of job candidates offer a "techno-MBA" specialty, making this unique group highly desirable to employers (Shelley, 1997, p. 30).

A motivated student might combine graduate degrees. Dora Anson, president of Susimo Cosmic Systems, sees the MS/MBA combination as ideal preparation for technical marketing (2004).

One disadvantage of a full-time graduate program is lost salary, compounded by school expenses. These costs must be weighed against the prospect of promotion and monetary rewards later in one's career.

AN OVERALL COMPARISON BY RELATIVE ADVANTAGE. Table 1 compares the four entry options on the basis of three criteria: immediate income, rate of advancement, and long-term potential.

Table 1 Relative Advantages Among Four Technical-Marketing Entry Options

Option	Relative Advantages		
	Early, immediate income	Greatest advancement in marketing	Long-term potential
Entry level, no experience	yes	yes	no
Training program	yes	yes	no
Practical experience	yes	no	yes
Graduate program	no	no	yes

FIGURE 24.4   An Analytical Report *(continued)*

CONCLUSION

Summary of Findings

Technical marketing and sales involves identifying, reaching, and selling the customer a product or service. Besides a solid technical background, the field requires motivation, communication skills, and interpersonal skills. This career offers job diversity and excellent income potential, balanced against hard work and relentless pressure to perform.

College graduates interested in this field confront four entry options: (1) direct entry with on-the-job training, (2) a formal training program, (3) prior experience in a technical specialty, and (4) graduate programs. Each option has benefits and drawbacks based on immediacy of income, rate of advancement, and long-term potential.

Interpretation of Findings

For graduates with a strong technical background and the right skills and motivation, technical marketing offers attractive career prospects. Anyone contemplating this field, however, needs to be able to enjoy customer contact and thrive in a highly competitive environment.

Those who decide that technical marketing is for them can choose various entry options:

- For hands-on experience, direct entry is the logical option.
- For sophisticated sales training, a formal program with a large company is best.
- For sharpening technical skills, prior work in one's specialty is invaluable.
- If immediate income is not vital, graduate school is an attractive option.

Recommendations

If your interests and abilities match the requirements, consider these suggestions:

1. To get a firsthand view, seek the advice and opinions of people in the field. You might begin by contacting professional organizations such as the Manufacturers' Agents National Association at www.manaonline.org or the Manufacturers' Representatives Educational Research Foundation at www.mrerf.org.
2. Before settling on an entry option, consider all its advantages and disadvantages and decide whether this option best coincides with your career goals. (Of course, you can always combine options during your professional life.)
3. When making any career decision, consider career counselor Phil Hawkins' advice: "Listen to your brain and your heart" (personal interview, February 11, 2004). Choose an option or options that offer not only professional advancement but also personal satisfaction.

REFERENCES

[The complete list of references is shown and discussed in Appendix A, page 713.]

FIGURE 24.4   An Analytical Report *(continued)*

24.3

Consider the cultural contexts of your analysis at <www.ablongman. com/lannonweb>

GUIDELINES for Reasoning through an Analytical Problem

Audiences approach an analytical report with this basic question:

| *Is this analysis based on sound reasoning?*

Whether your report documents a causal, comparative, or feasibility analysis (or some combination) you need to trace your line of reasoning so that readers can follow it clearly.

As you prepare your report, refer to the usability checklist on page 000 and observe the following guidelines:

For Causal Analysis

1. *Be sure the cause fits the effect.* Keep in mind that faulty causal reasoning is extremely common, especially when we ignore other possible causes or we confuse mere coincidence with causation.

2. *Make the links between effect and cause clear.* Identify the immediate cause (the one most closely related to the effect) as well as the distant cause(s) (the ones that precede the immediate cause). For example, the immediate cause of a particular airplane crash might be a fuel-tank explosion, caused by a short circuit in frayed wiring, caused by faulty design or poor quality control by the manufacturer. Discussing only the immediate cause often just scratches the surface of the problem.

3. *Clearly distinguish between possible, probable, and definite causes.* Unless the cause is obvious, limit your assertions by using *perhaps, probably, maybe, most likely, could, seems to, appears to,* or similar qualifiers that prevent you from making an insupportable claim.

For Comparative Analysis

1. *Rest the comparison on clear and definite criteria: costs, uses, benefits/drawbacks, appearance, results.* In evaluating the merits of competing items, identify your specific criteria and rank them in order of importance.

2. *Give each item balanced treatment.* Discuss points for each item in identical order.

3. *Support and clarify the comparison or contrast through credible examples.* Use research, if necessary, for examples that readers can visualize.

4. *Follow either a block pattern or a point-by-point pattern.* In the block pattern, first one item is discussed fully, then the next. Choose a block pattern when the overall picture is more important than the individual points.

   In the point-by-point pattern, one point about both items is discussed, then the next point, and so on. Choose a point-by-point pattern when specific points might be hard to remember unless placed side by side.

Block pattern	Point-by-point pattern
Item A	first point of A/first point of B, etc.
first point	
second point, etc.	
Item B	second point of A/second point of B, etc.
first point	
second point, etc.	

5. *Order your points for greatest emphasis.* Try ordering your points from least to most important or dramatic or useful or reasonable. Placing the most striking point last emphasizes it best.

6. *In an evaluative comparison ("X is better than Y"), offer your final judgment.* Base your judgment squarely on the criteria presented.

For Feasibility Analysis

1. *Consider the strength of supporting reasons.* Choose the best reasons for supporting the action or decision being considered—based on solid evidence.

2. *Consider the strength of opposing reasons.* Remember that people usually see only what they want to see. Avoid the temptation to overlook or downplay opposing reasons, especially for an action or decision that you have been promoting. Consider alternate points of view and examine all the evidence.

3. *Recommend a realistic course of action.* After weighing all the pros and cons, make your recommendation—but be prepared to backtrack if you discover that what seemed like the right course of action turns out to be the wrong one.

24.4

For more on usability testing visit <www.ablongman.com/lannonweb>

EXERCISE

 For more exercises, visit <www.ablongman.com/lannon>

Prepare an analytical report, using these guidelines:

a. Choose a subject for analysis from the list at the end of this exercise, from your major, or from a subject of interest.

b. Identify the problem or question so that you will know exactly what you are looking for.

c. Restate the main question as a declarative sentence in your statement of purpose.

d. Identify an audience—other than your instructor—who will use your information for a specific purpose.

e. Hold a private brainstorming session to generate major topics and subtopics.

f. Use the topics to make an outline based on the model outline in this chapter. Divide as far as necessary to identify all points of discussion.

g. Make a tentative list of all sources (primary and secondary) that you will investigate. Verify that adequate sources are available.

h. In a proposal memo to your instructor, describe the problem or question and your plan for analysis. Attach a tentative bibliography.

i. Use your working outline as a guide to research and observation. Evaluate sources and evidence, and interpret all evidence fully. Modify your outline as needed.

j. Submit a progress report to your instructor describing work completed, problems encountered, and work remaining.

k. Compose an audience and use profile. (Use the sample on page 000 as a model, along with the profile worksheet on page 68.)

l. Write the report for your stated audience. Work from a clear statement of purpose, and be sure that your reasoning is shown clearly. Verify that your evidence, conclusions, and recommendations are consistent. Be especially careful that your recommendations observe the critical-thinking guidelines in Figure 24.3.

m. After writing your first draft, make any needed changes in the outline and revise your report according to the revision checklist. Include all necessary supplements.

n. Exchange reports with a classmate for further suggestions for revision.

o. Prepare an oral report of your findings for the class as a whole.

Base your analysis on a question similar to these:

- What has and hasn't been done to protect a nearby port, nuclear plant, or chemical plant from sabotage or attack?
- How adequate is your area's evacuation plan?
- How can local hospitals reduce medical errors?
- Who are the biggest polluters in your area, and what can be done?
- How should your state deal with discarded computers and other ewaste?
- What can local schools do to stem the obesity epidemic?
- Is mass smallpox vaccination a good idea?
- How safe is our food supply, and what can be done to protect it?
- How should we deal with the Mad-Cow threat?
- Which gender is the more competitive, and what could this difference mean in the workplace?
- How will digital convergence change our work and our world?
- Which diets should be avoided, and why?
- How can future skyscrapers be made safer?

- Can wind power be profitable?
- Are efforts to deflect incoming asteroids worth the cost?
- Stem cell research: What progress has been made so far? Prospects?
- Can nuclear power plants be dismantled safely?
- What are the unintended consequences of banning DDT, or some other major policy decision?
- What are pros and cons of distance learning?
- Are treatments such as homeopathy or acupuncture feasible alternatives to traditional medicine?
- Are irradiated or genetically modified foods safe?
- What are the pros and cons of legalizing gambling in your state?
- Which should you buy: condominium or house?
- Should you move to a different part of the country or the world?
- How have budget cuts affected your public schools?
- Is police protection adequate in your area?
- How should people prepare for long-term job prospects in your field?
- In which fields are women paid less than men? Why?
- What are pros and cons of home birth (vs. hospital delivery).
- How can tourism be promoted in your area?
- Should you work before graduate school?

COLLABORATIVE PROJECTS

1. Divide into small groups. Choose a topic for group analysis—preferably, a campus issue—and brainstorm. Draw up a working outline that could be used as an analytical report on this subject.

2. Prepare a questionnaire based on your work above, and administer it to members of your campus community. Report the findings of your questionnaire and your conclusions and recommendations. (Review pages 154–61, on questionnaires and surveys.)

25

Front Matter and End Matter in Long Documents

A long document must be easily accessible and must accommodate users with different interests. Preceding the report is *front matter:* The cover, title page, letter of transmittal, table of contents, and abstract give summary information about the content of the document. Following the report (as needed) is *end matter:* The glossary, appendices, and list of works cited can either provide supporting data or help users understand technical sections. Users can refer to any of these supplements or skip them altogether, according to their needs.

COVER

Use a sturdy, plain cover with page fasteners. With the cover on, the open pages should lie flat. Use covers only for long documents.

Center the report title and your name four to five inches below the top of your page. (Many workplace reports include the company name and logo instead of the report author's name).

> THE FEASIBILITY OF A TECHNICAL MARKETING CAREER:
> AN ANALYSIS
> by
> Richard B. Larkin, Jr.

TITLE PAGE

The title page lists the report title, author's name, name of person(s) or organization to whom the report is addressed, and date of submission.

The title announces the report's purpose and subject by using descriptive words such as "analysis," "instructions," "proposal," "feasibility," "description," "progress." Do not number your title page but count it as page i of the front matter. Center the title and all other lines (Figure 25.1).

LETTER OF TRANSMITTAL

Include a letter of transmittal with any formal report or proposal addressed to a specific reader. As a gesture of courtesy, your letter might

What to include in a letter of transmittal

- acknowledge those who helped with the report
- thank the recipient for any special assistance
- refer to sections of special interest
- discuss the limitations of your study, or any problems gathering data
- discuss possible follow-up investigations
- offer personal (or off-the-record) observations
- suggest some special uses for the information
- urge the recipient to immediate action

The transmittal letter is tailored to a particular audience, as is Richard Larkin's in Figure 25.2. If a report is being sent to a number of people who are variously qualified and bear various relationships to the writer, individual letters of transmittal may vary.

Feasibility Analysis
of a Career
in Technical Marketing

for

Professor J. M. Lannon

Technical Writing Instructor

University of Massachusetts

North Dartmouth, Massachusetts

by

Richard B. Larkin, Jr.

English 266 Student

May 1, 20XX

FIGURE 25.1   Title Page for a Formal Report

165 Hammond Way
Hyannis, MA 02457
April 29, 20XX

John Fitton
Placement Director
University of Massachusetts
North Dartmouth, MA 02747

Dear Mr. Fitton:

Here is my report, Feasibility Analysis of a Career in Technical Marketing. In preparing
this report, I've learned a great deal about the requirements and modes of access to this
career, and I believe my information will help other students as well. Thank you for your
guidance and encouragement throughout this process.

Although committed to their specialties, some technical and science graduates seem
interested in careers in which they can apply their technical knowledge to customer and
business problems. Technical marketing may be an attractive choice of career for those
who know their field, who can relate to different personalities, and who are good
communicators.

Technical marketing is competitive and demanding, but highly rewarding. In fact, it is an
excellent route to upper-management and executive positions. Specifically, marketing
work enables one to develop a sound technical knowledge of a company's products, to
understand how these products fit into the marketplace, and to perfect sales techniques
and interpersonal skills. This is precisely the kind of background that paves the way to
top-level jobs.

I've enjoyed my work on this project, and would be happy to answer any questions.
Please phone me at 690-555-1122 anytime.

Sincerely,

Richard B. Larkin, Jr.

FIGURE 25.2   Letter of Transmittal for a Formal Report

How to prepare
the transmittal
letter

Begin your letter by referring to the user's original request, and introduce your report by name. Briefly review the reasons for your report or include a short abstract. Maintain a confident and positive tone. Indicate pride and satisfaction in your work. Avoid implied apologies, such as "I hope this report meets your expectations."

In the body section, include items from the above-mentioned list of possibilities (acknowledgments, special problems). Although your abstract or executive summary will summarize major findings, conclusions, and recommendations, your letter gives a brief and personal overview of the *entire project*.

End on a positive theme: "I believe that the data in this report are accurate, that they have been analyzed rigorously and impartially, and that the recommendations are sound." Express your willingness to answer questions or discuss findings. Show how the reader can get in touch quickly with the writer.

NOTE

For college reports, the letter of transmittal usually follows the title page and is bound as part of the report. For workplace reports, the letter usually is not bound in the report but is presented separately.

TABLE OF CONTENTS

Help readers find the information they're looking for by providing a table of contents. In designing your table of contents, follow these guidelines:

How to prepare a
table of contents

- List front matter (transmittal letter, abstract), numbering the pages with lowercase roman numerals. (The title page, though not listed, is counted page i.) Number glossary, appendix, and endnote pages with arabic numerals, continuing the page sequence of your report proper, in which page 1 is the first page of the report text.
- Include no headings in the table of contents not listed as headings or subheadings in the report; the report may, however, contain subheadings not listed in the table of contents.
- Phrase headings in the table of contents exactly as in the report.
- List headings at various levels in varying type styles and indention.
- Use *leader lines* (........) to connect headings to page numbers. Align rows of leader lines vertically, each above the other.

Figure 25.3 shows the table of contents for Richard Larkin's feasibility analysis.

With some word-processing programs you can generate a table of contents automatically provided that you have assigned styles or codes to all of the headings in your report. If your word-processing program does not have this feature, compose the table of contents by assigning page numbers to headings from your outline. Keep in mind, however, that not all levels of outline headings appear in your table of contents or your report. Excessive headings can fragment the discussion.

iii

CONTENTS

FIGURE 25.3 **Table of Contents for a Formal Report**

LIST OF TABLES AND FIGURES

Following the table of contents is a list of tables and figures, if needed. When a report has four or more visuals, place this table on a separate page. List the figures first, then the tables. Figure 25.4 shows the list of tables and figures in Larkin's report.

TABLES AND FIGURES

FIGURE 25.4   A List of Tables and Figures for a Formal Report

ABSTRACT OR EXECUTIVE SUMMARY

Reports are often read by many people: researchers, managers, executives, customers. For readers who are interested only in the big picture, the entire report may not be relevant. Many readers who don't have the time or willingness to read your entire report will consider the informative abstract the most useful part of the material you present. In addition to the Chapter 11 guidelines for summarizing information, follow these suggestions for preparing your abstract[1]:

How to prepare an informative abstract

- Make sure your abstract stands alone in terms of meaning.
- Write for a general audience. Readers of the abstract are likely to vary in expertise, perhaps more than those who read the report itself; therefore, adjust the vocabulary to suit the intended reader. When you send report copies to readers with varying levels of expertise, write a different summary for each type of reader.
- Add no new information. Simply present the report's highlights.
- Present your information in the following sequence:
  a. Identify the issue or need that led to the report.
  b. Offer the key facts, statistics, and findings—the material your reader *must* know.
  c. Include a condensed conclusion and recommendations, if any.

---

[1] My thanks to Professor Edith K. Weinstein for these suggestions.

The informative abstract in Figure 25.5 accompanies Richard Larkin's report.

NOTE   *This item can be called many different names including summary, abstract, informative abstract, executive summary, executive abstract, or report synopsis.*

ABSTRACT

The feasibility of technical marketing as a career is based on a college graduate's interests, abilities, and expectations, as well as on possible entry options.

Technical marketing is a feasible career for anyone who is motivated, who can communicate well, and who knows how to get along. Although this career offers job diversity and potential for excellent income, it entails almost constant travel, competition, and stress.

College graduates enter technical marketing through one of four options: entry-level positions that offer hands-on experience, formal training programs in large companies, prior experience in one's specialty, or graduate programs. The relative advantages and disadvantages of each option can be measured in resulting immediacy of income, rapidity of advancement, and long-term potential.

Anyone considering a technical marketing career should follow these recommendations:

• Speak with people who work in the field.
• Weigh the implications of each entry option carefully.
• Consider combining two or more options.
• Choose options for personal as well as professional benefits.

FIGURE 25.5   Informative Abstract

GLOSSARY

A glossary alphabetically lists specialized terms and their definitions. A glossary makes key definitions available to laypersons without interrupting technical readers. Use a glossary if your report contains more than five or six technical terms that may not be understood by all audience members. If fewer than five terms need defining, place them in the report introduction as working definitions, or use footnote definitions. If you use a separate glossary, announce its location: "(See the glossary at the end of this report)."

Follow these suggestions for preparing a glossary:

How to prepare a glossary

• Define all terms unfamiliar to an intelligent layperson. When in doubt, overdefining is safer than underdefining.

- Define all terms that have a special meaning in your report ("In this report, a small business is defined as . . .").
- Define all terms by giving their class and distinguishing features, unless some terms need expanded definitions.
- List all terms in alphabetical order. Highlight each term and use a colon to separate it from its definition.
- On first use, place an asterisk in the text by each item defined in the glossary.
- List your glossary and its first page number in the table of contents.

Figure 25.6 shows part of a glossary for a comparative analysis of two natural childbirth techniques, written by a nurse practitioner for expectant mothers and student nurses.

GLOSSARY

Analgesic: a medication given to relieve pain during the first stage of labor.

Cervix: the neck-shaped anatomical structure that forms the mouth of the uterus.

Dilation: cervical expansion occurring during the first stage of labor.

Episiotomy: an incision of the outer vaginal tissue, made by the obstetrician just before the delivery, to enlarge the vaginal opening.

First stage of labor: the stage in which the cervix dilates and the baby remains in the uterus.

Induction: the stimulating of labor by puncturing the membranes around the baby or by giving an oxytoxic drug (uterine contractant), or by doing both.

FIGURE 25.6 A Partial Glossary

APPENDICES

Add one or more appendices to your report if you have large blocks of material or other documents that are relevant but will bog readers down if placed in the middle of the report itself. (Page 600 shows an appendix to a budget proposal.) Items that belong in an appendix might include

What an appendix might include

- complex formulas
- details of an experiment
- interview questions and responses

- long quotations (one or more pages)
- maps
- material more essential to secondary readers than to primary readers
- photographs
- related correspondence (letters of inquiry, and so on)
- sample questionnaires and tabulated responses
- sample tests and tabulated results
- some visuals occupying more than one full page
- statistical or other measurements
- texts of laws and regulations

Do not stuff appendices with needless information or use them unethically for burying bad or embarrassing news that belongs in the report proper. In preparing your appendices, follow these suggestions:

How to prepare an appendix

- Include only relevant material.
- Use a separate appendix for each major item.
- Title each appendix clearly: "Appendix A: Projected Costs."
- Use appendices sparingly. Four or five appendices in a ten-page report indicates a poorly organized document.
- Limit an appendix to a few pages, unless more length is essential.
- Mention the appendix early in the introduction, and refer to it at appropriate points in the report: "(see Appendix A)."

Users should be able to understand your report without having to turn to the appendix. Distill essential facts from your appendix and place them in your report text.

IMPROPER REFERENCE	The whale population declined drastically between 1986 and 1997 (see Appendix B for details).
PROPER REFERENCE	The whale population declined by 16 percent from 1986 to 1997 (see Appendix B for statistical breakdown).

DOCUMENTATION

In the endnotes or works cited pages, list each of your outside references in alphabetical order or in the same numerical order as they are cited in the report proper. See Appendix A for discussion.

EXERCISES For more exercises, visit
<www.ablongman.com/lannon>

1. These titles are intended for investigative, research, or analytical reports. Revise each inadequate title to make it clear and accurate.

   a. The Effectiveness of the Prison Furlough Program in Our State
   b. Drug Testing on the Job
   c. The Effects of Nuclear Power Plants
   d. Woodburning Stoves
   e. Interviewing
   f. An Analysis of Vegetables (for a report assessing the physiological effects of a vegetarian diet)
   g. Wood as a Fuel Source
   h. Oral Contraceptives
   i. Lie Detectors and Employees

2. Prepare a title page, letter of transmittal (for a specific reader who can use your information in a definite way), table of contents, and informative abstract for a report you have written earlier.

3. Find a short but effective appendix in one of your textbooks, in a journal article in your field, or in a report from your workplace. In a memo to your instructor and classmates, explain how the appendix is used, how it relates to the main text, and why it is effective. Attach a copy of the appendix to your memo. Be prepared to discuss your evaluation in class.

26

Oral Presentations

We all need to present our ideas effectively in person. Oral presentations vary in style, range, complexity, and formality. They may include convention speeches, reports at national meetings, reports via teleconferencing networks, technical briefings for colleagues, and speeches to community groups. These talks may be designed to inform (to describe new government safety requirements), to persuade (to induce company officers to vote a pay raise), or to do both. The higher your status, on the job or in the community, the more you can expect to give oral presentations.

ADVANTAGES AND DRAWBACKS OF ORAL REPORTS

Advantages

Unlike written documents, oral presentations are truly interactive. In face-to-face communication, you can rely on body language, vocal tone, eye contact, and other elements of human chemistry—a likable personality can have a powerful effect on audience receptiveness. Also, oral presentations provide for give-and-take, which does not happen with traditional written documents. As you see how your audience reacts, you can adjust your presentation accordingly and answer questions immediately.

Drawbacks

In a written report you generally have plenty of time to think about what you're saying and how you're saying it, and to revise until the message is just right. For an oral report, one attempt is basically all you get, and all this pressure makes it easier to stumble. (People consistently rank fear of public speaking higher than fear of dying!) Also, an oral report is limited in the amount and complexity of information it can present. Readers of a written report can follow their own pace and direction, going back and forth, perhaps skimming some sections and studying others. In an oral presentation, you establish the pace and the information flow, thereby creating the risk of "losing" or boring the listeners.

AVOIDING PRESENTATION PITFALLS

An oral presentation is only the tip of a pyramid built from many earlier labors. But such presentations often serve as the concrete measure of your overall job performance. In short, your audience's only basis for judgment may be the brief moments during which you stand before them.

The podium or lectern can be a lonely and intimidating place. In the words of two experts, "most persons in most presentational settings do not perform well" (Goodall and Waagen 14–15). Despite the fact that they can help make or break a person's career, oral presentations often turn out to be boring, confusing, unconvincing, or too long. Many are delivered ineptly, with the presenter losing her or his place, fumbling through notes, apologizing for forgetting something, or generally seeming disorganized and unprofessional. Table 26.1 lists some of the things that go wrong.

	Speaker · · ·	Visuals * * *	Setting ▪▪▪
TABLE 26.1 **Common** **Pitfalls in Oral** **Presentations**	• makes no eye contact • seems like a robot • hides behind the lectern • speaks too softly/loudly • sways, fidgets, paces • rambles or loses her/his place • never gets to the point • fumbles with notes or visuals • has too much material	* are nonexistent * are hard to see * are hard to interpret * are out of sequence * are shown too rapidly * are shown too slowly * have typos/errors * are word-filled	▪ is too noisy ▪ is too hot or cold ▪ is too large or small ▪ is too bright for visuals ▪ is too dark for notes ▪ has equipment missing ▪ has broken equipment

Given such difficulties, how can any presenter display skill and confidence? By proceeding systematically through careful analysis, planning, and preparation.

PLANNING YOUR PRESENTATION

A successful presentation involves more than just getting up and talking. We have all sat through enough lectures and presentations during our student careers to know how to separate the excellent from the awful. The successful presenter knows how to forge a relationship with the listeners, how to establish rapport and persuade listeners their time has been well spent.

Analyze Your Listeners

Assess your listeners' needs, knowledge, concerns, level of involvement, and possible objections (Goodall and Waagen 16).

Many audiences include people with varied technical backgrounds. Unless you have a good idea of each person's background, speak to a general, heterogeneous audience, as in a classroom of mixed majors.

Work from an Explicit Purpose Statement

Formulate, on paper, a statement of purpose in two or three sentences. Why, exactly, are you speaking on this subject? Who are your listeners? What do you want the listeners to think, know, or do? (A solid purpose statement can also serve as the introduction to your presentation.)

Assume, for example, that you represent an environmental engineering firm that has completed a study of groundwater quality in your area. The organization that sponsored your study has asked you to give an oral version of your written report, titled "Pollution Threats to Local Groundwater," at a town meeting.

After careful thought, you settle on this purpose statement:

Purpose
statement

> *Purpose:*   By informing Cape Cod residents about the dangers to the Cape's freshwater supply posed by rapid population growth, this report is intended to increase local interest in the problem.

Now you are prepared to focus on the listeners and the speaking situation.

QUESTIONS FOR ANALYZING YOUR LISTENERS AND PURPOSE

- *Who are my listeners (strangers, peers, superiors, clients)?*
- *What is their attitude toward me or the topic (hostile, indifferent, needy, friendly)?*
- *Why are they here (they want to be here, they are forced to be here, they are curious)?*
- *What kind of presentation do they expect (brief, informal; long, detailed; lecture)?*
- *What do these listeners already know (nothing, a little, a lot)?*

- *What do they need or want to know (overview, bottom line, nitty gritty)?*
- *How large is their stake in this topic (about layoffs, new policies, pay raises)?*
- *Do I want to motivate, mollify, inform, instruct, or warn my listeners?*
- *What are their biggest concerns or objections about this topic?*
- *What do I want them to think, know, or do?*

Analyze Your Speaking Situation

The more you can discover about the circumstances, the setting, and the constraints for your presentation, the more deliberately you will be able to prepare.

Later parts of this chapter explain how to incorporate your answers to the questions in your preparation.

QUESTIONS FOR ANALYZING YOUR SPEAKING SITUATION

- *How much time will I have to speak?*
- *Will other people be speaking before or after me?*
- *How formal or informal is the setting?*
- *How large is the audience?*

- *How large is the room?*
- *How bright and adjustable is the lighting?*
- *What equipment is available?*
- *How much time do I have to prepare?*

Select an Appropriate Delivery Method

Your presentation's effectiveness will depend largely on *how* it connects with listeners. Different types of delivery create different connections.

THE MEMORIZED DELIVERY.   A memorized delivery seldom connects with listeners because the speaker is too busy reciting the lines and trying to remember everything. This type of delivery takes a long time to prepare, offers no chance for revision during the presentation, and spells disaster if you lose your place. Avoid this type of delivery in most workplace settings.

THE IMPROMPTU DELIVERY. An impromptu (off-the-cuff) delivery can be a natural way of connecting with listeners—but only when you really know your material, feel comfortable with your audience, and are in an informal speaking situation (group brainstorming, or a response to a question: "Tell us about your team's progress on the automation project"). Avoid impromptu deliveries for complex information—no matter how well you know the material.

Too many things can go wrong in an unplanned, spontaneous presentation: You might say something offensive or irrelevant; you might seem disorganized; you might not make sense. If you anticipate being called on, never assume "It's all in my head." Get your plan down on paper. If you have little warning, at least jot a few notes about what you want to say.

THE SCRIPTED DELIVERY. For a complex technical presentation, a conference paper, or a formal speech, you may want to read your material verbatim from a prepared script. Scripted presentations work well if you have many details to present, are talkative, or have a strict time limit (e.g., at a conference), or if this audience makes you nervous. Consider a scripted delivery when you want the content, organization, and style of your presentation to be as near perfect as possible.

Although a scripted delivery helps you control your material, it offers little chance for audience interaction and it can be boring.

If you *do* plan to read aloud, allow ample preparation time. Leave plenty of white space between lines and paragraphs. Rehearse until you are able to glance up from the script periodically without losing your place. Plan on roughly two minutes per double-spaced page.

26.1

For more on the role of Microsoft *PowerPoint* in presentations visit <www.ablongman.com/lannonweb>

THE EXTEMPORANEOUS DELIVERY. An extemporaneous delivery is carefully planned, practiced, and based on notes that keep you on track. In this natural way of addressing an audience, you glance at your material and speak in a conversational style. Extemporaneous delivery is based on key ideas in sentence or topic outline form, often projected as overhead transparencies or as slides generated from presentation software such as Microsoft *PowerPoint*™.

The dangers in extemporaneous delivery are that you might lose track of your material, forget something important, say something unclearly, or exceed your time limit. Careful preparation is the key.

Table 26.2 summarizes the various uses and drawbacks of the most common types of delivery. These need not be fixed, exclusive categories. In many instances, some combination of methods can be effective. For example, in an orientation for new employees, you might prefer the flexibility of an extemporaneous format but also read a brief passage aloud from time to time (e.g., excerpts from the company's formal code of ethics).

TABLE 26.2   A Comparison of Oral Presentation Methods

Delivery Method	* Main Uses*	• Main Drawbacks •
IMPROMPTU (inventing as you speak)	* in-house meetings * small, intimate groups * simple topics	• offers no chance to prepare • speaker might ramble • speaker might lose track
SCRIPTED (reading verbatim from a written work)	* formal speeches * large, unfamiliar groups * highly complex topics * strict time limit * cross-cultural audiences * highly nervous speaker	• takes a long time to prepare • speaker can't move around • limits human contact • can appear stiff and unnatural • might bore listeners • makes working with visuals difficult
EXTEMPORANEOUS (speaking from an outline of key points)	* face-to-face presentations * medium-sized, familiar groups * moderately complex topics * somewhat flexible time limit * visually based presentations	• speaker might lose track • speaker might leave something out • speaker might get tongue-tied • speaker might exceed time limit • speaker might fumble with notes, visuals, or equipment

PREPARING YOUR PRESENTATION

To stay in control and build confidence, plan the presentation systematically. We will assume here that your presentation is extemporaneous.

Research Your Topic

Do your homework. Be prepared to support each assertion, opinion, conclusion, and recommendation with evidence and reason. Check your facts for accuracy. Your audience expects to hear a knowledgeable speaker. Don't disappoint them.

Begin gathering material well ahead of time. Use summarizing techniques from Chapter 11 to identify and organize major points.

If your preparation is simply a spoken version of a written report, you need far less preparation. Simply expand your outline for the written report into a sentence outline.

Aim for Simplicity and Conciseness

Keep your presentation short and simple. Boil the material down to a few main points. Listeners' normal attention span is about twenty minutes. After that, they begin tuning out. Time yourself in practice sessions and trim as needed. (If the material requires a lengthy presentation, plan a short break, with refreshments if possible, about halfway through.)

Anticipate Audience Questions

Consider those parts of your presentation that listeners might question or challenge. You might need to clarify or justify information that is new, controversial, disappointing, or surprising.

Outline Your Presentation

26.2

Try alternative forms of outlining by storyboard in *PowerPoint* at <www.ablongman.com/lannonweb>

Presentation outline

Review Chapter 12 for organizing and outlining strategies. Each sentence in the following presentation outline is a topic sentence for a paragraph that a well-prepared speaker can develop in detail.

Pollution Threats to Local Groundwater

ARNOLD BORTHWICK

I. Introduction to the Problem
 A. Do you know what you are drinking when you turn on the tap and fill a glass?
 B. The quality of our water is good, but not guaranteed to last forever.
 C. Cape Cod's rapid population growth poses a serious threat to our freshwater supply. (Visual #1)
 D. Measurable pollution in some town water supplies has already occurred. (Visual #2)
 E. What are the major causes and consequences of this problem and what can we do about it? (Visual #3)

II. Description of the Aquifer
 A. The groundwater is collected and held in an aquifer.
  1. This porous rock formation creates a broad, continuous arch beneath the entire Cape. (Visual #4)
  2. The lighter freshwater flows on top of the heavier saltwater.
 B. This type of natural storage facility, combined with rapid population growth, creates potential for disaster.

III. Hazards from Sewage and Landfills
 A. With increasing population, sewage and solid waste from landfill dumps increasingly invade the aquifer.
 B. The Cape's sandy soil promotes rapid seepage of wastes into the groundwater.
 C. As wastes flow naturally toward the sea, they can invade the drawing radii of town wells. (Visual #5)

IV. Hazards from Saltwater Intrusion
 A. Increased population also causes overdraw on some town wells, resulting in saltwater intrusion. (Visual #6)
 B. Salt and calcium used in snow removal add to the problem by seeping into the aquifer from surface runoff.

V. Long-Term Environmental and Economic Consequences
    A. The environmental effects of continuing pollution of our water table will
        be far-reaching. (Visual #7)
        1. Drinking water will have to be piped in more than 100 miles from
            Quabbin Reservoir.
        2. The Cape's beautiful freshwater ponds will be unfit for swimming.
        3. Aquatic and aviary marsh life will be threatened.
        4. The Cape's sensitive ecology might well be damaged beyond repair.
    B. Such environmental damage would, in turn, spell economic disaster for
        Cape Cod's major industry—tourism.

VI. Conclusion and Recommendations
    A. This problem is becoming more real than theoretical.
    B. The conclusion is obvious: If the Cape is to survive ecologically and
        financially, we must take immediate action to preserve our *only* water
        supply.
    C. These recommendations offer a starting point for action. (Visual #8)
        1. Restrict population density in all Cape towns by creating larger building
            lot requirements.
        2. Keep strict watch on proposed high-density apartment and
            condominium projects.
        3. Create a committee in each town to educate residents about water
            conservation.
        4. Prohibit salt, calcium, and other additives in sand spread on snow-
            covered roads.
        5. Explore alternatives to landfills for solid waste disposal.
    D. Given its potential effects on our quality of life, such a crucial issue
        deserves the active involvement of every Cape resident.

Before practicing the delivery, transfer your presentation outline to 3" × 5" note-cards (one side only, each card numbered and perhaps color-coded), which you can hold in one hand and shuffle as needed. Or insert the outline pages in a loose-leaf binder for easy flipping. Type or print clearly, leaving enough white space to locate material at a glance.

Plan Your Visuals

Visuals increase listeners' interest, focus, understanding, and memory. Select visuals that will clarify and enhance your talk—without making you fade into the background.

DECIDE WHERE VISUALS WILL WORK BEST. Use visuals to emphasize a point and whenever *showing* would be more effective than just *telling*.

DECIDE WHICH VISUALS WILL WORK BEST.   Will you need numerical or prose tables, graphs, charts, graphic illustrations, computer graphics? How complex should these visuals be? Should they impress or simply inform? Use the visual planning sheet in Chapter 14 to guide your decisions.

DECIDE HOW MANY VISUALS ARE APPROPRIATE.   Prefer an array of lean and simple visuals that present material in digestible amounts to one or two overstuffed visuals that people end up staring at endlessly.

CREATE A STORYBOARD.   A presentation storyboard is a double-column format in which your discussion is outlined in the left column, aligned with the specific supporting visuals in the right column (Figure 26.1).

DECIDE WHICH VISUALS ARE ACHIEVABLE.   Fit each visual to the situation. The visuals you select will depend on the room, the equipment, and the production resources available.

Pollution Threats to Local Groundwater

1.0   Introduce the Problem

    1.1   Do you know what you are drinking when you turn on the tap and fill the glass?

    1.2   The quality of our water is good, but not guaranteed to last forever.

    1.3   Cape Cod's rapid population growth poses a serious threat to our freshwater supply.   *(slide: a line graph showing twenty-year population growth)*

    1.4   Measurable pollution in some town water supplies has already occurred.   *(slide: two side-by-side tables showing twenty-year increases in nitrate and chloride concentrations in three town wells)*

    1.5   What are the major causes and consequences of this problem and what can we do about it?   *(poster: a multicolored list that previews my five subtopics)*

2.0   Describe the Aquifer

    2.1   The groundwater is collected and held in an aquifer.

        2.1.1.   This porous rock formation creates a broad, continuous arch beneath the entire Cape.   *(slide: a cutaway view of the aquifer's geology)*

FIGURE 26.1   A Partial Storyboard

How large is the room and how is it arranged? Some visuals work well in small rooms, but not large ones, and vice versa. How well can the room be darkened? Which lights can be left on? Can the lighting be adjusted selectively? What size should visuals be, to be seen clearly by the whole room? (A smaller, intimate room is usually better than a room that is too big.)

What hardware is available (slide projector, opaque projector, overhead projector, film projector, videotape player, terminal with large-screen monitor)? What graphics programs are available? Which program is best for your purpose and listeners? How far in advance does this equipment have to be requested?

What resources are available for producing the visuals? Can drawings, charts, graphs, or maps be created as needed? Can transparencies (for overhead projection) be made or slides produced? Can handouts be typed and reproduced? Can multimedia displays be created?

SELECT YOUR MEDIA. Fit the medium to the situation. Which medium or combination is best for the topic, setting, and listeners? How fancy do listeners expect this to be? Which media are appropriate for this occasion?

- For a weekly meeting with colleagues in your department, scribbling on a blank transparency, chalkboard, or dry-erase markerboard might suffice.
- For interacting with listeners, you might use a chalkboard to record audience responses to your questions.
- For immediate orientation, you might begin with a poster listing key visuals/ ideas/themes to which you will refer repeatedly.
- For helping listeners take notes, absorb technical data, or remember complex material, you might distribute a presentation outline as a preview or provide handouts.
- For displaying and discussing written samples listeners bring in, you might use an opaque projector.
- For a presentation to investors, clients, or upper management, you might require polished and professionally prepared visuals, including computer graphics, such as an electronic slide presentation using *PowerPoint* software.

Figure 26.2 presents the various common media in approximate order of availability and ease of preparation.

Prepare Your Visuals

As you prepare visuals, focus on economy, clarity, and simplicity.

BE SELECTIVE. Use a visual only when it truly serves a purpose. Use restraint in choosing what to highlight with visuals. Try not to begin or end the presentation with a visual. At those times, listeners' attention should be focused on the presenter instead of the visual.

MAKE VISUALS EASY TO READ AND UNDERSTAND. Think of each visual as an image that flashes before your listeners. They will not have the luxury of studying the visual at leisure. Listeners need to know at a glance what they are looking at and what it means.

In addition to being able to *read* the visual, listeners need to *understand* it. Following are suggestions for achieving clarity.

When your material is extremely detailed or complex, distribute handouts to each listener.

GUIDELINES for Readable Visuals

- *Make visuals large enough to be read anywhere in the room.*
- *Don't cram too many words, ideas, designs, or type styles onto a visual.*
- *Keep wording and images simple.*
- *Boil your message down to the fewest words.*
- *Break things into small sections.*
- *Summarize with key words, phrases, or short sentences.*
- *Use 18–24 point type size and sans serif typeface (White,* Great Pages *80).*

GUIDELINES for Understandable Visuals

- *Display only one point per visual—unless previewing or reviewing (White 79).*
- *Give each visual a title that announces the topic.*
- *Use colour, sparingly, to highlight key words, facts, or the bottom line.*
- *Use the brightest colour for what is most important (White,* Great Pages *78–79).*
- *Label each part of a diagram or illustration.*
- *Proofread each visual carefully.*

LOOK FOR ALTERNATIVES TO WORD-FILLED VISUALS. Instead of just presenting overhead versions of printed pages, explore the full *visual* possibilities of your media. For example, anyone who tries to write a verbal equivalent of the visual message in Figure 26.3 will soon appreciate the power of images in relation to words alone. Whenever possible, use drawings, graphs, charts, photographs, and other visual representation discussed in Chapter 14.

Consider the Available Technology

Today's audiences expect communication that displays high visual quality. Using digital and video cameras, Web sites, and presentation software, you can create dynamic presentations that appeal to the listener's multiple senses. Using an automatic,

Whiteboard/Chalkboard

Uses
- simple, on-the-spot visuals
- recording audience responses
- informal settings
- small, well-lighted rooms

Tips
- copy long material in advance
- make it legible and visible to all
- use washable markers on markerboard
- speak to the listeners—not to the board

Poster

Uses
- overviews, previews, emphasis
- recurring themes
- formal or informal settings
- small, well-lighted rooms

Tips
- use 20" x 30" posterboard (or larger)
- use intense, washable colors
- keep each poster simple and uncrowded
- arrange/display posters in advance
- make each visible to the whole room
- point to what you are discussing

Flip Chart

Uses
- a sequence of visuals
- back-and-forth movement
- formal or informal settings
- small, well-lighted rooms

Tips
- use an easel pad and easel
- use intense, washable colors
- work from a storyboard
- check your sequence beforehand
- point to what you are discussing

Handouts

Uses
- present complex material
- help listeners follow along
- help listeners take notes
- help listeners remember

Tips
- staple or bind the packet
- number the pages
- try saving for the end
- if you must distribute up front, ask listeners to await instructions before reading the material

Opaque Projection

Uses
- direct display of paper documents
- samples listeners bring in
- pages from books, reports
- informal settings
- very small and dark rooms
- when spontaneity is more important than image quality

Tips
- allow projector to warm up
- don't shut off projector until you're done
- never stare at a bright bulb
- use a light source for viewing your notes
- use a laser or telescoping pointer
- place listeners close enough to see clearly

FIGURE 26.2   Selecting Media for Visual Presentations

Film and Video

Uses
- necessary display of moving images
- coordinated sound and visual images aid understanding and help persuade

Tips
- carefully introduce segment to be shown; tell viewers what to see or expect
- show only those segments needed to make your points
- if the segment is complex, or happens very quickly, play it more than once
- where appropriate, use video slow-motion replay
- always practice with the segment beforehand
- remember that the denser film image or DVD image can be projected on a larger screen to a larger audience than can VHS video

Overhead Projection

Uses
- on-the-spot or prepared visuals
- overlaid visuals[1]
- formal or informal settings
- small- or medium-sized rooms
- rooms needing to remain lighted

Tips
- use cardboard mounting frames for your acetate transparencies
- write discussion notes on each frame
- check your sequence beforehand
- turn projector off when not using it
- face the audience—not the screen
- point directly on the transparency
- use erasable color markers to highlight items as you discuss them

Slide Projection

Uses
- professional-quality visuals
- formal setting
- small or large dark rooms

Tips
- work from a storyboard
- check the slide sequence beforehand
- use a light source for viewing your notes
- use a laser or telescoping pointer

Computer Projection

Uses
- sophisticated charts, graphs, maps
- multimedia presentations
- formal settings
- small, dark rooms

Tips
- take lots of time to prepare/practice
- work from a storyboard
- check the whole system beforehand
- have a default plan in case something goes wrong

1. The overlay technique begins with one transparency showing a basic image, over which additional transparencies (with coordinated images, colors, labels, and so on) are added to produce an increasingly complex image.

FIGURE 26.2   Selecting Media for Visual Presentations *(continued)*

FIGURE 26.3 Images More Powerful Than Words

Source: "Distribution of the World's Water," as appeared in WWF Atlas of the Environment *by Geoffrey Lean and Don Hinrichsen. Copyright © 1994 Banson Marketing Ltd. Reprinted with permission.*

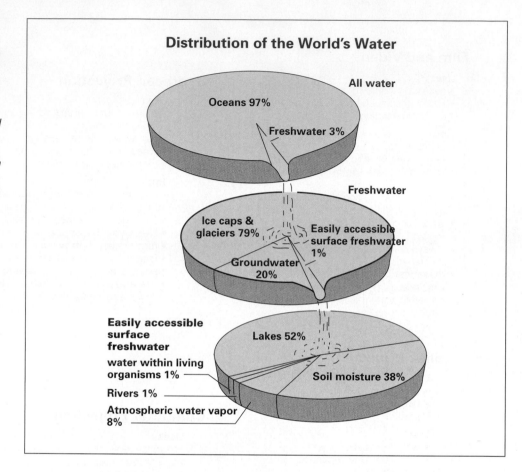

Distribution of the World's Water

remote-controlled transparency feeder and a pencil-sized laser pointer, you can deliver a smooth and elegant presentation. Despite the possibilities inherent in the technology, you are still responsible for a presentation that is well researched and professionally delivered.

Use *PowerPoint* or Other Software Wisely

Using presentation software such as Microsoft *PowerPoint* (Figure 26.4), you can produce professional-quality slides and then show them electronically:

A sampling of *PowerPoint's* design and display features

- Create slide designs in various colors, shading, and textures.
- Create drawings or graphs and import clip art, photographs, or other images.
- Create animated text and images: say, bullets that flash one-at-a-time on the screen or bars and lines on a graph that are highlighted individually, to emphasize specific characteristics of the data.

26.3

For models of *PowerPoint* presentations visit <www.ablongman.com/lannonweb>

- Create dynamic transitions between each slide, such as having one slide dissolve toward the right side of the screen as the following slide uncovers from the left.
- Amplify each slide with speaker notes that are invisible to the audience.
- Sort your slides into various sequences.
- Precisely time your entire presentation.
- Show your presentation directly on the computer screen or large-screen projector, online via the Web, as overhead transparencies, or as printed handouts. (Figure 26.5 shows printed versions of *PowerPoint* slides.)

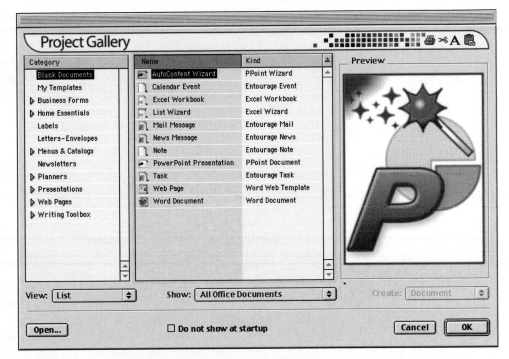

FIGURE 26.4 *PowerPoint* **Software's Opening Screen** From this gateway screen, even inexperienced users can explore topic categories (say, "Presenting a Technical Report," "Reporting Progress," or "Communicating Bad News"). Users can also find ideas for slide content and schemes for organizing their presentation. To get started, select "AutoContent Wizard."
Source: Reprinted by permission from Microsoft Corporation.

Because of its many features, *PowerPoint* has become the most widely used presentation software. In 2001, Microsoft estimated that some 30 million *PowerPoint* slide presentations were given daily (Parker 85). In a world in which images are everywhere, and electronic communication is the mode, *PowerPoint* is often regarded—rightly or wrongly—as "an indispensable corporate survival tool" (Nunberg 330).

The *PowerPoint* debate

PowerPoint advocates argue that bullet-style points help structure the story or the argument and help the presenter organize and stay on course. Critics argue that the mere content outline provided by the slides can oversimplify complex issues and that an endless list of bullets or animations, colors, and sounds can distract the audience from the message.

As an example of how overreliance on presentation slides can cloud the thinking process, consider the following scenario:

PowerPoint and the Space Shuttle *Columbia* Disaster

On February 1, 2003 the space shuttle *Columbia* burned up upon reentering the Earth's atmosphere. During launch a piece of insulating foam had broken off the shuttle and damaged the wing. (Page 15 describes the *Columbia* tragedy.)

During the days the *Columbia* was in orbit, NASA personnel tried to assess the damage and to recommend a course of action. Finally, it was decided that the damage did not seem serious enough to pose a significant threat, and reentry went ahead on schedule. (Lower-level suggestions that the shuttle fly close to a satellite that could have photographed the damage, for a clearer assessment, were overlooked and ultimately ignored by the final decision makers.)

The *Columbia* Accident Investigation Board concluded that a *PowerPoint* presentation to NASA officials had played a role in the disaster: Engineers presented their findings in a series of confusing and misleading slides that obscured errors in their own engineering analysis. Design expert Edward Tufte points out that one especially crucial slide was so crammed with data and bulleted points and so lacking in analysis that it was impossible to decipher accurately (8–9).

The Board's findings:

As information gets passed up an organization hierarchy, from people who do analysis to mid-level managers to high-level leadership, key explanations and supporting information are filtered out. In this context, it is easy to understand how a senior manager might read this *PowerPoint* slide and not realize that it addresses a life-threatening situation.

At many points during its investigation, the Board was surprised to receive similar presentation slides from NASA officials in place of technical reports. The Board views the endemic use of *PowerPoint* briefing slides instead of technical papers as an illustration of the problematic methods of technical communication at NASA. (Columbia Accident 191) ❏

In the end, technological tools are merely a *supplement* to your presentation; they are no *substitute* for the facts, ideas, examples, numbers, and interpretations that make up the clear and complete message audiences expect.

CHECK THE ROOM AND SETTING BEFOREHAND.   Make sure you have enough space, electrical outlets, and tables for your equipment. If you will be addressing a large

GUIDELINES for Using Presentation Software

- *Have a backup plan in case the system fails.* Be prepared to give the presentation without the software.

- *Prepare a handout.* In case of system failure, you can distribute a handout of your slides so that listeners can follow along. In any event, by not having to record each slide, the audience can devote its attention to what you are saying. If you save the handout for the end, tell your audience up-front.

- *Avoid slide overload (too many slides for the time allotted, too many words per slide).* Aim for no more than one content slide per minute, 7 lines per slide, 7 words per line. Use phrases instead of complete sentences. Keep bulleted lists grammatically parallel.

- *Don't let the medium obscure the message.* The audience should be focused on what you have to say, and not on the slide. Avoid colors and backgrounds that distract from the content. Avoid the whooshing and whiz-bang and other sounds unless absolutely necessary. Be conservative with any design and display feature.

- *Keep it simple but not simplistic.* Spice things up with a light dose of imported digital photos, charts, graphs, or diagrams, but avoid images that look so complex that they require detailed study.

- *Keep viewers oriented.* Open with a slide that previews the main topics in your presentation. Use divider slides as transitions from one topic to the next. Close with a "Conclusion" or "Questions?" slide.

- *Set a comfortable pace.* Present one idea per slide, bringing bullets on one at a time. Give viewers time to digest the data.

- *Avoid merely reciting the slides.* Instead, discuss each slide, with specific examples and details that round out the idea—but try not to digress or ramble.

NOTE *See also the general guidelines for preparing and presenting visuals (pages 663 and 672.)*

---

audience by microphone and plan to point to features on your visuals, be sure the microphone is movable. Pay careful attention to lighting, especially for chalkboards, flip charts, and posters. Don't forget a pointer if you need one.

Rehearse Your Delivery

Hold ample practice sessions to learn the geography of your report. Try to rehearse at least once in front of friends, or use a full-length mirror and a tape recorder. Assess your delivery from listeners' comments or from your taped voice (which will sound high to you). Use the evaluation sheet on page 676 as a guide.

If at all possible, rehearse using the actual equipment (overhead projector and so on) in the actual setting, to ensure that you have all you need and that everything works. Rehearsing a computer-projected presentation is essential.

DELIVERING YOUR PRESENTATION

You have planned and prepared carefully. Now consider the following simple steps to make your actual presentation enjoyable instead of terrifying.

Cultivate the Human Landscape

A successful presentation involves relationship building between presenter and audience.

GET TO KNOW YOUR AUDIENCE.   Try to meet some audience members before your presentation. We all feel more comfortable with people we know. Don't be afraid to smile.

DISPLAY ENTHUSIASM AND CONFIDENCE.   Nobody likes a speaker who seems half dead. Clean up verbal tics ("er," "ah," "uuh"). Overcome your shyness; research indicates that shy people are seen as less credible, trustworthy, likable, attractive, and knowledgeable.

BE REASONABLE.   Don't make your point at someone else's expense. If your topic is controversial (layoffs, policy changes, downsizing), decide how to speak candidly and persuasively with the least chance of offending anyone. For example, in your presentation about groundwater pollution, you don't want to attack the developers, since the building trade is a major producer of jobs, second only to tourism, on Cape Cod. Avoid personal attacks.

DON'T PREACH.   Speak like a person talking—not someone giving a sermon or the Gettysburg Address. Use *we, you, your, our,* to establish commonality with the audience. Avoid jokes or wisecracks.

Keep Your Listeners Oriented

Help your listeners to focus their attention and organize their understanding. Give them a map, some guidance, and highlights.

INTRODUCE YOUR TOPIC CLEARLY.   Open with a preview of your discussion:

A presentation
preview

> Today, I want to discuss *A, B,* and *C.*

Use visuals to highlight your main points and reveal your organization. For example, you might outline main points on a poster, a chalkboard, or a transparency, or hand out a presentation outline.

FOCUS ON LISTENERS' CONCERNS. Say something to establish immediate common ground. Show how your presentation has meaning for them, personally.

An appeal to listeners' concerns

> Do you know what you will be drinking when you turn on the tap?

Listeners who have a definite stake in the issues will be far more attentive and receptive.

PROVIDE EXPLICIT TRANSITIONS. Alert your listeners whenever you are preparing to switch gears:

Explicit transitions

> For my next point. . . .
> Turning now to my second point. . . .
> The third point I want to emphasize. . . .

Repeat key points or terms to keep them fresh in listeners' minds.

GIVE CONCRETE EXAMPLES. Good examples are informative and persuasive.

A concrete example

> Overdraw from town wells in Maloket and Tanford (two of our most rapidly growing towns) has resulted in measurable salt infusion at a yearly rate of 0.1 mg per liter.

Use examples that focus on listener concerns.

REVIEW AND INTERPRET. Last things are best remembered. Help listeners remember the main points:

A review of main points

> To summarize the dangers to our groundwater. . . .

Also, be clear about what this material means. Be emphatic about what listeners should be doing, thinking, or feeling:

An emphatic conclusion

> The conclusion is obvious: If the Cape is to survive, we must. . . .

Try to conclude with a forceful answer to this implied question from each listener: "What does this all mean to me personally?"

Manage Your Visuals

Presenting visuals effectively is a matter of good timing and careful management.

PREPARE EVERYTHING BEFOREHAND. If you plan to draw on a chalkboard or poster, do the drawings beforehand (in multicolors). Otherwise, listeners will be sitting idly while you draw away.

Prepare handouts if you want listeners to remember or study certain material. Distribute these *after* your talk. (You want the audience to be looking at and listening to you, instead of reading the handout.) Distribute handouts before or during the talk only if you want listeners to take notes—or if your equipment breaks down. When you do distribute handouts beforehand, ask listeners to await your instructions before they turn to a particular page.

Use transparency mounting frames for easy handling (the white cardboard frames allow you to number the transparencies and prepare notes for yourself on the frame).

Increasingly available are automatic transparency feeders with a remote control, making your transparency presentation work like a slide show. This device attaches easily to your overhead projector and allows you to reveal each point line by line.

ARRANGE EVERYTHING BEFOREHAND. Make sure you organize your media materials and the physical layout beforehand, to avoid fumbling during the presentation. Check your visual sequence against your storyboard.

FOLLOW A FEW SIMPLE GUIDELINES. Make your visuals part of a seamless presentation. Avoid listener distraction, confusion, and frustration by observing the following suggestions.

GUIDELINES for Presenting Visuals

- *Try not to begin with a visual.*
- *Try not to display a visual until you are ready to discuss it.*
- *Tell viewers what they should be looking for in the visual.*
- *Point to what is important.*
- *Stand aside when discussing a visual, so everyone can see it.*
- *Don't turn your back on the audience.*
- *After discussing the visual, remove it promptly.*
- *Switch off equipment that is not in use.*
- *Try not to end with a visual.*

Manage Your Presentation Style

Think about how you are moving, how you are speaking, and where you are looking. These are all elements of your personal style.

USE NATURAL MOVEMENTS AND REASONABLE POSTURES.   Move and gesture as you normally would in conversation, and maintain reasonable postures. Avoid foot shuffling, pencil tapping, swaying, slumping, or fidgeting.

ADJUST VOLUME, PRONUNCIATION, AND RATE.   With a microphone, don't speak too loudly. Without one, don't speak too softly. Be sure you can be heard clearly without shattering eardrums. Ask your audience about the sound and speed of your delivery after a few sentences.

Nervousness causes speakers to gallop along and mispronounce words. Slow down and pronounce clearly. Usually, a rate that seems a bit slow to you will be just right for listeners.

MAINTAIN EYE CONTACT.   Look directly into listeners' eyes. With a small audience, eye contact is one of your best connectors. As you speak, establish eye contact with as many listeners as possible. With a large group, maintain eye contact with those in the first rows. Establish eye contact immediately—before you even begin to speak—by looking around.

Manage Your Speaking Situation

Do everything you can to keep things running smoothly.

BE RESPONSIVE TO LISTENER FEEDBACK.   Assess listener feedback continually and make adjustments as needed. If you are laboring through a long list of facts or figures and people begin to doze or fidget, you might summarize. Likewise, if frowns, raised eyebrows, or questioning looks indicate confusion, skepticism, or indignation, you can backtrack with a specific example or explanation. By tuning in to your audience's reactions, you can avoid leaving them confused, hostile, or simply bored.

STICK TO YOUR PLAN.   Say what you came to say, then summarize and close— politely and on time. Don't punctuate your speech with digressions that pop into your head. Unless a specific anecdote was part of your original plan to clarify a point or increase interest, avoid excursions. We often tend to be more interested in what we have to say than our listeners are! Don't exceed your time limit.

LEAVE LISTENERS WITH SOMETHING TO REMEMBER.   Before ending, take a moment to summarize the major points and reemphasize anything of special importance. Are listeners supposed to remember something, have a different attitude, take a specific action? Let them know! As you conclude, thank your listeners.

26.4

For more on global communication visit <www.ablongman.com/ lannonweb>

ALLOW TIME FOR QUESTIONS AND ANSWERS.   At the very beginning, tell your listeners that a question-and-answer period will follow. Use the following suggestions for managing listener questions diplomatically and efficiently.

GUIDELINES for Managing Listener Questions

- *Announce a specific time limit* for the question period to avoid prolonged debates.
- *Listen carefully to each question.*
- *If you can't understand a question,* ask that it be rephrased.
- *Repeat every question,* to ensure that everyone hears it.
- *Be brief in your answers.*
- *If you need extra time to answer a question,* arrange for it after the presentation.
- *If anyone attempts lengthy debate,* offer to continue after the presentation.
- *If you can't answer a question,* say so and move on.
- *End the session with,* "We have time for one more question," or some such signal.

CONSIDER THIS Cross-Cultural Audiences May Have Specific Expectations

Imagine that you've been assigned to represent your company at an international conference or before international clients (e.g., of passenger aircraft or mainframe computers). As you plan and prepare your presentation, remain sensitive to various cultural expectations.

For example, some cultures might be offended by a presentation that gets right to the point without first observing formalities of politeness, well wishes, and the like.

Certain communication styles are welcomed in some cultures, but considered offensive in others. In southern Europe and the Middle East, people expect direct and prolonged eye contact as a way of showing honesty and respect. In Southeast Asia, this may be taken as a sign of aggression or disrespect (Gesteland 24). A sampling of the questions to consider:

- *Should I smile a lot or look serious? (Hulbert, "Overcoming" 42)*
- *Should I rely on expressive gestures and facial expressions?*

- *How loudly or softly, rapidly or slowly should I speak?*
- *Should I come out from behind the podium and approach the audience or keep my distance?*
- *Should I get right to the point or take plenty of time to lead into and discuss the matter thoroughly?*
- *Should I focus only on the key facts or on all the details and various interpretations?*
- *Should I be assertive in offering interpretations and conclusions, or should I allow listeners to reach their own conclusions?*
- *Which types of visuals and which media might or might not work?*
- *Should I invite questions from this audience, or would this be offensive?*

To account for language differences, prepare a handout of your entire script for distribution after the presentation, along with a copy of your visuals. This way, your audience will be able to study your material at their leisure.

EXERCISES For more exercises, visit
<www.ablongman.com/lannon>

1. In a memo to your instructor, identify and discuss the kinds of oral reporting duties you expect to encounter in your career.

2. Prepare an oral presentation for your class, based on your written long report. Develop a sentence outline and a storyboard that includes at least three visuals. If your instructor requests, create one or more of your presentation visuals using *PowerPoint* software. (For a step-by-step guide to getting started on *PowerPoint,* select "AutoContent Wizard" from the opening screen. See Figure 26.4.)

Practice your presentation with a tape recorder or a friend. Use the Peer Evaluation Checklist on page 676 to assess and refine your delivery.

3. Observe a lecture or speech, and evaluate it according to the Peer Evaluation Checklist. Write a memo to your instructor (without naming the speaker), identifying strong and weak areas and suggesting improvements.

4. In an oral presentation to the class, present your findings, conclusions, and recommendations from the analytical report assignment in Chapter 24.

Peer Evaluation Checklist for Oral Presentations

Presentation Evaluation for (name/topic)   _____

Content   *Comments*

☐ Stated a clear purpose.   _____
☐ Created interest in the topic.   _____
☐ Showed command of the material.   _____
☐ Supported assertions with evidence.   _____
☐ Used adequate and appropriate visuals.   _____
☐ Used material suited to this audience's   _____
   needs, knowledge, concerns, and interests.   _____
☐ Acknowledged opposing views.   _____
☐ Gave the right amount of information.   _____

Organization

☐ Began with a clear overview.   _____
☐ Presented a clear line of reasoning.   _____
☐ Moved from point to point effectively.   _____
☐ Stayed on course.   _____
☐ Used transitions effectively.   _____
☐ Avoided needless digressions.   _____
☐ Summarized before concluding.   _____
☐ Was clear about what the listeners   _____
   should think or do.

Style

☐ Dressed appropriately.
☐ Seemed confident, relaxed, and likable.   _____
☐ Seemed in control of the speaking situation.   _____
☐ Showed appropriate enthusiasm.   _____
☐ Pronounced, enunciated, and spoke well.   _____
☐ Used no slang whatsoever.   _____
☐ Used appropriate gestures, tone,   _____
   volume, and delivery rate.
☐ Had good posture and eye contact.   _____
☐ Interacted with the audience.   _____
☐ Kept the audience actively involved.   _____
☐ Answered questions concisely and convincingly.   _____

Overall professionalism: Superior _____ **Acceptable** _____ **Needs work** _____

Evaluator's signature: _____

PART VI

A Brief Handbook with Additional Sample Documents

Public Access versus Public Security on U.S. Web Sites

Some agencies of the U.S. government deleted or restricted public access to information on their Web sites in the wake of the September 11, 2001 terrorist attacks (Ruppe). The Environmental Protection Agency (EPA), for example, deleted "risk management plan" data from its site. Similarly, the Department of Transportation (DOT) decided to restrict access to its National Pipeline Mapping System, which includes data on locations where pipeline leaks might endanger drinking water.

Other agencies, including the Federal Aviation Administration (FAA) and the National Transportation Safety Board (NTSB) left their sites open and intact. The nonprofit group OMB Watch criticized the EPA and DOT decisions, citing federal laws mandating public access. But others, including *National Review Online* commentator Jonathan Adler, countered that EPA information in particular could help terrorists locate schools that might be endangered by chemical leaks. Ongoing debates pitting security versus access issues will no doubt continue to play out in both public and private sector Web design conferences for some time to come (Ruppe). ■

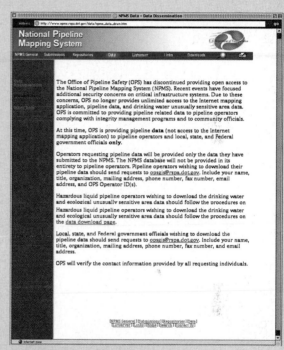

The Office of Pipeline Safety, U.S. Department of Transportation

Recording and Documenting Research Findings

TAKING NOTES

Many researchers take notes on a laptop computer, using electronic file programs or database management software that allows notes to be filed, shuffled, and retrieved by author, title, topic, date, or keywords. You can also take notes in a single word-processing file, then use the "find" command to locate notes quickly. Whether you use a computer or notecards, your notes should be easy to organize and reorganize.

GUIDELINES for Recording Research Findings

1. *Make a separate bibliography listing for each work you consult.* Record that work's complete entry (Figure A.1), using the citation format that will appear in your document. (See pages 688–716 for sample entries.) Record the information accurately so that you won't have to relocate a source at the last minute.

Record each bibliographic citation exactly as it will appear in your final report

> Pinsky, Mark A. The EMF Book: What You Should Know about Electromagnetic
>
> Fields, Electromagnetic Radiation, and Your Health. New York: Warner, 1995.

FIGURE A.1   Recording a Bibliographic Citation

When searching online, you can often print out the full bibliographic record for each work or save it to disk, thereby ensuring an accurate citation.

2. *Skim the entire work to locate relevant material.* Look over the table of contents and the index. Check the introduction for an overview or thesis. Look for informative headings.

3. *Go back and decide what to record.* Use a separate entry for each item.

4. *Be selective.* Don't copy or paraphrase every word. (See the guidelines for summarizing on page 199.)

5. *Record the item as a quotation or paraphrase.* When quoting others directly, be sure to record words and punctuation accurately. When restating material in your own words, preserve the original meaning and emphasis.

QUOTING THE WORK OF OTHERS

You must place quotation marks around all exact wording you borrow, whether the words were written, spoken (as in an interview or presentation), or appeared in electronic form. Even a single borrowed sentence or phrase, or a single word used in a special way, needs quotation marks, with the exact source properly cited. These sources include people with whom you collaborate.

Plagiarism is often unintentional

If your notes don't identify quoted material accurately, you might forget to credit the source. Even when this omission is unintentional, you face the charge of *plagiarism* (misrepresenting as your own the words or ideas of someone else). Possible consequences of plagiarism include expulsion from school, loss of a job, and a lawsuit.

The perils of buying plagiarized work online

It's no secret that any cheater can purchase reports, term papers, and other documents on the Web. But antiplagiarism Web sites, such as <plagiarism.org> now enable professors to cross-reference a suspicious paper against previously published material, flagging and identifying each plagiarized source.

GUIDELINES for Quoting the Work of Others

1. *Use a direct quotation only when absolutely necessary.* Sometimes a direct quotation is the only way to do justice to the author's own words—as in these instances:

   Expressions that warrant direct quotation

   > "Writing is a way to end up thinking something you couldn't have started out thinking" (Elbow 15).

   > Think of the topic sentence as "the one sentence you would keep if you could keep only one" (USAF Academy 11).

   Consider quoting directly for these purposes:

   Reasons for quoting directly

   - to preserve special phrasing or emphasis
   - to preserve precise meaning
   - to preserve the original line of reasoning
   - to preserve an especially striking or colorful example
   - to convey the authority and complexity of expert opinion
   - to convey the original's voice, sincerity, or emotional intensity

2. *Ensure accuracy.* Copy the selection word for word; record the exact page numbers; and double-check that you haven't altered the original expression in any way (Figure A.2).

3. *Keep the quotation as brief as possible.* For conciseness and emphasis, use *ellipses:* Use three spaced periods to indicate each omission within a single sentence. Add a fourth period to indicate each omission that includes the end of a sentence or multi-sentence sections of text.

   Ellipsis within and between sentences

   > Use three . . . periods to indicate each omission within a single sentence. Add a fourth period to indicate . . . the end of a sentence or . . . .

Place quotation marks around all directly quoted material

> Pinsky, Mark A. pp. 29–30.
>
> "Neither electromagnetic fields nor electromagnetic radiation cause cancer per se, most researchers agree. What they may do is promote cancer. Cancer is a multistage process that requires an 'initiator' that makes a cell or group of cells abnormal. Everyone has cancerous cells in his or her body. Cancer—the disease as we think of it—occurs when these cancerous cells grow uncontrollably."

FIGURE A.2   Recording a Quotation

The elliptical passage must be grammatical and must not distort the original meaning. (For additional guidelines, see page 771.)

4. *Use square brackets to insert your own clarifying comments or transitions.* To distinguish your words from those of your source, place them within brackets:

Brackets setting off the added words within a quotation

> "Job stress [in aircraft ground control] can lead to disaster."

5. *Embed quoted material in your sentences clearly and grammatically.* Introduce integrated quotations with phrases such as "Jones argues that," or "Gomez concludes that." More importantly, use a transitional phrase to show the relationship between the quoted idea and the sentence that precedes it:

An introduction that unifies a quotation with the discussion

> One investigation of age discrimination at select Fortune 500 companies found that "middle managers over age 45 are an endangered species" (Jablonski 69).

Your integrated sentence should be grammatical:

Quoted material integrated grammatically with the writer's words

> "The present farming crisis," Marx argues, "is a direct result of rampant land speculation" (41).

(For additional guidelines, see page 770.)

6. *Quote passages four lines or longer in block form.* Avoid relying on long quotations except in these instances:

Reasons for quoting a long passage

- to provide an extended example, definition, or analogy (see page 11)
- to analyze or discuss an idea or concept (see page 610)

Double-space a block quotation and indent the entire block ten spaces. Do not indent the first line of the passage, but do indent first lines of subsequent paragraphs three spaces. Do not use quotation marks.

7. *Introduce the quotation and discuss its significance.*

An introduction to quoted material

> Here is a corporate executive's description of some audiences you can expect to address:

8. *Cite the source of each quoted passage.*

Research writing is a process of independent thinking in which you work with the ideas of others in order to reach your own conclusions; unless the author's exact wording is essential, try to paraphrase, instead of quoting, borrowed material.

PARAPHRASING THE WORK OF OTHERS

Paraphrasing means more than changing or shuffling a few words; it means restating the original idea in your own words—sometimes in a clearer, more direct, and emphatic way—and giving full credit to the source.

Faulty paraphrasing is a form of plagiarism

To borrow or adapt someone else's ideas or reasoning without properly documenting the source is plagiarism. To offer as a paraphrase an original passage that is only slightly altered—even when you document the source—also is plagiarism. Equally unethical is offering a paraphrase, although documented, that distorts the original meaning.

GUIDELINES for Paraphrasing

1. *Refer to the author early in the paraphrase, to indicate the beginning of the borrowed passage.*
2. *Retain key words from the original, to preserve its meaning.*
3. *Restructure and combine original sentences for emphasis and fluency.*
4. *Delete needless words from the original, for conciseness.*
5. *Use your own words and phrases to clarify the author's ideas.*
6. *Cite (in parentheses) the exact source, to mark the end of the borrowed passage and to give full credit.*
7. *Be sure to preserve the author's original intent (Weinstein 3).*

Figure A.3 shows an entry paraphrased from Figure A.2. Paraphrased material is not enclosed within quotation marks, but it is documented to acknowledge your debt to the source.

Signal the beginning of the paraphrase by citing the author, and the end by citing the source.

Pinsky, Mark A.

Pinsky explains that electromagnetic waves probably do not directly cause cancer. However, they might contribute to the uncontrollable growth of those cancer cells normally present—but controlled—in the human body (29–30).

FIGURE A.3   Recording a Paraphrase

WHAT YOU SHOULD DOCUMENT

Document any insight, assertion, fact, finding, interpretation, judgment, or other "appropriated material that readers might otherwise mistake for your own" (Gibaldi and Achtert 155)—whether the material appears in published form or not. Specifically, you must document:

Sources that require documentation

- any source from which you use exact wording
- any source from which you adapt material in your own words
- any visual illustration: charts, graphs, drawings, or the like (see Chapter 14 for documenting visuals)

How to document a confidential source

In some instances, you might have reason to preserve the anonymity of unpublished sources: for example, to allow people to respond candidly without fear of reprisal (as with employee criticism of the company), or to protect their privacy (as with certain material from email inquiries or electronic newsgroups). You must still document the fact that you are not the originator of this material by providing a general acknowledgment in the text ("A number of employees expressed frustration with . . .") along with a general citation in your list of references or works cited ("Interviews with Polex employees, May 2004").

Common knowledge need not be documented

You don't need to document anything considered *common knowledge:* material that appears repeatedly in general sources. In medicine, for instance, it has become common knowledge that foods containing animal fat contribute to higher blood cholesterol levels. So in a report on fatty diets and heart disease, you probably would not need to document that well-known fact. But you would document information about how the fat/cholesterol connection was discovered, what subsequent studies have found (say, the role of saturated versus unsaturated fats), and any information for which some other person could claim specific credit. If the borrowed material can be found in only one specific source, not in multiple sources, document it. When in doubt, document the source.

HOW YOU SHOULD DOCUMENT

Cite borrowed material twice: at the exact place you use that material, and at the end of your document. Documentation practices vary widely, but all systems work almost identically: a brief reference in the text names the source and refers readers to the complete citation, which allows readers to retrieve the source.

Many disciplines, institutions, and organizations publish their own style guides or documentation manuals. Here are a few:

Style guides from various disciplines

Geographical Research and Writing
Style Manual for Engineering Authors and Editors
IBM Style Manual
NASA Publications Manual

This chapter illustrates citations and entries for three styles widely used for documenting sources in their respective disciplines:

- Modern Language Association (MLA) style, for the humanities
- American Psychological Association (APA) style, for the social sciences
- Council of Biology Editors (CBE) style, for the natural and applied sciences

Unless your audience has its own preference, any of these three styles can be adapted to most research writing. Use one style consistently throughout the document.

MLA DOCUMENTATION STYLE

Traditional MLA documentation of sources used superscript numbers (like this:[1]) in the text, followed by full references at the bottom of the page (footnotes) or at the end of the document (endnotes) and, finally, by a bibliography. But a more current form of documentation appears in the *MLA Handbook for Writers of Research Papers, 6th ed.* New York: Modern Language Association, 2003. Footnotes or endnotes are now used only to comment on material in the text or on sources or to suggest additional sources.

In current MLA style, in-text parenthetical references briefly identify the source(s). Full documentation then appears in a Works Cited section at the end of the document.

A parenthetical reference usually includes the author's surname and the exact page number(s) of the borrowed material:

> One notable study indicates an elevated risk of leukemia for children exposed to certain types of electromagnetic fields (Bowman et al. 59).

Readers seeking the complete citation for Bowman can refer easily to the Works Cited section, with entries listed alphabetically by author:

> Bowman, J. D., et al. "Hypothesis: The Risk of Childhood Leukemia Is Related
>   to Combinations of Power-Frequency and Static Magnetic Fields."
>   *Bioelectromagnetics* 16.1 (1995): 48–59.

This complete citation includes page numbers for the entire article.

MLA Parenthetical References

For clear and informative parenthetical references, observe these guidelines:

- If your discussion names the author, do not repeat the name in your parenthetical reference; simply give the page number(s):

Marginal notes:

Use this alternative to footnotes and bibliographies

Cite a source briefly in text and fully at the end

Parenthetical reference in the text

Full citation at document's end

How to cite briefly in text

| Citing page numbers only | Bowman et al. explain how their study indicates an elevated risk of leukemia for children exposed to certain types of electromagnetic fields (59). |

- If you cite two or more works in a single parenthetical reference, separate the citations with semicolons:

| Three works in a single reference | (Jones 32; Leduc 41; Gomez 293–94) |

- If you cite two or more authors with the same surnames, include the first initial in your parenthetical reference to each author:

| Two authors with identical surnames | (R. Jones 32)

(S. Jones 14–15) |

- If you cite two or more works by the same author, include the first significant word from each work's title, or a shortened version:

| Two works by one author | (Lamont, <u>Biophysics</u> 100–01)

(Lamont, <u>Diagnostic</u> *Tests* 81) |

- If the work is by an institutional or corporate author or if it is unsigned (that is, the author is unknown), use only the first few words of the institutional name or the work's title in your parenthetical reference:

| Institutional, corporate, or anonymous author | (American Medical Assn. 2)

("Distribution Systems" 18) |

To avoid distracting the reader, keep each parenthetical reference as brief as possible. (One method is to name the source in your discussion and to place only the page number(s) in parentheses.)

| Where to place a parenthetical reference | For a paraphrase, place the parenthetical reference *before* the closing punctuation mark. For a quotation that runs into the text, place the reference *between* the final quotation mark and the closing punctuation mark. For a quotation set off (indented) from the text, place the reference two spaces *after* the closing punctuation mark. |

MLA Works Cited Entries

| How to space and indent entries | The Works Cited list includes each source that you have paraphrased or quoted. In preparing the list, type the first line of each entry flush with the left margin. Indent the second and subsequent lines one-half inch. Use one character space after any period, comma, or colon. Double-space within and between each entry. |

| How to cite fully at the end | Following are examples of complete citations as they would appear in the Works Cited section of your document. Shown below each citation is its corresponding parenthetical reference as it would appear in the text. |

INDEX TO SAMPLE MLA WORKS CITED ENTRIES

Books

1. Book, single author
2. Book, two or three authors
3. Book, four or more authors
4. Book, anonymous author
5. Multiple books, same author
6. Book, one or more editors
7. Book, indirect source
8. Anthology selection or book chapter

Periodicals

9. Article, magazine
10. Article, journal with new pagination each issue
11. Article, journal with continuous pagination
12. Article, newspaper

Other Sources

13. Encyclopedia, dictionary, alphabetical reference
14. Report
15. Conference presentation
16. Interview, personally conducted
17. Interview, published

18. Letter, unpublished
19. Questionnaire
20. Brochure or pamphlet
21. Lecture
22. Government document
23. Document with corporate or foundation authorship
24. Map or other visual
25. Unpublished dissertation, report, or miscellaneous items

Electronic Sources

26. Reference database
27. Computer software
28. CD-ROM
29. Listserv
30. Usenet
31. Email
32. Home page for a course
33. Print article posted online
34. Real-time communication
35. Online abstract
36. General reference to a site

What to include in an MLA citation for a book

MLA WORKS CITED ENTRIES FOR BOOKS. Any citation for a book should contain the following information: author, title, editor or translator, edition, volume number, and facts about publication (city, publisher, date).

1. Book, Single Author—MLA

Kerzin-Fontana, Jane B. <u>Technology Management: A Handbook</u>. 3rd ed. Delmar,

     NY: American Management Assn., 2005.

Parenthetical reference: (Kerzin-Fontana 3–4)

Identify the state of publication by U.S. Postal Service abbreviations. If the city of publication is well known (Boston, Chicago), omit the state abbreviations. If several cities are listed on the title page, give only the first. For Canada, include the province abbreviation after the city. For all other countries include an abbreviation of the country name.

2. Book, Two or Three Authors—MLA

Aronson, Linda, Roger Katz, and Candide Moustafa. Toxic Waste Disposal

Methods. New Haven: Yale UP, 2004.

Parenthetical reference: (Aronson, Katz, and Moustafa 121–23)

Shorten publisher's names, as in "Simon" for Simon & Schuster, "GPO" for Government Printing Office, or "Yale UP" for Yale University Press. For page numbers with more than two digits, give only the final two digits for the second number if the first digit is identical.

3. Book, Four or More Authors—MLA

Santos, Ruth J., et al. Environmental Crises in Developing Countries. New York:

Harper, 2003.

Parenthetical reference: (Santos et al. 9)

"Et al." is the abbreviated form of the Latin "et alia," meaning "and others."

4. Book, Anonymous Author—MLA

Structured Programming. Boston: Meredith, 2005.

Parenthetical reference: (Structured 67)

5. Multiple Books, Same Author—MLA

Chang, John W. Biophysics. Boston: Little, 2002.

---. Diagnostic Techniques. New York: Radon, 1997.

Parenthetical references: (Chang, Biophysics 123–26), (Chang, Diagnostic 87)

When citing more than one work by the same author, do not repeat the author's name; simply type three hyphens followed by a period. List the works alphabetically by title.

6. Book, One or More Editors—MLA

Morris, A. J., and Louise B. Pardin-Walker, eds. Handbook of New Information

Technology. New York: Harper, 2003.

Parenthetical reference: (Morris and Pardin-Walker 34)

For more than three editors, name only the first, followed by "et al."

7. Book, Indirect Source—MLA

Kline, Thomas. Automated Systems. Boston: Rhodes, 1999.

Stubbs, John. White-Collar Productivity. Miami: Harris, 2004.

Parenthetical reference: (qtd. in Stubbs 116)

When your source (as in Stubbs, above) has quoted or cited another source, list each source in its appropriate alphabetical place in the Works Cited page. Use the name of the original source (here, Kline) in your text and precede your parenthetical reference with "qtd. in," or "cited in" for a paraphrase.

8. Anthology Selection or Book Chapter—MLA

Bowman, Joel P. "Electronic Conferencing." Communication and Technology:

Today and Tomorrow. Ed. Al Williams. Denton, TX: Assn. for Business

Communication, 1994, 123–42.

Parenthetical reference: (Bowman 129)

The page numbers in the complete citation are for the selection cited from the anthology.

What to include in an MLA citation for a periodical

MLA WORKS CITED ENTRIES FOR PERIODICALS. Give all available information in this order: author, article title, periodical title, volume or number (or both), date (day, month, year), and page numbers for the entire article—not just pages cited.

9. Article, Magazine—MLA

DesMarteau, Kathleen. "Study Links Sewing Machine Use to Alzheimer's

Disease." Bobbin Oct. 1994: 36–38.

Parenthetical reference: (DesMarteau 36)

No punctuation separates the magazine title and date. Nor is the abbreviation "p." or "pp." used to designate page numbers. If no author is given, list all other information:

"Distribution Systems for the New Decade." Power Technology Magazine 18

Oct. 2004: 18+.

Parenthetical reference: ("Distribution Systems" 18)

This article began on page 18 and continued on page 21. When an article does not appear on consecutive pages, give only the number of the first page, followed immediately by a plus sign. A three-letter abbreviation denotes any month spelled with five or more letters.

10. Article, Journal with New Pagination Each Issue—MLA

Thackman-White, Joan R. "Computer-Assisted Research." <u>American Librarian</u>

51.1 (2005): 3–9.

Parenthetical reference: (Thackman-White 4–5)

Because each issue for a given year will have page numbers beginning with "1," readers need the number of this issue. The "51" denotes the volume number; "1" denotes the issue number. Omit "The" or "A" or any other introductory article from a journal or magazine title.

11. Article, Journal with Continuous Pagination—MLA

Barnstead, Marion H. "The Writing Crisis." <u>Journal of Writing Theory</u> 12 (2004):

415–33.

Parenthetical reference: (Barnstead 415–16)

When page numbers continue from one issue to the next for the full year, readers won't need the issue number, because no other issue in that year repeats these same page numbers. (Include the issue number if you think it will help readers retrieve the article more easily.) The "12" denotes the volume number.

How to cite an abstract
If, instead of the complete work, you are citing merely an abstract found in a bound collection of abstracts, and not the full article, include the information on the abstracting service right after the information on the original article.

Barnstead, Marion H. "The Writing Crisis." <u>Journal of Writing Theory</u> 12 (2004):

415–33. <u>Rhetoric Abstracts</u> 67 (2005): item 1354.

If you are citing an abstract that appears before the printed article, add "Abstract," followed by a period, immediately after the original work's page number(s).

12. Article, Newspaper—MLA

Baranski, Vida H. "Errors in Technology Assessment." <u>Boston Times</u> 15 Jan.

2005, evening ed., sec. 2: 3.

Parenthetical reference: (Baranski 3)

When a daily newspaper has more than one edition, cite the edition after the date. Omit any introductory article in the newspaper's name (not *The Boston Times*). If no author is given, list all other information. If the newspaper's name does not include the city of publication, insert it, using brackets: *Sippican Sentinel* [Marion, MA].

What to include in
MLA citations for
a miscellaneous
source

MLA WORKS CITED ENTRIES FOR OTHER KINDS OF MATERIALS.   Miscellaneous sources range from unsigned encyclopedia entries to conference presentations to government publications. A full citation should give this information (as available): author, title, city, publisher, date, and page numbers.

13. Encyclopedia, Dictionary, Other Alphabetical Reference—MLA

"Communication." The Business Reference Book 2004.

Parenthetical reference: ("Communication")

Begin a signed entry with the author's name. For any work arranged alphabetically, omit page numbers in the citation and the parenthetical reference. For a well-known reference book, include only an edition (if stated) and a date. For other reference books, give the full publication information.

14. Report—MLA

Electrical Power Research Institute (EPRI). Epidemiologic Studies of Electric Utility
 Employees. (Report No. RP2964.5). Palo Alto, CA: EPRI, Nov. 1994.

Parenthetical reference: (Electrical Power Research Institute [EPRI] 27)

If no author is given, begin with the organization that sponsored the report.

For any report or other document with group authorship, as above, include the group's abbreviated name in your first parenthetical reference, and then use only that abbreviation in any subsequent reference.

15. Conference Presentation—MLA

Smith, Abelard A. "Radon Concentrations in Molded Concrete." First British
 Symposium in Environmental Engineering. London, 11–13 Oct. 2004.
 Ed. Anne Hodkins. London: Harrison, 2005. 106–21.

Parenthetical reference: (Smith 109)

The above example shows a presentation that has been included in the published proceedings of a conference. For an unpublished presentation, include the presenter's name, the title of the presentation, and the conference title, location, and date, but do not underline or italicize the conference information.

16. Interview, Personally Conducted—MLA

Nasser, Gamel. Chief Engineer for Northern Electric. Personal interview.
Rangeley, ME. 2 Apr., 2004.

Parenthetical reference: (Nasser)

17. Interview, Published—MLA

Lescault, James. "The Future of Graphics." Executive Views of Automation.
Ed. Karen Prell. Miami: Haber, 2005. 216–31.

Parenthetical reference: (Lescault 218)

The interviewee's name is placed in the entry's author slot.

18. Letter, Unpublished—MLA

Rogers, Leonard. Letter to the author. 15 May 2004.

Parenthetical reference: (Rogers)

19. Questionnaire—MLA

Taylor, Lynne. Questionnaire sent to 612 Massachusetts business executives.
14 Feb. 2004.

Parenthetical reference: (Taylor)

20. Brochure or Pamphlet—MLA

Investment Strategies for the 21st Century. San Francisco: Blount Economics
Assn., 2001.

Parenthetical reference: (*Investment*)

If the work is signed, begin with its author.

21. Lecture—MLA

Dumont, R. A. "Managing Natural Gas." Lecture. University of Massachusetts
at Dartmouth, 15 Jan. 2005.

Parenthetical reference: (Dumont)

If the lecture title is not known, write Address, Lecture, or Reading but do not use quotation marks. Include the sponsor and the location if they are available.

22. Government Document—MLA

> Virginia. Highway Dept. Standards for Bridge Maintenance. Richmond: Virginia
>
> > Highway Dept., 2004.

Parenthetical reference: (Virginia Highway Dept. 49)

If the author is unknown (as above), begin with the information in this order: name of the government, name of the issuing agency, document title, place, publisher, and date.

For any congressional document, identify the house of Congress (Senate or House of Representatives) before the title, and the number and session of Congress after the title:

> United States Cong. House, Armed Services Committee. Funding for the Military
>
> > Academies. 108th Congress, 2nd sess. Washington: GPO, 2004.

Parenthetical reference: (U.S. Cong. 41)

"GPO" is the abbreviation for the U.S. Government Printing Office.

For an entry from the *Congressional Record*, give only date and pages:

> Cong. Rec. 10 Mar. 2002: 2178–92.

Parenthetical reference: (Cong. Rec. 2184)

23. Document with Corporate or Foundation Authorship—MLA

> Hermitage Foundation. Global Warming Scenarios for the Year 2030.
>
> > Washington: Natl. Res. Council, 2002.

Parenthetical reference: (Hermitage Foun. 123)

24. Map or Other Visual—MLA

> Deaths Caused by Breast Cancer, by County. Map. Scientific American Oct.
>
> > 1995: 32D.

Parenthetical reference: (Deaths Caused)

If the creator of the visual is listed, give that name first. Identify the type of visual (Map, Graph, Table, Diagram) immediately following its title.

25. Unpublished Dissertation, Report, or Miscellaneous Items—MLA

> Author (if known). "Title." Sponsoring organization or publisher, date.

For any work that has group authorship (corporation, committee, task force), cite the name of the group or agency in place of the author's name.

MLA Works Cited Entries for Electronic Sources. Electronic sources include Internet sites, reference databases, CD-ROMs, computer software, and email. Any citation for an electronic source should allow readers to identify the original source (printed or electronic) and trace a clear path for retrieving the material. Provide all available information in the following order:

What to include in an MLA citation for an electronic source

1. Name of author or editor or creator of the electronic work or site.
2. Title of the document. For online postings, such as email discussion lists or newsgroups, give the title of the posting followed by the words "Online posting." For CD-ROM or software, give the title of the document or software followed by "CD-ROM" or "Diskette."
3. Publication information of the original printed version (as in the above entries), if such a version exists.
4. Information about the electronic publication, including the title of the site or database (as in "MEDLINE") and the date of the posting or the last update of the site. Name the sponsoring organization or provider of the CD-ROM (as in "ProQuest") or reference database service (as in "Dialog").
5. The date you accessed the source.
6. The full and accurate electronic address. For Internet sources, provide the complete URL (Uniform Resource Locator), enclosed in angle brackets (< >). For CD-ROM and database sources, give the document's retrieval number. Include page numbers only if the electronic document shows page numbers from the original print version. Include paragraph numbers only if they appear in the original Internet document.

NOTE *When a URL continues from one line to the next, break it only after a slash. Do not insert a hyphen.*

26. Reference Database—MLA

Sahl, J. D. "Power Lines, Viruses, and Childhood Leukemia." <u>Cancer Causes</u>

<u>Control</u> 6.1 (Jan. 1995): 83. <u>MEDLINE</u>. Online. 7 Nov. 2004. Dialog.

Parenthetical reference: (Sahl 83)

For entries with a printed equivalent, begin with publication information, then the database title (underlined or italicized), the "Online" designation to indicate the medium, and the service provider (or URL or email address) and the date of access. The access date is important because frequent updatings of databases can produce different versions of the material.

For entries with no printed equivalent, give the title and date of the work in quotation marks, followed by the electronic source information:

> Argent, Roger R. "An Analysis of International Exchange Rates for 1999."
>
> Accu-Data. Online. Dow Jones News Retrieval. 10 Jan. 2002.

Parenthetical reference: (Argent)

If the author is not known, begin with the work's title.

27. Computer Software—MLA

> Virtual Collaboration. Diskette. New York: Pearson, 2005.

Parenthetical reference: (*Virtual*)

Begin with the author's name, if known.

28. CD-ROM—MLA

> Canalte, Henry A. "Violent-Crime Statistics: Good News and Bad News." Law
>
> Enforcement Feb. 1995: 8. ABI/INFORM. CD-ROM. Proquest. Sept.
>
> 2004.

Parenthetical reference: (Canalte 8)

If the material is also available in print, begin with the information about the printed source, followed by the electronic source information: name of the data-base (underlined), CD-ROM designation, vendor name, and electronic publication date. If the material has no printed equivalent, list its author (if known) and title (in quotation marks), followed by the electronic source information.

For CD-ROM reference works and other material not routinely updated, give the title of the work, followed by the CD-ROM designation, place, electronic publisher, and date:

> Time Almanac. CD-ROM. Washington: Compact, 2004.

Parenthetical reference: (Time Almanac 74)

Begin with the author's name, if known.

29. Listserv—MLA

> Korsten, A. "Major Update of the WWWVL Migration and Ethnic Relations."
>
> 7 Apr. 1998. Online posting. ERCOMER News. 8 Apr. 2003.
>
> <www.ercomer.org/archive/ercomer-news/0002.html>.

Parenthetical reference: (Korsten)

Begin with the author's name (if known), followed by the title of the work (in quotation marks), publication date, the Online posting designation, title of discussion group (underlined), date of access, and the URL. The parenthetical reference includes no page number because none is given in an online posting.

30. Usenet—MLA

Dorsey, Michael. "Environmentalism or Racism." 25 Mar. 1998. Online posting.

1 Apr. 2002 <news:alt.org.sierra-club>.

Parenthetical reference: (Dorsey)

31. Email—MLA

Wallin, John Luther. "Frog Reveries." Email to the author. 12 Oct. 2004.

Parenthetical reference: (Wallin)

Cite personal email as you would printed correspondence. If the document has a subject line or title, enclose it in quotation marks.

For publicly posted email (say, a newsgroup or discussion list) include the address and date of access.

32. Home Page for a Course—MLA

Dumont, R. A. An Online Course in Technical Writing. Course Home Page. Fall

2004. Dept. of English, UMASS Dartmouth. 6 Jan. 2005.

<www.umassd.edu/englishdepartment.html.>.

Parenthetical reference: (Dumont)

Begin with the instructor's name and title of the course, followed by "Course Home Page," all without underlines or quotation marks. Then give course dates, the academic department, the school, your date of access, and the URL.

33. Print Article Posted Online—MLA

Jeffers, Anna D. "NAFTA's Effects on the U.S. Trade Deficit." Sultana Business

Quarterly 3.4 (2004): 65–74. April 2005.

<www.sol.org/sbc/2004vol3/jeffers2.html>.

Parenthetical reference: (Jeffers 66)

34. Real-Time Communication—MLA

Synchronous communication occurs in a "real-time" forum and MUDs (multi-user dungeons), MOOs (MUD object-oriented software), FTP (file transfer protocols), chatrooms, and instant messaging.

"Online Debate on Global Warming." 3 Apr. 2004. Frank Findle at EarthWatchMOO.

10 May 2004. <www.ab.liu/orb/globalwarm_3_4-04.htm>.

Parenthetical reference: ("Online Debate")

Begin with the type of communication (virtual conference, personal interview) and topic title followed by the posting date, name of communicator, name of forum, access date, and electronic address.

35. Online Abstract—MLA

Lane, Amanda D., et al. "The Promise of Microcircuits." Journal of

Nanotechnology 12.2 (2004). Abstract. 11 May 2004.

<http://www.jnt.org/abt/0105ab.htm>.

Parenthetical reference: (Lane et. al)

36. General Reference to a Site—MLA

When you are referring to a site in general instead of a specific document, include the address in your discussion and *not* in the list of Works Cited.

For the latest information about worldwide research in electromagnetic

radiation, go to Microwave News at <wwwmicrowavenews.com>.

MLA Sample Works Cited Page

Place your Works Cited section on a separate page at the end of the document. Arrange entries alphabetically by author's surname. When the author is unknown, list the title alphabetically according to its first word (excluding introductory articles). For a title that begins with a digit ("5," "6," etc.), alphabetize the entry as if the digit were spelled out.

The list of works cited in Figure A.4 accompanies the report on electromagnetic fields, pages 617–28. In the left margin, colored numbers refer to the elements discussed on the page preceding Figure A.4. Bracketed labels identify different types of sources cited.

Discussion of Figure A.4

1. Center the Works Cited title at the top of the page. Use one-inch margins. Double-space the entries, and order them alphabetically. Number works cited pages consecutively with text pages.
2. Indent five spaces for the second and subsequent lines of an entry.
3. Place quotation marks around article titles. Underline or italicize periodical or book titles. Capitalize the first letter of key words in all titles (also articles, prepositions, and conjunctions, but only if they come first or last). When an article skips pages in a publication, give only the first page number followed by a plus sign.
4. Do not cite a magazine's volume number, even if it is given.
5. For a CD-ROM database that is updated often (such as *ProQuest*), conclude your citation with the date of electronic publication.
6. For additional perspective beyond "establishment" viewpoints, examine "alternative" publications (such as the *Amicus Journal* and *In These Times*).
7. In citing an online database, include the date you accessed the source.
8. Use a period and one space to separate a citation's three major items (author, title, publication data). Skip one space after a comma or colon. Use no punctuation to separate magazine title and date.
9. Alphabetize hyphenated surnames according to the name that appears first.
10. Use the first author's name and "et al." for works with four or more authors or editors. When citing an abstract instead of the complete article, indicate this by inserting "Abstract" after the page numbers of the original.
11. For a journal with new pagination in each issue include the issue number after the volume number and separated by a period. For example, 26.4 would signify volume 26, issue 4. For page numbers of more than two digits, give only the final digits in the second number.
12. Use three-letter abbreviations for months with five or more letters.
13. For government reports, name the sponsoring agency and include all available information for retrieving the document.
14. When the privacy of the electronic source is not an issue (e.g., a library versus an email correspondent), include the electronic address in your entry.

Works Cited

Beardsley, Tim. "Say That Again?" <u>Scientific American</u> Dec. 1997: 20.

Broad, William J. "Cancer Fear Is Unfounded, Physicists Say." <u>New York Times</u> 14 May 1995:
A19.   *[newspaper article]*

Brodeur, Paul. "Annals of Radiation: The Cancer at Slater School." <u>New Yorker</u> 7 Dec. 1992: 86+.
                                                                                            [magazine article]

Cavanaugh, Herbert A. "EMF Study: Good News and Bad News." <u>Electrical World</u> Feb. 1995: 8.
        <u>ABI/INFORM</u>. CD-ROM. <u>ProQuest</u>. Sept. 2004.
                                                          [trade magazine article from CD-ROM database]

Dana, Amy, and Tom Turner. "Currents of Controversy." <u>Amicus Journal</u> Summer 1993: 29–32.
                                                                                        [alternative press]

de Jager, L., and L. de Bruyn. "Long-Term Effects of a 50 HZ Electric Field on the Life
        Expectancy of Mice." <u>Review of Environmental Health</u> 10.3 (1994): 221–24. <u>MEDLINE</u>.
        <u>DIALOG</u>. 8 May 2004 <www.dialog.com>.   *[database article]*

DesMarteau, Kathleen. "Study Links Sewing Machine Use to Alzheimer's Disease." <u>Bobbin</u>
        Oct. 1994: 36–38. <u>ABI/INFORM</u>. CD-ROM. ProQuest. Aug. 2004.

"Electrophobia: Overcoming Fears of EMFs." <u>University of California Wellness Letter</u> Nov. 1994:1.
                                                                                        [newsletter]

Goodman, E. M., B. Greenebaum, and M. T. Marron. "Effects of Electromagnetic Fields on Molecules
        and Cells." <u>International Review of Cytology</u> 158 (1995): 279–338. <u>MEDLINE</u>. <u>DIALOG</u>. 8 Mar.
        2004 <www.dialog.com>.

Goodman, Neville W. "The Media and The Power Line Scare." 23 Jan. 2004. 12 Mar. 2004
        <www.healthwatch-uk.org/nlett21.html#power>.   *[Web page]*

Gross, Liza. "Current Risks." <u>Sierra</u> May-June 1999: 30+.

Halloran-Barney, Marianne B. Energy Service Advisor for County Electric. Email to the author.
        3 Apr. 2004.   *[email inquiry]*

Henshaw, D. L., et.al. "Does Our Electricity Distribution System Pose a Serious Risk to Public
        Health?" (<u>Medical Hypotheses</u> 59.1 (2002): 39–51. Abstract. 22 April 2004
        <www.electric-fields.bris.ac.uk/MedHypoth.htm>.   *[online abstract]*

Jauchem, J. "Alleged Health Effects of Electromagnetic Fields: Misconceptions in the Scientific
        Literature." <u>Journal of Microwave Power and Electromagnetic Energy</u> 26.4 (1991): 189–95.
                                                                            [journal article from print source]

FIGURE A.4   A List of Works Cited (MLA Style)

Kirkpatrick, David. "Can Power Lines Give You Cancer?" Fortune 31 Dec. 1990: 80–85.

12 Lai, Henry, and Narendra D. Singh. "Magnetic-Field-Induced DNA Strand Breaks in Brain Cells of
the Rat." Environmental Health Perspectives 112.6 (2004): 687–694. Abstract. 22 Apr. 2004
<www.ehp.niehs.nih.gov/docs/2004/6355/abstract.html>.   *[Web page]*

Miltane, John. Chief Engineer for County Electric. Personal interview. 5 Apr. 2004.

Monmonier, Mark. Cartographies of Danger: Mapping Hazards in America. Chicago: U of Chicago P,
1997: 190.   *[book–one author]*

Moore, Taylor. "EMF Health Risks: The Story in Brief." EPRI Journal Mar./Apr. 1995: 7–17.

Moulder, John. "Electromagnetic Fields and Human Health." 21 Jan. 2001. Medical College of
Wisconsin. 10 Mar. 2004 <www.mcw.edu/gcrc/cop/powerlines-cancer-FAQ/>.

13 NIEHS EMF-RAPID Program Staff. Health Effects from Exposure to Power Line Frequency Electric
and Magnetic Fields (NIH Publication No. 99-4493). Research Triangle Park, NC: National
Institute of Environmental Health Sciences. 4 May 1999. 11 Mar. 2004
<www.niehs.nih.gov/emfrapid>.   *[govt. report posted online]*

Palfreman, John. "Apocalypse Not." Technology Review 24 April 1996: 24–33.

Pinsky, Mark A. The EMF Book: What You Should Know about Electromagnetic Fields,
Electromagnetic Radiation, and Your Health. New York: Warner, 1995.

Raloff, Janet. "Electromagnetic Fields May Damage Hearts." Science News 155.3 (1999): 70.

---. "Electromagnetic Fields May Trigger Enzymes." Science News 153.8 (1998): 119.

---. "EMFs' Biological Influences." Science News 153.2 (1998): 29–31.

Sivitz, Laura B. "Cells Proliferate in Magnetic Fields." Science News 158.18 (2000): 196–97.

Stix, Gary. "Are Power Lines a Dead Issue?" Scientific American Mar. 1998: 33–34.

14 "Strong Electric Fields Implicated in Major Leukemia Risk for Workers." Microwave News XX.3
(2000): 1–2. 15 Mar. 2004 <www.microwavenews.com>.   *[journal article posted online]*

Travis, R. C. "Melatonin and Breast Cancer: A Prospective Study" Journal of the National Cancer
Institute 96.6 (2004): 889-89. Abstract. 22 April 2004: <www.ncbi.nim.nih.gov:80/entrez/
query.fcgi?cmd=Retreive&db=pubmed&dopt=Abstract&List_ulds=15026473>.

Taubes, Gary. "Fields of Fear." Atlantic Monthly Nov. 1994: 94–108.

United States Environmental Protection Agency. EMF in Your Environment. Washington: GPO, 1992.

FIGURE A.4   A List of Works Cited (MLA Style) *Continued*

APA Documentation Style

One popular alternative to MLA style appears in the *Publication Manual of the American Psychological Association,* 5th ed., Washington: American Psychological Association, 2001. APA style is useful when writers wish to emphasize the publication dates of their references. A parenthetical reference in the text briefly identifies the source, date, and page number(s):

Reference cited in the text

> In one study, mice continuously exposed to an electromagnetic field tended to die earlier than mice in the control group (de Jager & de Bruyn, 1994, p. 224).

The full citation then appears in the alphabetical listing of "References," at the report's end:

Full citation at document's end

> de Jager, L., & de Bruyn, L. (1994). Long-term effects of a 50 Hz electric field
>
> on the life-expectancy of mice. *Review of Environmental Health,*
>
> *10*(3–4), 221–224.

Because it emphasizes the date, APA style (or some similar author-date style) is preferred in the sciences and social sciences, where information quickly becomes outdated.

APA Parenthetical References

How APA and MLA parenthetical references differ

APA's parenthetical references differ from MLA's (pages 686–87) as follows: The APA citation includes the publication date; a comma separates each item in the reference; and "p." or "pp." precedes the page number (which is optional in the APA system). When a subsequent reference to a work follows closely after the initial reference, the date need not be included. Here are specific guidelines:

- If your discussion names the author, do not repeat the name in your parenthetical reference; simply give the date and page numbers:

Author named in the text

> Researchers de Jager and de Bruyn explain that experimental mice exposed to
>
> an electromagnetic field tended to die earlier than mice in the control group
>
> (1994, p. 224).

When two authors of a work are named in the text, their names are connected by "and," but in a parenthetical reference, their names are connected by an ampersand, "&."

- If you cite two or more works in a single reference, list the authors in alphabetical order and separate the citations with semicolons:

Two or more works in a single reference	(Jones, 2004; Gomez, 2002; Leduc, 1999)

- If you cite a work with three to five authors, try to name them in your text, to avoid an excessively long parenthetical reference.

A work with three to five authors	Franks, Oblesky, Ryan, Jablar, and Perkins (2003) studied the role of electromagnetic fields in tumor formation.

In any subsequent references to this work, name only the first author, followed by "et al." (Latin abbreviation for "and others").

- If you cite two or more works by the same author published in the same year, assign a different letter to each work:

Two or more works by the same author in the same year	(Lamont, 2004a, p. 135) (Lamont, 2004b, pp. 67–68)

Other examples of parenthetical references appear with their corresponding entries in the following discussion of the reference list entries.

APA Reference List Entries

How to space and indent entries

The APA reference list includes each source you have cited in your document. In preparing the list of references, type the first line of each entry flush with the left margin. Indent the second and subsequent lines five character spaces (one-half inch). Skip one character space after any period, comma, or colon. Double-space within and between each entry.

Following are examples of complete citations as they would appear in the References section of your document. Shown immediately below each entry is its corresponding parenthetical reference as it would appear in the text. Note the capitalization, abbreviation, spacing, and punctuation in the sample entries.

What to include in an APA citation for a book

APA ENTRIES FOR BOOKS.   Any citation for a book should contain all applicable information in the following order: author, date, title, editor or translator, edition, volume number, and facts about publication (city and publisher).

1. Book, Single Author—APA

Kerzin-Fontana, J. B. (2005). *Technology management: A handbook* (3rd ed.).

Delmar, NY: American Management Association.

Parenthetical reference: (Kerzin-Fontana, 2005, pp. 3–4)

Use only initials for an author's first and middle name. Capitalize only the first word of a book's title and subtitle and any proper names. Identify a later edition in parentheses between the title and the period.

INDEX TO SAMPLE ENTRIES FOR APA REFERENCES

Books

1. Book, single author
2. Book, two to five authors
3. Book, six or more authors
4. Book, anonymous author
5. Multiple books, same author
6. Book, one to five editors
7. Book, indirect source
8. Anthology selection or book chapter

Periodicals

9. Article, magazine
10. Article, journal with new pagination each issue
11. Article, journal with continuous pagination
12. Article, newspaper

Other Sources

13. Encyclopedia, dictionary, alphabetical reference

14. Report
15. Conference presentation
16. Interview, personally conducted
17. Interview, published
18. Personal correspondence
19. Brochure or pamphlet
20. Lecture
21. Government document
22. Miscellaneous items

Electronic Sources

23. Online abstract
24. Print article posted online
25. Computer software or software manual
26. CD-ROM abstract
27. CD-ROM reference work
28. Personal email
29. Document from a university
30. Newsgroup, discussion list, online forum

---

2. Book, Two to Five Authors—APA

Aronson, L., Katz, R., & Moustafa, C. (2004). *Toxic waste disposal methods.*
New Haven: Yale University Press.

Parenthetical reference: (Aronson, Katz, & Moustafa, 2004)

Use an ampersand (&) before the name of the final author listed in an entry. As an alternative parenthetical reference, name the authors in your text and include date (and page numbers, if appropriate) in parentheses.

Give the publisher's full name (as in "Yale University Press") but omit the words "Publisher," "Company," and "Inc."

3. Book, Six or More Authors—APA

Fogle, S. T., et al. (2004). *Hyperspace technology.* Boston: Little, Brown.

Parenthetical reference: (Fogle et al., 2004, p. 34)

"Et al." is the Latin abbreviation for "et alia," meaning "and others."

4. Book, Anonymous Author—APA

Structured programming. (2005). Boston: Meredith Press.

Parenthetical reference: (*Structured programming*, 2005, p. 67)

In your list of references, place an anonymous work alphabetically by the first key word (not *The, A,* or *An*) in its title. In your parenthetical reference, capitalize all key words in a book, article, or journal title.

5. Multiple Books, Same Author—APA

Chang, J. W. (2002a). *Biophysics.* Boston: Little, Brown.

Chang, J. W. (2002b). *MindQuest.* Chicago: John Pressler.

Parenthetical references: (Chang, 2000a)

                    (Chang, 2000b)

Two or more works by the same author not published in the same year are distinguished by their respective dates alone, without the added letter.

6. Book, One to Five Editors—APA

Morris, A. J., & Pardin-Walker, L. B. (Eds.). (2003). *Handbook of new*

    information technology. New York: HarperCollins.

Parenthetical reference: (Morris & Pardin-Walker, 2003, p. 79)

For more than five editors, name only the first, followed by "et al."

7. Book, Indirect Source—APA

Stubbs, J. (2004). *White-collar productivity.* Miami: Harris.

Parenthetical reference: (cited in Stubbs, 2004, p. 47)

When your source (as in Stubbs, above) has cited another source, list only this second source in the References section, but name the original source in the text: "Kline's study (cited in Stubbs, 2004, p. 47) supports this conclusion."

8. Anthology Selection or Book Chapter—APA

Bowman, J. (1994). Electronic conferencing. In A. Williams (Ed.),

    Communication and technology: Today and tomorrow (pp. 123–142).

    Denton, TX: Association for Business Communication.

Parenthetical reference: (Bowman, 1994, p. 126)

The page numbers in the complete reference are for the selection cited from the anthology.

What to include
in an APA citation
for a periodical

APA ENTRIES FOR PERIODICALS. A citation for an article should give this information (as available), in order: author, publication date, article title (without quotation marks), volume or number (or both), and page numbers for the entire article—not just the page(s) cited.

9. Article, Magazine—APA

DesMarteau, K. (1994, October). Study links sewing machine use to

Alzheimer's disease. *Bobbin, 36,* 36–38.

Parenthetical reference: (DesMarteau, 1994, p. 36)

If no author is given, provide all other information. Capitalize the first word in an article's title and subtitle, and any proper nouns. Capitalize all key words in a periodical title. Italicize the periodical title, volume number, and commas (as shown above).

10. Article, Journal with New Pagination for Each Issue—APA

Thackman-White, J. R. (2005). Computer-assisted research. *American Library*

Journal, 51 (1), 3–9.

Parenthetical reference: (Thackman-White, 2005, pp. 4–5)

Because each issue for a given year has page numbers that begin at "1," readers need the issue number (in this instance, "1"). The "51" denotes the volume number, which is italicized.

11. Article, Journal with Continuous Pagination—APA

Barnstead, M. H. (2004). The writing crisis. *Journal of Writing Theory 12,*

415–433.

Parenthetical reference: (Barnstead, 2004, pp. 415–416)

The "12" denotes the volume number. When page numbers continue from issue to issue for the full year, readers won't need the issue number, because no other issue in that year repeats these same page numbers. (You can still include the issue number if you think it will help readers retrieve the article more easily.)

12. Article, Newspaper—APA

Baranski, V. H. (2005, January 15). Errors in technology assessment. *The*
Boston Times, p. B3.

Parenthetical reference: (Baranski, 2005, p. B3)

In addition to the year of publication, include the month and day. If the newspaper's name begins with "The," include it in your citation. Include "p." or "pp." before page numbers. For an article on nonconsecutive pages, list each page, separated by a comma.

What to include
in an APA citation
for a
miscellaneous
source

APA Entries for Other Sources. Miscellaneous sources range from unsigned encyclopedia entries to conference presentations to government documents. A full citation should give this information (as available): author, publication date, work title (and report or series number), page numbers (if applicable), city, and publisher.

13. Encyclopedia, Dictionary, Alphabetical Reference—APA

Communication. (2004). In *The business reference book.* Boston: Business
Resources Press.

Parenthetical reference: ("Communication," 2004)

For an entry that is signed, begin with the author's name and publication date.

14. Report—APA

Electrical Power Research Institute. (1994). *Epidemiologic studies of electric*
utility employees (Report No. RP2964.5). Palo Alto, CA: Author.

Parenthetical reference: (Electrical Power Research Institute [EPRI], 1994, p. 12)

If authors are named, list them first, followed by the publication date. When citing a group author, as above, include the group's abbreviated name in your first parenthetical reference, and use only that abbreviation in any subsequent reference. When the agency (or organization) and publisher are the same, list "Author" in the publisher's slot.

15. Conference Presentation—APA

Smith, A. A. (2003, March). Radon concentrations in molded concrete. In A.
Hodkins (Ed.), *First British Symposium on Environmental Engineering*
(pp. 106–121). London: Harrison Press, 2004.

Parenthetical reference: (Smith, 2004, p. 109)

In parentheses is the date of the presentation. The name of the symposium is a proper name, and so is capitalized. Following the publisher's name is the date of publication.

For an unpublished presentation, include the presenter's name, year and month, title of the presentation (italicized), and all available information about the conference or meeting: "Symposium held at. . . ." Do not italicize this last information.

16. Interview, Personally Conducted—APA

Parenthetical reference: (G. Nasser, personal interview, April 2, 2004)

This material is considered a nonrecoverable source, and so is cited in the text only, as a parenthetical reference. If you name the respondent in text, do not repeat the name in the citation.

17. Interview, Published—APA

Jable, C. K. (2004). The future of graphics [Interview with James Lescault]. In

K. Prell (Ed.), *Executive views of automation* (pp. 216–231). Miami:

Haber Press, 2005.

Parenthetical reference: (Jable, 2005, pp. 218–223)

Begin with the name of the interviewer, followed by the interview date and title (if available), the designation (in brackets), and the publication information, including the date.

18. Personal Correspondence—APA

Parenthetical reference: (L. Rogers, personal correspondence, May 15, 2004)

This material is considered nonrecoverable data, and so is cited in the text only, as a parenthetical reference. If you name the correspondent in text, do not repeat the name in the citation.

19. Brochure or Pamphlet—APA

This material follows the citation format for a book entry (page 703). After the title of the work, include the designation "Brochure" in brackets.

20. Lecture—APA

Dumont, R. A. (2005, January 15). *Managing natural gas.* Lecture presented at

the University of Massachusetts at Dartmouth.

Parenthetical reference: (Dumont, 2005)

If you name the lecturer in text, do not repeat the name in the citation.

21. Government Document—APA

> Virginia Highway Department. (2004). *Standards for bridge maintenance.*
>
> Richmond: Author.

Parenthetical reference: (Virginia Highway Department, 2004, p. 49)

If the author is unknown, present the information in this order: name of the issuing agency, publication date, document title, place, and publisher. When the issuing agency is both author and publisher, list "Author" in the publisher's slot.

For any congressional document, identify the house of Congress (Senate or House of Representatives) before the date.

> U.S. House Armed Services Committee. (2004). *Funding for the military*
>
> *academies.* Washington, DC: U.S. Government Printing Office.

Parenthetical reference: (U.S. House, 2004, p. 41)

22. Miscellaneous items (unpublished manuscripts, dissertations, and so on)—APA

> Author (if known). (Date of publication.) *Title of work.* Sponsoring organization
>
> or publisher.

For any work that has group authorship (corporation, committee, and so on), cite the name of the group or agency in place of the author's name.

APA ENTRIES FOR ELECTRONIC SOURCES.   When you cite sources in the References section of your document, identify the original source (printed or electronic) and give readers a path for retrieving the material. Provide all available information in the following order:

What to include in an APA citation for an electronic source

1. Author or editor or creator of the electronic work.
2. Date the work was published or was created electronically. For magazines and newspapers, include the month and day as well as the year. If the date of an electronic publication is not available, use "n.d."
3. Publication information of the original printed version (as in the above entries), if such a version exists. Follow this by designating the electronic medium (as in "[CD-ROM]") or the type of work (as in "[Abstract]" or "[Editorial]" or the like).
4. The word "Retrieved" followed by the date (month, day, year) you accessed the source.

5. Information about the electronic publication, including the title of the site or reference database (as in "MEDLINE") and the date of the posting or the last update of the site. Name the sponsoring organization or provider of the CD-ROM (as in "ProQuest") or database service (as in "Dialog").

6. The full and accurate electronic address. For Internet sources, provide the complete URL (Uniform Resource Locator). For CD-ROM and database sources, give the document's retrieval number. If the electronic version is identical to the original printed version, omit the URL and give only the publication information of the print version followed by "[Electronic version]."

23. Online Abstract—APA

Stevens, R. L. (2004). Cell phones and cancer rates [Abstract]. *Oncology*

Journal, 57(2), 41–43. Retrieved April 10, 2005, from Dialog database.

(MEDLINE item: AY 24598).

Parenthetical reference: (Stevens, 2004)

The above entry ends with a period. Only entries that close with a URL (as in entry no. 24, below) have no period at the end of the URL.

NOTE *If instead you are citing the entire article, retrieved from a full-text database, merely delete the "[Abstract]" from your citation.*

24. Print Article Posted Online—APA

Alley, R. A. (2003, January). Ergonomic influences on worker satisfaction.

Industrial Psychology 5(12). Retrieved April 8, 2004 from

www.psycharchives/index/indpsy/2003_1.html

Parenthetical reference: (Alley, 2003)

If the page numbers of the printed original were posted on the online source and if you were confident that the document's electronic version and print version were identical, you could omit the URL and insert "[Electronic version]" between the end of the article title and the period.

25. Computer Software or Software Manual—APA

Virtual collaboration [Computer software]. (2005). New York: Pearson.

Parenthetical reference: (Virtual, 2005)

For citing a manual, replace the "Computer software" designation in brackets with "Software manual."

26. CD-ROM Abstract—APA

Cavanaugh, H. (1995). An EMF study: Good news and bad news [CD-ROM].

Electrical World, 209(2), 8. Abstract retrieved April 7, 2002, from

ProQuest File: ABI/INFORM database (62-1498).

Parenthetical reference: (Cavanaugh, 1995)

The "8" in the entry above denotes the page number of this one-page article.

27. CD-ROM Reference Work—APA

Ecoterrorism [CD-ROM] (2004). *Ecological encyclopedia.* Washington: Redwood.

Parenthetical reference: (Ecoterrorism, 2004)

If the work on CD-ROM has a printed equivalent, APA currently prefers that it be cited in its printed form.

28. Personal Email—APA

Parenthetical reference: Fred Flynn (personal communication, May 10, 2005) provided these statistics.

Instead of being included in the list of references, personal email is cited directly in the text.

29. Document from a University—APA

Owens, P. (2003). *Internship guidelines.* Retrieved June 12, 2004, from

Cabrone College, Department of Communication Web site:

www.clayton.edu/comm/p-o.html

Parenthetical reference: (Owens, 2004)

30. Newsgroup, discussion list, online forum—APA

LaBarge, V. S. (2004, October 20). A cure for computer viruses. *Firewall*

DiscussionList. Retrieved December 15, 2004, from

www.srb/forums/frwl/webZ/m2237.html

Parenthetical reference: (LaBarge, 2004).

Although email is not included in the list of references, listserv, newsgroup, and forum postings are considered more retrievable.

APA Sample Reference List

APA's References section is an alphabetical listing (by author) equivalent to MLA's Works Cited section. Like Works Cited, the reference list includes only those works actually cited. (A bibliography usually would include background works or works consulted as well.) Unlike MLA style, APA style calls for only "recoverable" sources to appear in the reference list. Therefore, personal interviews, email messages, and other unpublished materials are cited in the text only.

The list of references in Figure A.5 accompanies the report on a technical marketing career, pages 631–38. In the left margin, colored numbers denote elements of Figure A.5 discussed below. Bracketed labels on the right identify different types of sources.

Discussion of Figure A.5

1. Center the References title at the top of the page. Use one-inch margins. Number reference pages consecutively with text pages. Include only recoverable data (material that readers could retrieve for themselves); cite personal interviews, email, and other personal correspondence parenthetically in the text only. See also item 8 in this list.

2. Double-space entries and order them alphabetically by author's last name (excluding *A*, *An*, or *The*). List initials only for authors' first and middle names. Write out names of all months. In student papers, indent the second and subsequent lines of an entry five spaces. In papers submitted for publication in an APA journal, the *first* line is indented instead.

3. Use the first key word in the title to alphabetize works whose author is not named.

4. Do not enclose article titles in quotation marks. Italicize periodical titles. Capitalize the first word in article or book titles and subtitles, and any proper nouns. Capitalize all key words in magazine or journal titles.

5. For more than one author or editor, use ampersands instead of spelling out "and."

6. Use italics for a journal's name, volume number, and the comma. Give the issue number in parentheses only if each issue begins on page 1. Do not include "p." or "pp." before journal page numbers (only before page numbers from a newspaper).

7. Omit punctuation from the end of an electronic address.

8. Treat an unpublished conference presentation as a recoverable source; include it in your list of references instead of only citing it parenthetically in your text.

1

References

2 Anson, D. (2004, March 12). *Engineering graduates and the job market.* Lecture presented at the University of Massachusetts at Dartmouth.   *[lecture]*

3 The answer to outsourcing. (2004, March 1). *BusinessWeek,* 136.

4 Average annual job openings, 2002–2012. (2004, Spring). *Occupation Report.* Retrieved April 24, 2004, from http://www.bls.gov/EMP   *[online report]*

Bureau of Labor Statistics, U.S. Department of Labor. (2004) Sales engineers. *Occupational Outlook Handbook,* 2004-2005 Edition. Retrieved April 17, 2004, from www.bls.gov/oco/ocos123.htm   *[reference work posted online]*

5 Cornelius, H., & Lewis, W. (1983). *Career guide for sales and marketing* (2nd ed.) New York: Monarch Press   *[book with two authors]*

Coy, P. (2004, March 22). The future of work. *Business Week,* 50–52.

Lok, C (2004, April). Where's my job? [Interview with Deborah Wince-Smith, President, Council on Competitiveness]. In *Technology review, 107*(3), 74–75.   *[published interview]*

Manufacturers' and wholesale sales representatives. (2000). In *Occupational outlook handbook 2000–2001 edition.* Washington DC: U.S. Department of Labor. Retrieved March 12, 2002, from http://www.bls.gov/oco/ocos119.htm   *[reference book posted online]*

6 Nelson, A. J. (2001, Fall) Sales engineers. *Occupational Outlook Quarterly, 44*(3), 20–24. Retrieved March 15, 2004, from http://www.bls.gov/opub/ooqhome.htm

[govt. periodical posted online—author named]

Shelley, K. J. (1997, Fall). A portrait of the M.B.A. *Occupational Outlook Quarterly, 41*(3), 26–33.

[govt. periodical—author named]

7 Technology Marketing Group, Inc. (1998). *Services for clients.* Retrieved March 18, 2004, from http://www.technology-marketing.com

8 Tolland, M. (2004, April). *Alternate careers in marketing.* Presentation at Electro '04 Conference in Boston.   *[unpublished conference presentation]*

The 2002–12 job outlook. (2004, Spring) *Occupational Outlook Quarterly. 45*(1), 9–43.

[govt. publication—no author named]

U.S. Department of Labor. (1997). *Tomorrow's jobs.* Washington, DC: Author.

FIGURE A.5   A List of References (APA Style)

CBE AND OTHER NUMERICAL DOCUMENTATION STYLES

In the numerical system, each work is assigned a number in order according to the first time it is cited. This same number is then used for any subsequent reference to that work. Numerical documentation is often used in the physical sciences (astronomy, chemistry, geology, physics) and in the applied sciences (mathematics, medicine, engineering, and computer science).

Particular disciplines have their own preferred documentation styles, described in manuals such as these:

A sampling of discipline-specific documentation manuals

- American Chemical Society, *The ACS Style Guide for Authors and Editors*
- American Institute for Physics, *AIP Style Guide*
- American Mathematical Society, *A Manual for Authors of Mathematical Papers*
- American Medical Association, *Manual of Style*

One widely consulted guide for numerical documentation is *Scientific Style and Format: The CBE Manual for Authors, Editors, and Publishers, 6th ed., 1994,* from the Council of Biology Editors. (In addition to its citation-sequence system for documentation, CBE offers a name-year system that basically duplicates the APA system described on pages 702–13.)

NOTE
In January 2000, the Council of Biology Editors changed its name to the Council of Science Editors because of its broadening membership. Until the Manual *is revised, CBE style will continue to be the preferred system for documenting sources in the sciences.*

CBE Numbered Citations

In one version of CBE style, a citation in the text appears as a superscript number immediately following the source to which it refers:

Numbered citations in the text

A recent study[1] indicates an elevated leukemia risk among children exposed to certain types of electromagnetic fields. Related studies[2-3] tend to confirm the EMF/cancer hypothesis.

When referring to two or more sources in a single note (as in "[2-3]" above) separate the numbers by a hyphen if they are in sequence and by commas but no space if they are out of sequence: ("[2,6,9]").

The full citation for each source then appears in the numerical listing of references at the end of the document.

REFERENCES

Full citations at
document's end

1. Baron, KL, et al. The electromagnetic spectrum. New York: Pearson; 2005. 476 p.

2. Klingman, JM. Nematode Infestation in Boreal Environments. J Entymol 2003;54:475-8.

CBE Reference List Entries

CBE's References section lists each source in the numerical order in which it was first cited. In preparing the list, which should be double-spaced, begin each entry on a new line. Type the number flush with the left margin, followed by a period and a space. Align subsequent lines directly under the first word of line one.

Following are examples of complete citations as they would appear in the References section for your document.

INDEX TO SAMPLE CBE ENTRIES

1. Book, single author
2. Book, multiple authors
3. Book, anonymous author
4. Book, one or more editors
5. Anthology selection or book chapter
6. Article, magazine
7. Article, journal with new pagination each issue
8. Article, journal with continuous pagination
9. Article, newspaper
10. Article, online source

CBE ENTRIES FOR BOOKS.   Any citation for a book should contain all available information in the following order: number assigned to the entry, author or editor, work title (and edition), facts about publication (place, publisher, date), and number of pages. Note the capitalization, abbreviation, spacing, and punctuation in the sample entries.

1. Book, Single Author—CBE

1. Kerzin-Fontana JB. Technology management: a handbook. 3rd ed. Delmar, NY: American Management Assn; 2005. 356p.

2. Book, Multiple Authors—CBE

2. Aronson L, Katz R, Moustafa C. Toxic waste disposal methods. New Haven: Yale Univ Pr; 2004. 316p.

3. Book, Anonymous Author—CBE

3. [Anonymous]. Structured programming. Boston: Meredith Pr; 2005. 267p.

4. Book, One or More Editors—CBE

4. Morris AJ, Pardin-Walker LB, editors. Handbook of new information technology. New York: Harper; 2003. 345p.

5. Anthology Selection or Book Chapter—CBE

5. Bowman JP. Electronic conferencing. In: Williams A, editor: Communication and technology: today and tomorrow. Denton, TX: Assn for Business Communication; 1994. p 123–42.

CBE ENTRIES FOR PERIODICALS. Any citation for an article should contain all available information in the following order: number assigned to the entry, author, article title, periodical title, date (year, month), volume and issue number, and inclusive page numbers for the article. Note the capitalization, abbreviation, spacing, and punctuation in the sample entries.

6. Article, Magazine—CBE

6. DesMarteau K. Study links sewing machine use to Alzheimer's disease. Bobbin 1994 Oct:36–8.

7. Article, Journal with New Pagination Each Issue—CBE

7. Thackman-White JR. Computer-assisted research. Am Library J 2005;51(1):3–9.

8. Article, Journal with Continuous Pagination—CBE

8. Barnstead MH. The writing crisis. J of Writing Theory 2004;12:415–33.

9. Article, Newspaper—CBE

9. Baranski VH. Errors in technology assessment. Boston Times 2005 Jan 15;Sect B: 33(col 2).

10. Article, Online Source—CBE

> 10. Alley RA. Ergonomic influences on worker satisfaction. Industrial Psychology [serial online] 2003 Jan;5(11). Available from: ftp.pub/journals/industrialpsychology/2003 via the INTERNET. Accessed 2004 Feb 10.

Citation for an article published online follows a similar format, with these differences: write "[article online]" between article title and publication date; after "Available from," give the URL followed by a period and your access information.

For more guidelines and examples, consult the *CBE Manual* or go to these sites: <www.wisc.edu/writing/Handbook/DocCBE.html> and <www.lib.ohio-state.edu/guides/cbegd.html>.

A Casebook of Sample Documents Illustrating the Writing Process

Every writing situation requires deliberate decisions for *working with the information* and for *planning, drafting,* and *revising* the document. Some of these decisions are illustrated in Figure B.1. Each writer approaches the process through a sequence of decisions that works best for *that* person. No stage of decisions is complete until *all* stages are complete.

CRITICAL THINKING IN THE WRITING PROCESS

The writing process is a *critical thinking* process: the writer makes a series of deliberate decisions in response to a situation. The actual "writing" (putting words on the page) is only a small part of the overall process—probably the least significant part.

In this section, we will follow one working writer through an everyday writing situation; we will see how he solves his unique information, persuasion, and ethics problems and how he collaborates to design a useful and efficient document.

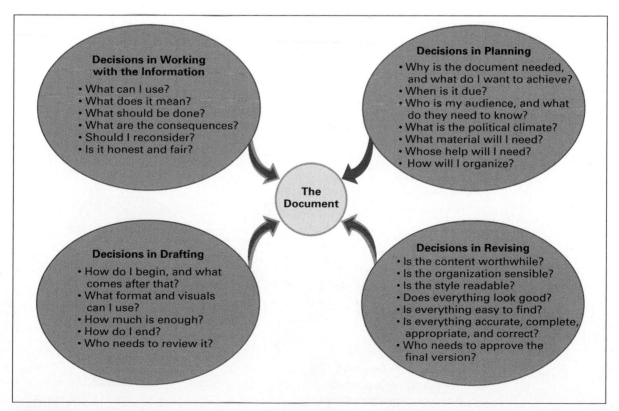

FIGURE B.1   Typical Decisions during the Writing Process

CASE 1—AN EVERYDAY WRITING SITUATION: THE EVOLUTION OF A SHORT REPORT

The company is Microbyte, maker of portable microcomputers. The writer is Glenn Tarullo (BS, Management; Minor: Computer Science). Glenn has been on the job three months as Assistant Training Manager for Microbyte's Marketing and Customer Service Division.

For three years, Glenn's boss, Marvin Long, has periodically offered a training program for new managers. Long's program combines an introduction to the company with instruction in management skills (time management, motivation, communication). Long seems satisfied with his two-week program but has asked Glenn to evaluate it and write a report as part of a company move to upgrade training procedures.

Glenn knows his report will be read by Long's boss, George Hopkins (Assistant Vice President, Personnel), and Charlotte Black (Vice President, Marketing, the person who devised the upgrading plan). Copies will go to other division heads, to the division's chief executive, and to Long's personnel file.

Glenn spends two weeks (Monday, October 3, to Friday, October 14) sitting in and taking notes on Long's classes. On October 14, the trainees evaluate the program. After reading these evaluations and reviewing his notes, Glenn concludes that the program was successful but could stand improvement. How can he be candid without harming or offending anyone (instructors, his boss, or guest speakers)? Figure B.2 depicts Glenn's problem.

Glenn is scheduled to present his report in conference with Long, Black, and Hopkins on Wednesday, October 19. Right after the final class (1 P.M., Friday, the 14th), Glenn begins work on his report.

FIGURE B.2
Glenn's
Fourfold
Problem

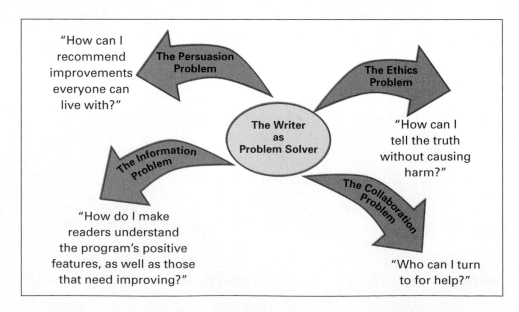

Working with the Information

Glenn spends half of Friday afternoon fretting over the details of his situation, the readers and other people involved, the political realities, constraints, and consequences. (He knows no love is lost between Long and Black, and he wants to steer clear of their ongoing conflict.) By 3 P.M., Glenn hasn't written a word. Desperate, he decides to write whatever comes to mind:

Glenn's first draft

> Although the October Management Training Session was deemed quite successful, several problems have emerged that require our immediate attention.
>
> - Too many of the instructors had poor presentation skills. A few never arrived on time. One didn't stick to the topic but rambled incessantly. Jones and Wells seemed poorly prepared. Instructors in general seemed to lack any clear objectives. Also, because too few visual aids were used, many presentations seemed colorless and apparently bored the trainees.
> - The trainees (all new people) were not at all cognizant of how the company was organized or functioned, so the majority of them often couldn't relate to what the speakers were talking about.
> - It is my impression that this was a weak session due to the fact that there were insufficient members (only five trainees). Such a small class makes the session a waste of time and money. For instance, Lester Beck, Senior Vice President of Personnel, came down to spend over one hour addressing only a handful of trainees. Another factor is that with fewer trainees in a class, less dialogue occurs, with people tending to just sit and get talked at.
> - Last but not least, executive speakers generally skirted the real issues, saying nothing about what it was really like to work here. They never really explained how to survive politically (e.g., never criticize your superior; never complain about the hard work or long hours; never tell anyone what you *really* think; never observe how few women are in executive or managerial positions, or how disorganized things seem to be). New employees shouldn't have to learn these things the hard way.
>
> In the final analysis, if these problems can be addressed immediately, it is my opinion we can look forward in the future to effectuating management training sessions of even higher quality than those we now have.

Glenn completes this draft at 5:10 P.M. Displeased with the results but not sure how to improve the piece, he asks an experienced colleague for advice and feedback. Blair Cordasco, a senior project manager, has collaborated with Glenn on several earlier projects. Cordasco agrees to study Glenn's draft over the weekend. Because of this document's sensitive nature, they decide to work on it face-to-face instead of transacting via email.

At 8:05 Monday morning, Cordasco reviews the document with Glenn. First, she points out obvious style problems: wordiness ("due to the fact that"), jargon ("effectuating"), triteness ("in the final analysis"), implied bias ("weak presentation," "skirted"), among others. Can you identify other style problems in Glenn's draft?

Cordasco points out other problems. The piece is disorganized, and even though Glenn is being honest, he isn't being particularly fair. The emphasis is too critical (making Glenn's boss look bad to his superiors), and the views are too subjective (no one is interested in hearing Glenn gripe about the company's political problems). Moreover, the report lacks persuasive force because it contains little useful advice for solving the problems he identifies. The tone is bossy and judgmental. Glenn is in no position to make this kind of *power connection* (see page 48). In this form, the report will only alienate people and harm Glenn's career. He needs to be more fair, diplomatic, and reasonable.

Planning the Document

Glenn realizes he needs to begin by focusing on his writing situation. His audience and use analysis goes like this[1]:

> I'd better decide *exactly* what my primary reader wants.
>
> Long requested the report, but only because Black developed the scheme for division-wide improvements. So I really have two primary readers: my boss and the big boss.
>
> My major question here: Am I including enough detail for all the bosses? The answer to this question will require answers to more specific questions:

> | ANTICIPATED READERS' QUESTIONS | What are we doing right, and how can we do it better? |
> | --- | --- |
> | | What are we doing wrong, and does it cost us money? |
> | | Have we left anything out, and does it matter? |
> | | How, specifically, can we improve the program, and how will those improvements help the company? |

> Because all readers have participated in these sessions (as trainees, instructors, or guest speakers), they don't need background explanations.
>
> I should begin with the *positive* features of the last session. Then I can discuss the problems and make recommendations. Maybe I can eliminate the bossy and judgmental tone by *suggesting improvements* instead of *criticizing weaknesses*. Also, I could be more persuasive by describing the *benefits* of my suggestions.

Glenn realizes that if he wants successful future programs, he can't afford to alienate anyone. After all, he wants to be seen as a loyal member of the company, yet preserve his self-esteem and demonstrate he is capable of making objective recommendations.

Now, I have a clear enough sense of what to do.

------

[1]Throughout this section, Glenn's analysis will address *all* the areas illustrated in the audience and use profile sheet (page 65).

STATEMENT OF PURPOSE	The purpose of my document is to provide my supervisor and interested executives with an evaluation of the workshop by describing its strengths, suggesting improvements, and explaining the benefits of these changes.

From this plan, I should be able to revise my first draft, but that first draft lacks important details. I should brainstorm to get *all* the details (including the *positive* ones) I want to include.

GLENN'S BRAINSTORMING LIST.   Glenn's first draft touched on several topics. Incorporating them into his brainstorming, he comes up with the following list.

1. better-prepared instructors and more visuals
2. on-the-job orientation *before* the training session
3. more members in training sessions
4. executive speakers should spell out qualities needed for success
5. beneficial emphasis on interpersonal communication
6. need follow-up evaluation (in six months?)
7. four types of training evaluations:
   a. trainees' reactions
   b. testing of classroom learning
   c. transference of skills to the job
   d. effect of training on the organization (high sales, more promotions, better-written reports)
8. videotaping and critiquing of trainee speeches worked well
9. acknowledge the positive features of the session
10. ongoing improvement ensures quality training
11. division of class topics into two areas was a good idea
12. additional trainees would increase classroom dialogue
13. the more trainees in a session, the less time and money wasted
14. instructors shouldn't drift from the topic
15. on-the-job training to give a broad view of the division
16. clear course objectives to increase audience interest and to measure the program's success
17. Marvin Long has done a great job with these sessions over the years

By 9:05 A.M., the office is hectic. Glenn puts his list aside to spend the day on work that has been piling up. Not until 4 P.M. does he return to his report.

Now what? I should delete whatever my audience already knows or doesn't need, or whatever seems unfair or insincere: 7 can go (this audience needs no lecture in training theory); 14 is too negative and critical—besides, the same idea is stated more positively in 4; 17 is obvious brown-nosing, and I'm in no position to make such grand judgments.

Maybe I can unscramble this list by arranging items within categories (strengths, suggested changes, and benefits) from my statement of purpose.

GLENN'S BRAINSTORMING LIST REARRANGED.   Notice here how Glenn discovers additional *content* (see italic type) while he's deciding about *organization*.

> **Strengths of the Workshop**
> - division of class topics into two areas was useful
> - emphasis on interpersonal communication
> - videotaping of trainees' oral reports, followed by critiques

Well, that's one category done. Maybe I should combine *suggested changes* with *benefits,* since I'll want to cover them together in the report.

> **Suggested Changes/Benefits**
> - more members per session would increase dialogue and use resources more efficiently
> - varied on-the-job experiences before the training sessions would give each member a broad view of the marketing division
> - executive speakers should spell out qualities required for success and *future sessions should cover professional behavior, to provide trainees with a clear guide*
> - follow-up evaluation in six months *by both supervisors and trainees would reveal the effectiveness of this training and suggest future improvements*
> - clear course objectives and more visual aids would increase *instructor efficiency* and audience interest

Now that he has a fairly sensible arrangement, Glenn can get this list into report form, even though he will probably think of more material to add as he works. Since this is *internal* correspondence, he uses a memo format.

Drafting the Document

Glenn produces a usable draft—one containing just about everything he wants to cover. (Sentences are numbered for later reference.)

A later draft

> [1]In my opinion, the Management Training Session for the month of October was somewhat successful. [2]This success was evidenced when most participants rated their training as "very good." [3]But improvements are still needed.
>
> [4]First and foremost, a number of innovative aspects in this October session proved especially useful. [5]Class topics were divided into two distinct areas. [6]These topics created a general-to-specific focus. [7]An emphasis on interpersonal communication skills was the most dramatic innovation. [8]This helped class members develop a better attitude toward things in general. [9]Videotaping of trainees' oral reports, followed by critiques, helped clarify strengths and weaknesses.
>
> [10]A detailed summary of the trainees' evaluations is attached. [11]Based on these and on my past observations, I have several suggestions.

- [12]All management training sessions should have a minimum of ten to fifteen members. [13]This would better utilize the larger number of managers involved and the time expended in the implementation of the training. [14]The quality of class interaction with the speakers would also be improved with a larger group.
- [15]There should be several brief on-the-job training experiences in different sales and service areas. [16]These should be developed prior to the training session. [17]This would provide each member with a broad view of the duties and responsibilities in all areas of the marketing division.
- [18]Executive speakers should take a few minutes to spell out the personal and professional qualities essential for success with our company. [19]This would provide trainees with a concrete guide to both general company and individual supervisors' expectations. [20]Additionally, by the next training session we should develop a presentation dealing with appropriate attitudes, manners, and behavior in the business environment.
- [21]Do a six-month follow-up. [22]Get feedback from supervisors as well as trainees. [23]Ask for any new recommendations. [24]This would provide a clear assessment of the long-range impact of this training on an individual's job performance.
- [25]We need to demand clearer course objectives. [26]Instructors should be required to use more visual aids and improve their course structure based on these objectives. [27]This would increase instructor quality and audience interest.
- [28]These changes are bound to help. [29]Please contact me if you have further questions.

Although now developed and organized, this version still is some way from the finished document. Glenn has to make further decisions about his style, content, arrangement, audience, and purpose.

Blair Cordasco offers to review the piece once again and to work with Glenn on a thorough edit.

Revising the Document

At 8:15 Tuesday morning, Cordasco and Glenn begin a sentence-by-sentence revision for worthwhile content, sensible organization, and readable style. Their discussion goes something like this:

Sentence 1 begins with a needless qualifier, has a redundant phrase, and sounds insulting ("somewhat successful"). Sentence 2 should be in the passive voice, to emphasize the training—not the participants. Also, 1 and 2 are choppy and repetitious, and should be combined.

ORIGINAL   In my opinion, the Management Training Session for the month of October was somewhat successful. This success was evidenced when most participants rated their training as "very good." (28 words)

> **REVISED**   The October Management Training Session was successful, with training rated "very good" by most participants. (15 words)[2]

Sentence 3 is too blunt. An orienting sentence should forecast content diplomatically. This statement can be candid without being so negative.

> **ORIGINAL**   But improvements are still needed.

> **REVISED**   A few changes—beyond the recent innovations—should result in even greater training efficiency.[3]

In sentence 4, "First and foremost" is trite, "aspects" is a clutter word, and word order needs changing to improve the emphasis (on innovations) and to lead into the examples.

> **ORIGINAL**   First and foremost, a number of innovative aspects in this October session proved especially useful.

> **REVISED**   Several program innovations were especially useful in this session.

In collaboration with his colleague, Glenn continues this editing and revising process on his report. Wednesday morning, after much revising and proofreading, Glenn prints out the final draft, shown in Figure B.3.

Glenn's final report is both informative and persuasive. But this document did not appear magically. Glenn made deliberate decisions about purpose, audience, content, organization, and style. He sought advice and feedback on every aspect of the document. Most importantly, he *spent time revising*.[4]

NOTE   *Writers work in different ways. Some begin by brainstorming. Some begin with an outline. Others simply write and rewrite. Some sketch a quick draft before thinking through their writing situation. Introductions and titles are often written last. Whether you write alone or collaborate in preparing a document, whether you are receiving feedback or providing it, no one step in the process is complete until the whole is complete. Notice, for instance, how Glenn sharpens his content and style while he organizes. Every document you write will require all these decisions, but you rarely will make them in the same sequence.*

No matter what the sequence, revision is a fact of life. It is the one constant in the writing process. When you've finished a draft, you have in a sense only begun. Sometimes you will have more time to compose than Glenn did, sometimes much less. Whenever your deadline allows, leave time to revise.

---

[2]Notice throughout how careful revision sharpens the writer's meaning while cutting needless words.
[3]This revision has more words, but also much more concrete and specific detail (in italic type). Completeness of information always takes priority over word count.
[4]A special thanks to Glenn Tarullo for his perseverance. I made his task doubly difficult by having him explain each of his decisions during this writing process.

:::. MICRO**BYTE**

October 19, 20XX

To: Marvin Long
From: Glenn Tarullo
Subject: October Management Training Program: Evaluation
          and Recommendations

Begins on a positive note, and cites evidence

States his claim

The October Management Training Session was successful, with training rated as "very good" by most participants. A few changes, beyond the recent innovations, should result in even greater training efficiency.

Workshop Strengths

Especially useful in this session were several program innovations:

Gives clear examples of "innovations"

—Dividing class topics into two areas created a general-to-specific focus: The first week's coverage of company structure and functions created a context for the second week's coverage of management skills.

—Videotaping and critiquing trainees' oral reports clarified their speaking strengths and weaknesses.

—Emphasizing interpersonal communication skills (listening, showing empathy, and reading nonverbal feedback) created a sense of ease about the group, the training, and the company.

Innovations like these ensure high-quality training. And future sessions could provide other innovative ideas.

Suggested Changes/Benefits

Cites the bases for his recommendations

Based on the trainees' evaluation of the October session (summary attached) and my observations, I recommend these additional changes:

—We should develop several brief (one-day) on-the-job rotations in different sales and service areas before the training session. These rotations would give each member a real-life view of duties and responsibilities throughout the company.

—All training sessions should have at least ten to fifteen members. Larger classes would make more efficient use of resources and improve class-speaker interaction.

FIGURE B.3   Glenn's Final Draft

Long, Oct. 19, 20XX, page 2

Supports each recommendation with convincing reasons

—We should ask instructors to follow a standard format (based on definite course objectives) for their presentations, and to use visuals liberally. These enhancements would ensure the greatest possible instructor efficiency and audience interest.

—Executive speakers should spell out personal and professional traits that are essential to success in our company. Such advice would give trainees a concrete guide to both general company and individual supervisor expectations. Also, by the next training session, we should assemble a presentation dealing with appropriate attitudes, manners, and behavior in the business environment.

—We should do a six-month follow-up of trainees (with feedback from supervisors as well as ex-trainees) to gain long-term insights, to measure the influence of this training on job performance, and to help design advanced training.

Closes by appealing to shared goals (efficiency and profit)

Inexpensive and easy to implement, these changes should produce more efficient training.

Copies: B. Hull, C. Black, G. Hopkins, J. Capilona, P. Maxwell, R. Sanders, L. Hunter

FIGURE B.3   Glenn's Final Draft *(continued)*

CASE 2—PREPARING A PERSONAL STATEMENT IN AN INTERNSHIP OR FELLOWSHIP APPLICATION

Applications for jobs, grants, scholarships, and graduate school typically require a personal statement that addresses these basic questions:

What readers of a personal statement want to know

- *Why should we select you?*
- *Why do you want this?*
- *What will you bring to this experience?*
- *How, exactly, do you plan to use this opportunity?*
- *What do you hope to gain?*

In a short essay, the candidate presents her/his best argument for being selected. Statements that stand out are the ones that make the final cut.

In the situation that follows, Mike Duval, a junior in marine biology, is applying for a prestigious and highly competitive summer research internship at a leading oceanographic institute. Application requirements include a personal statement. Before writing a word, Mike wisely decides to analyze and anticipate what, exactly, his audience is looking for.

AUDIENCE EXPECTATIONS IN A PERSONAL STATEMENT

About Content
- a brief but specific proposal for a research project
- some new and *significant* ideas about the research topic
- a summary of the writer's qualifications to undertake this project
- neither too much nor too little information
- answers to questions about *what*, *why*, *how*, and *when*

About Organization
- a distinct line of reasoning and a clear, sensible plan, consistent with the best scientific methods
- an introduction that offers brief background and justifies the need for the project
- a body section that outlines the scope, method, and sources for the proposed investigation
- a conclusion that describes the benefits of the research and encourages the reader's support

About Style
- a decisive tone with no hint of ambivalence
- at least a suggestion of enthusiasm
- an efficient style, in which nothing is wasted

Mike's technical writing instructor has invited him to bring in his best draft (shown in Figure B.4) for review by the entire class.

Personal Statement

Michael C. Duval

1   I want to study marine science during the summer at Woods Hole Oceanographic Institute as I hope to add to my background and understanding of the marine environment. Presently, I have been unable to conduct any full-time research projects due to the time factor involved and the responsibilities of a full semester's course load. However, your program is an opportunity to study any aspect of the marine environment largely independent of that time factor, except for summer limitations.

2   Textbooks cannot develop techniques; they can only present concepts. Therefore, one has to develop these techniques himself by actually doing the thinking, the designing, the manipulating, and the interpreting. Once such skills are perfected, they can be carried on for further application in graduate studies or in job situations. And this is an important aspect of the summer program: mastering skills necessary in any kind of research, and developing that "research frame-of-mind." As I plan to further my education by attending graduate school, I think this program will prove invaluable to me.

3   However, I would like to work a year or so before entering a graduate program so that I can observe senior researchers and understand the requirements of various positions. In this way, and by completing graduate school, I can become a more marketable person.

4   My research at Woods Hole could lead to continued work as the topic of my graduate thesis. Presently, I am interested in marine microbes and their interactions with invertebrates such as mollusks or crustaceans. Since little is known about these interactions, much attention must be given to this subject. Recent studies have found that some human pathogens are part of the indigenous bacterial fauna of the oyster and other similar shellfish, and can be introduced into the gastrointestinal system via direct consumption. This increases the need to understand and thus control such vectors of human disease.

FIGURE B.4   Mike's Best Early Draft

Discussion of Mike's First Draft (Figure B.4)

For the workshop on Mike's statement, the instructor asked the class to assume they were members of the committee screening fellowship applicants. Here is the summary of the class's critique:

1. An opening paragraph—especially in a competitive application—should grab the reader's attention and make the candidate stand out. Here, Mike opens with a self-evident observation followed by an unconvincing apology and then a rambling final sentence. Because the content is vague, readers have nothing concrete to *visualize*. Because the style is wordy, readers work harder than they should. Mike needs to paint a more vivid picture of who he is, why he is applying, and what he plans to do.

2. The second paragraph is where Mike should begin focusing on his proposed research topic, offering something new and significant. Instead, the content seems vague and abstract, lacking any real point or personality. Also the passive voice, excessive prepositions, and overall wordiness almost make the writer disappear. Mike needs to spell out his topic and show why it's important.

3. By this stage, Mike should be explaining what he can contribute to this fellowship experience instead of focusing only on the personal benefits he expects. In the audience's view, Mike's stated goal "to become a more marketable person" is hardly a persuasive reason. Mike needs to make a better case not only for what he hopes to gain but also for what he can offer—and he needs to project a likable *persona* (page 67) throughout.

4. The closing paragraph should leave readers with a clear and positive sense of this writer as a unique candidate who deserves to "make the cut." Instead, the paragraph continues the abstract theme about what the writer hopes to gain, then tells readers what they already know, and closes with a vague "this" statement that drowns the whole point of the essay. Mike needs to sum up his case clearly and emphatically.

Discussion of Mike's Final Draft (Figure B.5)

After additional revisions and class workshops, Mike produced the final draft shown in Figure B.5. This version is far more concise as well as more visual, offering concrete, persuasive support in a tone that is decisive. Mike's ideas are significant; his plan is clear and sensible; and his attitude is mature, realistic, and engaging. The persona suggests a writer who knows what he wants and how he can contribute.

Personal Statement

Michael C. Duval

Tells who he is, why he's applying, and what he plans to do

Please consider my application for a summer research internship. I am a marine biology major at southeastern Massachusetts University, interested in marine microbes and their interactions with mollusks and crustaceans. My plan is to attend graduate school, but I wish to work a year or so in the biological sciences before enrolling. In this way, I can recognize weak areas in my understanding of biology, and then take the appropriate graduate courses. After graduate studies, I plan to do research and advocate for the environment.

Describes his topic and explains its importance

I haven't yet had a chance to conduct full-time research; however, I am studying the indigenous bacterial fauna of the quahog (Mercenaria) as a semester project. I am particularly concerned with pathogenic interactions in shellfish, and their ultimate influence on public health through the transfer of dysentery and viral diseases such as herpes, hepatitis, and polio viruses.

Explains what he hopes to gain and what he can offer

Beyond the obvious prestige and resultant professional benefits of an internship at WHOI, I would anticipate more subtle and personal rewards. These include exposure to graduate-level work and practical applications of the research process—specifically the design, implementation, and interpretation of an experiment. My natural curiosity and determination to contribute to scientific understanding would, I think, be an asset to your program.

Mentions his work with a well-known professor

I enjoy working closely with my professors. For the past year, I have been working in a microbiological laboratory under Dr. Samuel Jennings. He, more than anyone, has helped me recognize my scientific abilities and deficiencies. I try to emulate his ways of thinking, incorporating that logic into my own problem solving. With his encouragement, I have learned to approach biology with the precision of a critical observer, the flexibility of an imaginative scientist, and the curiosity of a perennial student.

Sums up clearly and emphatically

I realize I have much to learn. Yet, I know enough to ask the kinds of questions that are socially and biologically significant. Are our methods of assessing microbial contamination in shellfish exacting enough? How can we approach a society that leans more toward reaction than prevention, and persuade its citizens that they must change their life-style in order to save their livelihood? A research internship at WHOI might prepare me to find answers to these questions.

FIGURE B.5   **Mike's Final Draft**

CASE 3—DOCUMENTS FOR THE COURSE PROJECT: A SEQUENCE CULMINATING IN THE FINAL REPORT

Mike Cabral, a communications major, works part-time as assistant to the production manager for *Megacrunch,* a computer magazine specializing in small-business applications.[5] (**Production** is the transforming of manuscripts into a published form.) Mike's writing instructor assigns a course project and encourages students to select topics from the workplace, when possible. Mike asks *Megacrunch* production manager Marcia White to suggest a research topic that might be useful to the magazine.

White outlines a problem she thinks needs careful attention: Now six years old, *Megacrunch* has enjoyed steady growth in sales volume and advertising revenue—until recently. In the past year alone, *Megacrunch* has lost $150,000 in subscriptions and one advertising account worth $60,000. White knows that most of these losses are caused by increasing competition (three competing magazines have emerged in 18 months). In response to these pressures, White and the executive staff have been exploring ways of reinvigorating the magazine—through added coverage and "hot" features, more appealing layouts and page design, and creative marketing. But White is concerned about another problem that seems partially responsible for the fall in revenues: too many errors are appearing in recent issues.

When *Megacrunch* first hit the shelves, errors in grammar or accuracy seemed rare. But as the magazine grew in complexity, errors increased. Recent issues contain misspellings, inaccurate technical details, unintelligible sentences and paragraphs, and scrambled source code (in sample programs).

White asks Mike if he's interested in researching the error problem and looking into quality-control measures. She feels that his three years' experience with the magazine qualifies him for the task. Mike accepts the assignment.

White cautions Mike that this topic is politically sensitive, especially to the editorial staff and to the investors. She wants to be sure that Mike's investigation doesn't merely turn up a lot of dirty laundry. Above all, White wants to preserve investor confidence—not to mention the morale of the editorial staff, who do a good job in a tough environment, plagued by impossible deadlines and constant pressure. White knows that offending people—even unintentionally—can be disastrous.

She therefore insists that all project documents express a supportive rather than critical point of view: "What could we be doing better?" instead of "What are we doing wrong?" Before agreeing to release the information from company files (complaint letters, notes from irate phone calls, and so on), White asks Mike to submit a proposal, in which the intent of this project is made absolutely clear.

---

[5]In the interest of privacy, the names of all people, publications, companies, and products in Mike's documents have been changed.

The Project Documents

Three types of documents lend shape and sequence to the research project: the **proposal,** which spells out the plan; the **progress report,** which keeps track of the investigation; and the **final report,** which analyzes the findings. This presentation shows how these documents function together in Mike Cabral's reporting process.

The Proposal Stage

Proposals offer plans for meeting needs. A proposal's primary audience consists of those who will decide whether to approve, fund, or otherwise support the project. Reviewers of a research proposal usually begin with questions like these:

- *What, exactly, do you intend to find out?*
- *Why is the question worth answering, or the problem worth solving?*
- *What benefits can we expect from this project?*

Once reviewers agree that the project is worthwhile, they will want to know all about the plan:

- *How, exactly, do you plan to do it?*
- *Is the plan realistic?*
- *Is the plan acceptable?*

Besides these questions, reviewers may have others: *How much will it cost? How long will it take? What makes you qualified to do it?* and so on. See Chapter 23 for more discussion and examples.

Mike Cabral knows his proposal will have only Marcia White (and possibly some executive board members) as the primary audience. The secondary audience is Mike's writing instructor, who also must approve the topic. And at some point, Mike's documents could find their way to his coworkers.

White already knows the background and she needs no persuading that the project is worthwhile, but she does expect a realistic plan before she will approve the project. (For his instructor, Mike attaches a short appendix [not shown here] outlining the background and his qualifications.) Also, Mike concentrates on his emphasis: He wants the proposal to be positive rather than critical, so as not to offend anyone. He therefore focuses on achieving *greater accuracy* rather than *fewer errors.* So that his instructor can approve the project, Mike submits the proposal by the semester's fourth week (Figure B.6).

The Progress Report Stage

The progress report keeps the audience up to date on the project's activities, new developments, accomplishments or setbacks, and timetable. Depending on the size and length of the particular project, the number of progress reports will vary. (Mike's course project will require only one.) The audience approaches any progress report with two big questions:

Rangeley Publishing Company

TO.:   Marcia White, Production Manager   September 26, 20XX
FROM: Mike Cabral, Production Assistant   *MC*
SUBJECT: Proposal for Studying Ways to Improve Quality Control at *Megacrunch*

Introduction

The growing number of grammatical, informational, and technical errors in each monthly issue of *Megacrunch* is raising complaints from authors, advertisers, and readers. Beyond compromising the magazine's reputation for accurate and dependable information, these errors—almost all of which seem avoidable—endanger our subscription and advertising revenues.

Summary of the Problem

Authors are complaining of errors in published versions of their articles. Software developers assert that errors in reviews and misinformation about products have damaged reputations and sales. For example, Osco Scientific, Inc. claims to have lost $150,000 in software sales because of an erroneous review in *Megacrunch*.

Although we continue to receive a good deal of "fan mail," we also receive letters speculating about whether *Megacrunch* has lost its edge as a leading resource for small-business users.

Proposed Study

I propose to examine the errors that most frequently recur in our publication, to analyze their causes, and to search for ways of improving quality.

Methods and Sources

In addition to close examination of recent *Megacrunch* issues and competing magazines, my primary data sources will include correspondence and other feedback now on file from authors, developers, and readers. I also plan telephone interviews with some of the above sources. In addition, interviews with our editorial staff should yield valuable insights and suggestions. As secondary material, books and articles on editing and writing can provide sources of theory and technique.

Conclusion

We should not allow avoidable errors to eclipse the hard work that has made *Megacrunch* the leading Cosmo resource for small-business users. I hope my research will help eliminate many such errors. With your approval, I will begin immediately.

Margin annotations:

Tells what the problem is, and what it means to the company

Further defines the problem and its effects

Describes the project and tells who will carry it out

Describes the research strategy

Encourages audience support by telling why the project will be beneficial

FIGURE B.6   Mike's Proposal

Rangeley Publishing Company

TO: Marcia White, Production Manager   November 6, 20XX
FROM: Mike Cabral, Production Assistant   *MC*
SUBJECT: Report of Progress on My Research Project: A Study of Ways for
            Improving Quality Control at *Megacrunch* Magazine

Work Completed

My topic was approved on September 28, and I immediately began both primary
and secondary research. I have since reviewed file letters from contributors and
readers, along with notes from phone conversations with various clients and from
interviews with *Megacrunch*'s editing staff. I have also surveyed the types and
frequency of errors in recent issues. Books and articles on writing and editing
round out my study. The project has moved ahead without complications. With my
research virtually completed, I have begun to interpret the findings.

Preliminary Interpretation of Findings

From my primary research and my own editing experience, I am developing a
focused idea of where some of the most avoidable problems lie and how they might
be solved. My secondary sources offer support for the solutions I expect to recommend,
and they suggest further ideas for implementing the recommendations. With a realistic
and efficient plan, I think we can go a long way toward improving our accuracy.

Work Remaining

So far, the project is on schedule. I plan to complete the interpretation of all findings
by the week of November 29, and then to organize, draft, and revise my final report
in time for the December 14 submission deadline.

Margin notes:

Tells what has been done so far, and what is now being done

Tells what has been found so far, and what it seems to mean

Assesses the project schedule; describes work remaining and gives a completion date

FIGURE B.7   Mike's Progress Report

- *Is the project moving ahead according to plan and schedule?*
- *If not, why not?*

The audience may have various subordinate questions as well. See pages 388-92 for more discussion and examples.

Mike designs his progress report for his boss *and* his instructor, and turns it in by the semester's tenth week (Figure B.7).

The Final Report Stage

The final report presents the results of the research project: findings, interpretations, and recommendations. This document answers questions like these:

- *What did you find?*
- *What does it all mean?*
- *What should we do?*

Depending on the topic and situation, of course, the audience will have specific questions as well. See Chapter 24 for discussion and examples.

During his research, Mike Cabral discovered problems over and above the published errors he had been assigned to investigate. For instance, after looking at competing magazines he decided that *Megacrunch* needed improved page design, along with a higher quality stock (the paper the magazine is printed on). He also concluded that a monthly section on business applications would help. But despite their usefulness, none of these findings or ideas was part of Mike's *original* assignment. White expected him to focus on these questions, specifically:

- *Which errors recur most frequently in our publication?*
- *Where are these errors coming from?*
- *What can we do to prevent them?*

Mike therefore decides to focus exclusively on the error problem. (He might later discuss those other issues with White—if the opportunity arises. But if *this* report were to include material that exceeds the assignment *and* the reader's expectations, Mike could end up appearing arrogant or presumptuous.)

Mike tries to give White only what she requested. He analyzes the problem and the causes, and then recommends a solution. Mike adapts the general outline on page 616 to shape the three major sections of his report: *introduction, findings and conclusions, and recommendations.* For the user's convenience and orientation, he includes the report supplements discussed in Chapter 25: *front matter* (title page, transmittal letter, table of contents, and informative abstract) and *end matter* (a Works Cited page and appendices [not shown here]).

After several revisions, Mike submits copies of the report to his boss and to his instructor (Figure B.8).

Virtually any long
report has a title
page

A forecasting title

Quality-Control Recommendations for *Megacrunch* Magazine

The primary user's
name, title, and
organization

Prepared for
Marcia S. White
Production Manager
Rangeley Publications

Author's name

by
Michael T. Cabral

Submission date

December 14, 20XX

FIGURE B.8   Mike's Final Report

A letter of transmittal usually accompanies a long report

Addressed to the primary user

Identifies the report's subject and describes its scope

Maintains a confident tone throughout
Gives an overview of entire project, and findings

Requests action

Invites follow-up

---

82 Stephens Road
Boca Grande, FL 08754
December 14, 20XX

Marcia S. White, Production Manager and Vice President
Rangeley Publications, Inc.
167 Dolphin Ave.
Englewood, FL 08567

Dear Ms. White:

Here is my report, recommending quality-control measures for Megacrunch magazine. The report briefly discusses the history of our quality-control problem, identifies the types of errors we are up against, analyzes possible causes, and recommends four realistic solutions.

My research confirmed exactly what you had feared. The problem is big and deeply rooted: our authors have legitimate complaints; our readers justifiably want information they can put to work; and developers and advertisers have the right to demand fair and complete representation. As a result of client dissatisfaction, competing magazines are gaining readers and authors at our expense.

To have an immediate effect on our quality-control problem, we should act now. Because of our limited budget, I have tried to recommend low-cost, high-return solutions. If you have other solutions in mind, I would be happy to research them for projected effectiveness and feasibility.

Sincerely,

Michael T. Cabral

Michael T. Cabral
Production Assistant
Rangeley Publications

FIGURE B.8   Mike's Final Report *(continued)*

Table of contents (reports with numerous visuals also have a table of figures)

Front matter (items that precede the report)

Heads and sub-heads from the report itself

All heads in the table of contents follow the exact phrasing of those in the report text

The various typefaces and indentations reflect the respective rank of various heads in the report

Each head listed in the table of contents is assigned a page number

End matter (items that follow the report)

iii

CONTENTS

FIGURE B.8   Mike's Final Report *(continued)*

iv

INFORMATIVE ABSTRACT

An investigation of the quality-control problem at Megacrunch magazine identifies the types of errors and their causes and recommends a solution.

Megacrunch suffers from the following avoidable errors:

- *Grammatical errors* are most frequent: misspellings, fragmented and jumbled sentences, misplaced punctuation, and so on.
- *Informational errors:* incorrect prices, products attributed to wrong companies, mismarked visuals, and so on.
- *Technical content errors* are less frequent, but the most dangerous: garbled source code, mismarked diagrams, misused technical terms, and so on.
- *Distortions of the author's original meaning:* introduced by editors who attempt to improve clarity and style.

The above errors seem to have the following causes:

- *Poor initial submissions from contributors* ignore basic rules of grammar, clarity, and organization.
- *Lack of structure in the editing cycle* allows for unrestrained and often excessive editing at all stages.
- *Lack of diversity in the editorial staff* leaves language specialists responsible for catching technical and informational errors.
- *Lack of communication with authors and advertisers* leaves the primary sources out of the production process.

On the basis of my findings, I offer four recommendations for improving quality control during the production process:

- *Expanded author's guide* that includes guidelines for effective use of active voice, visuals, direct address, audience analysis, and so on.
- *Five-stage editing cycle* that specifies everyone's duties at each stage. The cycle would require two additional staff members: a technical editor and a fact checker/typist for editorial changes.
- *A checklist for each stage,* to ensure consistent editing.
- *More communication with contributors* by exchanging galley proofs, increasing our use of the electronic network, and possibly using groupware.

The informative abstract summarizes the report's essential message (findings, conclusions, and recommendations). This is the one part of a long report read most often.

The summary stands alone in meaning—a kind of mini-report written for laypersons.

Busy audiences need to know quickly what is important. A summary gives them enough information to decide whether they should read the whole report, parts of it, or none of it.

FIGURE B.8   Mike's Final Report *(continued)*

1

INTRODUCTION

The reputation of *Megacrunch* magazine is jeopardized by grammatical, technical, and other errors appearing in each issue.

Megacrunch has begun to lose some long-time readers, advertisers, and authors. Although many readers continue to praise the usefulness of our information, complaints about errors are increasing and subscriptions are falling. Advertisers and authors increasingly point to articles or layouts in which excessive editing has been introduced, and some have taken their business and articles to competing magazines. One disgruntled subscriber sums up our problem by asking that we devote "more effort to publishing a magazine without the kinds of elementary error that distract readers from the content" (Grendel). This kind of complaint is typical of the sample letters in Appendix A.

Granted, complaints are inevitable—as can be seen in a quick review of "Letters to the Editor" in virtually any publication. But if *Megacrunch* is to withstand the competition and uphold its reputation as the leading resource for Cosmo applications in small business, we must minimize such complaints.

Such negative reactions to errors in our magazine should not be surprising: Roughly two decades of research have verified this basic fact: errors in a business document —especially errors in grammar and usage—frustrate readers, and cause them to mistrust the quality of the writing in general.

As two noted communication researchers point out, "business readers tend to read rapidly, for meaning. A perceived error in the writing trips them up. They might not fall down—that is, misunderstand or even quit reading—but they are discomfited, distracted, and even annoyed" (Gilsdorf and Leonard 459).

This report identifies the major errors that recur in our magazine, and investigates their causes. My data is compiled from interviews with our editorial staff, a review of complaint letters from authors and readers, and a spot-check for errors in the magazine itself. Books and articles on writing and editing provide theory and technique. The report concludes by recommending a four-part solution to our error problem.

Sidebar annotations (left margin):

This section tells what the report is about, why it was written, and how much it covers.

An overview of the problem and its effects on the magazine's revenues

The audience is referred to appendices for details that would interrupt the report flow.

Request for action

Citing expert opinion lends immediate credibility to the report.

Purpose and scope of report; overview of research methods and data sources

Because his primary audience knows the background, Mike keeps the introduction brief.

FIGURE B.8   Mike's Final Report (*continued*)

This section tells what was found and what it means.

The first subsection analyzes the problem; the second will examine causes.

Introduction to the problem, and a lead-in to the visual

A visual that illustrates parts of the problem

2

FINDINGS AND CONCLUSIONS

ELEMENTS OF THE PROBLEM

Errors in *Megacrunch* are limited to no single category. For example, some errors are tied to technical slip-ups, while others result from editors changing the author's intended meaning. My spot-check of *Megacrunch* 8.10, our most recent issue, revealed errors of the types listed in Table 1.

Table 1 Sample Errors

	Spot-check of *Megacrunch* 8.10	
Error Type	**As Published**	**Corrected Version**
mechanical	varity of software	variety of software
technical	Dos	DOS
informational	Deluxe Panel	DeluxePanel
grammatical	This will help	Editing will help
grammatical	. . . everyone helps for of a program's release date approaches. . . .	. . . everyone helps as a program's release date approaches . . . .
technical	ram	RAM
informational	cosmo	Cosmo
mechanical	We're back, now we will	We're back; now we will

My random analysis of only six pages identified errors in four categories: grammatical/mechanical, informational, technical, and distortions of meaning.

Errors in Grammar and Mechanics

Basic correctness is a given—and a problem—for any publication. According to one editing expert, errors such as sentence fragments, confused punctuation, and poor spelling "serve as evidence of ignorance or sloppiness" (Samson 10). Even worse, as another expert points out, "if the writing is bad, readers will often question the accuracy of the content as well" (Johnson-Rew 16). These assertions are borne out by the sampling of reader complaints in Appendix A.

One survey of college and workplace writing (Haswell 168) found that the average writer suffers from the following basic problems:

- Three to four words are misspelled in a memo-length piece.
- One of every ten sentences is a run-on or an "attachable sentence fragment."
- Every fifth possessive is incorrectly formed.

Discussion of the visual, and overview of the subsection

One part of the problem defined, with examples and effects of the problem

Citing authorities clarifies and supports the author's position.

FIGURE B.8   Mike's Final Report *(continued)*

Author interprets
and relates the
material to these
users.

Request for action
Author refutes
anticipated reader
assumptions about
a solution that
would be overly
simplistic

Another part of the
problem defined

Examples

Effects of the
problem

Other examples

The need for action

Another part of the
problem defined

Effects of the
problem

Examples

A vivid example,
as a visual

3

Given these findings, we should not be surprised to receive imperfect manuscripts. However, we must eliminate the slips of the so-called average writer before final copy goes to press.

Although spelling and grammar checkers can eliminate many such errors, these electronic editing tools can't spot them all. For example, spelling checkers cannot detect words that are spelled correctly but used incorrectly (*it's* for *its, effect* for *affect, there* for *their,* and so on). Nor can they detect typos that create the wrong word that happens to be correctly spelled (*fort* for *port, risk* for *disk, the* for *then,* and so on). Grammar checkers often suggest revisions that are too simplistic or that distort the intended meaning and emphasis (such as the advice to use smaller words or shorter sentences). In fact, one survey found that professional writers consider electronic editing tools somewhat helpful—but no substitute for the human role in editing and proofreading (Johnson-Rew 3–4).

Errors in Information Accuracy and Access

Beyond basic errors, we have published some inaccurate information. For instance, we sometimes attribute products to the wrong companies or we list incorrect prices. Inaccuracies of this kind infuriate readers, product developers, and suppliers. And a retraction printed in the magazine's subsequent issue has little impact once the damage has been done.

Besides inaccurate information, *Megacrunch* too often presents inaccessible information. Mismarked visuals, misplaced headings, and misnumbered references to pages and figures make the magazine hard to follow and to use selectively.

Technical Content Errors

Technical content errors seem to be one of our biggest problems. While some readers might raise a proverbial eyebrow over grammatical errors or skim over informational errors, technical content errors are more frustrating and incapacitating. On a page of text, a misplaced comma or a missing bracket can be irritating, but in a program listing, these same errors can render the program useless. Even worse, a misnamed or misnumbered pin or socket in a hardware diagram might cause users to inadvertently destroy their data or damage their hardware.

Some technical slips in *Megacrunch* have veered close to disaster. Consider, for example, the flawed diagram in Figure 1, from our 7.12 issue.

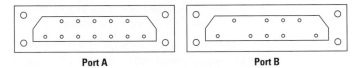

Port A   Port B

To connect the external drive, plug Cable Y into Port A

Figure 1   Partial View of the Port Panel on the AXL 100

FIGURE B.8 **Mike's Final Report** (*continued*)

4

Discussion of the
visual

Effects of the error

Interpretation—
what it all means
for the magazine

Another part of the
problem defined
Example

Interpretation

Citing an authority
to support the
interpretation
Another example

Effect of the error

Second subsection

Justification for
an analysis of
causes

Our published diagram instructed users to plug a 9-pin external-drive cable into Port A, a 12-pin modem port; the correct connection was to Port B. Ron Catabia, author of the article and respected tech wizard, explained the flaw in our reproduction of his diagram: "Had any users followed the instructions as printed, the read/write head on their external drive could have suffered permanent damage." Our lengthy correction printed one month later was in no way a sufficient response to an error of this importance. Nor could we placate an enraged and discredited author.

Such errors do little to encourage readers' perception of *Megacrunch* as the serious user's resource for the latest technical information.

Distortions of the Author's Original Meaning
Experts point out that editors are ethically obliged to "make only those changes that can be justified as assisting the reader while respecting the author's ownership of the work" (Allen and Voss 58). Most important in this regard is the need to preserve the meaning intended by the author. But our authors complain that, in our effort to increase clarity and readability, we distort their original, intended meaning. After reading the edited version of her article, one author insisted that "too often, edits changed what I had said to something I hadn't said—sometimes to the point of altering the facts" (Dimmersdale).

Editorial liberties inevitably alienate authors. Overzealous editors who set out to shorten a sentence or fine-tune a clause—while knowing nothing about the program being discussed—can distort the author's meaning. As one study confirms, "When reviewers [editors] criticize in areas outside their expertise, their misguided reviews are seen as an intrusion" (Barker 37).

The following excerpt typifies the distortions in recent issues of *Megacrunch*. Here, a seemingly minor edit (from "but" to "even") radically changes the meaning.

As submitted: A user can complete the Filibond program without ever having typed but a single command.

As published: A user can complete the Filibond program without ever having typed even a single command.

As the irate author later pointed out, "My intent was to indicate that a single command must be typed during the program run" (Klause).

This type of wholesale editing (more examples of which are shown in Appendix B) is a disservice to all parties: author, reader, and magazine.

CAUSES OF OUR EDITORIAL INACCURACY
Before devising a plan for dealing with our editing difficulties, we have to answer questions such as these:
- Where are these errors coming from?
- Can they be prevented?

FIGURE B.8   Mike's Final Report (continued)

5

Scope of this sub-
section, so that
users know what
to expect

Interviews with our editing staff, along with analysis of our editing practices and review of letters on file, uncovered the following causes: (1) poor initial submissions from contributors, (2) lack of structure in our editing cycle, (3) lack of diversity in our editing staff, and (4) lack of communication with the authors and advertisers.

Poor Initial Submissions from Contributors

First cause defined

Findings

Conclusion

Some contributors submit poorly written manuscripts. And so we edit heavily whenever "a submission otherwise deserves flat-out rejection," as one editor argues. Our editors claim that printing poor writing would be more damaging than the occasional editing excesses that now occur. Although editors can improve clarity and readability without in-depth knowledge of the subject, we often misinterpret the author's meaning. Clearer writing guidelines for authors would result in manuscripts needing less editing to begin with. Our single-page author's guide is inadequate.

Lack of Structure in Our Editing Cycle

Second cause
defined

Findings

Interpretation—
what it means

Findings

Interpretation
Conclusion

In our current editing cycle, the most thorough editors see an article repeatedly, as often as time allows. Various editors are free to edit heavily at all stages. And these editors are entirely responsible for judgments about grammatical, informational, and technical accuracy.

Although "having your best give their best" throughout the cycle seems a good idea, this approach leads to inconsistent editing and/or overediting. Some editors do a light editing job, choosing to preserve the original writing. Others prefer to "overhaul" the original. With light-versus-heavy editing styles entering the cycle randomly, errors slip by. As one editor noted, "Sometimes an article doesn't get a tough edit until the third or fourth reading. At that point, we have no time to review these last-minute changes" (*Megacrunch* editorial staff).

Any article heavily edited and rewritten in the final stages stands a chance of containing typographical and mechanical errors, some questionable sentence structures, inadvertent technical changes, and other problems that result from a "rough-and-tumble edit."

While some articles are edited inconsistently, others are overedited. Our editors tend to be vigilant in pursuit of clarity, conciseness, and tone. Unfortunately, they seem less vigilant about technical accuracy.

Instead of full-scale editing at all stages, we need a cycle that makes a manuscript progress from inadequate (or adequate) to excellent, through different levels of editorial attention. For example, a first edit should be thorough, but a final proofreading should be merely a fine-combing for typographical and mechanical errors.

FIGURE B.8 Mike's Final Report *(continued)*

6

Third cause defined and interpreted

Conclusion

Fourth cause defined and interpreted

Conclusion

This section tells what should be done.

Scope of this section, so that users know what to expect

Lead-in to first recommendation

The recommendation

Lead-in to second recommendation

Lack of Diversity in Our Editing Staff
The variety of errors suggests that our present staff alone cannot spot all problems. Strong writing backgrounds have not prepared our editors to recognize a jumbled line of programming code or a misquoted price. To snag all errors, we must hire technical specialists. We need both a technical editor and a fact checker, to pick up where current editors leave off.

Lack of Communication with Authors and Advertisers
Some of our editing troubles emerge from a gap between the meaning intended by contributors and the interpretation by editors. In the present system, contributors submit manuscripts without seeing any editorial changes until the published version appears. Along with an expanded author's guide, regular communication throughout the editing process (and perhaps the writing process as well) would involve contributors in developing the published piece, and thus make authors more responsible for their work.

RECOMMENDATIONS

To eliminate published inaccuracies, I recommend: (1) an expanded author's guide, (2) a five-stage editing cycle, (3) a checklist for each stage, and (4) improved communication with contributors.

EXPANDED AUTHOR'S GUIDE
The obvious way to limit editing changes would be to accept only near-perfect submissions. But as a technical resource we cannot afford to reject poorly written articles that are nonetheless technically valuable.

To reduce editing required on submissions, I recommend we expand our author's guide to include topics like these: audience analysis, use of direct address and active voice, principles of outlining and formatting, and use of visuals.

FIVE-STAGE EDITING CYCLE
In place of haphazard editing, I propose a progressive, five-stage cycle: Stages one through three would refine grammar, clarity, and readability. Two additional staff members, a fact checker and a technical editor, would check facts and technical accuracy in the final two stages. Figure 2 outlines responsibilities at each stage.

FIGURE B.8   Mike's Final Report (continued)

7

A CHECKLIST FOR EACH STAGE
To ensure a consistent focus throughout the editing cycle, our staff should collaborate immediately to develop a detailed checklist for each stage (Hansen 15). As a guide for editors as well as for contributors, these checklists would enhance communication among all parties.

Lead-in to final recommendation

IMPROVED COMMUNICATION WITH CONTRIBUTORS
The following measures would reduce errors caused by misunderstandings between contributors and editors.

First part of final recommendation

Author/Client Verification of Page Proofs
Two weeks before our deadline, we could send authors prepublication page proofs [which show the text as it will appear in published form]. Authors could check for technical errors or changes in meaning, and return proofs within five days.

Second part of final recommendation

Expanded Use of Our Electronic Mail Network
At any time during production, authors and editors could communicate through CompuServe by leaving questions and messages in one another's email boxes. (Virtually all our regular authors subscribe to CompuServe.) In addition, all parties could log onto CompuServe at one or more scheduled times daily to discuss the manuscript. The email alternative is cheaper than the telephone, eliminates "telephone tag," and could serve as a "hot line" for authors while they prepare a manuscript for submission.

Third part of final recommendation

Possible Use of Groupware
For unified and efficient collaboration, we should also explore the feasibility of groupware (software, such as *ForComment* or *MarkUp,* for collaborative editing online). With *MarkUp* software, for example, documents could be edited on-screen and copies of each stage of the revision cycle could be filed and accessed as needed.

Even though research indicates that editing is best done on hard copy pages rather than on a computer screen (Kaufer and Neuwirth 113–15), groupware would provide a record of all edits at all stages of the editing cycle. Such a record would be invaluable in helping us troubleshoot and refine our editing process.

A closing call for action

Taking these recommended steps will have an immediate impact on the quality of our magazine and on its prospects for long-term success in an increasingly competitive market.

FIGURE B.8   Mike's Final Report *(continued)*

8

The visual illustrates and summarizes the process being recommended.

Data sources appear directly below the visual.

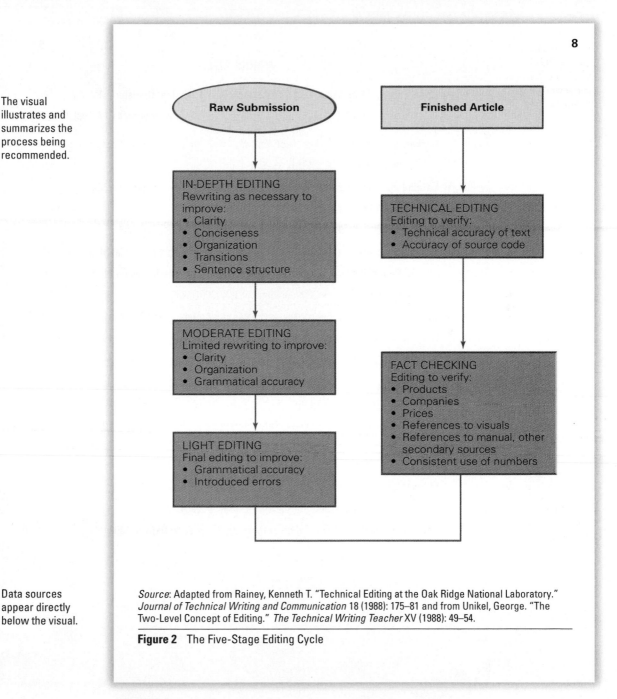

Source: Adapted from Rainey, Kenneth T. "Technical Editing at the Oak Ridge National Laboratory." *Journal of Technical Writing and Communication* 18 (1988): 175–81 and from Unikel, George. "The Two-Level Concept of Editing." *The Technical Writing Teacher* XV (1988): 49–54.

Figure 2   The Five-Stage Editing Cycle

FIGURE B.8   Mike's Final Report *(continued)*

This list of works cited clearly identifies each source cited in the report.

9

WORKS CITED

Allen, Lori, and Dan Voss. "Ethics for Editors: An Analytical Decision-Making Process." IEEE Transactions on Professional Communication 41.1 (1998): 58–65.

Barker, T. "Feedback in High-Tech Writing." Journal of Technical Writing and Communication 18.1 (1988): 35–51.

Catabia, R. Notes from author Catabia's phone conversation with the Managing Editor. 10 Mar. 2004.

Dimmersdale, O. Author's letter to the Managing Editor. 6 July 2004.

Gilsdorf, Jeanette, and Don Leonard. "Big Stuff, Little Stuff: A Dicennial Measurement of Executives' and Academics' Reactions to Questionable Usage Elements." The Journal of Business Communications 38.4 (2001): 439–75.

Grendel, M. L. Subscriber's letter to the Managing Editor. 14 May 2004.

Hansen, James B. "Editing Your Own Writing." Intercom Feb. 1997: 14–16.

Haswell, R. "Toward Competent Writing in the Workplace." Journal of Technical Writing and Communication 18.2 (1988): 161–72.

Johnson-Rew, Lois. Editing for Writers. Upper Saddle River, NJ: Prentice, 1999.

Kaufer, David S., and Chris Neuwirth. "Supporting Online Team Editing: Using Technology to Shape Performance and to Monitor Individual and Group Action." Computers and Composition 12 (1995): 113–24.

Klause, M. Author's letter to the Managing Editor. 11 Nov. 2004.

Megacrunch editing staff. Interviews. 12–14 Nov. 2004.

Samson, Donald C. Editing Technical Writing. New York: Oxford, 1993.

Specific names omitted from this citation, to protect in-house sources and to allow employees to speak candidly, without fear of reprisal

FIGURE B.8   Mike's Final Report (continued)

Appendix **C**

Editing for Grammar, Usage, and Mechanics

The rear endsheet of this book displays editing and revision symbols and corresponding page references. When your instructor marks a symbol on your paper, turn to the appropriate section for explanations and examples.

COMMON SENTENCE ERRORS

The following common sentence errors are easy to repair.

Sentence Fragment

frag

A sentence expresses a logically complete idea. Any complete idea must contain a subject and a verb and must not depend on another complete idea to make sense. Your sentence might contain several complete ideas, but it must contain at least one!

> [incomplete idea]   [complete idea]   [complete idea]
> Although Mary was injured, she grabbed the line, and she saved the boat.

Omitting some essential element (the subject, the verb, or another complete idea), leaves only a piece of a sentence—a *fragment*.

> Grabbed the line. [*a fragment because it lacks a subject*]
>
> Although Mary was injured. [*a fragment because—although it contains a subject and a verb—it needs to be joined with a complete idea to make sense*]
>
> Sam an electronics technician.

This last statement leaves the reader asking, "What about Sam the electronics technician?" The verb—the word that makes action happen—is missing. Adding a verb changes this fragment to a complete sentence.

SIMPLE VERB	Sam **is** an electronics technician.
VERB PLUS ADVERB	Sam, an electronics technician, **works hard.**
DEPENDENT CLAUSE, VERB, AND SUBJECTIVE COMPLEMENT	**Although he is well paid,** Sam, an electronics technician, **is not happy.**

Do not, however, mistake the following statement—which seems to contain a verb—for a complete sentence:

> Sam being an electronics technician.

Such "-ing" forms do not function as verbs unless accompanied by other verbs such as **is, was,** and **will be.**

> **Sam,** being an electronics technician, **was responsible for checking the circuitry.**

Likewise, the "*to* + verb" form (infinitive) is not a verb.

> **FRAGMENT**   To become an electronics technician.
>
> **COMPLETE**   To become an electronics technician, **Sam had to complete a two-year apprenticeship.**

Sometimes we inadvertently create fragments by adding subordinating conjunctions (**because, since, it, although, while, unless, until, when, where,** and others) to an already complete sentence.

> **Although** Sam is an electronics technician.

Such words subordinate the words that follow them; that is, they make the statement dependent on an additional idea, which must itself have a subject and a verb and be a complete sentence. (See also "Subordination"—pages 757–58.) We can complete the subordinate statement by adding an independent clause.

> **Although** Sam is an electronics technician, **he hopes to become an electrical engineer.**

NOTE   *Because the incomplete idea (dependent clause) depends on the complete idea (independent clause) for its meaning, you need only a* pause *(symbolized by a comma), not a* break *(symbolized by a semicolon).*

Acceptable Fragments

Not all fragments are unacceptable in all circumstances. For example, a fragmented sentence is acceptable in commands or exclamations because the subject ("you") is understood.

> **ACCEPTABLE**   Slow down.
> **FRAGMENTS**   Give me a hand.
>   Look out!

Also, questions and answers are sometimes expressed as incomplete sentences.

> **ACCEPTABLE**   How? By investing wisely.
> **FRAGMENTS**   When? At three o'clock.
>   Who? Bill.

In general, however, avoid fragments unless you have good reason to use one for special tone or emphasis.

Comma Splice

In a comma splice, two complete ideas (independent clauses) that should be *separated* by a period or a semicolon are incorrectly *joined* by a comma:

> Sarah did a great job, she was promoted.

You can choose among several possibilities for repair:

1. Substitute a period followed by a capital letter:

> Sarah did a great job. She was promoted.

2. Substitute a semicolon to signal a relationship between the two items:

> Sarah did a great job; she was promoted.

3. Use a semicolon with a connecting (conjunctive) adverb (a transitional word):

> Sarah did a great job; **consequently,** she was promoted.

4. Use a subordinating word to make the less important clause incomplete and thereby dependent on the other:

> **Because** Sarah did a great job, she was promoted.

5. Add a connecting word after the comma:

> Sarah did a great job, **and** she was promoted.

Your choice of construction will depend on the specific meaning or tone you intend to convey.

Run-On Sentence

The run-on sentence, a cousin to the comma splice, crams too many ideas together without needed breaks or pauses.

RUN-ON	The hourglass is more accurate than the waterclock for the water in a waterclock must always be at the same temperature in order to flow with the same speed since water evaporates it must be replenished at regular intervals thus not being as effective in measuring time as the hourglass.
REVISED	The hourglass is more accurate than the waterclock because water in a waterclock must always be at the same temperature to flow at the same speed. Also, water evaporates and must be replenished at regular intervals. These temperature and volume problems

make the waterclock less effective than the hourglass in measuring time.

Faulty Agreement—Subject and Verb

The subject should agree in number with the verb. Faulty agreement seldom occurs in short sentences, where subject and verb are not far apart: "Jack eat too much" instead of "Jack eats too much." But when the subject is separated from the verb by other words, we sometimes lose track of the subject-verb relationship.

> FAULTY   The lion's **share** of diesels **are** sold in Europe.

Although **diesels** is closest to the verb, the subject is **share,** a singular subject that needs a singular verb.

> REVISED   The lion's **share** of diesels **is** sold in Europe.

Agreement errors are easy to correct once subject and verb are identified.

> FAULTY   There **is** an estimated 29,000 **women** living in our city.
> REVISED   There **are** an estimated 29,000 **women** living in our city.
>
> FAULTY   **A system** of lines **extend** horizontally to form a grid.
> REVISED   **A system** of lines **extends** horizontally to form a grid.

A second problem with subject-verb agreement occurs with indefinite subject pronouns such as **each, everyone, anybody,** and **somebody.** They usually take a singular verb.

> FAULTY   **Each** of the crew members **were** injured during the storm.
> REVISED   **Each** of the crew members **was** injured during the storm.
>
> FAULTY   **Everyone** in the group **have** practiced long hours.
> REVISED   **Everyone** in the group **has** practiced long hours.

Collective nouns such as **herd, family, union, group, army, team, committee,** and **board** can call for a singular or plural verb, depending on your intended meaning. When denoting the group as a whole, use a singular verb.

> CORRECT   The **committee meets** weekly to discuss new business.
>   The editorial **board** of this magazine **has** high standards.

To denote individual members of the group, use a plural verb.

> CORRECT   The **committee disagree** on whether to hire Jim.
>   The editorial **board are** all published authors.

When two subjects are joined by **either . . . or** or **neither . . . nor,** the verb is singular if both subjects are singular and plural if both subjects are plural. If one subject is plural and one is singular, the verb agrees with the one closer to the verb.

CORRECT   Neither **John** nor **Bill works** regularly.
Either **apples** or **oranges are** good vitamin sources.
Either **Felix** or his **friends are** crazy.
Neither the **boys** nor their **father likes** the home team.

If, on the other hand, two subjects (singular, plural, or mixed) are joined by **both . . . and,** the verb will be plural.

CORRECT   **Both** Joe **and** Bill **are** resigning.

A single **and** between subjects makes for a plural subject.

Faulty Agreement—Pronoun and Referent

A pronoun must refer to a specific noun (its referent or antecedent), with which it must agree in gender and number.

CORRECT   **Jane** lost **her** book

The **students** complained that **they** had been treated unfairly.

When an indefinite pronoun such as **each, everyone, anybody, someone,** or **none** serves as the pronoun referent, the pronoun is singular.

CORRECT   **Anyone** can get **his** degree from that college.

Anyone can get **his** or **her** degree from that college.

Each candidate described **her** plans in detail.

Faulty Coordination

Give equal emphasis to ideas of equal importance by joining them, within simple or compound sentences, with coordinating conjunctions: **and, but, or, nor, for, so,** and **yet.**

> This course is difficult **but** worthwhile.
> My horse is old **and** gray.
> We must decide to support **or** reject the dean's proposal.

But do not confound your meaning by coordinating excessively.

EXCESSIVE COORDINATION   The climax in jogging comes after a few miles **and** I can no longer feel stride after stride **and** it seems as if I am floating **and** jogging becomes almost a reflex **and** my arms **and** legs continue to move **and** my mind no longer has to control their actions.

> **REVISED** The climax in jogging comes after a few miles when I can no longer feel stride after stride. By then I am jogging almost by reflex, nearly floating, my arms and legs still moving, my mind no longer having to control their actions.

Notice how the meaning becomes clear when the less important ideas (**nearly floating, arms and legs still moving, my mind no longer having**) are shown as dependent on, rather than equal to, the most important idea (**jogging almost by reflex**)—the idea that contains the lesser ones.

Avoid coordinating two or more ideas that cannot be sensibly connected:

> **FAULTY** John had a drinking problem **and** he dropped out of school.
>
> **REVISED** John's drinking problem depressed him so much that he couldn't study, so he quit school.
>
> **FAULTY** I was late for work **and** wrecked my car.
>
> **REVISED** Late for work, I backed out of the driveway too quickly, hit a truck, and wrecked my car.

Instead of *try and,* use *try to.*

> **FAULTY** I will try and help you.
>
> **REVISED** I will try to help you.

Faulty Subordination

Proper subordination shows that a less important idea is dependent on a more important idea. A dependent (or subordinate) clause in a sentence is signaled by a subordinating conjunction: **because, so, if, unless, after, until, since, while, as,** and **although.** Consider these complete ideas:

> Joe studies hard. He has severe math anxiety.

Because these ideas are expressed as simple sentences, they appear coordinate (equal in importance). But if you wanted to indicate your opinion of Joe's chances of succeeding in math, you would need a third sentence: **His disability probably will prevent him from succeeding,** or **His willpower will help him succeed.** To communicate the intended meaning concisely, combine the two ideas. Subordinate the one that deserves less emphasis and place the idea you want emphasized in the independent (main) clause.

sub

> Despite his severe math anxiety (*subordinate idea*), Joe studies hard (*independent idea*).

This first version suggests that Joe will succeed. Below, the subordination suggests the opposite meaning:

> Despite his diligent studying (*subordinate idea*), Joe has severe math anxiety (*independent idea*).

Do not coordinate when you should subordinate:

> **WEAK**   Television viewers can relate to an athlete they idolize and they feel obliged to buy the product endorsed by their hero.

Of the two ideas in the sentence above, one is the cause, the other the effect. Emphasize this relationship through subordination:

> **REVISED**   Because television viewers can relate to an athlete they idolize, they feel obliged to buy the product endorsed by their hero.

When combining several ideas within a sentence, decide which is most important, and subordinate the other ideas to it—do not merely coordinate:

> **FAULTY**   This employee is often late for work, and he writes illogical reports, and he is a poor manager, and he should be fired.

> **REVISED**   Because this employee is often late for work, writes illogical reports, and has poor management skills, **he should be fired.** (*The last clause is independent.*)

Faulty Pronoun Case

ca

A pronoun's case (nominative, objective, or possessive) is determined by its role in the sentence: as subject, object, or indicator of possession.

If the pronoun serves as the subject of a sentence (**I, we, you, she, he, it, they, who**), its case is *nominative.*

> **She** completed her graduate program in record time.
> **Who** broke the chair?

When a pronoun follows a version of the verb **to be** (a linking verb), it explains (complements) the subject, and so its case is nominative.

> The killer was **she.**
> The professor who perfected our new distillation process is **he.**

If the pronoun serves as the object of a verb or a preposition (**me, us, you, her, him, it, them, whom**), its case is *objective.*

OBJECT OF THE VERB	The employees gave **her** a parting gift.
OBJECT OF THE PREPOSITION	To **whom** do you wish to complain?

If a pronoun indicates possession (**my, mine, our, ours, your, yours, his, her, hers, its, their, whose**), its case is *possessive*.

> The brown briefcase is **mine.**
> Her offer was accepted.
> **Whose** opinion do you value most?

Here are some frequent errors in pronoun case:

FAULTY	**Whom** is responsible to **who?** [*The subject should be nominative and the object should be objective.*]
REVISED	**Who** is responsible to **whom?**
FAULTY	The debate was between Marsha and **I.** [*As object of the preposition, the pronoun should be objective.*]
REVISED	The debate was between Marsha and **me.**

Deleting the accompanying noun from the two latter examples reveals the correct pronoun case ("We . . . are accountable . . . ."; "A group of us . . . will fly . . . .").

Faulty Modification

Modifiers explain, define, or add detail to other words or ideas. Prepositional phrases, for example, usually define or limit adjacent words:

> the foundation **with the cracked wall**
> the repair job **on the old Ford**
> the journey **to the moon**

So do phrases with "-ing" verb forms:

> the student **painting the portrait**
> **Opening the door,** we entered quietly.

Phrases with "*to* + verb" form limit:

> **To succeed,** one must work hard.

Some clauses also limit:

> the person **who came to dinner**

Problems with ambiguity occur when a modifying phrase has no word to modify.

dgl

DANGLING MODIFIER   **Dialing the phone,** the cat ran out the open door.

The cat obviously did not dial the phone, but because the modifier **Dialing the phone** has no word to modify, the noun beginning the main clause (*cat*) seems to name the one who dialed the phone. Without any word to join itself to, the modifier *dangles*. Inserting a subject repairs this absurd message.

CORRECT   As **Joe** dialed the phone, the cat ran out the open door.

A dangling modifier can also obscure your meaning.

DANGLING MODIFIER   **After completing the student financial aid application form,** the Financial Aid Office will forward it to the appropriate state agency.

Who completes the form—the student or the financial aid office?

Here are some other dangling modifiers that make the message confusing, inaccurate, or downright absurd:

DANGLING MODIFIER   **While walking,** a cold chill ran through my body.

CORRECT   While **I** walked, a cold chill ran through my body.

DANGLING MODIFIER   Impurities have entered our bodies **by eating chemically processed foods.**

CORRECT   Impurities have entered our bodies by **our** eating chemically processed foods.

The order of adjectives and adverbs also affects the meaning of sentences.

I **often** remind myself of the need to balance my checkbook.
I remind myself of the need to balance my checkbook **often.**

Position modifiers to reflect your meaning.

MISPLACED MODIFIER   Joe typed another memo on our computer **that was useless.** (*Was the typewriter or the memo useless?*)

CORRECT   Joe typed another useless memo on our computer.

or

Joe typed another memo on our useless computer.

MISPLACED MODIFIER   She volunteered **immediately** to deliver the radioactive shipment. (*Volunteering immediately, or delivering immediately?*)

CORRECT	She immediately volunteered to deliver . . .

or

She volunteered to deliver immediately . . .

par

Faulty Parallelism

To reflect relationships among items of equal importance, express them in identical grammatical form:

CORRECT	We here highly resolve . . . that government **of the people, by the people, for the people** shall not perish from the earth.

Otherwise, the message would be garbled, like this:

FAULTY	We here highly resolve . . . that government **of the people, which the people created and maintain, serving the people** shall not perish from the earth.

If you begin the series with a noun, use nouns throughout the series; likewise for adjectives, adverbs, and specific types of clauses and phrases.

FAULTY	The new apprentice is **enthusiastic, skilled,** and **you can depend on her.**
CORRECT	The new apprentice is **enthusiastic, skilled,** and **dependable.** (*all subjective complements*)
FAULTY	In his new job, he felt **lonely** and **without a friend.**
CORRECT	In his new job, he felt **lonely** and **friendless.** (*both adjectives*)
FAULTY	She plans **to study** all this month and **on scoring well** in her licensing examination.
CORRECT	She plans **to study** all this month and **to score well** in her licensing examination. (*both infinitive phrases*)
FAULTY	She **sleeps** well and **jogs** daily, **as well as eating** high-protein foods.
CORRECT	She **sleeps** well, **jogs** daily, and **eats** high-protein foods. (*all verbs*)

Shift

Sentence Shifts

Shifts in point of view damage coherence. If you begin a sentence or paragraph with one subject or person, do not shift to another.

SHIFT IN PERSON	When **one** finishes such a great book, **you** will have a sense of achievement.

REVISED	When **you** finish such a great book, **you** will have a sense of achievement.
SHIFT IN NUMBER	**One** should sift the flour before **they** make the pie.
REVISED	**One** should sift the flour before **one** makes the pie. (*Or better: Sift the flour before making the pie.*)

Do not begin a sentence in the active voice and then shift to the passive voice.

SHIFT IN VOICE	**He** delivered the plans for the apartment complex, and the building site **was also inspected by him.**
REVISED	He **delivered** the plans for the apartment complex and also **inspected** the building site.

Do not shift tenses without good reason.

SHIFT IN TENSE	She **delivered** the blueprints, **inspected** the foundation, **wrote** her report, and **takes** the afternoon off.
REVISED	She **delivered** the blueprints, **inspected** the foundation, **wrote** her report, and **took** the afternoon off.

Do not shift from one verb mood to another (as from imperative to indicative mood in a set of instructions).

SHIFT IN MOOD	**Unscrew** the valve and then steel wool **should be used** to clean the fittings.
REVISED	**Unscrew** the valve and then **use** steel wool to clean the fittings.

 EFFECTIVE PUNCTUATION

Punctuation marks are like road signs and traffic signals. They govern reading speed and provide clues for navigation through a network of ideas.

End Punctuation

The three marks of end punctuation—period, question mark, and exclamation point—work like a red traffic light by signaling a complete stop.

 PERIOD.   A period ends a sentence and is the final mark in some abbreviations.

> Ms.   Assn.   Inc.

Periods serve as decimal points in numbers.

> $15.95
> 21.4%

QUESTION MARK.   A question mark follows a direct question.

> Where is the essay that was due today**?**

Do not use a question mark to end an indirect question.

FAULTY	Professor Grim asked if all students had completed the essay**?**
REVISED	Professor Grim asked if all students had completed the essay.

or

Professor Grim asked, "Did all students complete the essay**?**"

EXCLAMATION POINT.   Use an exclamation point only when expression of strong feeling is appropriate.

APPROPRIATE	Oh, no!
	Pay up!

Semicolon

Like a blinking red traffic light at an intersection, a semicolon signals a brief but definite stop.

SEMICOLONS SEPARATING INDEPENDENT CLAUSES.   Semicolons separate independent clauses (logically complete ideas) whose contents are closely related and that are not connected by a coordinating conjunction.

> The project was finally completed**;** we had done a good week's work.

The semicolon can replace the conjunction-comma combination that joins two independent ideas.

> The project was finally completed, and we were elated.
> The project was finally completed**;** we were elated.

The second version emphasizes the sense of elation.

SEMICOLONS USED WITH ADVERBS AS CONJUNCTIONS AND OTHER TRANSITIONAL EXPRESSIONS.   Semicolons must accompany conjunctive adverbs like **besides, otherwise, still, however, furthermore, moreover, consequently, therefore, on the other hand, in contrast,** or **in fact.**

The job is filled; **however**, we will keep your résumé on file.
Your background is impressive**; in fact**, it is the best among our applicants.

SEMICOLONS SEPARATING ITEMS IN A SERIES. When items in a series contain internal commas, semicolons provide clear separation between items.

I am applying for summer jobs in Santa Fe, New Mexico; Albany, New York; Montgomery, Alabama; and Moscow, Idaho.

Members of the survey crew were Juan Jimenez, a geologist; Hector Lightfoot, a surveyor; and Mary Shelley, a graduate student.

Colon

Like a flare in the road, a colon signals you to stop and then proceed, paying attention to the situation ahead. Usually a colon follows an introductory statement that requires a follow-up explanation.

We need this equipment immediately: a voltmeter, a portable generator, and three pairs of insulated gloves.

She is an ideal colleague: honest, reliable, and competent.

Except for salutations in formal correspondence (e.g., Dear Ms. Jones:) colons follow independent (logically and grammatically complete) statements.

FAULTY My plans include: finishing college, traveling for two years, and settling down in Sante Fe.

No punctuation should follow "include."
Colons can introduce quotations.

The supervisor's message was clear enough: "You're fired."

A colon can replace a semicolon between two related, complete statements when the second one explains or amplifies the first.

Pam's reason for accepting the lowest-paying job offer was simple: She had always wanted to live in the Northwest.

Comma

The comma is the most frequently used—and abused—punctuation mark. It works like a blinking yellow light, for which you slow down briefly without stopping. Never use a comma to signal a *break* between independent ideas.

COMMA AS A PAUSE BETWEEN COMPLETE IDEAS.　In a compound sentence in which a coordinating conjunction (**and, or, nor, for, but**) connects equal (independent) statements, a comma usually precedes the conjunction.

> This is an excellent course**,** but the work is difficult.

COMMA AS A PAUSE BETWEEN AN INCOMPLETE AND A COMPLETE IDEA.　A comma is usually placed between a complete and an incomplete statement in a complex sentence when the incomplete statement comes first.

> **Because he is a fat cat,** Jack diets often.
> **When he eats too much,** Jack gains weight.

When the order is reversed (complete idea followed by incomplete), the comma is usually omitted.

> Jack diets often **because he is a fat cat.**
> Jack gains weight **when he eats too much.**

Reading a sentence aloud should tell you whether to pause (and use a comma).

COMMAS SEPARATING ITEMS (WORDS, PHRASES, OR CLAUSES) IN A SERIES.　Use commas after items in a series, including the next-to-last item.

> **Helen, Joe, Marsha,** and **John** are joining us on the term project.
>
> He works hard **at home, on the job,** and even **during his vacation.**
>
> The new employee complained **that the hours were long, that the pay was low, that the work was boring, and that the supervisor was paranoid.**

Use no commas if **or** or **and** appears between all items in a series.

> She is willing to study in San Francisco or Seattle or even in Anchorage.

COMMAS SETTING OFF INTRODUCTORY PHRASES.　Infinitive, prepositional, or verbal phrases introducing a sentence are usually set off by commas, as are interjections.

INFINITIVE PHRASE	**To be or not to be,** that is the question.
PREPOSITIONAL PHRASE	**In Rome,** do as the Romans do.
PARTICIPIAL PHRASE	**Being fat,** Jack was slow at catching mice.
	Moving quickly, the army surrounded the enemy.
INTERJECTION	**Oh,** is that the verdict?

COMMAS SETTING OFF NONRESTRICTIVE ELEMENTS.   A *restrictive* phrase or clause modifies or defines the subject in such a way that deleting the modifier would change the meaning of the sentence.

> All students **who have work experience** will receive preference.

Without **who have work experience,** which *restricts* the subject by limiting the category **students,** the meaning would be entirely different: **All** students will receive preference.

Because this phrase is essential to the sentence's meaning, it is *not* set off by commas.

A *nonrestrictive* phrase or clause could be deleted without changing the sentence's meaning and *is* set off by commas.

> Our new manager, **who has only six weeks' experience,** is highly competent.

MODIFIER DELETED	Our new manager is highly competent.

> This house, **riddled with carpenter ants,** is falling apart.

MODIFIER DELETED	This house is falling apart.

COMMAS SETTING OFF PARENTHETICAL ELEMENTS.   Items that interrupt the flow of a sentence (such as **of course, as a result, as I recall,** and **however**) are called parenthetical and are enclosed by commas. They may denote emphasis, afterthought, clarification, or transition.

EMPHASIS	This deluxe model, **of course,** is more expensive.
AFTERTHOUGHT	Your essay, **by the way,** was excellent.
CLARIFICATION	The loss of my job was, **in a way,** a blessing.
TRANSITION	Our warranty, **however,** does not cover tire damage.

Direct address is parenthetical.

> Listen, **my children,** and you shall hear . . . .

A parenthetical expression at the beginning or the end of a sentence is set off by a comma.

> **Naturally,** we will expect a full guarantee.
> **My friends,** I think we have a problem.
> You've done a good job, **Jim.**
> **Yes,** you may use my name in your advertisement.

COMMAS SETTING OFF QUOTED MATERIAL. Quoted items within a sentence are set off by commas.

> The customer said, **"I'll take it,"** as soon as he laid eyes on our new model.

COMMAS SETTING OFF APPOSITIVES. An appositive, a word or words explaining a noun and placed immediately after it, is set off by commas when the appositive is nonrestrictive. (See page 766.)

> Martha Jones, **our new president,** is overhauling all personnel policies.
>
> Alpha waves, **the most prominent of the brain waves,** are typically recorded in a waking subject whose eyes are closed.
>
> Please make all checks payable to Sam Sawbuck, **school treasurer.**

COMMAS USED IN COMMON PRACTICE. Commas set off the day of the month from the year, in a date.

> May 10, 1989

Commas set off numbers in three-digit intervals.

> 11,215
> 6,463,657

They also set off street, city, and state in an address.

> Mail the bill to J. B. Smith, 18 Sea Street, Albany, Iowa 01642.

When the address is written vertically, however, the commas that would otherwise occur at the end of each address line are omitted.

> J. B. Smith
> 18 Sea Street
> Albany, Iowa 01642

Commas set off an address or date in a sentence.

> Room 3C, Margate Complex, is my summer address.
> June 15, 1987, is my graduation date.

Commas set off degrees and titles from proper nouns.

> Roger P. Cayer, M.D.
> Gordon Browne, Jr.
> Sandra Mello, Ph.D.

COMMAS USED INCORRECTLY. Avoid needless or inappropriate commas. Read a sentence aloud to identify inappropriate pauses.

FAULTY
The instructor told me, that I was late. [*separates the indirect from the direct object*]

The most universal symptom of the suicide impulse, is depression. [*separates the subject from its verb*]

This has been a long, difficult, semester. [*second comma separates the final adjective from its noun*]

John, Bill, and Sally, are joining us on the trip home. [*third comma separates the final subject from its verb*]

An employee, who expects rapid promotion, must quickly prove his or her worth. [*separates a modifier that should be restrictive*]

I spoke by phone with John, and Marsha. [*separates two nouns, linked by a coordinating conjunction*]

The room was, 18 feet long. [*separates the linking verb from the subjective complement*]

We painted the room, red. [*separates the object from its complement*]

Apostrophe

ap

Apostrophes indicate the possessive, a contraction, and the plural of numbers, letters, and figures.

APOSTROPHE INDICATING THE POSSESSIVE. At the end of a singular word, or of a plural word that does not end in *s*, add an apostrophe plus *s* to indicate the possessive. Single-syllable nouns that end in *s* take the apostrophe before an added *s*.

> The **people's** candidate won.
> The chainsaw was **Emma's.**
> The **women's** locker room burned.
> I borrowed **Chris's** book.

Do not add *s* to words that already end in *s* and have more than one syllable; add an apostrophe only.

> Aristophanes' death

Do not use an apostrophe to indicate the possessive form of either singular or plural pronouns.

> The book was **hers.**
> **Ours** is the best school in the county.
> The fault was **theirs.**

At the end of a plural word that ends in *s,* add an apostrophe only.

> the **cows'** water supply
> the **Jacksons'** wine cellar

At the end of a compound noun, add an apostrophe plus *s.*

> my **father-in-law's** false teeth

At the end of the last word in nouns of joint possession, add an apostrophe plus *s* if both own one item.

> **Joe and Sam's** lakefront cottage

Add an apostrophe plus *s* to both nouns if each owns specific items.

> **Joe's** and **Sam's** passports

APOSTROPHE INDICATING A CONTRACTION.   An apostrophe shows that you have omitted one or more letters in a phrase that is usually a combination of a pronoun and a verb.

> I'm   they're
> he's   you'd
> you're   who's

Don't confuse **they're** with **their** or **there.**

> FAULTY   there books
>   their now leaving
>   living their
>
> CORRECT   their books
>   they're now leaving
>   living there

Remember the distinction this way:

> Their friend knows they're there.

It's means "it is." **Its** is the possessive.

> It's watching its reflection in the pond.

Who's means "who is." **Whose** indicates the possessive.

> Who's interrupting whose work?

Other contractions are formed from the verb and the negative.

isn't	can't
don't	haven't
won't	wasn't

APOSTROPHE INDICATING THE PLURAL OF NUMBERS, LETTERS, AND FIGURES

> The **6's** on this new printer look like smudged **G's, 9's** are illegible, and the **%'s** are unclear.

Quotation Marks

Quotation marks set off the exact words borrowed from another speaker or writer. The period or comma at the end is placed within the quotation marks.

PERIODS AND COMMAS BELONG WITHIN QUOTATION MARKS

"Hurry up," Jack whispered.
Jack told Felicia, "I'm depressed."

The colon or semicolon is always placed outside quotation marks.

COLONS AND SEMICOLONS BELONG OUTSIDE QUOTATION MARKS

Our student handbook clearly defines "core requirements"; however, it does not list all the courses that fulfill the requirements.

When a question mark or exclamation point is part of a quotation, it belongs within the quotation marks, replacing the comma or period.

SOME PUNCTUATION BELONGS WITHIN QUOTATION MARKS

"Help!" he screamed.
Marsha asked John, "Can't we agree about anything?"

But if the question mark or exclamation point pertains to the attitude of the person quoting instead of the person being quoted, it is placed outside the quotation mark.

SOME PUNCTUATION BELONGS OUTSIDE QUOTATION MARKS

Why did Boris wink and whisper, "It's a big secret"?

Use quotation marks around titles of articles, paintings, book chapters, and poems.

CERTAIN TITLES BELONG WITHIN QUOTATION MARKS

The enclosed article, *"The Job Market for College Graduates,"* should provide some helpful insights.

But titles of books, journals, or newspapers should be underlined or italicized. Finally, use quotation marks (with restraint) to indicate irony.

QUOTATION MARKS TO INDICATE IRONY

She is some "friend"!

Ellipses

Three dots . . . indicate you have omitted material from a quotation. If the omitted words come at the end of the original sentence, a fourth dot indicates the period. (Also see page 682.)

> "Three dots . . . indicate . . . omitted . . . material . . . . A fourth dot . . . indicates the period . . . . "

Italics

In typing or longhand writing, indicate italics by underlining. On a word processor, use italic print for titles of books, periodicals, films, newspapers, and plays; for the names of ships; for foreign words or scientific names; sparingly, for emphasizing a word; and for indicating the special use of a word.

> The *Oxford English Dictionary* is a handy reference tool.
>
> The *Lusitania* sank rapidly.
>
> She reads the *Boston Globe* often.
>
> My only advice is *caveat emptor.*
>
> *Bacillus anthracis* is a highly virulent organism.
>
> *Do not* inhale these fumes under any circumstances!
>
> Our contract defines a *work-study student* as one who works a minimum of twenty hours weekly.

Parentheses

Use commas to set off parenthetical elements, dashes to give some emphasis to the material that is set off, and parentheses to enclose material that defines or explains the statement that precedes it.

> An anaerobic **(airless)** environment must be maintained for the cultivation of this organism.
>
> The cost of running our college has increased by 15 percent in one year **(see Appendix A for full cost breakdown).**
>
> This new calculator **(made by Ilco Corporation)** is perfect for science students.

Material between parentheses, like all other parenthetical material discussed earlier, can be deleted without harming the logical and grammatical structure of the sentence.

Brackets

Brackets in a quotation set off material that was not in the original quotation but is needed for clarification, such as an antecedent (or referent) for a pronoun.

> "She **[Amy]** was the outstanding candidate for the scholarship."

Brackets can enclose information taken from some other location within the context of the quotation.

> "It was in early spring **[April 2, to be exact]** that the tornado hit."

Use **sic** ("thus," or "so") when quoting an error from the original source.

> The assistant's comment was clear: "He don't **[sic]** want any."

Dashes

Dashes can be effective—if they are not overused. Parentheses deemphasize the enclosed material; dashes emphasize it.

> Have a good vacation—but watch out for sandfleas.
> Mary—a true friend—spent hours helping me rehearse.

TRANSITIONS

Transitional expressions **announce** relations between ideas. Words or phrases such as **for example, meanwhile,** and **however** work like bridges between thoughts.

Here is a paragraph in which these transitions are used to clarify the writer's line of thinking (emphasis added):

USING TRANSITIONS TO BRIDGE IDEAS

> Psychological and social problems of aging are too often aggravated by the final humiliation: poverty. One out of every three older Americans lives near or below the poverty level. **Meanwhile,** only one out of every nine younger adults lives in poverty. The American public assumes that Social Security and Medicare provide adequate support for the aged. These benefits alone, **however,** are rarely enough to raise an older person's living standards above the poverty level. **Moreover,** older people are the only group living in poverty whose population has recently increased rather than decreased. More and more of our aging citizens *thus* confront the prospect of living with less and less.

NOTE *Transitional expressions should be a limited option for achieving coherence. Use them sparingly, and only when a relationship is not already clear.*

Whole sentences can serve as transitions between paragraphs, and a whole paragraph can serve as a transition between sections of writing. Assume, for instance, that you work as a marketing intern for a stereo manufacturer. You have just completed a section of a memo on the advantages of the new AKS amplifier

COMMON TRANSITIONS AND THE RELATIONS THEY INDICATE

- **An addition:** *moreover, in addition, and, also*

  > I am majoring in naval architecture; **also,** I spent three years crewing on a racing yawl.

- **Results:** *thus, hence, therefore, accordingly, thereupon, as a result, and so, as a consequence*

  > Mary enjoyed all her courses; **therefore,** she worked especially hard last semester.

- **An example or illustration:** *for instance, to illustrate, namely, specifically*

  > Competition for part-time jobs is fierce; **for example,** 80 students applied for the clerk's job at Sears.

- **An explanation:** *in other words, simply stated, in fact*

  > Louise had a terrible semester; **in fact,** she flunked three courses.

- **A summary or conclusion:** *in closing, to conclude, to summarize, in brief, in summary, to sum up, all in all, on the whole, in retrospect, in conclusion*

  > Our credit is destroyed, our bank account is overdrawn, and our debts are piling up; **in short,** we are bankrupt.

- **Time:** *first, next, second, then, meanwhile, at length, later, now, the next day, in the meantime, in turn, subsequently*

  > Mow the ball field this morning; **then,** clean the dugouts.

- **A comparison:** *likewise, in the same way, in comparison*

  > Our reservoir is drying up because of the drought; **similarly,** water supplies in neighboring towns are dangerously low.

- **A contrast or alternative:** *however, nevertheless, yet, still, in contrast, otherwise, but, on the other hand, to the contrary, notwithstanding, conversely*

  > Felix worked hard; **however,** his grades remained poor.

and are now moving to a section on selling the idea to consumers. This next paragraph might link the two sections:

A TRANSITIONAL PARAGRAPH

> Because the AKS amplifier increases bass range by 15 percent, it should be installed as a standard item in all our stereo speakers. Tooling and installation adjustments, however, will add roughly $50 to the list price of each model. We must, therefore, explain the cartridge's long-range advantages to consumers. Let's consider ways of explaining these advantages.

Notice that this transitional paragraph *contains* transitional expressions as well.

EFFECTIVE MECHANICS

Correctness in abbreviation, hyphenation, capitalization, use of numbers, and spelling demonstrates your attention to detail.

Abbreviations

Avoid abbreviations in formal writing or in situations that might confuse your reader. When in doubt, write the word out.

Abbreviate some words and titles when they precede or immediately follow a proper name, but not military, religious, or political titles.

CORRECT	**Mr.** Jones
	Dr. Jekyll
	Raymond Dumont, **Jr.**
	Reverend Ormsby
	President Clinton

Abbreviate time designations only when they are used with actual times.

FAULTY	Plato lived sometime in the **BC** period.
	She arrived in the **a.m.**
CORRECT	400 **B.C.**
	5:15 **a.m.**

Abbreviate a unit of measurement only when it appears often in your report and is written out in full on first use. Use only abbreviations you are sure readers will understand. Abbreviate items in a visual aid only if you need to save space.

COMMON ABBREVIATIONS FOR REFERENCE IN MANUSCRIPTS

anon.	anonymous	fig.	figure
app.	appendix	i.e.	that is
b.	born	illus.	illustrated
©	copyright	jour.	journal
c., ca.	about (c. 1988)	l., ll.	line(s)
cf.	compare	ms., mss.	manuscript(s)
ch.	chapter	no.	number
col.	column	p., pp.	page(s)
d.	died	pt., pts.	part(s)
ed.	editor	rev.	revised or review
e.g.	for example	sec.	section
esp.	especially	sic.	thus, so (to cite an
et al.	and others		error in the quotation)
etc.	and so on	trans.	translation
ex.	example	vol.	volume
f. or ff.	the following page or pages		

Most dictionaries provide an alphabetical list of other abbreviations. For abbreviations in documentation of research sources, see pages 686–717.

COMMON ABBREVIATIONS FOR UNITS OF MEASUREMENT

AC	alternating current	kw	kilowatt
amp	ampere	kwh	kilowatt hour
Å	angstrom	l	liter
az	azimuth	lat	latitude
bbl	barrel	lb	pound
BTU	British Thermal Unit	lin	linear
C	Celsius	log	logarithm
cal	calorie	long	longitude
cc	cubic centimeter	m	meter
circ	circumference	max	maximum
cm	centimeter	mg	milligram
CPS	cycles per second	min	minute
cu ft	cubic foot	ml	milliliter
dB	decibel	mm	millimeter
DC	direct current	mo	month
dm	decimeter	mph	miles per hour
doz	dozen	oct	octane
DP	dewpoint	oz	ounce
F	Fahrenheit	psf	pounds per square foot
F	farad	psi	pounds per square inch
fbm	foot board measure	qt	quart
fl oz	fluid ounce	r	roentgen
FM	frequency modulation	rpm	revolutions per minute
freq	frequency	sec	second
ft	foot	sp gr	specific gravity
ft lb	foot pound	sq	square
gal	gallon	t	ton
GPM	gallons per minute	temp	temperature
gr	gram	tol	tolerance
hp	horsepower	ts	tensile strength
hr	hour	V	volt
in	inch	VA	volt ampere
IU	international unit	W	watt
J	joule	wk	week
ke	kinetic energy	WL	wavelength
kg	kilogram	yd	yard
km	kilometer	yr	year

Hyphen

Hyphens divide words at line breaks and join two or more words used as a single adjective if they precede the noun, but not if they follow it:

> com-
> puter
>
> the rough-hewn wood
>
> the all-too-human error
>
> The wood was rough hewn.
>
> The error was all too human.

Some other commonly hyphenated words:

- Most words that begin with the prefix *self-*. (Check your dictionary.)

> self-reliance
> self-discipline

- Combinations that might be ambiguous.

> re-creation [*a new creation*]
> recreation [*leisure activity*]

- Words that begin with *ex* only if *ex* means "past."

> ex-faculty member
> excommunicate

- All fractions, along with ratios that are used as adjectives and that precede the noun (but not those that follow it), and compound numbers from twenty-one through ninety-nine.

> a **two-thirds** majority
>
> In a **four-to-one** vote, the student senate defeated the proposal.
>
> The proposal was voted down **four to one.**
>
> **Thirty-eight** windows were broken.

Capitalization

Capitalize the first words of all sentences as well as titles of people, books, and chapters; languages; days of the weeks; the months; holidays; names of organizations or groups; races and nationalities; historical events; important documents; and names of structures or vehicles. In titles of books, films, and the like, capitalize the first word and all those following except articles or prepositions.

ITEMS THAT ARE	Joe Schmoe	Russian
CAPITALIZED	*A Tale of Two Cities*	Labor Day
	Protestant	Dupont Chemical Company
	Wednesday	Senator Barbara Boxer
	the *Queen Mary*	France
	the Statue of Liberty	The War of 1812

Do not capitalize the seasons (**spring, winter**) or general groups (**the younger generation, the leisure class**).

Capitalize adjectives that are derived from proper nouns.

> Chaucerian English

Capitalize titles preceding a proper noun (but not those following).

> State Senator Marsha Smith
> Marsha Smith, state senator

Capitalize words such as **street, road, corporation,** and **college** only when they accompany a proper noun.

> Bob Jones University
> High Street
> The Rand Corporation

Capitalize **north, south, east,** and **west** when they denote specific locations, not when they are simply directions.

> the South
> the Northwest
> Turn east at the next set of lights.

Use of Numbers

Numbers expressed in one or two words can be written out or written as numerals. Use numerals to express larger numbers, decimals, fractions, precise technical figures, or any other exact measurements.

543	2,800,357
$3\frac{1}{4}$	15 pounds of pressure
50 kilowatts	4,000 rpm

Use numerals for dates, census figures, addresses, page numbers, exact units of measurement, percentages, times with a.m. or p.m. designations, and monetary and mileage figures.

page 14	1:15 p.m.
18.4 pounds	9 feet
12 gallons	$15

Do not begin a sentence with a numeral. If your figure needs more than two words, revise your word order.

Six hundred students applied for the 102 available jobs.

The 102 available jobs brought 780 applicants.

Do not use numerals to express approximate figures, time not designated as a.m. or p.m., or streets named by numbers less than 100.

about seven hundred fifty

four fifteen

108 East Forty-Second Street

In contracts and other documents in which precision is vital, a number can be stated both in numerals and in words:

The tenant agrees to pay a rental fee of **three hundred seventy-five dollars ($375.00) monthly.**

Spelling

Take the time to use your dictionary for all writing assignments. When you read, note the spelling of words that give you trouble. Compile a list of troublesome words.

Works Cited

Abelman, Arthur F. "Legal Issues in Scholarly Publishing." *MLA Style Manual*. 2nd ed. New York: Modern Language Association, 1998: 30–57.

Adams, Gerald R., and Jay D. Schvaneveldt. *Understanding Research Methods*. New York: Longman, 1985.

Adler, Jerry. "For Humans, Evolution Ain't What It Used to Be." *Newsweek* 29 Sept. 1997: 17.

"Advertising and Marketing on the Internet." Sept. 2000. Online 8 pp. Federal Trade Commission. 17 Jan. 2001 <www.ftc.gov.bcp/conline/pubs/buspubs/ruleroad.htm>.

"Advisories on the Use of Medical Web Sites Issued." *Professional Ethics Report* [American Association for the Advancement of Science] XII.3 (Summer 1999): 2–3.

The Aldus Guide to Basic Design. Aldus Corporation, 1988.

Alleman, James E., and Brooke T. Mossman. "Asbestos Revisited." *Scientific American* July 1997: 70–75.

Allen, Lori, and Dan Voss. "Ethics for Editors: An Analytical Decision-Making Process". *IEEE Transactions on Professional Communication* 41.1 (Mar. 1998): 58–65.

American Psychological Association. *Publication Manual of the American Psychological Association*. 5th ed. Washington: Author, 2001.

"And the Winner of the Dubious-Study-of-Year Award Is . . . ." *University of California at Berkeley Wellness Letter* 14.6 (1998): 1+.

Anson, Chris M., and Robert A. Schwegler. *The Longman Handbook for Writers and Readers*, 2nd ed. New York: Longman, 2000.

"Any Alternative?" *The Economist* 1 Nov. 1997: 83–84.

Archee, Raymond K. "Online Intercultural Communication." *Intercom* Sept./Oct. 2003: 40–41.

"Are We in the Middle of a Cancer Epidemic?" *University of California at Berkeley Wellness Letter* 10.9 (1994): 4–5.

Armstrong, William H. "Learning to Listen." *American Educator* (Winter 1997–98): 24+.

Author's Guide. New York: Addison Wesley Longman, 1998.

Baker, Russ. "Surfer's Paradise." *Inc.* Nov. 1997: 57+.

Ball, Charles. "Figuring the Risks of Closer Runways." *Technology Review* Aug./Sept. 1996: 12–13.

Barbour, Ian. *Ethics in an Age of Technology*. New York: Harper, 1993.

Barnes, Shaleen. "Evaluating Sources Checklist." Information Literacy Project. 10 June 1997. Online. 23 June 1998 <www.2lib.umassd.edu/library2/INFOLIT/prop.html>.

Barnett, Arnold. "How Numbers Can Trick You." *Technology Review* Oct. 1994: 38–45.

Bashein, Barbara J., and M. Lynne Markus. "A Credibility Equation for IT Specialists." *Sloan Management Review* 38.4 (Summer 1997): 35–44.

Baumann, K. E., et al. "Three Mass Media Campaigns to Prevent Adolescent Cigarette Smoking." *Preventive Medicine* 17 (1988): 510–30.

Baumeister, Roy F. "Should Schools Try to Boost Self-Esteem?" *American Educator* (Summer 1996): 14+.

Bazerman, Max H., Kimberly P. Morgan, and George F. Loewenstein. "The Impossibility of Auditor Independence." *Sloan Management Review* 38.4 (Summer 1997): 89–94.

Beamer, Linda. "Learning Intercultural Communication Competence." *Journal of Business Communication* 29.3 (1992): 285–303.

Bedford, Marilyn S., and F. Cole Stearns. "The Technical Writer's Responsibility for Safety." *IEEE Transactions on Professional Communication* 30.3 (1987): 127–32.

Begley, Sharon. "Bad Days on the Lily Pad." *Newsweek* 13 July 1998: 67.

———. "Is Science Censored?" *Newsweek* 14 Sept. 1992: 63.

———. "Odds on the Greenhouse." *Newsweek* 1 Dec. 1997: 72.

Belkin, Lisa. "How Can We Save the Next Victim?" *New York Times Magazine* 15 June 1997: 28+.

Benson, Phillipa J. "Visual Design Consideration in Technical Publications." *Technical Communication* 32.4 (1985): 35–39.

Bernstein, Peter L. *Against the Gods: The Remarkable Story of Risk.* New York: Wiley, 1998.

Berry, Stephen R. "Scientific Information in the Electronic Era." *Professional Ethics Report* [American Association for the Advancement of Science] X.2 (Spring 1997): 1+.

Bjerklie, David. "E-Mail: The Boss Is Watching." *Technology Review* 14 Apr. 1993: 14–15.

Blaser, Martin J. "The Bacteria behind Ulcers." *Scientific American* Feb. 1996: 140+.

Blinder, Alan S., and Richard E. Quandt. "The Computer and the Economy." *Atlantic Monthly* Dec. 1997: 26–32.

Blum, Deborah. "Investigative Science Journalism." *Field Guide for Science Writers.* Eds. Deborah Blum and Mary Knudson. New York: Oxford, 1997. 86–93.

Bogert, Judith, and David Butt. "Opportunities Lost, Challenges Met: Understanding and Applying Group Dynamics in Writing Projects." *Bulletin of the Association for Business Communication* 53.2 (1990): 51–53.

Boiarsky, Carolyn. "Using Usability Testing to Teach Reader Response." *Technical Communication* 39.1 (1992): 100–02.

Bolles, Richard Nelson. *Job Hunting on the Internet.* 2nd ed. Berkeley, CA: Ten Speed, 1999.

Bosley, Deborah. "International Graphics: A Search for Neutral Territory." *INTERCOM* Aug./Sept. 1996: 4–7.

Boucher, Norman. "Back to the Everglades." *Technology Review* Aug./Sept. 1995: 24–35.

Boyd, Ruth-Anne. "Plain Language: Making It Work." *INTERCOM* Nov. 1997: 16–18.

Branscum, Deborah. "bigbrother@the.office.com." *Newsweek* 27 Apr. 1998: 78.

Brimelow, Peter. "Income Gap." *Forbes* 27 July 1998: 51.

Broad, William J. "NASA Budget Cuts Raise Concerns over Safety of Shuttle. *New York Times* 8 Mar. 1994, sec. B: 5+.

Brower, Vicki. "Ethics for Hire." *Technology Review* Mar./Apr. 1999: 25.

Brownell, Judi, and Michael Fitzgerald. "Teaching Ethics in Business Communication: The Effective/Ethical Balancing Scale." *Bulletin of the Association for Business Communication* 55.3 (1992): 15–18.

Bruhn, Mark J. "E-Mail's Conversational Value." *Business Communication Quarterly* 58.3 (1995): 43–44.

Bryan, John. "Down the Slippery Slope: Ethics and the Technical Writer as Marketer." *Technical Communication Quarterly* 1.1 (1992): 73–88.

Bureau of Labor Statistics. "Employee Tenure Summary." 19 Sept. 2002. *News.* 12 Feb. 2004 <http://stats.bls.gov/news.release/tenure.nr0.htm>.

Burger, Katrina. "Righteousness Pays." *Forbes* 22 Sept. 2000: 11.

Burghardt, M. David. *Introduction to the Engineering Profession.* New York: Harper, 1991.

Burnett, Rebecca E. "Substantive Conflict in a Cooperative Context: A Way to Improve the Collaborative Planning of Workplace Documents." *Technical Communication* 38.4 (1991): 532–39.

Busiel, Christopher, and Tom Maeglin. *Researching Online.* New York: Addison, 1998.

Byrd, Patricia, and Joy M. Reid. *Grammar in the Composition Classroom.* Boston: Heinle, 1998.

Caher, John M. "Technical Documentation and Legal Liability." *Journal of Technical Writing and Communication* 25.1 (1995): 5–10.

Carliner, Saul. "Demonstrating Effectiveness and Value: A Process for Evaluating Technical Communication Products and Services." *Technical Communication* 44.3 (1997): 252–65.

———. "Physical, Cognitive, and Affective: A Three-Part Framework for Information Design." *Technical Communication* 47.2 (2000): 561–76.

Caswell-Coward, Nancy. "Cross-Cultural Communication: Is It Greek to You?" *Technical Communication* 39.2 (1992): 264–66.

Chauncey, C. "The Art of Typography in the Information Age." *Technology Review* Feb./Mar. (1986): 26+.

Christians, C. G., et al. *Media Ethics: Cases and Moral Reasoning.* 2nd ed. White Plains, NY: Longman, 1978.

Cialdini, Robert B. "The Science of Persuasion." *Scientific American* Feb. 2001: 76–81.

Clark, Gregory. "Ethics in Technical Communication: A Rhetorical Perspective." *IEEE Transactions on Professional Communication* 30.3 (1987): 190–95.

Clark, Thomas. "Teaching Students to Enhance the Ecology of Small Group Meetings." *Business Communication Quarterly* 61.4 (Dec. 1998): 40–52.

———. "Teaching Students How to Write to Avoid Legal Liability." *Business Communication Quarterly* 60.3 (1997): 71–77.

Clement, David E. "Human Factors, Instructions, and Warnings, and Product Liability." *IEEE Transactions on Professional Communication* 30.3 (1987): 149–56.

Cochran, Jeffrey K., et al. "Guidelines for Evaluating Graphical Designs." *Technical Communication* 36.1 (1989): 25–32.

Coe, Marlana. *Human Factors for Technical Communicators.* New York: Wiley, 1996.

———. "Writing for Other Cultures: Ten Problem Areas." *INTERCOM* Jan. 1997: 17–19.

Cohn, Victor. "Coping with Statistics." *A Field Guide for Science Writers.* Eds. Deborah Blum and Mary Knudson. New York: Oxford, 1997. 102–09.

Cole-Gomolski B. "Users Loathe to Share Their Know-How." *Computerworld* 17 Nov. 1997: 6.

Columbia Accident Investigation Board [NASA]. *Report,* Volume 1. Washington, DC: GPO, 2003.

Columbia Accident Investigation Board Press Briefing. August 26, 2003. Transcript. 7 May 2004. <http://www.caib.us/events/press_briefings/20030826/transcript.html>.

Communication Concepts, Inc. "Electronic Media Poses New Copyright Issues." *Writing Concepts* ©. Reprinted in *INTERCOM* Nov. 1995: 13+.

Congressional Research Report. Washington, DC: GPO, 1990.

Conlin, Michelle. "And Now, the Just-in-Time Employee." *Business Week* 28 Aug. 2000: 169–70.

"Consequences of Whistle Blowing in Scientific Misconduct Reported." *Professional Ethics Report* [American Association for the Advancement of Science] IX.4 (Winter 1996): 2.

Consumer Product Safety Commission. *Fact Sheet No. 65.* Washington: GPO, 1989.

Cooper, Lyn O. "Listening Competency in the Workplace: A Model for Training." *Business Communication Quarterly* 60.4 (Dec. 1997): 75–84.

"Copyright Protection and Fair Use of Printed Information." *Addison Wesley Longman Author's Guide.* New York: Longman, 1998.

Corbett, Edward P. J. *Classical Rhetoric for the Modern Student,* 3rd ed. New York: Oxford, 1990.

Cortese, Amy. "Automatic Web Downloads—without the overload." *Business Week* 24 Nov. 1997: 152.

Cotton, Robert, ed. *The New Guide to Graphic Design.* Secaucus, NJ: Chartwell, 1990.

"Crime Spree." *Business Week* 9 Sept. 2002: 8.

Cronin, Mary J. "Knowing How Employees Use the Intranet Is Good Business." *Fortune* 21 July 1997: 103.

———. "Using the Web to Push Key Data to Decision Makers." *Fortune* 29 Sept. 1997: 254.

Crosby, Olivia. *Employment Interviewing.* Washington, DC: U.S. Department of Labor, 2000.

———. *Résumés, Applications, and Cover Letters.* Washington DC: U.S. Department of Labor, 1999.

Cross, Mary. "Aristotle and Business Writing: Why We Need to Teach Persuasion." *Bulletin of the Association for Business Communication* 54.1 (1991): 3–6.

Crossen, Cynthia. *Tainted Truth: The Manipulation of Fact in America.* New York: Simon, 1994.

Crumpton, Amy. "Secrecy in Science." *Professional Ethics Report* [American Association for the Advancement of Science] XII.1 (Winter 1999): 1+.

Curry, Jerome. "Trapping the Internet's Job Search Resources." *Business Communication Quarterly* 61.2 (1998): 100–06.

D'Aprix, Roger. "Related Thoughts." *Journal of Employee Communication Management* Nov./Dec. 1997: 66–70.

Daugherty, Shannon. "The Usability Evaluation: A Discount Approach to Usability Testing." *INTERCOM* Dec. 1997: 16–20.

Davenport, Thomas H. *Information Ecology.* New York: Oxford, 1997.

Debs, Mary Beth, "Collaborative Writing in Industry." In *Technical Writing: Theory and Practice.* Eds. Bertie E. Fearing and W. Keats Sparrow. New York: Modern Language Assn., 1989, 33–42.

———. "Recent Research on Collaborative Writing in Industry." *Technical Communication* 38.4 (1991): 476–85.

December, John. "An Information Development Methodology for the World Wide Web." *Technical Communication* 43.3 (1996): 369–75.

Desmond, Edward W. "How Your Data May Soon Seek You Out." *Fortune* 8 Sept. 1997: 149–50.

Detjen, Jim. "Environmental Writing." *A Field Guide for Science Writers.* Eds. Deborah Blum and Mary Knudson. New York: Oxford, 1997, 173–79.

Devlin, Keith. *Infosense: Turning Information into Knowledge.* New York: W. H. Freeman, 1999.

Dillard, James P., Denise H. Solomon, and Jennifer A. Samp. "Framing Social Reality: The Relevance of Relational Judgments." *Communication Research* 23.6 (1996): 703–22.

Dobnik, Verena. "Surgeons Who Play Videogames Err Less." 7 April 2004. Associated Press web story. 10 May 2004 <http://www.msnbc.msn.com/id/4685909/>.

Dombrowski, Paul M. "*Challenger* and the Social Contingency of Meaning: Two Lessons for the Technical Communication Classroom." *Technical Communication Quarterly* 1.3 (1992): 73–86.

Dorner, Deitrich. *The Logic of Failure.* Reading, MA: Addison, 1996.

Dowd, Charles. "Conducting an Effective Journalistic Interview." *INTERCOM* May 1996: 12–14.

Doyle, Rodger. "Amphibians and Risk." *Scientific American* Aug. 1998: 27.

Dragga, Sam, and Gwendolyn Gong. *Editing: The Design of Rhetoric.* Amityville, NY: Baywood, 1989.

Dulude, Jennifer. "The Web Marketing Handbook." Thesis. University of Massachusetts Dartmouth, 1997.

Dumont, R. A., and J. M. Lannon, *Business Communications.* 3rd ed. Glenview, IL: Scott, 1990.

"Earthquake Hazard Analysis for Nuclear Power Plants." *Energy and Technology Review* June 1984:8.

Easton, Thomas, and Stephan Herrara. "J&J's Dirty Little Secret." *Forbes* 12 Jan. 1998: 42–44.

Edelman, Rob. "Commentary on Prescription Drug Commercials." *Midday Magazine.* Albany, NY: WAMC Radio 5 Feb. 2001.

Elbow, Peter. *Writing without Teachers.* New York: Oxford, 1973.

"Electronic Mentors." *The Futurist* May 1992: 56.

Elias, Stephen. *Patent, Copyright, and Trademark.* Berkeley, CA: Nolo Press, 1997.

Elliot, Joel. "Evaluating Web Sites: Questions to Ask." 18 Feb. 1997. Online. List for Multimedia and New Technologies in Humanities Teaching. 9 Mar. 1997 <www.learnnc.org/documents/webeval.html>.

"Email Etiquette Revisited." *Manager's Legal Bulletin.* Ramsey, NJ: Alexander Hamilton Institute, 2000.

"Evaluating Internet-Based Information." May 1997. Online. Wolfgram Memorial Library, Widener University, PA. 17 Mar. 2001. <www.lme.mankato.msus.edu/class/629/wid.html>.

Evans, James. "Legal Briefs." *Internet World* Feb. 1998: 22.

———. "Whose Web Site Is It Anyway?" *Internet World* Sept. 1997: 46+.

Extejt, Marian M. "Teaching Students to Correspond Effectively Electronically." *Business Communication Quarterly* 61.2 (1998): 57–67.

Fackelmann, Kathleen. "Science Safari in Cyberspace." *Science News* 152.50 (1997): 397–98.

Facts and Figures about Cancer. Boston: Dana-Farber Cancer Institute, 1995.

"Fair Use." 10 June 1993. Online. 2 pp. United States Copyright Office, Library of Congress. 19 June 1998 <http://www.loc.gov/copyright>.

Farnham, Alan. "How Safe Are Your Secrets?" *Fortune* 8 Sept. 1997: 114–120.

"Fat Chance." *University of California at Berkeley Wellness Letter* 14.7 (1998): 2–3.

Fawcett, Heather. "*The New Oxford Dictionary* Project." *Technical Communication* 40.3 (1993): 379–82.

Felker, Daniel B., et al. *Guidelines for Document Designers.* Washington: American Institutes for Research, 1981.

Fineman, Howard, "The Power of Talk." *Newsweek* 8 Feb. 1993: 24–28.

Finkelstein, Leo, Jr. "The Social Implications of Computer Technology for the Technical Writer." *Technical Communication* 38.4 (1991): 466–73.

Fischman, Josh. "Who'll Pay for the Doc You Want?," *U.S. News & World Report* 18 Aug. 2003: 50.

Fisher, Anne. "Can I Stop Gay Bashing?" *Fortune* 7 July 1997: 205–06.

———. "I Didn't Spend Four Years in College to End Up as a Barista." *Fortune* May 12, 2003: 178.

———. "My Company Just Announced I May Be Laid Off. Now What?" *Fortune* 3 Mar. 2003.

———. "My Team Leader Is a Plagiarist." *Fortune* 27 Oct. 1997: 291–92.

———. "Truth and Consequences." *Fortune* 29 May 2000: 292.

Ford, Donna. "Phone Interviews: New Skills Required." *intercom* April 2002: 18–19.

Foster, Edward. "Why Users Beef about Documentation." *INTERCOM* Nov. 1998: 10.

Fox, Justin, " A Startling Notion—The Whole Truth," *Fortune* 24 Nov. 1997: 303.

Franke, Earnest A. "The Value of the Retrievable Technical Memorandum System to the Engineering Company." *IEEE Transactions on Professional Communication* 32.1 (Mar. 1989): 12–16.

Freundlich, Naomi. "When the Cure May Make You Sicker." *Business Week* 16 Mar. 1998: 14.

Friedland, Andrew J., and Carol L. Folt. *Writing Successful Science Proposals.* New Haven, CT: Yale UP, 2000.

"From Mir to Mars." *PBS Online.* 11 Nov. 1998 <www.PBS.org>.

Fugate, Alice E. "Mastering Search Tools for the Internet." *INTERCOM* Jan. 1998: 40–41.

———. "Wowing Them with Your Web Site." *INTERCOM* Nov. 2000: 33–35.

"Full Responsibility." 3 Nov. 2000. Online. 3 Oct. 2001 <www.ABC News.com>.

Gallagher, Leigh. "Isn't That Special?" *Forbes* 9 Mar. 1998: 39.

Gannon, Joseph P. "From GUI Guru to Web Weaver: Making the Transition." *INTERCOM* Dec. 1997: 21–25.

Garfield, Eugene, "What Scientific Journals Can Tell Us about Scientific Journals." *IEEE Transactions on Professional Communication* 16.4 (1973): 200–02.

Garner, Rochelle. "IS Newbies: Eager, Motivated, Clueless." *Computerworld* 1 Dec. 1997: 85–86.

Garten, Jeffrey E. "Globalism Doesn't Have to Be Cruel." *Business Week* 9 Feb. 1998: 26.

Gartaganis, Arthur. "Lasers." *Occupational Outlook Quarterly* Winter 1984: 22–26.

Gerstner, John. "Print Is Obsolete, but It Won't Go Away." *Journal of Employee Communication Management* Nov./Dec. 1997: 42–47.

Gesteland, Richard R. "Cross-Cultural Compromises." *Sky* May 1993: 20+.

Gibaldi, Joseph. *MLA Handbook for Writers of Research Papers.* 5th ed. New York: Modern Language Assn., 1998.

Gibaldi, Joseph, and Walter S. Achtert. *MLA Handbook for Writers of Research Papers.* 3rd ed. New York: Modern Language Assn., 1988.

Gibbs, W. Wayt. "The Price of Silence." *Scientific American* Nov. 1996: 15–16.

———. "Speech without Accountability." *Scientific American* Oct. 2000: 34+.

Gilbert, Nick, "1–800-ETHIC." *Financial World* 16 Aug. 1994: 20+.

Gilsdorf, Jeanette W. "Executives' and Academics' Perception of the Need for Instruction in Written Persuasion." *Journal of Business Communication* 23.4 (1986): 55–68.

———. "Write Me Your Best Case for . . ." *Bulletin of the Association for Business Communication* 54.1 (1991): 7–12.

Girill, T.R. "Technical Communication and Art." *Technical Communication* 31.2 (1984): 35.

———. "Technical Communication and Ethics." *Technical Communication* 34.3 (1987): 178–79.

———. "Technical Communication and Law." *Technical Communication* 32.3 (1985): 37.

Glassman, James K. "Dihydrogen Monoxide: It's a Killer." *Daily Hampshire Gazette* 22 Oct. 1997: 6.

Glidden, H. K. *Reports, Technical Writing and Specifications.* New York: McGraw, 1964.

Goby, Valerie P., and Lewis Justus Helen. "The Key Role of Listening in Business: A Study of the Singapore Insurance Industry." *Business Communication Quarterly* 63.2 (June 2000): 41–51.

Godin, Seth. "Blame It On Microsoft." 14 May 2004 <http://blog.fastcompany.com/archives/2003/09/29/blame_it_on_microsoft.html>.

Gogoi, Pallavi. "Teaching Men the Right Stuff." *Business Week* 20 Nov. 2000: 84.

Golen, Steven, et al. "How to Teach Ethics in a Basic Business Communications Class." *Journal of Business Communication* 22.1 (1985): 75–84.

Goodall, H. Lloyd, Jr., and Christopher L. Waagen. *The Persuasive Presentation.* New York: Harper, 1986.

Goodman, Danny. *Living at Light Speed.* New York: Random, 1994.

Goodman, Ellen. "Fear of Taxes Trumps Risk of Cancer." *Daily Hampshire Gazette* 24 June 1998: 6.

Goubil-Gambrell, Patricia. "Designing Effective Internet Assignments in Introductory Technical Communication Courses." *IEEE Transactions on Professional Communication* 39.4 (1996): 224–31.

Grant, Linda. "Where Did the Snap, Crackle, & Pop Go?" *Fortune* 4 Aug. 1997: 223+.

Grassian, Esther. "Thinking Critically about World Wide Web Resources." 20 Aug. 1997. UCLA College Library. 25 Oct. 1997 <www.library.ucla.edu/libraries/college/instruct/critical.htm>.

Graybill, Nina. "Freelancers and the Information Highway." *AWP Chronicle* Sept. 1997: 27.

Green, Heather. "Biotech: Can We Keep the Genie in the Bottle?" *Business Week* 2 Dec. 2002: 104.

Greenberg, Ilan. "Selling News Short." *Brill's Content* Mar. 2000: 64–65.

Gribbons, William M. "Organization by Design: Some Implications for Structuring Information." *Journal of Technical Writing and Communication* 22.1 (1992): 57–74.

Grice, Roger A. "Document Development in Industry." In *Technical Writing: Theory and Practice.* Eds. Bertie E. Fearing and W. Keats Sparrow. New York: Modern Language Assn. 1989, 27–32.

———. "Focus on Usability: Shazam!" *Technical Communication* 42.1 (1995): 131–33.

Grice, Roger A., and Lenore S. Ridgway. "Presenting Technical Information in Hypermedia Format: Benefits and Pitfalls." *Technical Communication Quarterly* 4.1 (1995): 35–46.

Griffin, Robert J. "Using Systematic Thinking to Choose and Evaluate Evidence." *Communicating Uncertainty: Media Coverage of New and Controversial Science.* Eds. Sharon Friedman, Sharon Dunwoody, and Carol Rogers. Mahwah, NJ: Erlbaum, 1999, 225–48.

Grimes, Brad. "Blazing a Paper Trail." *Fortune* 23 June 2003. [Advertising Supplement.]

Gross, Neil, "Between a Rock and a Hard Place." *Business Week* 20 Apr. 1998: 134+.

Grossman, Wendy M. "Downloading as a Crime." *Scientific American* Mar. 1998: 37.

Gurak, Laura J., and John M. Lannon. *A Concise Guide to Technical Communication.* New York: Longman, 2001.

———. *A Concise Guide to Technical Communication.* 2nd ed. New York: Longman, 2004.

Hafner, Kate, "Have Your Agent Call My Agent." *Newsweek* 27 Feb. 1995: 76–77.

Hall, Judith G. "Medicine on the Web: Finding the Wheat, Leaving the Chaff." *Technology Review* Mar./Apr. 1998: 60–61.

Halpern, Jean W. "An Electronic Odyssey." In *Writing in Nonacademic Settings.* Eds. Dixie Goswami and Lee Odell. New York: Guilford, 1985: 157–201.

Hamblen, Matt. "Volvo Taps AT&T for Global Net." *Computerworld* 1 Dec. 1997: 51+.

Hammett, Paula. "Evaluating Web Resources." 29 Mar. 1997. Ruben Salazar Library, Sonoma State University, 26 Oct. 1997 <www.libweb.sonoma.edu/resources/eval.html>.

"Handbooks." *The Employee Problem Solver.* Ramsey, NJ: Alexander Hamilton Institute, 2000.

Harcourt, Jules. "Teaching the Legal Aspects of Business Communication." *Bulletin of the Association for Business Communication* 53.3 (1990): 63–64.

Harris, Richard F. "Toxics and Risk Reporting." *A Field Guide for Science Writers.* Eds. Deborah Blum and Mary Knudson. New York: Oxford, 1997: 166–72.

Harris, Robert. "Evaluating Internet Research Sources." 17 Nov. 1997. Online, 23 June 1998 <www.sccu.edu/faculty/R_Harris/evalu8it.htm>.

Harrison, Bennett, "Don't Blame Technology This Time." *Technology Review* July 1997: 62.

Hart, Geoff. "Accentuate the Negative: Obtaining Effective Reviews through Focused Questions." *Technical Communication* 44.1 (1997): 52–57.

Hartley, James. *Designing Instructional Text.* 2nd ed. London: Kogan Page, 1985.

Haskin, David. "The Extranet Team Play." *Internet World* Aug. 1997: 57–60.

———. "Meetings without Walls." *Internet World* Oct. 1997: 53–60.

———. "A Push in the Right Direction." *Internet World* Sept. 1997: 75+.

Hauser, Gerald. *Introduction to Rhetorical Theory.* New York: Harper, 1986.

Hayakawa, S. I. *Language in Thought and Action.* 3rd ed. New York: Harcourt, 1972.

Hays, Robert. "Political Realities in Reader/Situation Analysis." *Technical Communication* 31.1 (1984): 16–20.

Hein, Robert G. "Culture and Communication." *Technical Communication* 38.1 (1991): 125–26.

Herper, Matthew. "Why Oscar Winners Live Longer." *Forbes* 7 July 2003: 12.

Hill-Duin, Ann. "Terms and Tools: A Theory and Research-Based Approach to Collaborative Writing." *Bulletin of the Association for Business Communication* 53.2 (1990): 45–50.

Hilligoss, Susan. *Visual Communication: A Writer's Guide.* New York: Longman, 1999.

Hilts, Philip J. "Web Sites Inconsistent on Health, Study Finds." *New York Times* 23 May 2001. 8 July 2001 <www.nytimes.com/2001/05/23/health/23 NET.html>.

Hodges, Mark. "Is Web Business Good Business?" *Technology Review* Aug./Sept. 1997: 22+.

Hoger, Elizabeth, James J. Cappel, and Mark A. Myerscough. "Navigating the Web with a Typology of Corporate Uses." *Business Communication Quarterly* 61.2 (1998): 39–47.

Hogge, Robert. Unpublished review of *Technical Writing.* 2nd ed.

Holler, Paul F. "The Challenge of Writing for Multimedia." *INTERCOM* July/Aug. 1995: 25.

Holloway, Marguerite. "Sounding Out Science." *Scientific American* Oct. 1996: 106–13.

Hollowitz, John C., and Donna Pawlowski. "The Development of an Ethical Integrity Interview for Pre-Employment Screening." *The Journal of Business Communication* 34.2 (1997): 203–19.

Holyoak, K. J. "Symbolic Connectionism." *Toward Third-Generation Theories of Expertise: Prospects and Limits.* Eds. K. A. Ericsson and J. Smith. New York: Cambridge UP, 1991. 331–35.

Hopkins-Tanne, Janice. "Writing Science for Magazines." *A Field Guide for Science Writers.* Eds. Deborah Blum and Mary Knudson. New York: Oxford, 1997:17–26.

Horgan, John. "Multicultural Studies." *Scientific American* Nov. 1996: 24+.

Horn, Robert E. *Visual Language: Global Communication for the 21st Century.* Bainbridge Island, WA: MacroVU, 1998.

Hornig-Priest, Susanna. "Popular Beliefs, Media, and Biotechnology." *Communicating Uncertainty: Media Coverage of New and Controversial Science.* Eds. Sharon Friedman, Sharon Dunwoody, and Carol Rogers. Mahwah, NJ: Erlbaum, 1999:95–112.

Horton, William. "Is Hypertext the Best Way to Document Your Product?" *Technical Communication* 38.1 (1991): 20–30.

———. "Mix Media, Not Metaphors." *Technical Communication* 41.4 (1994): 781–83.

Howard, Tharon. "Property Issue in E-Mail Research." *Bulletin of the Association for Business Communication* 56.2 (1993): 40–41.

Huff, Darrell. *How to Lie with Statistics.* New York: Norton, 1954.

Hugenberg, Lawrence W., Renee M. LaCivita, and Andra M. Lubanovic. "International Business and Training: Preparing for the Global Economy." *The Journal of Business Communication* 33.2 (1996): 205–22.

Hughes, Michael. "Rigor in Usability Testing." *Technical Communication* 46.4 (1999): 488–95.

Hulbert, Jack E. "Developing Collaborative Insights and Skills." *Bulletin of the Association for Business Communication* 57.2 (1994): 53–56.

———. "Overcoming Intercultural Communication Barriers." *Bulletin of the Association for Business Communication* 57.2 (1994): 41–44.

Humphreys, Donald S. "Making Your Hypertext Interface Usable." *Technical Communication* 40.4 (1993): 754–61.

Hunt, Kevin. "Establishing a Presence on the World Wide Web: A Rhetorical Approach." *Technical Communication* 46.4 (1996): 376–87.

Hutheesing, Nikhil. "What Are You Doing on That Porn Site?" *Forbes* 3 Nov. 1997: 368–69.

IBM Corporation, "IBM Solutions." Advertisement, 1997.

Imperato, Gina. "35 Ways to Land a Job Online." *Fast Company* Aug. 1998: 192–98.

"In Brief." *Scientific American* Oct. 1997: 28.

"International Copyright." July 2002. Online. United States Copyright Office, Library of Congress. 21 Mar. 2004 <www.loc.gov/copyright>.

Isaacs, Arlene B. "Tact Can Seal a Global Deal." *New York Times* 26 July 1997, sec. B: 43.

James-Catalano, P. "Fight for Privacy." *Internet World* Jan. 1997: 32+.

Jameson, Daphne A. "Using a simulation to Teach Intercultural Communication in Business Communication Courses." *Bulletin of the Association for Business Communication* 56.1 (1993): 3–11.

Janis, Irving L. *Victims of Groupthink: A Psychological Study of Foreign Policy Decisions and Fiascos.* Boston: Houghton, 1972.

Johannesen, Richard L. *Ethics in Human Communication,* 2nd ed. Prospect Heights, IL: Waveland, 1983.

Jones, Barbara. "Giving Women the Business." *Harper's Magazine* Dec. 1997: 47–58.

Journet, Debra. Unpublished review of *Technical Writing.* 3rd ed.

Kahneman, Daniel, and Amos Tversky. "Choices, Values, and Frames." *American Psychologist* 39.4 (1984): 342–47.

Kalb, Claudia, and Deborah Branscum. "Doctors Go Dot.Com." *Newsweek* 16 Aug. 1999: 65–66.

Kane, Kate. "Can You Perform under Pressure?" *Fast Company* Oct./Nov. 1997: 54+.

Kapoun, Jim. "Questioning Web Authority." *On Campus* Feb. 2000: 4.

Karaim, Reed. "The Invasion of Privacy." *Civilization* Oct./Nov. 1996: 70–77.

Kawasaki, Guy. "Get Your Facts Here." *Forbes* 23 Mar. 1998: 156.

———. "The Rules of E-Mail." *MACWORLD* Oct. 1995: 286.

Kelley-Reardon, Kathleen. *They Don't' Get It Do They? Communication in the Workplace—Closing the Gap between Women and Men.* Boston: Little, 1995.

Kelman, Herbert C. "Compliance, Identification, and Internalization: Three Processes of Attitude Change." *Journal of Conflict Resolution* 2 (1958): 51–60.

Keyes, Elizabeth. "Typography, Color, and Information Structure." *Technology Communication* 40.4 (1993): 638–54.

Kiely, Thomas. "The Idea Makers." *Technology Review* Jan. 1993: 33–40.

King, Ralph T. "Medical Journals Rarely Disclose Researchers' Ties." *Wall Street Journal* 2 Feb. 1999: B1+.

Kinik, Karina, "The Library That Never Closes." *Forbes ASAP* 19 Jan. 2000: 38.

Kipnis, David, and Stuart Schmidt. "The Language of Persuasion." *Psychology Today* Apr. 1985: 40–46. Rpt. in Raymond S. Ross, *Understanding Persuasion.* 3rd ed. Englewood Cliffs: Prentice, 1990.

Kirsh, Lawrence. "Take It from the Top." *MACWORLD* Apr. 1986: 112–15.

Kleimann, Susan D. "The Complexity of Workplace Review." *Technical Communication* 38.4 (1991): 520–26.

Kohl, John R., et al. "The Impact of Language and Culture on Technical Communication in Japan." *Technical Communication* 40.1 (1993): 62–72.

Koretz, Gene. "The New World of Work." *Business Week* 10 Jan. 2000: 36.

Kotulak, Ronald. "Reporting on Biology of Behavior." *A Field Guide for Science Writers.* Eds. Deborah Blum and Mary Knudson. New York: Oxford, 1997. 142–51.

Koudsi, Suzanne. "Actually, It Is Like Brain Surgery." *Fortune* 20 Mar. 2000: 233–34.

Kraft, Stephanie. "Whistleblower Bill's Holiday Adventures." *The Valley Advocate* [Northhampton, MA] 6 Jan. 1994: 5–6.

Krause, Tim. "Preparing an Online Résumé." *Business Communication Quarterly* 60.1 (1997): 159–61.

Kremers, Marshall, "Teaching Ethical Thinking in a Technical Writing Course." *IEEE Transactions on Professional Communication* 32.2 (1989): 58–61.

Lambe, Jennifer L. "Techniques for Successful SME Interviews." *INTERCOM* Mar. 2000: 30–32.

Lambert, Steve. *Presentation Graphics on the Apple® Macintosh.* Bellevue, WA: Microsoft, 1984.

Lang, Thomas A., and Michelle Secic. *How to Report Statistics in Medicine.* Philadelphia: American College of Physicians, 1997.

Larson, Charles U. *Persuasion: Perception and Responsibility.* 7th ed. Belmont, CA: Wadsworth: 1995.

Lavin, Michael R. *Business Information: How to Find it, How to Use It.* 2nd ed. Phoenix, AZ: Oryx, 1992.

"Learning to Love *PowerPoint.*" *Wired.* Sept. 2003. Archive. 10 May 2004 <http://www.wired.com/wired/archive/11.09/ppt1.html>.

Lederman, Douglas. "Colleges Report Rise in Violent Crime." *Chronicle of Higher Education* 3 Feb. 1995, sec. A: 5+.

Lee, Susan. "Death by Charcoal?" *Forbes* 25 Aug. 1997: 280.

Leki, Ilona. "The Technical Editor and the Non-native Speaker of English." *Technical Communication* 37.2 (1990): 148–52.

Lemonick, Michael. "The Evils of Milk?" *Times* 15 June 1998: 85.

Lenzer, Robert, and Carrie Shook. "Whose Rolodex Is It, Anyway?" *Forbes* 23 Feb. 1998: 100–04.

Le Vie, Donald S. "Résumés: You Can't Escape." *INTERCOM* Apr. 2000: 8–11.

Lewis, Howard L. "Penetrating the Riddle of Heart Attack." *Technology Review* Aug./Sept. 1997: 39–44.

Lewis, Kate Bohner. "Maybe Don't Take Two Aspirin." *Forbes* 23 May 1994: 222–23.

Lewis, Philip L., and N. L. Reinsch. "The Ethics of Business Communication." Proceedings of the American Business Communication Conference. Champaign, IL., 1981. In *Technical Communication and Ethics.* Eds. John R. Brockman and Fern Rook. Washington: Soc. for Technical Communication, 1989, 29–44.

Littlejohn, Stephen W., and David M. Jabusch. *Persuasive Transactions.* Glenview, IL: Scott, 1987.

Machlis, Sharon, "Surfing into a New Career as Webmaster." *Computerworld* 1 Dec. 1997: 45+.

MacKenzie, Nancy. Unpublished review of *Technical Writing.* 5th ed.

Mackin, John. "Surmounting the Barrier between Japanese and English Technical Documents." *Technical Communication* 36.4 (1989): 346–51.

Maeglin, Thomas. Unpublished review of *Technical Writing.* 7th ed.

Manning, Michael. "Hazard Communication 101." *INTERCOM* June 1998: 12–15.

Martin, Jeanette S., and Lillian H. Chaney. "Determination of Content for a Collegiate Course in Intercultural Business Communication by Three Delphi Panels." *Journal of Business Communication* 29.3 (1992): 267–83.

Martin, Justin. "Changing Jobs? Try the Net." *Fortune* 2 Mar. 1998: 205+.

———. "So, You Want to Work for the Best . . ." *Fortune* 12 Jan. 1998: 77–78.

Martin, Maurice, "Mars Needs Technical Communications." *INTERCOM* Jul./Aug. 2000: 3.

Matson, Eric. "(Search) Engines." *Fast Company* Oct./Nov. 1997: 249–52.

———. "The Seven Sins of Deadly Meetings." *Fast Company* Oct./Nov. 1997: 27–31.

Mayer, R. E. "When Less Is More: Meaningful Learning from Visual and Verbal Summaries of Science Textbook Lessons." *Journal of Educational Psychology* 88 (1996): 64–73.

McDonald, Kim A. "Covering Physics." *A Field Guide for Science Writers.* Eds. Deborah Blum and Mary Knudson. New York: Oxford, 1997. 188–95.

———. "Some Physicists Criticize Research Purporting to Show Links between Low-Level Electromagnetic Fields and Cancer." *Chronicle of Higher Education* 3 May 1991, sec. A: 5+.

McGuire, Gene. "Shared Minds: A Model of Collaboration." *Technical Communication* 39.3 (1992): 467–68.

McNair, Catherine. "New Technologies and Your Résumé." *INTERCOM* June 1997: 12–14.

Melymuka, Kathleen. "Not Another #$|&|$ Survey!" *Computerworld* 24 Nov. 1997: 82.

Menz, Mary. "Clip Art Comes of Age." *INTERCOM* May 1997: 4–8.

Merritt, Jennifer. "For MBAs, Soul-Searching 101." *Business Week* 16 Sept. 2002: 64–66.

———. "You Mean Cheating Is Wrong?" *Business Week* 9 Dec. 2002: 8.

Meyer, Benjamin D. "The ABCs of New-Look Publications." *Technical Communication* 33.1 (1986): 13–20.

Meyerson, Moe. "Grand Illusions." *Inc. Tech* 2 (1997): 35–36.

Microsoft Word User's Guide: Word Processing Program for the Macintosh, Version 5.0. Redmond, WA: Microsoft Corporation, 1992.

Miller, Julie. "Trade Journals." *A Handbook for Science Writers.* Eds. Deborah Blum and Many Knudson. New York: Oxford, 1997. 27–30.

Mirel, Barbara, Susan Feinberg, and Leif Allmendinger. "Designing Manuals for Active Learning Styles." *Technical Communication* 38.1 (1991): 75–87.

Mirsky, Steve. "Wonderful Town." *Scientific American* July 1996: 29.

"Misconduct Scandal Shakes German Science." *Professional Ethics Report* [American Assoc. for the Advancement of Science] X3 (Summer 1997): 2.

Mokhiber, Russell. "Crime in the Suites." *Greenpeace* May 1989: 14–16.

Monastersky, Richard. "Courting Reliable Science." *Science News* 153.16 (1998): 249–51.

———. "Do Clouds Provide a Greenhouse Thermostat?" *Science News* 142.16 (1992): 69.

Monmonier, Mark. *Cartographies of Danger: Mapping Hazards in America.* Chicago: U of Chicago P, 1997.

Morgan, Meg. "Patterns of Composing: Connections between Classroom and Workplace Collaborations." *Technical Communication* 38.4 (1991): 540–42.

Morgenson, Gretchen. "Would Uncle Sam Lie to You?" *Worth* Nov. 1994: 53+.

Morse, June. "Hypertext—What Can We Expect?" *INTERCOM* Feb. 1992: 6–7.

Munger, David. Unpublished review of *Technical Writing.* 7th ed.

Munger, Roger H. "Finding Proposal Money for Nonprofits." *INTERCOM* June 2001: 28–30.

Munter, Mary. "Meeting Technology: From Low-Tech to High-Tech. *Business Communication Quarterly* 61.2 (1998): 80–87.

Murphy, Kate. "Separating Ballyhoo from Breakthrough." *Business Week* 13 July 1998: 143.

Nakache, Patricia. "Is It Time to Start Bragging about Yourself? *Fortune* 27 Oct. 1997: 287–88.

Nantz, Karen S., and Cynthia L. Drexel. "Incorporating Electronic Mail with the Business Communication Course." *Business Communication Quarterly* 58.3 (1995): 45–51.

Neergaard, Lauran. "U.S. Adults Face 'Health Literacy' Crisis." Associated Press Wire story. April 8, 2004 <http://www.miami.com/mid/miamiherald/living/health/8389092.htm7lc>.

Nielsen, Jakob. "Be Succinct! (Writing for the Web)." 15 Mar. 1997. Alertbox. 8 Aug. 1998 <www.useit.com/alertbox/9719a.html>.

———. "Global Web: Driving the International Network Economy." Apr. 1998. Alertbox. 8 Aug. 1998 <www.useit.com/alertbox/9710a.html>.

———. "How Users Read on the Web." Oct. Alertbox. 8 Aug. 1998 <www.useit.com/alertbox/9710a.html>.

———. International Web Usability." Aug. 1996. Alertbox. 8 Aug. 1998 <www.useit.com/alertbox/9710a.html>.

———. "Inverted Pyramids in Cyberspace." June 1996. Alertbox. 8 Aug. 1998 <www.useit.com/alertbox/9710a.html>.

———. "Top Ten Web Design Mistakes of 2003." Alertbox. 12 May 2004 <www.uselt.com/alertbox/20031222.html>.

Nelson, Sandra J., and Douglas C. Smith. "Maximizing Cohesion and Minimizing Conflict in Collaborative Writing Groups." *Bulletin of the Association for Business Communication* 53.2 (1990): 59–62.

Nordenberg, Tamar. "Direct to You: TV Drug Ads That Make Sense." *FDA Consumer* Jan./Feb. 1998: 7–10.

Nunberg, G. "The Trouble with PowerPoint." *Fortune* 20 Dec. 1999: 330–34.

Nydell, Margaret K. *Understanding Arabs: A Guide for Westerners.* New York: Logan, 1987.

Office of Technology Assessment. *Harmful Non-Indigenous Species in the United States.* Washington, DC: GPO, 1993.

"On Line." *Chronicle of Higher Education* 21 Sept. 1992, sec. A: 29.

"Online Health Companies Announce New Set of Ethics and Privacy Guidelines." *Professional Ethics Report* [American Association for the Advancement of Science] XIII.2 (Spring 2000): 3–4.

Ornatowski, Cezar M. "Between Efficiency and Politics: Rhetoric and Ethics in Technical Writing." *Technical Communication Quarterly* 1.1 (1992): 91–103.

Ostrander, Elaine L. "Usability Evaluations: Rationale, Methods, and Guidelines." *INTERCOM* June 1999: 18–21.

Outing, Steve. "Does Your Site Contribute to Data Smog?" 28 May 1997. *Editor and Publisher Interactive.* 8 Aug. 1998 <www.mediainfo.com/ephome/news/newsshtm/stop/st052897.htm>.

Oxfeld, Jesse. "Analyze This." *Brill's Content* Mar. 2000: 105–06.

Parker, Ian. "Absolute *PowerPoint.*" *The New Yorker* 28 May 2001: 76–87.

Parrish, Deborah. "The Scientific Misconduct Definition and Falsification of Credentials." *Professional Ethics Report* [American Assoc. for the Advancement of Science] IX.4 (1996): 1+.

Parsons, Gerald M. Review of *Technical Writing.* 6th ed. *Journal of Technical Writing and Communication* 25.3 (1995): 322–24.

Pearce, C. Glenn, Iris W. Johnson, and Randolph T. Barker. "Enhancing the Student Listening Skills and Environment." *Business Communication Quarterly* 58.4 (Dec. 1995): 28–33.

Pender, Kathleen. "Dear Computer, I Need a Job." *Worth* Mar. 1995: 120–21.

"People, Performance, Profits." *Forbes* 20 Oct. 1997: 57.

"Performance Appraisal—Discrimination." *The Employee Problem Solver.* Ramsey, NJ: Alexander Hamilton Institute, 2000.

Perloff, Richard M. *The Dynamics of Persuasion.* Hillsdale, NJ: Erlbaum, 1993.

Peters, Tom. "The New Wired World of Work." *Business Week* 28 Aug. 2000: 172–74.

Petroski, Henry. *Invention by Design.* Cambridge, MA: Harvard UP, 1996.

Peyser, Marc, and Steve Rhodes. "When E-Mail Is Oops-Mail." *Newsweek* 16 Oct. 1995: 82.

Phillips, John I. *How to Think about Statistics.* New York: Freeman, 2000.

Pinelli, Thomas E., et al., "A Survey of Typography, Graphic Design, and Physical Media in Technical Reports." *Technical Communication* 32.2 (1986): 75–80.

Plumb, Carolyn, and Jan H. Spyridakis, "Survey Research in Technical Communication: Designing and Administering Questionnaires." *Technical Communication* 39.4 (1992): 625–38.

Pool, Robert. "When Failure Is Not an Option." *Technology Review* July 1997: 38–45.

Porter, James E. "Truth in Technical Advertising: A Case Study." *IEEE Transactions on Professional Communication* 33.3 (1987): 182–89.

Powell, Corey S. "Science in Court." *Scientific American* October 1997: 32+

Pugliano, Fiore. Unpublished review of *Technical Writing*, 5th ed.

Quible, Zane K. "Guiding Students in Finding Information on the Web." *Business Communication Quarterly* 62.3 (Sept. 1999): 57–70.

Raeburn, Paul. "Warning: Biotech Is Hurting Itself." *Business Week* 20 Dec. 1999: 78.

Raloff, Janet. "Chocolate Hearts: Yummy and Good Medicine?" *Science News* 157.12 (2000): 188–89.

Rao, Srikumar. "Diaper-Beer Syndrome." *Forbes* 9 Apr. 1998: 128.

Read Me First!: A Style Guide for the Computer Industry. Palo Alto, CA: Sun Microsystems Press, 2003.

Redish, Janice C., and David A. Schell. "Writing and Testing Instructions for Usability." *Technical Writing: Theory and Practice.* Eds. Bertie E. Fearing and W. Keats Sparrow. New York: Modern Language Assn., 1989. 61–71.

Redish, Janice C., et al. "Making Information Accessible to Readers." *Writing in Nonacademic Settings.* Eds. Lee Odell and Dixie Goswami. New York: Guilford, 1985.

Reichard, Kevin, "Web-Site Watchdogs." *Internet World* Dec. 1997: 106+.

Reinhardt, Andy. "From Gearhead to Grand High Poo-Bah." *Business Week* 28 Aug. 2000: 129–30.

Rensberger, Boyce. "Covering Science for Newspapers." *A Field Guide for Science Writers.* Eds. Deborah Blum and Mary Knudson. New York: Oxford, 1997. 7–16.

Research Triangle Institute. *Consequences of Whistleblowing for the Whistleblower in Misconduct in Science Cases.* (Report prepared for the Office of Research Integrity.) Washington: ORI, 1995.

Rifkin, William, and Brian Martin. "Negotiating Expert Status: Who Gets Taken Seriously." *IEEE Technology and Society Magazine* (Spring 1997): 30–39.

Riney, Larry A. *Technical Writing for Industry.* Englewood Cliffs: Prentice, 1989.

Ritzenthaler, Gary, and David H. Ostroff. "The Web and Corporate Communication: Potentials and Pitfalls." *IEEE Transactions on Professional Communication* 39.1 (1996): 16–20.

Rivers, William E. "Politics, Ethics, and Corporate Policy: U.S. Corporation's 1986 Position Papers on South Africa." *Journal of Business Communication* 37.4 (Oct. 2000): 369–407.

Robart, Kay. "Submitting Résumés via E-Mail." *INTERCOM* July/Aug. 1998: 13–14.

Robinson, Edward A. "Beware—Job Seekers Have No Secrets." *Fortune* 29 Dec. 1997: 285.

Rokeach, Milton. *The Nature of Human Values.* New York: Free, 1973.

Rosman, Katherine. "Finding Drug Ties at a Medical Mag." *Brill's Content* Mar. 2000: 100.

Ross, Philip E. "Enjoy It While It Lasts." *Forbes* 27 July 1998: 206.

———. "Lies, Damned Lies, and Medical Statistics." *Forbes* 14 Aug. 1995: 130–35.

Ross, Raymond S. *Understanding Persuasion.* 3rd ed. Englewood Cliffs: Prentice, 1990.

Ross-Flanigan, Nancy. "The Virtues (and Vices) of Virtual Collaboration." *Technology Review* Mar./Apr. 1998: 50–59.

Rottenberg, Annette T. *Elements of Argument,* 3rd ed. New York: St. Martin's, 1991.

Rowan, Katherine E. "Effective Explanation of Uncertain and Complex Science." *Communicating Uncertainty: Media Coverage of New and Controversial Science.* Eds. Sharon Friedman, Sharon Dunwoody, and Carol Rogers. Mahwah, NJ: Erlbaum, 1999. 201–23.

Rowland, D. *Japanese Business Etiquette: A Practical Guide to Success with the Japanese.* New York: Warner, 1985.

Ruggiero, Vincent R. *The Art of Thinking.* 3rd ed. New York: Harper, 1991.

———. *The Art of Thinking.* 5th ed. New York: Addison, 1998.

Ruhs, Michael A. "Usability Testing: A Definition Analyzed." *Boston Broadside* [Newsletter of the Soc. for Technical Communication] May/June 1992: 8+.

Ruppe, David. "Information Control." 4 Oct. 2001. Online. 5 Oct. 2001 <www.ABCNews.com>.

Sabath, Ann Marie. *Business Etiquette: 101 Ways to Conduct Business with Charm and Savvy.* Franklin Lakes, NJ: Career Press, 1998.

Samuelson, Robert J. "The Endless Paper Chase." *Newsweek* 1 Dec. 1997: 53.

———. "Merchants of Mediocrity." *Newsweek* 1 Aug. 1994: 44.

Savan, Leslie. "Truth in Advertising?" *Brill's Content* March 2000: 62+.

Schafer, Sarah. "Is Your Data Safe?" *Inc.* Feb. 1997: 93–97.

Schein, Edgar H. "How Can Organizations Learn Faster? The Challenge of Entering the Green Room." *Strategies for Success: Core Capabilities for Today's Managers.* Boston: Sloan Management Review Assoc., 1996. 34–39.

Schenk, Margaret T., and James K. Webster. *Engineering Information Resources.* New York: Decker, 1984.

Schrage, Michael. "Time for Face Time." *Fast Company* Oct./Nov. 1997: 232.

Schwartz, Leon. "Ideas and Trends: The Level of Discourse Continues to Slide." *New York Times* 28 Sept. 2003, late ed., sec. 4: 3.

Scott, James C. "Dear ???—Understanding British Forms of Address." *Business Communication Quarterly* 61.3 (1998): 50–61.

Scott, James C., and Diana J. Green. "British Perspectives on Organizing Bad-News Letters: Organizational Patterns Used by Major U.K. Companies." *Bulletin of the Association for Business Communication* 55.1 (1992): 17–19.

Seglin, Jeffrey L. "Would You Lie to Save Your Company?" *Inc.* July 1998: 53+.

Seligman, Dan. "Gender Mender." *Forbes* 6 Apr. 1998: 72+.

Selzer, Jack. "Composing Processes for Technical Discourse." *Technical Writing: Theory and Practice.* Eds. Bertie E. Fearing and W. Keats Sparrow. New York: Modern Language Assn., 1989. 43–50.

Senge, Peter M. "The Leader's New York: Building Learning Organizations." *Sloan Management Review* 32.1 (Fall 1990): 1–17.

Seppa, Nathan. "Broken Arms Way Up." *Science News* 164.14 (2003): 221.

Sharpe, Rochelle. "As Leaders, Women Rule." *Business Week* 20 Nov. 2000: 75+.

Shedroff, Nathan. "Information Interaction Design: A Unified Field Theory of Design." *Information Design.* Ed. Robert Jacobson. Cambridge, MA: MIT Press, 2000. 267–92.

Shenk, David. "Data Smog: Surviving the Information Glut." *Technology Review* May/June 1997: 18–26.

Sherblom, John C., Claire F. Sullivan, and Elizabeth C. Sherblom, "The What, the Whom, and the Hows of Survey Research," *Bulletin of the Association for Business Communication* 56:12 (1993): 58–64.

Sherif, Muzapher, et al. *Attitude and Attitude Change: The Social Judgment-Involvement Approach.* Philadelphia: Saunders, 1965.

Sittenfeld, Curtis. "Good Ways to Deliver Bad News." *Fast Company* Apr. 1999: 88+.

Sklaroff, Sara, and Michael Ash. "American Pie Charts." *Civilization* April/May 1997: 84–85.

Smart, Karl L., Matthew E. Whiting, and Kristen Bell DeTienne. "Assessing the Need for Printed and Online Documentation: A Study of Customer Preference and Use." *Journal of Business Communication* 38.3 (2001): 285–314.

Smith, Gary. "Eleven Commandments for Business Meeting Etiquette." *INTERCOM* Feb. 2000: 29.

Snyder, Joel. "Finding It on Your Own." *Internet World* June 1995: 89–90.

Sowell, Thomas. "Magic Numbers." *Forbes* 20 Oct. 1997: 120.

Specter, Michael. "Your Mail Has Vanished." *New Yorker* 16 Dec. 1999: 95–104.

Spencer, SueAnn. "Use Self-Help to Improve Document Usability." *Technical Communication* 43.1 (1996): 73–77.

Spragins, Ellyn E. "The Numbers Racket." *Newsweek* 5 May 1997: 77.

Spyridakis, Jan H. "Conducting Research in Technical Communication: The Application of True Experimental Design." *Technical Communication* 39.4 (1992): 607–24.

Spyridakis, Jan H., and Michael J. Wenger. "Writing for Human Performance: Relating Reading Research to Document Design." *Technical Communication* 39.2 (1992): 202–15.

St. Amant, Kirk R. "Resource and Strategies for Successful International Communication." *INTERCOM* Sept./Oct. 2000: 12–14.

Stanton, Mike. "Fiber Optics." *Occupational Outlook Quarterly* (Winter 1984): 27–30.

Stedman, Craig. "Data Mining for Fool's Gold." *Computerworld* 1 Dec. 1997: 1+.

Stemmer, John. "Citing Internet Sources." 4 Mar. 1997. Online. Political Science Research and Teaching List. 22 April 1997 <polpsrt@h—met.msu.edu>.

Stepanek, Marcia. "When in Beijing, Mum's the Word." *Business Week* 13 July 1998: 4.

Stevenson, Richard W. "Workers Who Turn in Bosses Use Law to Seek Big Rewards." *New York Times* 10 July 1989, sec. A: 7.

Stix, Gary. "Plant Matters: How Do You Regulate an Herb?" *Scientific American* Feb. 1998: 30+.

Stone, Peter H. "Forecast Cloudy: The Limits of Global Warming Models." *Technology Review* Feb./Mar. 1992: 32–40.

Stonecipher, Harry. *Editorial and Persuasive Writing.* New York: Hastings, 1979.

Sturges, David L. "Internationalizing the Business Communication Curriculum." "*Bulletin of the Association for Business Communication* 55.1 (1992): 30–39.

"Sunday Sermons." *Scientific American* Feb. 2003: 26.

Task Force on High-Performance Work and Workers. *Spanning the Chasm: Corporate and Academic Preparation to Improve Work-Force Preparation*. Report. Washington, DC: Business-Higher Education Forum, Jan. 1997.

Taubes, Gary. "Telling Time by the Second Hand." *Technology Review* May/June 1998: 76–78.

Taylor, John R. *Introduction to Error Analysis*. 2nd ed. Sausalito, CA: University Science Books, 1997.

Teague, John H. "Marketing on the World Wide Web." *Technical Communication* 42.2 (1995): 236–42.

Templeton, Brad. "10 Big Myths about Copyright Explained." 29 Nov. 1994. Online. 6 May 1995 <www.law/copyright/FAQ/myths/part1>.

"Testing Your Documents." 16 Apr. 2001. Online. *Plain English Network*. 4 May 2001 <www.plainlanguage.gov/howto/test.htm>.

Thatcher, Barry. "Cultural and Rhetorical Adaptation for South American Audiences." *Technical Communication* 46.2 (1999): 177–95.

"The Big Picture." *Business Week* 6 Nov. 2000: 14.

"The Safest Car May Be a Truck." *Fortune* 21 July 1997: 72.

Thrush, Emily A. "Bridging the Gap: Technical Communication in an Intercultural and Multicultural Society." *Technical Communication Quarterly* 2.3 (1993): 271–83.

Trafford, Abigail. "Critical Coverage of Public Health and Government." *A Field Guide for Science Writers*. Eds. Deborah Blum and Mary Knudson. New York: Oxford, 1997. 131–41.

Tufte, Edward R. *The Cognitive Style of PowerPoint*. Cheshire CT: Graphics Press, 2003.

Tullar, William, Paula Kaiser, and Pierre A. Balthazard. "Group Work and Electronic Meeting System: From Boardroom to Classroom." *Business Communication Quarterly* 61.4 (Dec. 1998): 53–65.

Turner, John R. "Online Use Raises New Ethical Issues." *INTERCOM* Sept. 1995: 5+.

Unger, Stephen H. *Controlling Technology: Ethics and the Responsible Engineer*. New York: Holt, 1982.

U.S. Air Force Academy. *Executive Writing Course*. Washington, DC: GPO, 1981.

U.S. Department of Commerce. *Statistical Abstract of the United States*. Washington, DC: GPO, 1994, 1997, 2000, 2003.

U.S. Department of Labor. *Tips for Finding the Right Job*. Washington, DC: GPO, 1993.

———. *Tomorrow's Jobs*. Washington, DC: GPO, 2000.

U.S. General Services Administration. *Your Rights to Federal Records*. Washington, DC: GPO, 1995.

"Using Icons as Communication." *Simply Stated* [Newsletter of the Document Design Center, American Institutes for Research] 75 (Sept./Oct. 1987): 1+.

van der Meij, Hans. "The ISTE Approach to Usability Testing." *IEEE Transactions on Professional Communication* 40.3 (1997): 209–23.

van der Meij, Hans, and John M. Carroll. "Principles and Heuristics for Designing Minimalist Instruction." *Technical Communication* 42.2 (1995): 243–61.

Van Pelt, William. Unpublished review of *Technical Writing*. 3rd ed.

Varchaver, Nicholas. "The Perils of E-mail." *Fortune* Feb. 17, 2003: 96–102.

Varner, Iris I., and Carson, H. Varner. "Legal Issues in Business Communications." *Journal of the American Association for Business Communication* 46.3 (1983): 31–40.

Vaughan, David K. "Abstracts and Summaries: Some Clarifying Distinctions." *Technical Writing Teacher* 18:2 (1991): 132–41.

Velotta, Christopher. "How to Design and Implement a Questionnaire." *Technical Communication* 38.3 (1991): 387–92.

Victor, David A. *International Business Communication*. New York: Harper, 1992.

"Vital Signs." *Internet World* Jan. 1998: 18.

"Vitamin C under Attack." *University of California, Berkeley Wellness Letter* 14.10 (1998): 1.

"Walking to Health." *Harvard Men's Watch* 2.12 (1998): 3–4.

Wallace, Bob. "Restaurant Franchiser Puts Internet on Menu." *Computerworld* 10 Nov. 1997: 12.

Wallich, Paul. "Not So Blind, After All." *Scientific American* May 1996: 20+.

Walter, Charles, and Thomas F. Marsteller. "Liability for the Dissemination of Defective Information." *IEEE Transactions on Professional Communication* 30.3 (1987): 164–67.

Wandycz, Katarzyna. "Damn Yankees." *Forbes* 10 March, 1997: 22–23.

Wang, Linda. "Veggies Prevent Cancer through Key Protein." *Science News* 159.12 (2001): 182.

Warshaw, Michael. "Have You Been House-Trained?" *Fast Company* Oct. 1998: 46+.

Weinstein, Edith K. Unpublished review of *Technical Writing*. 5th ed.

Weiss, Edmond H. *How to Write a Usable User Manual*. Philadelphia: ISI, 1985.

"Wellness Facts." *University of California, Berkeley Wellness Letter* 14.10 (1998): 1.

Weymouth, L. C. "Establishing Quality Standards and Trade Regulations for Technical Writing in World Trade." *Technical Communication* 37.2 (1990): 143–47.

White, Jan. *Color for the Electronic Age*. New York: Watson-Guptill, 1990.

———. *Editing by Design*. 2nd ed. New York: Bowker, 1982.

———. *Great Pages*. El Segundo, CA: Serif, 1990.

———. *Visual Design for the Electronic Age*. New York: Watson-Guptill, 1988.

Wickens, Christopher D. *Engineering Psychology and Human Performance*. 2nd ed. New York: Harper, 1992.

Wiggins, Richard. "The Word Electric." *Internet World* Sept. 1995: 31–34.

Wight, Eleanor, "How Creativity Turns Facts into Usable Information." *Technical Communication* 32.1 (1985): 9–12.

Wilkinson, Theresa A. "Defining Content for a Web Site." *INTERCOM* June 1998: 33–34.

Willett, Walter C., and Meir J. Stampfer. "Rebuilding the Food Pyramid." *Scientific American* Jan. 2003: 64–71.

Williams, Robert I. "Playing with Format, Style, and Reader Assumptions." *Technical Communication* 30.3 (1983): 11–13.

Wojahn, Patricia G. "Computer-Mediated Communication: The Great Equalizer between Men and Women?" *Technical Communication* 41.4 (1994): 747–51.

Woodhouse, E. J., and Dean Nieusma. "When Expert Advice Works and When It Does Not." *IEEE Technology and Society Magazine* Spring 1997: 23–29.

Wriston, Walter. *The Twilight of Sovereignty*. New York: Scribner's, 1992.

Writing User-Friendly Documents. Washington, DC: U.S. Bureau of Land Management, 2001.

Wurman, Richard Saul. *Information Anxiety*. New York: Doubleday, 1989.

Yen, Hope. "9/11 Panel: FAA Downplayed Suicide Hijacking." [Associated Press] *The Recorder* [Greenfield, MA] 28 Jan. 2004: 7.

Yoos, George. "A Revision of the Concept of Ethical Appeal." *Philosophy and Rhetoric* 12.4 (1979): 41–58.

Young, Patrick. "Writing Articles for Science Journals." *A Field Guide for Science Writers*. Eds. Deborah Blum and Mary Knudson. New York: Oxford, 1997. 110–16.

Zibell, Kristin J. "Usable Information through User-Centered Design." *INTERCOM* Dec. 1999: 12–14.

Zinsser, William. *On Writing Well*. New York: Harper, 1980.

Index

EDITING AND REVISION SYMBOLS

Symbol	Problem	Page*	Symbol	Problem	Page*
ab	wrong abbreviation	774	ital	italics	771
agr p	error in pronoun agreement	756	( ) /	parentheses	771
			. /	period	762
agr sv	error in subject-verb agreement	755	? /	question mark	763
			" / "	quotation	770
amb	ambiguous phrasing	245	; /	semicolon	763
av	active voice needed	250	pref	needless preface	256
bias	biased language	278	prep	needless preposition	257
ca	wrong pronoun case	758	pv	passive voice needed	252
cap	capital letter needed	776	qual	needless qualifier	260
cl	word adds clutter	759	red	redundant phrase	254
comb	sentences need to be combined	261	rep	needless repetition	254
			ref	faulty reference	245
cont	faulty contraction	775	ro	run-on sentence	754
coord	faulty coordination	756	sexist	sexist usage	280
cs	comma splice	754	shift	sentence shift	761
dgl	dangling modifier	760	short	short sent. needed	263
euph	euphemism	268	simple	simpler word needed	264
frag	sentence fragment	752	spec	specific word needed	271
jarg	needless jargon	266	sub	faulty subordination	757
mod	misplaced modifier	247	tel	telegraphic writing	246
neg	negative phrasing	259	th	faulty sent. opener	255
nom	nominalization	258	tone	inappropriate tone	274
offen	offensive usage	280	trans	transition needed	772
os	overstuffed sentence	253	trite	overused expression	268
over	overstatement	269	ts	faulty topic sentence	232
par	faulty parallelism	761	var	sent. variety needed	262
pct	faulty punctuation	762	w	too many words	254
ap/	apostrophe	773	wo	faulty word order	248
[ ] /	brackets	771	wv	weak verb	256
: /	colon	764	ww	wrong word	269
, /	comma	764	#	wrong use of numbers	777
— — /	dash	772	¶	new paragraph needed	230
... /	ellipses	771	¶ coh	para. lacks coherence	233
! /	exclamation point	763	¶ lgth	para. too long or short	235
— /	hyphen	776	¶ un	paragraph lacks unity	233

*Numbers refer to the first page of major discussion in the text.